Hormones in Neuroscience

Hormones in Neuroscience
The Thyroid Gland and Thyroid Hormones

Edited by

Colin R. Martin
Institute for Health and Wellbeing, University of Suffolk, Ipswich, United Kingdom

Vinood B. Patel
School of Life Sciences University of Westminster, London, United Kingdom

Victor R. Preedy
Department of Clinical Biochemistry, King's College Hospital, London, United Kingdom
Faculty of Life Sciences and Medicine, King's College London, London, United Kingdom
University of Suffolk, Ipswich, United Kingdom

Rajkumar Rajendram
Medical Protocol Department, King Abdulaziz Medical City, Ministry of National Guard Heath Affairs, Riyadh, Saudi Arabia

ELSEVIER
ACADEMIC PRESS
An imprint of Elsevier

Academic Press is an imprint of Elsevier
125 London Wall, London EC2Y 5AS, United Kingdom
50 Hampshire Street, 5th Floor, Cambridge, MA 02139, United States

ISBN: 978-0-443-30261-9

For information on all Academic Press publications visit our website at https://www.elsevier.com/books-and-journals

Publisher: Megan Ball
Acquisitions Editor: Megan McManus
Editorial Project Manager: Jahnuvi Chauhan
Production Project Manager: Kiruthiga devi
Cover Designer: Mark Rogers

Typeset by TNQ Tech

Contents

Disclaimer xv
Contributors xvii
Preface xxi

Section 1
Introductory chapters

1. Thyroid hormones and brain structural, molecular, and functional correlates

Asma Hallab

Introduction 3
Thyroid hormones and brain: Insights from structural neuroimaging studies 4
Thyroid hormones and brain structures: Gray matter analysis 4
Thyroid hormones and brain structures: White matter analysis 6
Thyroid hormones and brain: Insights from functional neuroimaging studies 6
Hypothyroidism and functional brain changes 6
Hyperthyroidism and functional brain changes 6
Thyroid hormones and brain: Insights from molecular neuroimaging studies 6
Thyroid function and glucose metabolism 6
Thyroid function and alzheimer's disease biomarkers 7
Other neuroimaging studies 7
Applications of the material 7
Conclusions 8
Mini-dictionary of terms 8
Summary points 8
References 9

2. Thyroid hormone signaling in neurodevelopment

Ehsan Noroozi, Alireza Rashidi, Hoda Borooghani, Maryam Eskandarpour, Mojan Ayati and Alaleh Alizadeh

Abbreviations 13
Introduction 13
Neurodevelopmental assessments within the thyroid-NDT-TAS 14
Regulation of TH action in the brain from development to adulthood 14
Tyrosine hydroxylase and excitatory glutamatergic neurons 14
TH and inhibitory GABAergic interneurons 14
TH and cholinergic systems 14
TH and dopaminergic systems 16
An overview of the signaling landscape of THs 16
Uncoupling of the genomic and non-genomic impacts of THs during development 18
Heterochronic signatures of THs in neurodevelopment 20
Neurodevelopmental consequences following IUGR 21
Cerebral TH signaling: Implication in IUGR 21
TH analogs: Therapeutic potential in the IUGR brain 22
Conclusion 22
Mini-dictionary of terms 23
Summary points 23
Declaration of AI use 23
References 24

3. Enzymes involved in thyroid hormone metabolism and status

Stanisław Surma and Kazimierz Ciechanowski

List of abbreviations: 27
Introduction 27
Thyroid hormones: Structure 27
Iodine: Biochemical transformations 28
NIS 28
Pendrin 28
Apical iodine transporter (AIT) 29
Wolff–chaikoff effect 29
Thyroid hormone synthesis 29
TPO 29
Iodotyrosine conjugation 30
Pathophysiological significance of TPO 30
Oxidases involved in thyroid hormone synthesis 30

Tg 31
Thyroid hormone metabolism 32
Thyroid hormone deiodination 32
Metabolism in the liver 33
Transport in serum 33
Transport of thyroid hormones into cells 35
Regulation of thyroid hormone secretion 35
TSH 36
Conclusions 36
Mini-dictionary of terms 36
Summary points 37
Other hormones, endocrine tissues, and pathways 37
Applications of the material 37
References 38

4. An overview of the consequences of subclinical hypothyroidism

Tiffany Mach and Oriana Hoi Yun Yu

Abbreviations 39
Introduction 39
Clinical assessment of subclinical hypothyroidism 40
Causes of subclinical hypothyroidism 40
Management of subclinical hypothyroidism 41
Subclinical hypothyroidism in children 41
Translational aspects of the chapter 41
Thyroid hormone and its physiologic role 41
Subclinical hypothyroidism and cardiovascular outcomes 41
Subclinical hypothyroidism and pregnancy outcomes 43
Subclinical hypothyroidism and renal outcomes 45
Subclinical hypothyroidism and cognitive outcomes 45
Subclinical hypothyroidism and effects on mood 45
Applications of the material 46
Mini-dictionary of terms 46
Summary points 46
References 47

5. The consequences of thyroidectomy: An overview new narrative

Bhuvnesh Deivakumar, Femi E. Ayeni, Saikrishna Ananthapadmanabhan, Mark Smith, Faruque Riffat and Niranjan Sritharan

List of abbreviations 51
Introduction 51
Early post-operative consequence 51
Hemorrhage 51
Endocrine and metabolic consequences 53
Peri-operative thyroid hormone derangements 53
Disruption to calcium homeostasis 53
Neurological consequence: Nerve injuries 54
Anatomical consideration & intraoperative nerve preservation 54
Functional consequences 55
Dysphonia 55
Dysphagia 56
Cosmetic and aesthetic consequence 56
Psychological consequence: Effect on quality of life 57
Thyroidectomy consideration in the pediatric population 57
Consequences of alternative thyroidectomy techniques 58
Other hormones and their pathways 59
Applications of the material 59
Conclusion 60
Mini-dictionary of terms 60
Summary points 60
References 60

6. Pregnancy and thyroid hormone status

Akiko Sankoda, Naoko Arata, Ai Yoshihara and Natsuko Watanabe

Abbreviations 63
Introduction 63
Maternal–fetal–placental thyroid physiology 64
Maternal thyroidal adaptation during pregnancy 64
The placenta as a thyroid-responsive endocrine organ 64
Epidemiology and pregnancy outcomes 65
Overt and subclinical hypothyroidism 65
Isolated hypothyroxinemia 65
Thyroid autoimmunity 65
Hyperthyroidism and gestational transient thyrotoxicosis 66
Diagnosis and screening in pregnancy 66
Biochemical assessment and pregnancy-specific reference ranges 66
Etiological classification 66
Universal versus targeted screening 67
Timing and follow-up 67
Management of thyroid dysfunction in pregnancy 67
Overt hypothyroidism 67
Subclinical hypothyroidism 67
Isolated hypothyroxinemia 67
Hyperthyroidism and gestational transient thyrotoxicosis 68

Thyroid autoimmunity in euthyroid women 68
Timing of intervention and clinical thresholds 68
Thyroid autoimmunity and reproductive outcomes 68
Epidemiology and reproductive risk 68
Thyroidal reserve and endocrine vulnerability 69
Immune dysregulation at the maternal–fetal interface 69
Evidence from interventional trials 69
Assisted reproduction and recurrent pregnancy loss 69
Preconception care and future directions 69
Rationale for preconception thyroid assessment 69
Preconception management strategies 70
Guideline perspectives: Convergence and divergence 70
Future directions 70
Mini-dictionary of terms 70
Summary points 71
Acknowledgments 71
References 71

Section 2
Focused areas of neurological pathology and physiology

7. Thyroid hormone transporters and the brain

Beatriz Muñoz-Falder and Soledad Bárez-López

Abbreviations 77
Introduction 77
MCT8 77
Structural and functional features of MCT8 77
Localization of MCT8 in the CNS 78
The pathophysiology of MCT8 mutations 80
OATP1C1 81
Structural and functional features of OATP1C1 81
Localization of OATP1C1 in the CNS 81
The pathophysiology of OATP1C1 mutations 83
Other TH transporters and partial redundancy 83
Contributions of TH transporters to TH brain availability 84
Identification of TH transporters at the brain barriers 84
Thyroid hormone transporters and brain development 86
Other hormones and their pathways 86
Applications of the material 87
Mini-dictionary of terms 87
Summary points 87
References 88

8. Thyroid hormones and neurodegenerative diseases

Mohammad Hossein Etemadi, Hanieh Shafizadeh, Mohammad Amin Karimi, Nazanin Mohammad Zadeh Kosari, Maryam Shafizadeh and Niloofar Deravi

List of abbreviations 91
Introduction 91
TH production, transport, and mechanisms of action 92
TH receptors and distribution in the brain 92
TH regulation and action in the brain 93
Sex-specific effects of TH in the brain 93
Oxidative stress, metabolism, aging, and TH-related neurodegeneration 93
Thyroid hormones in neurodegenerative and neuropsychiatric disorders 96
Thyroid function within the euthyroid range and cognitive risk 97
Overt hyperthyroidism and cognitive impairment 99
Subclinical hyperthyroidism and cognitive impairment 99
Overt hypothyroidism and dementia 99
Subclinical hypothyroidism and dementia 99
Thyroid dysfunction and Alzheimer's disease 101
Mini-dictionary of terms 102
Summary points 102
Disclosure of AI use 103
References 103

9. Thyroid hormones and Alzheimer's disease

Rakshit R. Sangolli, Ruturaj A. Kadu and Girdhari Lal Gupta

List of abbreviations 107
Introduction 107
Overview of Alzheimer's disease 107
Overview of thyroid hormones 107
Rationale for linking thyroid dysfunction with AD 108
Thyroid hormones: Biosynthesis, regulation, and function 108
Synthesis and secretion (T_3, T_4) 108
Hypothalamic-pituitary-thyroid axis 110
Roles in CNS development and function 110
Pathophysiology of Alzheimer's disease 111
Amyloid-β plaque formation 111
Tau hyperphosphorylation and neurofibrillary tangles 111
Neuroinflammation and oxidative stress 111
Neurodegeneration and cognitive decline 111

Thyroid dysfunction in aging and neurodegeneration 111
Hypothyroidism and subclinical hypothyroidism 111
Hyperthyroidism and subclinical hyperthyroidism 112
Epidemiological evidence linking thyroid disorders with AD 112
Mechanistic interplay between thyroid hormones and AD pathogenesis 112
Impact on amyloid-β precursor protein (APP) processing 112
Influence on tau phosphorylation 112
Thyroid hormones and mitochondrial function in neurons 112
Hormonal regulation of neuroinflammation and synaptic plasticity 112
Clinical and biomarker evidence 113
Thyroid hormone levels in AD patients 113
TSH, T_3, T_4 as potential biomarkers 113
Thyroid autoimmunity and Alzheimer's disease risk 113
Experimental and animal models 113
Transgenic AD models with altered thyroid function 114
Behavioral and cognitive assessments 114
Molecular insights from Preclinical studies 114
Therapeutic and preventive perspectives 115
Thyroid hormone replacement in AD 115
Challenges in hormonal therapy for neurodegeneration 115
Potential of modulating the HPT axis 116
Other hormones, endocrine tissues, and pathways: Translational perspectives 116
Hypothalamic–pituitary–adrenal (HPA) axis 116
Gonadal hormones and neuroprotection 116
Pancreatic hormones and brain metabolism 116
Adipose-derived hormones and inflammatory modulation 116
Gut–brain–endocrine axis 117
Integrated endocrine network in AD management 117
Current challenges and future directions 117
Gaps in knowledge 117
Future research strategies 117
Integrative approaches in neuroendocrinology 117
Conclusion 118
Applications of the material 118
Clinical screening and risk stratification 118
The process of creating biomarkers 118
Strategies for therapy 118
Interventions for prevention and lifestyle 118
Research that moves from the lab to the clinic and clinical trials 118
Personalized medicine strategies 119
Uses in education and training 119
Mini-dictionary 119
Summary points 119
References 119

10. Diseases of the thyroid gland – an overview

Stanisław Surma and Kazimierz Ciechanowski

List of abbreviations 123
Introduction 123
Hypothyroidism: Definitions and causes 124
Hypothyroidism: Clinical symptoms 125
Hypothyroidism: Diagnostic approach 126
Hypothyroidism: Treatment approach 127
Hyperthyroidism: Definitions and causes 127
Hyperthyroidism: Clinical symptoms 128
Hyperthyroidism: Diagnostic approach 129
Hyperthyroidism: Treatment 131
Management of thyroid storm 132
Management of thyroid-associated orbitopathy 132
Other thyroid diseases 133
Thyroid lesions 133
Thyroid cancers 134
Thyroiditis 135
Conclusions 136
Summary points 136
Mini-dictionary of terms 136
Other hormones, endocrine tissues, and pathways 136
Applications of the material 137
References 137

11. The thyroid hormone drives bidirectional communication along the brain-gut axis in vertebrate models

Dennis P. Jacob, Meenu Viswanath and M. C. Subhash Peter

Introduction 139
Brain-gut axis 139
TH and brain-gut axis 141
Other hormones, endocrine tissues, and pathways 142
TH and stress-responsive endocrine signaling 145
TR-R system and B-G axis 148
Applications of the material 149
Mini-dictionary terms 149
Summary Points 149
Acknowledgments 149
References 150

12. Scenarios of astrocyte dysfunction linked to thyroid hormone availability

Marina Guillén-Yunta, Víctor Valcárcel-Hernández and Ana Montero-Pedrazuela

List of abbreviations 155
Introduction 155
Astrocytes functions in health and disease 156
Astrocytes as mediators for thyroid hormone action in the CNS 156
Astrocytes are thyroid hormone target cells 157
Astrocyte defects associated to thyroid hormone alterations 158
Astrocyte functions in the neurovascular unit and blood-brain barrier 160
Neurovascular unit and blood-brain barrier alterations associated to astrocyte dysfunction 161
Astrocytes in neuroinflammation: Homeostatic role and reactive responses 162
Astrocyte dysfunction during neuroinflammation and thyroid hormone-related disorders 163
Other endocrine alterations 164
Other astroglial-like cells: Tanycytes and bergmann glia 164
Applications of the material: *Therapeutic perspectives* 165
Mini-dictionary of terms 165
Summary points 165
References 166

Section 3
Psychology and behavior as a consequence of changes in the endocrine status

13. Thyroid function and psychosis

Eloi Giné-Servén and Javier Labad

List of abbreviations 171
Introduction 171
Thyroid disease and psychosis 172
Hypothyroidism and myxedema psychosis 172
Hyperthyroidism and psychosis 172
Hashimoto's encephalopathy 172
Clinical implications and differential diagnosis 173
Thyroid abnormalities in schizophrenia and psychosis 173
Biological plausibility: Neurodevelopment and glia–neuron crosstalk 174
Cross-sectional and meta-analytic evidence on HPT-axis dysregulation 174
Prospective evidence and phenotype associations in early psychosis 174
Severity of psychopathology and dimensional correlates 175
Thyroid autoimmunity as an immunoendocrine marker 175
Longitudinal change in thyroid indices across the course of illness 175
Medication effects and other confounders 175
Clinical synthesis and forward look 176
Thyroid and cognition in psychosis 176
Cognitive dysfunction in thyroid disease 176
Thyroid hormones and cognition in psychosis 177
Potential mechanisms 177
Autoimmunity and cognition 178
Clinical implications 178
Therapeutic implications 178
Secondary psychoses 178
Adjunctive thyroid hormone therapy in primary psychoses 178
Risks and cautions 179
Future perspectives 179
Conclusions 179
Other hormones, endocrine tissues, and pathways 179
Hypothalamic–pituitary–adrenal (HPA) axis dysregulation 179
Gonadal steroids and sex differences 179
Metabolic hormones 180
Pineal hormones 180
Adipose and metabolic signaling 180
Integrative translational perspective 180
Mini-dictionary of terms 181
Summary points 181
Acknowledgments 181
References 182

14. Thyroid cancers and cognition

Nazan Dolu

Abbreviations 185
Introduction 185
Types of thyroid carcinoma 185
Methods used in the diagnosis of thyroid cancers 185
Methods used in thyroid cancer treatment 186
Thyroid cancers and cognition 186
Cognition before thyroid cancer treatment 186
The thyroidectomy effect on cognitive functions 187
RAI treatment effect on cognitive functions 188

TSH suppression therapy (ST) effect on cognitive functions 188
Systemic chemotherapy 190
Other hormones and pathways affecting cognition in thyroid cancer 190
Applications of the material 191
Mini-dictionary of terms 191
Summary points 191
Acknowledgments 191
References 191

15. The impact of thyroid hormones on learning and memory

Mohammad Rahmanian, Roshanak Ghorbani, Amirmohammad Jabbarpour, Mona Minadi, Mahdiyeh Nozad Varjovi and Andisheh Iranfar

Abbreviations 195
Introduction 195
Physiology and molecular basis 196
Developmental role in cognition 197
Adult brain: Learning, memory and neuroplasticity 199
Clinical implications 202
Other hormones, endocrine tissues, and pathways 205
Applications of the material 207
Mini-dictionary of terms 207
Summary points 208
Declaration of AI assistance 208
Conflicts of interest 208
References 208

16. Thyroid hormone signaling shaping fear memory: Implications for PTSD and other stress-related disorders

Olga Ponomareva, Radha Shah and Kerry J. Ressler

List of abbreviations 213
Introduction 213
Fear memory: Definitions and implications in disease 213
Thyroid signaling in fear learning: Insights from clinical studies 214
PTSD introduction and initial studies linking thyroid disorders 214
Findings from multicenter clinical studies 215
Emerging work: Thyroid signaling as a predictive PTSD biomarker 215
Thyroid signaling in fear learning: Insights from animal studies 215
Animal models of stress: Fear learning 215
Thyroid signaling in animal models of stress: Hippocampus 216
Thyroid signaling in animal models of stress: Amygdala 216
Other hormones and pathways 216
Applications of the material 217
Mini-dictionary of terms 217
Summary points 218
References 218
Further reading 220

Section 4
Treatment regimens modulating the thyroid status and effects on neurological tissues

17. Levothyroxine treatments in pregnancy and impact on cognition in offspring

Alice Batistuzzo and Miriam Oliveira Ribeiro

Abbreviations 223
Introduction 223
Thyroid organogenesis 223
Hormonal and metabolic adaptations affecting thyroid function during pregnancy 224
Hypothyroidism 225
Prevalence and clinical manifestations 226
Laboratory parameters of hypothyroidism in pregnancy 227
Treatment 227
Complications during pregnancy and effects on the offspring 228
Role of thyroid hormones in neurodevelopment 228
Complications in offspring resulting from gestational hypothyroidism in humans 228
Complications in offspring resulting from experimental gestational hypothyroidism in rodents 230
Possible strategies for stimulating development and cognition 231
Applications of the material 231
Clinical applications 231
Public health applications 231
Research applications 231
Conclusion 232
Mini-dictionary of terms 232
Summary Points 232
Use of AI acknowledgment 232
References 232

18. Nanomedicines with triiodothyronine and their effects on neurological tissues

Tanu Dixit, Sohini Chattopadhyay, Tan Thakur, Mansi Sawardekar and Selvan Ravindran

Introduction 235
Therapeutic potential and delivery of triiodothyronine in neurological applications 236
Role of triiodothyronine in brain development and function 236
Neurological disorders linked to T3 dysregulation 237
Pharmacokinetics and challenges in T3 delivery 237
Nanomedicine platforms for T3 delivery 238
Preclinical and clinical evidence 238
Applications of the material 239
Clinical potential of T3-loaded nanomedicines 239
Treatment of CNS hypothyroidism, brain injury, and demyelinating disorders 240
Future integration into personalized medicine and nanopharmacology 240
Mini-dictionary of terms 240
Summary points 241
References 241
Further reading 242

19. Hypothyroidism and improved spatial working memory

Satkarjit Kaur Jhandi and Ajaypal Singh

List of abbreviations 243
Introduction 243
Thyroid hormones and hypothyroidism 243
Hypothyroidism and working memory and neuropsychological symptoms 244
Levothyroxine (LT4) and spatial working memory 245
Screening of spatial working memory 246
Conclusion 246
Mini-dictionary of terms 246
Summary 246
References 247

20. Levothyroxine and effects on the brain: MRI and beyond

Sweta Sahu, Salma Younas, Muhammad Nafees, Soha Farooq, Rajkumar Rajendram and Victor R. Preedy

List of abbreviations 249
Introduction 249
Physiology of thyroid hormones and the brain 250
Normal thyroid hormone synthesis and regulation 250
Mechanisms of thyroid hormone action in the CNS 250
Key developmental and neurophysiological roles 251
Levothyroxine therapy: Pharmacology and clinical use 252
Pharmacokinetics and pharmacodynamics 252
Effects of levothyroxine on brain function 253
Cognitive outcomes 253
Mood and affective disorders 253
Neurodevelopmental implications (perinatal and pediatric) 255
Neurodegenerative disorders and the aging brain 255
Neuroimaging insights: MRI and advanced techniques 256
Structural MRI findings 256
Functional MRI (fMRI) studies 256
Diffusion tensor imaging (DTI) 257
MR spectroscopy 258
Beyond MRI: Multimodal and emerging techniques 258
Positron emission tomography (PET) and SPECT 258
Electroencephalography (EEG) and magnetoencephalography (MEG) 258
Biomarker studies (CSF, serum, and neurotrophic factors) 259
Clinical implications and case studies 259
Other hormones, endocrine tissues, and pathways 260
Applications of the material 261
Future directions 261
Conclusion 261
Mini-dictionary of terms 261
Summary points 262
References 262
Further reading 264

Section 5
Nonthyroid treatment regimens

21. Impact of epilepsy therapies on thyroid regulation: A narrative review

Anne Rochtus and Brigitte Decallonne

Abbreviations 269
Introduction 269

Regulation and metabolism of thyroid hormones 270
Thyroid hormone alterations in illness and hypothyroidism 270
The possible role of thyroid hormones in epilepsy 271
The impact of antiseizure medications and ketogenic diet on thyroid function 272
Valproate 273
Carbamazepine 274
Oxcarbazepine 274
Phenobarbital 275
Phenytoin 275
Lamotrigine 275
Gabapentin 275
Levetiracetam 276
Topiramate 276
Ketogenic diet 276
Polytherapy and intractable epilepsy 277
Practical guideline 277
Hypothyroxinemia in patients on ASMs 277
The clinical importance of mild hypothyroidism 277
Guidelines for monitoring and treatment 278
Monitoring recommendations 278
Conclusions 278
Mini-dictionary of terms 279
Summary points 280
Artificial intelligence 280
References 280

22. Hypothyroidism in patients treated with immune checkpoint inhibitors

N. Laghmouch, I. Meeusen and Th.B. Twickler

List of abbreviations 283
Introduction 283
ICI therapy, at a glance 283
Epidemiology of ICI-induced hypothyroidism 283
Prediction of hypothyroidism along with ICI therapy 284
Primary versus secondary hypothyroidism along with ICI therapy 285
Mechanisms of destructive thyroiditis related to ICI therapy 285
Screening for ICI-induced hypothyroidism 286
Treatment of ICI-induced hypothyroidism 287
Prognosis of ICI-induced hypothyroidism 287
Clinical approach 287
The clinical relevance of the diagnosis and treatment of ICI-induced hypothyroidism 287
Summary points 288
Conclusion 288
Mini dictionary 289
References 289

Section 6
Diseases of the thyroid or thyroid status and their impact on the brain and behavior

23. Focused diseases of the thyroid gland: Graves' hyperthyroidism

Michele Marinò and Elena Sabini

List of abbreviations 293
Mini-dictionary of terms 293
Summary points 294
Introduction 294
GH as a systemic endocrine disorder 294
Systemic consequences of hyperthyroidism 294
Neuroendocrine relevance of GH 294
Rationale and scope of the chapter 294
Epidemiology and clinical features of GH 294
Epidemiology 294
Clinical presentation 295
Disease management 295
Disease course and prognosis 295
Pathogenesis of GH 296
Autoantibody-mediated thyroid stimulation in GH 296
Adaptive immune dysregulation in GH 296
The TSH receptor–IGF-1 receptor signaling axis in GH 296
Fibroblast-driven tissue remodeling in extra-thyroidal manifestations of GH 296
GH and target tissues 297
Target tissues in GH 297
Skeletal and cardiac muscle 297
Pituitary gland 297
Liver 297
Bone 297
Thyroid gland 297
Skin, hair, and nails 297
Central nervous system 298

Effects of thyroid hormone excess on CNS 298
Thyroid hormone transport and local availability in the brain 298
Thyroid hormone receptors and regional brain vulnerability 298
TSH receptor signaling and autoimmune interactions in the brain 298
Structural, functional, and neurotransmitter alterations in the hyperthyroid brain 298
Neuropsychatric manifestation of GH 299
Other hormones and endocrine pathways 299
HPA axis and stress 299
HPG axis 300
Metabolic pathway 300
Somatotropin-IGF-1 axis 300
Applications of the material 301
Clinical implications 301
Research and translational perspectives 301
References 301

24. Graves' disease (autoimmune hyperthyroidism) and fatigue — understandable, identifiable, and manageable

Birgitta Johansson, Lars Rönnbäck and Helena Filipsson Nyström

Summary points 305
Understanding fatigue in Graves' disease 305
Identifying fatigue in Graves' disease 306
Managing fatigue in Graves' disease 308
References 308

Section 7
Diet and nutrition and their impact on the thyroid status and neuroscientific domains or behavior

25. The thyroid—brain axis in hyperthyroidism: Neurotoxic mechanisms and nutraceutical interventions

Mohammed A. Abdel-Rasol, Mariam A. Tammam, Yaser M. Hassan and Wael M. El-Sayed

List of abbreviations 313
Introduction 313
Neurological and cognitive effects of hyperthyroidism 313
Cognitive dysfunctions in hyperthyroidism 314
Mood and psychiatric disturbances 314
Congenital hyperthyroidism and neuro-developmental consequences 314
Structural and functional brain changes 316
Pathophysiological mechanisms of brain damage 316
Oxidative stress 316
Neuroinflammation 317
Excitotoxicity and neurotransmitter imbalance 317
Blood brain barrier disruption 318
Thyroid hormone receptor overactivation 319
Role of nutraceuticals in neuroprotection 320
Antioxidants 321
Vitamin E, vitamin C, and coenzyme Q10 321
Polyphenols 321
Anti-inflammatory agents 321
Adaptogens and neuroprotective herbs 321
Micronutrients 321
Gut—brain axis modulators 321
Limitations of nutraceuticals 321
Other hormones, endocrine tissues, and pathways 323
Crosstalk with the hypothalamic—pituitary—adrenal (HPA) axis 323
Interactions with sex hormones 323
Metabolic and pancreatic effects 323
Bone—endocrine interactions 323
Catecholaminergic and sympathetic pathways 323
Translational opportunities 324
Applications of the material 324
Clinical practice 324
Research and innovation 324
Public health and education 324
Therapeutic perspectives and future directions 324
Personalized nutraceutical strategies 324
Synergistic combinations 324
Vitamin E for mitochondrial protection 325
Gut—brain axis interventions 325
Microbiome modulation 325
Systems biology and AI-guided design 325
Biomarker-led trials 325
Summary points 325
Conclusion 325
Mini-dictionary of terms 326
Ethics approval 326
Availability of data and materials 326
Acknowledgments 326
Competing interests 326
Funding 326
Authors' contributions 326
References 326

26. Excess iodine intake and thyroid dysfunction

Hernando Vargas-Uricoechea

List of abbreviations 331
Introduction 331
Iodine metabolism 332
Iodine metabolism and thyroid hormone synthesis 332
Integrated physiological model between iodine metabolism and the thyroid hormones 335
Iodine-induced hyperthyroidism 335
Iodine-induced hypothyroidism 336
Excessive iodine intake and thyroid cancer 338
Amiodarone-induced thyroiditis (AIT) 339
Iodine intake and autoimmunity 340
Summary points 342
References 342

Section 8
Resources

27. Recommended resources for the study of hormones in neuroscience: The thyroid gland and thyroid hormones

Rajkumar Rajendram, M. Aadam Siraj, Vinood B. Patel and Victor R. Preedy

Introduction 347
Historical development of key mechanistic insights 347
Advancing the scientific fields of endocrinology and neuroscience 347
Other resources 354
Acknowledgments (in alphabetical order) 354
References 355

Index 357

Disclaimer

Every reasonable effort has been made to ensure that the data and information published are reliable. However, the publisher, editors, and authors cannot take responsibility for all of the materials' validity or any sequelae from their use. Neither the publishers nor the editors endorse any of the substances, treatments, interventions, protocols, or centers mentioned in the chapter contributions.

Contributors

Mohammed A. Abdel-Rasol, Department of Zoology, Faculty of Science, Ain Shams University, Cairo, Egypt

Alaleh Alizadeh, Faculty of Medicine, Mashhad Branch, Islamic Azad University, Mashhad, Iran

Saikrishna Ananthapadmanabhan, Department of Otolaryngology, Head and Neck Surgery, Nepean Hospital, Sydney, NSW, Australia

Naoko Arata, Women's Internal Medicine, Women's Health Center, National Center for Child Health and Development, Tokyo, Japan

Mojan Ayati, School of Medicine, Szeged University of Medical Sciences, Szeged, Hungary

Femi E. Ayeni, Department of Otolaryngology, Head and Neck Surgery, Nepean Hospital, Sydney, NSW, Australia; Nepean Institute of Academic Surgery, Nepean Clinical School, The University of Sydney, Sydney, NSW, Australia

Soledad Bárez-López, Instituto de Investigaciones Biomédicas Sols-Morreale, Consejo Superior de Investigaciones Científicas (CSIC), Universidad Autónoma de Madrid (UAM), Department of Neurological Diseases, Madrid, Spain

Alice Batistuzzo, Institute for Mind and Biology, Neuroscience Institute, Department of Psychology, University of Chicago, Chicago, IL, United States

Hoda Borooghani, School of Medicine, Iran University of Medical Sciences, Tehran, Iran

Sohini Chattopadhyay, Symbiosis School of Biological Sciences, Symbiosis International (Deemed University), Pune, Maharashtra, India

Kazimierz Ciechanowski, Department of Nephrology, Transplantology and Internal Medicine, Pomeranian Medical University of Szczecin, Szczecin, Poland

Brigitte Decallonne, Department of Endocrinology, University Hospital of Leuven, Leuven, Belgium

Bhuvnesh Deivakumar, Department of Otolaryngology, Head and Neck Surgery, Nepean Hospital, Sydney, NSW, Australia

Niloofar Deravi, School of Medicine, Shahid Beheshti University of Medical Sciences, Tehran, Iran

Tanu Dixit, Symbiosis School of Biological Sciences, Symbiosis International (Deemed University), Pune, Maharashtra, India

Nazan Dolu, Department of Physiology, School of Medicine, University of Istanbul Medipol, İstanbul, Türkiye

Wael M. El-Sayed, Department of Zoology, Faculty of Science, Ain Shams University, Cairo, Egypt

Maryam Eskandarpour, Department of General Practice, Dr Ozone Clinic, Dubai Healthcare City, Dubai, United Arab Emirates

Mohammad Hossein Etemadi, Student Research Committee, School of Medicine, Isfahan University of Medical Sciences, Isfahan, Iran

Soha Farooq, Department of Pharmacy, Shifa Tameer e Millat University, Islamabad, Pakistan

Roshanak Ghorbani, Universal Scientific Education and Research Network (USERN), Tehran, Iran

Eloi Giné-Servén, Department of Psychiatry, Hospital Universitario de Navarra, Pamplona, Spain; IdiSNA, Navarra Institute for Health Research, Pamplona, Spain

Marina Guillén-Yunta, Department of Neurological Diseases, Instituto de Investigaciones Biomédicas Sols-Morreale, Consejo Superior de Investigaciones Científicas-Universidad Autónoma de Madrid, Madrid, Spain; Department of Biosciences, Faculty of Biomedical and Health Sciences, Universidad Europea de Madrid, Madrid, Spain

Girdhari Lal Gupta, School of Pharmacy & Technology Management, SVKM's NMIMS Deemed-to-be University, Shirpur, Maharashtra, India

Asma Hallab, Department of Psychiatry and Psychotherapy, Otto-von-Guericke University of Medicine, University Hospital of Magdeburg, Magdeburg, Germany; Charité Universitätsmedizin Berlin, Freie Universitätsmedizin Berlin and Humboldt Universität Berlin, Berlin, Germany

Yaser M. Hassan, Biotechnology Program, Faculty of Science, Ain Shams University, Cairo, Egypt

Andisheh Iranfar, School of Pharmacy, Tehran University of Medical Sciences, Tehran, Iran

Amirmohammad Jabbarpour, Student Research Committee, Faculty of Medicine, Zanjan University of Medical Sciences, Zanjan, Iran

Dennis P. Jacob, Department of Medicine, Saint Francis Hospital, Hartford, CT, United States

Satkarjit Kaur Jhandi, Department of Physiotherapy, Khalsa College, Amritsar, Punjab, India

Birgitta Johansson, Institute of Neuroscience and Physiology, Department of Clinical Neuroscience, Sahlgrenska Academy, University of Gothenburg, Gothenburg, Sweden

Ruturaj A. Kadu, School of Pharmacy & Technology Management, SVKM's NMIMS Deemed-to-be University, Shirpur, Maharashtra, India

Mohammad Amin Karimi, School of Medicine, Shahid Beheshti University of Medical Sciences, Tehran, Iran

Javier Labad, Department of Mental Health and Addictions, Hospital Universitari de Mataró, Consorci Sanitari del Maresme, Universitat Autònoma de Barcelona, Mataró, Spain; Centro de Investigación biomédica en Red de Salud Mental (CIBERSAM), Instituto de Salud Carlos III, Madrid, Spain

N. Laghmouch, Antwerp University, Edegem, Belgium

Tiffany Mach, Department of Medicine, Endocrinology and Metabolism Residency Training Program, Montreal, QC, Canada

Michele Marinò, Department of Clinical and Experimental Medicine, Endocrinology Unit I, University of Pisa, Pisa, Italy

I. Meeusen, AZ Monica, Deurne, Belgium

Mona Minadi, Student Research Committee, Kermanshah University of Medical Sciences, Kermanshah, Iran

Nazanin Mohammad Zadeh Kosari, School of Medicine, Mashhad University of Medical Sciences, Mashhad, Iran

Ana Montero-Pedrazuela, Department of Neurological Diseases, Instituto de Investigaciones Biomédicas Sols-Morreale, Consejo Superior de Investigaciones Científicas-Universidad Autónoma de Madrid, Madrid, Spain

Beatriz Muñoz-Falder, Instituto de Investigaciones Biomédicas Sols-Morreale, Consejo Superior de Investigaciones Científicas (CSIC), Universidad Autónoma de Madrid (UAM), Department of Neurological Diseases, Madrid, Spain

Muhammad Nafees, Department of Pharmacy, University of Faisalabad, Faisalabad, Pakistan; Department of Pharmacy, Bahauddin Zakariya University, Multan, Pakistan

Ehsan Noroozi, Department of Internal Medicine, Shahid Labbafi Nezhad Hospital, Shahid Beheshti University of Medical Science, Tehran, Iran

Helena Filipsson Nyström, Institute of Medicine, Sahlgrenska Academy, University of Gothenburg, Gothenburg, Sweden; Department of Endocrinology, Sahlgrenska University Hospital, Gothenburg, Sweden; Wallenberg Center for Molecular and Translational Medicine, Gothenburg, Sweden; Gothenburg Centre for Person Centred Care, Gothenburg, Sweden

Vinood B. Patel, School of Life Sciences, University of Westminster, London, United Kingdom

M. C. Subhash Peter, Inter-University Centre for Evolutionary and Integrative Biology-iCEIB, School of Life Sciences, University of Kerala, Thiruvananthapuram, Kerala, India; Department of Zoology, University of Kerala, Thiruvananthapuram, Kerala, India; Sastrajeevan CIS-Life, F-17, Thiruvananthapuram, Kerala, India

Olga Ponomareva, McLean Hospital, Belmont, MA, United States

Victor R. Preedy, School of Life Course and Population Sciences, Faculty of Life Science and Medicine, King's College London, London, United Kingdom

Mohammad Rahmanian, Gastroenterology and Liver Diseases Research Center, Research Institute for Gastroenterology and Liver Diseases, Shahid Beheshti University of Medical Sciences, Tehran, Iran; Student Research Committee, School of Medicine, Shahid Beheshti University of Medical Sciences, Tehran, Iran

Rajkumar Rajendram, Research Unit, Department of Medical Education, College of Medicine, King Saud bin Abdulaziz University for Health Sciences, Riyadh, Saudi Arabia; Department of Medicine, King Abdulaziz Medical City, Ministry of National Guard Health Affairs, Riyadh, Saudi Arabia; Research Offices, King Abdullah International Medical Research Center, Riyadh, Saudi Arabia

Alireza Rashidi, Department of Pharmaceutics, School of Pharmacy and Novel Drug Delivery Systems Research Center, Isfahan University of Medical Sciences, Isfahan, Iran

Selvan Ravindran, Symbiosis School of Biological Sciences, Symbiosis International (Deemed University), Pune, Maharashtra, India

Kerry J. Ressler, McLean Hospital, Belmont, MA, United States

Miriam Oliveira Ribeiro, Human Developmental Sciences Graduate Program, Center for Biological Sciences and Health, Mackenzie Presbyterian University, Sao Paulo, Brazil

Faruque Riffat, Department of Otolaryngology, Head and Neck Surgery, Nepean Hospital, Sydney, NSW, Australia

Anne Rochtus, Department of Pediatrics, Section of Pediatric Endocrinology, University Hospital of Leuven, Leuven, Belgium

Lars Rönnbäck, Institute of Neuroscience and Physiology, Department of Clinical Neuroscience, Sahlgrenska Academy, University of Gothenburg, Gothenburg, Sweden

Elena Sabini, Department of Clinical and Experimental Medicine, Endocrinology Unit I, University of Pisa, Pisa, Italy

Sweta Sahu, Department of Medicine, JJM Medical College, Davanagere, Karnataka, India

Rakshit R. Sangolli, School of Pharmacy & Technology Management, SVKM's NMIMS Deemed-to-be University, Shirpur, Maharashtra, India

Akiko Sankoda, Women's Internal Medicine, Women's Health Center, National Center for Child Health and Development, Tokyo, Japan

Mansi Sawardekar, Symbiosis School of Biological Sciences, Symbiosis International (Deemed University), Pune, Maharashtra, India

Hanieh Shafizadeh, School of Medicine, Shahid Beheshti University of Medical Sciences, Tehran, Iran

Maryam Shafizadeh, School of Medicine, Iran University of Medical Sciences, Tehran, Iran

Radha Shah, University of Toledo, Toledo, OH, United States

Ajaypal Singh, Fortis Escort Hospital, Amritsar, Punjab, India

M. Aadam Siraj, Department of Clinical and Biomedical Sciences, University of Exeter Medical School, University of Exeter, Exeter, United Kingdom

Mark Smith, Department of Otolaryngology, Head and Neck Surgery, Nepean Hospital, Sydney, NSW, Australia

Niranjan Sritharan, Department of Otolaryngology, Head and Neck Surgery, Nepean Hospital, Sydney, NSW, Australia

Stanisław Surma, Department of Internal Medicine and Clinical Pharmacology, Medical University of Silesia Katowice, Poland

Mariam A. Tammam, Biotechnology Program, Faculty of Agriculture, Beni-Suef University, Beni-Suef, Egypt

Tan Thakur, Symbiosis School of Biological Sciences, Symbiosis International (Deemed University), Pune, Maharashtra, India

Th.B. Twickler, Antwerp University, Edegem, Belgium; AZ Monica, Deurne, Belgium

Víctor Valcárcel-Hernández, Laboratory Molecular Physiology and Adaptation, UMR 7221, Department Adaptations of Life, Muséum National d'Histoire Naturelle, Paris, France

Hernando Vargas-Uricoechea, Metabolic Diseases Study Group, Department of Internal Medicine, Universidad del Cauca, Popayán, Colombia

Mahdiyeh Nozad Varjovi, Student Research Committee, Faculty of Pharmacy, Tabriz University of Medical Sciences, Tabriz, Iran

Meenu Viswanath, Inter-University Centre for Evolutionary and Integrative Biology-iCEIB, School of Life Sciences, University of Kerala, Thiruvananthapuram, Kerala, India; Department of Zoology, University of Kerala, Thiruvananthapuram, Kerala, India

Natsuko Watanabe, Department of Internal Medicine, Ito Hospital, Tokyo, Japan

Ai Yoshihara, Department of Internal Medicine, Ito Hospital, Tokyo, Japan

Salma Younas, Department of Pharmacy, University of the Punjab, Lahore, Pakistan

Oriana Hoi Yun Yu, Department of Medicine, Division of Endocrinology and Metabolism, Jewish General Hospital, McGill University, Montreal, QC, Canada

Preface

The neuroendocrine system is a complex biological network of pathways that maintains the homeostatic milieu of neurological tissues. The importance of this system is highlighted by the numerous regulatory processes that modulate these pathways. These include the maintenance of circulating hormone levels, receptor concentrations, postreceptor responses, signaling pathways, and tissue effects. Of course, these simple abstractions belie the many layers of interactions among genes, epigenetic mechanisms, transcription, translation, hormonal turnover, feedback loops, and many other processes, which are interlinked with other hormones. Yet, diseases and surgical interventions may induce changes in neuroendocrine profiles that are adaptive and beneficial or maladaptive and harmful, causing further tissue and organ injury. Moreover, there are also connections among other modulators, such as development, aging, medications, diet, and nutrition. For example, dietary iodine is a component of thyroid hormones and iodine deficiency frequently occurs. Maternal iodine deficiency can cause neurological impairment in the offspring. Indeed, it has been argued that iodine deficiency is the biggest cause of mental retardation in the world. Some endocrine abnormalities also induce psychological disorders, such as anxiety and depression, and modulate the expression of emotions and behavior or affect the quality of life measures. One also needs to consider that some adaptive changes to the endocrine status are beneficial. For example, this occurs in trauma and critical care, where the mobilization of metabolic fuels is driven by endocrine stress responses.

To understand the links between hormones and neurological tissues, all the facets of homeostatic regulation must be considered. However, much of the available text that links hormones and neurological tissues is designed for the specialist or does not address the translational context. Moreover, much of the literature fails to address the behavioral consequences of altered endocrine gland pathology or the endocrine status. These deficiencies are addressed in the series *Hormones in Neuroscience*. While it embraces hormones and neurological tissues at its core, it also includes the causative and modulating influences as well as the consequential effects.

Hormones in Neuroscience: The Thyroid Gland and Thyroid Hormones adopts an illustrative ethos and has the following unique subsections:

- **Abstract** (published online)
- **Other Hormones, Endocrine Tissues, and Pathways**
- **Applications of the Material**
- **Mini-Dictionary of Terms**
- **Summary Points**

Other Hormones, Endocrine Tissues, and Pathways ensures that there is holistic coverage within the chapter, while the subsection **Applications of the Material** ensures that there is translational and cross-disciplinary information within the chapter. The **Mini-Dictionary of Terms** subsection is particularly useful for those who have expertise in other disciplines or who are novices to the area, and **Summary Points** encapsulates the entire chapter in succinct sentences, which is also useful for the novice.

In this volume, **The Thyroid Gland and Thyroid Hormones**, we have the following sections:

Section 1: Introductory chapters
Section 2: Focused areas of neurological pathology and physiology
Section 3: Psychology and behavior as a consequence of changes in the endocrine status
Section 4: Treatment regimens modulating the thyroid status and effects on neurological tissues
Section 5: Nonthyroid treatment regimens
Section 6: Diseases of the thyroid or thyroid status and their impact on the brain and behavior
Section 7: Diet and nutrition and their impact on the thyroid status and neuroscientific domains or behavior
Section 8: Resources

Hormones in Neuroscience: The Thyroid Gland and Thyroid Hormones is suitable for undergraduates entering university and beyond. Because of the translational and transdisciplinary nature of the book series, it will also be suitable for physicians, academics at the PhD level, researchers, university teachers, lecturers, and professors. Endocrinologists, pathologists, pharmacologists, healthcare professionals, neuroscientists, research scientists, behavioral scientists, psychologists, and psychiatrists will also benefit from this book. Clinical professionals working at the intersection of neurology and endocrinology will also find this useful, as will early career researchers, including postgraduate research students (e.g., Masters by research). Industrial heads or those leading biodiscovery groups in the pharmaceutical sector will also find these books very useful.

The editors
Professor Colin R. Martin
Professor Vinood B. Patel
Dr. Rajkumar Rajendram
Professor Victor R. Preedy

Section 1

Introductory chapters

Chapter 1

Thyroid hormones and brain structural, molecular, and functional correlates

Asma Hallab[1, 2]

[1]Department of Psychiatry and Psychotherapy, Otto-von-Guericke University of Medicine, University Hospital of Magdeburg, Magdeburg, Germany; [2]Charité Universitätsmedizin Berlin, Freie Universitätsmedizin Berlin and Humboldt Universität Berlin, Berlin, Germany

Introduction

The relationship between the brain and thyroid is fundamental and starts very early during the neurodevelopmental stages (Salas-Lucia, 2024). While the brain-located anterior pituitary gland controls thyroid function through the production, secretion, and regulation of thyroid stimulating hormone (TSH); thyroid hormones, namely free thyroxine (fT4) and free triiodothyronine (fT3), which are secreted by the thyroid gland and circulating in the blood, exert direct effects on brain structures and functions, depending on the developmental stages and underlying conditions (Endendijk et al., 2017; Ghassabian & Trasande, 2018). Thyroid hormones are particularly involved in regulating neurogenesis (Gothié et al., 2017), hippocampal synaptic plasticity (Salazar et al., 2017), brain-derived neurotrophic factor (BDNF) levels (Boustani et al., 2024), gliogenesis, and myelinations (Valcárcel-Hernández et al., 2022).

Variations in thyroid hormone levels within the normal reference ranges have not been associated with adverse neuropsychological or other health outcomes (Samuels et al., 2016a, 2016b). However, thyroid disorders are very common, with a global prevalence of overt hyperthyroidism ranging from 0.2%–1.4%, and subclinical hyperthyroidism from 0.7% to 1.4% (Lee & Pearce, 2023). Overt and subclinical hypothyroidism are even more frequent; as the prevalence of subclinical hypothyroidism alone might reach 10% of the adult population (Biondi et al., 2019). In some populations, depending on the geographic location and genetic predisposition, these rates might be up to two-fold higher (Jessy et al., 2024; Juárez-Cedillo et al., 2017).

A meta-analytical study found that approximately one-third of patients with hypothyroidism have neurocognitive impairment (Pankowski & Wytrychiewicz-Pankowska, 2025). In the Study in Aging and Dementia in Mexico (SADEM), overt hypothyroidism was associated with a 26% increase in the odds of cognitive impairment (Juárez-Cedillo et al., 2017). Findings from the Brazilian Longitudinal Study of Adult Health (ELSA-Brasil) suggested that, among middle-aged adults without overt thyroid disorders, low TSH levels were cross-sectionally associated with poorer executive function (Szlejf et al., 2018). On a longitudinal scale, although baseline TSH, fT3, and fT4 levels did not predict incident cognitive impairment, changes in fT4 levels over time - whether decrease or increase - were associated with cognitive decline in different domains (Gomes Gonçalves et al., 2024). These findings are consistent with another study in 85-year-old participants, in which only the longitudinal decrease in TSH levels predicted prospective cognitive decline (Gan et al., 2021).

Findings from the Rotterdam Study suggested an increased risk of dementia in individuals with higher and high-normal thyroid function (Chaker et al., 2016). This pattern was replicated in subsequent studies, in which both endogenous and iatrogenic thyrotoxicosis were associated with a higher risk of cognitive disorders (Adams et al., 2023).

Thyroid hormone levels might interact with underlying neuropsychological conditions, reflecting an interplay between the overall symptom severity and clinical outcomes. In patients with early-onset psychosis, fT4 levels were significantly associated with the global cognitive

Hormones in Neuroscience. https://doi.org/10.1016/B978-0-443-30261-9.00001-2

performance in those with mild to extreme psychosis severity (Saglam et al., 2024).

In a large multicohort study, thyroid dysfunction did not show any significant association with the global cognitive status (van Vliet et al., 2021). Other studies have also failed to demonstrate significant associations between thyroid dysfunction, particularly hypothyroidism, and cognitive impairment (Ojala et al., 2016; Pyun et al., 2022).

The conflicting results reported across different published studies might be attributable to the heterogeneity in the underlying populations (Sinha et al., 2024). One important biological confounder – sex – has been rarely explored in larger studies, yet it is attracting increasing attention in recent works, as the effects of thyroid hormones on brain structures and functions appear to be modulated in a sex-specific manner (Baksi & Pradhan, 2021; Endendijk et al., 2017; Winkler et al., 2016). A large population-based study showed that lower TSH levels are associated with informant-perceived anxiety symptoms in non-depressed middle-aged and older males but not females (Hallab, 2025a). Furthermore, in the same non-demented cohort, lower TSH levels were associated with poorer cognitive performance and higher odds of being diagnosed with concomitant mild cognitive impairment (Hallab, 2025b). Together, these studies highlighted the importance of the non-linearity and sex in the relationship between TSH levels and neuropsychological outcomes (Hallab, 2025a, 2025b).

The aim of this chapter is to review the current evidence on the association between thyroid hormones and brain structural, metabolic, and functional correlates, based on findings from recent neuroimaging studies using structural, functional, diffusion magnetic resonance imaging (MRI), MR-spectroscopy, as well as positron emission tomography (PET) techniques.

Graves' disease and Hashimoto's thyroiditis are specific forms of thyroid dysfunction, primarily related to autoimmune disorders (Burch & Cooper, 2015; Valencia-Sanchez et al., 2021), and will not be specifically covered in this review, beyond their contribution to hypo- and hyperthyroidism. This exclusion is mainly due to the complexity of their interactions with brain structures, which involve intricate autoimmune, hormonal, metabolic, genetic, and trans-synaptic mechanisms, increasing the likelihood of associated neurodegeneration and psychiatric disorders (Matsunaga et al., 2019; Naicker et al., 2019; Piekiełko-Witkowska et al., 2025; Wang et al., 2021; Wei et al., 2024; Yu et al., 2025). Similarly, the neuroimaging of thyroid-associated ophthalmopathy lies beyond the scope of this chapter (Zhang et al., 2024).

Thyroid hormones and brain: Insights from structural neuroimaging studies

Thyroid hormones and brain structures: Gray matter analysis

Hypothyroidism and brain structures

Compared to healthy controls, patients with subclinical hypothyroidism showed significantly lower gray matter volumes in the left precentral and postcentral gyri, as well as the left frontal orbital gyrus. On the right side, comparable differences were observed in the superior frontal gyrus, inferior occipital gyrus, and upper temporal pole (Zhang et al., 2021). In patients with subclinical hypothyroidism and comorbid depression, the total gray matter volume was significantly lower than that of healthy controls. The volume of the right middle frontal gyrus, in particular, was positively correlated with fT3 levels and cognitive scores (Digit Span Backward), implying an interplay between thyroid hormones, brain structures, mood, and cognition (Zhao et al., 2023). Untreated adult-onset overt hypothyroidism is generally associated with higher comorbid depression, anxiety, and cognitive impairment profiles. On a structural level, more gray matter regions seem to be involved. Compared to healthy controls, patients with overt hypothyroidism showed lower gray matter volumes in the orbital part of the right superior frontal gyrus, right superior temporal gyrus, left middle frontal gyrus, left dorsolateral superior frontal gyrus, and left supplementary motor area (Su et al., 2023). The bilateral cerebellar Crus I and left precentral gyrus had greater volumes in patients with overt hypothyroidism compared to healthy controls (Su et al., 2023). On a large population level, hypothyroidism was associated with lower total cerebellum and pallidum volumes (Chambers et al., 2021).

Hyperthyroidism and brain structures

Hyperthyroid patients showed reduced cortical gray matter volume and sulcal shallowing in the left hemisphere compared to healthy controls, whereas neither hypothalamic subfields nor limbic volumes showed significant differences between groups (Genç et al., 2024). In a bilateral volumetric analysis of a large cohort of non-demented middle-aged and older adults, and after excluding cases with TSH over 10 μIU/L, lower TSH levels were associated with lower gray matter volumes in the left hippocampus and left entorhinal cortex only in those with MCI, while no significant associations were observed between TSH and amygdala volumes (Hallab, 2026). In patients with Alzheimer's disease, higher thyroid hormone levels - particularly fT3 - showed a significant association with a greater annual decrease in both left and right hippocampal volumes. These findings contrasted with those observed in healthy controls and those with stable MCI, which demonstrated a longitudinal stability in their bilateral hippocampal volumes (Quinlan et al., 2022).

In premenopausal women with Graves' disease, amygdala and hippocampal volumes were bilaterally smaller compared to healthy controls. After treatment, these volumes increased significantly in size, reaching comparable sizes between groups, except for the left amygdala, which remained significantly smaller in the affected group. Only variations in fT3 levels correlated with the longitudinal differences in left-sided structure volumes (Holmberg et al., 2022).

Thyroid function and brain structures in psychiatric disorders

Patients with major depressive disorder exhibited lower serum TSH levels and higher cognitive impairment. These associations were mediated by reduced gray matter volume in the middle frontal gyrus, particularly in the right brain hemisphere (Zhao et al., 2021). Furthermore, among male patients with schizophrenia, those displaying aggressive behavior had higher TSH levels and lower brain volumes, especially in the left frontal pole, with a significant association between these two variables (Yu et al., 2024). In addition, in drug-naïve first episode psychosis patients, higher TSH levels were significantly associated with lower hippocampal volumes (Toll et al., 2023).

Thyroid function and small vessel disease

A few studies have examined the association between thyroid function and cerebral small vessel disease, but their findings have been inconsistent. Radiological markers of small vessel disease include confluent white matter hyperintensities, enlarged perivascular spaces, microbleeds, and lacunas. While subclinical hypothyroidism has been significantly associated with higher loads of small vessel disease biomarkers in stroke-free individuals (Chu et al., 2022) and in patients with minor stroke (Zhang et al., 2017), Mendelian randomization studies have not shown any causal relationship between cerebral small vessel disease and thyroid dysfunction (Tian et al., 2023).

Maternal thyroid dysfunction and the offspring's brain structures

Children born to women with gestational hypothyroidism show region-specific alterations in cortical thickness compared with those born to euthyroid mothers. Depending on the anatomical structure and hemisphere, both cortical thickening and thinning have been reported, and these structural changes correlate with the severity and timing of the maternal hypothyroidism (Lischinsky et al., 2016). In a large population-based study, maternal TSH levels during pregnancy showed a nonlinear association with both cortical surface area and gyrification on MRI in the offspring, particularly in the rostral middle frontal region (Mulder et al., 2025). In contrast with the previous studies, cortical thickness was not associated with maternal TSH or fT4 levels. At the subcortical level, only the caudate volume of the explored children showed a significant nonlinear association with the maternal fT4 during pregnancy. These results suggest that lower brain volumes are associated with both higher and lower extremes of maternal thyroid function during pregnancy (Mulder et al., 2025). Another study with a small sample size showed negative results in the association between maternal thyroid function during pregnancy and brain volumetric measures in the offspring (Scholz et al., 2025). However, within the same population, children expressed impaired white matter integrity in the inferior longitudinal fasciculus in association with maternal untreated suboptimal gestational thyroid function (McNabb et al., 2025).

Congenital hypothyroidism and brain structures

Both cortical thickening and cortical thinning across different cortical regions have been described in children with congenital hypothyroidism compared with typically developing controls. These structural anomalies correlate with the severity of hormone deficiency and might partly explain the unfavorable neurocognitive outcomes (Clairman et al., 2015). Abnormal cortical migration patterns observed in hypothyroid rodent models might help account for these findings in children with congenital hypothyroidism, as some regions might receive more neurons and others less than in physiological conditions (Cuevas et al., 2005).

At the subcortical level, children with congenital hypothyroidism exhibited reduced left hippocampal volume that did not correlate with age, in contrast to findings in healthy controls (Wheeler et al., 2011).

Thyroid function, interventions, and cerebral plasticity

An iatrogenically induced hyperthyroidism following an 8-week continuous administration of fT4 was associated, at the structural level, with an increase in gray matter volume in the left posterior cerebellum (lobule VI) and a decrease in gray matter volume in the bilateral visual cortex and the anterior cerebellum (lobules I-IV). At the neurocognitive level, only performance on the Attention Network Task was partially affected (Göbel et al., 2015).

The administration of glucocorticoid therapy in patients with thyroid-associated ophthalmopathy was associated with increased gray matter volumes in the left anterior cingulate gyrus and the medial superior frontal gyrus. The latter region had a significantly lower volume compared with healthy controls (Wu et al., 2024).

Thyroid hormones and brain structures: White matter analysis

Diffusion tensor imaging was used to assess the white matter integrity. In a large population-based study, associations between thyroid function in untreated individuals and diffusion tensor imaging biomarkers showed age-specific patterns. In older participants, higher fT4 values were associated with reduced white matter integrity (reflected by lower fractional anisotropy and higher mean diffusivity). In contrast, in younger participants, higher fT4 levels were associated with increased white matter integrity (Chaker et al., 2018).

Disruption of white matter integrity in relation to thyroid dysfunction has been reported across different age groups. In children with permanent congenital hypothyroidism, impaired white matter integrity has been associated with neurocognitive deficits, particularly lower mathematical scores (Perri et al., 2021). These white matter abnormalities were characterized by lower fractional anisotropy and radial diffusivity, and higher mean diffusivity and axial diffusivity, most prominently affecting the superior longitudinal fasciculi and uncinate fasciculi - major associative tracts - as well as motor tracts, namely corona radiata and internal capsules (Perri et al., 2021). Additional white matter tracts were affected in children with lower Intelligence Quotient and Perceptional Reasoning Index scores, including the cortico-spinal tracts, inferior longitudinal fasciculus, pontine crossing tracts, and medial lemniscus (Perri et al., 2021).

In subjects with adult-onset hypothyroidism, various abnormalities were observed in the morphology (and connectivity) of the anterior part of the corpus callosum, depending on whether the thyroid dysfunction was overt or subclinical (Chen et al., 2024).

Thyroid hormones and brain: Insights from functional neuroimaging studies

Hypothyroidism and functional brain changes

During the Stroop task-based functional MRI, patients with subclinical hypothyroidism showed impaired attention, accompanied by reduced activation in the dorsolateral prefrontal cortex and anterior cingulate cortex; regions primarily involved in attentional and cognitive control, respectively (Yin et al., 2021).

Resting-state MRI studies in patients with overt hypothyroidism have shown reduced functional connectivity across several networks compared with healthy controls (Singh et al., 2015). Affected networks included the right frontoparietal network, the medial visual network (lateral occipital gyrus, precuneus cortex, and cuneus), and the motor network (pre-/postcentral gyri, precuneus cortex, paracingulate gyrus, cingulate gyrus, and supramarginal gyrus). In contrast, neither the default mode network nor the left frontoparietal network showed differences between groups (Singh et al., 2015).

Hyperthyroidism and functional brain changes

Several functional MRI studies were dedicated to patients with thyroid-associated ophthalmopathy (Chen et al., 2021; Jiang et al., 2023; Liu et al., 2023, 2024). However, very few references have specifically examined the association between hyperthyroidism and brain activity using functional neuroimaging.

Hyperthyroid patients exhibited distinct patterns of abnormal resting-state functional connectivity within and between the salience network, default mode network, and executive control network, as revealed by seed-based analyses. While the internetwork connectivity between the default mode network and salience network was stronger in hyperthyroid patients compared to controls, the intranetwork connectivity within their salience network and executive control network was decreased (Liu et al., 2020). These directional imbalances suggest that hyperthyroidism might contribute to the disruption of salience network - default mode network anti-correlation, potentially driving attentional and cognitive symptoms, with insula hubs emerging as key reversible targets after treatment (Liu et al., 2020).

Task-based functional MRI was used to investigate the effects of induced thyrotoxicosis on memory and brain connectivity. Short-term thyroid hormone administration was associated with increased regional brain activity related to task difficulty and cognitive performance (Göbel et al., 2016). However, these results lacked robustness, and additional longitudinal interventional studies are required.

Thyroid hormones and brain: Insights from molecular neuroimaging studies

Thyroid function and glucose metabolism

A meta-analytical review of 18F-Fluorodeoxyglucose-Positron emission tomography (FDG-PET) studies in patients with thyroid dysfunction found that both hypothyroidism and hyperthyroidism might be associated with reduced regional perfusion and metabolic activity when compared to healthy controls (Pak et al., 2020). Regions of particular interest in hypothyroidism were the right inferior occipital gyrus and left cuneus, in addition to the left anterior cingulate. In patients with hyperthyroidism, lower FDG-uptake was observed in the right superior frontal gyrus compared to controls (Pak et al., 2020). Among patients who underwent thyroidectomy, those experiencing hypothyroidism without replacement therapy exhibit lower FDG-uptake in the right cuneus and right inferior occipital gyrus, in contrast with those achieving an euthyroid status under replacement therapy. Lower regional cerebral metabolic rates of glucose in the right cuneus were significantly associated with higher TSH levels (Jeong et al., 2017).

In non-demented patients, only in those with pathological amyloid-PET scans (PiB positive), higher TSH levels were associated with higher FDG-uptake in the frontal and temporoparietal lobes. In the same group, higher fT3 levels were associated with lower FDG-uptake in the temporoparietal lobe and posterior cingulate (Goto et al., 2022).

In patients with Alzheimer's disease, however, TSH levels and FDG-uptake were significantly correlated, particularly in the left frontal lobe and bilaterally in the anterior cingulate cortex. No significant correlations were found between fT3 or fT4 levels and regional cerebral glucose metabolism (Chiaravalloti et al., 2017).

The use of supraphysiological doses of L-T4 as an adjuvant treatment in patients with bipolar depression resulted in clinical improvement in depression scores and a reduction in the metabolic activity within the anterior limbic network, specifically in the left thalamus, left ventral striatum, right amygdala, right hippocampus, and right dorsal striatum (Bauer et al., 2016).

Thyroid function and alzheimer's disease biomarkers

Several studies have explored the potential association between thyroid hormone levels and an increased risk of Alzheimer's pathology in the brain, using either cerebrospinal fluid or plasma biomarkers (Chiaravalloti et al., 2018; Choi et al., 2017, 2020; Kimura et al., 2025; Quinlan et al., 2020). However, only one recent study has investigated this association specifically using advanced amyloid and tau PET neuroimaging modalities (Byeon et al., 2024). In particular, fT3 and fT4 levels appear to modulate the amyloid β-associated tau deposition in the brain, with stronger positive associations between concomitant amyloid β and tau uptake observed in individuals with higher thyroid hormone levels (Byeon et al., 2024).

Other neuroimaging studies

A few proton-magnetic resonance spectroscopy (^{1}H-MRS)-based studies have evaluated regional brain metabolism in association with thyroid function. Patients with hypothyroidism showed higher levels of choline compounds compared to healthy controls (Kumar et al., 2025). Significant differences were observed in the dorsolateral prefrontal cortex and posterior parietal cortex. Notably, choline levels in these regions did not improve after thyroxine supplementation and normalization of the thyroid function. Only in the posterior parietal cortex, significant correlations between TSH and choline levels were observed (Kumar et al., 2025). Choline levels in the posterior parietal cortex and dorsolateral prefrontal cortex showed a significant negative correlation with fT4 and were also correlated with worse cognitive scores in hypothyroid patients (Kumar et al., 2025). On the hippocampus level, patients with hypothyroidism showed lower glutamate/total creatine and myo-inositol/total creatine ratios compared to healthy controls; these alterations persisted even after treatment and clinical normalization of the thyroid function (Singh et al., 2016). In patients with first-episode major depressive disorder, serum TSH levels were significantly lower than in healthy controls, and the N-acetylaspartate/creatine ratio in the white matter of the right prefrontal lobe correlated positively with serum TSH levels (Jia et al., 2015).

In patients with thyroid eye disease, Choline and Glutamine levels in the optic radiation were higher than in healthy controls (Luo et al., 2023). Although ^{1}H-MRS did not reveal significant between-group differences in Glutamate levels, the use of Chemical Exchange Saturation Transfer (CEST) imaging detected higher Glutamate levels associated with thyroid disorder status compared to healthy controls (Luo et al., 2023).

Quantitative susceptibility mapping (QSM) is a noninvasive MRI technique with high sensitivity and specificity for visualizing and quantifying brain iron levels. In patients with thyroid-associated ophthalmopathy, QSM values were found to be increased in the bilateral caudate nucleus, left thalamus, left cuneus, left precuneus, right insula, and right middle frontal gyrus, which may partly account for the observed pathological cognitive scores (Hu et al., 2023).

Applications of the material

The available evidence is in favor of a lifelong active interplay between thyroid hormones and brain structure and function. While neonatal screening strategies have long been integrated into clinical practice to reduce risks associated with congenital hypothyroidism, current findings highlight the importance of ruling out thyroid dysfunction in patients with neurocognitive and psychiatric manifestations, particularly in advanced ages. Patients with lower thyroid hormone levels might significantly benefit from a hormone supplementation therapy. This might include associated improvement in depression symptoms, suicidality, and brain fog. Patients under thyroid hormone therapy might need regular monitoring to prevent iatrogenic effects associating hyperthyroidism to cognitive impairment and anxiety symptoms. Future research needs to further explore the association between thyroid hormones and Alzheimer's disease. In addition to the preventive aspect associated with maintaining thyroid hormones within age-adapted ranges, understanding these interactions might promote complementary therapeutic strategies to mitigate the modifiable risk factors of dementia. Furthermore, explaining thyroid-brain interactions might be a valuable educational tool to enhance the awareness of patients toward the neuropsychological symptoms associated with thyroid dysfunction and promote their adherence to therapeutic options.

Researchers are encouraged to refine and validate advanced imaging methods and analytical strategies to optimize the sensitivity and specificity of neuroimaging tools for detecting brain structural and functional changes associated with variations in thyroid hormone levels. This includes improving spatial and temporal resolution, integrating multimodal techniques such as structural MRI, functional MRI, diffusion imaging, and PET, and applying sophisticated quantitative approaches, including machine learning and network analysis, to better characterize subtle thyroid-related alterations in neural circuits. By tailoring imaging protocols to endocrine and developmental variables, and by carefully correlating imaging findings with hormonal profiles and cognitive or behavioral measures, these efforts might clarify brain–thyroid relationships and ultimately support earlier diagnosis, more precise risk stratification, and targeted therapeutic interventions.

Finally, this chapter highlights the need for a multidisciplinary collaborative approach between neurologists, psychiatrists, endocrinologists, pediatricians/geriatricians, and primary care providers to optimize the management of patients at risk.

Conclusions

There is increasing evidence of an association between thyroid function and the structural, metabolic, and functional organization of the brain. However, most published studies were of a cross-sectional observational nature, and only a limited number had a longitudinal design, allowing an objective assessment of dynamic changes associated with normal versus pathological thyroid function. Randomized clinical trials are scarce, limiting the interpretability of the current findings in causal frameworks. Most of the studies included a small number of patients, increasing the heterogeneity of the results and limiting the generalizability of the findings. It is therefore important to interpret the findings of the current evidence with caution and critically question the studies' limitations and generalizability. Larger population-based and interventional studies using neuroimaging follow-up are needed to tackle these limitations and allow a better interpretation of potential causal associations between thyroid hormones and brain structures.

Mini-dictionary of terms

Structural MRI: A noninvasive MRI technique that produces high-resolution, static images of brain anatomy, allowing the differentiation between gray matter, white matter, and cerebrospinal fluid. Advanced processing allows the measurement of volume, thickness, and shape of brain structures.

Diffusion MRI: An MRI technique that measures the diffusion of water molecules in tissue. Advanced processing gives information on white matter microstructure and the structural connectivity of fiber tracts in the brain.

Task-based functional MRI: A functional MRI method that measures changes in brain activity while a person performs a specific task, identifying regions whose blood flow and activation proportionally increase in response to that task.

Resting-state MRI: A form of functional MRI acquired while the person is not performing any task, used to assess spontaneous brain activity and functional connectivity between regions at rest.

18F-Fluorodeoxyglucose-Positron emission tomography (FDG-PET): A PET imaging technique based on the administration of a radioactive glucose analog (18F-FDG) and quantification of its uptake to estimate regional brain glucose metabolism.

Amyloid-PET and Tau-PET: Amyloid-PET uses tracers to detect beta-amyloid plaques; Tau-PET visualizes neurofibrillary tau tangles.

Proton-magnetic resonance spectroscopy (1H-MRS): A non-invasive MRI-based method to measure brain metabolite concentrations (e.g., choline, creatine); surrogate biomarkers of neuronal integrity and pathology.

Quantitative susceptibility mapping (QSM): is a non-invasive MRI technique with high sensitivity and specificity for visualizing and quantifying brain iron levels.

Summary points

- Thyroid dysfunction might be associated with impaired cognitive function and concomitant changes in the brain architecture.

- Both hyper- and hypothyroidism might be associated with lower cerebral gray matter volume, particularly in limbic structures.
- Thyroid dysfunction might be associated with impaired white matter integrity.
- Both hyper- and hypothyroidism might be associated with impaired regional perfusion and metabolic activity.
- Thyroid dysfunction might be associated with dysregulation of the functional cerebral connectivity and task-related regional activation.
- Advanced neuroimaging methods, exploring amyloid, tau, and iron deposition, are gaining increasing interest in the thyroid-brain field.

References

Adams, R., Oh, E. S., Yasar, S., Lyketsos, C. G., & Mammen, J. S. (2023). Endogenous and exogenous thyrotoxicosis and risk of incident cognitive disorders in older adults. *JAMA Internal Medicine, 183*(12), 1324–1331.

Baksi, S., & Pradhan, A. (2021). Thyroid hormone: Sex-dependent role in nervous system regulation and disease. *Biology of Sex Differences, 12*(1), 25.

Bauer, M., Berman, S., Stamm, T., Plotkin, M., Adli, M., Pilhatsch, M., et al. (2016). Levothyroxine effects on depressive symptoms and limbic glucose metabolism in bipolar disorder: A randomized, placebo-controlled positron emission tomography study. *Molecular Psychiatry, 21*(2), 229–236.

Biondi, B., Cappola, A. R., & Cooper, D. S. (2019). Subclinical hypothyroidism: A review. *Journal of the American Medical Association (JAMA), 322*(2), 153–160.

Boustani, A., Rashidy-Pour, A., Bozorgi, H., Vafaei, A. A., & Raise-Abdullahi, P. (2024). Mild exercise plus levothyroxine ameliorate deficits of spatial navigation, anxiety profile, and hippocampal BDNF in hypothyroid male offspring rats. *Brain and behavior, 14*(7), Article e3614.

Burch, H. B., & Cooper, D. S. (2015). Management of graves disease: A review. *Journal of the American Medical Association (JAMA)., 314*(23), 2544–2554.

Byeon, J. H., Byun, M. S., Yi, D., Jung, J. H., Sohn, B. K., Chang, Y. Y., et al. (2024). Moderation of thyroid hormones for the relationship between amyloid and tau pathology. *Alzheimer's Research & Therapy, 16*(1), 164.

Chaker, L., Cremers, L. G. M., Korevaar, T. I. M., de Groot, M., Dehghan, A., Franco, O. H., et al. (2018). Age-dependent association of thyroid function with brain morphology and microstructural organization: Evidence from brain imaging. *Neurobiology of Aging, 61*, 44–51.

Chaker, L., Wolters, F. J., Bos, D., Korevaar, T. I., Hofman, A., van der Lugt, A., et al. (2016). Thyroid function and the risk of dementia: The rotterdam study. *Neurology, 87*(16), 1688–1695.

Chambers, T., Anney, R., Taylor, P. N., Teumer, A., Peeters, R. P., Medici, M., et al. (2021). Effects of thyroid status on regional brain volumes: A diagnostic and genetic imaging study in UK biobank. *The Journal of Clinical Endocrinology and Metabolism, 106*(3), 688–696.

Chen, C., Cao, J., Zhang, T., Zhang, H., Shi, Q., Li, X., et al. (2024). Alterations in corpus callosum subregions morphology and functional connectivity in patients with adult-onset hypothyroidism. *Brain Research, 1840*, Article 149110.

Chen, W., Wu, Q., Chen, L., Zhou, J., Chen, H. H., Xu, X. Q., et al. (2021). Aberrant brain voxel-wise resting state fMRI in patients with thyroid-associated ophthalmopathy. *Journal of Neuroimaging, 31*(4), 773–783.

Chiaravalloti, A., Barbagallo, G., Ricci, M., Martorana, A., Ursini, F., Sannino, P., et al. (2018). Brain metabolic correlates of CSF tau protein in a large cohort of alzheimer's disease patients: A CSF and FDG PET study. *Brain Research, 1678*, 116–122.

Chiaravalloti, A., Ursini, F., Fiorentini, A., Barbagallo, G., Martorana, A., Koch, G., et al. (2017). Functional correlates of TSH, fT3 and fT4 in alzheimer disease: A F-18 FDG PET/CT study. *Scientific Reports, 7*(1), 6220.

Choi, H. J., Byun, M. S., Yi, D., Sohn, B. K., Lee, J. H., Lee, J. Y., et al. (2017). Associations of thyroid hormone serum levels with in-vivo alzheimer's disease pathologies. *Alzheimer's Research & Therapy, 9*(1), 64.

Choi, B. W., Kim, S., Kang, S., Won, K. S., Yi, H. A., & Kim, H. W. (2020). Relationship between thyroid hormone levels and the pathology of alzheimer's disease in euthyroid subjects. *Thyroid, 30*(11), 1547–1555.

Chu, M., Cai, Y., Zhong, J., Qian, Y., Cen, Y., Dou, M., et al. (2022). Subclinical hypothyroidism is associated with basal ganglia enlarged perivascular spaces and overall cerebral small vessel disease load. *Quantitative Imaging in Medicine and Surgery, 12*(2), 1475–1483.

Clairman, H., Skocic, J., Lischinsky, J. E., & Rovet, J. (2015). Do children with congenital hypothyroidism exhibit abnormal cortical morphology? *Pediatric Research, 78*(3), 286–297.

Cuevas, E., Ausó, E., Telefont, M., Morreale de Escobar, G., Sotelo, C., & Berbel, P. (2005). Transient maternal hypothyroxinemia at onset of corticogenesis alters tangential migration of medial ganglionic eminence-derived neurons. *European Journal of Neuroscience, 22*(3), 541–551.

Endendijk, J. J., Wijnen, H. A. A., Pop, V. J. M., & van Baar, A. L. (2017). Maternal thyroid hormone trajectories during pregnancy and child behavioral problems. *Hormones and Behavior, 94*, 84–92.

Göbel, A., Heldmann, M., Göttlich, M., Dirk, A.-L., Brabant, G., & Münte, T. F. (2015). Effect of experimental thyrotoxicosis on brain gray matter: A voxel-based morphometry study. *European Thyroid Journal, 4*(1_Suppl. 1), 113–118.

Göbel, A., Heldmann, M., Göttlich, M., Dirk, A. L., Brabant, G., & Münte, T. F. (2016). Effect of mild thyrotoxicosis on performance and brain activations in a working memory task. *PLoS One, 11*(8), Article e0161552.

Gan, E. H., Jagger, C., Yadegarfar, M. E., Duncan, R., & Pearce, S. H. (2021). Changes in serum thyroid function predict cognitive decline in the very old: Longitudinal findings from the newcastle 85+ study. *Thyroid, 31*(8), 1182–1191.

Genç, B., Aslan, K., Avcı, U., İncesu, L., & Günbey, H. P. (2024). Opposing effects of thyroid hormones on hypothalamic subunits and limbic structures in hyperthyroidism patients: A comprehensive volumetric study. *Journal of Neuroendocrinology, 36*(3), Article e13369.

Ghassabian, A., & Trasande, L. (2018). Disruption in thyroid signaling pathway: A mechanism for the effect of endocrine-disrupting chemicals on child neurodevelopment. *Frontiers in Endocrinology, 9*, 204.

Gomes Gonçalves, N., Szlejf, C., Lotufo, P. A., Bensenor, I. M., & Suemoto, C. K. (2024). Association between subclinical thyroid dysfunction and cognitive decline: Findings from the ELSA-Brasil study. *The Journals of Gerontology. Series A, Biological Sciences and Medical Sciences, 79*(8).

Gothié, J. D., Demeneix, B., & Remaud, S. (2017). Comparative approaches to understanding thyroid hormone regulation of neurogenesis. *Molecular and Cellular Endocrinology, 459*, 104–115.

Goto, M., Kimura, N., & Matsubara, E. (2022). Association of serum thyroid hormone levels with positron emission tomography imaging in non-demented older adults. *Psychogeriatrics, 22*(3), 373–381.

Hallab, A. (2025a). Sex-modulated association between thyroid stimulating hormone and informant-perceived anxiety in non-depressed older adults: Prediction models and relevant cutoff value. *Scientific Reports, 15*(1), 2526.

Hallab, A. (2025b). Thyroid-stimulating hormone and cognitive impairment in non-depressed non-demented multiethnic middle-aged and older US adults: Assessing sex-specific risk prediction. *Archives of Gerontology and Geriatrics, 2*(3).

Hallab, A., & Alzheimer's Disease Neuroimaging Initiative (ADNI). (2026). Association of hippocampus, entorhinal cortex, and amygdala with thyroid function: a bilateral volumetric analysis. *Thyroid Research, 19*(1), 8. https://doi.org/10.1186/s13044-026-00289-4

Holmberg, M., Malmgren, H., Heckemann, R. A., Johansson, B., Klasson, N., Olsson, E., et al. (2022). A longitudinal study of medial temporal lobe volumes in graves disease. *The Journal of clinical endocrinology and metabolism, 107*(4), 1040–1052.

Hu, H., Zhou, J., Fang, W., Chen, H. H., Jiang, W. H., Pu, X. Y., et al. (2023). Increased brain iron in patients with thyroid-associated ophthalmopathy: A whole-brain analysis. *Frontiers in Endocrinology, 14*, Article 1268279.

Jeong, H. S., Choi, E. K., Song, I. U., Chung, Y. A., Park, J. S., & Oh, J. K. (2017). Differences in brain glucose metabolism during preparation for (131)I ablation in thyroid cancer patients: Thyroid hormone withdrawal versus recombinant human thyrotropin. *Thyroid, 27*(1), 23–28.

Jessy, A. S., Sandhya, G., Monisha, S., Sundarakumar, J. S., Stezin, A., & Issac, T. G. (2024). Prevalence of hypothyroidism in older adults and its association with cognition: A cross-sectional study from a South Indian ageing urban cohort. *Brain communications, 6*(6), Article fcae391.

Jia, Y., Zhong, S., Wang, Y., Liu, T., Liao, X., & Huang, L. (2015). The correlation between biochemical abnormalities in frontal white matter, hippocampus and serum thyroid hormone levels in first-episode patients with major depressive disorder. *Journal of Affective Disorders, 180*, 162–169.

Jiang, M., Zhang, H., Liu, Y., Wu, S., Qu, J., Tang, Y., et al. (2023). Abnormal regional spontaneous neural activity and functional connectivity in thyroid-associated ophthalmopathy patients with different activity: A resting-state fMRI study. *Frontiers in Neurology, 14*, Article 1199251.

Juárez-Cedillo, T., Basurto-Acevedo, L., Vega-García, S., Sánchez-Rodríguez Martha, A., Retana-Ugalde, R., Juárez-Cedillo, E., et al. (2017). Prevalence of thyroid dysfunction and its impact on cognition in older mexican adults: (SADEM study). *Journal of Endocrinological Investigation, 40*(9), 945–952.

Kimura, N., Sasaki, K., Masuda, T., Ataka, T., Matsumoto, M., Kitamura, M., et al. (2025). Machine learning models for dementia screening to classify brain amyloid positivity on positron emission tomography using blood markers and demographic characteristics: A retrospective observational study. *Alzheimer's Research & Therapy, 17*(1), 25.

Kumar, M., Singh, S., Rana, P., D'Souza, M., Kumaran, S. S., Sekhri, T., et al. (2025). Neurometabolite and cognitive changes in hypothyroid patients in response to treatment: In-vivo (1)H MRS study. *Neurochemistry International, 182*, Article 105915.

Lee, S. Y., & Pearce, E. N. (2023). Hyperthyroidism: A review. *JAMA, 330*(15), 1472–1483.

Lischinsky, J. E., Skocic, J., Clairman, H., & Rovet, J. (2016). Preliminary findings show maternal hypothyroidism may contribute to abnormal cortical morphology in offspring. *Frontiers in Endocrinology, 7*, 16.

Liu, P., Luo, B., Feng, Y., Chen, L., Zhao, Y. L., Wang, Q. X., et al. (2023). Aberrant spontaneous brain activity in patients with thyroid-associated ophthalmopathy with and without optic neuropathy: A resting-state functional MRI study. *European Radiology, 33*(11), 7981–7991.

Liu, B., Wen, L., Ran, Q., Zhang, S., Hu, J., Gong, M., et al. (2020). Dysregulation within the salience network and default mode network in hyperthyroid patients: A follow-up resting-state functional MRI study. *Brain Imaging and Behavior, 14*(1), 30–41.

Liu, H., Zhong, Y. L., & Huang, X. (2024). Specific static and dynamic functional network connectivity changes in thyroid-associated ophthalmopathy and it predictive values using machine learning. *Frontiers in Neuroscience, 18*, Article 1429084.

Luo, L., Zhang, L., Huang, H., Guan, J., Zhang, X., Lin, Y., et al. (2023). 3.0 T multi-parametric MRI reveals metabolic and microstructural abnormalities in the posterior visual pathways in patients with thyroid eye disease. *Frontiers in Neuroscience, 17*, Article 1306364.

Matsunaga, A., Ikawa, M., Kawamura, Y., Kishitani, T., Yamamura, O., Hamano, T., et al. (2019). Serial brain MRI changes related to autoimmune pathophysiology in hashimoto encephalopathy with anti-NAE antibodies: A case-series study. *Journal of the Neurological Sciences, 406*, Article 116453.

McNabb, C. B., Scholz, A., Bloomfield, L., Bhargava, R., Hales, C., Dayan, C. M., et al. (2025). Controlled antenatal thyroid screening study III: Effects of gestational thyroid status on brain microstructure. *The Journal of Clinical Endocrinology and Metabolism, 110*(12), 3322–3330.

Mulder, T. A., Muetzel, R. L., Peeters, R. P., Tiemeier, H., & Korevaar, T. I. M. (2025). Maternal thyroid function during pregnancy and early adolescent regional differences in cerebral gray matter morphology. *The Journal of Clinical Endocrinology and Metabolism, 111*, 572–579.

Naicker, M., Abbai, N., & Naidoo, S. (2019). Bipolar limbic expression of auto-immune thyroid targets: Thyroglobulin and thyroid-stimulating hormone receptor. *Metabolic Brain Disease, 34*(5), 1281–1298.

Ojala, A. K., Schalin-Jäntti, C., Pitkälä, K. H., Tilvis, R. S., & Strandberg, T. E. (2016). Serum thyroid-stimulating hormone and cognition in older people. *Age and Ageing, 45*(1), 155–157.

Pak, K., Kim, M., Kim, K., Kim, B. H., Kim, S. J., & Kim, I. J. (2020). Cerebral glucose metabolism and cerebral blood flow in thyroid dysfunction: An activation likelihood estimation meta-analysis. *Scientific Reports, 10*(1), 1335.

Pankowski, D., & Wytrychiewicz-Pankowska, K. (2025). Prevalence, hormonal correlates, severity, and neural basis of neurocognitive impairment in patients with hypothyroidism: Systematic review and meta-analyses. *Alzheimer's & Dementia, 21*(11), Article e70924.

Perri, K., De Mori, L., Tortora, D., Calevo, M. G., Allegri, A. E. M., Napoli, F., et al. (2021). Cognitive and white matter microstructure development in congenital hypothyroidism and familial thyroid disorders. *The Journal of Clinical Endocrinology and Metabolism, 106* (10), e3990–e4006.

Piekiełko-Witkowska, A., Duda, M. K., Bogusławska, J., & Mackiewicz, U. (2025). The impact of autoimmune thyroid disease on cognitive and psychiatric disorders: Focus on clinical, pre-clinical and molecular studies. *European Thyroid Journal, 14*(3).

Pyun, J. M., Park, Y. H., & Kim, S. (2022). Subclinical hypothyroidism and cognitive impairment. *Journal of Alzheimer's Disease, 88*(2), 757–762.

Quinlan, P., Horvath, A., Eckerström, C., Wallin, A., & Svensson, J. (2020). Altered thyroid hormone profile in patients with alzheimer's disease. *Psychoneuroendocrinology, 121*, Article 104844.

Quinlan, P., Horvath, A., Eckerström, C., Wallin, A., & Svensson, J. (2022). Higher thyroid function is associated with accelerated hippocampal volume loss in alzheimer's disease. *Psychoneuroendocrinology, 139*, Article 105710.

Saglam, Y., Ermis, C., Tanyolac, D., Oz, A., Turan, S., Korkmaz, H. A., et al. (2024). The association between plasma thyroxine levels and neurocognitive impairment in early-onset schizophrenia and other psychosis spectrum disorders. *Progress In Neuro-Psychopharmacology & Biological Psychiatry, 131*, Article 110940.

Salas-Lucia, F. (2024). Mapping thyroid hormone action in the human brain. *Thyroid, 34*(7), 815–826.

Salazar, P., Cisternas, P., Codocedo, J. F., & Inestrosa, N. C. (2017). Induction of hypothyroidism during early postnatal stages triggers a decrease in cognitive performance by decreasing hippocampal synaptic plasticity. *Biochimica et Biophysica Acta (BBA) - Molecular Basis of Disease, 1863*(4), 870–883.

Samuels, M. H., Kaimal, R., Waring, A., Fink, H. A., Yaffe, K., Hoffman, A. R., et al. (2016). Thyroid function variations within the reference range do not affect quality of life, mood, or cognitive function in community-dwelling older men. *Thyroid, 26*(9), 1185–1194.

Samuels, M. H., Kolobova, I., Smeraglio, A., Niederhausen, M., Janowsky, J. S., & Schuff, K. G. (2016). Effect of thyroid function variations within the laboratory reference range on health status, mood, and cognition in levothyroxine-treated subjects. *Thyroid, 26* (9), 1173–1184.

Scholz, A., McNabb, C. B., Bloomfield, L., Bhargava, R., Hales, C., Dayan, C. M., et al. (2025). Controlled antenatal thyroid screening study III: Effects of gestational thyroid status on adolescent brain morphology. *The Journal of Clinical Endocrinology and Metabolism, 110*(4), e1094–e1102.

Singh, S., Kumar, M., Modi, S., Kaur, P., Shankar, L. R., & Khushu, S. (2015). Alterations of functional connectivity among resting-state networks in hypothyroidism. *Journal of Neuroendocrinology, 27*(7), 609–615.

Singh, S., Rana, P., Kumar, P., Shankar, L. R., & Khushu, S. (2016). Hippocampal neurometabolite changes in hypothyroidism: An in vivo (1) H magnetic resonance spectroscopy study before and after thyroxine treatment. *Journal of Neuroendocrinology, 28*(9).

Sinha, S. H., Zietlow, K., & Papaleontiou, M. (2024). Thyroid function and cognitive decline: A narrative review. *Endocrine Practice, 30* (11), 1113–1118.

Su, W., Zhao, L., Bao, S., Qin, R., Cao, J., Tian, J., et al. (2023). Alterations in gray matter morphology and functional connectivity in adult patients with newly diagnosed untreated hypothyroidism. *Thyroid, 33*(7), 791–803.

Szlejf, C., Suemoto, C. K., Santos, I. S., Lotufo, P. A., Haueisen Sander Diniz, M. F., Barreto, S. M., et al. (2018). Thyrotropin level and cognitive performance: Baseline results from the ELSA-Brasil study. *Psychoneuroendocrinology, 87*, 152–158.

Tian, Y., Yao, D., Jin, A., Wang, M., Pan, Y., Wang, Y., et al. (2023). Thyroid function in causal relation to MRI markers of cerebral small vessel disease: A mendelian randomization analysis. *The Journal of Clinical Endocrinology and Metabolism, 108*(9), 2290–2298.

Toll, A., Blanco-Hinojo, L., Berge, D., Manzano, A., El Abidi, K., Perez-Solà, V., et al. (2023). Relationship between thyroid-stimulating hormone, BDNF levels, and hippocampal volume in antipsychotic-naïve first-episode psychosis patients. *Frontiers in Psychiatry, 14*, Article 1301714.

Valcárcel-Hernández, V., López-Espíndola, D., Guillén-Yunta, M., García-Aldea, Á., López de Toledo Soler, I., Bárez-López, S., et al. (2022). Deficient thyroid hormone transport to the brain leads to impairments in axonal caliber and oligodendroglial development. *Neurobiology of Disease, 162*, Article 105567.

Valencia-Sanchez, C., Pittock, S. J., Mead-Harvey, C., Dubey, D., Flanagan, E. P., Lopez-Chiriboga, S., et al. (2021). Brain dysfunction and thyroid antibodies: Autoimmune diagnosis and misdiagnosis. *Brain Communications, 3*(2), Article fcaa233.

van Vliet, N. A., van Heemst, D., Almeida, O. P., Åsvold, B. O., Aubert, C. E., Bae, J. B., et al. (2021). Association of thyroid dysfunction with cognitive function: An individual participant data analysis. *JAMA Internal Medicine, 181*(11), 1440–1450.

Wang, N., Sun, Y., Yang, H., Xu, Y., Cai, Y., Liu, T., et al. (2021). Hashimoto's thyroiditis induces hippocampus-dependent cognitive alterations by impairing astrocytes in euthyroid mice. *Thyroid, 31*(3), 482–493.

Wei, Q., Zhang, H., Guan, H., Song, X., & Zhou, H. (2024). Brain alteration of autoimmune thyroid disease: Neuropsychiatric impact, neuroimaging insights, and neurobiological implications. *Frontiers in Neuroendocrinology, 75*, Article 101159.

Wheeler, S. M., Willoughby, K. A., McAndrews, M. P., & Rovet, J. F. (2011). Hippocampal size and memory functioning in children and adolescents with congenital hypothyroidism. *The Journal of Clinical Endocrinology and Metabolism, 96*(9), E1427–E1434.

Winkler, A., Weimar, C., Jöckel, K. H., Erbel, R., Dragano, N., Broecker-Preuss, M., et al. (2016). Thyroid-stimulating hormone and mild cognitive impairment: Results of the heinz nixdorf recall study. *Journal of Alzheimer's Disease, 49*(3), 797–807.

Wu, Q., Zhou, J., Fang, W., Jiang, W. H., Pu, X. Y., Chen, H. H., et al. (2024). Structural and functional brain changes after glucocorticoid therapy in thyroid-associated ophthalmopathy. *The Journal of Clinical Endocrinology and Metabolism, 109*(3), 649–658.

Yin, J., Xie, L., Luo, D., Huang, J., Guo, R., Zheng, Y., et al. (2021). Changes of structural and functional attention control networks in subclinical hypothyroidism. *Frontiers in Behavioral Neuroscience, 15*, Article 725908.

Yu, H., Li, Z., Gao, X., Liu, X., Cui, W., Li, N., et al. (2025). Multi-omics data integration reveals novel genes related to autoimmune hypothyroidism in the brain: A molecular basis for the brain-thyroid axis. *Progress In Neuro-Psychopharmacology & Biological Psychiatry, 136*, Article 111239.

Yu, T., Pei, W., Zhang, X., & Deng, C. (2024). Associations between thyroid hormones levels and gray matter volume of frontal lobe involved into violence in male schizophrenia patients. *Neuropsychiatric Disease and Treatment, 20*, 2169–2175.

Zhang, H., Liu, Y., Zhang, Z., Jiang, M., Tao, X., Lee, X. N., et al. (2024). Neuroimaging in thyroid eye disease: A systematic review. *Autoimmunity Reviews, 23*(12), Article 103667.

Zhang, X., Xie, Y., Ding, C., Xiao, J., Tang, Y., Jiang, X., et al. (2017). Subclinical hypothyroidism and risk of cerebral small vessel disease: A hospital-based observational study. *Clinical Endocrinology, 87*(5), 581–586.

Zhang, Y., Yang, Y., Tao, B., Lv, Q., Lui, S., & He, L. (2021). Gray matter and regional brain activity abnormalities in subclinical hypothyroidism. *Frontiers in Endocrinology, 12*, Article 582519.

Zhao, S., Du, Y., Zhang, Y., Wang, X., Xia, Y., Sun, H., et al. (2023). Gray matter reduction is associated with cognitive dysfunction in depressed patients comorbid with subclinical hypothyroidism. *Frontiers in Aging Neuroscience, 15*, Article 1106792.

Zhao, S., Xia, Y., Huang, Y., Zou, H., Wang, X., Chen, Z., et al. (2021). The correlation between thyroid function, frontal gray matter, and executive function in patients with major depressive disorder. *Frontiers in Endocrinology, 12*, Article 779693.

Chapter 2

Thyroid hormone signaling in neurodevelopment

Ehsan Noroozi[1], Alireza Rashidi[2], Hoda Borooghani[3], Maryam Eskandarpour[4], Mojan Ayati[5] and Alaleh Alizadeh[6]

[1]Department of Internal Medicine, Shahid Labbafi Nezhad Hospital, Shahid Beheshti University of Medical Science, Tehran, Iran; [2]Department of Pharmaceutics, School of Pharmacy and Novel Drug Delivery Systems Research Center, Isfahan University of Medical Sciences, Isfahan, Iran; [3]School of Medicine, Iran University of Medical Sciences, Tehran, Iran; [4]Department of General Practice, Dr Ozone Clinic, Dubai Healthcare City, Dubai, United Arab Emirates; [5]School of Medicine, Szeged University of Medical Sciences, Szeged, Hungary; [6]Faculty of Medicine, Mashhad Branch, Islamic Azad University, Mashhad, Iran

Abbreviations

ACh Acetylcholine
AChE Acetylcholine-Esterase
ADHD Attention-Deficit/Hyperactivity Disorder
AHDS Allan-Herndon-Dudley Syndrome
AOP Adverse Outcome Pathway
ASD Autism Spectrum Disorders
ChAT Choline-O-Acetyltransferase
CNS Central Nervous System
D1 Deiodinase 1
D2 Deiodinase 2
D3 Deiodinase 3
DITPA 3,5-Diiodothyropropionic Acid
E/I Excitation/Inhibition
IUGR Intrauterine Growth Restriction
MCT8 Monocarboxylate Transporter 8
OATP1C1 Organic Anion-Transporting Polypeptide 1C1
PI3K Phosphoinositide 3-Kinase
PTU 6-Propyl-2-Thiouracil
RLS Restless Legs Syndrome
ROS Reactive Oxygen Species
T2 3,5-Diiodothyronine
T3 Triiodothyronine
T4 Thyroxine
TETRAC 3′,5′,3,5-Tetraiodothyroacetic Acid
TH Thyroid Hormone
Thyroid-NDT-TAS Thyroid-Neurodevelopmental Toxicity-Test Assessment Strategy
TR Thyroid Receptor
TREs Thyroid Hormone Response Elements
TRIAC 3′,3,5-Triiodothyroacetic Acid
TRα Thyroid Receptor Alpha
TRβ Thyroid Receptor Beta
TRα1 Thyroid Receptor Alpha 1
TRβ1 Thyroid Receptor Beta 1
TRβ2 Thyroid Receptor Beta 2
TSH Thyroid-Stimulating Hormone

Introduction

Thyroid hormone (TH) is essential for normal brain development because it affects many important processes, ranging from neuronal differentiation to synaptogenesis. However, TH functions as a timing signal that synchronizes and aligns the parallel growth of several neuronal systems and related glial cells rather than actively triggering specific processes (Anderson et al., 2003; Alcaide Martin & Mayerl, 2023). The need for TH changes at different developmental stages. It has been observed that TH is most likely to be present between the late stages of pregnancy and weeks three and four after birth in rodents, and in the first few months after birth in humans. A lack of TH during this critical period has a significant impact on brain development by disrupting the developmental order, which causes permanent structural and functional abnormalities (Williams, 2008).

The fetal thyroid gland only becomes functional in the second trimester. This reliance on maternal TH highlights how susceptible the developing brain is to iodine deficiency or maternal hypothyroidism, both of which have been strongly linked to serious neurodevelopmental consequences such as cretinism, cerebral palsy, and deaf-mutism (Ishii et al., 2021; Leto et al., 2016).

At the molecular level, THs function through both genomic and non-genomic mechanisms. Thyroid receptor

Hormones in Neuroscience. https://doi.org/10.1016/B978-0-443-30261-9.00002-4

alpha and beta (TRα and TRβ) are nuclear TH receptors. The active hormone (T3) binds to them, after which these receptors use TH response elements (TREs) to control the transcription of genes that respond to TH. THs also participate in non-genomic signaling pathways in the cytoplasm, for example, Phosphoinositide 3-kinase (PI3K), Protein kinase B (Akt), and Mitogen-activated protein kinase (MAPK) pathways. These help cells respond quickly (Aranda, 2025a). Within cells, transporters and deiodinase enzymes (D1, D2, D3) precisely regulate TH activity. These enzymes modify the amount of T3 and thyroxine (T4) that are available in the brain and other systems (Dentice & Salvatore, 2019).

Neurodevelopmental assessments within the thyroid-NDT-TAS

The Thyroid-Neurodevelopmental Toxicity-Test Assessment Strategy (Thyroid-NDT-TAS) is a justifiable, science-based test procedure for detecting and identifying adverse neurodevelopmental effects that may be caused by maternal interference with THs. The Thyroid-NDT-TAS framework systematically evaluates the impact of thyroid hormone disruptions on neurodevelopment, as outlined in the assessment pipeline (Fig. 2.1). Since THs, especially thyroxine (T4), are crucial to fetal development, this test procedure is designed to recognize even slight and persistent neurological effects resulting from chemical interference with the endocrine system (Zoeller & Rovet, 2004).

This procedure assesses three general domains of neurodevelopmental function: cognition, behavior, and brain spatial organization (Table 2.1).

Regulation of TH action in the brain from development to adulthood

TRs are found at different stages of cerebral development. Thyroid receptor alpha 1 (TRα1) is expressed early in life and plays a significant role in the early development of neural cells. Conversely, thyroid receptors beta 1 and 2 (TRβ1 and TRβ2) are expressed later and are largely linked to the tissues controlling sensory functions, such as the auditory and visual systems (Bianco & Kim, 2006).

Tyrosine hydroxylase and excitatory glutamatergic neurons

Excitatory glutamatergic neurons, including the pyramidal neurons found in the cortex, are crucial for learning, memory, and higher intellectual functions. Recent studies indicate that THs play a critical role in guiding the proper development and long-term maintenance of these neurons throughout life (Alcaide Martin & Mayerl, 2023).

During early development, THs direct cells as they extend, migrate into position, and organize. Inside the cell, receptors like TH receptor alpha 1 (TRα1) regulate gene expression, ensuring normal development. A shortage of THs due to maternal hypothyroidism or deficiencies in transporters or receptors can result in abnormal brain structure, reduced neuron populations, and a thinner cortex (Alcaide Martin & Mayerl, 2023; Patel et al., 2011).

TH and inhibitory GABAergic interneurons

THs also regulate the inhibitory system of the brain through the promotion of the development of GABAergic interneurons, as needed for the regulation of the excitation–inhibition (E/I) balance (Bernal, 2024). TH influences the expression of key genes that lead to the differentiation and integration of these interneurons into local circuits during the development of neurons (Talhada et al., 2019).

TH and cholinergic systems

Cholinergic neurons are important in preserving the balance between inhibition and E/I in the central nervous system (CNS) and are involved in modulating several processes such as sleep, movement, cognition, memory, and learning (Ahmed et al., 2019; Okada et al., 2022). There is a robust association between cholinergic function and TH status (Alcaide Martin & Mayerl, 2023).

Although there are obvious connections between TH and cholinergic function, the exact mechanisms through which TH influences the development and physiological activity of cholinergic neurons remain unclear. Cholinergic neurons are mainly located in the basal forebrain and brainstem of adults. Cholinergic basal forebrain neurons project to the hippocampus and cortex, where they regulate learning and memory, particularly via the septo-hippocampal pathway. The majority of these neurons are derived from the medial ganglionic eminence, preoptic area, and septal regions, with some coming from the pallium. THs regulate the generation of Nkx2.1, a determinant of cholinergic neuron development, and as a consequence, learning and memory.

These observations offer significant anatomical evidence coupling TH availability to the activity of cholinergic neurons and lend support to the notion of a close cooperation between the thyroid and cholinergic systems during development and adulthood. TRα1 involvement is implicated by a few reports, and THs can also indirectly influence cholinergic neuron maintenance via nerve growth factor mechanisms (Ahmed et al., 2019; Magno et al., 2017).

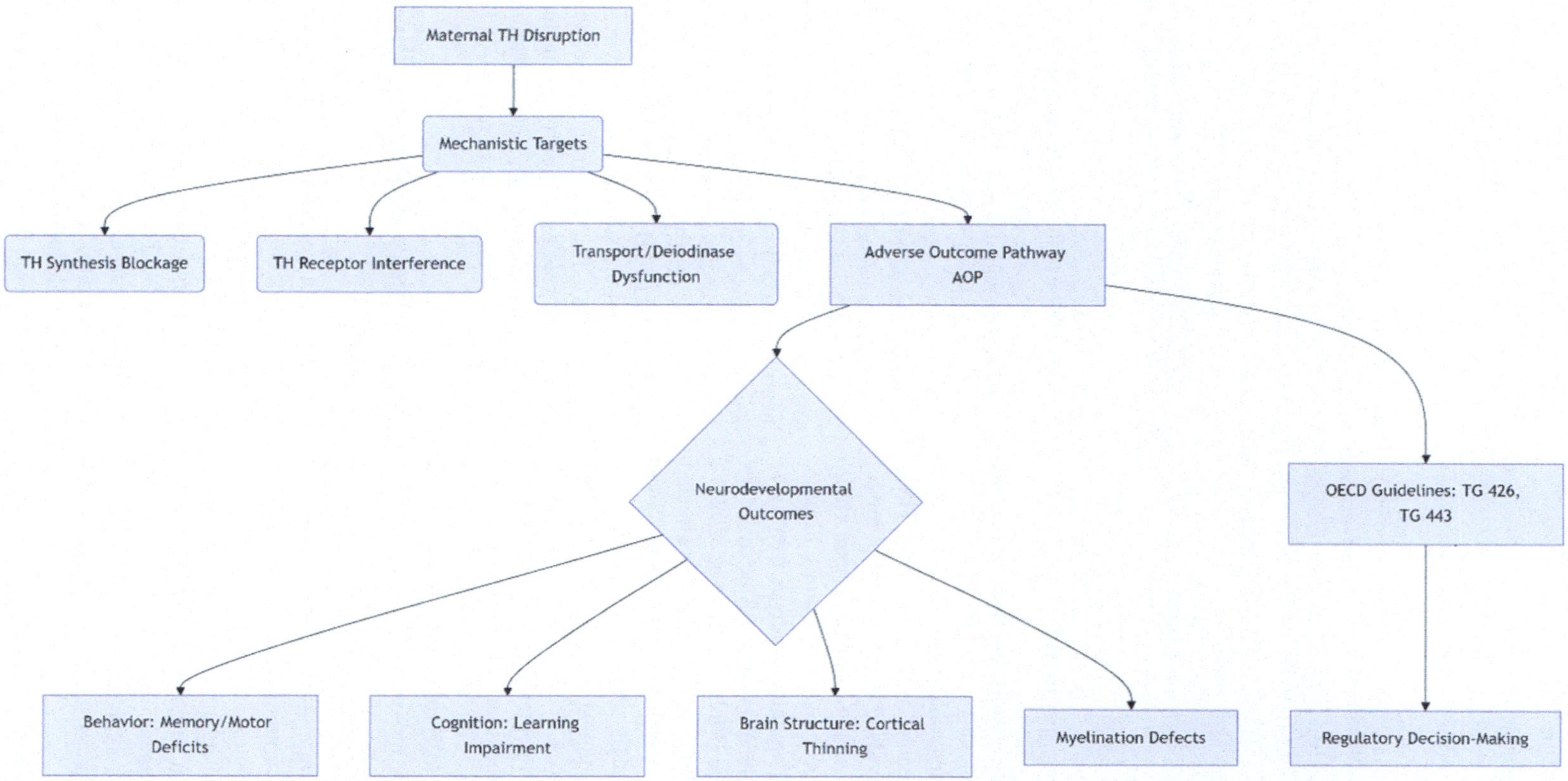

FIGURE 2.1 **Thyroid-NDT-TAS neurodevelopmental assessment Pipeline.** Illustrates the stepwise evaluation of maternal thyroid hormone (TH) disruption on brain development. Key components: **TH disruption targets**: TH synthesis blockage (e.g., deiodinase inhibition), receptor interference (TRα/TRβ), or transport dysfunction (MCT8/OATP1C1). **Adverse outcome pathway (AOP)**: Connects molecular disruptions to behavioral (memory/motor deficits), cognitive (learning impairment), and structural (cortical thinning, myelination defects) outcomes. **OECD guidelines**: TG 426 (Developmental neurotoxicity study) and TG 443 (Extended one-generation study) inform regulatory decisions.

TABLE 2.1 Thyroid-NDT-TAS framework – key features and applications.

Component	Description	Purpose
Neuro developmental testing	Assesses behavior, cognition, and brain structure (e.g., memory, motor skills, startle response)	Detect subtle neurodevelopmental impairments caused by thyroid disruption
OECD guidelines	Uses standardized test protocols like OECD TG 426 and TG 443	Ensures scientific validity and international acceptance
Mechanistic mapping (AOP)	Traces biological pathways from molecular disruption to behavioral outcomes	Links chemical exposure to specific thyroid-related neurodevelopmental deficits
Imaging & histology	Analyzes brain structure: Cortical layering, hippocampal organization, myelination	Identifies structural biomarkers of thyroid hormone deficiency
Regulatory application	Integrates multiple levels of evidence to inform chemical safety assessments	Helps distinguish thyroid-specific neurotoxicity from general toxicity

This table outlines the components of the Thyroid-NDT-TAS framework, including neurodevelopmental testing; OECD guidelines, mechanistic mapping via AOP; imaging/histology, and regulatory applications; designed to detect and characterize thyroid hormone-related neurodevelopmental deficits.

TH and dopaminergic systems

Dopaminergic neurons that are broadly functional across the adult forebrain are characteristically concentrated in specific areas like the olfactory bulb, hypothalamus, substantia nigra pars compacta, and the ventral tegmental area. These dopaminergic neurons are involved in the control of movement, cognition, mood, and behavior, and have been implicated in a variety of disorders like Parkinson's disease, schizophrenia, and ADHD. Dopamine synthesis from tyrosine requires the activity of the enzymes tyrosine hydroxylase and DOPA decarboxylase (Mohamadian et al., 2023).

Dopamine levels in the brain are significantly affected by THs. Thyroid dysfunction is linked to various movement disorders. If not accurately diagnosed, hypothyroidism may manifest with symptoms that mimic idiopathic Parkinson's disease, and thus a delay in proper diagnosis may occur, and treatment strategies may be complicated. Hyperthyroidism, on the other hand, is more frequently associated with hyperkinetic movement disorders like tremor, whereas its relation to chorea and dystonia is less common. The severity of the movement disorder tends to parallel the level of THs. Despite the fact that exact pathophysiological mechanisms are not well understood, they can include dopaminergic dysregulation, increased receptor sensitivity, and autoimmune processes that have particular relevance in the context of diseases like Graves' disease and Hashimoto's encephalopathy. Hence, thyroid function should be evaluated in patients with new-onset, unusual, or refractory movement disorders since thyroid-related movement disorders typically respond to treatment using appropriate thyroid medications (Schneider et al., 2023).

The diverse roles of thyroid hormones in regulating glutamatergic, GABAergic, cholinergic, and dopaminergic neurons are summarized in a class diagram (Fig. 2.2).

Developmentally, mDA neurons start forming from progenitors along the ventral midline of the neural tube in mice from embryonic day 8.5–12.5 with the assistance of morphogens like Sonic hedgehog and Fibroblast Growth Factor 8. Recent studies show that TH is a critical extrinsic factor regulating the development of mDA neurons and exerts direct influence on neural precursor cells. TH/TRα1 signaling increases Otx2, which enhances the expression of Nurr1 and Ngn2 both of which are required for the generation of mDA neurons. Knockdown of Otx2 eliminates this effect, and TH signaling also regulates TRPC1 calcium channels, which helps in neuronal differentiation. Nonetheless, the exact mechanisms of T3 delivery to mDA progenitors and the involvement of TH transporters and deiodinases during brain development are still unclear (Chen et al., 2015; Sillitoe & Vogel, 2008). The specific effects of thyroid hormone disruptions on various neuron types, including glutamatergic, GABAergic, cholinergic, and dopaminergic systems, are detailed in a comprehensive summary (Table 2.2).

An overview of the signaling landscape of THs

THs use both genomic and non-genomic mechanisms to regulate physiological processes. The principal hormone released from the thyroid gland, thyroxine (T4), is activated to T3 by 5′-deiodinase. T3 regulates gene expression by binding in a cell-specific manner to thyroid receptors (TRs) TRα and TRβ. Negative feedback is mediated when T3 inhibits the expression of genes that generate TH in thyroid and pituitary cells.

This accelerates the production of ROS in mitochondria. In addition, thyronamines decarboxylated metabolites of TH enhance ROS production by inhibiting mitochondrial complex III. While excessive ROS can be

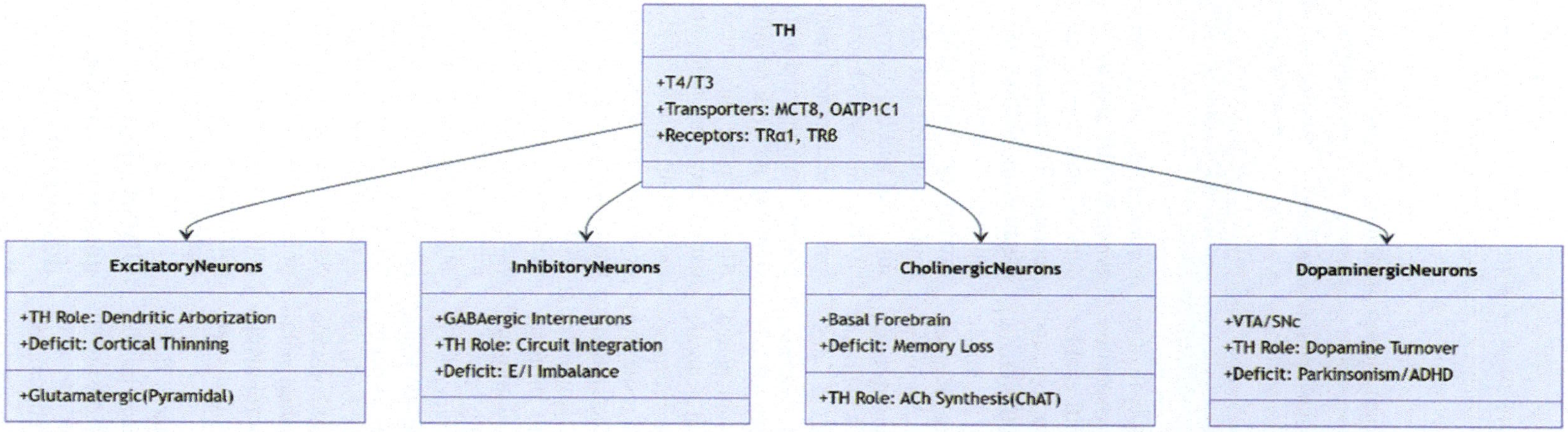

FIGURE 2.2 TH regulation of neuron Subtypes. Class diagram mapping TH actions on: **Glutamatergic neurons**: Pyramidal cell dendritogenesis (TRα1/MCT8-dependent); deficits → cortical thinning. **GABAergic interneurons**: Nkx2.1-mediated differentiation; deficits → excitation/inhibition (E/I) imbalance. **Cholinergic neurons**: Acetylcholine (ACh) synthesis in basal forebrain; deficits → memory loss. **Dopaminergic neurons**: Dopamine turnover in substantia nigra pars compacta (SNc)/ventral tegmental area (VTA); deficits → motor/cognitive disorders.

TABLE 2.2 Effects of thyroid hormones (THs) on different types of neurons.

Neuron type	Role of thyroid hormones	Effects of TH disruption
Glutamatergic (excitatory)	Guides growth, positioning, and maturation via TRα1 and transporters (MCT8, OATP1C1)	Reduced neuron numbers, disorganized brain architecture, cortical thinning
GABAergic (inhibitory)	Supports differentiation and local circuit integration	Fewer and less effective inhibitory neurons → excitatory/inhibitory (E/I) imbalance
Cholinergic	Regulates acetylcholine production and function, drives development from basal forebrain regions	Memory and learning deficits, motor control impairments
Dopaminergic	Modulates dopaminergic gene expression and neurotransmitter levels	Movement disorders, Parkinson-like symptoms, ADHD-like behaviors

This table summarizes the roles of thyroid hormones in the development and function of glutamatergic, GABAergic; cholinergic, and dopaminergic neurons; highlighting specific disruptions caused by thyroid hormone deficiencies, such as cortical thinning; E/I imbalance, and motor/cognitive impairments.

detrimental, temporary bursts of ROS are necessary to alter signaling pathways like PI3K/Akt through redox-sensitive proteins such as PTEN. This brief oxidative shift enables important cellular functions like brain development and neuronal differentiation to occur (Lee et al., 2002; Özsoy et al., 2021; Vujovic et al., 2023).

Recent research suggests that the distinction between genomic and non-genomic TH effects may be oversimplified. Thyroid hormones exert their effects through both genomic and non-genomic pathways, which are contrasted in the schematic representation of their signaling mechanisms (Fig. 2.3). Tedeschi et al. described TH signaling as a multifaceted, integrated network wherein genomic and non-genomic mechanisms are often paired instead of segregated. THs trigger their classical genomic actions by interactions of TRα and TRβ nuclear receptors with TREs, while simultaneously inducing rapid responses by integrin αvβ3 at the plasma membrane. The integrin pathway, activated mainly by T4, immediately stimulates PI3K/AKT and MAPK signaling pathways to promote processes including cellular proliferation and tumorigenesis. TH receptor isoforms also display shuttling between the nucleus and cytoplasm, increasing the accuracy of TH responses in different cellular contexts. Transport proteins, including MCT8, help to drive the entry of THs into cells, while post-translational modifications influence both the localization and the activity of TRs, and as a result, provide another regulatory dimension to the TH signaling paradigm. Genomic and non-genomic interplay cooperatively creates the wide array of tissue-specific effects that characterize THs (Tedeschi et al., 2021).

In addition, TH signaling is seemingly biphasic: rapid, ROS-mediated non-genomic actions that enhance growth and mitochondrial function, and slower, TR-mediated genomic responses that engage antioxidant defenses to restore cellular homeostasis. Additionally, the elimination of TRs can correct developmental problems caused by hypothyroidism, supporting the hypothesis that TR-mediated genomic effects typically counteract non-genomic TH actions. This integrated signaling system demonstrates the importance of both genomic and non-genomic TH pathways in fine-tuning development and metabolism (Morte et al., 2002). A hypothyroidism-induced E/I imbalance, arising from glutamatergic hyper-excitability and GABAergic hypoactivity, underlies neurodevelopmental disorders that are detectable via structural and functional biomarkers (Fig. 2.4).

THs are well known for their role in metabolism, growth, and mitochondrial control. They also exhibit potent immunomodulatory effects that connect their signaling network to inflammation and disease processes. An increasing amount of evidence indicates that THs, particularly via their nuclear receptors, play a role in modulating immunological responses and retarding the development of inflammation-associated disorders. THs have been shown to modulate key inflammatory pathways, such as NF-κB, p38MAPK, and JAK/STAT signaling, with significant implications for conditions like sepsis, cancer, and COVID-19. T3 binds to nuclear TRs, notably TRβ, inhibiting pro-inflammatory cytokine production, inducing anti-inflammatory macrophage morphologies, and suppressing NF-κB and STAT3 transcription. This genomic control balances non-genomic, integrin αvβ3-mediated actions that can augment inflammation and stimulate cancer growth. These observations emphasize the dual function of THs in immunological homeostasis and suggest potential therapeutic applications that target TR-mediated pathways to decrease inflammation and tumor development (Lasa & Contreras-Jurado, 2022).

Uncoupling of the genomic and non-genomic impacts of THs during development

Novel work in the developing mouse brain demonstrates that T3 selectively activates nuclear TRα1 in neural stem

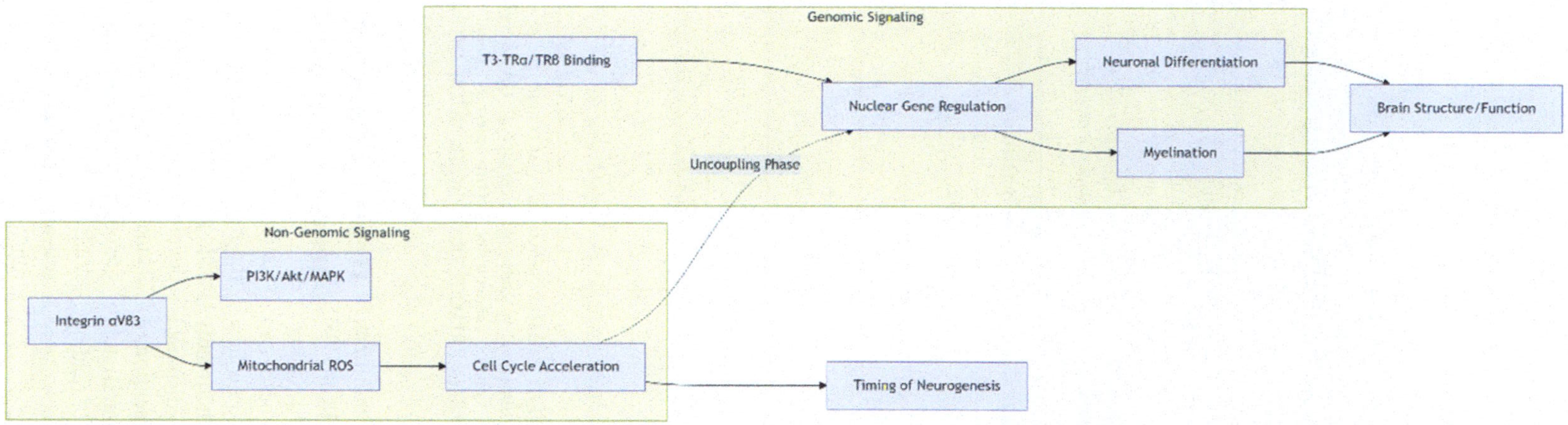

FIGURE 2.3 Dual TH signaling mechanisms in brain development. Highlights contrasting pathways: **Genomic signaling**: Slow, nuclear receptor (TRα1/TRβ)-dependent gene regulation (e.g., NeuroD, Otx2) driving neuronal differentiation and myelination. **Non-Genomic signaling**: Rapid integrin αVβ3-mediated PI3K/Akt/MAPK activation and mitochondrial T2-induced reactive oxygen species (ROS) production, accelerating cell cycle progression. **Uncoupling phase**: Developmental stages where non-genomic ROS signaling dominates over genomic TR actions.

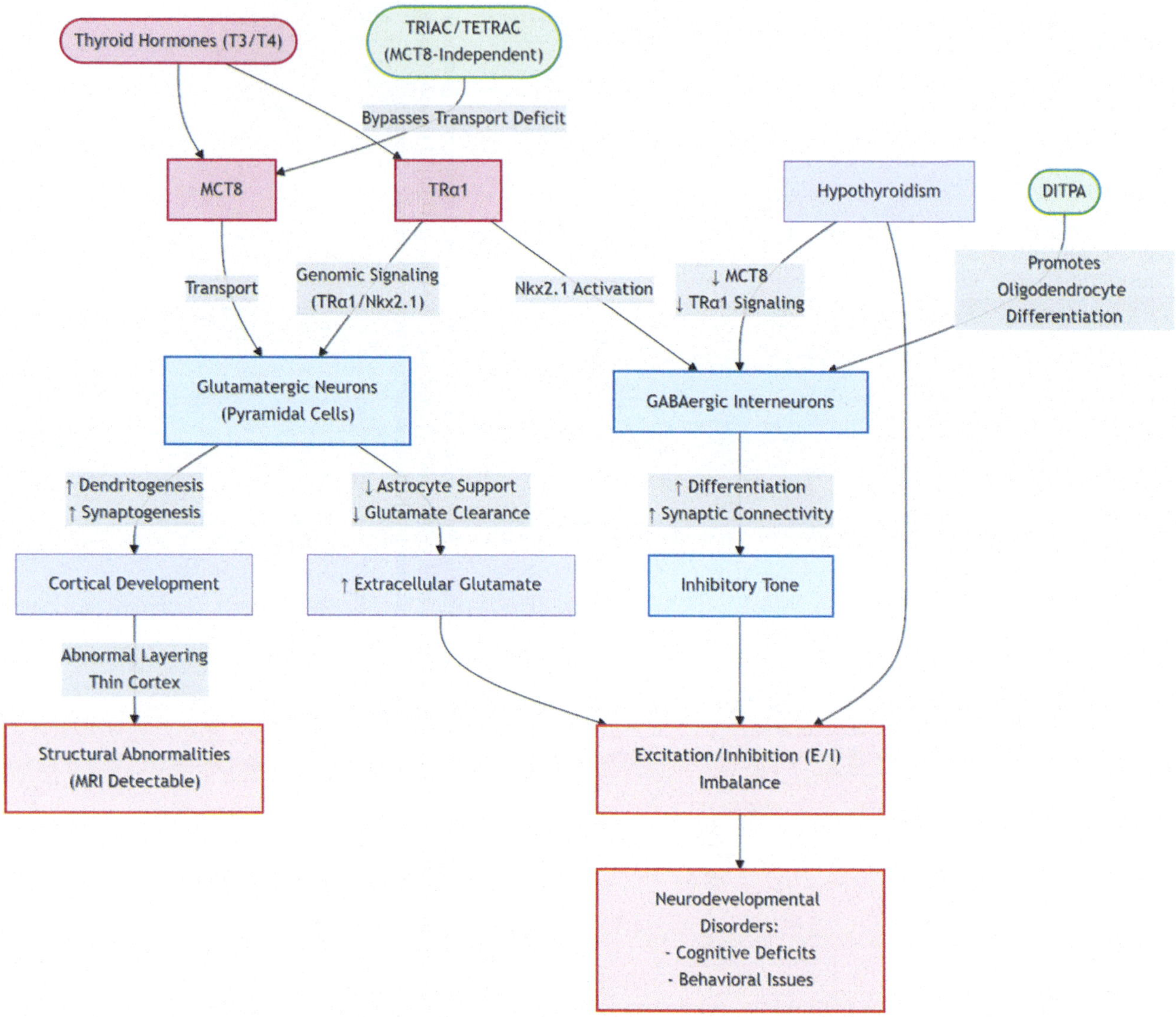

FIGURE 2.4 TH dysregulation of excitation/inhibition (E/I) Balance. Mechanisms and consequences: **Glutamatergic pathway**: MCT8/TRα1-dependent dendritic growth ↑ synaptogenesis; astrocyte dysfunction → glutamate clearance ↓. **GABAergic pathway**: Nkx2.1 activation ↑ interneuron differentiation; hypothyroidism (↓ MCT8/TRα1) → synaptic connectivity ↓. **Outcomes**: E/I imbalance → cognitive/behavioral deficits (e.g., ADHD); cortical layering defects (MRI-detectable). **Therapies**: TRIAC/TETRAC (bypass MCT8); DITPA (Wnt-Notch inhibition).

cells (NSCs) to advance their neurogenic commitment. Concurrently, it engages non-genomic PI3K Akt mTOR signaling in Tbr2+ intermediate progenitors (IPs), changing their metabolic and survival pathways. These two distinct mechanisms in separate cell types illustrate how genomic and non-genomic TH actions can be functionally uncoupled within the same tissue to regulate disparate facets of neurodevelopmental progression. The mention of the "T3-PI3K Akt mTOR signaling axis" in neural stem cells was incorporated into this sentence for ease of reading (Nappi et al., 2021).

Heterochronic signatures of THs in neurodevelopment

Heterochrony is the alteration of ontogeny due to changes in the rate or timing of developmental processes, and places THs as heterochronic regulators of dual function for brain development. Dual outcome following the influence of THs stems from a dynamic reciprocal interaction between genomic (receptor-mediated) and non-genomic (mitochondrial) processes that are critically responsible for the rate of self-organization and subsequent expression of form and function in the developing brain (Zhou et al., 2024).

Mitochondrial non-genomic actions of THs are primarily cell cycle accelerants. TH activation of mitochondria shortens G1 phase duration, accelerating progress and "winding developmental time forward" at the cellular level (Gorospe et al., 2023). At the population level, this shorter G1 duration enhances proliferative capacity and enhances synchronicity of cycling neural progenitors (Molina et al., 2022). The synchronized, accelerated cycling lowers the probability of differentiation through two major mechanisms: the rapid passage through the G1 phase increases the resistance to differentiation stimuli, and synchronized cycling limits differentiation-sensitive G1 phases to short temporal windows. The T2 acts on mitochondrial respiratory chains directly by binding to subunit Va of cytochrome *c* oxidase, thus abolishing ATP's allosteric inhibition (Senese et al., 2018) and increasing the yield of ROS, which subsequently facilitates neuronal differentiation and brain development.

In contrast, TR-mediated genomic effects oppose non-genomic effects (Tambones & le Maire, 2024). Genomic action, mediated largely by L-T3 binding to nuclear TRα and TRβ isoforms (Aranda, 2025a,2025b; Zhu & Cheng, 2018), results in cell cycle arrest and prime cycling cells for differentiation, for example, upregulate pro-neural transcription factor NeuroD during cerebellar maturation (Vujovic & Farahani, 2025). Hypothyroidism significantly slows brain development due to reduced pace of cell self-organization (Vujovic & Farahani, 2025), resulting in reduced bilateral cerebellar and pallidum volumes (Chambers et al., 2021), reduced neuron density within neurogenic regions, disrupted oligodendrocyte maturation, impaired intrahemispheric commissure formation, and delayed developmental milestones (Salazar et al., 2019) including subplate disappearance and external granular layer regression (Bernal, 2000).

Neurodevelopmental consequences following IUGR

Intrauterine growth restriction (IUGR) is defined by limited fetal growth and development, most commonly caused by a compromised supply of oxygen and nutrients via the placenta, which is typically a consequence of uteroplacental vascular insufficiency (Calek et al., 2023) that compromises 25%–30% of IUGR incidences.

The compromised oxygen and nutrition supply has a significant impact on the growth of the fetal brain (Ramirez Zegarra et al., 2022), with loss of both gray and white matter structure and integrity. Magnetic resonance imaging studies demonstrate reduced cerebral cortical gray matter and intracranial volumes of IUGR preterm newborns compared to appropriate-for-gestational-age controls (Hansen et al., 2022). Hippocampal gray matter volume loss, deranged cortical development patterns (Brosch et al., 2022) like deeper fissures and smaller brain volumes have been described. At the neuronal level, the animal models display neuronal cell death, dysynchronous migration patterns, reduced dendritic extension and branching, and aberrant synaptic pruning in cerebral cortex (Wixey et al., 2019), hippocampus, and cerebellum (Sakai, 2020).

Aside from gray matter alterations, IUGR results in extreme disturbance of white matter tracts development, for example, reduced myelination and connectivity, that are implicated in long-term neurodevelopmental compromise (Fung, 2023). MRI changes exhibit retarded myelination (Barkovich & Deon, 2016), reduced fractal dimension of white matter indicating reduced structural complexity, and disturbed connectivity patterns. Reduced fractional anisotropy, indicative of white matter complexity, has been noted in several tracts (Rimol et al., 2019).

Such profound structural and cellular brain deficiencies render IUGR babies susceptible to adverse neurodevelopmental effects, including cognitive, motor, and behavioral dysfunction (Sacchi et al., 2021). Human studies consistently report associations of IUGR with poorer motor function, cognition, language (Hartkopf et al., 2018; Murray et al., 2015), vision, and hearing outcomes. Longitudinal investigations reveal greater likelihood of learning disability (Ji et al., 2021), memory deficit, attention deficit, and executive function deficit. Animal models also exhibit the same cognitive, motor, and behavioral dysfunction like persistent motor coordination defects, gait disturbance, and spatial memory deficits. No effective therapy yet exists to prevent or reverse IUGR; clinical management is largely reliant on antenatal monitoring and optimal timing of delivery (Mutamba et al., 2022). This calls for urgent preclinical and clinical studies to evolve safe and effective treatments that avert long-term neurodevelopmental effects.

Cerebral TH signaling: Implication in IUGR

Cerebral TH signaling plays a critical role in normal brain development and growth, regulating fundamental processes including neuronal maturation and migration, neuronal and glial cell differentiation, myelinization, corticogenesis, and synaptogenesis (de Souza, 2024). Of particular importance, TH signaling is disrupted in IUGR and can make a significant contribution to neurodevelopmental delay (Chincarini et al., 2024).

THs primarily exist as thyroxine (T4), the precursor, and T3, the biologically active form (Carvalho & Dupuy, 2017). Although less than 1% of circulating TH is unbound and free to enter cells, cell uptake is largely mediated by specific thyroid hormone transmembrane transporters. Of these, MCT8 is the most active and is the sole transporter that is known to transport THs exclusively (Tan et al., 2025). CNS MCT8 is critical for the transport of THs across blood-brain and blood-cerebrospinal fluid barriers (Guillén-Yunta et al., 2023) to facilitate the entry of T3 into neurons (Tan et al., 2025). Approximately 75% of T3 entry into neurons from various regions of the brain is mediated by MCT8 (Salas-Lucia & Bianco, 2022; Tan et al., 2025). Intracellular conversion of T4 to active T3 is done by D2 (Murakami, 2016), and it binds to nuclear TH receptors to regulate gene expression in brain development (Sabatino et al., 2021).

The critical role of MCT8 is also evident in Allan-Herndon-Dudley syndrome (AHDS), caused by mutations in the SLC16A2 gene encoding MCT8 (Beheshti et al., 2022). AHDS typically present with elevated serum free T3, low free T4, and severe neurodevelopmental abnormalities like hypotonia, spasticity, dystonia, and delayed brain maturation with hypomyelination and reduced white matter volume (Beheshti et al., 2022; Rodrigues et al., 2014). These neurologic abnormalities resemble those in cerebral hypothyroidism and IUGR (Beheshti et al., 2022).

Most significantly, MCT8 mRNA expression is markedly reduced in 24–28 weeks gestation postmortem human IUGR fetal cortical plates, a prominent neuropathological characteristic (Graffunder et al., 2024). Other supportive evidence includes exceedingly low circulating free T3 and T4 in IUGR fetuses, inversely related maternal free T4 levels and birthweight, and reduced TH receptor isoform expression in cerebral cortex and cerebellum. With reduced MCT8 expression and impaired TH uptake in IUGR, conventional TH therapies will be ineffective (Hartley et al., 2019; Mutamba et al., 2023) and even cause severe thyrotoxicosis, highlighting the creation of new therapeutic approaches bypassing MCT8 transport deficiency (Chincarini et al., 2024).

TH analogs: Therapeutic potential in the IUGR brain

Given dysregulated TH signaling and decreased MCT8 expression in IUGR fetal brains, TH analogs are potential therapeutic agents (Chincarini et al., 2024). The chief benefit of these analogs is binding endogenous TH receptors with avoidance of MCT8 requirements for cellular uptake, so transport deficiency seen in IUGR is overcome while systemic side effects like peripheral thyrotoxicosis from traditional TH replacement are evaded (van Geest et al., 2021).

There are three major TH analogs that have been investigated as drugs: 3′,3,5-triiodothyroacetic acid (TRIAC), 3′,5′,3,5-tetraiodothyroacetic acid (TETRAC), and 3,5-diiodothyropropionic acid (DITPA) (Senese et al., 2019). The iodoamino acids are T3 and T4 derivatives (Köhrle, 2019).

TRIAC and TETRAC are metabolites of T3 and T4 deamination and oxidative decarboxylation, respectively, and TETRAC is a pro-hormone of TRIAC (Ayers et al., 2012). TRIAC clinical trials in children and adult patients with congenital MCT8 deficiency have shown an alleviation of peripheral thyrotoxicosis (Chen et al., 2023). There were mild but temporary side effects of increased perspiration and irritability, but these resolved with the withdrawal of treatment (Groeneweg et al., 2019). A follow-up trial is currently evaluating the long-term neurodevelopmental benefits in male infants with MCT8 deficiency. Preclinical evaluation of MCT8/OATP1C1 double-knockout and MCT8-knockout mouse models shows the potential of TRIAC and TETRAC to stimulate T3-dependent gene expression and neuronal differentiation (Chen et al., 2023).

DITPA is an MCT8-independent TH receptor agonist. In MCT8-negative cell lines examined in vitro, DITPA treatment increased oligodendrocyte differentiation transcription factors and myelin protein gene expression. DITPA has greater potency compared to T3 for inhibiting Wnt-Notch signaling pathways that usually inhibit oligodendrocyte differentiation. Clinical and preclinical studies of DITPA determine safety profiles (Lee et al., 2017). In children lacking MCT8, oral DITPA restored peripheral thyroid function and suppressed hypermetabolism. Most strikingly, in two children, brain imaging demonstrated that abnormally delayed myelination at 9 months was normalized by 47 months after treatment with DITPA. Maternal DITPA administration in MCT8-knockout mice demonstrated placental and fetal tissue crossing, neuronal cell entry, and TH-dependent regulation of gene expression, ameliorating cerebral hypothyroidism without peripheral thyrotoxic effects (Ferrara et al., 2014). Despite encouraging outcomes in MCT8-deficient conditions, comprehensive preclinical and clinical investigations with appropriate IUGR animal models are needed to ascertain long-term efficacy and safety of such analogs before clinical trials in vulnerable IUGR individuals (Chincarini et al., 2024; Visser, 2021).

Conclusion

The chapter reviewed a large body of evidence demonstrating that even subtle disruptions of TH signaling can have profound long-term neurodevelopmental consequences, which can result from various causes including environmental toxins, genetic defects, or developmental disorders such as IUGR. Several excitatory, inhibitory, cholinergic, and dopaminergic systems are complexly affected, with frequent dysfunctions specific to particular circuits. The effects go beyond a single neuronal type.

Notably, the hormonal actions involved are mediated by both non-genomic and genetic pathways, operating in a time-dependent fashion that can at times produce opposing effects at critical periods of brain development. Detectable changes in brain anatomy, that is, cortical thinning, suboptimal hippocampal organization, or changes in myelination, are overt manifestations of such early hormonal disruptions. Standard treatments currently in use with TH replacement may not always provide beneficial outcomes, especially when cellular transport mechanisms (e.g., MCT8) are defective, as can happen with IUGR or in various genetic diseases. This makes identifying TH analogs that do not require MCT8, overcome biological barriers, and support brain development without causing systemic side effects an urgent requirement.

In sum, the identification of TH signaling as a critical, time-regulated influence on brain development presents opportunities for new research, improved diagnostics, and more effective therapeutic interventions. Looking ahead, the investigation of methods to manipulate the timing of brain development using targeted therapies should take high priority—holding out encouraging possibilities for improving outcomes in at-risk populations.

Mini-dictionary of terms

- **Adverse Outcome Pathway (AOP)**: An organized concept that connects a molecular initiating event, for example, interference with hormone signaling, to harmful biological effects like defective brain development or behavioral impairment.
- **Cholinergic Neurons**: Nerve cells that utilize acetylcholine as a neurotransmitter, important for activities such as memory, learning, and motor control, and regulated by hormone signaling in brain development.
- **Excitation/Inhibition (E/I) Balance**: The balance between excitatory (e.g., glutamatergic) and inhibitory (e.g., GABAergic) neural activity, crucial for normal brain function and deranged by hormonal imbalances.
- **Genomic Signaling**: A process in which hormones attach to nuclear receptors in order to control gene expression, cell differentiation, and brain development on longer time scales.
- **Intrauterine Growth Restriction (IUGR)**: A condition of restricted fetal growth caused by inadequate supply of nutrients and oxygen, usually resulting in neurodevelopmental disabilities.
- **Monocarboxylate Transporter 8 (MCT8)**: A protein that mediates the passage of hormones through cell membranes, essential for the supply of active hormones to neurons in developing brains.
- **Non-Genomic Signaling**: Fast cell responses to hormones that take place outside the nucleus, including stimulation of signaling pathways or mitochondrial function, influencing cell development and metabolism.
- **Nucleophagy**: A type of autophagy that breaks down nuclear entities, momentarily decreasing genomic activity in cell differentiation so that rapid hormone-stimulated processes can take over.
- **Thyroid Hormone Analogs**: Synthetic compounds, such as TRIAC and DITPA, that aim to reproduce hormone actions and circumvent defective transport mechanisms for the treatment of neurodevelopmental abnormalities.

Summary points

- The thyroid hormone signaling pathways control the essential processes of neurodevelopment, such as differentiation, migration, myelination, and synaptogenesis of neurons, via genomic and non-genomic actions.
- Intrauterine growth restriction (IUGR) interferes with cerebral thyroid hormone signaling, primarily through decreased expression of monocarboxylate transporter 8 (MCT8), which is associated with neurodevelopmental disorders such as cortical thinning and hypomyelination.
- The Thyroid-NDT-TAS system screens neurodevelopmental toxicity by quantifying cognition, behavior, and brain structure using standard OECD guidelines to detect thyroid hormone disruption.
- Thyroid hormone analogs, such as TRIAC and DITPA, hold promise for the therapy of IUGR-related brain defects through bypassing defective MCT8 transport and restoring hormone-dependent gene expression.
- Maternal hypothyroidism in pregnancy is associated with structural changes in the brain, decreased IQ, and a higher risk of neurological disorders such as epilepsy, autism, and ADHD in the child.
- Non-genomic effects of thyroid hormone, mediated via mitochondrial reactive oxygen species and integrin signaling, accelerate cell cycle progression and neural progenitor synchronization during development.
- Dysregulation of thyroid hormone signaling affects a number of neuronal systems, including glutamatergic, GABAergic, cholinergic, and dopaminergic neurons, and leads to cognitive, motor, and behavioral defects.
- The genomic actions of thyroid hormones, which are mediated by nuclear receptors, promote cellular differentiation but may antagonize non-genomic actions, thus creating a dynamic balance that is essential for the timing of brain development.

Declaration of AI use

During the preparation of this work, the authors used ChatGPT, from OpenAI, to improve readability and correct grammar errors. After using this tool/service, the authors reviewed and edited the content as needed and take full responsibility for the content of the publication.

References

Ahmed, N. Y., Knowles, R., & Dehorter, N. (2019). New insights into cholinergic neuron diversity. *Frontiers in Molecular Neuroscience, 12*, 204.

Alcaide Martin, A., & Mayerl, S. (2023). Local thyroid hormone action in brain development. *International Journal of Molecular Sciences, 24*(15), Article 12352. https://doi.org/10.3390/ijms241512352

Anderson, G. W., Schoonover, C. M., & Jones, S. A. (2003). Control of thyroid hormone action in the developing rat brain. *Thyroid, 13*(11), 1039–1056.

Aranda, A. (2025a). Thyroid hormone action by genomic and non-genomic molecular mechanisms. In C. Contreras-Jurado (Ed.), *Thyroid hormones: Methods and protocols* (pp. 17–34). New York, NY: Springer US.

Aranda, A. (2025b). Thyroid hormone action by genomic and non-genomic molecular mechanisms. *Methods in Molecular Biology, 2876*, 17–34. https://doi.org/10.1007/978-1-0716-4252-8_2

Ayers, S., Baxter, J. D., & Webb, P. (2012). *Chapter 9: Thyroid hormone structure-function relationships.*

Barkovich, A. J., & Deon, S. (2016). Hypomyelinating disorders: An MRI approach. *Neurobiology of Disease, 87*, 50–58. https://doi.org/10.1016/j.nbd.2015.10.015

Beheshti, R., Aprile, J., & Lee, C. (2022). Allan-herndon-dudley syndrome: A novel pathogenic variant of the SLC16A2 gene. *Cureus, 14*(1), Article e21771. https://doi.org/10.7759/cureus.21771

Bernal, J. (2000). Thyroid hormones in brain development and function. In K. R. Feingold, S. F. Ahmed, B. Anawalt, M. R. Blackman, A. Boyce, G. Chrousos, E. Corpas, W. W. de Herder, K. Dhatariya, K. Dungan, J. Hofland, S. Kalra, G. Kaltsas, N. Kapoor, C. Koch, P. Kopp, M. Korbonits, C. S. Kovacs, W. Kuohung, et al. D. P. Wilson (Eds.), *Endotext*. South Dartmouth (MA): Copyright © 2000-2025, MDText.com, Inc.

Bernal, J. (2024). *Thyroid and brain: Understanding the actions of thyroid hormones in brain development and function*. Bentham Science Publishers.

Bianco, A. C., & Kim, B. W. (2006). Deiodinases: Implications of the local control of thyroid hormone action. *Journal of Clinical Investigation, 116*(10), 2571–2579. https://doi.org/10.1172/jci29812

Brosch, K., Stein, F., Schmitt, S., Pfarr, J.-K., Ringwald, K. G., Thomas-Odenthal, F., … Kircher, T. (2022). Reduced hippocampal gray matter volume is a common feature of patients with major depression, bipolar disorder, and schizophrenia spectrum disorders. *Molecular Psychiatry, 27*(10), 4234–4243. https://doi.org/10.1038/s41380-022-01687-4

Calek, E., Binder, J., Palmrich, P., Eibensteiner, F., Thajer, A., Kainz, T., … Binder, C. (2023). Effects of intrauterine growth restriction (IUGR) on growth and body composition compared to constitutionally small infants. *Nutrients, 15*(19). https://doi.org/10.3390/nu15194158

Carvalho, D. P., & Dupuy, C. (2017). Thyroid hormone biosynthesis and release. *Molecular and Cellular Endocrinology, 458*, 6–15. https://doi.org/10.1016/j.mce.2017.01.038

Chambers, T., Anney, R., Taylor, P. N., Teumer, A., Peeters, R. P., Medici, M., … Rees, D. A. (2021). Effects of thyroid status on regional brain volumes: A diagnostic and genetic imaging study in UK biobank. *The Journal of Clinical Endocrinology and metabolism, 106*(3), 688–696. https://doi.org/10.1210/clinem/dgaa903

Chen, C., Ma, Q., Chen, X., Zhong, M., Deng, P., Zhu, G., … Zhang, K. (2015). Thyroid hormone-Otx2 signaling is required for embryonic ventral midbrain neural stem cells differentiated into dopamine neurons. *Stem Cells and Development, 24*(15), 1751–1765.

Chen, J., Salveridou, E., Liebmann, L., Sundaram, S. M., Doycheva, D., Markova, B., … Mayerl, S. (2023). Triac treatment prevents neurodevelopmental and locomotor impairments in thyroid hormone transporter Mct8/Oatp1c1 deficient mice. *International Journal of Molecular Sciences, 24*(4), 3452. Retrieved from https://www.mdpi.com/1422-0067/24/4/3452.

Chincarini, G., Walker, D. W., Wong, F., Richardson, S. J., Cumberland, A., & Tolcos, M. (2024). Thyroid hormone analogues: Promising therapeutic avenues to improve the neurodevelopmental outcomes of intrauterine growth restriction. *Journal of Neurochemistry, 168*(9), 2335–2350. https://doi.org/10.1111/jnc.16124

de Souza, J. S. (2024). Chapter ten - thyroid hormone biosynthesis and its role in brain development and maintenance. In R. Donev (Ed.), *Advances in protein chemistry and structural biology* (Vol 142, pp. 329–365). Academic Press.

Dentice, M., & Salvatore, D. (2019). Deiodination and peripheral metabolism of thyroid hormone. In M. Luster, L. H. Duntas, & L. Wartofsky (Eds.), *The thyroid and its diseases: A comprehensive guide for the clinician* (pp. 61–72). Cham: Springer International Publishing.

Ferrara, A. M., Liao, X. H., Gil-Ibáñez, P., Bernal, J., Weiss, R. E., Dumitrescu, A. M., & Refetoff, S. (2014). Placenta passage of the thyroid hormone analog DITPA to male wild-type and Mct8-deficient mice. *Endocrinology, 155*(10), 4088–4093. https://doi.org/10.1210/en.2014-1085

Fung, C. M. (2023). Effects of intrauterine growth restriction on embryonic hippocampal dentate gyrus neurogenesis and postnatal critical period of synaptic plasticity that govern learning and memory function. *Frontiers in Neuroscience, 17*, Article 1092357. https://doi.org/10.3389/fnins.2023.1092357

Gorospe, C. M., Carvalho, G., Herrera Curbelo, A., Marchhart, L., Mendes, I. C., Niedźwiecka, K., & Wanrooij, P. H. (2023). Mitochondrial membrane potential acts as a retrograde signal to regulate cell cycle progression. *Life Science Alliance, 6*(12), Article e202302091. https://doi.org/10.26508/lsa.202302091 (%J Life Science Alliance).

Graffunder, A. S., Bresser, A. A. J., Fernandez Vallone, V., Megges, M., Stachelscheid, H., Kühnen, P., & Opitz, R. (2024). Spatiotemporal expression of thyroid hormone transporter MCT8 and THRA mRNA in human cerebral organoids recapitulating first trimester cortex development. *Scientific Reports, 14*(1), 9355. https://doi.org/10.1038/s41598-024-59533-2

Groeneweg, S., Peeters, R. P., Moran, C., Stoupa, A., Auriol, F., Tonduti, D., … Visser, W. E. (2019). Effectiveness and safety of the tri-iodothyronine analogue triac in children and adults with MCT8 deficiency: An international, single-arm, open-label, phase 2 trial. *Lancet Diabetes Endocrinol, 7*(9), 695–706. https://doi.org/10.1016/s2213-8587(19)30155-x

Guillén-Yunta, M., Valcárcel-Hernández, V., García-Aldea, Á., Soria, G., García-Verdugo, J. M., Montero-Pedrazuela, A., & Guadaño-Ferraz, A. (2023). Neurovascular unit disruption and blood–brain barrier leakage in MCT8 deficiency. *Fluids and Barriers of the CNS, 20*(1), 79. https://doi.org/10.1186/s12987-023-00481-w

Hansen, T. M., Muthulingam, J. A., Brock, B., Drewes, A. M., Juhl, A., Vorum, H., … Frøkjær, J. B. (2022). Reduced gray matter brain volume and cortical thickness in adults with type 1 diabetes and neuropathy. *Neuroscience Research, 176*, 66–72. https://doi.org/10.1016/j.neures.2021.10.002

Hartkopf, J., Schleger, F., Keune, J., Wiechers, C., Pauluschke-Froehlich, J., Weiss, M., … Kiefer-Schmidt, I. (2018). Impact of intrauterine growth restriction on cognitive and motor development at 2 years of age. *Frontiers in Physiology, 9*, 1278. https://doi.org/10.3389/fphys.2018.01278

Hartley, M. D., Banerji, T., Tagge, I. J., Kirkemo, L. L., Chaudhary, P., Calkins, E., … Scanlan, T. S. (2019). Myelin repair stimulated by CNS-selective thyroid hormone action. *JCI Insight, 4*(8). https://doi.org/10.1172/jci.insight.126329

Ishii, S., Amano, I., & Koibuchi, N. (2021). The role of thyroid hormone in the regulation of cerebellar development. *Endocrinology and Metabolism, 36*(4), 703–716.

Ji, L., Zhao, Q., Gu, H., Chen, Y., Zhao, J., Jiang, X., & Wu, L. (2021). Effect of executive function on event-based prospective memory for different forms of learning disabilities. *Frontiers in Psychology, 12* (2021). https://doi.org/10.3389/fpsyg.2021.528883

Köhrle, J. (2019). The colorful diversity of thyroid hormone metabolites. *European Thyroid Journal, 8*(3), 115–129. https://doi.org/10.1159/000497141

Lasa, M., & Contreras-Jurado, C. (2022). Thyroid hormones act as modulators of inflammation through their nuclear receptors. *Frontiers in Endocrinology, 13*, Article 937099.

Lee, J. Y., Kim, M. J., Deliyanti, D., Azari, M. F., Rossello, F., Costin, A., … Petratos, S. (2017). Overcoming monocarboxylate transporter 8 (MCT8)-deficiency to promote human oligodendrocyte differentiation and myelination. *EBioMedicine, 25*, 122–135. https://doi.org/10.1016/j.ebiom.2017.10.016

Lee, S.-R., Yang, K.-S., Kwon, J., Lee, C., Jeong, W., & Rhee, S. G. (2002). Reversible inactivation of the tumor suppressor PTEN by H2O2. *Journal of Biological Chemistry, 277*(23), 20336–20342.

Leto, K., Arancillo, M., Becker, E. B., Buffo, A., Chiang, C., Ding, B., … Hatten, M. E. (2016). Consensus paper: Cerebellar development. *The Cerebellum, 15*, 789–828.

Magno, L., Barry, C., Schmidt-Hieber, C., Theodotou, P., Häusser, M., & Kessaris, N. (2017). NKX2-1 is required in the embryonic septum for cholinergic system development, learning, and memory. *Cell Reports, 20*(7), 1572–1584.

Mohamadian, M., Fallah, H., Ghofrani-Jahromi, Z., Rahimi-Danesh, M., Shokouhi Qare Saadlou, M.-S., & Vaseghi, S. (2023). Mood and behavior regulation: Interaction of lithium and dopaminergic system. *Naunyn-Schmiedeberg's Archives of Pharmacology, 396*(7), 1339–1359.

Molina, A., Bonnet, F., Pignolet, J., Lobjois, V., Bel-Vialar, S., Gautrais, J., … Agius, E. (2022). Single-cell imaging of the cell cycle reveals CDC25B-induced heterogeneity of G1 phase length in neural progenitor cells. *Development, 149*(11). https://doi.org/10.1242/dev.199660

Morte, B., Manzano, J., Scanlan, T., Vennström, B., & Bernal, J. (2002). Deletion of the thyroid hormone receptor α1 prevents the structural alterations of the cerebellum induced by hypothyroidism. *Proceedings of the National Academy of Sciences, 99*(6), 3985–3989.

Murakami, M. (2016). Deiodinase and brain development. In N. Koibuchi, & P. M. Yen (Eds.), *Thyroid hormone disruption and neurodevelopment* (pp. 23–31). New York, NY: Springer New York.

Murray, E., Fernandes, M., Fazel, M., Kennedy, S. H., Villar, J., & Stein, A. (2015). Differential effect of intrauterine growth restriction on childhood neurodevelopment: A systematic review. *BJOG: an International Journal of Obstetrics and Gynaecology, 122*(8), 1062–1072. https://doi.org/10.1111/1471-0528.13435

Mutamba, A. K., He, X., & Wang, T. (2022). Therapeutic advances in overcoming intrauterine growth restriction induced metabolic syndrome. *Frontiers in Pediatrics, 10*, Article 1040742. https://doi.org/10.3389/fped.2022.1040742

Mutamba, A. K., He, X., & Wang, T. (2023). Therapeutic advances in overcoming intrauterine growth restriction induced metabolic syndrome. *Frontiers in Pediatrics, 10*(2022). https://doi.org/10.3389/fped.2022.1040742

Nappi, A., Murolo, M., Sagliocchi, S., Miro, C., Cicatiello, A. G., Di Cicco, E., … Stornaiuolo, M. (2021). Selective inhibition of genomic and non-genomic effects of thyroid hormone regulates muscle cell differentiation and metabolic behavior. *International Journal of Molecular Sciences, 22*(13), 7175.

Özsoy, Ş., Vujovic, F., Simonian, M., Valova, V., Hunter, N., & Farahani, R. M. (2021). Cannibalized erythroblasts accelerate developmental neurogenesis by regulating mitochondrial dynamics. *Cell Reports, 35*(1).

Okada, K., Hashimoto, K., & Kobayashi, K. (2022). Cholinergic regulation of object recognition memory. *Frontiers in Behavioral Neuroscience, 16*, Article 996089.

Patel, J., Landers, K., Li, H., Mortimer, R. H., & Richard, K. (2011). Thyroid hormones and fetal neurological development. *%J Journal of Endocrinology, 209*(1), 1–8. https://doi.org/10.1530/joe-10-0444

Ramirez Zegarra, R., Dall'Asta, A., & Ghi, T. (2022). Mechanisms of fetal adaptation to chronic hypoxia following placental insufficiency: A review. *Fetal Diagnosis and Therapy, 49*(5–6), 279–292. https://doi.org/10.1159/000525717

Rimol, L. M., Botellero, V. L., Bjuland, K. J., Løhaugen, G. C. C., Lydersen, S., Evensen, K. A. I., … Skranes, J. (2019). Reduced white matter fractional anisotropy mediates cortical thickening in adults born preterm with very low birthweight. *NeuroImage, 188*, 217–227. https://doi.org/10.1016/j.neuroimage.2018.11.050

Rodrigues, F., Grenha, J., Ortez, C., Nascimento, A., Morte, B., M-Belinchón, M., … Colomer, J. (2014). Hypotonic male infant and MCT8 deficiency - A diagnosis to think about. *BMC Pediatrics, 14* (1), 252. https://doi.org/10.1186/1471-2431-14-252

Sabatino, L., Vassalle, C., Del Seppia, C., & Iervasi, G. (2021). Deiodinases and the three types of thyroid hormone deiodination reactions. *Endocrinology and Metabolism, 36*(5), 952–964. https://doi.org/10.3803/EnM.2021.1198

Sacchi, C., O'Muircheartaigh, J., Batalle, D., Counsell, S. J., Simonelli, A., Cesano, M., … Nosarti, C. (2021). Neurodevelopmental outcomes following intrauterine growth restriction and very preterm birth. *The Journal of Pediatrics, 238*, 135–144. e110. https://doi.org/10.1016/j.jpeds.2021.07.002

Sakai, J. (2020). How synaptic pruning shapes neural wiring during development and, possibly, in disease. *Proceedings of the National Academy of Sciences of the United States of America, 117*(28), 16096–16099.

Salas-Lucia, F., & Bianco, A. C. (2022). T3 levels and thyroid hormone signaling. *Frontiers in Endocrinology, 13*, Article 1044691. https://doi.org/10.3389/fendo.2022.1044691

Salazar, P., Cisternas, P., Martinez, M., & Inestrosa, N. C. (2019). Hypothyroidism and cognitive disorders during development and adulthood: Implications in the central nervous system. *Molecular Neurobiology, 56*(4), 2952–2963. https://doi.org/10.1007/s12035-018-1270-y

Schneider, S. A., Tschaidse, L., & Reisch, N. (2023). Thyroid disorders and movement disorders—Asystematic review. *Movement Disorders Clinical Practice, 10*(3), 360–368.

Senese, R., Cioffi, F., Petito, G., Goglia, F., & Lanni, A. (2019). Thyroid hormone metabolites and analogues. *Endocrine, 66*(1), 105–114. https://doi.org/10.1007/s12020-019-02025-5

Senese, R., de Lange, P., Petito, G., Moreno, M., Goglia, F., & Lanni, A. (2018). 3,5-Diiodothyronine: A novel thyroid hormone metabolite and potent modulator of energy metabolism. *Frontiers in Endocrinology, 9*, 427. https://doi.org/10.3389/fendo.2018.00427

Sillitoe, R. V., & Vogel, M. W. (2008). Desire, disease, and the origins of the dopaminergic system. *Schizophrenia Bulletin, 34*(2), 212–219.

Talhada, D., Santos, C. R. A., Gonçalves, I., & Ruscher, K. (2019). Thyroid hormones in the brain and their impact in recovery mechanisms after stroke. *Frontiers in Neurology, 10*(2019). https://doi.org/10.3389/fneur.2019.01103

Tambones, I., & le Maire, A. (2024). Structural insights into thyroid hormone receptors. *Endocrinology, 166*(1). https://doi.org/10.1210/endocr/bqae154

Tan, J., Xiao, Y., Kong, F., Qian, J., Zhu, A., & Yan, C. (2025). Structural insights into thyroid hormone transporter MCT8. *Nature Communications, 16*(1), 2958. https://doi.org/10.1038/s41467-025-58131-8

Tedeschi, L., Vassalle, C., Iervasi, G., & Sabatino, L. (2021). Main factors involved in thyroid hormone action. *Molecules, 26*(23), 7337.

van Geest, F. S., Gunhanlar, N., Groeneweg, S., & Visser, W. E. (2021). Monocarboxylate transporter 8 deficiency: From pathophysiological understanding to therapy development. *Frontiers in Endocrinology, 12*(2021). https://doi.org/10.3389/fendo.2021.723750

Visser, W. E. (2021). Therapeutic applications of thyroid hormone analogues. *Annales d'Endocrinologie, 82*(3), 170–172. https://doi.org/10.1016/j.ando.2020.03.004

Vujovic, F., & Farahani, R. M. (2025). Thyroid hormones and brain development: A focus on the role of mitochondria as regulators of developmental time. *Cells, 14*(3), 150. https://doi.org/10.3390/cells14030150

Vujovic, F., Shepherd, C. E., Witting, P. K., Hunter, N., & Farahani, R. M. (2023). Redox-mediated rewiring of signalling pathways: The role of a cellular clock in brain health and disease. *Antioxidants, 12*(10), 1873.

Williams, G. R. (2008). Neurodevelopmental and neurophysiological actions of thyroid hormone. *Journal of Neuroendocrinology, 20*(6), 784–794.

Wixey, J. A., Lee, K. M., Miller, S. M., Goasdoue, K., Colditz, P. B., Tracey Bjorkman, S., & Chand, K. K. (2019). Neuropathology in intrauterine growth restricted newborn piglets is associated with glial activation and proinflammatory status in the brain. *Journal of Neuroinflammation, 16*(1), 5. https://doi.org/10.1186/s12974-018-1392-1

Zhou, J. L., Hong, W., & Kao, J. C. J. A. (2024). *Reciprocal reward influence encourages cooperation from self-interested agents. abs/2406.01641.*

Zhu, X., & Cheng, S.-y. (2018). Thyroid hormone nuclear receptors and molecular actions. In A. Belfiore, & D. LeRoith (Eds.), *Principles of endocrinology and hormone action* (pp. 233–257). Cham: Springer International Publishing.

Zoeller, R. T., & Rovet, J. (2004). Timing of thyroid hormone action in the developing brain: Clinical observations and experimental findings. *Journal of Neuroendocrinology, 16*(10), 809–818. https://doi.org/10.1111/j.1365-2826.2004.01243.x

Chapter 3

Enzymes involved in thyroid hormone metabolism and status

Stanisław Surma[1] and Kazimierz Ciechanowski[2]

[1]Department of Internal Medicine and Clinical Pharmacology, Medical University of Silesia Katowice, Poland; [2]Department of Nephrology, Transplantology and Internal Medicine, Pomeranian Medical University of Szczecin, Szczecin, Poland

List of abbreviations

DIO1–3 thyroid hormone deiodinases
DUOX1/2 NADPH oxidases in the thyroid gland
MCT8/MCT10 monocarboxylate transporters
MIT/DIT mono- and diiodotyrosine
NIS sodium-iodide symporter
OATP organic anion transporters
rT_3 reverse triiodothyronine
T_3 triiodothyronine
T_4 thyroxine
TBG thyroxine-binding globulin
Tg thyroglobulin
TPO thyroperoxidase
TRH thyrotropin-releasing hormone
TSH thyroid-stimulating hormone
TTR transthyretin

Introduction

Thyroid hormones have a multifaceted mechanism of action, playing a crucial role in regulating metabolic processes, differentiation, and growth. In fact, thyroid hormones affect most body tissues. The biosynthesis and secretion of thyroxine (T_4) and triiodothyronine (T_3) are precisely regulated by a negative feedback system that involves the hypothalamus, pituitary, and thyroid gland. The metabolic function of the thyroid gland, its growth, and development are regulated by thyroid-stimulating hormone (TSH), secreted by the anterior pituitary gland (Cicatiello et al., 2018; Köhrle, 2023; Ortiga-Carvalho et al., 2016).

TSH stimulation of the TSH receptor (TSH-R) increases intracellular cyclic adenosine monophosphate (cAMP) levels and stimulates protein kinase A (PKA)-dependent metabolic pathways. TSH stimulates the transcription of numerous genes for thyroid proteins, including the sodium-iodide symporter (NIS), thyroglobulin (Tg), and thyroid peroxidase (TPO). Iodine supplied with food is absorbed in the stomach (NOTE! Iodine is not absorbed through the respiratory tract) and small intestine and transported to the thyroid gland, where it is actively taken up into the NIS and then stored in follicular cells. Iodide ions are oxidized by TPO in the presence of hydrogen peroxide (H_2O_2) and attached to the tyrosine residues of Tg. This results in the formation of 3-monoiodotyrosine (MIT) and 3,5-diiodotyrosine (DIT) residues, which are the substrate for the formation of T_4 and T_3 (via conjugation). Th is stored in the colloid of thyroid follicular cells. Within the apical membrane, the colloid undergoes pinycytosis and is transported into the cell, where, in conjunction with lysosomes, its contents, including Tg, are digested. The released T_4 and T_3 are secreted into the blood and transported to target tissues by carrier proteins (thyroxine-binding globulin - TGB, transthyretin - TTR and albumin) (Cicatiello et al., 2018; Köhrle, 2023).

Thyroid hormones: Structure

Thyroid hormones are tyrosine derivatives containing iodine. Their molecules consist of two aromatic rings connected by an ether bridge (–O–). The tyrosine residues are linked in such a way that they are positioned obliquely, which is crucial for their biological activity. During hormone synthesis, iodine is sequentially incorporated into positions 3 and 5 of the thyroglobulin tyrosine ring, leading to the formation of mono- and diiodotyrosine (MIT and DIT). Subsequently, the conjugation of two DIT molecules produces thyroxine (T_4), while the combination of MIT with DIT leads to the synthesis of triiodothyronine (T_3). The thyroid gland primarily produces T_4, which is converted in peripheral tissues to the more biologically active T_3 (D'Aurizio et al., 2023).

Hormones in Neuroscience. https://doi.org/10.1016/B978-0-443-30261-9.00003-6

Iodine: Biochemical transformations

Iodine constitutes approximately 65% of the mass of a thyroxine molecule and 58% of the mass of triiodothyronine. This element is supplied to the body primarily through the diet in the form of organic and inorganic compounds. Before absorption in the gastrointestinal tract, iodine contained in food is reduced to the iodide form (I^-). Both iodides and iodinated amino acids (such as T_4 and T_3) are well absorbed in the small intestine. Absorption is rapid and almost complete. In healthy individuals, serum iodine concentrations are low, and excess iodine is excreted primarily by the kidneys (Köhrle, 2023; Zhang et al., 2024).

The average daily iodine requirement for adults is 150–200 μg. The thyroid gland absorbs only a portion of the iodine supplied, while the remaining amount circulates throughout the body and is eliminated in the urine. Iodine stored in the thyroid gland is primarily found in thyroglobulin and provides a sufficient supply for several weeks of normal hormone synthesis. To ensure adequate iodine intake, it is recommended to use table salt fortified with this element and consume foods naturally rich in iodine, such as saltwater fish, dairy products, and mineral water (Köhrle, 2023; Zhang et al., 2024).

Thyroid follicular cells are characterized by the ability to actively uptake iodide from the blood, leading to a significant increase in its concentration within the gland compared to plasma. This process occurs thanks to the NIS, located in the basolateral membrane of thyrocytes, which utilizes the sodium gradient generated by the Na^+/K^+-ATPase pump. Iodides are then transported into the lumen of the thyroid follicle via the iodide transporter located in the apical membrane of the cells (Köhrle, 2023; Zhang et al., 2024).

NIS

The NIS is a membrane protein located in the basolateral part of thyroid follicular cells. Its function is the active transport of iodide from the blood into the thyrocytes, a process that is strictly dependent on the presence of sodium ions. The gene encoding NIS is located on chromosome 19 and is responsible for the synthesis of a protein of approximately 85 kDa, composed of 643 amino acids. Structurally, NIS consists of 13 domains spanning the cell membrane (Gadisi et al., 2025; Ravera et al., 2022).

The mechanism of action of the symporter involves the simultaneous movement of iodide and sodium ions in the same direction, along the electrochemical sodium gradient. This transport requires two Na^+ ions for one I^- ion and is indirectly dependent on the activity of the sodium-potassium pump (Na^+/K^+-ATPase), which maintains a low intracellular sodium concentration. The energy required for NIS function comes from ATP hydrolysis, which is used to maintain an ionic gradient. The symporter exhibits high selectivity for iodide but can also transport other anions with similar chemical properties (Gadisi et al., 2025; Ravera et al., 2022).

Under normal conditions, iodide transport activity and NIS expression are regulated primarily by TSH. This hormone increases both protein synthesis and its functional activity. The process of iodine accumulation in the thyroid gland is also subject to feedback mechanisms, dependent on iodine concentration in the blood and within follicular cells (Gadisi et al., 2025; Ravera et al., 2022).

Abnormalities in the NIS gene, including point mutations, deletions, or altered protein localization within the cell, can lead to impaired iodide transport into thyrocytes. Such abnormalities result in reduced thyroid hormone synthesis, which can clinically manifest as hypothyroidism. Numerous variants of NIS mutations have been described, varying in the degree of their impact on the symporter's transport function (Gadisi et al., 2025; Ravera et al., 2022).

NIS expression is not limited to the thyroid gland. This protein has also been detected in other tissues, such as the salivary glands, gastric mucosa, mammary gland, ovaries, and placenta. In these organs, NIS participates in iodide transport, but its regulation and physiological significance differ from those observed in the thyroid. In some tumors, including thyroid cancers of varying degrees of differentiation, abnormal NIS expression is observed, which has significant diagnostic and therapeutic implications (Gadisi et al., 2025; Ravera et al., 2022).

Pendrin

After iodide enters thyrocytes, its further transport into the follicle lumen is mediated by pendrin. This membrane protein, located in the apical part of follicular cells, facilitates the passage of I^- ions into the colloid, where thyroid hormone synthesis occurs. Pendrin has a highly hydrophobic structure, adapted to function in the cell membrane (Kopp et al., 2008; Szybinski, 2017; Tamma & Dossena, 2022; Targovnik et al., 2017).

The gene encoding pendrin (PDS) is located on chromosome 7 and is responsible for the synthesis of a protein of approximately 115 kDa, composed of approximately 780 amino acids. Pendrin belongs to the family of anion transporters and may also participate in the transport of other inorganic ions. Experimental studies indicate that this protein plays a key role in the proper iodination of thyroglobulin and the effective synthesis of thyroid hormones (Kopp et al., 2008; Szybinski, 2017; Tamma & Dossena, 2022; Targovnik et al., 2017).

Pendrin activity is controlled by thyrotropin and thyroglobulin. TSH stimulates the expression of this protein, while high iodine concentrations lead to its inhibition.

In Graves' disease, elevated levels of pendrin are observed, whereas in most thyroid cancers, its expression is decreased. Furthermore, in neoplastic lesions, this protein often exhibits abnormal localization, remaining in the cytoplasm instead of the cell membrane (Kopp et al., 2008; Szybinski, 2017; Tamma & Dossena, 2022; Targovnik et al., 2017).

Alterations in the PDS gene cause Pendred syndrome, an autosomal recessive genetic disorder. This condition is characterized by congenital deafness, partial impairment of iodine utilization by the thyroid gland, and the presence of goiter, which occurs in approximately half of patients. Genetic analysis has revealed over 90 different mutations, most often leading to loss of protein function or its retention in the cytoplasm (Kopp et al., 2008; Szybinski, 2017; Tamma & Dossena, 2022; Targovnik et al., 2017).

In recent years, autoantibodies against pendrin have been detected in the serum of patients with autoimmune thyroid disease in selected populations, including Japan. However, their role in pathogenesis and clinical significance have not been clearly defined. In addition to the thyroid, pendrin is also found in other organs, such as the kidneys, middle ear, and lungs (Kopp et al., 2008; Szybinski, 2017; Tamma & Dossena, 2022; Targovnik et al., 2017).

Apical iodine transporter (AIT)

AIT is a protein consisting of approximately 610 amino acids and exhibiting nearly 46% sequence similarity to the sodium-iodide symporter. It belongs to the same family of transporters. Functional studies suggest that AIT participates in the passive movement of iodide ions from the thyrocytes to the lumen of the thyroid follicle (Kopp et al., 2008; Szybinski, 2017; Tamma & Dossena, 2022; Targovnik et al., 2017).

Wolff–chaikoff effect

The most important factor regulating iodine accumulation in the thyroid gland, alongside TSH, is iodine itself. The function of the thyroid gland depends on its availability in the diet and on the concentration of iodide ions in the serum. Low I^- concentrations stimulate iodine uptake, while excess iodine—especially when the supply significantly exceeds daily requirements—inhibits further accumulation of the element (Khudair et al., 2025).

Increased iodine supply leads to an increase in intrathyroidal iodine concentration, which results in suppression of the gland's hormonal activity. This process involves reduced iodide uptake, reduced hydrogen peroxide production, inhibition of iodine organification, and decreased synthesis and secretion of thyroid hormones. This mechanism, independent of TSH, is referred to as the Wolff-Chaikoff effect (Khudair et al., 2025).

Thyroid suppression is transient and typically lasts 25–50 hours. Prolonged high iodine concentrations lead to a glandular adaptation known as "escape" from the Wolff-Chaikoff effect, which involves a decrease in intrathyroidal iodine concentration and a gradual restoration of hormone synthesis. This adaptation is associated with a decrease in NIS mRNA expression and a shortened half-life of this transporter. Both mechanisms are part of the precise autoregulation of the thyroid gland, protecting it from the toxic effects of excess iodine while simultaneously enabling the synthesis of T_3 and T_4 under physiological conditions (Khudair et al., 2025).

Thyroid hormone synthesis

The process of producing thyroid hormones is shown in Fig. 3.1.

TPO

Thyroid peroxidase plays a key role in the biosynthesis of thyroid hormones and is essential for the proper functioning of the gland. It is a glycoprotein located in the apical membrane of follicular cells, facing the follicle lumen, where essential steps in hormone synthesis occur (Zhou et al., 2025).

The gene encoding TPO is located on chromosome 2p25, and its expression is controlled by thyroid-specific transcription factors such as TTF-1, TTF-2, and Pax-8. Alternative mRNA splicing leads to the generation of several protein variants, most of which, in shorter forms, lack enzymatic activity due to the absence of the region encoded by exons 8–10. Inactive TPO variants do not integrate into the membrane and are rapidly degraded (Koibuchi, 2018; Zhou et al., 2025).

The TPO protein consists of 933 amino acids, the vast majority of which is the extracellular domain, containing the catalytic center with a heme molecule. The remaining structure consists of a transmembrane segment and a short cytoplasmic segment. TPO domains exhibit structural similarity to myeloperoxidase, complement proteins, and epidermal growth factor. TPO synthesis occurs on polysomes of the endoplasmic reticulum, after which the protein undergoes glycosylation in the Golgi apparatus and is transported to the apical pole of the cell along with Tg (Koibuchi, 2018; Zhou et al., 2025).

TPO catalyzes:

- the oxidation of iodide ions to reactive forms,
- the incorporation of iodine into thyroglobulin tyrosine residues (organization),

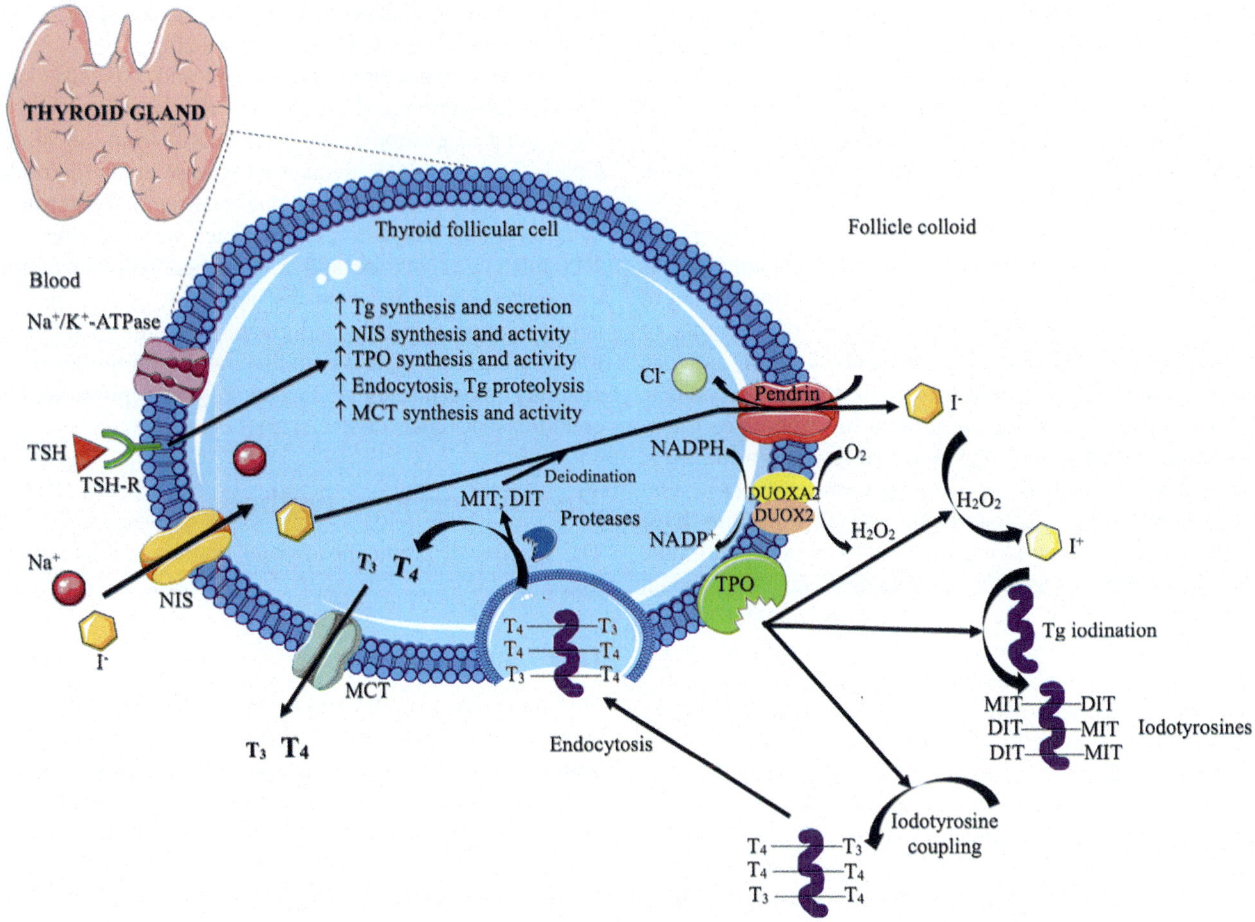

FIGURE 3.1 Synthesis and secretion of thyroid hormones.

- the conjugation of iodotyrosines, leading to the formation of thyroglobulin-bound T_3 and T_4.

Iodotyrosine conjugation

The conjugation process is a multi-step process that includes an oxidative phase, a non-oxidative phase, and a degradative phase. The effectiveness of the reaction depends on the structural integrity of thyroglobulin and the appropriate arrangement of iodotyrosine residues. During thyroglobulin proteolysis, thyroid hormone molecules and dehydroalanine are released (Koibuchi, 2018; Zhou et al., 2025).

Pathophysiological significance of TPO

TPO is one of the three main thyroid autoantigens involved in autoimmune processes. Antibodies directed against TPO play an important role in the pathogenesis of autoimmune thyroid diseases (Koibuchi, 2018; Zhou et al., 2025).

The vast majority of patients with autoimmune thyroid diseases, such as Hashimoto's thyroiditis and Graves' disease, have autoantibodies directed against thyroid peroxidase. Therefore, anti-TPO antibodies are a key serological marker for these conditions. Thyroid peroxidase is also one of the best-known thyroid autoantigens, and the structures recognized by autoantibodies are primarily located within the A and B domains of the protein (Koibuchi, 2018; Zhou et al., 2025).

Mutations in the gene encoding thyroid peroxidase are among the most common causes of congenital hypothyroidism resulting from defects in thyroid hormone synthesis. These disorders are inherited in an autosomal recessive manner and lead to impaired iodine organization (Koibuchi, 2018; Zhou et al., 2025).

Oxidases involved in thyroid hormone synthesis

The processes catalyzed by thyroid peroxidase require the presence of hydrogen peroxide, which serves as a secondary substrate for the reaction. H_2O_2 in the thyroid is produced by specialized membrane enzymes belonging to the NADPH oxidase family—DUOX1 and DUOX2. These enzymes are located in the apical membrane of thyrocytes, and their activity depends on the presence of calcium ions. Thyrotropin (TSH) stimulates DUOX, thereby increasing hydrogen peroxide production, while high iodine concentrations within the thyroid gland lead to inhibition of H_2O_2 synthesis (Koibuchi, 2018; Zhou et al., 2025).

Tg

Thyroglobulin is the main iodine-containing protein in the thyroid gland and is the primary substrate for thyroid hormone synthesis. It is a glycoprotein that accounts for approximately 75%–80% of the total thyroid protein pool. It performs two essential biological functions: it participates directly in the production of the thyroid hormones T_3 and T_4 and serves as their storage site within the follicular colloid (Coscia et al., 2020; Koibuchi, 2018; Zhou et al., 2025).

The gene encoding thyroglobulin is located on chromosome 8, in the 8q24.2–q24.3 region, and is one of the largest genes in humans. The gene's large size translates into a complex protein structure, enabling the proper iodination of tyrosine residues and the efficient formation of thyroid hormones (Coscia et al., 2020; Koibuchi, 2018; Zhou et al., 2025).

The Tg gene spans approximately 300,000 base pairs and contains 48 exons. Its expression product is a protein composed of approximately 2760 amino acids. Transcription of the TG gene is regulated by thyroid-specific factors and stimulated by TSH. The synthesis of thyroglobulin monomers occurs in the endoplasmic reticulum, where they undergo folding, dimerization, and post-translational modifications, including glycosylation. Sugar residues constitute over 10% of the molecule's mass and play a crucial role in both thyroid hormone synthesis and protein immunoreactivity (Coscia et al., 2020; Koibuchi, 2018; Zhou et al., 2025).

Mature thyroglobulin is transported in secretory vesicles to the apical pole of thyrocytes, from where it is secreted into the lumen of the thyroid follicle (Coscia et al., 2020; Koibuchi, 2018; Zhou et al., 2025).

Tg is the largest protein produced by the thyroid gland, with a mass of approximately 660 kDa. It exists as a dimer composed of two subunits of approximately 330 kDa each. It is the main component of follicular colloid and serves as a prohormone, also serving as a storage site for thyroid hormones (Coscia et al., 2020; Koibuchi, 2018; Zhou et al., 2025).

The Tg molecule contains 132 tyrosine residues, but only a small fraction (approximately 25–40) are iodinated. These residues are arranged to facilitate efficient conjugation reactions. During these reactions, one tyrosine residue donates an iodinated phenyl group to the other, leading to the formation of iodothyronine, while dehydroalanine remains in the donor site (Coscia et al., 2020; Koibuchi, 2018; Zhou et al., 2025).

The TG structure is divided into four key hormone-producing regions (A, B, C, and D), with domain A responsible for the synthesis of triiodothyronine. A single thyroglobulin molecule can contain four to six thyroid hormone molecules. In addition to its precursor function, thyroglobulin participates in the regulation of thyroid function, influencing the activity of transcription factors and indirectly modulating the expression of thyroid genes (Coscia et al., 2020; Koibuchi, 2018; Zhou et al., 2025).

Besides the predominant soluble form (19S), heavier soluble fractions (27 and 37S) and insoluble forms also occur, accounting for approximately one-third of the total Tg pool (Coscia et al., 2020; Koibuchi, 2018; Zhou et al., 2025).

For thyroid hormones to exert biological activity, they must be released from the thyroglobulin molecule and transported to their site of action. The first step in this process is the endocytosis of thyroglobulin from the lumen of the thyroid follicle (Coscia et al., 2020; Koibuchi, 2018; Zhou et al., 2025).

Colloid internalization can occur via both micropinocytosis and macropinocytosis. In micropinocytosis, tiny depressions form in the apical membrane of thyrocytes, leading to the formation of vesicles containing thyroglobulin, a process regulated by TSH. Macropinocytosis, on the other hand, involves the invagination of larger membrane fragments, which encompass significant amounts of colloid and form droplets that are then transported to lysosomes (Coscia et al., 2020; Koibuchi, 2018; Zhou et al., 2025).

After internalization, endocytic vesicles fuse into endosomes, where initial sorting of thyroglobulin occurs. Endosomes then fuse with lysosomes containing proteolytic enzymes such as cathepsins D, B, and L. These enzymes lead to the gradual degradation of thyroglobulin and the release of thyroid hormones (Coscia et al., 2020; Koibuchi, 2018; Zhou et al., 2025).

Research indicates that Tg degradation occurs in two stages: initially, selective proteolysis leads to the release of T_3 and T_4, followed by complete protein degradation, resulting in the formation of inactive iodotyrosines, peptides, and free amino acids. Highly iodinated thyroglobulin is preferentially degraded, while weakly iodinated molecules can recirculate and be retransported to the apical or basal membrane of thyrocytes. This mechanism is responsible for the presence of thyroglobulin in the blood (Coscia et al., 2020; Koibuchi, 2018; Zhou et al., 2025).

Some Tg can be degraded within the lumen of the thyroid follicle, allowing for the maintenance of an adequate pool of this protein in the colloid and the preservation of its structure necessary for biological functions (Coscia et al., 2020; Koibuchi, 2018; Zhou et al., 2025).

Thyroid hormone metabolism

The primary thyroid hormone acting on target tissues is T_3, a physiological ligand for nuclear receptors. Over 80% of T_3 is produced extrathyroidally, primarily in the liver, kidneys, and skeletal muscle, through enzymatic deiodination of T_4 (Sabatino et al., 2021).

Removal of the iodine atom from the 5′ position of the outer ring of T_4 produces biologically active T_3, while deiodination at the 5′ position of the inner ring results in the synthesis of an inactive metabolite, rT_3. Under physiological conditions, approximately one-third of T_4 is converted to T_3, a similar fraction to rT_3, and the remaining fraction is converted to other metabolites (Sabatino et al., 2021).

Thyroid hormone deiodination

T_3 is formed by the enzymatic removal of an iodine atom from the outer ring of thyroxine, while deiodination occurring in the inner ring leads to the synthesis of rT_3, which is biologically inactive. These processes are key to regulating the amount of active thyroid hormones in the body (Sabatino et al., 2021).

Deiodination reactions are catalyzed by three types (izoforms) of enzymes belonging to the deiodinase family: D1, D2, and D3 (Table 3.1) (Sabatino et al., 2021).

These are selenoproteins containing a selenocysteine residue in their active center, essential for the proper conduct of the reaction. Deiodinases differ in their tissue

TABLE 3.1 Characterization of deiodinases (DIO).

	DIO1	DIO2	DIO3
Subcellular location	Cell membrane	ER membrane	Cell membrane
Principal tissue location	Liver, kidney, thyroid	BAT, skeletal muscle, heart, CNS	CNS, skin, placenta
Deiodination site in THs	Outer/inner ring	Outer ring	Inner ring
Reaction	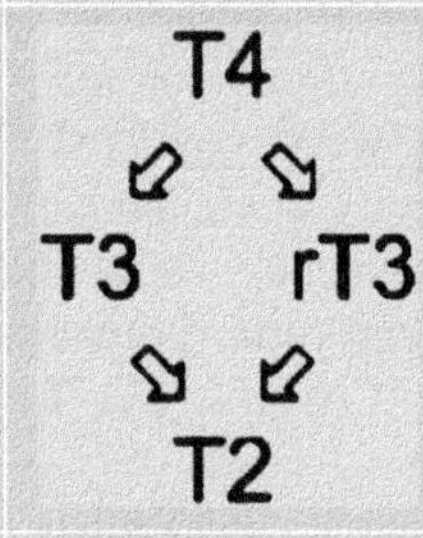	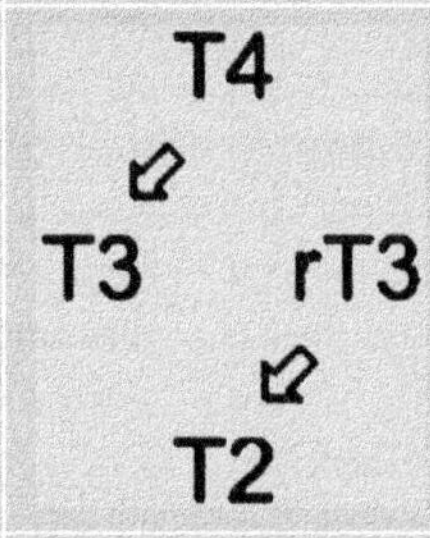	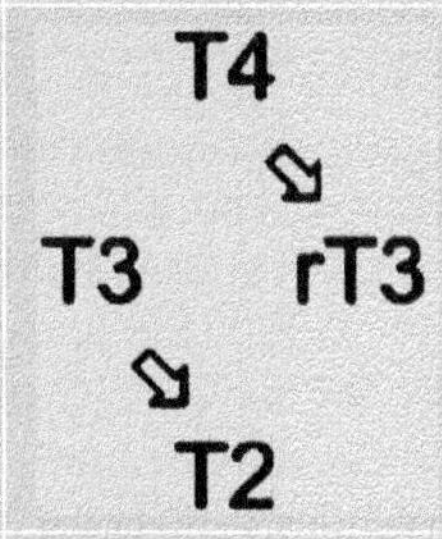
Substrate preference	rT3 > T4 > T3	T4 > rT3	T3 > T4
Thiol cofactor	Stimulatory	Stimulatory	Stimulatory
Propylthiouracil	Inhibitory	No effect	No effect
Km for main substrate	1–2 μM	1–2 nM	~40 nM
Kinetic mechanism	Ping-pong	Sequential	Sequential
Half-life	>12 hours	40 minutes	12 hours
Key biological role	• Contributes to systemic thyroid hormone homeostasis. • Peripheral conversion of T4→T3 and clearance of rT3. • Functions as a recycling/ detoxification enzyme in liver and kidney.	• Major source of local intracellular T3 production in target tissues. • Critical for hypothalamic–pituitary feedback regulation. • Supports adaptive thermogenesis in brown adipose tissue.	• Main inactivating enzyme terminating thyroid hormone action (anti-T3 safeguard). • Protects fetus and placenta from excessive thyroid hormone exposure. • Regulates brain thyroid hormone levels and prevents overactivation of T3 signaling.
Hypothyroidism	Decrease	Increase	Decrease
Hyperthyroidism	Increase	Decrease	Increase

Based on information from Luongo C, Dentice M, Salvatore D. Deiodinases and their intricate role in thyroid hormone homeostasis. Nature Reviews. Endocrinology. 2019;15(8):479-488. https://doi.org/10.1038/s41574-019-0218-2; Peeters, R.P., & Visser, T.J. (2000). Metabolism of thyroid hormone [Updated 2017 Jan 1]. In Feingold, K.R., Adler, R.A., & Ahmed, S.F., et al. (Eds.), *Endotext*. South Dartmouth (MA): MDText.com, Inc. Available from https://www.ncbi.nlm.nih.gov/books/NBK285545/; Sabatino L, Vassalle C, Del Seppia C, Iervasi G. Deiodinases and the three types of thyroid hormone deiodination reactions. Endocrinology and Metabolism. 2021;36(5):952-964. https://doi.org/10.3803/EnM.2021.1198.

localization, substrate specificity, and physiological function, enabling precise control of thyroid hormone concentrations in both the systemic circulation and individual tissues (Sabatino et al., 2021).

These enzymes are not only responsible for the production of T_3 but also participate in the inactivation of thyroid hormones, which helps maintain hormonal homeostasis. Deiodinase activity can be regulated locally, independently of serum hormone concentrations (Sabatino et al., 2021).

Type 1 deiodinase (D1) catalyzes the removal of iodine from both the outer and inner rings of thyroid hormones. It is responsible not only for the activation of T_4 to T_3 but also for the degradation of the hormones through conversion to rT_3 and other metabolites. The highest activity of D1 is observed primarily in the liver, kidneys, and thyroid gland, and to a lesser extent in skeletal muscle (Sabatino et al., 2021).

This enzyme plays a crucial role in the regulation of plasma T_3 concentration. Its activity increases in hyperthyroidism and decreases in iodine deficiency and hypothyroidism. D1 is also sensitive to selenium availability, and its deficiency leads to impaired thyroid hormone deiodination (Sabatino et al., 2021).

Type 2 deiodinase (D2) removes an iodine atom exclusively from the outer ring of thyroxine, leading to the formation of biologically active triiodothyronine. This enzyme is found primarily in the central nervous system, pituitary gland, brown adipose tissue, skeletal muscle, and the heart (Sabatino et al., 2021).

D2 plays a key role in maintaining adequate T_3 concentrations in tissues that are particularly sensitive to thyroid hormones. In brown adipose tissue, D2 plays an important role in thyroid hormone-dependent thermogenesis. In the pituitary gland, it enables precise regulation of TSH secretion, regardless of fluctuations in circulating hormone levels. D2 expression increases in hypothyroidism, promoting increased local T_3 production, while it decreases in hyperthyroidism (Sabatino et al., 2021).

Type 3 deiodinase (D3) removes iodine exclusively from the inner ring of iodothyronine. Its action leads to the inactivation of thyroid hormones by converting thyroxine (T_4) to reverse triiodothyronine (rT_3) and triiodothyronine (T_3) to diiodothyronine (T_2). This enzyme is particularly active in the brain, placenta, and fetal tissues, where its expression is significantly higher than in adults. D3 activity increases in hyperthyroidism and decreases in hypothyroidism, which is a protective mechanism that stabilizes serum thyroid hormone levels (Luongo et al., 2019; Sabatino et al., 2021).

The metabolism of thyroid hormones under the influence of deiodinases is shown in Fig. 3.2.

Metabolism in the liver

Thyroid hormones undergo not only deiodination in the liver but also numerous enzymatic transformations, such as conjugation, deamination, and decarboxylation. Modifications involving the hydroxyl group and the alanine side chain are particularly well understood. These processes increase the aqueous solubility of metabolites, facilitating their excretion from the body via bile and urine (Cicatiello et al., 2018; Gnocchi et al., 2016).

The hydroxyl group of thyroid hormones can undergo conjugation with sulfates and glucuronic acid. These reactions occur via enzymes such as sulfotransferases and glucuronyltransferases. The concentrations of sulfate derivatives of thyroid hormones (e.g., T_4S, T_3S, or rT_3S) in serum and urine are very low because these compounds are rapidly degraded by deiodination (Cicatiello et al., 2018; Gnocchi et al., 2016).

Glucuronides of thyroid hormones, on the other hand, are excreted primarily via bile. Some of them can be degraded by bacterial β-glucuronidase in the small intestine, and the released thyroxine can be reabsorbed into the circulation, which is part of the enterohepatic circulation.

The amino and carboxyl groups of iodothyronine also undergo metabolic changes. Under the influence of L-amino acid decarboxylases and monoamine oxidases, derivatives such as tetracyloacetic acid (TETRAC) and triiodoacetic acid (TRIAC) are formed, which can then participate in glucuronic acid conjugation reactions (Cicatiello et al., 2018; Gnocchi et al., 2016).

Transport in serum

Thyroid hormones are hydrophobic and poorly soluble in water. For this reason, they are transported in the bloodstream primarily bound to carrier proteins. Only a small percentage of hormones are present in free form—approximately 0.03% of T_4 and 0.3% of T_3—and only this fraction is biologically available to peripheral tissues (Braun & Schweizer, 2018; Janssen & Janssen, 2017).

Thyroxine has a long half-life (approximately 7 days) and high plasma concentrations, which result from its strong binding to transport proteins (Braun & Schweizer, 2018; Janssen & Janssen, 2017).

Hormones released from thyroglobulin bind noncovalently in serum to three major transport proteins:

- thyroxine-binding globulin (TBG),
- transthyretin (TTR),
- albumin.

In addition to these, other molecules, such as HDL lipoproteins, also play a minor role. Albumin reaches the highest concentration of the carrier proteins, many times higher than TTR and TBG (Braun & Schweizer, 2018; Janssen & Janssen, 2017).

However, TBG has the greatest affinity for thyroid hormones, binding iodothyronine much more strongly than other proteins. It is estimated that approximately 75% of thyroxine is bound to TBG, approximately 20% to TTR, while albumin binds only about 5% of T_4 (Braun & Schweizer, 2018; Janssen & Janssen, 2017).

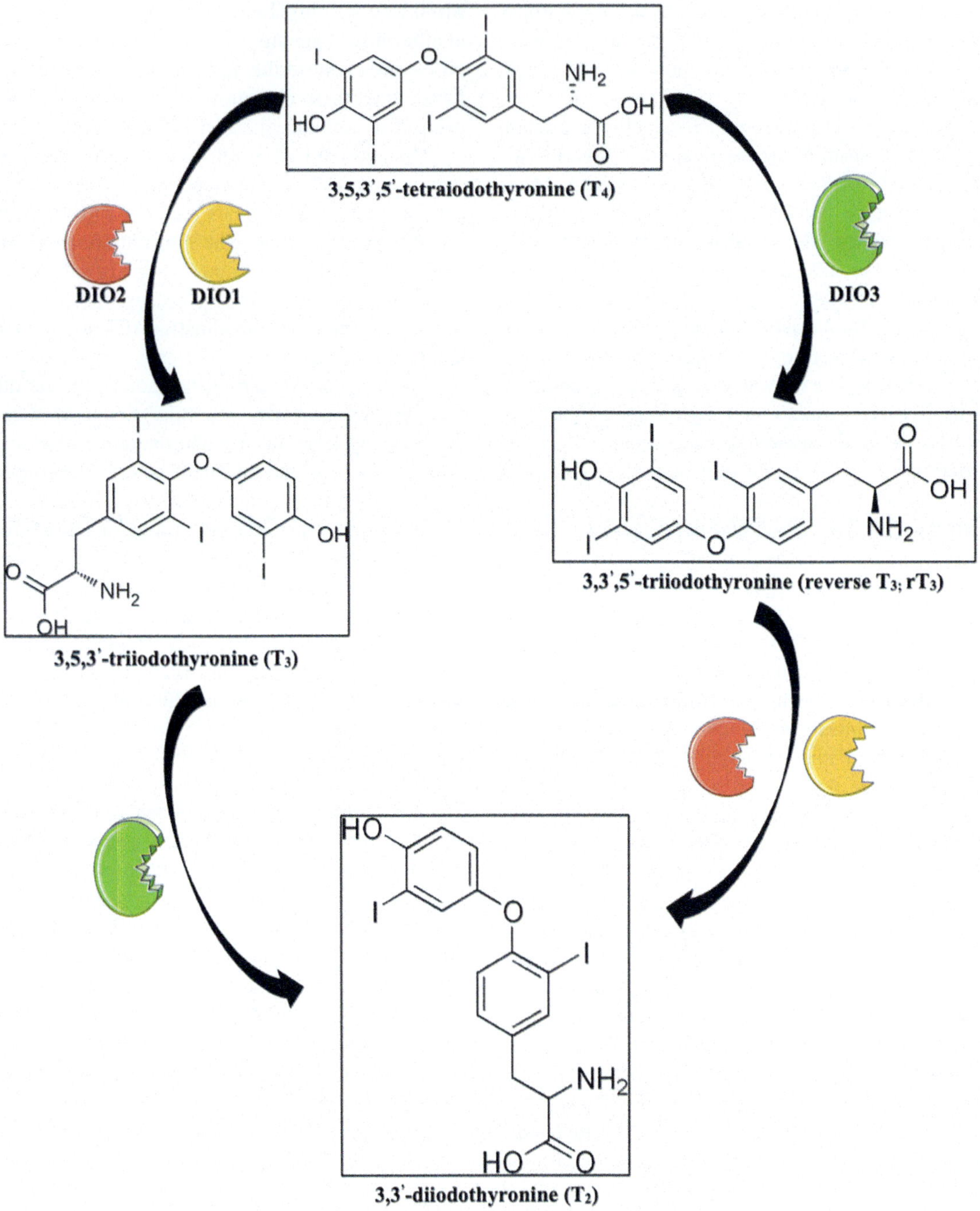

FIGURE 3.2 Deiodinase (DIO) thyroid hormone action.

TBG is an acidic glycoprotein of approximately 54 kDa, synthesized in the liver as a single polypeptide chain. It consists of 415 amino acids, and its gene is located on chromosome Xq22.2, near the binding sequence of the hepatic transcription factor HNF-1. This protein belongs to the SERPIN family (Braun & Schweizer, 2018; Janssen & Janssen, 2017).

The TBG molecule contains four sugar residues, which constitute approximately 20% of its mass and are essential for its proper structure. It has a single binding site for T_4, T_3, and rT_3, with the highest affinity for thyroxine. TBG can exist in high- and low-affinity conformations, allowing for both binding and release of hormones. The half-life of TBG is approximately 5 days (Braun & Schweizer, 2018; Janssen & Janssen, 2017).

Serum TBG concentration in healthy individuals ranges from 1.1 to 2.1 mg/dL, and its changes significantly impact total circulating thyroid hormone levels (Braun & Schweizer, 2018; Janssen & Janssen, 2017).

Transthyretin is a liver-synthesized protein of approximately 55 kDa, composed of four identical subunits of 127 amino acids each. The TTR gene is located on chromosome 18q11.2–q12.1, and its expression depends on the transcription factors HNF-1, -2, -3, and -4 (Braun & Schweizer, 2018; Janssen & Janssen, 2017).

TTR has two binding sites for thyroid hormones, although physiologically only one is usually occupied. The half-life of this protein is approximately 2 days. TTR's affinity for T_4 is approximately tenfold greater than for T_3. Under normal conditions, TTR transports approximately 20% of circulating thyroxine and less than 5% of T_3 (Braun & Schweizer, 2018; Janssen & Janssen, 2017).

Due to the dominant role of TBG in thyroid hormone binding, even significant changes in TTR concentration have little effect on free thyroxine levels (Braun & Schweizer, 2018; Janssen & Janssen, 2017).

Albumin is a plasma protein of approximately 66.5 kDa, synthesized in the liver. Composed of 585 amino acids, it is the most abundant serum protein and also serves as a nonspecific carrier of thyroid hormones (Braun & Schweizer, 2018; Janssen & Janssen, 2017).

Albumin has many binding sites for various compounds, but only one of them has the ability to bind thyroid hormones, primarily T_4 and, to a lesser extent, T_3. Albumin's affinity for iodothyronine is significantly weaker than that of TBG or transthyretin, despite its very high serum concentration of approximately 40 g/L. Albumin's half-life in circulation averages 10–15 days (Braun & Schweizer, 2018; Janssen & Janssen, 2017).

Due to albumin's ability to bind numerous hydrophobic substances, such as drugs and other steroid hormones, its role in thyroid hormone transport is nonspecific. Albumin constitutes approximately half of all plasma proteins, but its role as a carrier of T_4 and T_3 is limited. Even significant changes in its concentration do not significantly affect thyroid hormone levels in the blood (Braun & Schweizer, 2018; Janssen & Janssen, 2017).

Transport of thyroid hormones into cells

The biological activity of thyroid hormones depends largely on the concentration of free T_3 within the cell, which can bind to nuclear receptors. Thyroid hormones cross cell membranes not only by diffusion but also require the participation of specialized membrane transporters. In recent years, several groups of proteins with high affinity for iodothyronine have been identified, including (Bernal et al., 2015):

- the sodium-dependent taurocholate transporter (NTCP),
- sodium-independent organic anion transporters (OATPs),
- L-amino acid transporters (LAT1 and LAT2),
- monocarboxylate transporters, such as MCT8 and MCT10.

OATP proteins belong to a large family of membrane transporters that transport amphipathic organic compounds, including thyroid hormones. They are present in many organs, including the liver, kidneys, lungs, placenta, small intestine, and brain. Some isoforms exhibit high specificity for T_4 and T_3 and tissue-restricted expression (Bernal et al., 2015).

For example, OATP1B1 and OATP1B3 are found primarily in the liver, while OATP1C1 is characteristic of the central nervous system. No clear link has yet been demonstrated between OATP mutations and human diseases (Bernal et al., 2015).

Thyroid hormones are derivatives of amino acids, so their transport into cells can also occur via amino acid transporters. A special role in this process is attributed to the LAT1 and LAT2 transporters, which participate in the cellular uptake of iodothyronine (Bernal et al., 2015).

The most important specific thyroid hormone transporters include MCT8 and MCT10. They are widely distributed throughout the body and are found in the liver, kidneys, thyroid, heart, placenta, brain, skeletal muscle, and lungs, among other organs. MCT8 expression in the brain is particularly important for the proper functioning of the nervous system (Bernal et al., 2015).

MCT8 primarily transports T_4, T_3, and rT_3, but does not transport the sulfated metabolites of these hormones. MCT10 is also a functionally active thyroid hormone transporter, although its precise physiological role in various tissues has not yet been fully elucidated (Bernal et al., 2015).

In recent years, numerous mutations in the MCT8 gene have been described, leading to a severe neurological syndrome associated with impaired thyroid hormone transport into neurons—Allan-Herndon-Dudley syndrome (AHDS). However, no specific mutation in the MCT10 gene has yet been identified (Bernal et al., 2015).

Regulation of thyroid hormone secretion

Thyroid growth and function are controlled by a complex feedback system within the hypothalamic-pituitary-thyroid axis (Fig. 3.3).

The most important factor stimulating the gland is thyrotropin (TSH), secreted by the pituitary gland. The synthesis and release of TSH is stimulated by the hypothalamic hormone thyrotropin-releasing hormone (TRH) (Mullur et al., 2014; Hershman & Beck-Peccoz, 2023).

The gene encoding the precursor of TRH (prepro-TRH) is located on chromosome 3q21.1. TRH secretion in the hypothalamus is inhibited by negative feedback by thyroid hormones, particularly T_3, which can originate both from the blood and from local conversion of T_4 in the hypothalamic tissues (Mullur et al., 2014; Hershman & Beck-Peccoz, 2023).

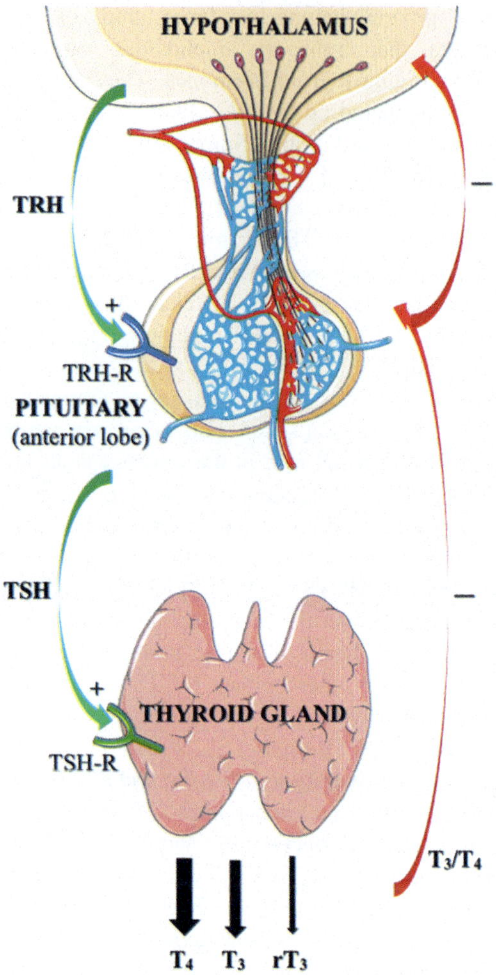

FIGURE 3.3 Regulation of the thyroid hormone axis. *Based on information from Cheng, X., Zhang, H., Guan, S., Zhao, Q., & Shan, Y. (2023). Receptor modulators associated with the hypothalamus-pituitary-thyroid axis.* Frontiers in Pharmacology, 14, *Article 1291856. https://doi.org/10.3389/fphar.2023.1291856. PMID: 38111381; PMCID: PMC10725963.*

TRH reaches the anterior pituitary via the portal circulation, where it binds to receptors on thyrotrophs and stimulates the synthesis and release of TSH. Increased thyroid hormone and TSH levels lead to inhibition of TRH gene expression and decreased thyrotropin-releasing hormone secretion. Furthermore, thyroid hormones limit TSH production by reducing the number of TRH receptors in the pituitary. The interplay of these mechanisms ensures stable thyroid hormone levels in the body (Mullur et al., 2014; Hershman & Beck-Peccoz, 2023).

TSH

TSH is a glycoprotein pituitary hormone and has a heterodimeric structure—it consists of two subunits: α and β, noncovalently linked. The α subunit is also common to other hormones, such as LH, FSH, and hCG, while the β subunit is responsible for the biological specificity of TSH (Mullur et al., 2014; Hershman & Beck-Peccoz, 2023).

The α chain contains approximately 92 amino acids, and the β chain approximately 118. The gene encoding the α subunit is located on chromosome 6, while the β subunit is located on chromosome 1. TSH is glycosylated, and the presence of sugar residues is essential for the hormone's normal biological activity. The synthesis and secretion of TSH are inhibited by high concentrations of thyroid hormones (Mullur et al., 2014; Hershman & Beck-Peccoz, 2023).

The release of TRH and TSH is controlled primarily by a negative feedback mechanism dependent on thyroid hormones. Experimental data indicate, however, that short regulatory loops may also exist in addition to the classic hypothalamic-pituitary-thyroid axis model. For example, TSH secreted by the pituitary can inhibit both TRH secretion in the hypothalamus and its own production within the pituitary. The physiological significance of these additional mechanisms has not yet been fully elucidated, although it is hypothesized that they may support the maintenance of hormonal homeostasis and the pulsatile nature of TSH secretion (Mullur et al., 2014; Hershman & Beck-Peccoz, 2023).

Conclusions

This chapter discusses the key steps in the synthesis, storage, transport, and metabolism of thyroid hormones. The importance of iodine as an essential component of T3 and T4 is presented, along with the role of the transporters NIS and pendrin in its uptake and translocation within the thyroid gland. The functions of thyroglobulin as a precursor protein and thyroperoxidase, which catalyzes iodination and tyrosine conjugation, are described. Particular attention is paid to deiodinases, which regulate hormone activation and inactivation in peripheral tissues. The importance of the hypothalamic-pituitary-thyroid axis in maintaining hormonal balance and the Wolff-Chaikof effect is also emphasized.

Mini-dictionary of terms

- NIS (sodium-iodide symporter) – a protein enabling the active transport of iodide into thyrocytes.
- Pendrin – an apical transporter that transports iodine into the lumen of the thyroid follicle.
- Thyroperoxidase (TPO) – an enzyme catalyzing the oxidation of iodine and the iodination of thyroglobulin.
- Deiodinases (D1–D3) – enzymes that convert T4 into active or inactive metabolites.
- Wolff–Chaikoff effect – a transient inhibition of thyroid hormone synthesis in the presence of excess iodine.

Summary points

- The thyroid gland's most important role is the production and release of T_4 and T_3 hormones, which are essential for proper body growth, cell maturation, and the regulation of numerous metabolic pathways.
- A key component of thyroid hormone molecules is iodine, which constitutes a significant portion of the mass of thyroxine. This element is supplied in the diet, absorbed in the gastrointestinal tract, and transported to the thyroid gland as iodide ions. Their uptake into thyrocytes is actively mediated by the Na^+/I^- symporter present in the basement membrane of the cells. The iodides are then transported into the lumen of the thyroid follicle by pendrin, located in the apical membrane.
- Thyrocytes produce two essential proteins essential for hormone synthesis: thyroglobulin (TG) and thyroid peroxidase (TPO). Thyroglobulin is a large glycoprotein secreted into the colloid, containing tyrosine residues that can be iodinated. It serves as the direct precursor of thyroid hormones and their storage site.
- Thyroid peroxidase, a membrane enzyme, catalyzes the oxidation of iodide ions to reactive forms and their incorporation into the tyrosine residues of thyroid hormones (TG). Hydrogen peroxide, which is provided by the enzymes DUOX1 and DUOX2, is essential for these reactions. The resulting mono- and diiodotyrosines are then conjugated, leading to the synthesis of thyroglobulin-bound T_4 and T_3.
- When the body's demand for thyroid hormones increases, thyroglobulin is endocytosed from the colloid into the thyrocytes. Endocytic vesicles fuse with lysosomes, where TG is proteolytically degraded, and the released hormones enter the bloodstream.
- Thyroid hormones do not circulate in free form in plasma; they are transported by carrier proteins, primarily TBG, transthyretin, and albumin, which enable their stable transport to target tissues.
- The thyroid gland primarily produces thyroxine (T_4), which has relatively low biological activity and acts as a prohormone. Active triiodothyronine (T_3) is largely formed in peripheral tissues through the deiodination of T_4. These processes are catalyzed by D1, D2, and D3 deiodinase enzymes, which contain selenocysteine at their active site. In addition to deiodination, thyroid hormones may also undergo other metabolic transformations, such as conjugation or deamination.
- For hormones to exert their biological effects, they must be transported into cells. This is accomplished by specialized membrane transporters, including: OATPs and the monocarboxylate transporters MCT8 and MCT10.

Other hormones, endocrine tissues, and pathways

Thyroid hormone metabolism is part of a broader network of endocrine regulation. The hypothalamic-pituitary-thyroid (HPT) axis interacts with other hormonal axes, such as the adrenal axis (HPA). Cortisol can influence the conversion of T4 to T3 by modulating deiodinase activity, as observed in stress and chronic diseases.

Peripheral tissues such as the liver, kidneys, and brown adipose tissue act as "endocrine effector organs" where thyroid hormones are activated or inactivated. As with steroid hormones, conjugation processes (glucuronidation, sulfation) are important, increasing the solubility of metabolites and enabling their excretion.

Hormonal interactions also involve leptin and insulin, which modulate the effect of thyroid hormones on energy metabolism. Moreover, membrane transporters (OATP, MCT8) resemble systems that transport other hormones and drugs, which is important in pharmacokinetics.

This chapter demonstrates that the thyroid gland does not operate in isolation – it is part of a complex system in which enzymes, transporters, and other hormones work together to maintain the body's metabolic homeostasis.

Applications of the material

Understanding the enzymes and transporters involved in thyroid hormone metabolism is crucial in clinical practice. Knowledge of NIS function allows for better interpretation of diagnostics and radioiodine therapy in thyroid cancer, where the expression of this symporter determines treatment efficacy. Mutations in the NIS or pendrin gene can lead to congenital hypothyroidism and Pendred syndrome, demonstrating the importance of genetic testing in endocrinology.

TPO is one of the main autoantigens in autoimmune thyroid diseases – determining anti-TPO antibodies is the basis for diagnosing Hashimoto's and Graves' thyroiditis. Deiodinases, on the other hand, play a crucial role in the local regulation of T3 availability, which is important in critical illnesses, malnutrition, and liver disorders, for example.

Transporters such as MCT8 are important in neurology – their defects lead to Allan-Herndon-Dudley syndrome. This material also has pharmacological applications: drugs (e.g., propylthiouracil) affect deiodinases, and excess iodine can induce the Wolff–Chaikoff effect. This chapter therefore bridges the gap between biochemistry and translational medicine.

References

Bernal J, Guadaño-Ferraz A, Morte B. Thyroid hormone transporters–functions and clinical implications. Nature Reviews. Endocrinology. July 2015;11(7):406-417. https://doi.org/10.1038/nrendo.2015.66. Epub May 5, 2015. Erratum in: Nat rev endocrinol. September 2015;11(9):506. https://doi.org/10.1038/nrendo.2015.113. Erratum in: Nat Rev Endocrinol. December 2015;11(12):690. https://doi.org/10.1038/nrendo.2015.186. PMID: 25942657.

Braun, D., & Schweizer, U. (2018). Thyroid hHormone tTransport and tTransporters. *Vitam Horm, 106*, 19–44. https://doi.org/10.1016/bs.vh.2017.04.005. Epub 2017 Jun 12. PMID: 29407435.

Cheng, X., Zhang, H., Guan, S., Zhao, Q., & Shan, Y. (2023). Receptor modulators associated with the hypothalamus-pituitary-thyroid axis. *Frontiers in Pharmacology,* 14, Article 1291856. https://doi.org/10.3389/fphar.2023.1291856. PMID: 38111381; PMCID: PMC10725963.

Cicatiello, A. G., Di Girolamo, D., & Dentice, M. (2018). Metabolic effects of the intracellular regulation of thyroid hormone: Old players, new concepts. *Frontiers in Endocrinology, 9*, 474. https://doi.org/10.3389/fendo.2018.00474

Coscia, F., Taler-Verčič, A., Chang, V. T., Sinn, L., O'Reilly, F. J., Izoré, T., Renko, M., Berger, I., Rappsilber, J., Turk, D., & Löwe, J. (2020). The structure of human thyroglobulin. *Nature, 578*(7796), 627–630. https://doi.org/10.1038/s41586-020-1995-4

D'Aurizio, F., Kratzsch, J., Gruson, D., Petranović Ovčariček, P., & Giovanella, L. (2023). Free thyroxine measurement in clinical practice: How to optimize indications, analytical procedures, and interpretation criteria while waiting for global standardization. *Critical Reviews in Clinical Laboratory Sciences, 60*(2), 101–140. https://doi.org/10.1080/10408363.2022.2121960. Epub 2022 Oct 13. PMID: 36227760.

Gadisi, R. P., Naicker, M., & Naidoo, S. (2025). The extra-thyroidal distribution of sodium iodide symporter. *Frontiers in Endocrinology, 16*, Article 1567405. https://doi.org/10.3389/fendo.2025.1567405. PMID: 40678322; PMCID: PMC12267006.

Gnocchi, D., Steffensen, K. R., Bruscalupi, G., & Parini, P. (2016). Emerging role of thyroid hormone metabolites. *Acta Physiologica, 217*(3), 184–216. https://doi.org/10.1111/apha.12648. Epub 2016 Jan 24. PMID: 26748938.

Hershman, J. M., & Beck-Peccoz, P. (2023). Discoveries around the hypothalamic-pituitary-hypothalamic-pituitary-thyroid axis. *Thyroid, 33*(7), 785–790. https://doi.org/10.1089/thy.2022.0258

Janssen, S. T., & Janssen, O. E. (2017). Directional thyroid hormone distribution via the blood stream to target sites. *Molecular and Cellular Endocrinology, 458*, 16–21. https://doi.org/10.1016/j.mce.2017.02.037. Epub 2017 Feb 28. PMID: 28257828.

Köhrle, J. (2023). Selenium, iodine and iron-essential trace elements for thyroid hormone synthesis and metabolism. *International Journal of Molecular Sciences, 24*(4), 3393. https://doi.org/10.3390/ijms24043393. PMID: 36834802; PMCID: PMC9967593.

Khudair, A., Khudair, A., Niinuma, S. A., Habib, H., & Butler, A. E. (2025). Beyond thyroid dysfunction: The systemic impact of iodine excess. *Frontiers in Endocrinology, 16*, Article 1568807. https://doi.org/10.3389/fendo.2025.1568807

Koibuchi, N. (2018). Molecular mechanisms of thyroid hormone synthesis and secretion. In A. Belfiore, & D. LeRoith (Eds.), *Principles of endocrinology and hormone action. Endocrinology*. Cham: Springer. https://doi.org/10.1007/978-3-319-44675-2_5

Kopp, P., Pesce, L., & Solis-S, J. C. (2008). Pendred syndrome and iodide transport in the thyroid. *Trends in Endocrinology and Metabolism : TEM, 19*(7), 260–268. https://doi.org/10.1016/j.tem.2008.07.001

Luongo, C., Dentice, M., & Salvatore, D. (2019). Deiodinases and their intricate role in thyroid hormone homeostasis. *Nature Reviews. Endocrinology, 15*(8), 479–488. https://doi.org/10.1038/s41574-019-0218-2. PMID: 31160732.

Mullur, R., Liu, Y. Y., & Brent, G. A. (2014). Thyroid hormone regulation of metabolism. *Physiological Reviews, 94*(2), 355–382. https://doi.org/10.1152/physrev.00030.2013. PMID: 24692351; PMCID: PMC4044302.

Ortiga-Carvalho, T. M., Chiamolera, M. I., Pazos-Moura, C. C., & Wondisford, F. E. (2016). Hypothalamus-pituitary-thyroid axis. *Comprehensive Physiology, 6*(3), 1387–1428. https://doi.org/10.1002/cphy.c150027. PMID: 27347897.

Peeters, R. P., & Visser, T. J. (2000). Metabolism of thyroid hormone [Updated 2017 Jan 1]. In K. R. Feingold, R. A. Adler, S. F. Ahmed, et al. (Eds.), *Endotext*. South Dartmouth (MA): MDText.com, Inc. Available from https://www.ncbi.nlm.nih.gov/books/NBK285545/.

Ravera, S., Nicola, J. P., Salazar-De, S. G., Sigworth, F. J., Karakas, E., Amzel, L. M., Bianchet, M. A., & Carrasco, N. (2022). Structural insights into the mechanism of the sodium/iodide symporter. *Nature, 612*(7941), 795–801. https://doi.org/10.1038/s41586-022-05530-2. Epub 2022 Dec 14. PMID: 36517601; PMCID: PMC10501339.

Sabatino, L., Vassalle, C., Del Seppia, C., & Iervasi, G. (2021). Deiodinases and the three types of thyroid hormone deiodination reactions. *Endocrinology and Metabolism, 36*(5), 952–964. https://doi.org/10.3803/EnM.2021.1198. Epub 2021 Oct 21. PMID: 34674502; PMCID: PMC8566136.

Szybinski, Z. (2017). Role of iodine in metabolism. *Recent Patents on Endocrine. Metabolic & Immune Drug Discovery, 10*(2), 123–126. https://doi.org/10.2174/1872214811666170119110618

Tamma, G., & Dossena, S. (2022). Functional interplay between CFTR and pendrin: Physiological and pathophysiological relevance. *Frontiers in Bioscience (Landmark Edition), 27*(2), 75. https://doi.org/10.31083/j.fbl2702075. PMID: 35227018.

Targovnik, H. M., Citterio, C. E., & Rivolta, C. M. (2017). Iodide handling disorders (NIS, TPO, TG, IYD). *Best Practice & Research. Clinical Endocrinology & Metabolism, 31*(2), 195–212. https://doi.org/10.1016/j.beem.2017.03.006. Epub 2017 Apr 4. PMID: 28648508.

Zhang, L., Shang, F., Liu, C., & Zhai, X. (2024). The correlation between iodine and metabolism: A review. *Frontiers in Nutrition, 11*, Article 1346452. https://doi.org/10.3389/fnut.2024.1346452. PMID: 38567251; PMCID: PMC10985161.

Zhou, C., Liu, W., Zheng, J., Wu, Q., & Ai, Z. (2025). Molecular mechanisms of thyroid hormone signaling in thyroid cancer: Oncogenesis, progression, and therapeutic implications. *Biomedicines, 13* (10), 2552. https://doi.org/10.3390/biomedicines13102552. PMID: 41153831; PMCID: PMC12561264.

Chapter 4

An overview of the consequences of subclinical hypothyroidism

Tiffany Mach[1] and Oriana Hoi Yun Yu[2]

[1]*Department of Medicine, Endocrinology and Metabolism Residency Training Program, Montreal, QC, Canada;* [2]*Department of Medicine, Division of Endocrinology and Metabolism, Jewish General Hospital, McGill University, Montreal, QC, Canada*

Abbreviations

ACOG American College of Obstetricians and Gynaecologists
Anti-TPOAb Anti-thyroid peroxidase antibodies
ATA American Thyroid Association
CKD Chronic kidney disease
eGFR Estimated glomerular filtration rate
fT4 Free thyroxine
HDL High-density lipoprotein
LDL Low-density lipoprotein
LT4 Levothyroxine
MASH Metabolic dysfunction-associated steatohepatitis
MASLD Metabolic dysfunction-associated steatotic liver disease
NHANES National Health and Nutrition Examination Survey
OR Odds ratio
RCT Randomized clinical trial
SCH Subclinical hypothyroidism
T3 Triiodothyronine
TBG Thyroxine-binding globulin
THR Thyroid hormone receptor
TSH Thyroid stimulating hormone/Thyrotropin

Introduction

Subclinical hypothyroidism (SCH) is a biochemical diagnosis characterized by an elevated serum thyroid-stimulating hormone (TSH) (also known as thyrotropin) with free thyroxine (fT4) levels that are within the population reference range. The hypothalamic-pituitary-thyroid axis is exquisitely sensitive such that a slight decrease in fT4 can trigger a disproportionate rise in TSH levels, leading to transient elevations in TSH while the fT4 levels remain within the reference range. The higher incidence of SCH is associated with increasing age, female sex, higher body-mass index and suboptimal iodine intake (Cooper & Biondi, 2012; Peeters, 2017). Due to routine thyroid function testing, SCH has become a common incidental finding, particularly in older adults.

Most people diagnosed with SCH are asymptomatic, although some may experience mild hypothyroid symptoms (Biondi et al., 2019). Despite the often subtle presentation, thyroid hormones play a crucial role in numerous physiological pathways and influence a wide range of organ systems. These include the cardiovascular, central and autonomic nervous systems, musculoskeletal, respiratory, and gastrointestinal systems, as well as regulation of the basal metabolic rate. In the cardiovascular system, thyroid hormones increase heart rate, contractility and cardiac output (Inoue et al., 2020). Within the nervous system, they support cognitive function, mood stability, and affect reflexes and autonomic regulation. They also stimulate the respiratory drive and maintain normal ventilation, while in the gastrointestinal system, they enhance gut motility and digestive function. Furthermore, thyroid hormones are key regulators of basal metabolic rate and thermogenesis, thus dictating energy expenditure and body temperature. Overall, thyroid hormones are essential for maintaining homeostasis and normal physiological function throughout the body.

Clinical trials showed that treatment of SCH among people age 65 years and over did not improve the quality of life compared to people given placebo (Mooijaart et al., 2019; Stott et al., 2017). Observational studies showed that treatment of SCH may be associated with a decrease in the risk of cardiovascular outcomes (Razvi et al., 2012; Rodondi et al., 2010; Yu et al., 2024). Despite the role of thyroid hormone in cognitive function, current evidence suggests that SCH is not associated with cognitive impairment among the elderly but may have cognitive effects among younger individuals. One meta-analysis

Hormones in Neuroscience. https://doi.org/10.1016/B978-0-443-30261-9.00004-8

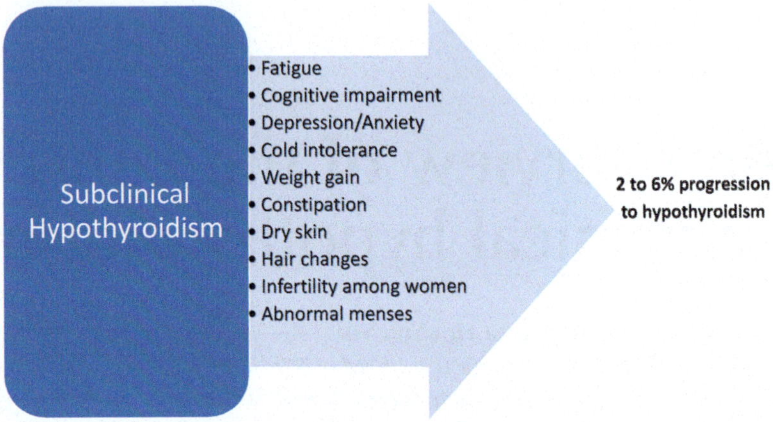

FIGURE 4.1 Clinical symptoms of subclinical hypothyroidism.

showed that SCH was associated with cognitive impairment among people younger than 75 years of age, whereas this association was not observed among people older than 75 years of age (Pasqualetti et al., 2015). SCH has been shown to be associated with an increased risk of having chronic kidney disease (Chonchol et al., 2008). More recently, there is evidence that SCH is associated with an increased risk of metabolic dysfunction-associated steatotic liver disease (MASLD) (Kizivat et al., 2020). However, whether treatment of SCH affects the risk of developing MASLD has not been studied. During pregnancy, SCH has been shown to be associated with adverse pregnancy outcomes, including increased risk of pregnancy loss and neonatal death (Maraka et al., 2016). Treatment of SCH was shown to decrease the risk of pregnancy loss and increase live birth rates (Bein et al., 2021; Leng et al., 2022). However, this evidence has not been consistent across studies (Dong et al., 2020; Maraka et al., 2016). In people who are not undergoing pregnancy, there is uncertainty on whether treatment of SCH results in any clinical benefits, with some recommendations on treating SCH when the TSH level increases beyond 10–20 mIU/L, presence of thyroid autoimmunity, or in the presence of hypothyroid symptoms (Bekkering et al., 2019; Biondi et al., 2019).

Clinical assessment of subclinical hypothyroidism

SCH is a prevalent condition, affecting 5%–10% of the population globally (Pearce et al., 2013; Surks et al., 2004). SCH is a biochemical finding that occurs during routine laboratory testing and requires at least two measures more than 3 months apart to confirm the diagnosis as approximately 50% of these initial findings revert to normal range spontaneously (Meyerovitch et al., 2007). A recent study investigated the incidence and determinants of spontaneous TSH normalization in adults aged 65 years and older with persistent SCH (van der Spoel et al., 2024). Among individuals with at least one elevated TSH measurement, 60.8% experienced spontaneous normalization of their TSH levels without intervention after a median follow-up of 1 year (van der Spoel et al., 2024). In certain cases, SCH can progress to overt hypothyroidism, where TSH levels continue to rise and fT4 levels decline below the normal range. The annual risk of progression to overt hypothyroidism is approximately 2%–6%, particularly in the presence of autoimmune thyroid disease (Peeters, 2017) and as such, treatment is warranted in these cases. Therefore, SCH can be seen as a mild or early form of thyroid failure.

Given that the fT4 levels are within normal range in people with SCH, most people do not have symptoms. However, as SCH is a form of mild hypothyroidism, people with this condition may experience mild symptoms of hypothyroidism (Fig. 4.1). These symptoms are expected to resolve once SCH is treated (i.e., a normal TSH level is attained with thyroid hormone replacement).

Causes of subclinical hypothyroidism

Causes of SCH are similar to those that result in hypothyroidism. The most common cause is from autoimmunity, where people develop anti-thyroid peroxidase antibodies (anti-TPOAb), inhibiting thyroid hormone production and ultimately causing thyroid cellular death and thyroid atrophy. Certain medications used to treat other disorders may affect thyroid function leading to SCH and these include amiodarone, lithium, and immunotherapy used in cancer treatments. Finally, a history of surgery where a part of the thyroid is excised, or irradiation treatments to the neck area where the thyroid is inadvertently exposed to radiation can also result in partial loss of thyroid function leading to SCH.

Management of subclinical hypothyroidism

Treatment of SCH consists of levothyroxine (LT4) replacement to normalize TSH levels. LT4 monotherapy is the standard of care to treat SCH as recommended by major guidelines and needs to be taken when fasting with water to ensure adequate absorption. Certain foods, including milk and caffeine, can affect the absorption of LT4. As such, beverages, aside from water, and foods should be consumed at least 30 minutes after LT4 is taken (Taylor et al., 2024). To determine if the LT4 dose is adequate to treat SCH, the TSH level needs to be repeated at least 6 weeks after LT4 is initiated (Taylor et al., 2024). LT4 is rarely associated with side effects except that it requires regular laboratory testing and surveillance to ensure that there is no over replacement as this can lead to an increased risk of arrhythmias and development of osteoporosis.

Subclinical hypothyroidism in children

The prevalence of SCH in children is much lower compared to adults with a reported prevalence of 1.7% (Salerno et al., 2016). SCH among children, from age 8 to 18 years, is commonly caused by autoimmune disease due to the presence of anti-TPOAb (Metwalley & Farghaly, 2021). In children with no evidence of autoimmune disease, other causes for SCH include genetic mutations, such as TSH-receptor or dual oxidase two mutations, genetic syndromes, such as Down or Turner syndrome, obesity, iron deficiency, nephrotic syndrome, iodine deficiency or excess, ionizing radiation or medication use, such as amiodarone, interferon-α, or antiepileptic medications (i.e., valproic acid, phenytoin, etc) (Metwalley & Farghaly, 2021). Clinical symptoms from SCH in children are similar to that in adults. SCH has not been shown to affect growth in children. However, SCH has been shown to be associated with cardiovascular risk factors, such as a higher risk of hypertension, having abnormal lipid profiles, with lower HDL levels and a higher triglyceride to HDL-C ratio. SCH was also shown to be associated with cognitive function in terms of overall cognitive scores, attention, and verbal affluency (Metwalley & Farghaly, 2021).

The TSH cutoff to determine when treatment of SCH is warranted has been controversial. In general, treatment with LT4 is recommended for children with TSH levels above 10 mU/L. For children with TSH levels above the limit of normal but less than 10 mU/L, treatment is recommended if there is the presence of a goiter, clinical evidence of hypothyroidism, presence of thyroid nodules or if they have an underlying condition that would likely result in overt hypothyroidism (i.e., known thyroid autoimmunity,genetic mutations or syndromes). Finally, if SCH is medication-induced, LT4 treatment is recommended until the medication causing SCH can be stopped. Currently, there has not been studies assessing long-term effects of SCH in children.

Translational aspects of the chapter

Thyroid hormone and its physiologic role

Thyroid hormone plays important physiologic roles across various organs: thyroid hormone has been shown to play direct roles in cell signaling in the central nervous system, retina, heart, kidney, liver, and bone (Mullur et al., 2014). The thyroid hormone activates thyroid hormone receptors (THR) which are present as two isoforms: THR α and β. The THR isoforms are expressed in different tissues and play important physiological roles in various organs, affecting development, growth, and metabolism (Table 4.1) (Anyetei-Anum et al., 2018).

More specifically, the thyroid hormone has been shown to have a direct role in glucose and lipid metabolism in the liver (Mullur et al., 2014). While SCH is a form of mild hypothyroidism, its effects on human health has been studied, with evidence showing that SCH is associated with a higher risk of cardiovascular outcomes when the TSH level rises above 10 mU/L (Rodondi et al., 2010). The clinical consequences of SCH in other body systems have been inconsistent (Fig. 4.2). Below, we discuss the evidence to date on SCH and the effects on other systems and whether treatment of SCH provides any benefits.

Subclinical hypothyroidism and cardiovascular outcomes

Thyroid hormone has been shown to affect gene expression and enzyme activity in cardiac myocytes and vascular smooth muscle (Danzi & Klein, 2014). Thyroid hormone, via its active form, triiodothyronine (T3), affects cellular pathways in cardiac myocytes which increase cardiac inotropy and chronotropy, and decrease systemic vascular resistance (Klein & Ojamaa, 2001) (Fig. 4.3). Thus, hypothyroidism can result in bradycardia, mild hypertension with a narrowed pulse pressure, decrease in cardiac contractility, and increase in systemic vascular resistance. Hypothyroidism also increases cholesterol levels and results in weight gain which may increase the risk of developing cardiovascular disease if hypothyroidism remains untreated (Klein & Ojamaa, 2001). SCH is a mild form of hypothyroidism, which may result in these cardiovascular changes but to a lesser extent compared to overt hypothyroidism.

SCH has been shown to be associated with an increased risk of coronary heart disease and cardiovascular-related mortality, specifically when the TSH level is above 10 mIU/L (Rodondi et al., 2010). Prior observational studies showed that treatment of SCH is associated with a reduced risk of major cardiovascular adverse events and

TABLE 4.1 Thyroid hormone receptor isoforms: Site of activity and functions.

Isoform	Site of activity	Role/Function
α	Central nervous system	Regulates energy metabolism, protein synthesis, and neurotransmitter function in neurons and glial cells
	Heart	Heart rate and cardiac cellular metabolism
	Gastrointestinal tract	Gastric motility and digestive activities
	Bone	Increases bone resorption via effects in osteoblast cell lines and osteoclasts
β1	Liver	Affects hepatic gluconeogenesis, lipid metabolism, fatty acid synthesis, and bile acid synthesis
	Kidney	Affects sodium and water homeostasis, renal hemodynamics, blood flow and glomerular filtration rate
	Inner ear	Required for inner ear development, needed for hearing
β2	Retina	Required for development of cone differentiation and development in the retina
	Cochlea	Helps in development of cell types which are needed for hearing
	Hypothalamus	Regulates energy homeostasis and appetite

cardiovascular related mortality (Razvi et al., 2012; Yu et al., 2024). Randomized clinical trials (RCT) conducted to date were conducted in participants 65 years and older. These RCTs found that LT4 treatment of SCH did not affect the risk of major cardiovascular events or cardiovascular-related mortality (Mooijaart et al., 2019; Stott et al., 2017). However, these trials were underpowered to assess these outcomes. An RCT conducted among people with SCH and

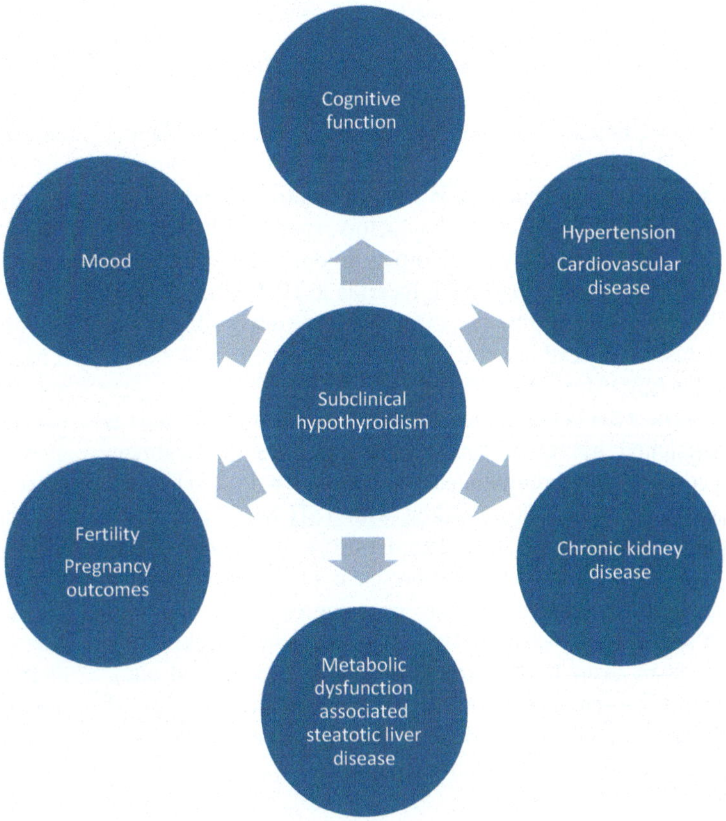

FIGURE 4.2 Clinical effects associated with subclinical hypothyroidism.

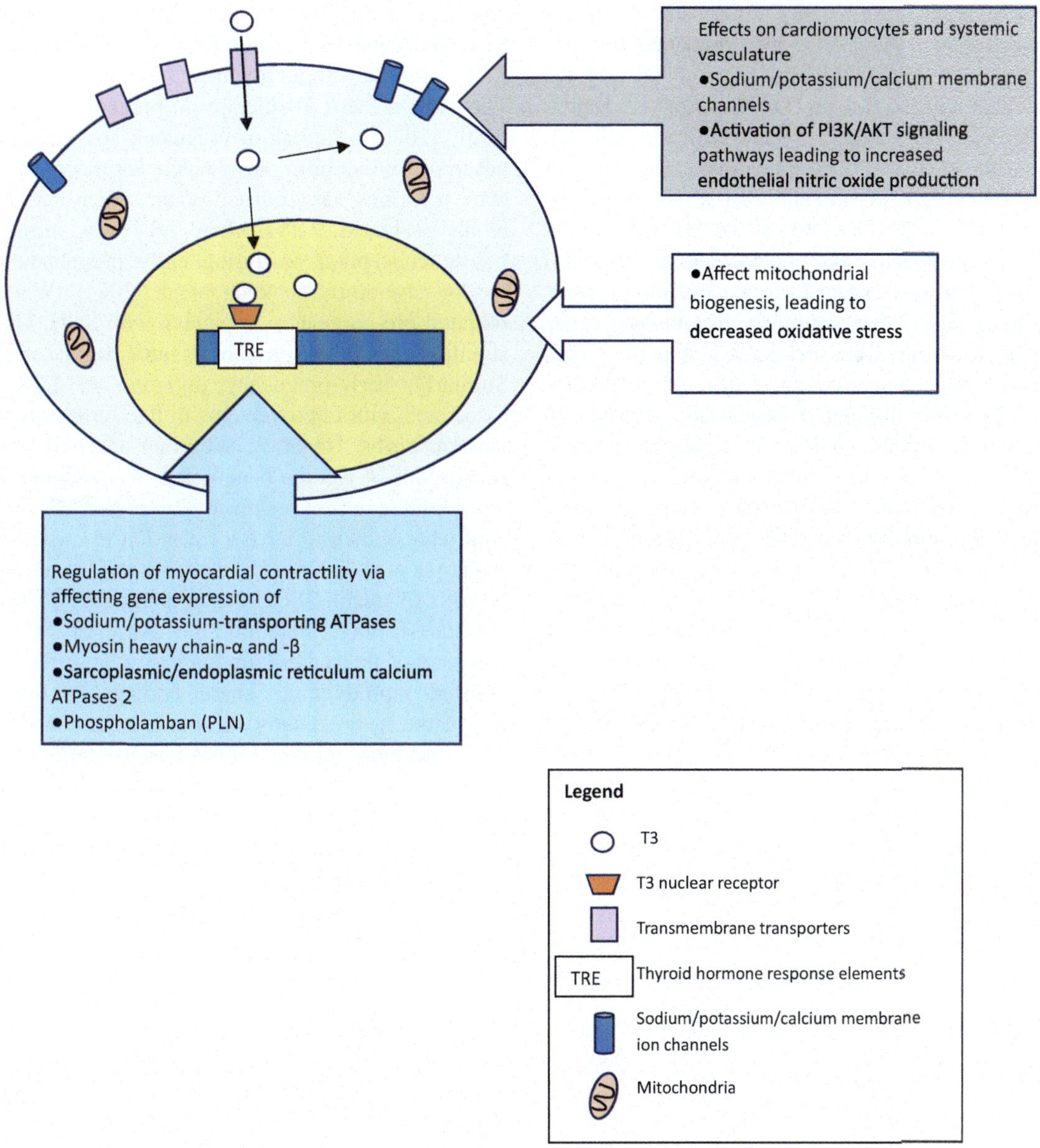

FIGURE 4.3 T3 effects in cardiac myocytes. T3 affects cardiomyocytes and systemic vasculature, via effects on gene expression and mitochondrial activity.

acute myocardial infarction found that LT4 treatment compared to placebo did not improve ventricular ejection fraction after 52 weeks (Jabbar et al., 2020). Overall, there is no strong evidence supporting treatment of SCH to decrease the risk of cardiovascular events or cardiovascular-related mortality. However, few RCTs have been conducted to address this issue and there is a paucity of studies conducted among people younger than 65 years of age.

As such, in the absence of a thyroid autoimmune condition, the American Thyroid Association (ATA) guidelines recommend initiation of LT4 among people with SCH when the TSH level is above 10 mIU/L. For people with SCH and TSH levels above the upper limit of normal but less than 10 mIU/L, treatment should be considered if there is evidence of thyroid autoimmune disease, atherosclerotic cardiac disease, heart failure, other risk factors for cardiovascular disease, or if there are symptoms of hypothyroidism (Garber et al., 2012). Future RCTs are required to confirm whether there are cardiovascular benefits with long-term treatment of SCH.

Subclinical hypothyroidism and pregnancy outcomes

SCH affects 15% of pregnancies in the United States and 14% of pregnancies in Europe (Maraka et al., 2016). Due to the possible effects on maternal-fetal outcomes, the treatment of SCH has been a source of debate and guidelines from different societies are also reflecting this uncertainty. The 2017 ATA guidelines recommend LT4 for pregnant women with SCH if the TSH exceeds pregnancy-specific reference ranges (or >4 mIU/L if unavailable), especially with positive TPO antibodies, and suggest initiating treatment at lower thresholds in high-risk groups (Alexander et al., 2017). In contrast, the 2020 American College of Obstetricians and Gynecologists (ACOG) advise against treatment except for overt hypothyroidism.

During pregnancy, the maternal thyroid undergoes adaptive changes to meet the metabolic demands of both the mother and fetus. Therefore, thyroid requirements increase by 30%–50% through estrogen-induced rises in thyroxine-binding globulin (TBG), hCG-mediated stimulation of the TSH receptor, increased renal iodine clearance and fetal transfer of iodine. SCH occurs when the metabolic demands of the body exceed the thyroid reserve, thus leading to an elevated TSH with normal free T4. Common causes include autoimmune thyroiditis, iodine deficiency and underlying thyroid disease. Even though the free T4 remains within normal limits, SCH represents an early sign of thyroid failure, which may influence placental function and fetal development, especially given that the fetal thyroid relies solely on maternal free T4 until mid-gestation.

The effect of LT4 on fertility and pregnancy outcomes in SCH have been evaluated in several RCTs, however the results have been controversial. An updated systematic review and meta-analysis by Provinciatto et al. in 2024 analyzed 11 RCTs involving 2749 pregnant women with SCH and compared outcomes between those treated with LT4 and those without treatment (Provinciatto et al., 2024). They found that LT4 therapy in pregnant women significantly reduced the risk of pregnancy loss, however it did not improve live birth rates, which was a consistent finding across various subgroups. On the other hand, Sankoda et al. evaluated RCTs where LT4 was initiated either before conception or during early pregnancy (until 20 weeks gestation) (Sankoda et al., 2024). When LT4 is initiated pre-pregnancy in women with SCH, LT4 did not significantly improve live birth rates or miscarriage rates. Similarly, early-pregnancy initiation of LT4 was not associated with improvements in live birth, miscarriage or preterm births. However, subgroup analyses showed that two groups of women benefited from treatment with LT4, those with recurrent pregnancy losses and TSH >4 mIU/L, who were shown to have a reduction in miscarriage rates (Sankoda et al., 2024). The inconsistency of findings on the benefits of treating SCH during pregnancy may be due to multiple factors, including the cause of SCH and the presence of thyroid autoimmunity, or infertility, requiring assisted reproduction. These findings in the literature reflect the heterogeneity in opinions for the treatment of SCH and the importance of tailoring treatment decisions. Rather than adopting a universal approach for all patients, the evidence seems to support a risk-based strategy based on their TSH levels and risk factors for adverse pregnancy outcomes (Table 4.2).

TABLE 4.2 Summary of various society guidelines for the treatment of subclinical hypothyroidism in pregnancy.

Society/Guideline	Year	Treat SCH in pregnancy?
American Thyroid Association (ATA)	2017	• Initiate treatment if TPOAb positive and if TSH level above the pregnancy-specific upper limit. • Treat if TSH ≥10 mIU/L regardless of antibody status. • Consider treatment if TPOAb positive with TSH >2.5 mIU/L but below or equal to the upper limit of the reference range. • Consider treatment if TPOAb negative with TSH level above the upper limit of normal but <10 mIU/L.
American College of Obstetricians and Gynaecologists (ACOG)	2020	• Routine treatment of all women with subclinical hypothyroidism is not recommended given that evidence does not show improved neurocognitive outcomes or consistent obstetric benefit. • Individualize treatment and management of SCH.
European Thyroid Association (ETA)	2014	• Initiate treatment if TSH ≥10 mIU/L. • Initiate treatment for SCH before or during pregnancy, especially if TPOAb positive or if TSH above pregnancy-specific upper limit.
Endocrine Society (ES)	2012	• Initiate treatment for women with SCH, especially when TPOAb positive. • Consider treatment if TSH <10 mIU/L. Treatment decision should be individualized according to antibody status and clinical context.

TPO Ab, thyroid peroxidase antibody; *TSH*, thyroid stimulating hormone.

Subclinical hypothyroidism and renal outcomes

The prevalence of SCH has been shown to be higher at approximately 18% among people with chronic kidney disease (CKD) (Chonchol et al., 2008). Furthermore, higher TSH levels are associated with a lower estimated glomerular filtration rate (eGFR) (Gopinath et al., 2013). Thyroid and renal function have interlinked pathways where uremia affects the response of TSH to thyrotropin-releasing hormone (TRH) and CKD affects iodine metabolism, causing iodine retention and inducing hypothyroidism via the Wolff-Chaikoff effect (Spahia et al., 2023). In animal studies, hypothyroidism was shown to affect renal structure and function. Rats with thyroidectomy-induced hypothyroidism were found to have decreased single nephron glomerular filtration rate, renal plasma flow, and glomerular transcapillary pressure (Falk et al., 1991; Spahia et al., 2023). Thyroid hormones may alter renal function via its effect on cardiac function and vascular changes, affecting renal blood flow (Agahi et al., 2024).

Studies showed that SCH is associated with an increased risk of adverse renal outcomes (Urgatz & Razvi, 2023). While there is a paucity of studies assessing whether treatment of SCH is associated with a decreased risk of adverse renal outcomes and among people with SCH, the results from studies conducted so far have been inconsistent. In one small RCT, involving 32 participants, LT4 treatment was found to reduce proteinuria among people with SCH and CKD compared to placebo (Black-aller et al., 2021). Similarly, in one retrospective study, people with stage 2–4 CKD and SCH, treated with LT4 had a significantly reduced risk of developing a 50% reduction in eGFR and end stage renal disease compared to people not given LT4 treatment (Shin et al., 2012). This finding contrasts with another retrospective study which did not find a reduced risk of developing a 50% decrease in eGFR or an increased risk of developing end stage renal disease among people with SCH given LT4 versus those not treated (Yu et al., 2025). Another observational study assessed whether LT4 treatment of SCH affected kidney function among people with CKD over a period of 2 years. In this study, LT4 was not associated with a change in eGFR or progression of CKD. However, LT4 treatment was associated with a shorter duration of hospitalization for CKD-related issues (Hennessey et al., 2021).

SCH may be associated with an increased risk of renal dysfunction. However, there is insufficient evidence showing whether treatment of SCH lowers the risk of renal dysfunction.

Subclinical hypothyroidism and cognitive outcomes

Thyroid hormones have direct effects on various areas of the brain. In animal studies, hypothyroidism has been shown to affect the hippocampal neurogenesis, dentate gyrus and hippocampal functions, which may reverse with treatment. Thyroid hormones have also been shown to affect neurotransmitter activity in the brain, namely acetylcholine and monoamines in the hippocampus and frontal cortex (Eslami-Amirabadi & Sajjadi, 2021). Hypothyroidism may affect cognitive function by decreasing cerebral blood flow and cerebral glucose metabolism. Hypothyroidism increases the risk of developing a hypercoagulable state and lipid or blood pressure abnormalities which may increase the risk of stroke, resulting in cognitive impairment (Sinha et al., 2024). From these effects, people with hypothyroidism develop cognitive abnormalities in the area of attention, concentration, verbal memory and executive function, which may reverse with treatment (Eslami-Amirabadi & Sajjadi, 2021).

The literature on the effects of SCH on cognitive function has been inconsistent. Three observational studies did not find an association between SCH and cognitive dysfunction (de Jongh et al., 2011; Roberts et al., 2006; van Vliet et al., 2021), whereas in a systematic review and meta-analysis study, cognitive dysfunction was found to be associated with SCH among people younger than 75 years of age, whereas there was no association between SCH and cognitive dysfunction among people older than 75 years of age (Pasqualetti et al., 2015). The largest RCT conducted in people aged 65 years and older did not find that treatment of SCH improved executive cognitive function, measured using the letter-digit coding test (Stott et al., 2017). Finally, two small RCTs did not find that treatment of SCH changed cognitive function (Jorde et al., 2006; Parle et al., 2010). Overall, there is inconsistent evidence showing whether SCH affects cognitive function and whether treatment of SCH confers any cognitive benefits.

Subclinical hypothyroidism and effects on mood

While overt hypothyroidism is known to affect mood, including causing a major-depressive disorder, or melancholia, the effects of SCH on mood has been controversial. Thyroid function screening is recommended for patients who have depressive or anxiety symptoms. Some studies found an association between SCH and having symptoms of depression or having a major depressive disorder. This association was found more commonly among younger or middle-aged adults with SCH and the evidence for the association between SCH and depression related symptoms of diagnosis have not been shown among the elderly (Joffe et al., 2013). In a systematic review and meta-analysis including pooled data of 12,315 individuals, SCH was found to be associated with an increased risk of depression among people with SCH (relative risk 2.35; 95% confidence interval: 1.84–3.02) (Loh et al., 2019). In this study, treatment with LT4 was not associated with improvement in depression symptoms among people with SCH and concomitant depression. RCTs in elderly participants with SCH did not find that LT4 treatment affected mood or improve overall quality of life (Mooijaart et al., 2019; Stott et al., 2017).

Subclinical hypothyroidism and metabolic dysfunction-associated steatotic liver disease

The thyroid hormone is known to affect glucose and lipid metabolism in the liver. Thyroid hormones (i.e., T4 and T3), activate THR, which leads to hepatic fatty acid oxidation and decreases hepatocyte fatty acids (Lanni et al., 2005). Hypothyroidism has been found to be associated with an increased risk of developing metabolic dysfunction-associated steatotic liver disease (MASLD) independent of several metabolic risk factors, suggesting a causal mechanism by which thyroid function impacts MASLD pathogenesis (Hatziagelaki et al., 2022). A study using data from the Third National Health and Nutrition Examination Survey (NHANES III) showed that individuals with SCH had a significantly higher risk of developing MASLD compared to people with normal thyroid function (odds ratio [OR]: 1.62; 95% CI: 1.13–2.33) (Chen et al., 2023). A recent Mendelian randomization study also showed a dose-response relationship between TSH levels and risk of developing MASLD (Q4 vs. Q1: OR: 1.29; 95% CI: 1.10–1.52), whereby higher TSH levels were associated with an increasing risk of developing MASLD (Fan et al., 2022). Overall, these studies suggest a direct link between thyroid hormone function and liver metabolism. As such, there is an interest in developing treatments targeting thyroid hormone pathways in hepatocytes to reverse MASLD. Indeed, recently a phase 3 RCT was conducted using a THRβ agonist, resmetirom, and showed that this agent improved liver fibrosis by one stage among participants with metabolic dysfunction-associated steatohepatitis (MASH) (Harrison et al., 2024).

Currently, there has not been any study that determines whether treatment of SCH lowers the risk of developing MASLD or whether treatment of this condition among people with pre-existing MASLD decreases the risk of developing adverse liver outcomes. Further research is required in this area to inform clinicians whether treating SCH lowers the risk of developing MASLD and whether SCH treatment is warranted among people with known MASLD or MASH.

Applications of the material

Is treatment of subclinical hypothyroidism necessary?

SCH affects multiple body systems but whether treatment of this condition leads to any clinical benefits have been controversial. The few RCTs conducted in this area did not include people younger than age 65 years and there is a lack of studies on the clinical benefits of treating SCH in the younger population. Most major guidelines recommend treatment of SCH when the TSH level is beyond 10 mIU/L, when patients have symptoms of hypothyroidism, or when there is an underlying cause of hypothyroidism identified such as autoimmune disease. The guidelines and recommendations are summarized in Table 4.3.

Mini-dictionary of terms

- **Acetylcholine** A neurotransmitter required for neuron activity in the central and peripheral nervous systems.
- **Dual oxidase 2** An enzyme that produces hydrogen peroxide required for the organification of iodine by the thyroid peroxidase enzyme, resulting in the production of thyroid hormone.
- **Immunotherapy** A biological medication that affects the immune system to recognize and attack cancer cells.
- **Metabolic dysfunction-associated steatotic liver disease** A diagnosis formerly known as non-alcoholic fatty liver disease whereby there is increased adipose tissue around the liver with metabolic abnormalities resulting in increased inflammation in the liver, resulting in fibrosis and ultimately cirrhosis.
- **Wolff-Chaikoff** A term used to describe the effect of large quantity of iodine causing inhibition of thyroid hormone biosynthesis, resulting in hypothyroidism.

Summary points

- Subclinical hypothyroidism (SCH) is defined as an elevated TSH with normal free thyroxine (fT4) levels.
- SCH is associated with an increased risk of cardiovascular disease and mortality when TSH >10 mIU/L, however randomized trials have not shown clear benefit with levothyroxine (LT4) therapy.
- In children, treatment is generally reserved for TSH >10 mIU/L, or in the presence of a goiter, hypothyroid symptoms, thyroid autoimmunity, or high-risk conditions.
- In pregnancy, SCH is associated with miscarriages and adverse perinatal outcomes, however treatment initiation remains inconsistent among trials; guidelines vary widely with their recommendations.
- SCH prevalence is higher among patients with chronic kidney disease and can worsen renal function, however evidence for renal benefit of LT4 treatment is inconclusive.
- SCH may affect cognition in individuals less than 75 years old, however RCTs show no improvement in cognitive outcomes with LT4 therapy.

TABLE 4.3 Guideline recommendations on treatment of subclinical hypothyroidism.

Society/Guideline	Year	Treatment recommendations
American Thyroid Association (ATA) and American Association of Clinical Endocrinologists (AACE)	2012	• To initiate treatment if TPOAb positive, TSH level >10 mIU/L, patient at risk of heart failure or cardiovascular mortality.
European Thyroid Association (ETA)	2013	• To initiate treatment if TSH ≥10 mIU/L, if there is a diffuse or nodular goiter, or if there are symptoms of hypothyroidism among people <65 years of age. If TSH elevated but <10 mIU/L, in people older than 80 years of age, to monitor and follow. • To consider treatment among people >80 years if TSH ≥10 mIU/L, symptoms of hypothyroidism or has an increased vascular risk.
National Institute for Health and Care Excellence	2019	• To treat if TSH above upper limit of normal but <10 mIU/L and hypothyroid symptoms among people <65 years of age. • To treat if the TSH is confirmed >10 mIU/L on 2 separate occasions and 3 months apart. • To consider treatment if evidence of underlying cause of hypothyroidism and for people with symptoms of hypothyroidism.
Korean Thyroid Association	2023	• To treat if TSH above the upper limit of normal but <10 mIU/L and patient has symptoms of hypothyroidism, progressive hypothyroidism, TPO Ab positive, presence of goiter or history of thyroidectomy. • To treat SCH if there is a history of coronary artery disease or heart failure and to consider treatment if there is dyslipidemia. • To treat if TSH >10 mIU/L in patients <70 years of age, and to monitor and consider treatment if clinically indicated for patients >70 years of age with TSH >10 mIU/L.

SCH, subclinical hypothyroidism; *TSH*, thyroid stimulating hormone; *TPO Ab*, thyroid, peroxidase antibody.

- Observational studies suggest an association between SCH and metabolic dysfunction-associated steatotic liver disease (MASLD), but no studies have assessed whether LT4 affects liver outcomes.
- Overall, the decision to treat SCH should be individualized, considering the TSH level, age, symptoms, comorbidities, thyroid antibody status and pregnancy considerations.

References

Agahi, S., Amouzegar, A., Honarvar, M., Azizi, F., & Mehran, L. (2024). Interrelationship between thyroid hormones and reduced renal function, a review article. *Thyroid Research, 17*(1), 14. https://doi.org/10.1186/s13044-024-00201-y

Alexander, E. K., Pearce, E. N., Brent, G. A., Brown, R. S., Chen, H., Dosiou, C., Grobman, W. A., Laurberg, P., Lazarus, J. H., Mandel, S. J., Peeters, R. P., & Sullivan, S. (2017). 2017 Guidelines of the American Thyroid Association for the diagnosis and management of thyroid disease during pregnancy and the postpartum. *Thyroid, 27*(3), 315–389. https://doi.org/10.1089/thy.2016.0457

Anyetei-Anum, C. S., Roggero, V. R., & Allison, L. A. (2018). Thyroid hormone receptor localization in target tissues. *Journal of Endocrinology, 237*(1), R19–r34. https://doi.org/10.1530/joe-17-0708

Bein, M., Yu, O. H. Y., Grandi, S. M., Frati, F. Y. E., Kandil, I., & Filion, K. B. (2021). Levothyroxine and the risk of adverse pregnancy outcomes in women with subclinical hypothyroidism: A systematic review and meta-analysis. *BMC Endocrine Disorders, 21*(1), 34. https://doi.org/10.1186/s12902-021-00699-5

Bekkering, G. E., Agoritsas, T., Lytvyn, L., Heen, A. F., Feller, M., Moutzouri, E., Abdulazeem, H., Aertgeerts, B., Beecher, D., Brito, J. P., Farhoumand, P. D., Singh Ospina, N., Rodondi, N., van Driel, M., Wallace, E., Snel, M., Okwen, P. M., Siemieniuk, R., Vandvik, P. O., … Vermandere, M. (2019). Thyroid hormones treatment for subclinical hypothyroidism: A clinical practice guideline. *British Medical Journal, 365*. https://doi.org/10.1136/bmj.l2006

Biondi, B., Cappola, A. R., & Cooper, D. S. (2019). Subclinical hypothyroidism: A review. *Journal of the American Medical Association, 322*(2), 153–160. https://doi.org/10.1001/jama.2019.9052

Blackaller, G. N., Chávez-Iñiguez, J. S., Carreón-Bautista, E. E., González-Torres, F. J., Villareal-Contreras, M., Barrientos Avalos, J. R., Aguilera, P. M., Rosales, F. R., José Antonio, T. M., Gómez Fregoso, J. A., Michel Gonzalez, J. I., & García-García, G.

(2021). A pilot trial on the effect of levothyroxine on proteinuria in patients with advanced CKD. *Kidney International Reports, 6*(1), 110–119. https://doi.org/10.1016/j.ekir.2020.10.016

Chen, Y. L., Tian, S., Wu, J., Li, H., Li, S., Xu, Z., Liang, X. Y., Adhikari, V. P., Xiao, J., Song, J. Y., Ma, C. Y., She, R. L., Li, Z. X., Wu, K. N., & Kong, L. Q. (2023). Impact of thyroid function on the prevalence and mortality of metabolic dysfunction-associated fatty liver disease. *Journal of Clinical Endocrinology & Metabolism, 108*(7), e434–e443. https://doi.org/10.1210/clinem/dgad016

Chonchol, M., Lippi, G., Salvagno, G., Zoppini, G., Muggeo, M., & Targher, G. (2008). Prevalence of subclinical hypothyroidism in patients with chronic kidney disease. *Clinical Journal of the American Society of Nephrology, 3*(5), 1296–1300. https://doi.org/10.2215/cjn.00800208

Cooper, D. S., & Biondi, B. (2012). Subclinical thyroid disease. *Lancet, 379*(9821), 1142–1154. https://doi.org/10.1016/s0140-6736(11)60276-6

Danzi, S., & Klein, I. (2014). Thyroid disease and the cardiovascular system. *Endocrinology and Metabolism Clinics of North America, 43*(2), 517–528. https://doi.org/10.1016/j.ecl.2014.02.005

de Jongh, R. T., Lips, P., van Schoor, N. M., Rijs, K. J., Deeg, D. J., Comijs, H. C., Kramer, M. H., Vandenbroucke, J. P., & Dekkers, O. M. (2011). Endogenous subclinical thyroid disorders, physical and cognitive function, depression, and mortality in older individuals. *European Journal of Endocrinology, 165*(4), 545–554. https://doi.org/10.1530/eje-11-0430

Dong, A. C., Morgan, J., Kane, M., Stagnaro-Green, A., & Stephenson, M. D. (2020). Subclinical hypothyroidism and thyroid autoimmunity in recurrent pregnancy loss: A systematic review and meta-analysis. *Fertility and Sterility, 113*(3), 587–600.e581. https://doi.org/10.1016/j.fertnstert.2019.11.003

Eslami-Amirabadi, M., & Sajjadi, S. A. (2021). The relation between thyroid dysregulation and impaired cognition/behaviour: An integrative review. *Journal of Neuroendocrinology, 33*(3). https://doi.org/10.1111/jne.12948

Falk, S. A., Buric, V., Hammond, W. S., & Conger, J. D. (1991). Serial glomerular and tubular dynamics in thyroidectomized rats with remnant kidneys. *American Journal of Kidney Diseases, 17*(2), 218–227. https://doi.org/10.1016/S0272-6386(12)81132-2

Fan, H., Liu, Z., Zhang, X., Wu, S., Shi, T., Zhang, P., Xu, Y., Chen, X., & Zhang, T. (2022). Thyroid stimulating hormone levels are associated with genetically predicted nonalcoholic fatty liver disease. *Journal of Clinical Endocrinology & Metabolism, 107*(9), 2522–2529. https://doi.org/10.1210/clinem/dgac393

Garber, J. R., Cobin, R. H., Gharib, H., Hennessey, J. V., Klein, I., Mechanick, J. I., Pessah-Pollack, R., Singer, P. A., & Woeber, K. A. (2012). Clinical practice guidelines for hypothyroidism in adults: Cosponsored by the American Association of clinical endocrinologists and the American Thyroid Association. *Endocrine Practice, 18*(6), 988–1028. https://doi.org/10.4158/EP12280.GL

Gopinath, B., Harris, D. C., Wall, J. R., Kifley, A., & Mitchell, P. (2013). Relationship between thyroid dysfunction and chronic kidney disease in community-dwelling older adults. *Maturitas, 75*(2), 159–164. https://doi.org/10.1016/j.maturitas.2013.03.009

Harrison, S. A., Bedossa, P., Guy, C. D., Schattenberg, J. M., Loomba, R., Taub, R., Labriola, D., Moussa, S. E., Neff, G. W., Rinella, M. E., Anstee, Q. M., Abdelmalek, M. F., Younossi, Z., Baum, S. J., Francque, S., Charlton, M. R., Newsome, P. N., Lanthier, N., Schiefke, I., … Ratziu, V. (2024). A phase 3, randomized, controlled trial of resmetirom in NASH with liver fibrosis. *New England Journal of Medicine, 390*(6), 497–509. https://doi.org/10.1056/NEJMoa2309000

Hatziagelaki, E., Paschou, S. A., Schön, M., Psaltopoulou, T., & Roden, M. (2022). NAFLD and thyroid function: Pathophysiological and therapeutic considerations. *Trends in Endocrinology & Metabolism, 33*(11), 755–768. https://doi.org/10.1016/j.tem.2022.08.001

Hennessey, J. V., Weir, M. R., Soni-Brahmbhatt, S., Duan, Y., & Gossain, V. V. (2021). Effect of levothyroxine on kidney function in chronic kidney disease with subclinical hypothyroidism in US veterans: A retrospective observational cohort study. *Advances in Therapy, 38*(2), 1185–1201. https://doi.org/10.1007/s12325-020-01589-3

Inoue, K., Ritz, B., Brent, G. A., Ebrahimi, R., Rhee, C. M., & Leung, A. M. (2020). Association of subclinical hypothyroidism and cardiovascular disease with mortality. *JAMA Network Open, 3*(2). https://doi.org/10.1001/jamanetworkopen.2019.20745

Jabbar, A., Ingoe, L., Junejo, S., Carey, P., Addison, C., Thomas, H., Parikh, J. D., Austin, D., Hollingsworth, K. G., Stocken, D. D., Pearce, S. H. S., Greenwood, J. P., Zaman, A., & Razvi, S. (2020). Effect of levothyroxine on left ventricular ejection fraction in patients with subclinical hypothyroidism and acute myocardial infarction: A randomized clinical trial. *Journal of the American Medical Association, 324*(3), 249–258. https://doi.org/10.1001/jama.2020.9389

Joffe, R. T., Pearce, E. N., Hennessey, J. V., Ryan, J. J., & Stern, R. A. (2013). Subclinical hypothyroidism, mood, and cognition in older adults: A review. *International Journal of Geriatric Psychiatry, 28*(2), 111–118. https://doi.org/10.1002/gps.3796

Jorde, R., Waterloo, K., Storhaug, H., Nyrnes, A., Sundsfjord, J., & Jenssen, T. G. (2006). Neuropsychological function and symptoms in subjects with subclinical hypothyroidism and the effect of thyroxine treatment. *Journal of Clinical Endocrinology & Metabolism, 91*(1), 145–153. https://doi.org/10.1210/jc.2005-1775

Kizivat, T., Maric, I., Mudri, D., Curcic, I. B., Primorac, D., & Smolic, M. (2020). Hypothyroidism and nonalcoholic fatty liver disease: Pathophysiological associations and therapeutic implications. *Journal of Clinical and Translational Hepatology, 8*(3), 347–353. https://doi.org/10.14218/jcth.2020.00027

Klein, I., & Ojamaa, K. (2001). Thyroid hormone and the cardiovascular system. *New England Journal of Medicine, 344*(7), 501–509. https://doi.org/10.1056/nejm200102153440707

Lanni, A., Moreno, M., Lombardi, A., de Lange, P., Silvestri, E., Ragni, M., Farina, P., Baccari, G. C., Fallahi, P., Antonelli, A., & Goglia, F. (2005). 3,5-diiodo-L-thyronine powerfully reduces adiposity in rats by increasing the burning of fats. *FASEB Journal, 19*(11), 1552–1554. https://doi.org/10.1096/fj.05-3977fje

Leng, T., Li, X., & Zhang, H. (2022). Levothyroxine treatment for subclinical hypothyroidism improves the rate of live births in pregnant women with recurrent pregnancy loss: A randomized clinical trial. *Gynecological Endocrinology, 38*(6), 488–494. https://doi.org/10.1080/09513590.2022.2063831

Loh, H. H., Lim, L. L., Yee, A., & Loh, H. S. (2019). Association between subclinical hypothyroidism and depression: An updated systematic review and meta-analysis. *BMC Psychiatry, 19*(1), 12. https://doi.org/10.1186/s12888-018-2006-2

Maraka, S., Ospina, N. M., O'Keeffe, D. T., Espinosa De Ycaza, A. E., Gionfriddo, M. R., Erwin, P. J., Coddington, C. C., Stan, M. N.,

Murad, M. H., & Montori, V. M. (2016). Subclinical hypothyroidism in pregnancy: A systematic review and meta-analysis. *Thyroid, 26*(4), 580–590. https://doi.org/10.1089/thy.2015.0418

Metwalley, K. A., & Farghaly, H. S. (2021). Subclinical hypothyroidism in children: Updates for pediatricians. *Annals of Pediatric Endocrinology & Metabolism, 26*(2), 80–85. https://doi.org/10.6065/apem.2040242.121

Meyerovitch, J., Rotman-Pikielny, P., Sherf, M., Battat, E., Levy, Y., & Surks, M. I. (2007). Serum thyrotropin measurements in the community: Five-year follow-up in a large network of primary care physicians. *Archives of Internal Medicine, 167*(14), 1533–1538. https://doi.org/10.1001/archinte.167.14.1533

Mooijaart, S. P., Du Puy, R. S., Stott, D. J., Kearney, P. M., Rodondi, N., Westendorp, R. G. J., den Elzen, W. P. J., Postmus, I., Poortvliet, R. K. E., van Heemst, D., van Munster, B. C., Peeters, R. P., Ford, I., Kean, S., Messow, C. M., Blum, M. R., Collet, T. H., Watt, T., Dekkers, O. M., … Gussekloo, J. (2019). Association between levothyroxine treatment and thyroid-related symptoms among adults aged 80 years and older with subclinical hypothyroidism. *Journal of the American Medical Association, 322* (20), 1977–1986. https://doi.org/10.1001/jama.2019.17274

Mullur, R., Liu, Y. Y., & Brent, G. A. (2014). Thyroid hormone regulation of metabolism. *Physiological Reviews, 94*(2), 355–382. https://doi.org/10.1152/physrev.00030.2013

Parle, J., Roberts, L., Wilson, S., Pattison, H., Roalfe, A., Haque, M. S., Heath, C., Sheppard, M., Franklyn, J., & Hobbs, F. D. (2010). A randomized controlled trial of the effect of thyroxine replacement on cognitive function in community-living elderly subjects with subclinical hypothyroidism: The Birmingham elderly thyroid study. *Journal of Clinical Endocrinology & Metabolism, 95*(8), 3623–3632. https://doi.org/10.1210/jc.2009-2571

Pasqualetti, G., Pagano, G., Rengo, G., Ferrara, N., & Monzani, F. (2015). Subclinical hypothyroidism and cognitive impairment: Systematic review and meta-analysis. *Journal of Clinical Endocrinology & Metabolism, 100*(11), 4240–4248. https://doi.org/10.1210/jc.2015-2046

Pearce, S. H., Brabant, G., Duntas, L. H., Monzani, F., Peeters, R. P., Razvi, S., & Wemeau, J. L. (2013). 2013 ETA guideline: Management of subclinical hypothyroidism. *European Thyroid Journal, 2*(4), 215–228. https://doi.org/10.1159/000356507

Peeters, R. P. (2017). Subclinical hypothyroidism. *New England Journal of Medicine, 376*(26), 2556–2565. https://doi.org/10.1056/NEJMcp1611144

Provinciatto, H., Moreira, M. V. B., Neves, G. R., De Freitas, L. R., Mitsui, H. C., Zhang, J. M. F., & Araujo Júnior, E. (2024). Levothyroxine for subclinical hypothyroidism during pregnancy: An updated systematic review and meta-analysis of randomized controlled trials. *Archives of Gynecology and Obstetrics, 309*(6), 2387–2393. https://doi.org/10.1007/s00404-024-07512-3

Razvi, S., Weaver, J. U., Butler, T. J., & Pearce, S. H. (2012). Levothyroxine treatment of subclinical hypothyroidism, fatal and nonfatal cardiovascular events, and mortality. *Archives of Internal Medicine, 172*(10), 811–817. https://doi.org/10.1001/archinternmed.2012.1159

Roberts, L. M., Pattison, H., Roalfe, A., Franklyn, J., Wilson, S., Hobbs, F. D., & Parle, J. V. (2006). Is subclinical thyroid dysfunction in the elderly associated with depression or cognitive dysfunction? *Annals of Internal Medicine, 145*(8), 573–581. https://doi.org/10.7326/0003-4819-145-8-200610170-00006

Rodondi, N., den Elzen, W. P. J., Bauer, D. C., Cappola, A. R., Razvi, S., Walsh, J. P., Åsvold, B. O., Iervasi, G., Imaizumi, M., Collet, T.-H., Bremner, A., Maisonneuve, P., Sgarbi, J. A., Khaw, K.-T., Vanderpump, M. P. J., Newman, A. B., Cornuz, J., Franklyn, J. A. Westendorp, R. G. J., … . (2010). Subclinical hypothyroidism and the risk of coronary heart disease and mortality. *Journal of the American Medical Association, 304*(12), 1365–1374. https://doi.org/10.1001/jama.2010.1361

Salerno, M., Capalbo, D., Cerbone, M., & De Luca, F. (2016). Subclinical hypothyroidism in childhood - Current knowledge and open issues. *Nature Reviews Endocrinology, 12*(12), 734–746. https://doi.org/10.1038/nrendo.2016.100

Sankoda, A., Suzuki, H., Imaizumi, M., Yoshihara, A., Kobayashi, S., Katai, M., Hamada, K., Hidaka, Y., Yoshihara, A., Nakamura, H., Kubota, S., Kakita-Kobayashi, M., Iwase, A., Sugiyama, T., Ota, E., & Arata, N. (2024). Effects of levothyroxine treatment on fertility and pregnancy outcomes in subclinical hypothyroidism: A systematic review and meta-analysis of randomized controlled trials. *Thyroid, 34* (4), 519–530. https://doi.org/10.1089/thy.2023.0546

Shin, D. H., Lee, M. J., Kim, S. J., Oh, H. J., Kim, H. R., Han, J. H., Koo, H. M., Doh, F. M., Park, J. T., Han, S. H., Yoo, T.-H., & Kang, S.-W. (2012). Preservation of renal function by thyroid hormone replacement therapy in chronic kidney disease patients with subclinical hypothyroidism. *The Journal of Clinical Endocrinology & Metabolism, 97*(8), 2732–2740. https://doi.org/10.1210/jc.2012-1663

Sinha, S. H., Zietlow, K., & Papaleontiou, M. (2024). Thyroid function and cognitive decline: A narrative review. *Endocrine Practice, 30* (11), 1113–1118. https://doi.org/10.1016/j.eprac.2024.07.013

Spahia, N., Rroji, M., Barbullushi, M., & Spasovski, G. (2023). Subclinical hypothyroidism, kidney, and heart from normal to uremic milieu. *Metabolic Syndrome and Related Disorders, 21*(8), 415–425. https://doi.org/10.1089/met.2023.0057

Stott, D. J., Rodondi, N., Kearney, P. M., Ford, I., Westendorp, R. G. J., Mooijaart, S. P., Sattar, N., Aubert, C. E., Aujesky, D., Bauer, D. C., Baumgartner, C., Blum, M. R., Browne, J. P., Byrne, S., Collet, T.-H., Dekkers, O. M., den Elzen, W. P. J., Du Puy, R. S., Ellis, G., … Gussekloo, J. (2017). Thyroid hormone therapy for older adults with subclinical hypothyroidism. *New England Journal of Medicine, 376* (26), 2534–2544. https://doi.org/10.1056/NEJMoa1603825

Surks, M. I., Ortiz, E., Daniels, G. H., Sawin, C. T., Col, N. F., Cobin, R. H., Franklyn, J. A., Hershman, J. M., Burman, K. D., Denke, M. A., Gorman, C., Cooper, R. S., & Weissman, N. J. (2004). Subclinical thyroid disease: Scientific review and guidelines for diagnosis and management. *Journal of the American Medical Association, 291*(2), 228–238. https://doi.org/10.1001/jama.291.2.228

Taylor, P. N., Medici, M. M., Hubalewska-Dydejczyk, A., & Boelaert, K. (2024). Hypothyroidism. *The Lancet, 404*(10460), 1347–1364. https://doi.org/10.1016/S0140-6736(24)01614-3

Urgatz, B., & Razvi, S. (2023). Subclinical hypothyroidism, outcomes and management guidelines: A narrative review and update of recent literature. *Current Medical Research and Opinion, 39*(3), 351–365. https://doi.org/10.1080/03007995.2023.2165811

van der Spoel, E., van Vliet, N. A., Poortvliet, R. K. E., Du Puy, R. S., den Elzen, W. P. J., Quinn, T. J., Stott, D. J., Sattar, N., Kearney, P. M., Blum, M. R., Alwan, H., Rodondi, N., Collet, T. H., Westendorp, R. G. J., Ballieux, B. E., Jukema, J. W., Dekkers, O. M.,

Gussekloo, J., Mooijaart, S. P., & van Heemst, D. (2024). Incidence and determinants of spontaneous normalization of subclinical hypothyroidism in older adults. *Journal of Clinical Endocrinology & Metabolism, 109*(3), e1167–e1174. https://doi.org/10.1210/clinem/dgad623

van Vliet, N. A., van Heemst, D., Almeida, O. P., Åsvold, B. O., Aubert, C. E., Bae, J. B., Barnes, L. E., Bauer, D. C., Blauw, G. J., Brayne, C., Cappola, A. R., Ceresini, G., Comijs, H. C., Dartigues, J. F., Degryse, J. M., Dullaart, R. P. F., van Eersel, M. E. A., den Elzen, W. P. J., Ferrucci, L., ... Trompet, S. (2021). Association of thyroid dysfunction with cognitive function: An individual participant data analysis. *JAMA Internal Medicine, 181* (11), 1440–1450. https://doi.org/10.1001/jamainternmed.2021.5078

Yu, O. H. Y., Filliter, C., Filion, K. B., Platt, R. W., Grad, R., Rad, M. V., & Renoux, C. (2025). Levothyroxine treatment for subclinical hypothyroidism and risk of adverse renal outcomes: A population-based cohort study. *Canadian Medical Association Journal, 197*(26), E744–e753. https://doi.org/10.1503/cmaj.250356

Yu, O. H. Y., Filliter, C., Filion, K. B., Platt, R. W., Grad, R., & Renoux, C. (2024). Levothyroxine treatment of subclinical hypothyroidism and the risk of adverse cardiovascular events. *Thyroid.* https://doi.org/10.1089/thy.2024.0227

Chapter 5

The consequences of thyroidectomy: An overview new narrative

Bhuvnesh Deivakumar[1], Femi E. Ayeni[1,2], Saikrishna Ananthapadmanabhan[1], Mark Smith[1], Faruque Riffat[1] and Niranjan Sritharan[1]

[1]*Department of Otolaryngology, Head and Neck Surgery, Nepean Hospital, Sydney, NSW, Australia;* [2]*Nepean Institute of Academic Surgery, Nepean Clinical School, The University of Sydney, Sydney, NSW, Australia*

List of abbreviations

ACTH Adrenocorticotropic hormone
CRH Corticotropin releasing hormone
EBSLN External Branch of the Superior Laryngeal Nerve
HCG Human Chorionic Gonadotropin
HPA Hypothalamic Pituitary Axis
IONM Intra-operative nerve monitoring
MDADI MD Anderson Dysphagia Index
RLN Recurrent Laryngeal Nerve
SIS Swallow Impairment Score
VHI Voice Handicap Index

Introduction

Thyroid pathologies are relatively common among the general population. Although some of these disorders are managed medically, thyroidectomy remains the most widely performed head and neck surgery, serving therapeutic and curative indications in benign and malignant conditions (Stefanou et al., 2022).

Historically, thyroidectomy was considered a highly dangerous procedure with severe mortality risk secondary to infection and hemorrhage, and was actively discouraged (Stefanou et al., 2022). The pioneering work of Theodore Kocher, often regarded as the father of modern thyroid surgery, refined the technique with meticulous attention to anatomy and surgical principles. His essential tenets emphasized a comprehensive understanding of the surgical anatomy of the thyroid gland, gentle handling of the proximal key structures, and rigorous hemostatic control. These principles significantly reduced the mortality risk in the 1900s and remain relevant to contemporary practice (Kapre, 2020; Stefanou et al., 2022).

Early post-operative consequence

Hemorrhage

Post-operative hemorrhage often manifests and is clinically recognized in the immediate postoperative period. The rich vascular supply and high blood perfusion of the thyroid gland make bleeding a constant risk factor with variable bleeding risk encountered across different thyroid pathologies. Although the risk of post-thyroidectomy hemorrhage is small, at 0.7%–2.1% (Promberger et al., 2012), the potential for immediate life-threatening consequences, including airway obstruction from an expanding hematoma in the surgical bed, remains a point of contention in the debate over same-day discharge following thyroidectomy (Leyre et al., 2008). Intraoperatively, bleeding or difficult-to-control bleeding may obscure anatomic landmarks and nearby critical structures, increasing the risk of inadvertent injuries to the recurrent laryngeal nerve or parathyroid glands.

Postoperatively, an expanding hematoma causes veno-lymphatic obstruction resulting in laryngopharyngeal edema and airway obstruction – in the setting of respiratory distress and clinical airway compromise, this is an imminent problem requiring urgent evacuation (Iliff et al., 2021). In time-critical situations, intubation should be performed after opening the neck and evacuating the hematoma, with practitioners also cognizant of the potential need to secure front-of-neck airway access (Fig. 5.1).

The perioperative risk of hemorrhage is influenced by patient and disease factors. Pathologies such as Graves' disease, toxic adenoma, and thyroid malignancies have increased bleeding tendencies. In addition, several patient factors such as advanced age, male gender, coagulopathies,

Hormones in Neuroscience. https://doi.org/10.1016/B978-0-443-30261-9.00005-X

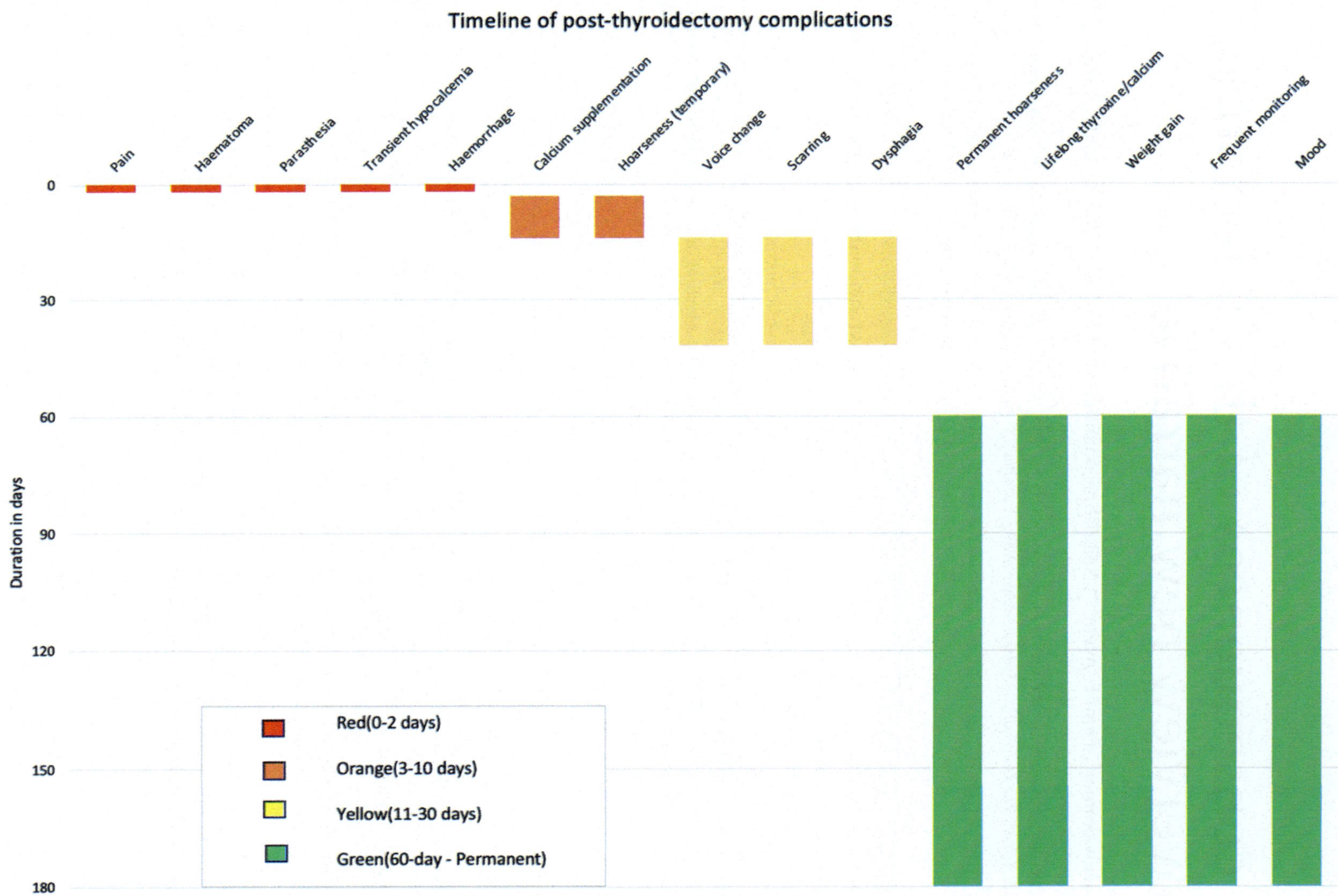

FIGURE 5.1 Timeline of post-thyroidectomy complications.

use of anti-thrombotic agents, or pathologies complicated by retrosternal extension or requiring neck dissection or revision surgery increase the risk of bleeding (Liu et al., 2017; Wojtczak et al., 2018).

With multiple potential sources of perioperative bleeding, early identification of hemorrhage is critical. Bleeding in the postoperative period may occur due to residual thyroid tissue, unsealed vessels, oozing from skin or muscle, or loosely tied knots. Surgical advancements, including novel vessel-sealing energy devices and hemostatic agents, have been welcomed in practice (Materazzi et al., 2017). Intraoperatively, various surgical techniques have been described to aid in the early identification of hemorrhage. The Valsalva maneuver can help identify potential sites of bleeding (Morton et al., 2011), while the use of the Trendelenburg tilt further assists in the assessment of hemostasis (Moumoulidis et al., 2010).

In contrast to popular belief, contemporary evidence does not support the routine insertion of a neck drain in thyroidectomy, as there is no observed benefit in reducing the rate of postoperative hematomas (Maroun et al., 2020; Materazzi et al., 2017). Similarly, the use of topical hemostatic agents also demonstrates variable results. While they are an adjunct in achieving intraoperative hemostasis, thereby decreasing operating times, drain outputs, and duration of hospital stay, there is inadequate evidence to support a reduction in postoperative bleeding and hematomas (Khadra et al., 2018).

Post-operatively, close monitoring during the first 6 hours of surgery, maintenance of systolic blood pressure less than 150 mm Hg (Morton et al., 2011), nursing education around assessment of increasing neck circumference associated with dyspnea and early identification of an unusual amount of blood in the suction drain can assist in minimizing the risk of, and early detection of, post-operative bleeding and hematoma (Wojtczak et al., 2018).

Endocrine and metabolic consequences

Peri-operative thyroid hormone derangements

Partial and total thyroidectomy can lead to endocrine disruptions, most commonly hypothyroidism and acute or chronic hypoparathyroidism, which can result in hypocalcaemia.

The risk of intra-operative thyroid manipulation and the resultant rare complication of a thyroid storm or thyrotoxic crisis is the most feared complication in thyrotoxic patients undergoing thyroidectomy. Thyroid storm is characterized by sinus tachycardia, severe hypertension and can lead to cardiac failure and death. This under-reported and under researched complication highlights the importance of pre-operative assessment and treatment of thyrotoxicity with methimazole and potassium iodide essential, as recommended by the American Thyroid Association Guidelines (de Mul et al., 2021; Stefanou et al., 2022).

Hypothyroidism is an inevitable consequence following total thyroidectomy. Studies suggest that it also occurs in 29% of patients following hemithyroidectomy (Li et al., 2020). Hemithyroidectomy is a prevalent choice in unilateral goiter, single toxic adenoma, and low-risk papillary thyroid carcinoma, offering a safer alternative to total thyroidectomy with less severe consequences. However, high preoperative TSH indicates a reduced thyroid reserve function, along with factors such as older age, female, and Hashimoto's thyroiditis, further increasing the risk of hypothyroidism (Li et al., 2020). Thyroid pathologies with high post-operative risk of hypothyroidism, ongoing need for thyroxine supplementation, and risk of disease recurrence contralaterally and possible reoperation make studious pre-operative planning, including consideration of total thyroidectomy and appropriate post-operative planning and follow-up, crucial (Li et al., 2020).

Disruption to calcium homeostasis

Hypocalcemia is the most common endocrine complication following thyroidectomy, with reported incidences ranging from 25% for transient hypocalcemia and <2% for permanent hypocalcemia (Edafe et al., 2014; Qin et al., 2021). The parathyroid glands play a critical role in calcium homeostasis and are especially vulnerable to injury during thyroidectomy due to their small size, proximity to the thyroid gland, and variable anatomy. The inferior parathyroid glands in particular, are at a greater risk of devascularization, due to their reliance on the inferior thyroid artery and the inadvertent trauma during the surgery (Jeong et al., 2016; Zhou et al., 2016).

Ligature of the inferior thyroid artery is therefore performed as close as possible to the thyroid gland capsule to preserve parathyroid blood supply. A central neck dissection further increases the risk of incidental parathyroidectomy due to the extensive dissection required in the region which may lead to hypoparathyroidism (Edafe et al., 2014; Jeong et al., 2016; Stefanou et al., 2022). Intraoperative factors such as trauma, inadvertent excision, anatomical positioning of the parathyroid gland, and disruption to its vascular supply remain the primary contributors to post-operative hypocalcemia.

Strategies such as the use of autofluorescence to identify parathyroid glands, careful documentation of congested parathyroid glands and their viability, and selective auto transplantation aid in the early identification of vulnerable patients who will require supplementation, while minimizing intraoperative risks (Stefanou et al., 2022; Wiseman et al., 2021).

Beyond surgical factors, certain patient and disease factors contribute to the risk. Patients with pre-operative thyrotoxicosis have a higher incidence of postoperative hypocalcemia compared to euthyroid patients (Fields et al., 2023). The under-researched phenomenon of hungry bone syndrome partly explains this in the context of thyroidectomy. Thyrotoxicosis is associated with high bone turnover, with the catabolic effects of triiodothyronine favoring bone resorption.

This can lead to thyrotoxic osteodystrophy, hypercalcemia, and reactive hypoparathyroidism. Postoperatively, the reactive biochemical hypoparathyroidism in previously thyrotoxic patients induces transient hypocalcaemia (Fields et al., 2023). In addition, the abrupt post-operative osteoblast-mediated bone formation promotes skeletal calcium uptake, resulting in biochemical hypocalcaemia. This increases the risk of muscular spasms, including life-threatening laryngospasms, and warrants routine calcium monitoring following thyroidectomy (Fields et al., 2023; Karunakaran et al., 2018).

Biochemical predictors of hypocalcemia have been well defined. Pre-surgical assessment of serum PTH, calcium and Vitamin D levels help stratify the risk of post-operative hypocalcemia in the vulnerable population (Ankitha et al., 2021). Post-operative PTH measured at 3–4 hours, attributing to the known short-half-life of PTH, has shown high sensitivity and specificity in predicting hypocalcemia (Metere et al., 2021).

Alternatively, a PTH gradient comparing PTH in the immediate post-operative period and day one post-surgery values has also been suggested to be a valuable marker to guide ongoing calcium management (Garg et al., 2021). In lieu of the higher incidence of this complication, routine calcium augmentation following thyroidectomy has been widely advocated in all patients to reduce the incidence of symptomatic hypocalcemia and its associated morbidity (Verma et al., 2020).

Neurological consequence: Nerve injuries

Preservation of both the recurrent laryngeal nerve (RLN) and the external branch of the superior laryngeal nerve (ESBLN) is essential in thyroid surgery to effectively maintain vocal cord function. Preoperative assessment of voice and vocal cord function using naso-laryngoscopy aids in risk stratification and counseling, especially in neoplastic disease states with suspicion of nerve involvement (Fig. 5.2).

Unilateral injury to the recurrent laryngeal nerve can cause hoarseness in voice, cough, and micro aspirations. In contrast, bilateral injuries may result in glottal obstruction leading to dyspnea, respiratory failure or death. The prevalence of these injuries' ranges from 6%–10% for a transient RLN injury – typically defined as resolution of symptoms within a year, and 1%–1.5% for permanent injuries (Jeannon et al., 2009; Stefanou et al., 2022).

Anatomical consideration & intraoperative nerve preservation

A key principle to safe surgery is the thorough knowledge of the surgical anatomy of RLN and EBSLN, including awareness of uncommon anatomic variants. The RLN branches from the vagus nerve at the level of the subclavian artery on the right and at the level of the aorta on the left. The nerves then course through the tracheoesophageal groove proximal to thyroid gland and the inferior thyroid artery. It serves as the main motor nerve to all intrinsic laryngeal muscles. The cricothyroid muscle, responsible for a high-pitched voice, is an exception, and is supplied by the external branch of the superior laryngeal nerve (EBSLN).

The EBSLN is also vulnerable to injury during thyroid surgery due to its variable course and proximity to the superior thyroid vessels (Chandrasekhar et al., 2013; Stefanou et al., 2022). Attention must be paid to infrequent anatomic variants including extra-laryngeal branching patterns of the RLN, the presence of a non-recurrent laryngeal nerve, and anastomoses between the RLN and the internal and external branches of the superior laryngeal nerve (Chandrasekhar et al., 2013). Injury to these nerves can occur during surgery due to traction, contusion, crush, burn, disruption to the vascular supply, or transection. The risk is higher in pathologies such as thyroid cancer with extra-thoracic extension or multinodular goiter where the nerve may already be stretched or involved by disease, making postoperative paralysis more likely (Affleck et al., 2003; Stefanou et al., 2022).

Nerve monitoring is pivotal in thyroid surgery, particularly since the shift toward the practice of direct visualization of the RLN intra-operatively to prevent injury. Intraoperative nerve monitoring (IONM), a prevalent adjunctive method that utilizes electrical stimulation to identify the recurrent laryngeal nerve during thyroid surgery has been widely accepted and integrated into standards of care despite mixed evidence (Duong et al., 2021). Several studies have evidently demonstrated that the use of a sterile nerve stimulator not only aids in identification of the RLN, but also strongly correlates with the prevention of transient and permanent RLN injury, aiding in neural mapping during dissection, identification of neural injury and prognostication, especially in neuropraxic injuries (Bai & Chen, 2018; Chandrasekhar et al., 2013; Duong et al., 2021)). While other studies report no significant benefit with the use of IONM (Higgins et al., 2011).

Ligation of the superior thyroid vessels closer to the thyroid capsule remains the gold standard in the preservation of EBSLN. However, the varied anatomical course of EBSLN warrants the use of IONM in its identification and not be limited to the RLN (Chandrasekhar et al., 2013; Stefanou et al., 2022). Management strategies depend on

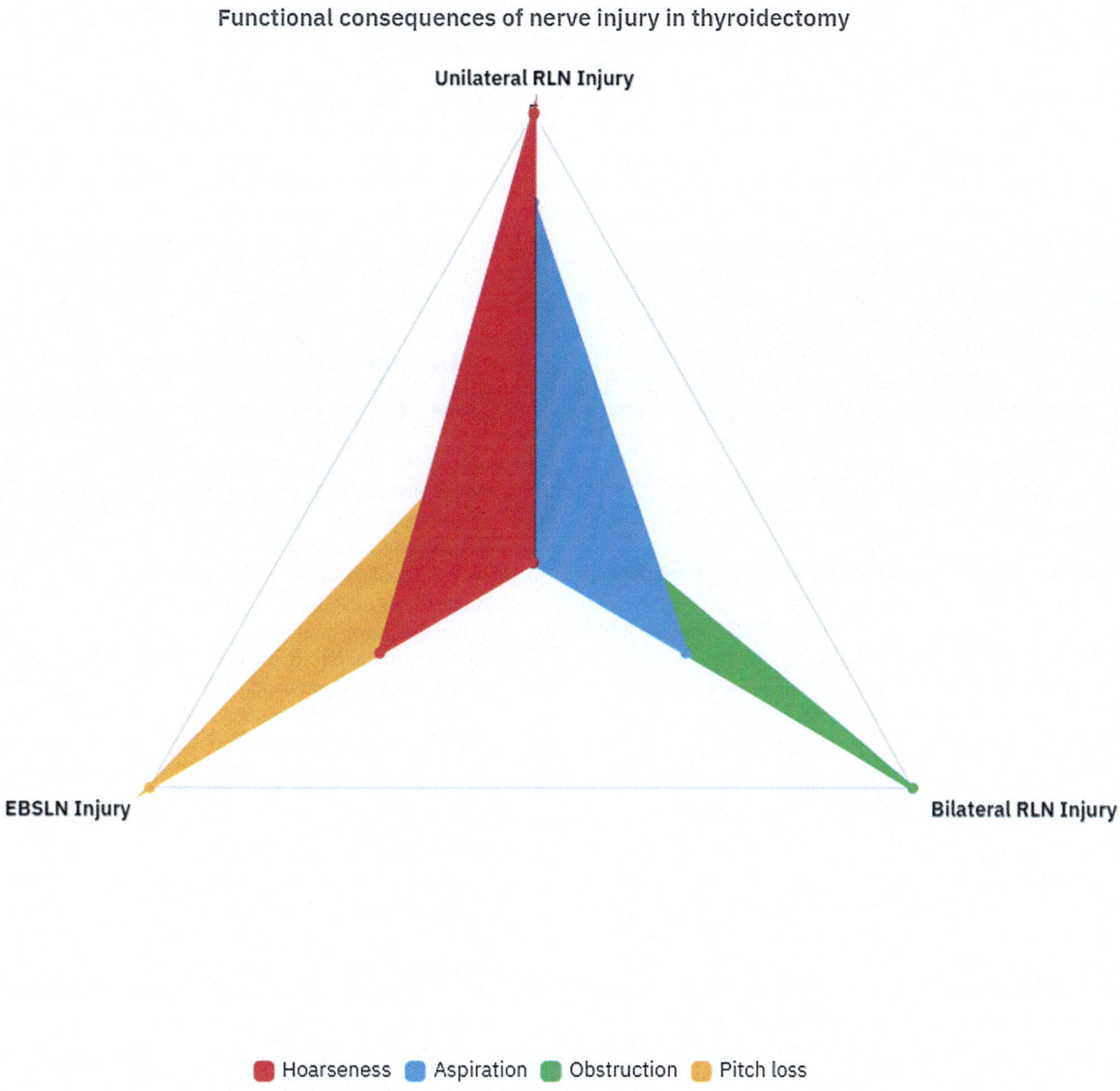

FIGURE 5.2 Functional consequences of nerve injury in thyroidectomy.

the type of injury and its impact on the patient's quality of life. Treatments include patient education, voice therapy, voice augmentation, and reinnervation of the transected nerve. In severe cases with bilateral nerve injury, tracheostomy is performed for airway protection (Chandrasekhar et al., 2013).

Functional consequences

Dysphonia

Dysphonia may occur following thyroidectomy even in the absence of the anticipated recurrent laryngeal nerve injury (Kaur et al., 2021). Although several hypotheses suggest that this is likely due to laryngeal edema, intubation trauma, or strap muscle damage resulting in temporary malfunction or vocal cord folding, the mechanism is still not fully understood. Patient subjectively reports temporary or persistent hoarseness, upper pitch loss, difficulty speaking out loud and vocal fatigue, post-surgery (Stojadinovic et al., 2002).

Subjective assessment of voice impairment using various parameters of voice quality such as fundamental frequency, maximum frequency, and maximum phonation length, pre and post operatively identified decline in voice performance in the immediate post-operative period. These studies, however, also identified the general inclination of these subjective scores at a 3-month period toward their pre-operative scoring highlighting the temporary nature of these symptoms (Kaur et al., 2021; Marchese et al., 2021).

Although thyroidectomy is one of the most commonly performed neck procedures worldwide, it carries the risk of voice dysfunction, that can be particularly debilitating. This emphasizes the imperative of addressing material risk during shared decision making, especially in high impact population. Surgeons typically quote voice related implications associated with thyroidectomy to be 1%–4% despite the evidence suggesting otherwise as identified by a qualitative study characterizing pre-operative consultations (Kao et al., 2025). This narrow estimate reflects the focus on nerve–related injury, thereby overlooking the broader spectrum of non-nerve related dysphonia frequently reported by patients.

Dysphagia

Thyroidectomy is often regarded as the procedure to resolve preoperative dysphagic symptoms in various thyroid pathologies, particularly substernal goiters, which may lead to swallowing difficulties. Dysphagia may be frequently reported pre-operatively amongst patients with benign and malignant thyroid pathologies, making it a common surgical indication. Ironically, these symptoms prevail even in the absence of an obvious superior laryngeal nerve or recurrent laryngeal nerve injury post-operatively, however transient (Vardaxi et al., 2022). This phenomenon has been explained by several hypotheses including secondary to orotracheal intubation, adhesions of the skin flap, changes to esophageal motility, and the unevaluable injury to the extrinsic peri thyroid neural plexus (Scerrino et al., 2017).

Post-operative swallowing symptoms are frequently reported irrespective of the thyroid pathology, both subjectively and objectively, with no statistical significance between benign and malignant thyroid pathologies or endoscopic and open thyroidectomies (Vardaxi et al., 2022). Although the prevalence of dysphagia peaks in the immediate post-operative period, this generally improves substantially within a few months (Vardaxi et al., 2022).

Postoperative dysphonia and dysphagia can have psychosocial consequences, impacting quality of life, daily social interactions, functional independence, medication compliance, and nutritional intake. The assessment of these patients can include fibreoptic laryngoscopy, fluoroscopic swallow studies, or esophageal manometry. Instrumental diagnostic tools and screening questionnaires including the Voice Handicap Index (VHI), Swallow Impairment Score (SIS), and the MD Anderson Dysphagia Index (MDADI) can supplement assessment (Scerrino et al., 2017; Đanić Hadžibegović, 2020).

Iatrogenic injury resulting in complications such as esophageal perforation is rare. However, the risk increases with poor surgical technique or extensive malignancy with pathological invasion of the upper aerodigestive tract. Adequate preoperative evaluation using contrast imaging can aid in assessment, early surgical planning, and prognostication. Nevertheless, esophageal injuries can result in stricture formation leading to long-term dysphagia, requiring further extensive interventions such as endoscopic dilation when identified early or esophagectomy and reconstruction in severe cases (Peng et al., 2014).

Cosmetic and aesthetic consequence

Visible scar in the neck region can have significant cosmetic and psychological impact. Atypical scar response can not only be aesthetically displeasing but also cause emotional distress, disabling the patient both functionally and psychosocially (Hong et al., 2024). This makes meticulous surgical technique essential during skin closure to achieve optimal cosmetic outcome.

Advanced surgical techniques implemented over the last few decades have minimized the incision length from a 10 cm Kocher incision to as little as 15 mm in video assisted thyroidectomy procedures. Despite the emergence of minimally invasive techniques, including robotic surgeries via intra-oral and post-auricular approaches, the conventional cervical incision via the Kocher's technique remains the most widely adopted method for thyroidectomy (Hong et al., 2024).

Certain patient factors such as Afro-Caribbean descent and a personal or family history of hypertrophic scarring or keloid can amplify scarring response. Since these innate tendencies are difficult to modify, perioperative strategies are important especially in certain high-risk populations (Hong et al., 2024). Intra-operative use of subcuticular sutures for skin closure provides favorable esthetic outcomes (Chung et al., 2021). The use of non-ablative fractional laser, intra-lesional steroid injection, and scar revision in the post-operative period can help treat hypertrophic or keloid scars.

Another factor impacting cosmesis is the development of seromas. Seromas are postoperative collections of serous fluid within the surgical field. The likelihood of a seroma post-operatively typically ranges from 1.3% to 7% and its clinical significance largely depends on its size (Ramouz et al., 2017). While small seromas often resolve spontaneously within 6–8 weeks, larger ones usually require aspiration. Notably, a single aspiration increases the likelihood of subsequent aspirations. The presence of a seroma can compromise cosmetic outcomes by increasing the risk of infection, neck swelling, flap necrosis, and thereby increasing the need for repeated interventions (Ramouz et al., 2017).

Psychological consequence: Effect on quality of life

Beyond physical symptoms, thyroid hormone deficiency is closely linked to anxiety and depression following thyroid surgery, significantly affecting psychological well-being (Huynh et al., 2020; Shin et al., 2016). Hypothyroidism is an inevitable consequence following total thyroidectomy. Hypothyroidism notably also occurs in 7%–35% of patients following other less extensive thyroid procedures, such as lobectomy and hemithyroidectomy (Kapre, 2020). Hypothyroidism, and the resultant fall in the basal metabolic rate manifest clinically as weight gain, persistent fatigue, and reduced energy levels.

These consequences make prompt achievement of optimal thyroid hormone replacement in the immediate post-operative period crucial. The 2013 European Thyroid Association guidelines recommend titrating levothyroxine replacement to maintain serum TSH within the reference range of 0.4–4.0uIU/ml, with an advised target range to be (0.4–2.5uIU/ml) (Yaniv et al., 2022). However, an euthyroid state across all tissues is unlikely to be captured by a single measure of serum TSH concentration, underscoring the limitations of titrating levothyroxine supplementation solely based on biochemical values rather than patient-reported symptoms and well-being.

Patients undergoing thyroidectomy may require lifelong thyroxine replacement. Hyperthyroidism is a metabolic state that increases bone turnover, particularly the cortical bone of the hip and forearm (Hung et al., 2018). Post-operatively, the use of exogenous levothyroxine mimics this metabolic state and increases the risk of post-thyroidectomy osteoporosis following total, sub total or partial thyroidectomy (Hung et al., 2018). However, achieving an optimal level of thyroid hormone level remains a significant clinical challenge. Despite various thyroxine supplementation strategies, including fixed dosing, weight-based dosing, and lean body mass dosing, only one-third of the patients achieve a euthyroid state at their first post-operative review (Chiardi et al., 2025).

In contrast to the risk of osteoporosis, in metabolic states of inadequate supplementation, patients may continue to experience weight gain, anxiety, and depression despite normal TSH levels. This suggests that these deficiencies may extend beyond biochemical targets, requiring frequent clinical monitoring. Individuals who fail to consistently achieve a clinical euthyroid state report poorer quality of life compared to their counterparts, irrespective of the extent of surgery performed (Yaniv et al., 2022). These observations highlight the need for individualized dose titration strategies that account for specific factors such as age, sex, BMI, and lean body mass.

Conversely, inadequate thyroid hormone replacement following total thyroidectomy and, in select cases of lobectomy, may lead to bradycardia, dyslipidemia, and associated accelerated atherosclerosis. These contrasting needs in post-operative care highlight the importance of individualized titration and long-term surveillance. The balance in maintaining euthyroidism can significantly impair one's quality of life post-operatively and should be factored into surgical decision-making (Chiardi et al., 2025; Yaniv et al., 2022).

Thyroidectomy consideration in the pediatric population

Thyroid surgery is becoming an increasingly utilized treatment modality amongst the pediatric population. Thyrotoxicosis by Grave's disease refractory to medical therapy to achieve remission is the most common indication for thyroidectomy in children. Additional indications include benign pathologies such as large goiters and thyroid nodules, differentiated thyroid malignancies, and prophylaxis in familial syndromes predisposing to medullary thyroid carcinoma (Angeliki et al., 2024). Surgical management of thyroid pathologies in children, however, presents distinct challenges due to the peculiar anatomical variation compared with adults, in addition to other consequences discussed earlier.

Children are susceptible to post-operative complications such as transient or permanent hypocalcemia, transient or permanent hypothyroidism, and are at an increased risk of recurrent laryngeal nerve (RLN) injuries compared with adults (Angeliki et al., 2024). This increased risk arises from the difficulty in identifying RLN due to its smaller caliber in children and potentially being obscured by ectopic cervical thymus tissue. Furthermore, the smaller thyroid volume and softer laryngeal structures in children add to the technical complexity of the surgery (Zhang et al., 2022). Hypoparathyroidism is the most common complication following thyroid surgery in children, as in adults. Post-operative management of hypocalcemia in children, however, is often more challenging due to difficulty in achieving medication compliance with oral calcium and Vitamin D supplementation. This frequently necessitates extended inpatient admissions requiring intravenous calcium supplementation (Julie et al., 2008).

Total thyroidectomy is the most common thyroid procedure performed in the pediatric population (Angeliki et al., 2024). The resultant acquired hypothyroidism is of particular concern in this group, as it may contribute to growth delay, short stature and discordant puberty, if not adequately managed. This highlights the importance of prompt initiation of thyroxine therapy following surgery and close specialist monitoring thereafter (Bhattacharyya & Singh, 2023).

Thyroidectomy in children remains an effective intervention, however, carries unique risks and consequences.

To mitigate complications, the procedure should be performed by an experienced high-volume surgeon, with the support of adjuncts such as Intra-operative nerve monitoring (Angeliki et al., 2024; Zhang et al., 2022).

Consequences of alternative thyroidectomy techniques

Remote access endoscopic and robotic thyroidectomy procedures via cervical, axillary, and post-auricular approaches are alternatives to conventional open cervical thyroidectomy (Kapre, 2020). These newer alternatives are particularly appealing to patients who prioritize better cosmetic outcomes (Fig. 5.3).

In addition to having minimally visible scarring, these procedures provide a magnified surgical field, making it easier to identify the RLN and parathyroid gland (Kapre, 2020). This is aimed at reducing common post-operative complications in theory. While these techniques offer advantages to the conventional surgical approach, they introduce a distinct set of risks that differ from those associated with the traditional open cervical approach.

Endoscopic thyroidectomy utilizes CO_2 insufflation through a smaller incision to enhance the visualization of the surgical field. This provides better cosmesis; however, CO_2 insufflation increases the risks of subcutaneous emphysema, hypercarbia, and pneumomediastinum (Kapre, 2020). Alternatively, gasless endoscopic procedures maintain a clear surgical view in the absence of CO_2 insufflation and subsequent risks. However, this is not a minimally invasive surgery as it requires an extensive skin flap elevation and a wider dissection area (Kapre, 2020).

Robotic thyroidectomy is effective in improving cosmesis, but it is associated with an increased risk of intraoperative complications when performed by surgeons with low procedural volume. Although cosmetic results are enhanced, there is an increased risk of transient RLN injury and longer operative times compared to the conventional approach (Kapre, 2020; de Vries et al., 2021). These risks are largely due to the steep learning curve, increasing the likelihood of other severe complications such as tracheal injury, hematoma, and CO_2 embolism. Additionally, robotic access via the trans-axillary, post-auricular, and trans oral approaches also raise the risk of additional complications such as facial and chest wall paresthesia, injury to the brachial plexus, marginal mandibular nerve, and mental nerve. This highlights the need to reserve these procedures for carefully selected patients (Kapre, 2020).

However, the risk of other unusual complications, especially during the steep learning curve, necessitates an

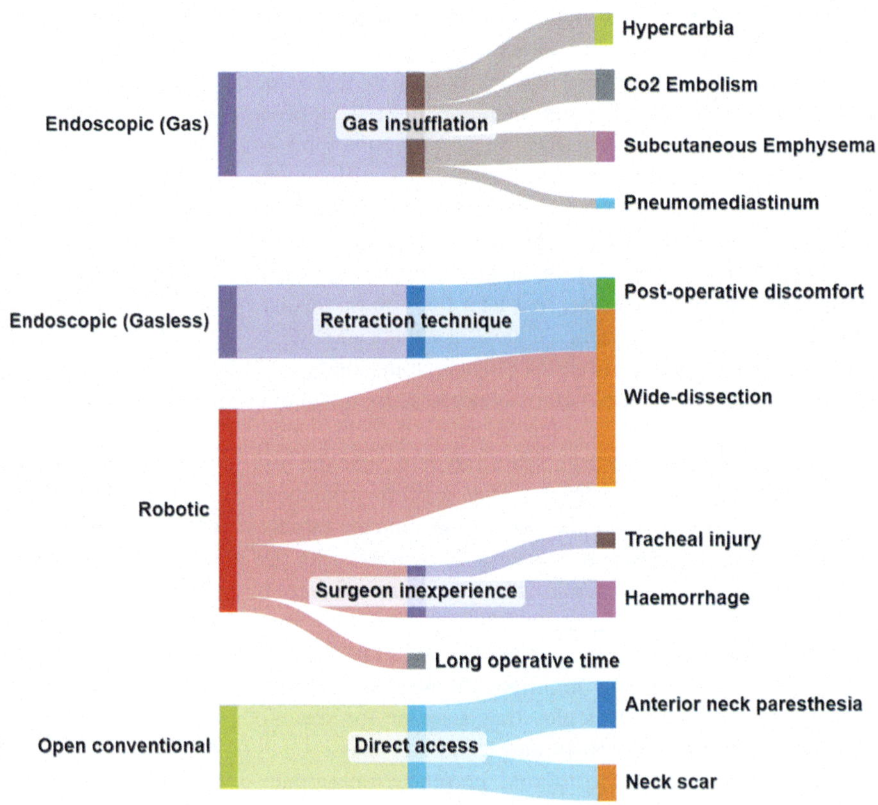

FIGURE 5.3 Consequences in alternative thyroidectomy.

appropriate institutional training program with adequate volume for surgeons. Furthermore, with patient safety taken into consideration, they should be reserved for selective individuals, and the possibility of converting to an open procedure in an emergency should always be discussed preoperatively (Kapre, 2020).

Other hormones and their pathways

The risk of intraoperative thyroid manipulation and the resultant thyroid storm or thyrotoxic crisis is the most feared complication in thyrotoxic patients undergoing thyroidectomy. Thyroid storm is characterized by sinus tachycardia, severe hypertension, and can lead to cardiac failure and death. This under-researched complication highlights the importance of pre-operative assessment and treatment of thyrotoxicity with methimazole and potassium iodide, which is essential, as recommended by the American Thyroid Association Guidelines (de Mul et al., 2021; Stefanou et al., 2022).

In the postoperative period, calcium dysregulation is the most frequently addressed issue; however, hypomagnesemia and its link to the post-thyroidectomy hypocalcaemia should also be considered. Emerging evidence suggests a correlation between low magnesium and impaired PTH secretion, which may compound the risk of hypocalcemia. This reiterates the importance of perioperative electrolyte monitoring, as such abnormalities can prolong hospital stay and delay recovery (Chincholikar & Ambiger, 2018; Radhakrishnan et al., 2023).

Total thyroidectomy is performed with a curative intent in patients with differentiated thyroid carcinoma. The trophic influence of thyroid-stimulating hormone on residual thyroid tissue makes suppressive levothyroxine therapy essential to prevent recurrence. While the cardiovascular consequences of long-term TSH suppression, such as atrial fibrillation, cerebrovascular accidents, and heart failure (Pucci et al., 2025), are well recognized, minimal attention has been paid to its indirect effect on other endocrine pathways. Exogenous thyroxine supplementation increases the hepatic clearance of cortisol, promoting a compensatory increase in secretion of CRH and ACTH. This adaptive dysregulated HPA axis response may impair stress tolerance and contribute to depressive or cognitive symptoms in patients following thyroidectomy (Lizcano & Rodríguez, 2011).

Thyroid surgery, including hemithyroidectomy, is relatively common in young women of reproductive age, making consideration of fertility and pregnancy particularly relevant. Pregnancy has a marked impact on thyroid physiology. Increased secretion of human chorionic gonadotropin (hCG) in pregnancy stimulates thyroid hormone production (T3, T4), while elevated estrogen levels during pregnancy increase thyroid-binding globulin concentrations. This physiological change makes normal pregnant females clinically euthyroid or slightly hyperthyroid, with transient suppression of serum TSH levels in the first trimester (Yap et al., 2023).

Pregnant women who have previously undergone total thyroidectomy are entirely dependent on exogenous levothyroxine. The increased metabolic and hormonal demand of gestation necessitates higher doses of thyroid hormone in these women with reduced thyroidal reserve, predisposing them to uncontrolled hypothyroidism (Alexander et al., 2017). Overt hypothyroidism following thyroid surgery is associated with a greater need for fertility treatments as well as higher rates of miscarriage and placental abruption compared to women with hypothyroidism from other causes (Cohen et al., 2010).

Applications of the material

A thorough pre-operative assessment by the surgeon, anesthesiologist, and endocrinologist is essential for risk stratification and counseling. Clinicians play a crucial role in obtaining informed consent from patients by clearly explaining the procedure and discussing the expected complications and potential red flag symptoms, which facilitates an early identification of minor setbacks before they progress into major complications. Precise technique and meticulous surgical dissection, supplemented by thorough knowledge of normal and pathological surgical anatomy, as well as other relevant anatomical variations, are critical to minimizing surgeon-led intraoperative setbacks. Intraoperative nerve monitoring has been popularized as an important adjunct in neural mapping, safe dissection around the nerve, and an essential aid in prognostication where nerve trauma is suspected. Early detection of post-operative complications is paramount. Vigilant monitoring for neck hematoma in the immediate post-operative period is critical, as an expanding neck hematoma can result in airway compromise. This limits the same-day discharge of patients following thyroidectomy. A highly sensitive post-operative PTH level guides the prediction of hypocalcemia, the most common endocrine complication following thyroidectomy. Dysphagia and dysphonia can occur in the absence of nerve injury, often transient due to laryngeal edema or strap muscle manipulation. Post-operative seromas are common and will require multiple fine needle aspirations for ongoing symptomatic and cosmetic outcomes. This emphasizes the need for patient education and the expectant lifestyle modifications. Following thyroidectomy, a multidisciplinary specialist care follow-up is essential in managing endocrine, metabolic, and psychological consequences to improve patient satisfaction peri-operatively. Special populations, including pregnant patients and children, require tailored perioperative planning addressing the unique physiological variation and anatomical complexity associated with thyroidectomy.

Conclusion

Thyroidectomy is a common endocrine procedure with a well-recognized complication profile that occurs with a low incidence. Despite its established safety, the post-operative complications can have a high morbidity and mortality risk, imposing long-term repercussions that impact quality of life and psychological well-being. Nevertheless, thyroidectomy remains an essential procedure in the management of thyroid disease and requires thorough perioperative care and careful planning to reduce complications.

Mini-dictionary of terms

Dysphagia: Difficulty or discomfort whilst swallowing.
Dysphonia: Impaired voice secondary to disruption to quality, pitch or volume.
Hematoma: A localized extra-vascular collection of blood following trauma or surgery.
Hemi-thyroidectomy: Surgical removal of one half of the thyroid gland.
Pneumomediastinum: Presence of air in the central compartment of chest.
Seroma: Collection of serous fluid in the post-operative surgical field.

Summary points

- Modern-day thyroidectomy, although a frequently performed endocrine procedure, is still associated with immediate, intermediate, and long-term consequences.
- Post-operative hemorrhage is an emergent complication since a rapidly expanding neck hematoma can cause airway compromise.
- Hypocalcemia is the prevalent endocrine complication, and post-operative PTH is a reliable predictor that guides safe early discharge.
- Intraoperative nerve monitoring is a key surgical adjunct in preventing injury to the recurrent laryngeal nerve and superior laryngeal nerve.
- Long-term specialist care and dose adjustments are essential to address post-operative hormonal, metabolic, and psychological consequences.

References

Affleck, B. D., Swartz, K., & Brennan, J. (2003). Surgical considerations and controversies in thyroid and parathyroid surgery. *Otolaryngologic Clinics of North America, 36*(1), 159–187. https://doi.org/10.1016/s0030-6665(02)00135-4

Alexander, E. K., Pearce, E. N., Brent, G. A., Brown, R. S., Chen, H., Dosiou, C., Grobman, W. A., Laurberg, P., Lazarus, J. H., Mandel, S. J., Peeters, R. P., & Sullivan, S. (2017). 2017 Guidelines of the American thyroid association for the diagnosis and management of thyroid disease during pregnancy and the postpartum. *Thyroid, 27*(3), 315–389. https://doi.org/10.1089/thy.2016.0457

Angeliki, V., Angeliki, C., Georgios, T., Menni, A., Stergianna, N., Giza, S., Tzitiridou-Chatzopoulou, M., Sohail, B., Assimina, G.-T., Michalopoulos, A., & Papavramidis, T. S. (2024). Thyroidectomy in children and adolescents: A systematic review. *Gland Surgery, 13*(5), 697–712. https://doi.org/10.21037/gs-24-16

Ankitha, R., Reddy, A. T., Dalal, P., Rastatter, J. C., Josefson, J. L., Samis, J., Beestrum, M., Tian, Y., & Raval, M. V. (2021). Hypocalcemia prevention and management after thyroidectomy in children: A systematic review. *Journal of Pediatric Surgery, 56*(3), 526–533. https://doi.org/10.1016/j.jpedsurg.2020.08.032

Bai, B., & Chen, W. (2018). Protective effects of intraoperative nerve monitoring (IONM) for recurrent laryngeal nerve injury in thyroidectomy: Meta-analysis. *Scientific Reports, 8*(1). https://doi.org/10.1038/s41598-018-26219-5

Bhattacharyya, S. S., & Singh, A. (2023). Acquired hypothyroidism in children. *Indian Journal of Pediatrics, 90*(10), 1025–1029. https://doi.org/10.1007/s12098-023-04578-w

Chandrasekhar, S. S., Randolph, G. W., Seidman, M. D., Rosenfeld, R. M., Angelos, P., Barkmeier-Kraemer, J., Benninger, M. S., Blumin, J. H., Dennis, G., Hanks, J., Haymart, M. R., Kloos, R. T., Seals, B., Schreibstein, J. M., Thomas, M. A., Waddington, C., Warren, B., & Robertson, P. J. (2013). Clinical practice guideline. *Otolaryngology–Head and Neck Surgery, 148*(6_Suppl. 1), S1–S37. https://doi.org/10.1177/0194599813487301

Chiardi, I., Croce, L., Caccavale, P., Bertini, J., Coperchini, F., Magri, F., Cappelli, C., Trimboli, P., & Rotondi, M. (2025). *Establishing the adequate levothyroxine dose after total thyroidectomy: A systematic review with meta-analysis*. PubMed. https://doi.org/10.1210/clinem/dgaf417

Chincholikar, S., & Ambiger, S. (2018). Association of hypomagnesemia with hypocalcemia after thyroidectomy. *Indian Journal of Endocrinology and Metabolism, 22*(5), 656. https://doi.org/10.4103/ijem.ijem_599_17

Chung, J., Kim, D., Cheon, J., Yoon, J., Baek, S., Jung, K., Yoon, E., & Park, S. (2021). Current protocol for aesthetic scar management in thyroid surgery. *The Laryngoscope, 131*(7). https://doi.org/10.1002/lary.29441

Cohen, N., Levy, A., Arnon, W., & Sheiner, E. (2010). Perinatal outcomes in post-thyroidectomy pregnancies. *Gynecological Endocrinology, 27*(5), 314–318. https://doi.org/10.3109/09513590.2010.491166

Đanić Hadžibegović, A. (2020). Thyroidectomy-related swallowing difficulties: Review of the literature. *Acta Clinica Croatica, 59*(Suppl 1), 38–49. 10.20471/acc.2020.59.s1.05.

de Mul, N., Damstra, J., Nieveen van Dijkum, E. J. M., Fischli, S., Kalkman, C. J., Schellekens, W.-J. M., & Immink, R. V. (2021). Risk of perioperative thyroid storm in hyperthyroid patients: A systematic review. *British Journal of Anaesthesia, 127*(6), 879–889. https://doi.org/10.1016/j.bja.2021.06.043

de Vries, L. H., Aykan, D., Lodewijk, L., Damen, J. A. A., Borel Rinkes, I. H. M., & Vriens, M. R. (2021). Outcomes of minimally invasive thyroid surgery - A systematic review and meta-analysis. *Frontiers in Endocrinology, 12*, 719397. https://doi.org/10.3389/fendo.2021.719397

Duong, W., Grigorian, A., Farzaneh, C., Elfenbein, D., Yamamoto, M., Rosenbaum, K., Lekawa, M., & Nahmias, J. (2021). Nerve monitoring decreases recurrent laryngeal nerve injury risk for neoplasm-related thyroidectomy. *The American Journal of Surgery, 223*(5), 918–922. https://doi.org/10.1016/j.amjsurg.2021.10.013

Edafe, O., Antakia, R., Laskar, N., Uttley, L., & Balasubramanian, S. P. (2014). Systematic review and meta-analysis of predictors of post-thyroidectomy hypocalcaemia. *British Journal of Surgery, 101*(4), 307–320. https://doi.org/10.1002/bjs.9384

Fields, T., Ramonell, K., Fazendin, J., Gillis, A., Zmijewski, P., Porterfield, J., Chen, H., & Lindeman, B. (2023). Postoperative hypocalcemia in hyperthyroid patients: The parathyroids aren't always to blame. *Journal of Surgical Research, 288*, 202–207. https://doi.org/10.1016/j.jss.2023.03.001

Garg, S., Mishra, A. K., Singh, K. R., Enny, L. E., & Ramakant, P. (2021). PTH gradient as a predictor of post thyroidectomy hypocalcemia. *Indian Journal of Endocrinology and Metabolism, 25*(4), 332–336. https://doi.org/10.4103/ijem.ijem_797_20

Higgins, T. S., Gupta, R., Ketcham, A. S., Sataloff, R. T., Wadsworth, J. T., & Sinacori, J. T. (2011). Recurrent laryngeal nerve monitoring versus identification alone on post-thyroidectomy true vocal fold palsy: A meta-analysis. *The Laryngoscope, 121*(5), 1009–1017. https://doi.org/10.1002/lary.21578

Hong, N., Sheng, B., & Yu, P. (2024). Early postoperative interventions in the prevention and management of thyroidectomy scars. *Frontiers in Physiology, 15*. https://doi.org/10.3389/fphys.2024.1341287

Hung, C.-L., Yeh, C.-C., Sung, P.-S., Hung, C.-J., Muo, C.-H., Sung, F.-C., Jou, I.-M., & Tsai, K.-J. (2018). Is partial or total thyroidectomy associated with risk of long-term osteoporosis: A nationwide population-based study. *World Journal of Surgery, 42*(9), 2864–2871. https://doi.org/10.1007/s00268-018-4573-2

Huynh, C. N., Pearce, J. V., Kang, L., & Celi, F. S. (2020). Weight gain after thyroidectomy: A systematic review and meta-analysis. *The Journal of Clinical Endocrinology & Metabolism*. https://doi.org/10.1210/clinem/dgaa754

Iliff, H. A., El-Boghdadly, K., Ahmad, I., Davis, J., Harris, A., Khan, S., Lan-Pak-Kee, V., O'Connor, J., Powell, L., Rees, G., & Tatla, T. S. (2021). Management of haematoma after thyroid surgery: Systematic review and multidisciplinary consensus guidelines from the difficult airway society, the British association of endocrine and thyroid surgeons and the British association of otorhinolaryngology, head and neck surgery. *Anaesthesia, 77*(1). https://doi.org/10.1111/anae.15585

Jeannon, J.-P., Orabi, A. A., Bruch, G. A., Abdalsalam, H. A., & Simo, R. (2009). Diagnosis of recurrent laryngeal nerve palsy after thyroidectomy: A systematic review. *International Journal of Clinical Practice, 63*(4), 624–629. https://doi.org/10.1111/j.1742-1241.2008.01875.x

Jeong, N. C., Park, W. S., & Sun, Y. M. (2016). Predictors and risk factors of hypoparathyroidism after total thyroidectomy. *International Journal of Surgery, 34*, 47–52. https://doi.org/10.1016/j.ijsu.2016.08.019

Julie, A. S., Tuggle, C. T., Wang, T. Y., Thomas, D., Boudourakis, L., Rivkees, S. A., & Roman, S. A. (2008). Clinical and economic outcomes of thyroid and parathyroid surgery in children. *The Journal of Clinical Endocrinology and Metabolism, 93*(8), 3058–3065. https://doi.org/10.1210/jc.2008-0660

Kao, D. D., Jensen, C. B., Bacon, E., Hogikyan, N. D., Roman, B. R., & Pitt, S. C. (2025). Communication of voice-related complications in thyroidectomy: A qualitative analysis. *Otolaryngology–Head and Neck Surgery, 172*(5), 1560–1569. https://doi.org/10.1002/ohn.1162

Kapre, M. L. (2020). *Thyroid surgery principles and practice*. Milton: Taylor & Francis Group.

Karunakaran, P., Maharajan, C., Ramalingam, S., & Rachmadugu, S. (2018). Is hungry bone syndrome a cause of postoperative hypocalcemia after total thyroidectomy in thyrotoxicosis? a prospective study with bone mineral density correlation. *Surgery, 163*(2), 367–372. https://doi.org/10.1016/j.surg.2017.09.008

Kaur, J., Singh, G., Davessar, J. L., & Bhatia, R. (2021). Impact of thyroid surgery on voice: A prospective study. *Indian Journal of Otolaryngology and Head & Neck Surgery, 74*(S3), 5094–5097. https://doi.org/10.1007/s12070-021-02787-3

Khadra, H., Bakeer, M., Hauch, A., Hu, T., & Kandil, E. (2018). Hemostatic agent use in thyroid surgery: A meta-analysis. *Gland Surgery, 7*(S1), S34–S41. https://doi.org/10.21037/gs.2018.03.02

Leyre, P., et al. (2008). Does the risk of compressive hematoma after thyroidectomy authorize 1-day surgery? *Langenbeck's Archives of Surgery, 393*(5), 733–737. Available at:.

Li, Z., Qiu, Y., Fei, Y., Xing, Z., Zhu, J., & Su, A. (2020). Prevalence of and risk factors for hypothyroidism after hemithyroidectomy: A systematic review and meta-analysis. *Endocrine, 70*(2), 243–255. https://doi.org/10.1007/s12020-020-02410-5

Liu, J., Sun, W., Dong, W., Wang, Z., Zhang, P., Zhang, T., & Zhang, H. (2017). Risk factors for post-thyroidectomy haemorrhage: A meta-analysis. *European Journal of Endocrinology, 176*(5), 591–602. https://doi.org/10.1530/eje-16-0757

Lizcano, F., & Rodríguez, J. S. (2011). Thyroid hormone therapy modulates hypothalamo-pituitary-adrenal axis. *Endocrine Journal, 58*(2), 137–142. https://doi.org/10.1507/endocrj.k10e-369

Marchese, M. R., Galli, J., D'Alatri, L., D'Amore, A., Sionne, F., Trozzi, L., Gaetano, P., Bellantone, R., & Lombardi, C. P. (2021). Neck complaints before and after uncomplicated thyroidectomy: Prevalence, postoperative outcome and relationships with thyroid weight and reflux like symptoms. *Endocrine, 73*(1), 98–106. https://doi.org/10.1007/s12020-020-02568-y

Maroun, C. A., El Asmar, M., Park, S.-J., El Asmar, M. L., Zhu, G., Gourin, C. G., Fakhry, C., Dhillon, V., Tufano, R. P., Russell, J. O., & Mandal, R. (2020). Drain placement in thyroidectomy is associated with longer hospital stay without preventing hematoma. *The Laryngoscope, 130*(5), 1349–1356. https://doi.org/10.1002/lary.28269

Materazzi, G., Ambrosini, C. E., Fregoli, L., De Napoli, L., Frustaci, G., Matteucci, V., Papini, P., Bakkar, S., & Miccoli, P. (2017). Prevention and management of bleeding in thyroid surgery. *Gland Surgery, 6*(5), 510–515. https://doi.org/10.21037/gs.2017.06.14

Metere, A., Biancucci, A., Natili, A., Intini, G., & Graves, C. E. (2021). PTH after thyroidectomy as a predictor of post-operative hypocalcemia. *Diagnostics, 11*(9), 1733. https://doi.org/10.3390/diagnostics11091733

Morton, R. P., Mak, V., Moss, D., Ahmad, Z., & Sevao, J. (2011). 'Risk of bleeding after thyroid surgery: Matched pairs analysis'. *Journal of Laryngology & Otology, 126*(3), 285–288. https://doi.org/10.1017/s0022215111001460

Moumoulidis, I., Martinez Del Pero, M., Brennan, L., & Jani, P. (2010). 'Haemostasis in head and neck surgical procedures: Valsalva manoeuvre versus trendelenburg tilt'. *Annals of the Royal College of Surgeons of England, 92*(4), 292–294. Available at:.

Peng, H., Wang, S. J., & Li, W. (2014). Rare complication after thyroidectomy-cervical esophageal stenosis: A case report and literature review. *World Journal of Surgical Oncology, 12*(1), 308. https://doi.org/10.1186/1477-7819-12-308

Promberger, R., et al. (2012). Risk factors for postoperative bleeding after thyroid surgery. *British Journal of Surgery, 99*(3), 373–379. Available at:.

Pucci, M., Marilena, C., Ghezzi, F., Iannotta, M., Gifuni, M., Ruberto, S., Crescenzo, F., Esposito, R., & Biondi, B. (2025). Cardiovascular risk after total thyroidectomy in patients with differentiated thyroid carcinoma undergoing levothyroxine replacement monotherapy. *Frontiers in Endocrinology, 16*. https://doi.org/10.3389/fendo.2025.1659736

Qin, Y., Sun, W., Wang, Z., Dong, W., He, L., Zhang, T., & Zhang, H. (2021). A meta-analysis of risk factors for transient and permanent hypocalcemia after total thyroidectomy. *Frontiers in Oncology, 10*. https://doi.org/10.3389/fonc.2020.614089

Radhakrishnan, A., Chauhan, I., Thakur, J. S., Azad, R., & Dhadwal, M. (2023). Hypomagnesemia in post thyroidectomy patient: An underestimated entity. *Indian Journal of Otolaryngology and Head & Neck Surgery, 75*(3), 1986–1991. https://doi.org/10.1007/s12070-023-03777-3

Ramouz, A., Seyed, Z. R., Faeze, D., Faraji, E., & Rouhani, S. (2017). Predisposing factors for seroma formation in patients undergoing thyroidectomy: Cross-sectional study. *Annals of Medicine and Surgery, 23*, 8–12. https://doi.org/10.1016/j.amsu.2017.09.001

Scerrino, G., Tudisca, C., Bonventre, S., Raspanti, C., Picone, D., Porrello, C., Paladino, N. C., Vernuccio, F., Cupido, F., Cocorullo, G., Lo Re, G., & Gulotta, G. (2017). Swallowing disorders after thyroidectomy: What we know and where we are. A systematic review. *International Journal of Surgery, 41*, S94–S102. https://doi.org/10.1016/j.ijsu.2017.03.078

Shin, Y.-W., Choi, Y. M., Kim, H. S., Kim, D.-J., Jo, H. J., O'Donnell, B. F., Jang, E. K., Kim, T. Y., Shong, Y. K., Hong, J. P., & Kim, W. B. (2016). Diminished quality of life and increased brain functional connectivity in patients with hypothyroidism after total thyroidectomy. *Thyroid, 26*(5), 641–649. https://doi.org/10.1089/thy.2015.0452

Stefanou, C. K., Papathanakos, G., Stefanou, S. K., Tepelenis, K., Kitsouli, A., Barbouti, A., Tsoumanis, P., Kanavaros, P., & Kitsoulis, P. (2022). Surgical tips and techniques to avoid complications of thyroid surgery. *Innovative Surgical Sciences, 7*(3–4), 115–123. https://doi.org/10.1515/iss-2021-0038

Stojadinovic, A., Shaha, A. R., Orlikoff, R. F., Nissan, A., Kornak, M.-F., Singh, B., Boyle, J. O., Shah, J. P., Brennan, M. F., & Kraus, D. H. (2002). Prospective functional voice assessment in patients undergoing thyroid surgery. *Annals of Surgery, 236*(6), 823–832. https://doi.org/10.1097/00000658-200212000-00015

Vardaxi, C., Tsetsos, N., Koliastasi, A., Poutoglidis, A., Sapalidis, K., Triaridis, S., & Printza, A. (2022). Swallowing disorders after thyroidectomy: A systematic review and meta-analysis. *European Archives of Oto-Rhino-Laryngology,279(9)*, 4213–4227. https://doi.org/10.1007/s00405-022-07386-8

Verma, H., Arun, P., Sharan, R., Manikantan, K., & Jain, P. (2020). Risk of hypocalcemia and the need to augment calcium supplementation after total thyroidectomy. *Indian Journal of Surgical Oncology, 13* (1), 7–10. https://doi.org/10.1007/s13193-020-01098-3

Wiseman, S. M., Saleh, N., Amir, T., Eshraghi, P., Jama, R., & Saleh, S. (2021). Parathyroid identification during thyroid and parathyroid operations: A pilot study evaluating a novel low cost autofluorescence based device. *The American Journal of Surgery, 221*(6), 1150–1158. https://doi.org/10.1016/j.amjsurg.2021.03.005

Wojtczak, B., et al. (2018). Consequences of bleeding after thyroid surgery – Analysis of 7805 operations performed in a single center. *Archives of Medical Science, 14*(2), 329–335. Available at:.

Yaniv, D., Vainer, I., Amir, I., Robenshtok, E., Hirsch, D., Watt, T., Hilly, O., Shkedy, Y., Shpitzer, T., Bachar, G., Feinmesser, R., & Mizrachi, A. (2022). Quality of life following lobectomy versus total thyroidectomy is significantly related to hypothyroidism. *Journal of Surgical Oncology, 126*(4), 640–648. https://doi.org/10.1002/jso.26983

Yap, Y. W., Onyekwelu, E., & Alam, U. (2023). Thyroid disease in pregnancy. *Clinical Medicine, 23*(2), 125–128. https://doi.org/10.7861/clinmed.2023-0018

Zhang, D., Sun, H., Hoon, Y. K., Pino, A., Patroniti, S., Frattini, F., Impellizzeri, P., Romeo, C., Gregory, W. R., Wu, C.-W., Gianlorenzo, D., & Fama, F. (2022). Optimal monitoring technology for pediatric thyroidectomy. *Cancers, 14*(11). https://doi.org/10.3390/cancers14112586

Zhou, H. Y., He, J. C., & McHenry, C. R. (2016). Inadvertent parathyroidectomy: Incidence, risk factors, and outcomes. *Journal of Surgical Research, 205*(1), 70–75. https://doi.org/10.1016/j.jss.2016.06.019

Chapter 6

Pregnancy and thyroid hormone status

Akiko Sankoda[1], Naoko Arata[1], Ai Yoshihara[2] and Natsuko Watanabe[2]

[1]*Women's Internal Medicine, Women's Health Center, National Center for Child Health and Development, Tokyo, Japan;* [2]*Department of Internal Medicine, Ito Hospital, Tokyo, Japan*

Abbreviations

FT4 Free thyroxine
HCG Human chorionic gonadotropin
IH Isolated hypothyroidism
LT4 levothyroxine
OH Overt hypothyroidism
SCH Subclinical hypothyroidism
TBG Thyroxine-binding globulin
TgAb Thyroglobulin antibodies
TPOAb Thyroid peroxidase antibody
TRAb Thyrotropin receptor antibody
TSH Thyroid-stimulating hormone

Introduction

Thyroid disorders are among the most common endocrine conditions affecting women of reproductive age. Overt hypothyroidism occurs in approximately 0.3%–0.5% of pregnant women, while subclinical hypothyroidism is observed in 3%–4%. In addition, thyroid peroxidase antibody (TPOAb) positivity is present in approximately 10% of pregnant women, many of whom are euthyroid at conception (Alexander et al., 2017; Dong & Stagnaro-Green, 2019). These conditions are clinically relevant because early embryonic and placental development depends largely on maternal thyroxine until the fetal thyroid becomes functionally mature at mid-gestation.

Pregnancy is accompanied by profound physiological adaptations of the maternal thyroid axis designed to meet the increased hormonal demands of the developing fetus and placenta. Estrogen-induced upregulation of thyroxine-binding globulin (TBG) leads to an expansion of circulating total thyroid hormone pools, necessitating increased thyroid hormone production to maintain free hormone concentrations. In parallel, rising levels of human chorionic gonadotropin (HCG) during early pregnancy exert thyrotropic stimulation via partial agonism of the TSH receptor, resulting in the characteristic first-trimester suppression of serum TSH. Additional contributors include increased renal iodine clearance and enhanced placental thyroid hormone metabolism. Together, these pregnancy-specific alterations shift the pregnancy-specific reference distribution of TSH and free thyroxine (FT4), rendering fixed nonpregnant reference ranges inappropriate for the assessment of thyroid function during gestation (Burrow et al., 1994; Korevaar et al., 2017).

The placenta plays an important role in mediating the maternal–fetal thyroid hormone axis and functions as a thyroid-responsive endocrine interface, rather than serving merely as a passive conduit for hormone transfer (Chan et al., 2009; Forhead & Fowden, 2014; Huang et al., 2003; Kilby et al., 2005). Thyroid hormone transporters, deiodinases, and thyroid hormone receptors are differentially expressed across trophoblast subtypes and decidual tissues, allowing tight regulation of fetal thyroid hormone exposure. Through these mechanisms, thyroid hormones influence trophoblast proliferation, syncytialization, extravillous trophoblast invasion, and placental angiogenesis. Dysregulation of thyroid hormone availability—whether due to hypothyroidism, hyperthyroidism, iodine imbalance, or autoimmune thyroid disease—may therefore impair placental development and contribute to adverse pregnancy outcomes such as hypertensive disorders, fetal growth restriction, and preterm birth.

Interpretation of maternal thyroid function during pregnancy requires trimester-specific—and ideally cohort-specific—reference ranges derived from the 2.5th to 97.5th percentiles of appropriately defined reference populations. Both the American Thyroid Association (ATA) and the European Thyroid Association (ETA) emphasize that nonpregnant diagnostic thresholds (e.g., TSH $\geq$ 4.0 mIU/L) should not be applied indiscriminately in pregnancy, as physiological changes in hCG stimulation, TBG expansion, and iodine kinetics substantially alter the normal distributions of TSH and FT4. Within this framework,

Hormones in Neuroscience. https://doi.org/10.1016/B978-0-443-30261-9.00006-1

recent individual patient data (IPD) meta-analyses including more than 80,000 pregnancies from multinational cohorts have demonstrated continuous, non-linear associations between maternal TSH and FT4 concentrations and adverse obstetric outcomes, including hypertensive disorders of pregnancy, small-for-gestational-age birth, and preterm delivery—even within conventionally normal reference ranges (Toloza et al., 2022; Derakhshan et al., 2020; Korevaar et al., 2019). These dose–response relationships underscore the need to balance a continuum-based understanding of thyroid-related risk with the practical requirement for clinically actionable thresholds in pregnancy. Contemporary guideline-based management therefore relies on pregnancy-specific reference ranges and outcome-informed cutoffs to guide screening, monitoring, and treatment decisions, while avoiding both under-recognition of clinically relevant risk and unnecessary intervention in women with physiological thyroid adaptations.

Maternal–fetal–placental thyroid physiology

Maternal thyroidal adaptation during pregnancy

Pregnancy represents a unique endocrine state in which maternal thyroid physiology undergoes coordinated adaptations to support placental function and fetal development. These changes begin early in gestation and are particularly critical during the first trimester, when fetal organogenesis and placental development are highly dependent on maternal thyroid hormone supply. The maternal–fetal–placental thyroid axis therefore functions as an integrated system encompassing maternal hormone production, placental transport and metabolism, and the gradual maturation of the fetal thyroid gland (Glinoer, 1997).

Maternal thyroid hormone requirements increase by approximately 30%–50% during pregnancy. Rising estrogen concentrations induce hepatic synthesis and reduced clearance of thyroxine-binding globulin (TBG), resulting in expansion of circulating total thyroid hormone pools and necessitating increased thyroid hormone production to maintain adequate free thyroxine (FT4) concentrations (Burrow et al., 1994; Korevaar et al., 2017). In parallel, human chorionic gonadotropin (hCG) exerts a thyrotropic effect through partial agonism of the TSH receptor. Peak hCG concentrations at around 10th gestational week lead to physiological suppression of serum TSH, thereby enhancing thyroid hormone availability during a critical window of placental growth and early fetal development (Glinoer, 2001; Zimmermann & Boelaert, 2015). Pregnancy is also characterized by marked changes in iodine kinetics. Increased renal blood flow and glomerular filtration enhance urinary iodine excretion, while transplacental iodine transfer and increased thyroid hormone synthesis further raise iodine requirements. Consequently, both iodine deficiency and iodine excess may disrupt maternal thyroid hormone availability, underscoring iodine status as an important modifier of thyroid physiology during gestation.

Although the fetal thyroid gland begins structural development early in gestation, it does not reach functional maturity until approximately 18–20 weeks. Iodine uptake and limited hormone synthesis can be detected from 10 to 12 weeks; however, fetal thyroid hormone production remains insufficient during early gestation. As a result, the fetus relies almost entirely on maternal thyroxine during the period critical for neuronal migration, cortical organization, and early brain development (Forhead & Fowden, 2014).

Maternal FT4 concentrations in early pregnancy are associated with offspring neurocognitive outcomes, reinforcing the biological importance of adequate maternal thyroid hormone availability before mid-gestation. These observations highlight early pregnancy as a particularly vulnerable window in which relatively subtle disturbances in maternal thyroid function may exert disproportionate effects on placental development and fetal neurodevelopment (de Escobar et al., 2004; Jansen et al., 2019; Thompson et al., 2018).

The placenta as a thyroid-responsive endocrine organ

The placenta contributes to the regulation of the maternal–fetal thyroid hormone axis and acts as a thyroid-responsive endocrine interface, rather than functioning solely as a passive conduit for hormone transfer (Kilby et al., 2005; Chan et al., 2009; Chen et al., 2022; Huang et al., 2003). Across different trophoblast subtypes and decidual tissues, thyroid hormone transporters, deiodinases, and receptors are expressed in a cell-specific manner, permitting fine-tuned control of thyroid hormone transfer to the fetus. Placental deiodinases are central to this regulation. Type 3 deiodinase is highly expressed in the placenta and inactivates thyroxine and triiodothyronine, thereby protecting the fetus from excessive thyroid hormone exposure, whereas type 2 deiodinase locally generates bioactive triiodothyronine to support placental metabolic demands and trophoblast differentiation. Thyroid hormone receptors are functionally expressed in trophoblast subtypes, and thyroid hormones influence trophoblast proliferation, syncytialization, extravillous trophoblast invasion, and placental angiogenesis. Altered regulation of thyroid hormone signaling, whether related to thyroid dysfunction, iodine status, or autoimmune processes, has the potential to adversely affect placental maturation and is associated with an increased risk of complications such as hypertensive disorders, impaired fetal growth, and preterm delivery.

Maternal thyroid adaptation, placental regulation of hormone transport and metabolism, and delayed fetal thyroid maturation together form a tightly interconnected system. Within this system, relatively small deviations in maternal thyroid hormone availability may have stage-specific vulnerability during early gestation, providing a mechanistic basis for epidemiological associations between maternal thyroid dysfunction and adverse pregnancy outcomes.

Epidemiology and pregnancy outcomes

Maternal thyroid dysfunction is common in pregnancy and encompasses a spectrum ranging from overt thyroid disease to subclinical and immunological abnormalities. Epidemiological studies over the past 2 decades have consistently demonstrated associations between maternal thyroid status and adverse obstetric and neonatal outcomes, although the strength and clinical implications of these associations vary according to disease severity, timing of exposure, and population characteristics.

Overt and subclinical hypothyroidism

Overt hypothyroidism affects approximately 0.3%–0.5% of pregnant women and is clearly associated with increased risks of miscarriage, preeclampsia, placental abruption, preterm birth, and impaired neurodevelopment in offspring. These associations form the basis for universal agreement regarding the need for prompt diagnosis and treatment (Alexander et al., 2017).

Subclinical hypothyroidism (SCH), defined by elevated TSH with normal FT4 concentrations using pregnancy-specific reference ranges, is considerably more prevalent, affecting 3%–4% of pregnancies. The relationship between SCH and pregnancy outcomes has been more heterogeneous. Earlier cohort studies suggested associations with miscarriage, hypertensive disorders of pregnancy, preterm birth, and small-for-gestational-age (SGA) infants, although results varied widely depending on TSH thresholds, iodine status, and the presence of thyroid autoimmunity (Casey et al., 2005; Lazarus et al., 2014).

Recent large-scale individual patient data (IPD) meta-analyses have provided greater clarity by harmonizing definitions and adjusting for confounding factors. These analyses demonstrate non-linear, dose–response relationships between maternal TSH concentrations and adverse pregnancy outcomes, with increased risks observed even within ranges previously considered normal (Toloza et al., 2022; Korevaar et al., 2019; Derakhshan et al., 2020). Importantly, the magnitude of risk appears modest at the population level but becomes clinically relevant at the extremes of the pregnancy-specific distribution, supporting a risk continuum rather than a dichotomous disease model.

Isolated hypothyroxinemia

Isolated hypothyroxinemia (IH), defined as low maternal free thyroxine (FT4) concentrations in the presence of normal TSH using pregnancy-specific reference ranges, represents a distinct biochemical phenotype whose clinical relevance remains incompletely understood. Reported prevalence ranges from 1% to 2% in iodine-sufficient populations but varies substantially according to assay methodology, reference range definition, gestational timing, and iodine status. Previous studies have most consistently linked IH in early pregnancy to adverse neurodevelopmental outcomes in offspring, including impaired cognitive performance, altered cortical morphology, and reduced gray matter volume in childhood (Pop et al., 1999; Levie et al., 2018; Korevaar et al., 2016). These findings are biologically plausible given the fetus's near-complete dependence on maternal thyroxine before mid-gestation and the established role of thyroid hormones in neuronal migration and synaptogenesis. In contrast, associations between IH and obstetric outcomes such as hypertensive disorders of pregnancy, gestational diabetes, fetal growth restriction, and preterm birth have been less consistent, suggesting substantial effect modification by population characteristics and residual confounding (Tehrani et al., 2021). Importantly, randomized controlled trials evaluating levothyroxine supplementation initiated after the first trimester have failed to demonstrate improvements in neurocognitive or obstetric outcomes. (Lazarus et al., 2012; Casey et al., 2017). This temporal mismatch between the presumed critical window of thyroid hormone action and the timing of intervention likely explains the discordance between observational and interventional evidence and cautions against routine treatment of IH outside research settings.

Thyroid autoimmunity

Thyroid autoimmunity, most identified by TPOAb positivity, affects approximately 8%–14% of pregnant women, many of whom are biochemically euthyroid at conception. Epidemiological studies and meta-analyses have consistently demonstrated associations between TPOAb positivity and increased risks of miscarriage and preterm birth, even after adjustment for maternal thyroid function (Prummel & Wiersinga, 2004; Thangaratinam et al., 2011; Consortium on et al., 2019).

Several mechanisms have been proposed to explain these associations. TPOAb-positive women exhibit reduced thyroidal reserve and a diminished capacity to meet the increased hormonal demands of pregnancy, predisposing them to gestational hypothyroxinemia (Glinoer et al., 1991; Korevaar et al., 2017; Alexander et al., 2017). In addition, thyroid autoimmunity may reflect broader immune dysregulation at the maternal–fetal interface, including altered natural killer cell activity, impaired regulatory T-cell function, and pro-inflammatory cytokine profiles that adversely affect implantation and placentation (Lu et al., 2018; Saito et al., 2010; Tańska et al., 2023). Despite these associations, randomized trials of levothyroxine therapy in euthyroid TPOAb-positive women—including TABLET and T4LIFE—have not demonstrated reductions in miscarriage or preterm birth (Dhillon-Smith et al., 2019; van Dijk et al., 2022). These findings argue against a simple hormone-deficiency model and support the interpretation of TPOAb positivity as a risk marker rather than a direct therapeutic target. Therefore, routine levothyroxine treatment in euthyroid TPOAb-positive women is not recommended, favoring individualized risk stratification and close biochemical monitoring.

Hyperthyroidism and gestational transient thyrotoxicosis

Maternal hyperthyroidism during pregnancy comprises a spectrum ranging from overt autoimmune disease to physiological, hCG-mediated thyrotoxicosis. Although less prevalent than hypothyroid disorders, hyperthyroid states confer etiology-specific maternal and fetal risks that depend on the magnitude and persistence of thyroid hormone excess.

Overt hyperthyroidism, most commonly due to Graves' disease, affects approximately 0.1%–0.4% of pregnancies and is clearly associated with increased risks of miscarriage, hypertensive disorders of pregnancy, placental abruption, preterm birth, low birth weight, and fetal loss when inadequately controlled (Laurberg et al., 2009). In addition, transplacental transfer of maternal thyrotropin receptor antibodies (TRAb) may result in fetal or neonatal thyrotoxicosis, even in women with biochemically controlled disease. (van Dijk et al., 2018) Antithyroid drugs should be titrated to maintain maternal FT4 concentrations in the upper pregnancy-specific reference range while minimizing fetal exposure (Alexander et al., 2017; Luton et al., 2005).

In contrast, **gestational transient thyrotoxicosis (GTT/GTH)** represents a self-limited, hCG-mediated condition characterized by suppressed TSH with elevated or high-normal FT4 concentrations in early pregnancy, in the absence of thyroid autoimmunity. It is frequently associated with hyperemesis gravidarum and multiple gestations. GTT typically resolves spontaneously by midgestation as hCG concentrations decline, and most observational studies indicate no increased risk of adverse obstetric outcomes when thyrotoxicosis is mild and transient. Accordingly, antithyroid drug therapy is generally not indicated, and management is supportive with biochemical monitoring (Glinoer, 1997; Goodwin et al., 1992).

Diagnosis and screening in pregnancy

Accurate diagnosis and appropriate screening for thyroid dysfunction during pregnancy are essential. Physiological changes unique to pregnancy profoundly alter thyroid hormone dynamics, rendering nonpregnant diagnostic thresholds inappropriate and complicating interpretation of biochemical tests. Contemporary approaches therefore emphasize trimester-specific reference ranges, etiological classification, and risk-stratified screening strategies.

Biochemical assessment and pregnancy-specific reference ranges

Interpretation of thyroid function tests during pregnancy requires trimester-specific—and ideally population- and assay-specific—reference ranges derived from healthy pregnant cohorts. Both the American Thyroid Association (ATA) and the European Thyroid Association (ETA) emphasize that fixed nonpregnant thresholds are inappropriate in pregnancy (Alexander et al., 2017; Lazarus et al., 2014). When local pregnancy-specific reference ranges are unavailable, the ATA recommends a pragmatic approach deducting approximately 0.5 mIU/L from nonpregnant TSH reference limits. The definition of TSH and FT4 reference intervals substantially influences the estimated prevalence of gestational thyroid dysfunction (Dong & Stagnaro-Green, 2019; Osinga et al., 2022).

Etiological classification

Beyond biochemical diagnosis, accurate etiological classification is critical for guiding management. Differentiating autoimmune thyroid disease from physiological or transient conditions has direct therapeutic implications.

TPOAb testing identifies autoimmune thyroid disease and provides prognostic information even in euthyroid women. TPOAb positivity is associated with an increased risk of progression to hypothyroidism during pregnancy and postpartum, reflecting limited thyroidal reserve under increased hormonal demand. In the setting of suppressed TSH, distinction between Graves' disease and gestational transient thyrotoxicosis (GTT) is essential. Clinical features such as goiter, ophthalmopathy, and a history of autoimmune thyroid disease, together with measurement of thyrotropin receptor antibodies (TRAb), support the diagnosis of Graves' disease. In contrast, GTT is typically characterized by absent autoantibodies, mild and transient biochemical abnormalities, and spontaneous resolution as hCG levels decline. TRAb measurement is particularly important in women with current or past Graves' disease, including those who have undergone total thyroidectomy or radioactive iodine therapy, as transplacental antibody transfer may result in fetal or neonatal thyrotoxicosis, independent of maternal thyroid function at the time of assessment.

Universal versus targeted screening

Whether universal thyroid screening should be implemented during pregnancy remains controversial. Universal screening increases detection of biochemical thyroid abnormalities but raises concerns regarding overdiagnosis and unnecessary intervention. Both the American Thyroid Association (ATA) and the European Thyroid Association (ETA) currently recommend a targeted, risk-based screening strategy. This approach prioritizes testing in women with established risk factors for thyroid dysfunction, including a personal or family history of thyroid disease, autoimmune disorders, infertility or recurrent pregnancy loss, prior preterm delivery, iodine deficiency, multiple gestation, and exposure to medications affecting thyroid function (Alexander et al., 2017; Lazarus et al., 2014).

However, the effectiveness of risk-based screening has been increasingly questioned. In a recent systematic review encompassing a broad range of observational studies, Liu et al. critically evaluated the performance of currently recommended risk factors and demonstrated that a substantial proportion of pregnant women with abnormal thyroid function tests would not be identified using existing risk-based criteria alone (Liu et al., 2025). These findings underscore an inherent limitation of risk-based screening strategies: while improving specificity, they may fail to capture a meaningful subset of women with clinically relevant thyroid dysfunction.

Timing and follow-up

Given the dynamic nature of thyroid physiology during pregnancy, timing of assessment is critical. Initial thyroid function testing is ideally performed in early pregnancy or preconception, particularly in high-risk women. In TPOAb-positive or borderline cases, repeat testing during mid- and late gestation is recommended to detect evolving dysfunction. Postpartum follow-up is also essential, as women with thyroid autoimmunity are at increased risk of postpartum thyroiditis and long-term hypothyroidism. Recognition of pregnancy as a "stress test" for thyroid function underscores the importance of longitudinal assessment beyond delivery.

Management of thyroid dysfunction in pregnancy

Management of thyroid dysfunction during pregnancy requires balancing the biological continuum of thyroid-related risk with the need for clinically actionable treatment thresholds. While overt thyroid disease clearly warrants intervention, the management of milder biochemical abnormalities remains nuanced and must be guided by etiology, gestational timing, symptom burden, and available evidence on maternal and fetal outcomes.

Overt hypothyroidism

Overt hypothyroidism in pregnancy is unequivocally associated with adverse obstetric and neurodevelopmental outcomes and represents a clear indication for treatment. Levothyroxine (LT4) therapy should be initiated promptly upon diagnosis, with the goal of restoring maternal euthyroidism as early as possible. Women with pre-existing hypothyroidism typically require a 30%–50% increase in LT4 dose during early pregnancy, reflecting increased thyroid hormone requirements. Close biochemical monitoring is essential to ensure adequate control.

Subclinical hypothyroidism

Management of subclinical hypothyroidism (SCH) remains an area of ongoing debate. Observational studies suggest associations between SCH and adverse pregnancy outcomes, particularly at higher TSH concentrations; however, randomized controlled trials have yielded inconsistent results regarding the benefits of LT4 therapy (Provinciatto et al., 2024; Sankoda et al., 2024). Accordingly, individualized decision-making is recommended, taking into account risk factors such as gestational age, TPOAb positivity, degree of TSH elevation, and a history of recurrent miscarriage. Importantly, the potential benefits of treatment must be weighed against the absence of clear evidence for improved outcomes when therapy is initiated after the first trimester.

Importantly, emerging evidence suggests that excessive maternal thyroid hormone exposure may adversely affect fetal neurodevelopment. Large population-based cohort studies have demonstrated non-linear, often inverted U-shaped, associations between maternal FT4 concentrations and offspring cognitive outcomes, indicating potential harm at both low and high extremes (Korevaar et al., 2016; Hales et al., 2020; Hales et al., 2018).

Isolated hypothyroxinemia

Isolated hypothyroxinemia (IH), defined by low FT4 concentrations with normal TSH, presents a particular management challenge. Although the association between IH and adverse neurodevelopmental outcomes has been shown (Korevaar et al., 2016; Pop et al., 2003; Henrichs et al., 2010), randomized trials of LT4 supplementation initiated in mid-gestation have failed to demonstrate benefit (Casey et al., 2017; Lazarus et al., 2012). Consequently, routine treatment of IH is not recommended by current guidelines. Management should focus on identifying and correcting potential contributing factors, including iodine insufficiency, assay-related issues, and non-thyroidal illness. Monitoring of thyroid function during pregnancy may be appropriate in selected cases, particularly when IH is detected early or coexists with thyroid autoimmunity.

Hyperthyroidism and gestational transient thyrotoxicosis

Overt hyperthyroidism, most commonly due to Graves' disease, requires active management to minimize maternal and fetal risk. Antithyroid drugs (ATDs) remain the mainstay of therapy, with propylthiouracil (PTU) generally preferred in the first trimester due to concerns regarding methimazole-associated embryopathy (Andersen et al., 2013). Therapy should be titrated to maintain maternal FT4 concentrations in the upper pregnancy-specific reference range, avoiding overtreatment and fetal hypothyroidism. Measurement of thyrotropin receptor antibodies (TRAb) is recommended in women with current or past Graves' disease to assess the risk of fetal or neonatal thyrotoxicosis and guide fetal surveillance. In contrast, gestational transient thyrotoxicosis (GTT) is a self-limited, hCG-mediated condition that typically does not require ATD therapy. Management is supportive, focusing on symptom control and biochemical monitoring. Careful differentiation between GTT and Graves' disease is essential to avoid unnecessary treatment (Alexander et al., 2017).

Thyroid autoimmunity in euthyroid women

Euthyroid women with thyroid autoimmunity represent a large and clinically important subgroup. While TPOAb positivity is associated with increased risks of miscarriage and preterm birth (Thangaratinam et al., 2011), randomized trials have not demonstrated benefit from routine LT4 therapy in euthyroid TPOAb-positive women (Dhillon-Smith et al., 2019; van Dijk et al., 2022). Universal treatment in this population is not recommended, favoring close biochemical surveillance throughout pregnancy. Recognition of pregnancy as a physiological "stress test" for the thyroid supports ongoing monitoring, as TPOAb-positive women are at increased risk of developing hypothyroidism during gestation or postpartum (Korevaar et al., 2017).

Timing of intervention and clinical thresholds

A recurring theme across thyroid disorders in pregnancy is the importance of timing. Evidence suggests that the critical window during which maternal thyroid hormone availability influences fetal neurodevelopment lies in early gestation, often before clinical detection and intervention are feasible (Casey et al., 2017; Hales et al., 2020; Hales et al., 2018). This temporal constraint likely explains the discrepancy between observational associations and negative results from interventional trials initiated after the first trimester. Accordingly, contemporary management strategies emphasize early identification of overt disease, judicious use of treatment thresholds informed by pregnancy-specific reference distributions, and avoidance of indiscriminate therapy in women with physiological adaptations or mild biochemical abnormalities.

Thyroid autoimmunity and reproductive outcomes

Thyroid autoimmunity, most commonly identified by the presence of TPOAb or thyroglobulin antibodies (TgAb), represents the most prevalent autoimmune condition affecting women of reproductive age. Unlike overt thyroid dysfunction, thyroid autoimmunity often exists in a biochemically euthyroid state, yet has consistently been associated with adverse reproductive outcomes. This paradox—strong epidemiological associations in the absence of clear therapeutic benefit—has positioned thyroid autoimmunity as a complex biological modifier of reproductive risk rather than a straightforward endocrine disorder.

Epidemiology and reproductive risk

Thyroid autoimmunity is detected in approximately 8%–14% of pregnant women, with prevalence varying by iodine status, age, and assay methodology. Large observational studies and meta-analyses have demonstrated that TPOAb positivity is associated with a 1.5–2.0-fold increased risk of miscarriage and a modest but reproducible increase in preterm birth risk, even among women who are euthyroid at conception (Thangaratinam et al., 2011; Chen & Hu, 2011). These associations extend across spontaneous conceptions and assisted reproductive technologies (ART), suggesting that thyroid autoimmunity exerts its effects early in gestation, potentially at the level of implantation and early placentation. Importantly, the magnitude of risk attributable to thyroid autoimmunity is generally smaller than that associated with overt thyroid dysfunction but remains clinically relevant at the population level due to the high prevalence of antibody positivity (Poppe et al., 2008; Thangaratinam et al., 2011).

Thyroidal reserve and endocrine vulnerability

One mechanistic explanation for the observed associations is reduced thyroidal reserve in antibody-positive women. Thyroid autoimmunity reflects chronic lymphocytic infiltration and subclinical glandular damage, which may limit the capacity of the thyroid to adapt to the increased hormonal demands of pregnancy. Pregnancy thus functions as a physiological "stress test," unmasking latent insufficiency and predisposing susceptible women to gestational hypothyroxinemia or progressive TSH elevation later in gestation (Glinoer, 1997). Emerging evidence further suggests that thyroid autoimmunity attenuates the normal thyroidal response to hCG stimulation in early pregnancy. Attenuated hCG-mediated TSH suppression and FT4 augmentation have been observed in TPOAb-positive women, potentially narrowing the margin of endocrine adaptation during the critical window of placental development (Pekonen et al., 1988; Korevaar et al., 2017).

Immune dysregulation at the maternal–fetal interface

Beyond endocrine vulnerability, thyroid autoimmunity may serve as a surrogate marker for broader immune dysregulation. Successful implantation and placentation require a finely tuned balance between immune tolerance and controlled inflammation at the maternal–fetal interface. TPOAb-positive women have been shown to exhibit alterations in systemic immune profiles and, potentially, immune regulation at the maternal–fetal interface, including increased Th1 and Th17 polarization, impaired regulatory T-cell (Treg) function, and altered natural killer (NK) cell activity (Mor & Cardenas, 2010; Iyidir et al., 2015; Marazuela et al., 2006; Kim et al., 2011). These immune perturbations may disrupt trophoblast invasion, spiral artery remodeling, and early placental development, thereby increasing susceptibility to pregnancy loss and placental-mediated complications. Importantly, these mechanisms are not directly corrected by levothyroxine supplementation, providing a biological rationale for the limited efficacy of hormone-based interventions in euthyroid antibody-positive women.

Evidence from interventional trials

Randomized controlled trials evaluating levothyroxine therapy in euthyroid TPOAb-positive women—including large studies such as TABLET and T4LIFE—have consistently failed to demonstrate reductions in miscarriage or preterm birth (Dhillon-Smith et al., 2019; van Dijk et al., 2022). These findings argue strongly against a simple hormone-deficiency model and suggest that thyroid autoimmunity-associated reproductive risk is not primarily mediated by subtle abnormalities in circulating thyroid hormone concentrations. Notably, some earlier small trials suggested benefit in selected subgroups, contributing to historical enthusiasm for treatment. However, these studies were limited by small sample sizes, heterogeneous inclusion criteria, and inconsistent outcome definitions (Negro et al., 2006; Negro et al., 2005; Kim et al., 2011). Contemporary guideline recommendations are going to be shifted away from routine treatment toward careful surveillance and individualized risk assessment.

Assisted reproduction and recurrent pregnancy loss

Thyroid autoimmunity has been extensively studied in the context of ART and recurrent pregnancy loss (RPL). While TPOAb positivity is associated with lower implantation rates and higher miscarriage rates in some ART cohorts, treatment with levothyroxine has not consistently improved live birth rates (Poppe et al., 2002; Busnelli et al., 2022; Leng et al., 2022). These findings reinforce the concept that thyroid autoimmunity reflects an underlying immunological or endometrial vulnerability rather than a modifiable endocrine defect. In women with RPL, thyroid autoimmunity should prompt thorough evaluation for coexisting autoimmune conditions, including antiphospholipid antibody syndrome, and consideration of multidisciplinary management strategies, rather than reflexive thyroid hormone therapy (RPL et al., 2023).

Preconception care and future directions

The recognition that thyroid-related influences on pregnancy outcomes often preceded clinical detection during gestation has increasingly shifted attention toward the preconception period. At the same time, evolving clinical recommendations reflect a growing appreciation of biological complexity, diagnostic uncertainty, and the limitations of hormone-based interventions. This section integrates current perspectives on preconception care with contemporary guideline positions and outlines future directions for research and clinical practice.

Rationale for preconception thyroid assessment

Evidence accumulated over the past decade indicates that critical windows for thyroid hormone–dependent processes—including implantation, placentation, and early neurodevelopment—occur in very early pregnancy, often before routine antenatal screening is performed. Observational studies linking maternal thyroid function in the first trimester to miscarriage risk and offspring neurodevelopment strongly suggest that optimization of thyroid status after pregnancy recognition may be insufficient to fully mitigate risk. Preconception care therefore represents a unique opportunity to identify overt thyroid disease, correct modifiable risk factors, and counsel women at increased risk of thyroid dysfunction during pregnancy. This approach is particularly relevant for women with known thyroid disease, thyroid autoimmunity, infertility, recurrent pregnancy loss, or planned assisted reproductive technologies (ART), as well as those residing in iodine-deficient regions.

Preconception management strategies

In women with overt hypothyroidism, preconception counseling should emphasize the need for early dose adjustment of levothyroxine upon confirmation of pregnancy and prompt biochemical monitoring. Achieving euthyroidism prior to conception remains a cornerstone of risk reduction. For euthyroid women with thyroid autoimmunity, current evidence does not support routine levothyroxine therapy before conception. Instead, preconception identification of antibody positivity should inform anticipatory guidance, including the likelihood of increased thyroid hormone requirements during pregnancy and the need for close surveillance. Importantly, this strategy avoids unnecessary treatment while acknowledging heightened vulnerability under the physiological stress of gestation. Iodine nutrition represents another critical component of preconception care. Both iodine deficiency and excess may impair thyroid adaptation during pregnancy, and assessment of dietary intake or regional iodine sufficiency should be considered as part of broader preconception counseling.

Guideline perspectives: Convergence and divergence

Contemporary guidelines from the American Thyroid Association (ATA) and the European Thyroid Association (ETA) increasingly converge on several key principles (Alexander et al., 2017; Lazarus et al., 2014). These include the use of pregnancy-specific reference ranges, avoidance of fixed nonpregnant diagnostic thresholds, and a cautious approach to the treatment of subclinical thyroid dysfunction.

However, notable areas of divergence remain. Recommendations regarding screening strategies, treatment thresholds for subclinical hypothyroidism (SCH), and the management of euthyroid antibody-positive women differ in emphasis, reflecting varying interpretations of an incomplete evidence base. Importantly, both guidelines explicitly acknowledge the lack of high-quality evidence supporting universal screening or routine treatment of mild biochemical abnormalities.

Rather than signaling uncertainty, these nuanced positions reflect an evolving paradigm that prioritizes individualized risk assessment over uniform intervention. The shift away from aggressive treatment strategies underscores growing recognition of the potential harms associated with overtreatment, particularly with respect to fetal neurodevelopment. In this context, the ATA guidelines are expected to undergo further revision, particularly concerning the treatment of SCH and TPOAb–positive women, in response to recent high-quality randomized controlled trials that have failed to demonstrate clear benefits of routine levothyroxine therapy on pregnancy and offspring outcomes.

Future directions

Future research priorities include refinement of preconception risk stratification tools that integrate endocrine, immunological, and placental biomarkers; development of population- and assay-specific reference ranges; and identification of clinically meaningful subphenotypes within heterogeneous groups such as subclinical hypothyroidism and thyroid autoimmunity. Longitudinal studies beginning in the preconception period will be essential to clarify causal pathways and optimal timing of intervention. In parallel, advances in systems biology and precision medicine may enable individualized approaches that move beyond one-size-fits-all thresholds. From a clinical standpoint, future guideline updates are likely to further emphasize early identification of overt disease, judicious use of treatment in selected high-risk scenarios, and avoidance of indiscriminate intervention. Preconception care will increasingly serve as the foundation upon which safe and effective management of thyroid-related reproductive risk is built. Moving forward, integration of biological insight with pragmatic clinical decision-making will be essential to optimize maternal and fetal health while minimizing unintended harm.

Mini-dictionary of terms

- **Autoimmune thyroid disease.**

An inflammatory thyroiditis in which immune-mediated processes target components of the hormone-producing gland, often detectable through circulating antibodies, including Graves' Disease and Hashimoto Disease.

- **Gestational transient thyrotoxicosis.**

A self-limited thyrotoxicosis in early pregnancy caused by high concentrations of placental hormones.

- **Human chorionic gonadotropin.**

A placental hormone that rises sharply in early pregnancy, peaking around 10th gestational week, and exerts stimulatory effects on maternal endocrine pathways, contributing to characteristic first-trimester biochemical changes.

- **Isolated hypothyroxinemia.**

A biochemical pattern defined by low circulating freethyroxine levels despite of normal TSH levels, most commonly observed in early pregnancy and influenced by iodine status.

- **Pregnancy-specific reference range.**

A statistically derived reference interval used to interpret thyroid function during pregnancy, reflecting physiological differences from the non-pregnant state and variation across gestational stages.

Summary points

- Maternal thyroid physiology undergoes coordinated adaptations during pregnancy to meet increased hormonal demands required for placental development and early fetal organogenesis.
- Because fetal thyroid hormone production is insufficient until mid-gestation, early fetal development depends largely on maternal hormone availability during the first trimester.
- Pregnancy-specific reference ranges are essential for accurate interpretation of thyroid function tests, as physiological changes shift biochemical distributions across gestation.
- Large individual participant data meta-analyses demonstrate continuous, non-linear associations between maternal thyroid function markers and adverse pregnancy outcomes, even within conventionally normal ranges.
- Overt thyroid dysfunction is clearly associated with miscarriage, hypertensive disorders of pregnancy, preterm birth, and impaired fetal growth, warranting prompt diagnosis and treatment.
- Subclinical thyroid dysfunction and isolated hypothyroxinemia show heterogeneous associations with pregnancy and neurodevelopmental outcomes, with randomized trials failing to demonstrate consistent benefit from routine treatment.
- Thyroid autoimmunity is common in pregnancy and is associated with increased risks of miscarriage and preterm birth, despite the absence of clear benefit from levothyroxine therapy in euthyroid women.
- The placenta actively regulates maternal–fetal thyroid hormone transfer and metabolism, contributing to stage-specific vulnerability during early gestation.
- Preconception assessment and early pregnancy monitoring would guide individualized management while avoiding unnecessary intervention.

Acknowledgments

During the preparation of this work the authors used ChatGPT 5.2, from OpenAI, strictly to improve readability and language usage. Under no circumstances has it been used to draw insights from data, conclude or produce recommendations. After using this tool/service, the authors reviewed and edited the content as needed and take full responsibility for the content of the publication.

References

Alexander, E. K., Pearce, E. N., Brent, G. A., Brown, R. S., Chen, H., Dosiou, C., & Sullivan, S. (2017). 2017 guidelines of the American thyroid association for the diagnosis and management of thyroid disease during pregnancy and the postpartum. *Thyroid, 27*(3), 315–389. https://doi.org/10.1089/thy.2016.0457

Andersen, S. L., Olsen, J., Wu, C. S., & Laurberg, P. (2013). Birth defects after early pregnancy use of antithyroid drugs: A Danish nationwide study. *The Journal of Clinical Endocrinology & Metabolism, 98*(11), 4373–4381.

Burrow, G. N., Fisher, D. A., & Larsen, P. R. (1994). Maternal and fetal thyroid function. *New England Journal of Medicine, 331*(16), 1072–1078.

Busnelli, A., Beltratti, C., Cirillo, F., Bulfoni, A., Lania, A., & Levi-Setti, P. E. (2022). Impact of thyroid autoimmunity on assisted reproductive technology outcomes and ovarian reserve markers: An updated systematic review and meta-analysis. *Thyroid, 32*(9), 1010–1028.

Casey, B. M., Dashe, J. S., Wells, C. E., McIntire, D., Leveno, K. J., & Cunningham, F. G. (2005). Subclinical hypothyroidism and pregnancy outcomes. *Obstetrics & Gynecology, 106*(1), 198–199.

Casey, B. M., Thom, E. A., Peaceman, A. M., Varner, M. W., Sorokin, Y., Hirtz, D. G., … Saade, G. (2017). Treatment of subclinical hypothyroidism or hypothyroxinemia in pregnancy. *New England Journal of Medicine, 376*(9), 815–825.

Casey, B. M., Thom, E. A., Peaceman, A. M., Varner, M. W., Sorokin, Y., Hirtz, D. G., … VanDorsten, J. P. (2017). Treatment of subclinical hypothyroidism or hypothyroxinemia in pregnancy. *New England Journal of Medicine, 376*(9), 815–825. https://doi.org/10.1056/NEJMoa1606205

Chan, S. Y., Vasilopoulou, E., & Kilby, M. D. (2009). The role of the placenta in thyroid hormone delivery to the fetus. *Nature Clinical Practice Endocrinology & Metabolism, 5*(1), 45–54.

Chen, L., & Hu, R. (2011). Thyroid autoimmunity and miscarriage: A meta-analysis. *Clinical endocrinology, 74*(4), 513–519.

Chen, Z., van der Sman, A. E., Groeneweg, S., de Rooij, L. J., Visser, W. E., Peeters, R. P., & Meima, M. E. (2022). Thyroid hormone transporters in a human placental cell model. *Thyroid, 32*(9), 1129–1137.

Consortium on, T., Pregnancy-Study Group on Preterm, B., Korevaar, T. I. M., Derakhshan, A., Taylor, P. N., Meima, M., … Peeters, R. P. (2019). Association of thyroid function test abnormalities and thyroid autoimmunity with preterm birth: A systematic review and meta-analysis. *The Journal of the American Medical Association, 322*(7), 632–641. https://doi.org/10.1001/jama.2019.10931

de Escobar, G. M., Obregón, M.a. J., & del Rey, F. E. (2004). Maternal thyroid hormones early in pregnancy and fetal brain development. *Best Practice & Research Clinical Endocrinology & Metabolism, 18*(2), 225–248.

Derakhshan, A., Peeters, R. P., Taylor, P. N., Bliddal, S., Carty, D. M., Meems, M., … Korevaar, T. I. M. (2020). Association of maternal thyroid function with birthweight: A systematic review and individual-participant data meta-analysis. *Lancet Diabetes Endocrinol, 8*(6), 501–510. https://doi.org/10.1016/s2213-8587(20)30061-9

Dhillon-Smith, R. K., Middleton, L. J., Sunner, K. K., Cheed, V., Baker, K., Farrell-Carver, S., … Coomarasamy, A. (2019). Levothyroxine in women with thyroid peroxidase antibodies before conception. *New England Journal of Medicine, 380*(14), 1316–1325. https://doi.org/10.1056/NEJMoa1812537

Dong, A. C., & Stagnaro-Green, A. (2019). Differences in diagnostic criteria mask the true prevalence of thyroid disease in pregnancy: Asystematic review and meta-analysis. *Thyroid, 29*(2), 278–289. https://doi.org/10.1089/thy.2018.0475

Forhead, A. J., & Fowden, A. L. (2014). Thyroid hormones in fetal growth and prepartum maturation. *Journal of Endocrinology, 221*(3), R87–R103.

Glinoer, D. (1997). The regulation of thyroid function in pregnancy: Pathways of endocrine adaptation from physiology to pathology. *Endocrine Reviews, 18*(3), 404–433.

Glinoer, D. (2001). Pregnancy and iodine. *Thyroid, 11*(5), 471–481.

Glinoer, D., Soto, M. F., Bourdoux, P., Lejeune, B., Delange, F., Lemone, M., … Nayer, P. D. (1991). Pregnancy in patients with mild thyroid abnormalities: Maternal and neonatal repercussions. *The Journal of Clinical Endocrinology & Metabolism, 73*(2), 421–427.

Goodwin, T. M., Montoro, M., & Mestman, J. H. (1992). Transient hyperthyroidism and hyperemesis gravidarum: Clinical aspects. *American Journal of Obstetrics and Gynecology, 167*(3), 648–652.

Hales, C., Taylor, P. N., Channon, S., McEwan, K., Thapar, A., Langley, K., & Ludgate, M. (2020). Controlled antenatal thyroid screening II: Effect of treating maternal suboptimal thyroid function on child behavior. *The Journal of Clinical Endocrinology & Metabolism, 105*(3). https://doi.org/10.1210/clinem/dgz098

Hales, C., Taylor, P. N., Channon, S., Paradice, R., McEwan, K., Zhang, L., & Ludgate, M. (2018). Controlled antenatal thyroid screening II: Effect of treating maternal suboptimal thyroid function on child cognition. *The Journal of Clinical Endocrinology & Metabolism, 103*(4), 1583–1591. https://doi.org/10.1210/jc.2017-02378

Henrichs, J., Bongers-Schokking, J. J., Schenk, J. J., Ghassabian, A., Schmidt, H. G., Visser, T. J., … Jaddoe, V. V. (2010). Maternal thyroid function during early pregnancy and cognitive functioning in early childhood: The generation R study. *The Journal of Clinical Endocrinology & Metabolism, 95*(9), 4227–4234.

Huang, S. A., Dorfman, D. M., Genest, D. R., Salvatore, D., & Larsen, P. R. (2003). Type 3 iodothyronine deiodinase is highly expressed in the human uteroplacental unit and in fetal epithelium. *The Journal of Clinical Endocrinology & Metabolism, 88*(3), 1384–1388.

Iyidir, O. T., Degertekin, C. K., Sonmez, C., Yucel, A. A., Erdem, M., Akturk, M., & Ayvaz, G. (2015). The effect of thyroid autoimmunity on T-cell responses in early pregnancy. *Journal of Reproductive Immunology, 110*, 61–66.

Jansen, T. A., Korevaar, T. I., Mulder, T. A., White, T., Muetzel, R. L., Peeters, R. P., & Tiemeier, H. (2019). Maternal thyroid function during pregnancy and child brain morphology: A time window-specific analysis of a prospective cohort. *The lancet Diabetes & endocrinology, 7*(8), 629–637.

Kilby, M., Barber, K., Hobbs, E., & Franklyn, J. (2005). Thyroid hormone action in the placenta. *Placenta, 26*(2–3), 105–113.

Kim, C. H., Ahn, J. W., Kang, S. P., Kim, S. H., Chae, H. D., & Kang, B. M. (2011). Effect of levothyroxine treatment on in vitro fertilization and pregnancy outcome in infertile women with subclinical hypothyroidism undergoing in vitro fertilization/intracytoplasmic sperm injection. *Fertility and Sterility, 95*(5), 1650–1654. https://doi.org/10.1016/j.fertnstert.2010.12.004

Kim, N. Y., Cho, H. J., Kim, H. Y., Yang, K. M., Ahn, H. K., Thornton, S., … Kwak-Kim, J. (2011). Thyroid autoimmunity and its association with cellular and humoral immunity in women with reproductive failures. *American journal of reproductive immunology, 65*(1), 78–87.

Korevaar, T. I. M., Derakhshan, A., Taylor, P. N., Meima, M., Chen, L., Bliddal, S., … Peeters, R. P. (2019). Association of thyroid function test abnormalities and thyroid autoimmunity with preterm birth: A systematic review and meta-analysis. *The Journal of the American Medical Association, 322*(7), 632–641. https://doi.org/10.1001/jama.2019.10931

Korevaar, T. I. M., Medici, M., Visser, T. J., & Peeters, R. P. (2017). Thyroid disease in pregnancy: New insights in diagnosis and clinical management. *Nature Reviews Endocrinology, 13*(10), 610–622. https://doi.org/10.1038/nrendo.2017.93

Korevaar, T. I., Muetzel, R., Medici, M., Chaker, L., Jaddoe, V. W., de Rijke, Y. B., … Tiemeier, H. (2016). Association of maternal thyroid function during early pregnancy with offspring IQ and brain morphology in childhood: A population-based prospective cohort study. *The lancet Diabetes & endocrinology, 4*(1), 35–43.

Korevaar, T. I., Steegers, E. A., Pop, V. J., Broeren, M. A., Chaker, L., de Rijke, Y. B., … Peeters, R. P. (2017). Thyroid autoimmunity impairs the thyroidal response to human chorionic gonadotropin: Two population-based prospective cohort studies. *The Journal of Clinical Endocrinology & Metabolism, 102*(1), 69–77. https://doi.org/10.1210/jc.2016-2942

Laurberg, P., Bournaud, C., Karmisholt, J., & Orgiazzi, J. (2009). Management of Graves' hyperthyroidism in pregnancy: Focus on both maternal and fetal thyroid function, and caution against surgical thyroidectomy in pregnancy. *European Journal of Endocrinology, 160*(1), 1–8.

Lazarus, J. H., Bestwick, J. P., Channon, S., Paradice, R., Maina, A., Rees, R., … George, L. M. (2012). Antenatal thyroid screening and childhood cognitive function. *New England Journal of Medicine, 366*(6), 493–501.

Lazarus, J. H., Bestwick, J. P., Channon, S., Paradice, R., Maina, A., Rees, R., ... Wald, N. J. (2012). Antenatal thyroid screening and childhood cognitive function. *New England Journal of Medicine, 366*(6), 493–501. https://doi.org/10.1056/NEJMoa1106104

Lazarus, J., Brown, R. S., Daumerie, C., Hubalewska-Dydejczyk, A., Negro, R., & Vaidya, B. (2014). 2014 European thyroid association guidelines for the management of subclinical hypothyroidism in pregnancy and in children. *European Thyroid Journal, 3*(2), 76–94.

Leng, T., Li, X., & Zhang, H. (2022). Levothyroxine treatment for subclinical hypothyroidism improves the rate of live births in pregnant women with recurrent pregnancy loss: A randomized clinical trial. *Gynecological Endocrinology, 38*(6), 488–494. https://doi.org/10.1080/09513590.2022.2063831

Levie, D., Korevaar, T. I., Bath, S. C., Dalmau-Bueno, A., Murcia, M., Espada, M., ... Tiemeier, H. (2018). Thyroid function in early pregnancy, child IQ, and autistic traits: A meta-analysis of individual participant data. *The Journal of Clinical Endocrinology & Metabolism, 103*(8), 2967–2979.

Liu, Y., Osinga, J. A. J., Maraka, S., Bliddal, S., Alexander, E. K., Dosiou, C., ... Korevaar, T. I. M. (2025). Risk factors for thyroid function test abnormalities during pregnancy: A systematic review of the literature to validate current risk factors and identify novel ones. *Thyroid.* https://doi.org/10.1089/thy.2024.0743

Lu, H., Huang, Y., Xin, H., Hao, C., & Cui, Y. (2018). The expression of cytokines IFN-γ, IL-4, IL-17A, and TGF-β1 in peripheral blood and follicular fluid of patients testing positive for anti-thyroid autoantibodies and its influence on in vitro fertilization and embryo transfer pregnancy outcomes. *Gynecological Endocrinology, 34*(11), 933–939.

Luton, D., Le Gac, I., Vuillard, E., Castanet, M., Guibourdenche, J., Noel, M., ... Schlageter, M.-H. (2005). Management of Graves' disease during pregnancy: The key role of fetal thyroid gland monitoring. *The Journal of Clinical Endocrinology & Metabolism, 90*(11), 6093–6098.

Marazuela, M., Garcia-Lopez, M. A., Figueroa-Vega, N., de la Fuente, H., Alvarado-Sánchez, B., Monsiváis-Urenda, A., ... Gonzalez-Amaro, R. (2006). Regulatory T cells in human autoimmune thyroid disease. *The Journal of Clinical Endocrinology & Metabolism, 91*(9), 3639–3646.

Mor, G., & Cardenas, I. (2010). The immune system in pregnancy: A unique complexity. *American journal of reproductive immunology, 63*(6), 425–433.

Negro, R., Formoso, G., Mangieri, T., Pezzarossa, A., Dazzi, D., & Hassan, H. (2006). Levothyroxine treatment in euthyroid pregnant women with autoimmune thyroid disease: Effects on obstetrical complications. *The Journal of Clinical Endocrinology & Metabolism, 91*(7), 2587–2591. https://doi.org/10.1210/jc.2005-1603

Negro, R., Mangieri, T., Coppola, L., Presicce, G., Casavola, E. C., Gismondi, R., ... Hassan, H. (2005). Levothyroxine treatment in thyroid peroxidase antibody-positive women undergoing assisted reproduction technologies: A prospective study. *Human Reproduction, 20*(6), 1529–1533. https://doi.org/10.1093/humrep/deh843

Osinga, J. A. J., Derakhshan, A., Palomaki, G. E., Ashoor, G., Mannisto, T., Maraka, S., ... Korevaar, T. I. M. (2022). TSH and FT4 reference intervals in pregnancy: A systematic review and individual participant data meta-analysis. *The Journal of Clinical Endocrinology & Metabolism, 107*(10), 2925–2933. https://doi.org/10.1210/clinem/dgac425

Pekonen, F., Alfthan, H., Stenman, U.-H., & Ylikorkala, O. (1988). Human chorionic gonadotropin (hCG) and thyroid function in early human pregnancy: Circadian variation and evidence for intrinsic thyrotropic activity of hCG. *The Journal of Clinical Endocrinology & Metabolism, 66*(4), 853–856.

Pop, V. J., Brouwers, E. P., Vader, H. L., Vulsma, T., Van Baar, A. L., & De Vijlder, J. J. (2003). Maternal hypothyroxinaemia during early pregnancy and subsequent child development: A 3-year follow-up study. *Clinical endocrinology, 59*(3), 282–288.

Pop, V. J., Kuijpens, J. L., van Baar, A. L., Verkerk, G., van Son, M. M., de Vijlder, J. J., ... Vader, H. L. (1999). Low maternal free thyroxine concentrations during early pregnancy are associated with impaired psychomotor development in infancy. *Clinical endocrinology, 50*(2), 149–155.

Poppe, K., Glinoer, D., Van Steirteghem, A., Tournaye, H., Devroey, P., Schiettecatte, J., & Velkeniers, B. (2002). Thyroid dysfunction and autoimmunity in infertile women. *Thyroid, 12*(11), 997–1001.

Poppe, K., Velkeniers, B., & Glinoer, D. (2008). The role of thyroid autoimmunity in fertility and pregnancy. *Nature Clinical Practice Endocrinology & Metabolism, 4*(7), 394–405.

Provinciatto, H., Moreira, M. V. B., Neves, G. R., De Freitas, L. R., Mitsui, H. C., Zhang, J. M. F., & Araujo Júnior, E. (2024). Levothyroxine for subclinical hypothyroidism during pregnancy: An updated systematic review and meta-analysis of randomized controlled trials. *Archives of Gynecology and Obstetrics, 309*(6), 2387–2393.

Prummel, M. F., & Wiersinga, W. M. (2004). Thyroid autoimmunity and miscarriage. *European Journal of Endocrinology, 150*(6), 751–755.

Rpl, E. G. G.o., Bender Atik, R., Christiansen, O. B., Elson, J., Kolte, A. M., Lewis, S., ... Quenby, S. (2023). ESHRE guideline: Recurrent pregnancy loss: An update in 2022. *Human reproduction open, 2023*(1), Article hoad002.

Saito, S., Nakashima, A., Shima, T., & Ito, M. (2010). Th1/Th2/Th17 and regulatory t-cell paradigm in pregnancy. *American journal of reproductive immunology, 63*(6), 601–610.

Sankoda, A., Suzuki, H., Imaizumi, M., Yoshihara, A., Kobayashi, S., Katai, M., ... Arata, N. (2024). Effects of levothyroxine treatment on fertility and pregnancy outcomes in subclinical hypothyroidism: A systematic review and meta-analysis of randomized controlled trials. *Thyroid, 34*(4), 519–530. https://doi.org/10.1089/thy.2023.0546

Tańska, K., Gietka-Czernel, M., Glinicki, P., & Kozakowski, J. (2023). Thyroid autoimmunity and its negative impact on female fertility and maternal pregnancy outcomes. *Frontiers in Endocrinology, 13*, Article 1049665.

Tehrani, F. R., Nazarpour, S., & Behboudi-Gandevani, S. (2021). Isolated maternal hypothyroxinemia and adverse pregnancy outcomes: A systematic review. *Journal of Gynecology Obstetrics and Human Reproduction, 50*(7), Article 102057.

Thangaratinam, S., Tan, A., Knox, E., Kilby, M. D., Franklyn, J., & Coomarasamy, A. (2011). Association between thyroid autoantibodies and miscarriage and preterm birth: Meta-analysis of evidence. *British Medical Journal, 342*.

Thompson, W., Russell, G., Baragwanath, G., Matthews, J., Vaidya, B., & Thompson-Coon, J. (2018). Maternal thyroid hormone insufficiency during pregnancy and risk of neurodevelopmental disorders in offspring: A systematic review and meta-analysis. *Clinical endocrinology, 88*(4), 575–584.

Toloza, F. J. K., Derakhshan, A., Mannisto, T., Bliddal, S., Popova, P. V., Carty, D. M., … Maraka, S. (2022). Association between maternal thyroid function and risk of gestational hypertension and pre-eclampsia: A systematic review and individual-participant data meta-analysis. *Lancet Diabetes Endocrinol, 10*(4), 243–252. https://doi.org/10.1016/S2213-8587(22)00007-9

van Dijk, M. M., Smits, I. H., Fliers, E., & Bisschop, P. H. (2018). Maternal thyrotropin receptor antibody concentration and the risk of fetal and neonatal thyrotoxicosis: A systematic review. *Thyroid, 28* (2), 257–264.

van Dijk, M. M., Vissenberg, R., Fliers, E., van der Post, J. A. M., van der Hoorn, M. P., de Weerd, S., … Goddijn, M. (2022). Levothyroxine in euthyroid thyroid peroxidase antibody positive women with recurrent pregnancy loss (T4LIFE trial): A multicentre, randomised, double-blind, placebo-controlled, phase 3 trial. *Lancet Diabetes Endocrinol, 10*(5), 322–329. https://doi.org/10.1016/s2213-8587(22)00045-6

Zimmermann, M. B., & Boelaert, K. (2015). Iodine deficiency and thyroid disorders. *The lancet Diabetes & endocrinology, 3*(4), 286–295.

Section 2

Focused areas of neurological pathology and physiology

Chapter 7

Thyroid hormone transporters and the brain

Beatriz Muñoz-Falder and Soledad Bárez-López
Instituto de Investigaciones Biomédicas Sols-Morreale, Consejo Superior de Investigaciones Científicas (CSIC), Universidad Autónoma de Madrid (UAM), Department of Neurological Diseases, Madrid, Spain

Abbreviations

AHDS Allan-Herndon-Dudley syndrome
BBB Blood-brain barrier
CNS Central nervous system
CSF Cerebrospinal fluid
DIO2 Type 2 deiodinase
DIO3 Type 3 deiodinase
ER Estrogen receptor
Km Michaelis–Menten constant
LAT1 L-type amino acid transporter 1
LAT2 L-type amino acid transporter 2
MCT10 Monocarboxylate transporter 10
MCT8 Monocarboxylate transporter 8
OATP1A2 Organic anion-transporting polypeptide 1A2
OATP1C1 Organic anion-transporting polypeptide 1C1
RNA-seq RNA sequencing
rT3 Reverse triiodothyronine
SLCO/OATP Solute carrier organic anion transporter family
T3 Triiodothyronine
T4 Thyroxine
TH Thyroid hormone
TM Transmembrane domain
TRα Thyroid hormone receptor alpha
TRβ Thyroid hormone receptor beta
TSH Thyroid-stimulating hormone

Introduction

Thyroid hormones (THs) play an essential role in the developing and adult central nervous system by the modulation of gene expression patterns. The essential role of THs in brain development and functioning was recognized long before the molecular mechanisms governing their cellular entry were known. For decades, it was widely believed that the THs thyroxine (T4) and triiodothyronine (T3) crossed cell membranes by passive diffusion. Under this model, intracellular TH levels were assumed to mirror circulating concentrations, and regional specificity of TH action in the brain was attributed primarily to differences in deiodinase activity or nuclear receptor distribution.

Beginning in the late 1970s, emerging experimental evidence challenged the classical diffusion-based model of TH uptake in neural tissue. Studies in neurons and astrocytes demonstrated saturable, temperature-dependent uptake kinetics that could be competitively inhibited, which are hallmarks of carrier-mediated transport for T3 and T4 (Hennemann et al., 2001; Kastellakis & Valcana, 1989). Importantly, the kinetic properties differed between cell types, suggesting the existence of distinct, brain-specific transporters for THs rather than simple passive diffusion. These findings highlighted that local TH availability in the central nervous system (CNS) is regulated not only by circulating levels but also by selective cellular transport.

The implications for the brain were profound and triggered an intensive search for the molecular identity of TH transporters expressed in the CNS. By the early 2000s, this search culminated in the identification of several transporter families capable of importing iodothyronines, among which the monocarboxylate transporter 8 (MCT8) (Friesema et al., 2003) and the organic anion-transporting polypeptide 1C1 (OATP1C1) (Pizzagalli et al., 2002) stand out for their pathophysiological roles.

MCT8

Structural and functional features of MCT8

MCT8 is encoded by *SLC16A2* on chromosome Xq13.2. In humans, the *SLC16A2* gene comprises two transcriptional start sites, resulting either in an MCT8 protein of 613 amino acids (known as the 'long' isoform) or 539 amino acids (known as the 'short' isoform). Although both

Hormones in Neuroscience. https://doi.org/10.1016/B978-0-443-30261-9.00007-3

isoforms exhibit very similar functional properties, the short isoform is considered the physiologically relevant form, as has higher interspecies homology and is the only one detected in human tissues to date (van Geest et al., 2021).

MCT8 is a transmembrane protein that contains 12 α-helical transmembrane domains, consistent with models of the solute carrier 16 (SLC16) family. Recent structural analysis via cryo-electron microscopy have confirmed that human MCT8 adopts the canonical Major Facilitator Superfamily fold: each monomer comprises two six-helix bundles (TM1–6 and TM7–12), and the structure reveals a central transport channel with binding pockets that accommodate the iodinated rings and carboxyl group of T3 and a semi-hydrophilic pocket for the 5′-iodine atom of T4, providing a molecular basis for substrate specificity (Tan et al., 2025; Tassinari et al., 2025). MCT8 exists as a monomerer or homodimer in humans, with the latter displaying greater resistance to denaturing conditions (Visser et al., 2009).

Functionally, MCT8 transports T3, T4 and reverse T3 (rT3) (Friesema et al., 2003). Transport by MCT8 is saturable, stereospecific, and competitively inhibitable, independent of sodium or proton gradients, consistent with facilitated diffusion. Moreover, uptake of T3, T4, and rT3 follows Michaelis–Menten kinetics, with apparent Km values of approximately 4.0 μM for T3, 4.7 μM for T4 and and 2.2 μm for rT3 in Xenopus oocytes (Friesema et al., 2003).

Localization of MCT8 in the CNS

MCT8 is highly expressed throughout the brain (Fig. 7.1). Immunolabeling studies have established a detailed map of MCT8 expression across multiple cellular types of the mammalian CNS (Table 7.1). The most consistently documented sites of expression are the interfaces forming the brain's barrier systems (Fig. 7.2). In both fetal and adult human tissue, MCT8 is robustly detected in endothelial cells throughout anterior and posterior brain regions, confirming its key role in mediating TH transport across the blood-brain barrier (BBB) (López-Espíndola et al., 2019; Roberts et al., 2008). A second major barrier location is the choroid plexus: MCT8 is present in ependymal cells in juvenile and adult mice (Mayerl et al., 2014) and in fetal and adult human choroid plexus and ventricular ependyma (Chan et al., 2014; López-Espíndola et al., 2019). In addition, tanycytes of the hypothalamus express MCT8 in juvenile and adult mice (Wirth et al., 2009) as well as in the adult and fetal human hypothalamus (Alkemade et al., 2006; López-Espíndola et al., 2019), suggesting a conserved role in local regulation of TH signaling in neuroendocrine circuits. More recently, MCT8 has also been identified in pericytes from adult humans and macaques within basal ganglia vasculature, motor cortex capillaries, and venules (Wang et al., 2023a, 2023b). Fetal human leptomeningeal cells additionally express MCT8, extending its presence to the meningeal barrier system (López-Espíndola et al., 2019).

Immunolabeling studies in developing human brain show strong MCT8 expression in neural progenitor populations of the ventricular and subventricular zones (López-Espíndola et al., 2019), indicating an early requirement for TH transport during neurogenesis. In the neuronal lineage, MCT8 has been detected in multiple regions and stages. In mice, neurons in the cortex, hippocampus, and cerebellum express MCT8 from fetal to adult life (Wirth et al., 2009). Human studies demonstrate expression in the cortex, cortical plate, marginal zone,

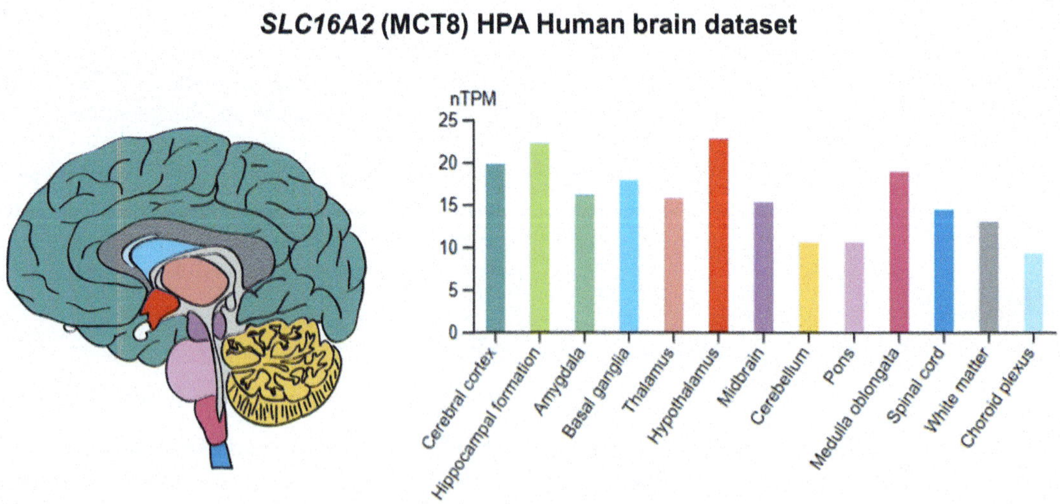

FIGURE 7.1 Expression of *SLC16A2* (MCT8) across 13 human brain regions. Bar plots show normalized RNA expression levels (nTPM), with each bar color-coded to match the corresponding anatomical region in the brain diagram. The figure was assembled using graphical outputs obtained from the human protein atlas (*HPA*) (www.proteinatlas.org; accessed February 01, 2025).

TABLE 7.1 Summary of previous studies of MCT8 expression in the brain of different mammalian species by immunolabeling.

	Cell type	Species	Distribution	References
Brain barriers	**Endothelial cells**	Fetal and adult human	Brain (various anterior and posterior brain regions)	Sugiyama et al., 2003; López-Espíndola et al., 2019
	Ependymal cells	Fetal, juvenile to adult mice	Choroid plexus	Mayerl et al., 2014
		Fetal and adult human	Choroid plexus, ventricles	López-Espíndola et al., 2019; Chan et al., 2014
	Tanycytes	Juvenile to adult mice	Hypothalamus	Wirth et al., 2009
		Fetal and adult human	Hypothalamus	López-Espíndola et al., 2019; Alkemade et al., 2006
	Pericytes	Adult human, adult macaque	Basal ganglia + motor thalamus vasculature; motor cortex capillaries and venules	Wang et al., 2023a; 2023b
	Leptomeningeal cells	Fetal human	Cerebral barriers	López-Espíndola et al., 2019
Neuronal lineage	**Neural cells**	Fetal human	Subventricular zone and ventricular zone	Lopez-Espíndola et al., 2019
	Cajal–Retzius cells	Fetal human	Cortex	López-Espíndola et al., 2019
	Neurons	Fetal, juvenile to adult mice	Cortex	Wirth et al., 2009
		Fetal and adult human	Cortex, cortical plate, marginal zone, hippocampus, hypothalamus	Chan et al., 2014
	Pyramidal neurons	Juvenile mouse	Cortex	Wirth et al., 2009
		Fetal human	Hippocampus	Wirth et al., 2009
		Adult human, adult macaque	Motor cortex	Wang et al., 2023b
	Medium-sized spiny neurons (MSNs)	Adult human, adult macaque	Striatum (direct and indirect basal ganglia circuits)	Wang et al., 2023b
	GABAergic interneurons	Adult human, adult macaque	Striatal microcircuit interneurons; Motor cortex	Wang et al., 2023a; 2023b
	Projection neurons	Adult human, adult macaque	GPe, GPi, STN, SNr, Motor thalamus	Wang et al., 2023b
		Adult human	Nucleus basalis of Meynert	Wang et al., 2023b
	Granule cells	Adult mouse	Hippocampus	Wirth et al., 2009
		Fetal human	Hippocampus, subiculum	López-Espíndola et al., 2019
	Purkinje cells	Adult mouse	Cerebellum	Wirth et al., 2009
Glial lineage	**Radial glia cells**	Fetal human	Cortex, cortical plate, hippocampus, ventricular zone, subventricular zone	López-Espíndola et al., 2019
	Astrocytes	Fetal and adult human	Temporal lobe cortex, cerebral barriers	López-Espíndola et al., 2019
	Glia	Fetal and adult human	Hypothalamus, midline raphe	López-Espíndola et al., 2019
	Folliculostellate cells	Adult human	Anterior pituitary	Alkemade et al., 2006

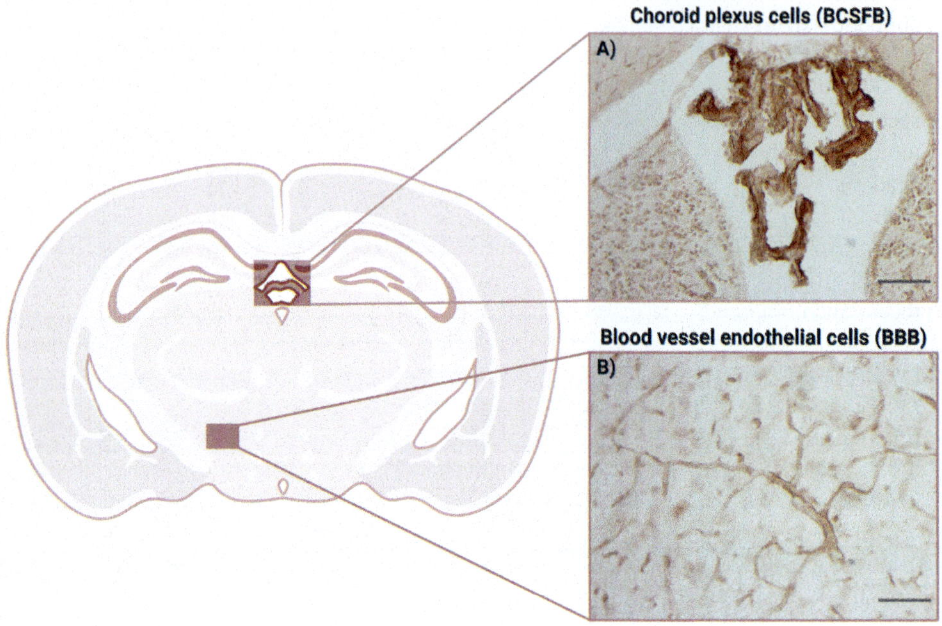

FIGURE 7.2 MCT8 expression at brain barriers. Representative mouse brain sections immunostained for MCT8 show strong transporter expression at key brain barrier interfaces, including the endothelial cells of the blood-brain barrier (*BBB*) and the epithelial cells of the choroid plexus, forming the blood-cerebrospinal fluid barrier (*BCSFB*). These patterns highlight the prominent role of MCT8 in regulating thyroid hormone transport into the central nervous system. Scale bar A = 100 μm and in B = 50 μm. *Created in BioRender. Montero, A. (2025) https://BioRender.com/a8gid9c.*

hippocampus, and hypothalamus during both fetal development and adulthood (Chan et al., 2014). Specific neuronal subtypes identified by immunolabeling include cortical pyramidal neurons in juvenile mice (Wirth et al., 2009), pyramidal neurons in the fetal human hippocampus (Wirth et al., 2009), and adult human and macaque pyramidal neurons of the motor cortex (Wang et al., 2023b). Additional neuronal classes expressing MCT8 include medium-sized spiny neurons of the striatum (Wang et al., 2023b), GABAergic interneurons in striatal and motor cortical microcircuits (Wang et al., 2023a, 2023b), projection neurons of the external and internal globus pallidus, subthalamic nucleus, substantia nigra pars reticulata, and motor thalamus (Wang et al., 2023b) and neurons of the nucleus basalis of Meynert (Wang et al., 2023b). Within the hippocampal formation, granule cells express MCT8 in adult mice (Wirth et al., 2009) and in fetal human hippocampus and subiculum (López-Espíndola et al., 2019). Purkinje cells of the adult mouse cerebellum also show robust labeling (Wirth et al., 2009). In fetal human cortex, Cajal-Retzius cells express MCT8, consistent with a role in early cortical patterning (López-Espíndola et al., 2019).

Glial-lineage expression is also well documented. In fetal human brain, MCT8 is present in radial glia of the cortex, cortical plate, hippocampus, ventricular zone, and subventricular zone (López-Espíndola et al., 2019). Astrocytes in both fetal and adult human temporal cortex and cerebral barriers also exhibit MCT8 immunoreactivity (López-Espíndola et al., 2019). Additional glial expression has been reported in the hypothalamus and midline raphe of fetal and adult human tissue (López-Espíndola et al., 2019). Outside the CNS parenchyma, MCT8 is expressed in folliculostellate cells of the adult human anterior pituitary (Alkemade et al., 2006).

Together, the available immunolabeling evidence reveals a broad yet anatomically coherent expression pattern for MCT8. The transporter is enriched at CNS barrier interfaces (including endothelial cells, ventricular and choroid plexus ependymal cells, tanycytes and pericytes) while also being present in diverse neuronal and glial populations across development and adulthood. This distribution underscores the central role of MCT8 in regulating access, distribution, and cellular utilization of TH within the brain.

The pathophysiology of MCT8 mutations

Mutations in MCT8 result in the Allan-Herndon-Dusdley syndrome (AHDS). Patients display characteristic alterations in serum TH levels, including elevated T3, reduced T4 and rT3, and slightly increased or normal thyroid stimulating hormone levels. In parallel with these endocrine abnormalities, AHDS patients exhibit a complex neurological phenotype (Dumitrescu et al., 2004; Friesema et al., 2004). Although brain impairments have been documented in postmortem fetal samples (López-Espíndola et al., 2014), symptoms are not detectable at birth and Apgar scores are usually normal. Within the first 6 months of life, however, the syndrome becomes clinically evident. Patients begin to show global developmental delay and profound intellectual disability (IQ < 30), central hypotonia that later progresses to spasticity, dystonic movements, muscular hypoplasia, and, in some cases, epileptic seizures (Masnada et al., 2022; Remerand et al., 2019).

The clinical manifestation of AHDS spans a wide spectrum of severity. While some patients achieve limited motor and communicative abilities, such as holding their head or establishing simple interactions, others present with far more severe incapacitating phenotypes (Masnada et al., 2022). Despite shared core features, these phenotypical differences underscore the complexity of AHDS. A recent and comprehensive integrative study by Groeneweg et al., (2025) has now established a clear relationship between the functional impact of MCT8 variants and the severity of multiple disease manifestations. Their loss-of-function classification system, categorizing variants as mild, moderate, or severe, accurately predicted survival as well as key neurodevelopmental (such as motor skills, seizure occurrence), anthropometric (such as body weight, head circumference), thyrotoxic (such as premature atrial complexes, sex hormone binding globulin levels) features, and circulating TH concentrations (Groeneweg et al., 2025). These findings could have direct clinical implications by enabling treatments to be tailored to patients with MCT8 mutations, according to the specific functional impact of their variant.

OATP1C1

Structural and functional features of OATP1C1

OATP1C1 is encoded by the *SLCO1C1* gene, located on chromosome 12p12.1. It belongs to the solute carrier organic anion transporter family (SLCO/OATP), and the human protein contains 12 α-helical transmembrane domains, consistent with the structural architecture of the OATP family and other Major Facilitator Superfamily transporters. Recent cryo-electron microscopy has now resolved the human OATP1C1 structure in complex with T4 and T3 (Ge et al., 2025), providing insights into its transport mechanism. Structurally, OATP1C1 adopts an outward-open conformation characterized by a central amphipathic cavity that contains a specialized T4-binding pocket.

Functionally, OATP1C1 acts as a high-affinity transporter of T4 and rT3, operating through sodium-independent facilitated diffusion, consistent with the transport mechanism of other OATPs (van der Deure et al., 2008). In vitro transport assays using COS-1 cells or Xenopus oocytes expressing OATP1C1 demonstrated robust T4 uptake, weaker rT3 transport, and minimal T3 or T3-sulfate uptake (Pizzagalli et al., 2002). Further studies expressing the rat OATP1C1 ortholog in embryonic kidney 293 cells provided kinetic insight, demonstrating that T4 is transported with a Km of 0.18 μM, whereas T3 uptake is ~six-fold lower. Moreover, the authors showed that OATP1C1 can also transport other amphipathic organic anions such as 17β-estradiol-d-17β-glucuronide (Km = 10 μM), cerivastatin (Km = 1.3 μM), and troglitazone sulfate (Km = 0.76 μM) (Sugiyama et al., 2003).

Localization of OATP1C1 in the CNS

MCT8 is highly expressed throughout the brain (Fig. 7.3). Immunolabeling studies across mammalian species reveal a complex, developmentally regulated, and cell-type-specific distribution of OATP1C1 in the CNS (Table 7.2).

In fetal and juvenile mice, OATP1C1 is prominently expressed in endothelial cells of brain capillaries,

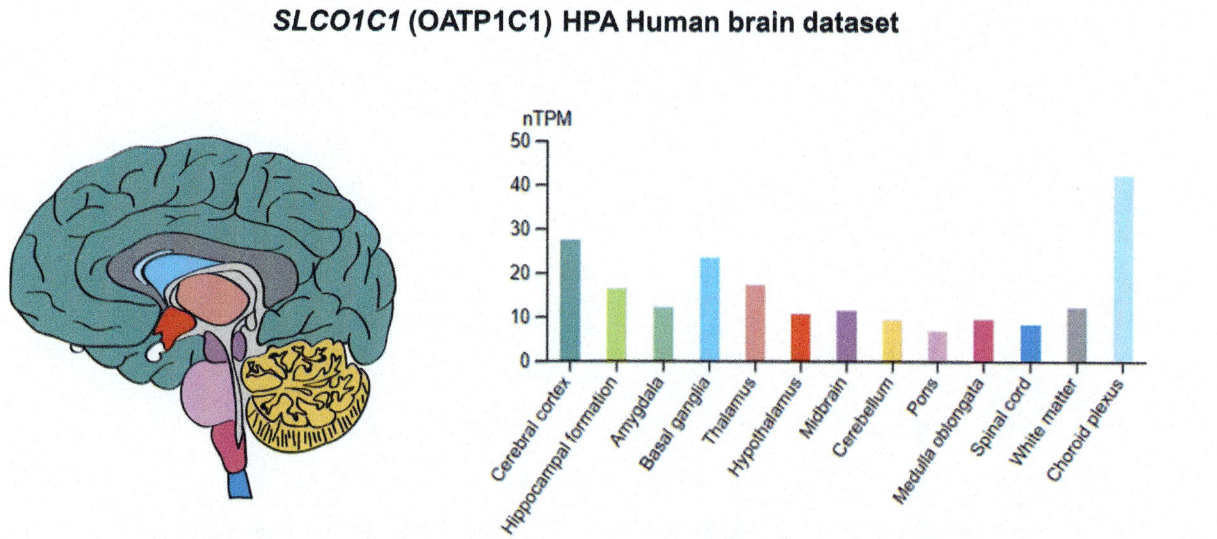

FIGURE 7.3 Expression of *SLCO1C1* (OATP1C1) across 13 human brain regions. Bar plots show normalized RNA expression levels (nTPM), with each bar color-coded to match the corresponding anatomical region in the brain diagram. The figure was assembled using graphical outputs obtained from the human protein atlas (*HPA*) (www.proteinatlas.org; accessed February 01, 2025).

TABLE 7.2 Summary of previous studies of OAP1C1 expression in the brain of different mammalian species by immunolabeling.

	Cell type	Species	Distribution	References
Brain barriers	**Endothelial cells**	Fetal and juvenile mouse	Brain capillary endothelial cells (blood-brain barrier, BBB)	Sugiyama et al., 2003; López-Espíndola et al., 2019
		Adult human, adult macaque	Motor cortex	Wang et al., 2023b
	Ependymal cells	Fetal and adult human	Choroid plexus, cerebral barriers	López-Espíndola et al., 2019
	Tanycytes	Fetal human	Hypothalamus	López-Espíndola et al., 2019
Neuronal lineage	**Neural cells**	Fetal human	Third ventricle, hypothalamus	López-Espíndola et al., 2019
	Cajal–retzius Cells	Fetal human	Cortex, cortical plate, subplate	López-Espíndola et al., 2019
	Neurons	Fetal, infant and adult human	Hypothalamus, subventricular zone, ventricular zone, intermediate zone	Alkemade et al., 2011; López-Espíndola et al., 2019
	Pyramidal neurons	Adult human, adult macaque	Motor cortex	Wang et al., 2023b
	Medium-sized spiny neurons (MSNs)	Adult human, adult macaque	Striatum (direct and indirect basal ganglia circuits)	Wang et al., 2023b
	GABAergic interneurons	Adult human, adult macaque	Striatal microcircuit interneurons; Motor cortex	Wang et al., 2023b
	Projection neurons	Adult human, adult macaque	Globus pallidus externus, globus pallidus internus, subthalamic nucleus, substantia nigra pars reticulata, motor thalamus	Wang et al., 2023b
		Adult human	Nucleus basalis of Meynert	Wang et al., 2023b
	Immature neurones	Fetal human	Cortex, brainstem, raphe, hypothalamus	López-Espíndola et al., 2019
Glial lineage	**Radial glia cells**	Fetal human	Cortex	López-Espíndola et al., 2019
	Astrocytes	Adult human, adult macaque	Hupothalamus, motor cortex (inside the *Corpora amylacea* complexes)	Alkemade et al., 2011; Wang et al., 2023b

indicating a role in TH handling at the BBB (López-Espíndola et al., 2019; Roberts et al., 2008; Sugiyama et al., 2003). In the adult primate motor cortex, endothelial immunoreactivity has also been documented in humans and macaques (Wang et al., 2023b), extending this barrier-associated expression to adulthood and across species. Beyond the endothelium, OATP1C1 is present in additional epithelial and barrier-associated cell types in the human brain. Fetal and adult human ependymal cells of the choroid plexus exhibit clear immunostaining (López-Espíndola et al., 2019). Tanycytes express OATP1C1 in fetal human tissue, indicating a conserved distribution across multiple CNS barrier systems (López-Espíndola et al., 2019).

Human fetal studies reveal widespread OATP1C1 expression during early neurodevelopment. Neural cells of the ventricular, subventricular, and intermediate zones express OATP1C1, as do cells of the third ventricle and hypothalamus (López-Espíndola et al., 2019). Specialized developmental populations, including Cajal–Retzius cells of the cortex, cortical plate, and subplate, also display immunoreactivity (López-Espíndola et al., 2019). Immature neurons in the fetal cortex, brainstem, raphe, and hypothalamus show robust labeling as well (López-Espíndola et al., 2019).

In the adult CNS, the most comprehensive neuronal mapping comes from recent primate studies. In the human and macaque motor cortex, OATP1C1 is expressed in long-projection pyramidal neurons (Wang et al., 2023b). Additional work in adult humans and macaques demonstrates OATP1C1 in medium-sized spiny neurons of the striatum, in multiple classes of striatal and cortical GABAergic interneurons, and in projection neurons across the basal ganglia and thalamic motor circuits, including the external and internal globus pallidus, subthalamic nucleus, substantia nigra pars reticulata, and motor thalamus (Wang et al., 2023b). Expression is also detected in neurons of the nucleus basalis of Meynert (Wang et al., 2023b). Moreover, OATP1C1 has also been detected in neuronal populations within several hypothalamic nuclei (Alkemade et al., 2011). Collectively, these findings firmly establish OATP1C1 as a transporter present in diverse excitatory and inhibitory neuronal subtypes in the adult primate brain.

During human fetal development, OATP1C1 is abundantly expressed in radial glia of the cortex (López-Espíndola et al., 2019). In the adult primate CNS, astrocytes in the human and macaque motor cortex exhibit clear OATP1C1 immunoreactivity (Alkemade et al., 2011; Wang et al., 2023b). Notably, OATP1C1 labeling is also found in astrocytes located within corpora amylacea complexes in the human motor cortex (Wang et al., 2023b). At present, no immunolabeling evidence has documented expression in oligodendrocytes, microglia, or oligodendrocyte precursor cells.

Across species and developmental stages, inmunolabeling evidence confirms OATP1C1 expression in endothelial cells, choroid-plexus ependyma, tanycytes, radial glia, multiple classes of fetal neural and neuronal precursor populations, pyramidal neurons, medium-sized spiny neurons, GABAergic interneurons, projection neurons of basal ganglia–thalamic circuits, and astrocytes. This distribution highlights OATP1C1 as a transporter with roles extending beyond classical barrier-mediated TH entry, contributing to TH handling within both neuronal and glial networks across the human and nonhuman primate brain.

The pathophysiology of OATP1C1 mutations

To date, only a single loss-of-function mutation in *SLCO1C1* has been clearly linked to severe neurological disease. Strømme et al., (2018) reported a patient homozygous for the D252N mutation, who exhibited progressive neurodegeneration, spasticity, cold intolerance, and cognitive decline. Functional studies revealed that this mutant transporter fails to reach the plasma membrane and shows a near-complete loss of T4 transport capacity (Strømme et al., 2018).

In contrast, common *SLCO1C1* polymorphisms in humans have been associated with milder clinical features, such as fatigue and depressive symptoms in population cohorts (van der Deure et al., 2008). However, these variants have not been shown to produce the profound neurological phenotype observed with the D252N mutation, and evidence linking them to objective neurocognitive deficits remains limited.

Overall, these findings highlight that while OATP1C1 function is critical for maintaining central TH homeostasis, only severe loss-of-function mutations appear sufficient to induce overt neurological disease, whereas more common variants may contribute to subtle or subclinical phenotypes.

Other TH transporters and partial redundancy

Several additional transporters are expressed in the brain and may contribute to TH uptake in specific cells or regions:

- **MCT10 (*SLC16A10*)**: MCT10 is a transporter primarily of aromatic amino acids (tryptophan, tyrosine), but it also mediates low-affinity transport of iodothyronines, although its affinity for T3 and T4 is lower than that of MCT8 (Friesema et al., 2008). MCT10 is found in neurons, mature oligodendrocytes and microglia (Alkemade et al., 2011; Braun et al., 2011; Müller & Heuer, 2014).
- **LAT1 (*SLC7A5*) and LAT2 (*SLC7A8*)**: LAT1 (*SLC7A5*) and LAT2 (*SLC7A8*) are sodium-independent

L-type amino acid transporters that operate as heterodimers with the heavy chain 4F2hc (CD98) (Ritchie et al., 1999). Their primary substrates are large neutral amino acids, but both transporters can also transport T4 and T3, and have a *K*m in the μM range, LAT1 being a more efficient T3 transporter than LAT2 (Friesema et al., 2001). LAT1 is expressed in neurons, astrocytes, oligodendrocyte precursors and endothelial cells of the BBB and LAT2 in neurons, in a lesser extent than LAT1, astrocytes and highly expressed in microglia (Boado et al., 1999; Braun et al., 2011). Although *Lat1* mRNA is expressed in microglia LAT1 protein has not been detected in this cell type, also this is the case with *Lat2* mRNA in endotelial cells (Braun et al., 2011).

- **OATP1A2 (*SLCO1A2*)**: A broad-specificity organic anion transporter expressed in brain endothelial cells and neurons. Although capable of T4 transport in vitro, there is no experimental evidence of compensatory TH transport in vivo when OATP1C1 or MCT8 are absent (Bernal et al., 2015; Ianculescu et al., 2010).

Contributions of TH transporters to TH brain availability

Since the pioneering work of Crantz et al., (1982), it has been known that, in the postnatal brain, only about 20%–25% of T3 originates directly from the blood circulation, while local conversion of T4 to T3 accounts for approximately 80% of nuclear T3 receptor occupancy in the cerebral cortex (Crantz et al., 1982).

Subsequent studies in adult hypothyroid rats revealed that continuous infusion of T3 at relatively low doses was sufficient to normalize T3 concentrations in peripheral tissues such as the liver, kidney, heart, and lung. However, much higher doses were required to restore normal T3 levels in the cerebral cortex. In contrast, administration of T4 normalized T3 content in the cerebral cortex at lower doses than those needed to achieve normalization in peripheral tissues. Moreover, T3 concentrations in the cerebral cortex remained within the physiological range even when the administered T4 dose increased (Escobar-Morreale et al., 1999).

These findings indicate that T4 serves as the primary source of T3 in the brain, and that local conversion of T4 to T3 is a tightly regulated process, preventing excessive T3 accumulation. Years later, it was demonstrated that DIO2 is primarily expressed in astrocytes (Guadaño-Ferraz et al., 1997), leading to the current model in which the main source of T3 in the brain derives from T4 deiodination within astrocytes. The newly generated T3 is then made available to surrounding neural cells, where it exerts its genomic effects (Fig. 7.4).

However, the identity of the TH transporters mediating this brain T3 availability remained unknown for many years thereafter.

Identification of TH transporters at the brain barriers

The first insights into the identity of the TH transporters at the brain barriers derived from the generation of mouse models for the AHDS. The first animal models, MCT8 knock-out (MCT8 KO) mice, faithfully replicated the syndrome's endocrine aspects but failed to recapitulate the psychomotor or histopathological brain defects (Dumitrescu et al., 2006; Trajkovic et al., 2007). This discrepancy led researchers to propose the existence of compensatory mechanisms, particularly the presence of alternative TH transporters that are abundant in mice but scarce in humans, and one candidate was OATP1C1.

Roberts et al. demonstrated that MCT8 is expressed in the cerebral microvessels of humans, mice, and rats, whereas OATP1C1 is present in the cerebral microvessels of mice and rats but largely absent in humans (Sugiyama et al., 2003; Trajkovic et al., 2007). These findings suggested that TH transport across the BBB is primarily mediated by MCT8 in humans, while in rodents, both MCT8 and OATP1C1 contribute to this process (Fig. 7.4). Subsequent studies confirmed this: mice deficient for both TH transporters, MCT8 and OATP1C1 (MCT8/OATP1C1 double knock-out, DKO), exhibited alterations in peripheral TH homeostasis similar to those observed in MCT8 KO mice, and uptake of both T3 and T4 into the brain was severely reduced. Importantly, MCT8/OATP1C1 DKO mice displayed neurological impairments and pronounced locomotor abnormalities (Mayerl et al., 2014), highlighting the complementary roles of these transporters in rodents. Similarly, mice deficient for both MCT8 and DIO2 (MCT8/DIO2 DKO) showed comparable alterations in peripheral TH homeostasis, strongly reduced cerebral T3 and T4 uptake, and neurological and locomotor deficits (Bárez-López et al., 2019). These findings confirmed the TH brain availability model, suggesting that in rodents, circulating T4 enters the brain through both MCT8 and OATP1C1, is then converted to T3 by DIO2 in astrocytes, and subsequently becomes available to other neural cells. A portion of T3 can also be transported directly across the BBB via MCT8. In contrast, in humans, this process relies almost exclusively on MCT8, with no compensatory contribution from other TH transporters (Fig. 7.4).

Evidence indicated that MCT8 is present and essential at the BBB, however, the relative contributions of the BBB versus the neuronal plasma membrane in restricting T3 access to target neural cells remained unclear. Pioneering work by Ceballos et al. addressed this question by comparing the effects of low doses of T4 and T3 in the

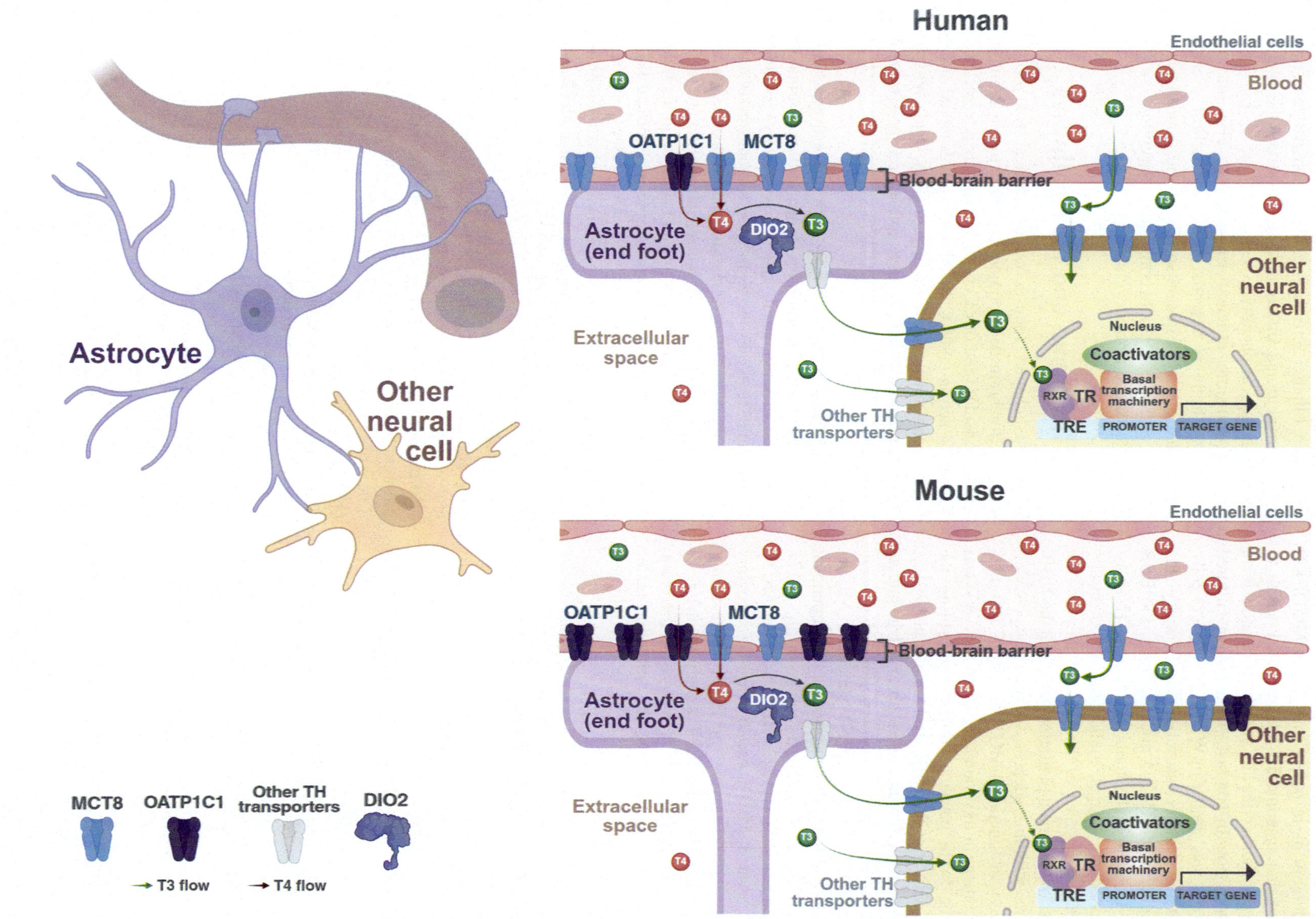

FIGURE 7.4 Comparative mechanisms of thyroid hormone transport in the human and mouse brain. Diagram illustrating the comparative mechanisms of thyroid hormone transport in the human and mouse brain. Schematic representation of how circulating thyroid hormones (T4 and T3) cross the blood-brain barrier (*BBB*) and are processed by astrocytes and other neural cells. In the primary pathway, T4 enters the brain via MCT8 or OATP1C1 transporters. A major species difference lies in OATP1C1 expression at the BBB: In humans (top), OATP1C1 levels are relatively low, making MCT8 the predominant route for thyroid hormone entry, whereas in mice (bottom), OATP1C1 is highly expressed. Once inside the brain, astrocytic DIO2 converts T4 to the active hormone T3, which is then released via thyroid hormone transporters and taken up by neighboring neural cells, where it enters the nucleus to regulate target gene expression. Additionally, T3 can cross the BBB directly through MCT8 to reach the brain parenchyma for use by target neural cells. *Created in BioRender. Montero, A. (2025) https://BioRender.com/4u8210j.*

brain of wild-type (Wt) and MCT8 KO mice (Ceballos et al., 2009). They found that, compared with Wt animals, T4 was considerably more potent than T3 in MCT8 KO mice, indicating that systemic T3 has restricted access to neurons in vivo, whereas the T3 locally generated from systemic T4 does not. In contrast, T3 action in cultured cerebellar neurons was similar in Wt and MCT8 KO cells, suggesting that the primary restriction for T3 entry into neural target cells in MCT8 KO mice occurs at the BBB rather than at the neuronal membrane. Other TH transporters expressed in target cells could potentially compensate for the absence of MCT8, but they do not appear to overcome the main limitation present at the BBB.

In accordance with this, adeno-associated virus (AAV)-mediated delivery of functional MCT8 to MCT8 KO mice increased TH content in the brain only when the AAV was delivered to the brain barriers, further highlighting the critical role of MCT8 at the BBB (Iwayama et al., 2016).

Years later, similar findings were observed in human samples. Using patient-specific induced pluripotent stem cells (iPSCs), Vatine et al., (2017) generated MCT8-deficient neural cells that exhibited normal TH-dependent neuronal properties and maturation (Vatine et al., 2017). However, when an iPSC-based BBB model of MCT8 deficiency was created, MCT8 was shown to be necessary for the transport of THs across the BBB. These results indicate that, in humans with MCT8 mutations, the BBB represents the main restriction to TH entry into the brain, suggesting that strategies to bypass the limitation at the BBB and deliver active TH directly to the brain could be a viable therapeutic approach.

Thyroid hormone transporters and brain development

THs play an essential role during brain development, orchestrating key processes such as neuronal migration, maturation, myelination, and synaptogenesis (Bernal, 2000).

As outlined above, the TH transporters MCT8 and OATP1C1 are expressed in multiple neural and supporting cell types during prenatal life. Among these, MCT8 appears to be particularly essential, as demonstrated by the profound neurodevelopmental abnormalities observed in individuals with MCT8 deficiency. Studies of MCT8-deficient human fetuses have revealed delayed cortical and cerebellar development, defective myelination, loss of parvalbumin expression, abnormal calbindin-D28k levels, impaired axonal maturation, and reduced biochemical differentiation of Purkinje cells. These findings indicate that neurological damage in MCT8 deficiency is diffuse and begins prenatally, even when early postnatal examinations appear normal (López-Espíndola et al., 2014). Recently, cerebral organoids derived from patient iPSCs lacking MCT8 (Salas-Lucia et al., 2024) exhibit impaired T3 uptake, blunted activation of T3-responsive genes, and defects in neuronal differentiation, mirroring the human neuropathological phenotype.

In conclusion, efficient transport of THs into the developing brain is essential from the earliest prenatal stages, and a deeper understanding of these mechanisms remains critical for uncovering how disruptions in TH availability shape neurodevelopment.

Other hormones and their pathways

While this chapter focuses on the critical role of TH transport for brain development and function, it is important to consider how other endocrine signals intersect with T3 and T4 action in the CNS. THs exert widespread effects on neuronal differentiation, myelination, and synaptogenesis, processes that are also influenced by estrogens, testosterone, insulin, and metabolic hormones such as ghrelin.

In particular, sex steroids such as estrogens and testosterone play essential roles in neural tissue formation, phenotypic differentiation, and neuroplasticity. Estradiol, for example, regulates the expression of genes involved in synaptic organization, myelination, neurotransmission, and intracellular signaling. Estrogen receptors (ER) are widely distributed across the developing and adult brain, including pyramidal neurons of the hippocampus and cortical layers II–VI, where both ERα and ERβ are present in nuclei, dendritic spines, and astrocytes. Functional studies further support a developmental role: pharmacological blockade of estrogen receptors alters the number and distribution of neurons and glial cells in the medial prefrontal cortex, and ERβ-knockout mice show defects in cortical lamination, suggesting that ERβ contributes to cytoskeletal dynamics and neuronal orientation during corticogenesis. Estradiol also promotes proliferation of embryonic neural stem cells through ERα and ERβ-dependent mechanisms, although this proliferative effect appears restricted to early developmental windows (Denley et al., 2018). Despite extensive work, the precise biochemical pathways through which estradiol shapes cortical architecture and neuronal identity remain incompletely understood.

Interestingly, evidence suggests that THs and estrogens can mutually modulate one another's activity. One proposed mechanism involves the ability of thyroid and estrogen receptors to bind overlapping hormone-response elements on shared target genes, potentially competing for or coordinating transcriptional control (Vasudevan et al., 2002). Such cross-talk likely represents a molecular mechanism that integrates environmental and endocrine cues, ultimately fine-tuning gene regulatory programs that drive brain development and influence adult neural function.

In summary, THs do not act in isolation in the brain; rather, they function within an intricate endocrine network in which multiple hormones converge on shared cellular targets and signaling pathways. Understanding these interactions is crucial for elucidating the pathophysiology of neurodevelopmental disorders and cognitive decline, and it highlights the importance of therapeutic strategies that consider multiple hormonal axes simultaneously.

Applications of the material

This chapter reviews how THs access the developing brain, the cellular expression of TH transporters, and the consequences of impaired transport, which has significant translational implications for understanding and treating neurodevelopmental disorders. Disorders such as AHDS, caused by loss-of-function mutations in the MCT8 transporter, demonstrate how restricted TH entry into the fetal and infant brain results in profound and irreversible neurological disability. As such, insights into the timing, regulation, and mechanisms of TH transport can directly inform therapeutic strategies.

A critical application of this knowledge is the development of TH-based treatments that bypass the BBB when MCT8 function is compromised. Several TH analogs have emerged as promising candidates because they can enter the brain independently of MCT8. These analogs have shown partial correction of metabolic abnormalities in patients and in some cases prevention or recovery of neurological impairments in animal models (Grijota-Martínez et al., 2020). Understanding which cell types express TH transporters during specific developmental windows allows researchers to better predict where and when these analogs must act to meaningfully influence neurodevelopment.

Another important translational avenue involves therapeutic strategies designed to restore or replace transporter function, such as gene therapy approaches using AAV vectors to introduce functional MCT8 into the brain endothelium. Recent preclinical studies demonstrate that re-establishing TH transport into the CNS improves myelination and neuronal maturation, even when treatment is initiated postnatally (Liao et al., 2022). These findings underscore the need to more precisely define the temporal trajectory of TH-dependent developmental processes in order to identify therapeutic windows during which intervention can still yield meaningful improvement.

Finally, mapping the precise developmental stages at which TH deprivation leads to specific structural and functional abnormalities has direct clinical relevance. It helps clinicians determine when treatment must begin to prevent irreversible injury, but also where recovery may still be possible late in development. These insights also extend beyond AHDS, informing the management of maternal hypothyroidism, preterm birth, and other conditions associated with impaired fetal TH delivery.

In summary, understanding how THs reach the developing brain directly enables the design of therapies that either restore physiological transport, bypass the BBB, or act downstream to correct impaired gene regulation. These advances bring the field closer to effective treatments for AHDS and other neurodevelopmental disorders rooted in disrupted TH signaling.

Mini-dictionary of terms

- **Allan-Herndon-Dudley Syndrome.** A rare X-linked neurodevelopmental disorder caused by mutations in the MCT8 transporter, leading to severe intellectual and motor impairment.
- **Blood-brain barrier.** A selective physical and chemical barrier formed by endothelial cells, astrocytic end-feet, and another specialized cell type (pericytes), that regulates the passage of molecules from the bloodstream into the brain.
- **Deiodinases.** Enzymes that activate or inactivate THs, controlling the local availability of active hormones in tissues.
- **Monocarboxylate Transporter 8.** A transmembrane protein responsible for T4 and T3 transport across cell membranes.
- **Organic anion-transporting polypeptide 1C1.** A transmembrane protein responsible for T4 transport across cell membranes.
- **Radial Glia Cells.** Multipotent progenitor cells in the developing brain that give rise to neurons and glia and guide neuronal migration.
- **Tanycytes.** Specialized ependymal glia located in the median eminence and basal hypothalamus.

Summary points

- Brain access to THs depends on specific transporters, mainly MCT8 and OATP1C1.
- Disruption of these transporters alters local T3 brain availability.
- MCT8 deficiency causes diffuse prenatal brain injury, leading to severe and persistent neurodevelopmental impairment.
- Advances in human fetal tissue studies and transgenic models have improved understanding of how TH transporter expression shapes regional hormone availability in the developing brain.
- Insights into TH transport mechanisms are informing therapeutic strategies, including BBB-independent delivery, TH analogs and gene therapy.

During the preparation of this work, the authors used ChatGPT3.5 to correct grammar errors. After using this tool, the authors reviewed and edited the content as needed and take full responsibility for the publication's content.

References

Alkemade, A., Friesema, E. C., Kalsbeek, A., Swaab, D. F., Visser, T. J., & Fliers, E. (2011). Expression of thyroid hormone transporters in the human hypothalamus. *The Journal of clinical endocrinology and metabolism, 96*(6), E967–E971. https://doi.org/10.1210/jc.2010-2750

Alkemade, A., Friesema, E. C., Kuiper, G. G., Wiersinga, W. M., Swaab, D. F., Visser, T. J., & Fliers, E. (2006). Novel neuroanatomical pathways for thyroid hormone action in the human anterior pituitary. *European Journal of Endocrinology, 154*(3), 491–500. https://doi.org/10.1530/eje.1.02111

Bárez-López, S., Grijota-Martínez, C., Ausó, E., Fernández-de Frutos, M., Montero-Pedrazuela, A., & Guadaño-Ferraz, A. (2019). Adult mice lacking Mct8 and Dio2 proteins present alterations in peripheral thyroid hormone levels and severe brain and motor skill impairments. *Thyroid, 29*(11), 1669–1682. https://doi.org/10.1089/thy.2019.0068

Bernal, J. (2000). Thyroid hormones in brain development and function. In K. R. Feingold, S. F. Ahmed, B. Anawalt, M. R. Blackman, A. Boyce, G. Chrousos, E. Corpas, W. W. de Herder, K. Dhatariya, K. Dungan, J. Hofland, S. Kalra, G. Kaltsas, N. Kapoor, C. Koch, P. Kopp, M. Korbonits, C. S. Kovacs, W. Kuohung, & D. P. Wilson (Eds.), *Endotext*. MDText.com, Inc.

Bernal, J., Guadaño-Ferraz, A., & Morte, B. (2015). Thyroid hormone transporters–functions and clinical implications. *Nature reviews. Endocrinology, 11*(7), 406–417. https://doi.org/10.1038/nrendo.2015.66

Boado, R. J., Li, J. Y., Nagaya, M., Zhang, C., & Pardridge, W. M. (1999). Selective expression of the large neutral amino acid transporter at the blood-brain barrier. *Proceedings of the National Academy of Sciences of the United States of America, 96*(21), 12079–12084. https://doi.org/10.1073/pnas.96.21.12079

Braun, D., Kinne, A., Bräuer, A. U., Sapin, R., Klein, M. O., Köhrle, J., & Schweizer, U. (2011). Developmental and cell type-specific expression of thyroid hormone transporters in the mouse brain and in primary brain cells. *Glia, 59*(3), 463–471. https://doi.org/10.1002/glia.21116

Ceballos, A., Belinchon, M. M., Sanchez-Mendoza, E., Grijota-Martinez, C., Dumitrescu, A. M., Refetoff, S., & Bernal, J. (2009). Importance of monocarboxylate transporter 8 for the blood-brain barrier-dependent availability of 3,5,3'-triiodo-L-thyronine. *Endocrinology, 150*(5), 2491–2496. https://doi.org/10.1210/en.2008-1616

Chan, S. Y., Hancox, L. A., Martín-Santos, A., Loubière, L. S., Walter, M. N., González, A. M., & Kilby, M. D. (2014). MCT8 expression in human fetal cerebral cortex is reduced in severe intrauterine growth restriction. *Journal of Endocrinology, 220*(2), 85–95. https://doi.org/10.1530/joe-13-0400

Crantz, F. R., Silva, J. E., & Larsen, P. R. (1982). An analysis of the sources and quantity of 3,5,3'-triiodothyronine specifically bound to nuclear receptors in rat cerebral cortex and cerebellum. *Endocrinology, 110*(2), 367–375. https://doi.org/10.1210/endo-110-2-367

Denley, M. C. S., Gatford, N. J. F., Sellers, K. J., & Srivastava, D. P. (2018). Estradiol and the development of the cerebral cortex: An unexpected role? *Frontiers in Neuroscience, 12*, 245. https://doi.org/10.3389/fnins.2018.00245

Dumitrescu, A. M., Liao, X. H., Best, T. B., Brockmann, K., & Refetoff, S. (2004). A novel syndrome combining thyroid and neurological abnormalities is associated with mutations in a monocarboxylate transporter gene. *The American Journal of Human Genetics, 74*(1), 168–175. https://doi.org/10.1086/380999

Dumitrescu, A. M., Liao, X. H., Weiss, R. E., Millen, K., & Refetoff, S. (2006). Tissue-specific thyroid hormone deprivation and excess in monocarboxylate transporter (mct) 8-deficient mice. *Endocrinology, 147*(9), 4036–4043. https://doi.org/10.1210/en.2006-0390

Escobar-Morreale, H. F., Obregón, M. J., Escobar del Rey, F., & Morreale de Escobar, G. (1999). Tissue-specific patterns of changes in 3,5,3'-triiodo-L-thyronine concentrations in thyroidectomized rats infused with increasing doses of the hormone. which are the regulatory mechanisms? *Biochimie, 81*(5), 453–462. https://doi.org/10.1016/s0300-9084(99)80095-9

Friesema, E. C., Docter, R., Moerings, E. P., Verrey, F., Krenning, E. P., Hennemann, G., & Visser, T. J. (2001). Thyroid hormone transport by the heterodimeric human system L amino acid transporter. *Endocrinology, 142*(10), 4339–4348. https://doi.org/10.1210/endo.142.10.8418

Friesema, E. C., Ganguly, S., Abdalla, A., Manning Fox, J. E., Halestrap, A. P., & Visser, T. J. (2003). Identification of monocarboxylate transporter 8 as a specific thyroid hormone transporter. *Journal of Biological Chemistry, 278*(41), 40128–40135. https://doi.org/10.1074/jbc.M300909200

Friesema, E. C., Grueters, A., Biebermann, H., Krude, H., von Moers, A., Reeser, M., & Visser, T. J. (2004). Association between mutations in a thyroid hormone transporter and severe X-linked psychomotor retardation. *Lancet, 364*(9443), 1435–1437. https://doi.org/10.1016/s0140-6736(04)17226-7

Friesema, E. C., Jansen, J., Jachtenberg, J. W., Visser, W. E., Kester, M. H., & Visser, T. J. (2008). Effective cellular uptake and efflux of thyroid hormone by human monocarboxylate transporter 10. *Molecular Endocrinology, 22*(6), 1357–1369. https://doi.org/10.1210/me.2007-0112

Ge, Y., Dou, T., Nguyen, T. U., Yadav, G. P., Wensel, T. G., Jiang, J., & Huang, P. (2025). Structural insights into brain thyroid hormone transport via MCT8 and OATP1C1. *Cell, 188*(20), 5576–5588. https://doi.org/10.1016/j.cell.2025.06.032. e5517.

Grijota-Martínez, C., Bárez-López, S., Gómez-Andrés, D., & Guadaño-Ferraz, A. (2020). MCT8 deficiency: The road to therapies for a rare disease. *Frontiers in Neuroscience, 14*, 380. https://doi.org/10.3389/fnins.2020.00380

Groeneweg, S., van Geest, F. S., Martín, M., Dias, M., Frazer, J., Medina-Gomez, C., & Visser, W. E. (2025). Mapping variants in thyroid hormone transporter MCT8 to disease severity by genomic, phenotypic, functional, structural and deep learning integration. *Nature Communications, 16*(1), 2479. https://doi.org/10.1038/s41467-025-56628-w

Guadaño-Ferraz, A., Obregón, M. J., St Germain, D. L., & Bernal, J. (1997). The type 2 iodothyronine deiodinase is expressed primarily in glial cells in the neonatal rat brain. *Proceedings of the National Academy of Sciences of the United States of America, 94*(19), 10391–10396. https://doi.org/10.1073/pnas.94.19.10391

Hennemann, G., Docter, R., Friesema, E. C., de Jong, M., Krenning, E. P., & Visser, T. J. (2001). Plasma membrane transport of thyroid hormones and its role in thyroid hormone metabolism and bioavailability. *Endocrine Reviews, 22*(4), 451–476. https://doi.org/10.1210/edrv.22.4.0435

Ianculescu, A. G., Friesema, E. C., Visser, T. J., Giacomini, K. M., & Scanlan, T. S. (2010). Transport of thyroid hormones is selectively inhibited by 3-iodothyronamine. *Molecular Biosystems, 6*(8), 1403–1410. https://doi.org/10.1039/b926588k

Iwayama, H., Liao, X. H., Braun, L., Bárez-López, S., Kaspar, B., Weiss, R. E., & Refetoff, S. (2016). Adeno associated virus 9-Based gene therapy delivers a functional monocarboxylate transporter 8, improving thyroid hormone availability to the brain of Mct8-deficient mice. *Thyroid, 26*(9), 1311–1319. https://doi.org/10.1089/thy.2016.0060

Kastellakis, A., & Valcana, T. (1989). Characterization of thyroid hormone transport in synaptosomes from rat brain. *Molecular and Cellular Endocrinology, 67*(2–3), 231–241. https://doi.org/10.1016/0303-7207(89)90213-x

López-Espíndola, D., García-Aldea, Á., Gómez de la Riva, I., Rodríguez-García, A. M., Salvatore, D., Visser, T. J., & Guadaño-Ferraz, A. (2019). Thyroid hormone availability in the human fetal brain: Novel entry pathways and role of radial glia. *Brain structure & function, 224*(6), 2103–2119. https://doi.org/10.1007/s00429-019-01896-8

López-Espíndola, D., Morales-Bastos, C., Grijota-Martínez, C., Liao, X. H., Lev, D., Sugo, E., & Guadaño-Ferraz, A. (2014). Mutations of the thyroid hormone transporter MCT8 cause prenatal brain damage and persistent hypomyelination. *The Journal of clinical endocrinology and metabolism, 99*(12), E2799–E2804. https://doi.org/10.1210/jc.2014-2162

Liao, X. H., Avalos, P., Shelest, O., Ofan, R., Shilo, M., Bresee, C., & Vatine, G. D. (2022). AAV9-MCT8 delivery at juvenile stage ameliorates neurological and behavioral deficits in a mouse model of MCT8-deficiency. *Thyroid, 32*(7), 849–859. https://doi.org/10.1089/thy.2022.0034

Müller, J., & Heuer, H. (2014). Expression pattern of thyroid hormone transporters in the postnatal mouse brain. *Frontiers in Endocrinology, 5*, 92. https://doi.org/10.3389/fendo.2014.00092

Masnada, S., Sarret, C., Antonello, C. E., Fadilah, A., Krude, H., Mura, E., & Tonduti, D. (2022). Movement disorders in MCT8 deficiency/allan-herndon-dudley syndrome. *Molecular genetics and metabolism, 135*(1), 109–113. https://doi.org/10.1016/j.ymgme.2021.12.003

Mayerl, S., Müller, J., Bauer, R., Richert, S., Kassmann, C. M., Darras, V. M., & Heuer, H. (2014). Transporters MCT8 and OATP1C1 maintain murine brain thyroid hormone homeostasis. *Journal of Clinical Investigation, 124*(5), 1987–1999. https://doi.org/10.1172/jci70324

Pizzagalli, F., Hagenbuch, B., Stieger, B., Klenk, U., Folkers, G., & Meier, P. J. (2002). Identification of a novel human organic anion transporting polypeptide as a high affinity thyroxine transporter. *Molecular Endocrinology, 16*(10), 2283–2296. https://doi.org/10.1210/me.2001-0309

Remerand, G., Boespflug-Tanguy, O., Tonduti, D., Touraine, R., Rodriguez, D., Curie, A., & Sarret, C. (2019). Expanding the phenotypic spectrum of allan-herndon-dudley syndrome in patients with SLC16A2 mutations. *Developmental Medicine and Child Neurology, 61*(12), 1439–1447. https://doi.org/10.1111/dmcn.14332

Ritchie, J. W., Peter, G. J., Shi, Y. B., & Taylor, P. M. (1999). Thyroid hormone transport by 4F2hc-IU12 heterodimers expressed in xenopus oocytes. *Journal of Endocrinology, 163*(2), R5–R9. https://doi.org/10.1677/joe.0.163r005

Roberts, L. M., Woodford, K., Zhou, M., Black, D. S., Haggerty, J. E., Tate, E. H., & Zerangue, N. (2008). Expression of the thyroid hormone transporters monocarboxylate transporter-8 (SLC16A2) and organic ion transporter-14 (SLCO1C1) at the blood-brain barrier. *Endocrinology, 149*(12), 6251–6261. https://doi.org/10.1210/en.2008-0378

Salas-Lucia, F., Escamilla, S., Bianco, A. C., Dumitrescu, A., & Refetoff, S. (2024). Impaired T3 uptake and action in MCT8-deficient cerebral organoids underlie allan-herndon-dudley syndrome. *JCI Insight, 9*(7). https://doi.org/10.1172/jci.insight.174645

Strømme, P., Groeneweg, S., Lima de Souza, E. C., Zevenbergen, C., Torgersbråten, A., Holmgren, A., & Visser, T. J. (2018). Mutated thyroid hormone transporter OATP1C1 associates with severe brain hypometabolism and juvenile neurodegeneration. *Thyroid, 28*(11), 1406–1415. https://doi.org/10.1089/thy.2018.0595

Sugiyama, D., Kusuhara, H., Taniguchi, H., Ishikawa, S., Nozaki, Y., Aburatani, H., & Sugiyama, Y. (2003). Functional characterization of rat brain-specific organic anion transporter (Oatp14) at the blood-brain barrier: High affinity transporter for thyroxine. *Journal of Biological Chemistry, 278*(44), 43489–43495. https://doi.org/10.1074/jbc.M306933200

Tan, J., Xiao, Y., Kong, F., Qian, J., Zhu, A., & Yan, C. (2025). Structural insights into thyroid hormone transporter MCT8. *Nature Communications, 16*(1), 2958. https://doi.org/10.1038/s41467-025-58131-8

Tassinari, M., Tanzi, G., Maggiore, F., Groeneweg, S., van Geest, Freund, M. E. T., Stavast, C. J., Boniardi, I., Pasqualato, S., Visser, W. E., & Coscia, F. (2025). Molecular mechanism of thyroxine transport by monocarboxylate transporters. *Nature Communications, 16*(1), 4493. https://doi.org/10.1038/s41467-025-59751-w

Trajkovic, M., Visser, T. J., Mittag, J., Horn, S., Lukas, J., Darras, V. M., & Heuer, H. (2007). Abnormal thyroid hormone metabolism in mice lacking the monocarboxylate transporter 8. *Journal of Clinical Investigation, 117*(3), 627–635. https://doi.org/10.1172/jci28253

van der Deure, W. M., Hansen, P. S., Peeters, R. P., Kyvik, K. O., Friesema, E. C., Hegedüs, L., & Visser, T. J. (2008). Thyroid hormone transport and metabolism by organic anion transporter 1C1 and consequences of genetic variation. *Endocrinology, 149*(10), 5307–5314. https://doi.org/10.1210/en.2008-0430

van Geest, F. S., Gunhanlar, N., Groeneweg, S., & Visser, W. E. (2021). Monocarboxylate transporter 8 deficiency: From pathophysiological understanding to therapy development. *Frontiers in Endocrinology, 12*, Article 723750. https://doi.org/10.3389/fendo.2021.723750

Vasudevan, N., Ogawa, S., & Pfaff, D. (2002). Estrogen and thyroid hormone receptor interactions: Physiological flexibility by molecular specificity. *Physiological Reviews, 82*(4), 923–944. https://doi.org/10.1152/physrev.00014.2002

Vatine, G. D., Al-Ahmad, A., Barriga, B. K., Svendsen, S., Salim, A., Garcia, L., & Svendsen, C. N. (2017). Modeling psychomotor retardation using iPSCs from MCT8-Deficient patients indicates a prominent role for the blood-brain barrier. *Cell Stem Cell, 20*(6), 831–843.e835. https://doi.org/10.1016/j.stem.2017.04.002

Visser, W. E., Philp, N. J., van Dijk, T. B., Klootwijk, W., Friesema, E. C., Jansen, J., & Visser, T. J. (2009). Evidence for a homodimeric structure of human monocarboxylate transporter 8. *Endocrinology, 150*(11), 5163–5170. https://doi.org/10.1210/en.2009-0699

Wang, Y., Wang, T., Montero-Pedrazuela, A., Guadaño-Ferraz, A., & Rausell, E. (2023a). Thyroid hormone transporters MCT8 and

OATP1C1 are expressed in pyramidal neurons and interneurons in the adult motor cortex of human and macaque brain. *International Journal of Molecular Sciences, 24*(4). https://doi.org/10.3390/ijms24043207

Wang, T., Wang, Y., Montero-Pedrazuela, A., Prensa, L., Guadaño-Ferraz, A., & Rausell, E. (2023b). Thyroid hormone transporters MCT8 and OATP1C1 are expressed in projection neurons and interneurons of basal ganglia and motor thalamus in the adult human and macaque brains. *International Journal of Molecular Sciences, 24*(11). https://doi.org/10.3390/ijms24119643

Wirth, E. K., Roth, S., Blechschmidt, C., Hölter, S. M., Becker, L., Racz, I., & Schweizer, U. (2009). Neuronal 3',3,5-triiodothyronine (T3) uptake and behavioral phenotype of mice deficient in Mct8, the neuronal T3 transporter mutated in allan-herndon-dudley syndrome. *Journal of Neuroscience, 29*(30), 9439–9449. https://doi.org/10.1523/jneurosci.6055-08.2009

Chapter 8

Thyroid hormones and neurodegenerative diseases

Mohammad Hossein Etemadi[1], Hanieh Shafizadeh[2], Mohammad Amin Karimi[2], Nazanin Mohammad Zadeh Kosari[3], Maryam Shafizadeh[4] and Niloofar Deravi[2]

[1]*Student Research Committee, School of Medicine, Isfahan University of Medical Sciences, Isfahan, Iran;* [2]*School of Medicine, Shahid Beheshti University of Medical Sciences, Tehran, Iran;* [3]*School of Medicine, Mashhad University of Medical Sciences, Mashhad, Iran;* [4]*School of Medicine, Iran University of Medical Sciences, Tehran, Iran*

List of abbreviations

Aβ Amyloid beta
AMPAR α-amino-3-hydroxy-5-methyl-4-isoxazolepropionic acid receptor
ATP Adenosine triphosphate
c-Abl Abelson non-receptor tyrosine kinase
DIO1/DIO2/DIO3 or D1/D2/D3 Iodothyronine deiodinase 1/2/3
ER Endoplasmic reticulum
ETC Electron transport chain
FT3/fT3 Free triiodothyronine
FT4/fT4 Free thyroxine
GPx Glutathione peroxidase
GSK-3 Glycogen synthase kinase 3
LTP Long-term potentiation
LXR Liver X receptor
MAPK Mitogen-activated protein kinase
MCT8 / MCT10 Monocarboxylate transporter 8/10
MMSE Mini-Mental State Examination
NF-κB Nuclear factor kappa B
NMDAR N-methyl-D-aspartate receptor
OA Okadaic acid
OATP1C1 Organic anion-transporting polypeptide 1C1
RXR Retinoid X receptor
SLC16A2 Solute carrier family 16 member 2
SLCO1C1 Solute carrier organic anion transporter family member 1C1
SOD Superoxide dismutase
T3 Triiodothyronine
T4 Thyroxine
TH/THs Thyroid hormone / thyroid hormones
THR or TR Thyroid hormone receptor
THRα/THRβ or TRα/TRβ Thyroid hormone receptor alpha/beta
TRE Thyroid hormone response element
TRH Thyrotropin-releasing hormone
TSH Thyroid-stimulating hormone
TrxR Thioredoxin reductase
TTR Transthyretin

Introduction

The brain is functionally integrated with multiple peripheral systems and does not operate in isolation. One important interface is the endocrine system, which communicates with the brain through hormones that regulate emotion, cognition, neurodevelopment, and other central nervous system functions. Hormones influence the function and plasticity of neurons by acting as signaling compounds that cross the blood-brain barrier. To preserve homeostasis and promote adaptive behaviors, the brain and endocrine system must communicate dynamically (de Souza, 2024).

Thyroid hormones (THs), including triiodothyronine (T3) and thyroxine (T4), are synthesized in thyroid follicular cells through iodination of tyrosyl residues in thyroglobulin, using iodine obtained from the diet. They are released into the bloodstream in response to thyroid-stimulating hormone (TSH), with T3 representing the more biologically active form.

THs play a crucial role in regulating physiological processes by influencing oxygen consumption and the metabolism of proteins, carbohydrates, lipids, and vitamins. They regulate gene expression by interacting with thyroid hormone receptors and nuclear transcription factors (de Souza, 2024). By altering protein expression within pathways, THs also impact the production and breakdown of various hormones and growth factors. Thus, THs influence broader endocrine signaling through their effects on hormone production, metabolism, and growth-factor pathways. Beyond their metabolic and growth-related

Hormones in Neuroscience. https://doi.org/10.1016/B978-0-443-30261-9.00008-5

functions, they are crucial for brain development from the embryonic period through to adulthood (de Souza, 2024; Samuels, 2014).

The pathophysiology of the TH-related changes in the cerebral tissue is diverse and includes changes in neurotransmission, alterations in neuronal or glial cell gene expression, blood-brain barrier (BBB) dysfunction, increased risk of cerebrovascular disease, and infrequently cerebral inflammatory disease in the setting of autoimmune thyroid disease. Thyroid dysfunction may cause notable mood and cognitive disturbances, such as anxiety, depression, irritability, executive function impairments, and potentially reversible dementia (Jurado-Flores et al., 2022; Samuels, 2014).

Several studies show that thyroid dysfunction and variations within the normal range of thyroid function, including serum TSH levels, are associated with changes in cognitive performance (Kim et al., 2022; Santhanam et al., 2022; Thyroid).

Given the reported associations between thyroid dysfunction and cognitive or neurodegenerative outcomes, appropriate screening and treatment of thyroid disease may help identify potentially reversible cognitive symptoms and reduce avoidable neurological morbidity (Jurado-Flores et al., 2022; Samuels, 2014; Thyroid).

In this chapter, we aim to further discuss the actions of thyroid hormones in the context of neurodegenerative diseases.

TH production, transport, and mechanisms of action

The biosynthesis of thyroid hormones T3 and T4 is stimulated by thyroid-stimulating hormone (TSH), which is secreted by the anterior pituitary gland and acts on thyroid follicular cells. Three deiodinase enzymes, including types 1, 2, and 3, are involved in the regulation of thyroid hormones.

Type 2 deiodinase primarily activates thyroid hormone by converting T4 to T3, whereas type 3 deiodinase inactivates thyroid hormones by converting T4 to reverse T3 and T3 to T2. Type 1 deiodinase contributes to peripheral thyroid hormone metabolism and may have activating or inactivating roles depending on tissue context (Braun & Schweizer, 2018).

Although T3 and T4 have lipophilic properties, their cellular uptake is largely mediated by specific membrane transporters. Their transport relies on specialized membrane carriers, primarily from two key families: monocarboxylate transporter 8 (MCT8), monocarboxylate transporter 10 (MCT10), and organic anion-transporting polypeptide 1C1 (OATP1C1). Notably, inactivation of the MCT8 gene in humans can lead to Allan-Herndon-Dudley syndrome, a severe neurodevelopmental disorder characterized by significant neurological impairments due to disrupted thyroid hormone transport into the brain (Dumitrescu et al., 2004). Thyroid hormones exert both genomic and non-genomic effects. Genomic TH action is primarily mediated through binding of T3 to thyroid hormone receptors (THRs), mainly THRα and THRβ, which are ligand-dependent nuclear receptors. THRs are ligand-dependent nuclear receptors that regulate transcription after binding thyroid hormone. THRs typically bind thyroid hormone response elements (TREs), often as heterodimers with retinoid X receptors (RXRs), thereby regulating transcription of target genes (Phan et al., 2010). The TRE sequence consists of two consensus half sites, AGGT/ACA, arranged either as direct repeat, palindrome, or inverted palindrome. The RXR binds to 5′ half-site while THR binds 3′ half-site. In the absence of TH, corepressors associate with THR to inhibit gene transcription. The binding of TH to the THR facilitates conformational change of THR, dissociation of corepressors from THR, and recruitment of coactivators, and this, in turn, drives gene transcription (Chatonnet et al., 2013). The non-genomic effects of THs include energy production, oxygen consumption, activation of the sodium-hydrogen exchanger, and increased intracellular acidity; however, the exact mechanisms are still not well understood. Non-genomic action of TH is considered to be important for brain development as T4 has been shown to alter actin polymerization and neural migration. It is also suggested that activation of protein kinase Akt and endothelial nitric oxide synthase via T3 non-genomic action in rat brain could contribute to neuroprotective effects of TH (Hiroi et al., 2006). Further investigation of TH non-genomic action in brain development will help to understand molecular mechanisms of TH in sex-specific regulation of neuronal signaling. The presence of crosstalk between genomic and non-genomic TH activities has also been proposed, indicating that the molecular mechanisms of TH action are diverse and complex.

TH receptors and distribution in the brain

Thyroid hormones exert many of their effects by binding intracellular nuclear receptors, particularly THRα and THRβ, which regulate gene transcription. These receptors are part of the nuclear receptor family and consist of two distinct forms: alpha and beta types. Although these two receptors share certain similarities, patients with mutations in either receptor display markedly different clinical characteristics (Cheng et al., 2010). Patients with THRβ mutations and Thrb-mutant mice may show goiter, hearing impairment, and dysregulation of the hypothalamic–pituitary–thyroid (HPT) axis, whereas Thra mutations in mice have been associated with increased mortality, reduced fertility, and growth impairment (Ortiga-Carvalho et al., 2014). In mammals, various forms of thyroid hormone receptors have been recognized. For example, in humans, three variants of the alpha receptor (1, 2, and 3) and three of the beta receptors (1, 2, and 3) are listed in the biological sequence repository. The alpha variant is broadly distributed across brain regions, while the beta variant is predominantly localized in the deeper, subcortical layers. The regulation of thyroid hormone activity strongly depends on the expression patterns of both receptor types.

TH regulation and action in the brain

T3 and T4 produced by the thyroid gland enter the circulation and are distributed to peripheral tissues, including the brain. Both T4 and T3 can cross the BBB. The TH transporters solute carrier family 16 member 2 (SLC16A2/MCT8) and solute carrier organic anion transporter family member 1C1 (SLCO1C1/OATP1C1) are expressed in BBB endothelial cells (Roberts et al., 2008). OATP1C1 is a T4 transporter, whereas MCT8 can transport both T3 and T4. From BBB endothelial cells, T4 is transported to astrocytes via OATP1C1, where it is converted to T3 by iodothyronine deiodinase 2 (DIO2). Research has shown that the maturation of specific regions in the brain depends on the activity of enzymes that transform the hormone T4 into its more biologically active form, T3. This conversion process plays a fundamental role in the proper development of those brain structures. DIO1 activity is mainly detected in the mouse cerebellum and is largely absent from other brain regions, making DIO2 and DIO3 the major iodothyronine deiodinases in the brain. Deiodinases are membrane-bound proteins. DIO2 is mainly localized to the endoplasmic reticulum, whereas DIO1 and DIO3 are primarily membrane-associated enzymes with catalytic domains oriented toward the cytosol. DIO2 is mainly expressed in astrocytes, whereas DIO3 is predominantly expressed in neurons (Dentice et al., 2005, 2007).

Higher THRα than THRβ expression in the human fetal brain may contribute to the prominent neurodevelopmental consequences of impaired THRα-mediated signaling.

Interestingly, experimental data suggest that some adverse neurodevelopmental effects of hypothyroidism are related to reduced circulating T3 availability rather than isolated loss of either THRα or THRβ signaling. T3 plays a pivotal role in the maturation of microglial cells, and it also promotes their migration and phagocytic activity (Gil-Ibañez et al., 2013).

Sex-specific effects of TH in the brain

Available clinical and experimental evidence suggests that TH signaling may have sex-dependent effects on brain maturation and neuropsychiatric vulnerability, potentially through interactions with gonadal steroids. However, the magnitude and clinical relevance of these effects likely vary by developmental stage, thyroid state, and disease context. Male and female brains respond differently to these hormones, especially during key developmental windows such as gestation and adolescence, a variation that may stem from hormonal interplay between thyroid hormones and gonadal steroids like estrogen and testosterone. This interaction shapes molecular patterns of gene activity, neural connectivity, and overall brain architecture in a sex-dependent fashion. Research from both clinical observations and animal studies reveals sex-specific presentations of neurological conditions associated with thyroid imbalance. For example, women are disproportionately affected by hypothyroidism-related mood disturbances, while men may exhibit increased susceptibility to certain psychiatric disorders linked to altered thyroid function. Recognizing these disparities could pave the way for more individualized approaches to treating thyroid-related neuropsychiatric illnesses (Baksi & Pradhan, 2021) (Table 8.1, Fig. 8.1).

Oxidative stress, metabolism, aging, and TH-related neurodegeneration

Reactive oxygen species (ROS) include chemically reactive oxygen-containing molecules such as superoxide anion ($O_2\bullet-$), hydrogen peroxide (H_2O_2), and hydroxyl radical (•OH) (Sies et al., 2022). These reactive species are mostly produced as a byproduct of mitochondrial oxidative metabolism, especially in the electron transport chain activities. In the mitochondrial electron transport chain (ETC), electron leakage is known to occur primarily at complexes I and III, which allow the generation of ROS during aerobic respiration. ROS are also critical at manageable concentrations for physiological signaling, such as redox-mediated signaling, kinase, and transcription factor activation. However, when ROS production exceeds antioxidant capacity, these species can indiscriminately oxidize critical biological molecules, including lipids, proteins, and nucleic acids. These oxidative modifications are associated with structural damage and loss of function and therefore to the development and/or progression of several pathological conditions.

The concentration of intracellular ROS is tightly regulated by an enzymatic antioxidant network, which is composed of superoxide dismutase (SOD), catalase, glutathione peroxidase (GPx), and thioredoxin reductase (TrxR) (Villanueva et al., 2013). An imbalance between ROS production and enzymatic detoxification culminates in oxidative stress, which may reduce mitochondrial function and activate pro-inflammatory signaling pathways. ROS are endogenously formed by-products of normal cellular physiology. THs influence intracellular ROS levels partly by increasing basal metabolic rate. Increased mitochondrial oxidative phosphorylation enhances ATP production; however, increased metabolic

TABLE 8.1 Overview of thyroid hormone (TH) functions, mechanisms, and effects on brain health.

Aspect	Description	Key mechanisms	Implications for brain health
TH production and transport	THs (T3, T4) released by thyroid gland under TSH stimulation; transported by MCT8, MCT10, OATP1C1 across BBB; deiodinases (D1/D2 activate, D3 inactivate).	Genomic: Bind to THRα/β receptors, heterodimerize with RXR, and regulate gene expression via TREs. Non-genomic: Regulate actin polymerization, Akt activation, nitric oxide synthase.	Essential for neurodevelopment, neuronal migration, and synaptic plasticity; MCT8 mutations can cause Allan–Herndon–Dudley syndrome.
TH receptors and distribution	THRα (extensive distribution, more in fetal brain) and THRβ (subcortical layers); isoforms vary in region.	Ligand-binding-induced conformational changes, coactivator recruitment for transcription; sex-specific regulation by gonadal steroids.	Mutations cause distinctive phenotypes (e.g., THRα: Dwarfism; THRβ: Goiter, HPT axis deregulation); required for maturation of microglia and neurogenesis.
TH regulation in brain	Conversion of T4 to T3 in astrocytes by DIO2; TH is inactivated by DIO3 in neurons; transporters on BBB (MCT8, OATP1C1).	Astrocyte-neuron interaction; upregulation of THRα in fetal brain; T3 enhances microglial migration/phagocytosis.	Maintains homeostasis; disruptions disrupt neurotransmission, BBB integrity, gene expression; linked to cognitive/mood disturbances.
Sex-specific effects	Females are more vulnerable to mood/psychiatric issues; interactions with estrogen/testosterone during development/adolescence.	Gene expression patterns and neural networks may differ by sex; clinical variations (e.g., women: Depression secondary to hypothyroidism).	Requires individualized, sex-specific therapeutic approaches; women are at higher risk of emotional disturbances at menopause.
Oxidative stress, metabolism, and aging	THs regulate ROS via mitochondrial ETC; hyperthyroidism increases ROS/oxidative stress; hypothyroidism impairs antioxidants.	Stimulates metabolism, electron leakage in complexes I/III; mitochondrial disease and ROS build-up exacerbated by aging.	Part of neurodegeneration (e.g., Alzheimer's, Parkinson's); THRP overexpression causes neuronal death; protective in some contexts.

Akt, protein kinase B; *BBB*, blood–brain barrier; *D1*, type 1 deiodinase; *D2*, type 2 deiodinase; *D3*, type 3 deiodinase; *DIO2*, iodothyronine deiodinase 2; *DIO3*, iodothyronine deiodinase 3; *ETC*, electron transport chain; *HPT*, hypothalamic–pituitary–thyroid; *MCT8*, monocarboxylate transporter 8; *MCT10*, monocarboxylate transporter 10; *OATP1C1*, organic anion-transporting polypeptide 1C1; *ROS*, reactive oxygen species; *RXR*, retinoid X receptor; *T3*, triiodothyronine; *T4*, thyroxine; *TH*, thyroid hormone; *THR*, thyroid hormone receptor; *THRα*, thyroid hormone receptor alpha; *THRβ*, thyroid hormone receptor beta; *THRP*, thyroid hormone-responsive protein; *THs*, thyroid hormones; *TRE*, thyroid hormone response element; *TSH*, thyroid-stimulating hormone.

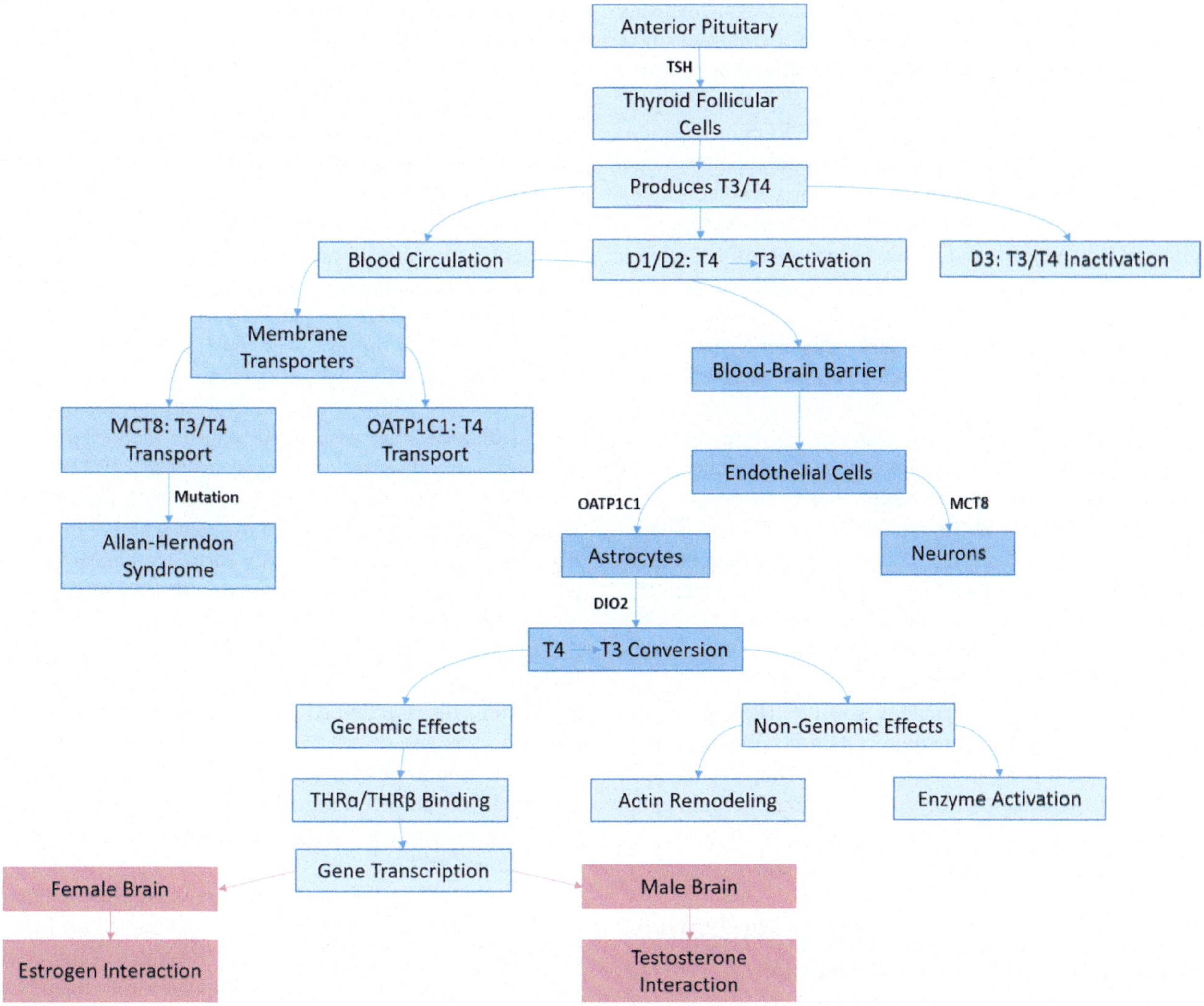

FIGURE 8.1 Pathway of thyroid hormone (T3/T4) production, transport, and mechanisms of action in the brain, with emphasis on sex-specific effects. *D1*, type 1 deiodinase; *D2*, type 2 deiodinase; *D3*, type 3 deiodinase; *DIO2*, iodothyronine deiodinase 2; *MCT8*, monocarboxylate transporter 8; *OATP1C1*, organic anion-transporting polypeptide 1C1; *T3*, triiodothyronine; *T4*, thyroxine; *THRα*, thyroid hormone receptor alpha; *THRβ*, thyroid hormone receptor beta; *TSH*, thyroid-stimulating hormone.

activity may also increase electron leakage from the electron transport chain, ultimately resulting in increased ROS generation (Villanueva et al., 2013). Although TH-dependent regulation plays an important role in redox homeostasis, persistent hyperthyroidism may shift cellular redox balance toward a pro-oxidant state. Conversely, hypothyroidism may reduce ROS production but may also impair antioxidant defenses, particularly in older adults (Macvanin et al., 2023; Marcocci et al., 2012; Villanueva et al., 2013). Therefore, thyroid hormone–controlled metabolism is an important element for cellular redox homeostasis.

Aging is characterized by a gradual decrease in mitochondrial function and increased oxidative stress. With aging, cumulative ROS production can damage electron transport chain components, reducing mitochondrial efficiency and creating a positive feedback loop that further increases ROS generation. This chronic oxidative damage to cellular constituents, including mitochondrial DNA, proteins, and lipids, damages and deregulates these cellular components, which appears to be a key feature of age-related cellular dysfunction and is increasingly associated with a variety of neurodegenerative disorders.

THs exert broad regulatory effects in the central nervous system, including effects on neuronal development, energy metabolism, and redox homeostasis (Alcaide Martin & Mayerl, 2023; Głombik et al., 2021). In the senescent brain, alterations in TH signaling have also been increasingly linked to increased susceptibility to age-related neurodegenerative diseases, mainly due to mechanisms implicating mitochondrial damage and oxidative stress (Mooradian & Haas, 2025). Thyroid dysfunction can substantially alter brain redox homeostasis. Hyperthyroid states enhance cellular respiration and ROS generation, whereas hypothyroidism may impair antioxidant defense and reduce neuronal resilience, particularly in older adults (Villanueva et al., 2013). These redox imbalances may impair BBB integrity and disrupt neurotransmitter release cascades, both of which are closely regulated by THs (Mooradian & Haas, 2025).

Experimental evidence suggests that abnormal activation of c-Abl signaling may contribute to neuronal injury and cell death in neurodegenerative disease models. This cytotoxic effect may occur through a non-canonical pathway involving c-Abl, suggesting an alternative TH-related mechanism of neuronal injury (Mooradian & Haas, 2025; Motaln & Rogelj, 2023). Moreover, altered TH levels may affect the course of neurodegenerative diseases, such as Parkinson's disease and Alzheimer's disease (Charoenngam et al., 2022; Li & Liu, 2024). Subclinical TH changes have been associated with cognitive alterations in older adults, but this effect is age dependent and disease context specific (Rieben et al., 2016). Interestingly, under some conditions, fluctuations of TH can be neutral or protective, emphasizing the intricate and context-dependent effects of TH actions on brain aging and pathology (Mooradian & Haas, 2025) (Fig. 8.2).

Thyroid hormones in neurodegenerative and neuropsychiatric disorders

Thyroid hormones including T4 and T3 play a vital role in regulating the function of the central nervous system, and disturbances in T3 and T4 levels have been associated with neurodegenerative and neuropsychiatric disorders (Villanueva et al., 2013). In hyperthyroid states, increased basal metabolic rate may promote ROS overproduction and exacerbate oxidative stress. In contrast, hypothyroidism may weaken cellular antioxidant defenses and increase vulnerability to cumulative brain injury.

In contrast, hypothyroidism, by reducing metabolic activity, weakens the cell's antioxidant defenses and makes the brain more vulnerable to cumulative damage (Mooradian & Haas, 2025). Disorders in TH metabolism have been associated with structural neuronal changes, reduced hippocampal volume, and neurotransmitter disruption, which may contribute to cognitive and neuropsychiatric disorders. (Zhang, Wang, et al., 2024).

In Alzheimer's disease, impaired thyroid hormone signaling has been associated with amyloid accumulation, tau hyperphosphorylation, and cognitive dysfunction (Peng et al., 2024). Brain imaging studies have shown that higher free triiodothyronine (FT3) and free thyroxine (FT4) levels are associated with increased amyloid and tau deposition in the brain (Byeon et al., 2024). This interaction may accelerate the process of neurodegeneration and lead to decreased memory and learning function (AlAnazi et al., 2023). Transthyretin (TTR), a transporter of T4 and T3 in the brain, has been proposed to have a protective role against amyloid accumulation, and TTR stabilization may represent a potential strategy for modifying Alzheimer's disease-related pathology; however, further clinical evidence is needed (Corino et al., 2025).

In Parkinson's disease, THs may affect dopaminergic neuron vulnerability by influencing inflammatory and oxidative stress pathways. Meta-analyses have shown that thyroid disorders are associated with an increased risk of developing the disease (Charoenngam et al., 2022).

In ALS, endocrine dysfunction has been discussed in relation to disease progression; however, whether disrupted TH signaling contributes directly to motor neuron degeneration requires further disease-specific evidence (Peng et al., 2024). Population-based analyses have suggested that thyroid disorders, particularly hypothyroidism, may be associated with increased risks of Lewy body dementia, Parkinson's disease, and Alzheimer's disease (Toraih et al., 2025).

In addition to neurodegenerative diseases, THs play an important role in mood regulation through their effects on neurotransmitters such as serotonin and dopamine (Baksi & Pradhan, 2021).

Hypothyroidism is associated with decreased serotonin levels and an increased risk of depression, while hyperthyroidism can lead to irritability and anxiety (Zeng et al., 2024). Sex differences in the brain response to THs have also been reported, with women appearing more susceptible to mood disorders associated with thyroid dysfunction (Valcárcel-Hernández et al., 2024).

At the cellular level, the non-receptor kinase Abelson non-receptor tyrosine kinase (c-Abl) plays an important role in regulating cellular stress responses, DNA repair, and programmed cell death (Motaln & Rogelj, 2023).

Abnormal activation of this kinase has been observed in diseases such as Alzheimer's disease, Parkinson's disease, and ALS, and its inhibition may have protective effects against neuronal degeneration. c-Abl-related signaling may intersect with pathways influenced by TH signaling, but the extent and clinical relevance of this interaction require further investigation (Mooradian & Haas, 2025). Recent evidence suggests that thyroid dysfunction during menopause may be associated with increased risks of depression and anxiety, potentially related to changes in T3 and T4 levels (Mintziori et al., 2024).

In neurogenic brain regions such as the hippocampus, THs are involved in regenerative processes by regulating neural stem cell differentiation and stimulating the

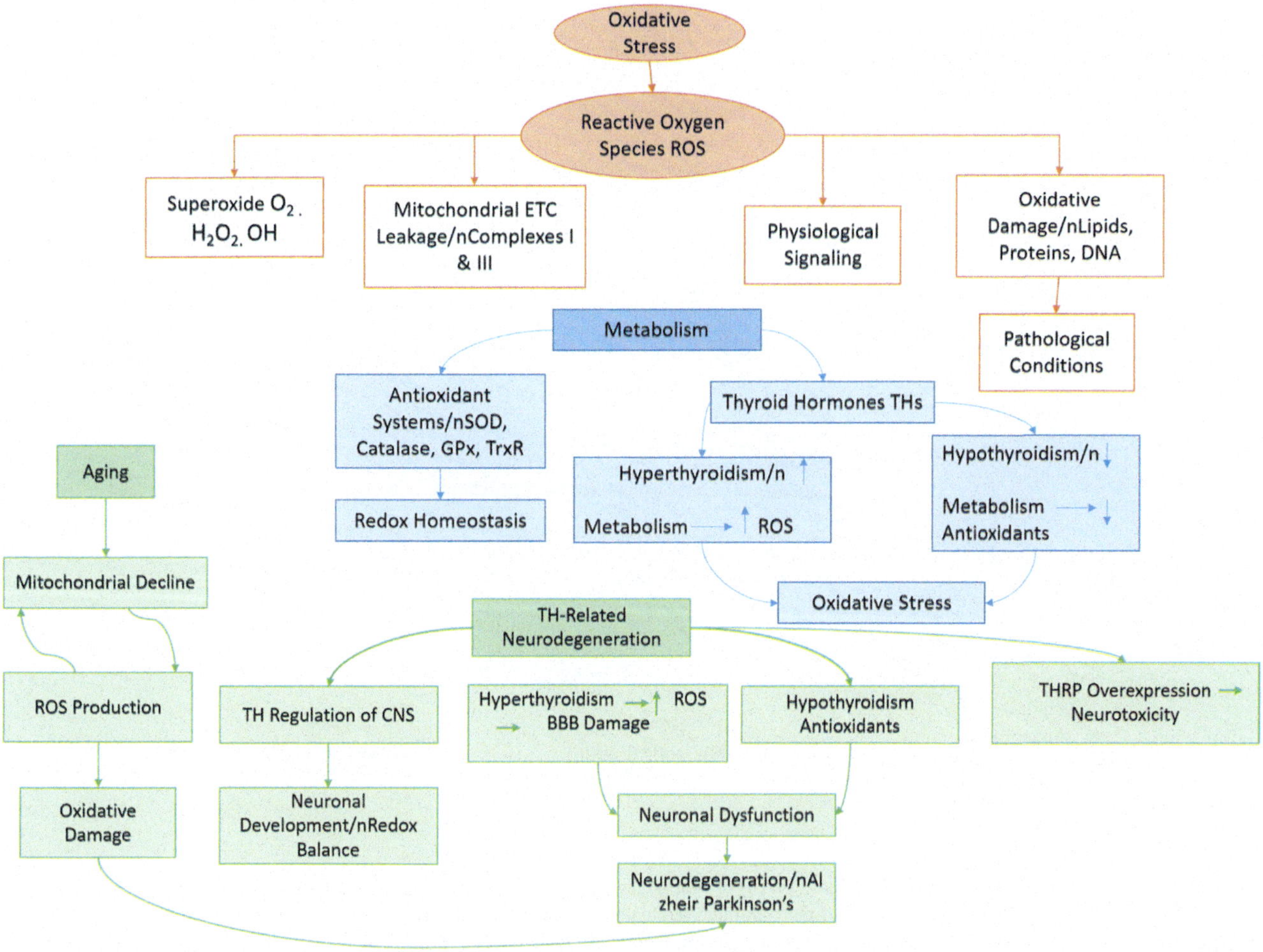

FIGURE 8.2 The interaction between reactive oxygen species (ROS), metabolism, thyroid hormones (THs), aging, and TH-related neurodegeneration is crucial for understanding the pathogenesis of neurodegenerative disorders, such as Alzheimer's disease and Parkinson's disease. ROS, generated by mitochondrial electron transport chain (ETC) leakage, cause oxidative damage, while THs control metabolism and ROS production. Aging-related mitochondrial decline increases ROS production and oxidative damage. *BBB*, blood–brain barrier; *CNS*, central nervous system; *ETC*, electron transport chain; *GPx*, glutathione peroxidase; *H_2O_2*, hydrogen peroxide; *O_2•−*, superoxide anion; *•OH*, hydroxyl radical; *ROS*, reactive oxygen species; *SOD*, superoxide dismutase; *TH*, thyroid hormone; *THRP*, thyroid hormone-responsive protein; *THs*, thyroid hormones; *TrxR*, thioredoxin reductase.

production of new neurons. In conditions of brain injury or ischemia, stimulation of thyroid pathways can help activate repair mechanisms and accelerate neuronal regeneration (Valcárcel-Hernández et al., 2024). Additionally, autoimmune thyroid disorders such as Graves' disease and Hashimoto's thyroiditis have been associated with increased risks of cognitive and psychiatric disorders (Piekiełko-Witkowska et al., 2025). In animal models, activation of thyroid hormone receptor signaling in the hippocampus has been associated with increased neurogenesis and improved spatial memory, suggesting a potential role for TH pathways in brain repair.

Hormone receptors such as thyroid hormone receptors and liver X receptors (LXRs) are involved in regulating brain function and may contribute to neuroprotection. These receptors may contribute to neuroprotective pathways by regulating cholesterol homeostasis and supporting neuronal health. In recent studies, the role of THs-responsive proteins such as THRP in neuronal degeneration has been highlighted, which may act independently of classical pathways such as c-Abl (Mooradian & Haas, 2025). Recent studies have shown that thyroid hormone receptors, which belong to the nuclear receptor family, can interact with inflammation-related transcription factors and may exert anti-inflammatory effects in the brain through inhibition of nuclear factor kappa B (NF-κB) signaling (Salehipour et al., 2023) (Fig. 8.3).

Thyroid function within the euthyroid range and cognitive risk

Cognitive impairment can occur in overt hypothyroidism and may be at least partially reversible with levothyroxine

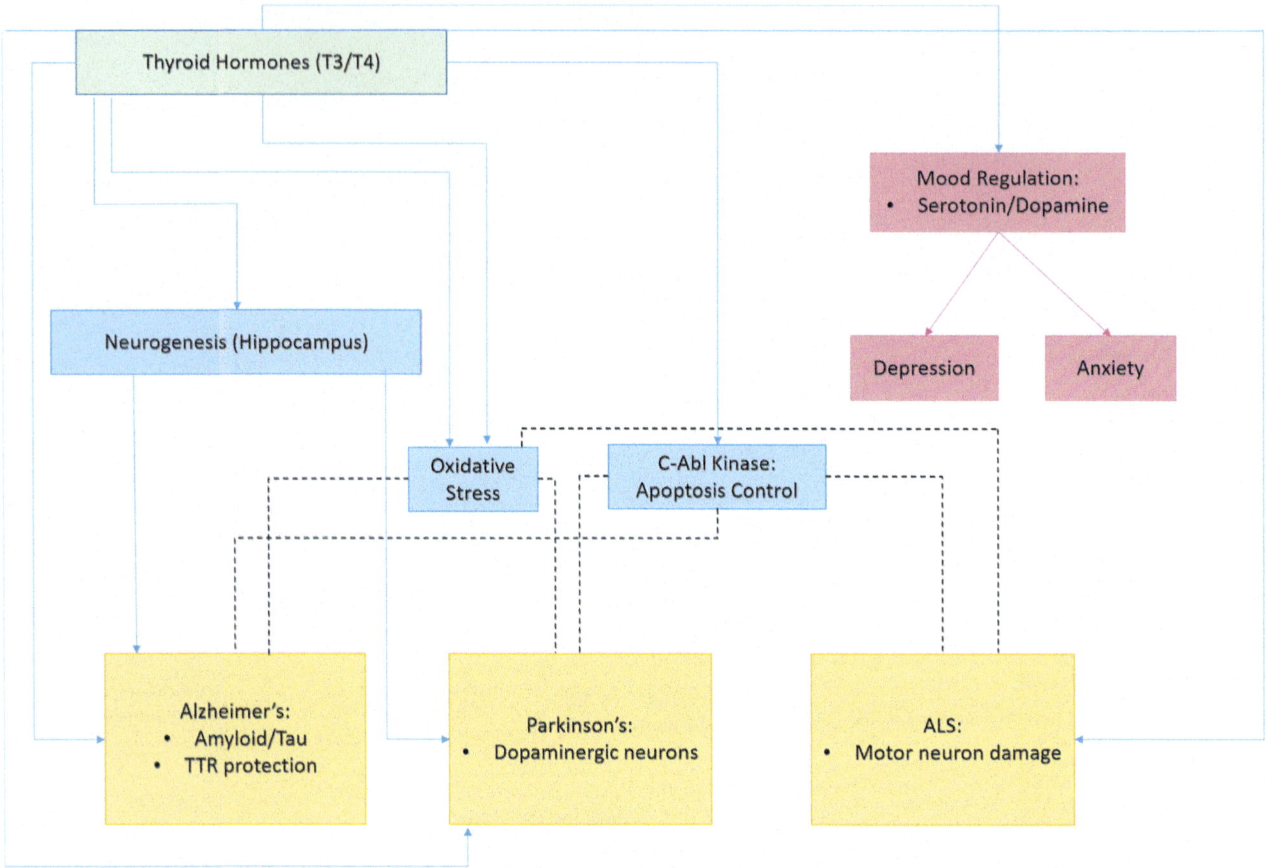

FIGURE 8.3 Multifaceted role of thyroid hormones (T3/T4) in neurological health and disease. The figure illustrates how THs affect brain function, including neurogenesis, as well as the modulation of mood through serotonin and dopamine pathways, and the potential connection to diseases such as Alzheimer's, Parkinson's, and ALS. Dysfunction is linked to neuropsychiatric disorders like depression and anxiety. *ALS*, amyotrophic lateral sclerosis; *c-Abl*, Abelson non-receptor tyrosine kinase; *T3*, triiodothyronine; *T4*, thyroxine; *TTR*, transthyretin.

treatment in selected patients (Mooradian & Haas, 2025; Rieben et al., 2016). Cognitive impairment can occur in overt hypothyroidism and may be at least partially reversible with levothyroxine treatment in selected patients; cognitive impairment has also been reported in hyperthyroidism.

The appropriate reference range for thyroid function remains debated, particularly in older adults. Research has shown that aging is associated with a rightward shift in both median TSH levels and the normal TSH distribution, which may represent an adaptive physiological response (Sinha et al., 2024). Older adults may experience reduced thyroid hormone activity, leading to lower basal metabolic rate. The pituitary gland may increase TSH secretion to compensate, potentially helping to prevent catabolism (Li & Liu, 2024; Mullur et al., 2014).

In the Cardiovascular Health Study, which included 2843 U.S. adults aged 65 years and older, no overall association was found between thyroid function tests, including TSH, free T4, and total T3, and dementia. However, individuals with TSH levels in the highest quartile of the normal range had significantly lower dementia rates than those in the lowest quartile (Cappola et al., 2015). Similarly, the Rotterdam Study, which included 9446 participants, found that higher TSH levels near the upper limit of the reference range were associated with approximately half the risk of developing dementia compared with lower TSH levels (Chaker et al., 2016).

This finding may be explained by an adaptive physiological shift whereby elevated TSH levels help preserve hippocampal volume and synaptic plasticity which are the key factors in maintaining cognitive function as well as decreasing inflammation through modulation of NF-κB pathways and enhancing neuroprotective mechanisms. Additionally, higher TSH might also be associated with lower free T4 levels, which have been associated with less tau hyperphosphorylation and misfolded protein deposition, thus slowing neurodegenerative processes like those seen in Alzheimer's disease (Quinlan et al., 2020). In addition, the Framingham Study reported a U-shaped association between TSH and Alzheimer's disease risk in women, with both low and high TSH tertiles associated with more than twice the risk compared with the middle tertile; this pattern was not observed in men (Tan et al., 2008). In contrast, the Osteoporotic Fractures in Men Study, which followed 539 euthyroid older men, found no significant link between thyroid hormone levels and cognitive function. Although these data suggest a potential protective association between higher-normal TSH levels and dementia risk, further research is needed to determine optimal thyroid function targets in aging populations.

Overt hyperthyroidism and cognitive impairment

Overt hyperthyroidism affects approximately 1%–3% of older adults, primarily due to thyroid hormone medication use. Graves' disease, multinodular goiter, and toxic adenomas are common causes (Samuels, 2000).

Overt hyperthyroidism is linked to a higher risk of all-cause dementia, particularly with prolonged periods of suppressed TSH levels, according to population-based registry cohorts like the Danish National Patient Registry and OPENTHYRO registry (Folkestad et al., 2020).

Thyroid hormones may accelerate ROS production in neurons and suppress antioxidant mechanisms, resulting in oxidative damage to cellular structures, including mitochondria, and potentially contributing to neuronal apoptosis (Villanueva et al., 2013). Elevated serum thyroid hormone levels may accelerate amyloid beta (Aβ)-dependent tau accumulation in the brain. These findings emphasize the potential cognitive risks of overt hyperthyroidism in older adults and highlight the importance of careful management to avoid overtreatment and minimize harm.

Subclinical hyperthyroidism and cognitive impairment

Subclinical hyperthyroidism affects ~3–8% of older adults, with many cases due to thyroid hormone therapy. Several studies have found that subclinical hyperthyroidism is associated with an increased risk of dementia (Aubert et al., 2017; Rieben et al., 2016).

Hypotheses were proposed to explain this relationship. First, given that several thyroid hormone target genes are implicated in neurogenesis, genetic susceptibility may partly explain the association (Kapoor et al., 2015).

Second, reduced brain choline may contribute to cognitive dysfunction; in a pilot study of untreated Graves' disease, the brain choline/creatine ratio decreased and returned to normal after treatment.

On the other hand, a large multicohort analysis of 74,565 adults found no association between subclinical hyperthyroidism and cognitive function (van Vliet et al., 2021). Due to uncertain evidence, further studies with longer follow-up periods and sensitive cognitive assessments are needed to clarify the impact of subclinical hyperthyroidism on cognition.

Overt hypothyroidism and dementia

In the general U.S. population, untreated overt hypothyroidism is estimated to affect approximately 0.33% of individuals, whereas the prevalence of treated hypothyroidism is substantially higher (Diab et al., 2019; Zhang, Zhao et al., 2024).

Studies have shown an association between hypothyroidism and cognitive impairment in older adults, with memory being the primary affected area. Large cohort analyses from the Danish National Patient Registry and OPENTHYRO cohort suggest a slightly increased risk of dementia (Thvilum et al., 2021).

Prolonged untreated thyroid hormone deficiency may contribute to dementia risk through several molecular pathways. By changing the regulation of genes encoding amyloid-β precursor proteins and upsetting the distribution of Aβ, low thyroid hormone levels have been connected to the pathogenesis of Alzheimer's disease. The hippocampus and neurogenesis, which are essential for learning, memory, and overall cognitive function, are also influenced by low thyroid hormone availability. Over time, these changes may increase the risk of dementia and accelerate neurodegenerative processes.

A population-based study in Mexico involving 1750 participants found that overt hypothyroidism was significantly associated with mild cognitive impairment (MCI), as assessed by the Mini-Mental State Examination (MMSE) (Juárez-Cedillo et al., 2017). On the other hand, a meta-analysis of more than one million participants found no significant association between overt hypothyroidism and cognitive impairment or dementia after accounting for vascular comorbidities (Ye et al., 2022). Given the mixed evidence, careful management of hypothyroidism in older adults remains important in clinical practice to avoid both overtreatment and undertreatment and to support cognitive well-being (Table 8.2) (Thvilum et al., 2021).

Subclinical hypothyroidism and dementia

Subclinical hypothyroidism happens more commonly than overt hypothyroidism, with prevalence rates reported between 3.8% and 20% (Diab et al., 2019).

Large multicohort analyses found no consistent relationship between subclinical hypothyroidism and cognitive function (van Vliet et al., 2021). A meta-analysis of 13 studies found an increased risk of cognitive impairment

TABLE 8.2 Thyroid dysfunction links to neurodegenerative/psychiatric disorders and cognitive impairment.

Thyroid state	Associated disorders/risks	Key mechanisms/effects	Treatment considerations and limitations
Hyperthyroidism (Overt/subclinical)	AD (amyloid/tau deposition, memory loss); PD (injury to dopaminergic neurons); anxiety/irritability; excess dementia risk (40%–50% increased).	Elevated ROS, protein misfolding, mitochondrial oxidative damage; disrupts neurotransmitter systems, hippocampal plasticity; promotes Aβ/tau formation.	Antithyroid drugs reduce symptoms, improve synaptic plasticity/memory; do not fully reverse chronic brain damage; avoid overtreatment in elderly.
Hypothyroidism (Overt/subclinical)	AD (loss of executive function/memory, atrophy of the hippocampus); PD/ALS (motor dysfunction, Lewy body dementia); depression (low serotonin/dopamine); risk increased in women.	Reduced antioxidant defense, neurotransmitter disruption (e.g., glutamate/cholinergic), endoplasmic reticulum stress; compromises neurogenesis, increases vulnerability to damage.	Levothyroxine replacement alleviates symptoms, re-establishes circuits; not sufficient for full return of cognition; not recommended solely for cognitive improvement in older adults with subclinical hypothyroidism
Euthyroid range variations	Protection from dementia with increased TSH (upper normal); U-shaped risk in women (high or low TSH increases AD risk).	Maintains hippocampal volume, reduces inflammation (NF-κB modulation), inhibits tau hyperphosphorylation.	Adaptive in aging (with antioxidant protection); further research needed to achieve best TSH targets; no direct correlation in some studies (e.g., in men).
Specific to Alzheimer's disease	Low FT3 levels are associated with an approximately twofold higher AD risk; altered TSH/TRH; hyperthyroidism reduces dendritic spines/LTP; hypothyroidism disrupts glutamate/cholinergic transmission, cerebral perfusion.	GSK-3 activation, tau hyperphosphorylation, Aβ deposition; compensatory upregulation of THR in hippocampus; hypoperfusion promotes pathological proteins.	Hormone replacement/antithyroid treatment improves cognition/synaptic function; may alleviate symptoms but may not fully reverse pre-existing cognitive deficits; targets such as TTR stabilization for prevention.
General neuropsychiatric links	Mood disorders (anxiety/depression); autoimmune thyroid disease, including Graves' disease and Hashimoto's thyroiditis increase cognitive/psychiatric risks; c-Abl-mediated stress and apoptotic signaling may contribute to neuronal injury	Disrupts serotonin/dopamine; inflammation through NF-κB; regenerative actions in hippocampus (new neuron formation).	Personalized treatments targeting TH regulation, sex-specific strategies, potential of THR/LXR agonists in neuroprotection against AD/PD/ALS.

AD, Alzheimer's disease; *ALS*, amyotrophic lateral sclerosis; *Aβ*, amyloid beta; *c-Abl*, Abelson non-receptor tyrosine kinase; *FT3*, free triiodothyronine; *GSK-3*, glycogen synthase kinase 3; *LTP*, long-term potentiation; *LXR*, liver X receptor; *NF-κB*, nuclear factor kappa B; *PD*, Parkinson's disease; *ROS*, reactive oxygen species; *TH*, thyroid hormone; *THR*, thyroid hormone receptor; *TRH*, thyrotropin-releasing hormone; *TSH*, thyroid-stimulating hormone; *TTR*, transthyretin.

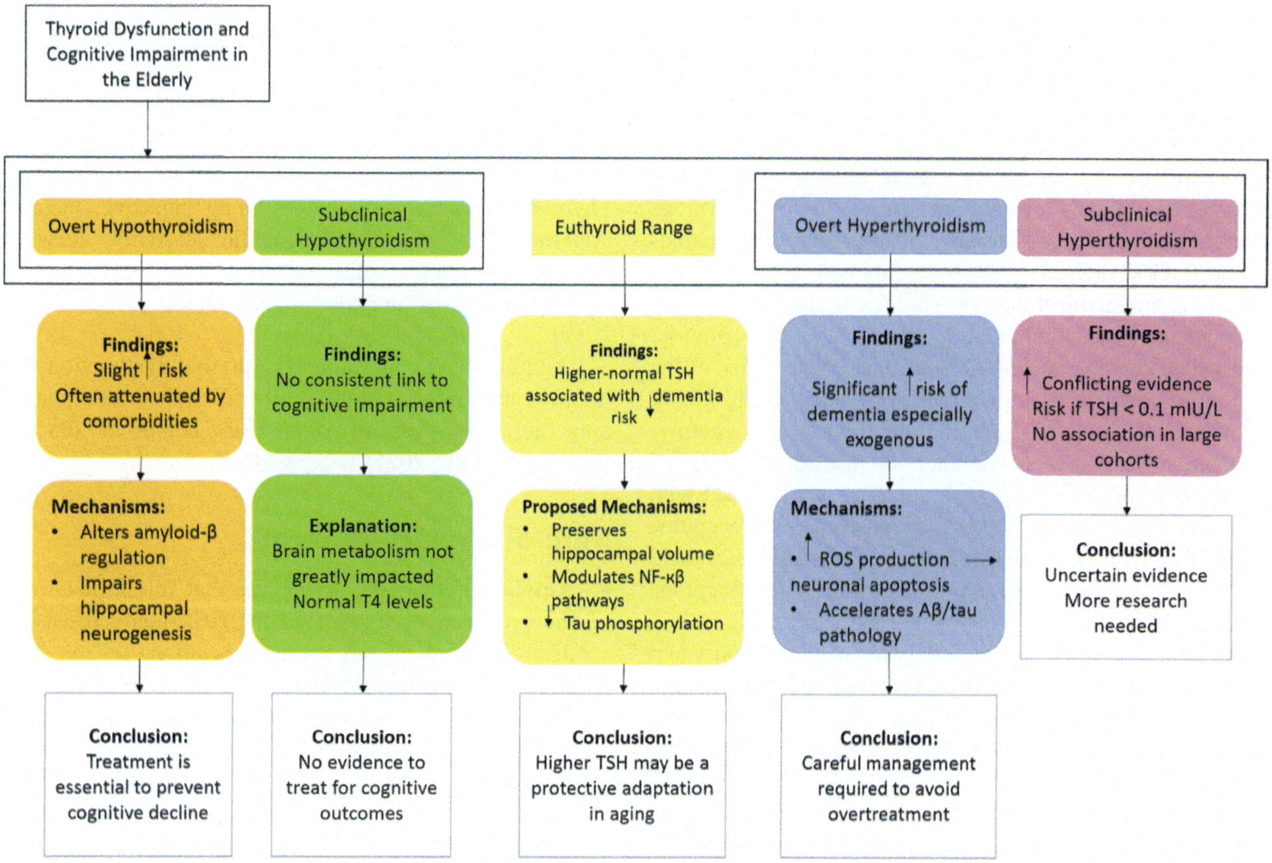

FIGURE 8.4 Schematic flowchart integrating thyroid function and cognitive risk in older adults. Overt thyroid dysfunction may be associated with increased dementia risk through pathways such as oxidative stress and impaired neurogenesis, whereas subclinical thyroid dysfunction has shown inconsistent associations. Borderline-normal TSH levels in the euthyroid range may represent an adaptive protective mechanism of aging. *Aβ*, amyloid beta; *NF-κB*, nuclear factor kappa B; *ROS*, reactive oxygen species; *T4*, thyroxine; *TSH*, thyroid-stimulating hormone.

among individuals younger than 75 years, whereas this association was not consistently observed in older age groups (Pasqualetti et al., 2015). This may be explained by the fact that thyroid hormone levels remain within the reference range and brain metabolism may not be substantially affected. Functional magnetic resonance imaging (fMRI) findings suggest that treatment may normalize specific patterns of brain activity related to working memory and executive function (Zhu et al., 2006).

Functional magnetic resonance imaging (fMRI) findings suggest that treatment may normalize specific patterns of brain activity related to working memory and executive function (Fig. 8.4).

Thyroid dysfunction and Alzheimer's disease

The relationship between thyroid dysfunction and Alzheimer's disease is complex and multifaceted, with multiple pathways and sometimes contradictory results.

A 2006 cohort study by de Jong et al. did not support the proposed association between hyperthyroidism and increased Alzheimer's disease risk, suggesting that thyroid status may have limited influence on AD pathogenesis in that cohort (de Jong et al., 2006). In contrast, Quinlan et al. reported an inverse association between low serum FT3 and Alzheimer's disease incidence, with individuals in the lowest FT3 quartile having more than a twofold higher risk than those in the highest quartile (Quinlan et al., 2019).

Experimental models have been utilized to investigate potential mechanisms. In a rat model of Alzheimer's disease induced by intrahippocampal okadaic acid injection, Ren et al. reported GSK-3 activation, pathological tau hyperphosphorylation, oxidative stress, Aβ neurotoxicity, increased cerebral FT3, FT4, TSH, and TRH levels, and increased hippocampal THR expression compared with controls (Ren et al., 2023). This may indicate a compensatory mechanism or dysregulation of central thyroid hormone pathways in AD pathology.

Other clinical findings also support disrupted thyroid hormone dynamics in AD. Chen et al. indicated that in one study, the circadian rhythm of serum TSH was disrupted and was significantly lower in Alzheimer's disease patients compared to controls (Chen et al., 2013). In addition, with the development of Alzheimer's disease, the brain's ability to transport thyroid hormones may also decrease (Tang et al., 2019). These results indicate that systemic and central nervous system regulation of thyroid hormone in AD is disrupted. The neurological impact of thyroid dysfunction is also supported by studies of thyroid hormone excess and deficiency. Studies reveal that excess thyroid hormone treatment in young rats strongly inhibits long-term potentiation (LTP), damages cognitive function, and produces drastic neurochemical and morphological alterations in the hippocampal formation (Pavlides et al., 1991). Hyperthyroid individuals exhibit decreased functional connectivity of the hippocampus with the cerebral cortex (Zhang et al., 2014). Experimental studies using murine models of hyperthyroidism have shown a reduction in mature hippocampal dendritic spines. The structural alteration is associated with impaired function of the critical postsynaptic receptors, α-amino-3-hydroxy-5-methyl-4-isoxazole propionic acid (AMPAR) and N-methyl-D-aspartate (NMDAR) receptors, resulting in impaired excitatory synaptic transmission and potential defects in cognitive functions such as memory and learning (Zhu et al., 2022). These findings suggest that hyperthyroidism may impair cognitive processes, including memory and learning, although the extent and reversibility of these changes may vary across patients. Further, downregulation of TRH levels suppresses mitogen-activated protein kinase (MAPK) signaling, thereby reducing inhibitory control of GSK-3 activity and promoting tau hyperphosphorylation. Hypothyroidism may also exert harmful effects through distinct mechanisms. Hypothyroidism has also been shown to disrupt glutamate signaling and impair cholinergic neuronal cells, as reflected by reduced acetyltransferase activity and acetylcholine content.

Furthermore, these neurons exhibit aberrant electrophysiological responses, including increased excitation thresholds and depolarized resting membrane potentials (Patel et al., 1987; Wang et al., 2015).

Hypothyroidism can increase oxidative stress via endoplasmic reticulum stress and damage the hippocampus (Torres-Manzo et al., 2018). Downstream consequences may include impaired cerebral perfusion and reduced protein clearance, which can promote aggregation of pathological proteins typical of Alzheimer's disease, including β-amyloid and hyperphosphorylated tau (Fukui et al., 2001; Park et al., 2019).

Notably, these pathways also represent therapeutic targets. Recent neuroimaging studies suggest that antithyroid therapy may be associated with partial recovery of cognitive performance and brain functional connectivity in selected patients with hyperthyroidism-related cognitive impairment; however, these changes may not fully normalize after treatment. In a clinicoradiological case report of hyperthyroid dementia, cognitive improvement after beta-blocker and methimazole therapy was accompanied by gradual improvement of regional tracer-uptake abnormalities on single-photon emission computed tomography (SPECT) findings (Kumar et al., 2022; Liu et al., 2017) https://pubmed.ncbi.nlm.nih.gov/34905237/(14). Effective treatment of hypothyroidism may improve cognitive symptoms in selected patients. Thyroid hormone replacement may improve memory impairment by enhancing synaptic plasticity, reducing neuroinflammation, and restoring key hippocampal signaling pathways in experimental models (Chaalal et al., 2019).

Overall, although treatment of thyroid dysfunction may improve neurological symptoms and clinical outcomes, it may not fully restore premorbid cognitive function (Knopman et al., 2021) (Fig. 8.5).

Mini-dictionary of terms

- **Amyloid-beta (Aβ):** A peptide that forms plaques in the brain and is implicated in Alzheimer's disease, a disorder characterized by Aβ deposition.
- **Blood–Brain Barrier (BBB):** A specialized endothelial barrier that regulates the movement of molecules and cells between the bloodstream and the CNS.
- **Cognitive decline:** Gradual deterioration in cognitive domains such as memory, learning, and executive function, commonly observed in dementia.
- **Free Radicals:** Molecules or atoms with unpaired electrons that can oxidize lipids, proteins, and DNA, contributing to oxidative stress.
- **Glucocorticoids:** Steroid hormones that control metabolism and the response to stress, frequently disrupted in neurodegenerative diseases.
- **Mild Cognitive Impairment (MCI):** Cognitive impairment greater than expected for age and education level but not severe enough to substantially impair daily activities or independent function.
- **Mitochondrial dysfunction:** Impaired mitochondrial energy production in neurons, which may increase vulnerability to cellular dysfunction and death.
- **Neuroinflammation:** The activation of glial cells and the production of pro-inflammatory cytokines in the CNS.
- **Proteostasis:** The balance of protein synthesis, folding, trafficking, and degradation; disruption of proteostasis contributes to protein-aggregation diseases.
- **Synaptic plasticity:** The ability of synapses to strengthen or weaken over time, which is essential for learning and memory.

Summary points

- Thyroid dysfunction has been associated with an increased risk of neurodegenerative outcomes.

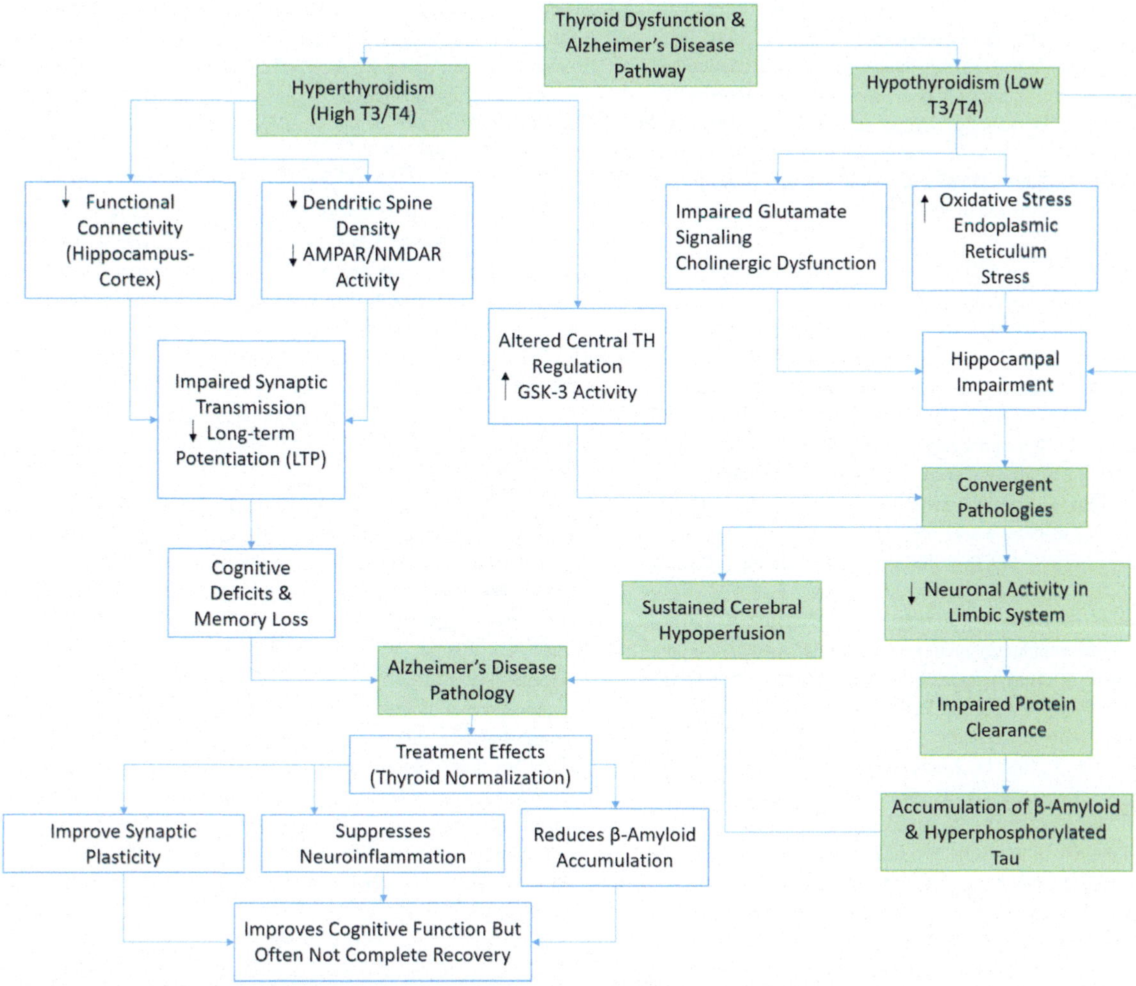

FIGURE 8.5 The diagram illustrates proposed pathways through which thyroid dysfunction may contribute to Alzheimer's disease-related pathology, with hyperthyroidism and hypothyroidism initiating intracellular cascades. Restoring euthyroidism can improve cognitive outcomes, but may not completely restore premorbid cognitive ability. *AMPAR*, α-amino-3-hydroxy-5-methyl-4-isoxazolepropionic acid receptor; *Aβ*, amyloid beta; *GSK-3*, glycogen synthase kinase 3; *LTP*, long-term potentiation; *NMDAR*, N-methyl-D-aspartate receptor; *T3*, triiodothyronine; *T4*, thyroxine; *TH*, thyroid hormone.

- Oxidative stress and mitochondrial dysfunction may link thyroid hormone deficiency to neuronal vulnerability.
- The association between subclinical thyroid dysfunction and subsequent cognitive decline or dementia remains uncertain.
- Thyroid hormone status and AD are related in clinical and cohort studies.
- Parkinson's disease pathology may interact with TH imbalance and oxidative stress pathways.
- Potential therapeutic approaches include antioxidant strategies and careful correction of TH abnormalities.
- Early diagnosis of thyroid disorders may prevent neurodegenerative consequences.
- Advanced methods, including fMRI, mass spectrometry, and transcriptomics, may provide further insight into thyroid–brain interactions.

Disclosure of AI use

During the preparation of this chapter, DeepSeek-R1 was used to improve readability and language. All scientific content, interpretations, and references were reviewed and verified by the authors.

References

AlAnazi, F. H., Al-Kuraishy, H. M., Alexiou, A., Papadakis, M., Ashour, M. H. M., Alnaaim, S. A., Elhussieny, O., Saad, H. M., & Batiha, G. E. (2023). Primary hypothyroidism and alzheimer's disease: A tale of two. *Cellular and Molecular Neurobiology, 43*(7), 3405–3416. https://doi.org/10.1007/s10571-023-01392-y

Alcaide Martin, A., & Mayerl, S. (2023). Local thyroid hormone action in brain development. *International Journal of Molecular Sciences, 24* (15), Article 12352. https://www.mdpi.com/1422-0067/24/15/12352.

Aubert, C. E., Bauer, D. C., da Costa, B. R., Feller, M., Rieben, C., Simonsick, E. M., Yaffe, K., Rodondi, N., & Health, A. B. C. S. (2017). The association between subclinical thyroid dysfunction and dementia: The health, aging and body composition (health ABC) study. *Clinical Endocrinology, 87*(5), 617–626. https://doi.org/10.1111/cen.13458

Baksi, S., & Pradhan, A. (2021). Thyroid hormone: Sex-dependent role in nervous system regulation and disease. *Biology of Sex Differences, 12*(1), 25.

Braun, D., & Schweizer, U. (2018). Thyroid hormone transport and transporters. *Vitamins and hormones, 106*, 19–44.

Byeon, J. H., Byun, M. S., Yi, D., Jung, J. H., Sohn, B. K., Chang, Y. Y., Kong, N., Jung, G., Ahn, H., Lee, J. Y., Lee, Y. S., Kim, Y. K., & Lee, D. Y. (2024). Moderation of thyroid hormones for the relationship between amyloid and tau pathology. *Alzheimer's Research & Therapy, 16*(1), 164. https://doi.org/10.1186/s13195-024-01534-4

Cappola, A. R., Arnold, A. M., Wulczyn, K., Carlson, M., Robbins, J., & Psaty, B. M. (2015). Thyroid function in the euthyroid range and adverse outcomes in older adults. *The Journal of Clinical Endocrinology & Metabolism, 100*(3), 1088–1096. https://doi.org/10.1210/jc.2014-3586

Chaalal, A., Poirier, R., Blum, D., Laroche, S., & Enderlin, V. (2019). Thyroid hormone supplementation restores spatial memory, hippocampal markers of neuroinflammation, plasticity-related signaling molecules, and β-amyloid peptide load in hypothyroid rats. *Molecular Neurobiology, 56*(1), 722–735.

Chaker, L., Wolters, F. J., Bos, D., Korevaar, T. I. M., Hofman, A., Van Der Lugt, A., Koudstaal, P. J., Franco, O. H., Dehghan, A., Vernooij, M. W., Peeters, R. P., & Ikram, M. A. (2016). Thyroid function and the risk of dementia: The rotterdam study. *Neurology, 87*(16), 1688–1695. https://doi.org/10.1212/WNL.0000000000003227

Charoenngam, N., Rittiphairoj, T., Ponvilawan, B., & Prasongdee, K. (2022). Thyroid dysfunction and risk of parkinson's disease: A systematic review and meta-analysis [systematic review]. *Frontiers in Endocrinology, 13*. https://doi.org/10.3389/fendo.2022.863281

Chatonnet, F., Guyot, R., Benoît, G., & Flamant, F. (2013). Genome-wide analysis of thyroid hormone receptors shared and specific functions in neural cells. *Proceedings of the National Academy of Sciences, 110*(8), E766–E775.

Chen, J.-M., Huang, C.-Q., Ai, M., & Kuang, L. (2013). Circadian rhythm of TSH levels in subjects with alzheimer's disease (AD). *Aging Clinical and Experimental Research, 25*(2), 153–157.

Cheng, S.-Y., Leonard, J. L., & Davis, P. J. (2010). Molecular aspects of thyroid hormone actions. *Endocrine Reviews, 31*(2), 139–170.

Corino, C., Aimo, A., Luigetti, M., Ciccone, L., Ferrari Chen, Y. F., Panichella, G., Musetti, V., Castiglione, V., Vergaro, G., Emdin, M., & Franzini, M. (2025). Tetrameric transthyretin as a protective factor against alzheimer's disease. *Molecular Neurobiology, 62*(3), 2945–2954. https://doi.org/10.1007/s12035-024-04442-8

de Jong, F. J., den Heijer, T., Visser, T. J., de Rijke, Y. B., Drexhage, H. A., Hofman, A., & Breteler, M. M. (2006). Thyroid hormones, dementia, and atrophy of the medial temporal lobe. *The Journal of Clinical Endocrinology & Metabolism, 91*(7), 2569–2573.

de Souza, J. S. (2024). Chapter ten - thyroid hormone biosynthesis and its role in brain development and maintenance. In R. Donev (Ed.), *Advances in protein chemistry and structural biology* (Vol. 142., pp. 329–365). Academic Press. https://www.sciencedirect.com/science/article/pii/S1876162323001360.

Dentice, M., Bandyopadhyay, A., Gereben, B., Callebaut, I., Christoffolete, M. A., Kim, B. W., Nissim, S., Mornon, J.-P., Zavacki, A. M., & Zeöld, A. (2005). The hedgehog-inducible ubiquitin ligase subunit WSB-1 modulates thyroid hormone activation and PTHrP secretion in the developing growth plate. *Nature Cell Biology, 7*(7), 698–705.

Dentice, M., Luongo, C., Huang, S., Ambrosio, R., Elefante, A., Mirebeau-Prunier, D., Zavacki, A. M., Fenzi, G., Grachtchouk, M., & Hutchin, M. (2007). Sonic hedgehog-induced type 3 deiodinase blocks thyroid hormone action enhancing proliferation of normal and malignant keratinocytes. *Proceedings of the National Academy of Sciences, 104*(36), 14466–14471.

Diab, N., Daya, N. R., Juraschek, S. P., Martin, S. S., McEvoy, J. W., Schultheiß, U. T., Köttgen, A., & Selvin, E. (2019). Prevalence and risk factors of thyroid dysfunction in older adults in the community. *Scientific Reports, 9*(1), Article 13156. https://doi.org/10.1038/s41598-019-49540-z

Dumitrescu, A. M., Liao, X.-H., Best, T. B., Brockmann, K., & Refetoff, S. (2004). A novel syndrome combining thyroid and neurological abnormalities is associated with mutations in a monocarboxylate transporter gene. *The American Journal of Human Genetics, 74*(1), 168–175.

Folkestad, L., Brandt, F., Lillevang-Johansen, M., Brix, T. H., & Hegedüs, L. (2020). Graves' disease and toxic nodular goiter, aggravated by duration of hyperthyroidism, are associated with alzheimer's and vascular dementia: A registry-based long-term follow-up of two large cohorts. *Thyroid: Official Journal of the American Thyroid Association, 30*(5), 672–680. https://doi.org/10.1089/thy.2019.0672

Fukui, T., Hasegawa, Y., & Takenaka, H. (2001). Hyperthyroid dementia: Clinicoradiological findings and response to treatment. *Journal of the Neurological Sciences, 184*(1), 81–88.

Głombik, K., Detka, J., & Budziszewska, B. (2021). Hormonal regulation of oxidative phosphorylation in the brain in health and disease. *Cells, 10*(11), 2937. https://www.mdpi.com/2073-4409/10/11/2937.

Gil-Ibañez, P., Morte, B., & Bernal, J. (2013). Role of thyroid hormone receptor subtypes α and β on gene expression in the cerebral cortex and striatum of postnatal mice. *Endocrinology, 154*(5), 1940–1947.

Hiroi, Y., Kim, H.-H., Ying, H., Furuya, F., Huang, Z., Simoncini, T., Noma, K., Ueki, K., Nguyen, N.-H., & Scanlan, T. S. (2006). Rapid nongenomic actions of thyroid hormone. *Proceedings of the National Academy of Sciences, 103*(38), 14104–14109.

Juárez-Cedillo, T., Basurto-Acevedo, L., Vega-García, S., Sánchez-Rodríguez Martha, A., Retana-Ugalde, R., Juárez-Cedillo, E., Gonzalez-Melendez Roberto, C., & Escobedo-de-la-Peña, J. (2017). Prevalence of thyroid dysfunction and its impact on cognition in older Mexican adults: (SADEM study). *Journal of Endocrinological Investigation, 40*(9), 945–952. https://doi.org/10.1007/s40618-017-0654-6

Jurado-Flores, M., Warda, F., & Mooradian, A. (2022). Pathophysiology and clinical features of neuropsychiatric manifestations of thyroid disease. *Journal of the Endocrine Society, 6*(2), Article bvab194. https://doi.org/10.1210/jendso/bvab194

Kapoor, R., Fanibunda, S. E., Desouza, L. A., Guha, S. K., & Vaidya, V. A. (2015). Perspectives on thyroid hormone action in adult neurogenesis. *Journal of Neurochemistry, 133*(5), 599–616. https://doi.org/10.1111/jnc.13093

Kim, J. H., Lee, H. S., Kim, Y. H., Kwon, M. J., Kim, J.-H., Min, C. Y., Yoo, D. M., & Choi, H. G. (2022). The association between thyroid diseases and alzheimer's disease in a national health screening cohort in Korea. *Frontiers in Endocrinology, 13*, Article 815063. https://doi.org/10.3389/fendo.2022.815063

Knopman, D. S., Amieva, H., Petersen, R. C., Chételat, G., Holtzman, D. M., Hyman, B. T., Nixon, R. A., & Jones, D. T. (2021). Alzheimer disease. *Nature Reviews Disease Primers, 7*(1), 33.

Kumar, M., Singh, S., Rana, P., Modi, S., Sekhri, T., Kanwar, R., … Khushu, S. (2022). Brain functional connectivity in patients with hyperthyroidism after anti-thyroid treatment. *Journal of Neuroendocrinology, 34*(1), Article e13075.

Li, Z., & Liu, J. (2024). Thyroid dysfunction and alzheimer's disease, a vicious circle. *Frontiers in Endocrinology, 15*. https://doi.org/10.3389/fendo.2024.1354372 [Review].

Liu, B., Ran, Q., Liu, D., Zhang, S., & Zhang, D. (2017). Changes in resting-state cerebral activity in patients with hyperthyroidism: a short-term follow-up functional MR imaging study. *Scientific reports, 7*(1), 10627.

Macvanin, M. T., Gluvic, Z., Zafirovic, S., Gao, X., Essack, M., & Isenovic, E. R. (2023). The protective role of nutritional antioxidants against oxidative stress in thyroid disorders. *Frontiers in Endocrinology, 13*(2022). https://doi.org/10.3389/fendo.2022.1092837 [Review].

Marcocci, C., Leo, M., & Altea, M. A. (2012). Oxidative stress in graves' disease. *European Thyroid Journal, 1*(2), 80–87. https://doi.org/10.1159/000337976

Mintziori, G., Veneti, S., Poppe, K., Goulis, D. G., Armeni, E., Erel, C. T., Fistonić, I., Hillard, T., Hirschberg, A. L., Meczekalski, B., Mendoza, N., Mueck, A. O., Simoncini, T., Stute, P., van Dijken, D., Rees, M., Duntas, L., & Lambrinoudaki, I. (2024). EMAS position statement: Thyroid disease and menopause. *Maturitas, 185*, Article 107991. https://doi.org/10.1016/j.maturitas.2024.107991

Mooradian, A. D., & Haas, M. J. (2025). Role of thyroid hormone in neurodegenerative disorders of older people. *Cells, 14*(2). https://doi.org/10.3390/cells14020140

Motaln, H., & Rogelj, B. (2023). The role of c-Abl tyrosine kinase in brain and its pathologies. *Cells, 12*(16), 2041. https://www.mdpi.com/2073-4409/12/16/2041.

Mullur, R., Liu, Y.-Y., & Brent, G. A. (2014). Thyroid hormone regulation of metabolism. *Physiological Reviews, 94*(2), 355–382. https://doi.org/10.1152/physrev.00030.2013

Ortiga-Carvalho, T. M., Sidhaye, A. R., & Wondisford, F. E. (2014). Thyroid hormone receptors and resistance to thyroid hormone disorders. *Nature Reviews Endocrinology, 10*(10), 582–591.

Park, J.-H., Hong, J.-H., Lee, S.-W., Ji, H. D., Jung, J.-A., Yoon, K.-W., Lee, J.-I., Won, K. S., Song, B.-I., & Kim, H. W. (2019). The effect of chronic cerebral hypoperfusion on the pathology of alzheimer's disease: A positron emission tomography study in rats. *Scientific Reports, 9*(1), Article 14102.

Pasqualetti, G., Pagano, G., Rengo, G., Ferrara, N., & Monzani, F. (2015). Subclinical hypothyroidism and cognitive impairment: systematic review and meta-analysis. *The Journal of Clinical Endocrinology & Metabolism, 100*(11), 4240–4248.

Patel, A. J., Hayashi, M., & Hunt, A. (1987). Selective persistent reduction in choline acetyltransferase activity in basal forebrain of the rat after thyroid deficiency during early life. *Brain Research, 422*(1), 182–185.

Pavlides, C., Westlind-Danielsson, A., Nyborg, H., & McEwen, B. (1991). Neonatal hyperthyroidism disrupts hippocampal LTP and spatial learning. *Experimental Brain Research, 85*(3), 559–564.

Peng, Y., Zhu, L., Bai, Q., Wang, L., & Li, Q. (2024). Associations between thyroid hormones and cognitive impairment in patients with parkinson's disease. *eNeuro, 11*(10). https://doi.org/10.1523/eneuro.0239-24.2024

Phan, T. Q., Jow, M. M., & Privalsky, M. L. (2010). DNA recognition by thyroid hormone and retinoic acid receptors: 3, 4, 5 rule modified. *Molecular and Cellular Endocrinology, 319*(1–2), 88–98.

Piekiełko-Witkowska, A., Duda, M. K., Bogusławska, J., & Mackiewicz, U. (2025). The impact of autoimmune thyroid disease on cognitive and psychiatric disorders: Focus on clinical, pre-clinical and molecular studies. *European Thyroid Journal, 14*(3). https://doi.org/10.1530/etj-24-0406

Quinlan, P., Horvath, A., Eckerström, C., Wallin, A., & Svensson, J. (2020). Altered thyroid hormone profile in patients with alzheimer's disease. *Psychoneuroendocrinology, 121*, Article 104844. https://doi.org/10.1016/j.psyneuen.2020.104844

Quinlan, P., Horvath, A., Wallin, A., & Svensson, J. (2019). Low serum concentration of free triiodothyronine (FT3) is associated with increased risk of alzheimer's disease. *Psychoneuroendocrinology, 99*, 112–119.

Ren, B., Ma, J., Tao, M., Jing, G., Han, S., Zhou, C., Wang, X., & Wang, J. (2023). The disturbance of thyroid-associated hormone and its receptors in brain and blood circulation existed in the early stage of mouse model of alzheimer's disease. *Aging (Albany NY), 15*(5), 1591.

Rieben, C., Segna, D., da Costa, B. R., Collet, T.-H., Chaker, L., Aubert, C. E., Baumgartner, C., Almeida, O. P., Hogervorst, E., Trompet, S., Masaki, K., Mooijaart, S. P., Gussekloo, J., Peeters, R. P., Bauer, D. C., Aujesky, D., & Rodondi, N. (2016). Subclinical thyroid dysfunction and the risk of cognitive decline: A meta-analysis of prospective cohort studies. *The Journal of Clinical Endocrinology and Metabolism, 101*(12), 4945–4954. https://doi.org/10.1210/jc.2016-2129

Roberts, L. M., Woodford, K., Zhou, M., Black, D. S., Haggerty, J. E., Tate, E. H., Grindstaff, K. K., Mengesha, W., Raman, C., & Zerangue, N. (2008). Expression of the thyroid hormone transporters monocarboxylate transporter-8 (SLC16A2) and organic ion transporter-14 (SLCO1C1) at the blood-brain barrier. *Endocrinology, 149*(12), 6251–6261.

Salehipour, A., Dolatshahi, M., Haghshomar, M., & Amin, J. (2023). The role of thyroid dysfunction in alzheimer's disease: A systematic review and meta-analysis. *The journal of prevention of Alzheimer's disease, 10*(2), 276–286. https://doi.org/10.14283/jpad.2023.20

Samuels, M. H. (2000). Hyperthyroidism in aging. In K. R. Feingold, S. F. Ahmed, B. Anawalt, M. R. Blackman, A. Boyce, G. Chrousos, E. Corpas, W. W. de Herder, K. Dhatariya, K. Dungan, J. Hofland, S. Kalra, G. Kaltsas, N. Kapoor, C. Koch, P. Kopp, M. Korbonits, C. S. Kovacs, W. Kuohung, et al.D. P. Wilson (Eds.), *Endotext*. MDText.com, Inc. http://www.ncbi.nlm.nih.gov/books/NBK278986/.

Samuels, M. H. (2014). Thyroid disease and cognition. *Endocrinology and Metabolism Clinics of North America, 43*(2), 529–543. https://doi.org/10.1016/j.ecl.2014.02.006

Santhanam, P., Nath, T., Lindquist, M. A., & Cooper, D. S. (2022). Relationship between TSH levels and cognition in the young adult: An analysis of the human connectome project data. *The Journal of Clinical Endocrinology and Metabolism, 107*(7), 1897–1905. https://doi.org/10.1210/clinem/dgac189

Sies, H., Belousov, V. V., Chandel, N. S., Davies, M. J., Jones, D. P., Mann, G. E., Murphy, M. P., Yamamoto, M., & Winterbourn, C. (2022). Defining roles of specific reactive oxygen species (ROS) in cell biology and physiology. *Nature Reviews Molecular Cell Biology, 23*(7), 499–515. https://doi.org/10.1038/s41580-022-00456-z

Sinha, S. H., Zietlow, K., & Papaleontiou, M. (2024). Thyroid function and cognitive decline: A narrative review. *Endocrine Practice, 30* (11), 1113–1118. https://doi.org/10.1016/j.eprac.2024.07.013

Tan, Z. S., Beiser, A., Vasan, R. S., Au, R., Auerbach, S., Kiel, D. P., Wolf, P. A., & Seshadri, S. (2008). Thyroid function and the risk of alzheimer disease: The framingham study. *Archives of Internal Medicine, 168*(14), 1514–1520. https://doi.org/10.1001/archinte.168.14.1514

Tang, X., Li, Z., Zhang, W., & Yao, Z. (2019). Nitric oxide might be an inducing factor in cognitive impairment in alzheimer's disease via downregulating the monocarboxylate transporter 1. *Nitric Oxide, 91*, 35–41.

Thvilum, M., Brandt, F., Lillevang-Johansen, M., Folkestad, L., Brix, T. H., & Hegedüs, L. (2021). Increased risk of dementia in hypothyroidism: A danish nationwide register-based study. *Clinical Endocrinology, 94*(6), 1017–1024. https://doi.org/10.1111/cen.14424

, Thyroid hormones are associated with longitudinal cognitive change in an urban adult population - PubMed. https://pubmed.ncbi.nlm.nih.gov/26329688/.

Toraih, E. A., Siddiqui, S., Siddiqui, S., Shirini, K., Elfezzani, N., Abdelmaksoud, A., Elshazli, R. M., Hussein, M. H., Elmorsy, E. M., & Fawzy, M. S. (2025). Thyroid disorders as a risk factor for neurodegenerative proteinopathies: A large-scale propensity score-matched analysis. *Neuroepidemiology, 1–13*. https://doi.org/10.1159/000545369

Torres-Manzo, A. P., Franco-Colín, M., Blas-Valdivia, V., Pineda-Reynoso, M., & Cano-Europa, E. (2018). Hypothyroidism causes endoplasmic reticulum stress in adult rat hippocampus: A mechanism associated with hippocampal damage. *Oxidative Medicine and Cellular Longevity, 2018*(1), Article 2089404.

Valcárcel-Hernández, V., Mayerl, S., Guadaño-Ferraz, A., & Remaud, S. (2024). Thyroid hormone action in adult neurogliogenic niches: The known and unknown. *Frontiers in Endocrinology, 15*, Article 1347802. https://doi.org/10.3389/fendo.2024.1347802

van Vliet, N. A., van Heemst, D., Almeida, O. P., Åsvold, B. O., Aubert, C. E., Bae, J. B., Barnes, L. E., Bauer, D. C., Blauw, G. J., Brayne, C., Cappola, A. R., Ceresini, G., Comijs, H. C., Dartigues, J.-F., Degryse, J.-M., Dullaart, R. P. F., van Eersel, M. E. A., den Elzen, W. P. J., Ferrucci, L., & Thyroid Studies, C. (2021). Association of thyroid dysfunction with cognitive function: An individual participant data analysis. *JAMA Internal Medicine, 181*(11), 1440–1450. https://doi.org/10.1001/jamainternmed.2021.5078

Villanueva, I., Alva-Sánchez, C., & Pacheco-Rosado, J. (2013). The role of thyroid hormones as inductors of oxidative stress and neurodegeneration. *Oxidative Medicine and Cellular Longevity, 2013*, Article 218145. https://doi.org/10.1155/2013/218145

Wang, F., Zeng, X., Zhu, Y., Ning, D., Liu, J., Liu, C., Jia, X., & Zhu, D. (2015). Effects of thyroxine and donepezil on hippocampal acetylcholine content, acetylcholinesterase activity, synaptotagmin-1 and SNAP-25 expression in hypothyroid adult rats. *Molecular Medicine Reports, 11*(2), 775–782.

Ye, Y., Wang, Y., Li, S., Guo, J., Ding, L., & Liu, M. (2022). Association of hypothyroidism and the risk of cognitive dysfunction: A meta-analysis. *Journal of Clinical Medicine, 11*(22), 6726. https://doi.org/10.3390/jcm11226726

Zeng, Y., Cao, S., & Yang, H. (2024). No causal relationship between thyroid function and parkinson's disease: A bidirectional mendelian randomization study. *Neurological Sciences, 45*(4), 1481–1487. https://doi.org/10.1007/s10072-023-07176-y

Zhang, W., Liu, X., Zhang, Y., Song, L., Hou, J., Chen, B., He, M., Cai, P., & Lii, H. (2014). Disrupted functional connectivity of the hippocampus in patients with hyperthyroidism: Evidence from resting-state fMRI. *European journal of Radiology, 83*(10), 1907–1913.

Zhang, X., Wang, Y., Wang, H., & Zhang, X. (2024). Trends in prevalence of thyroid dysfunction and its associations with mortality among US participants, 1988-2012. *The Journal of Clinical Endocrinology and Metabolism, 109*(2), e657–e666. https://doi.org/10.1210/clinem/dgad558

Zhang, T., Zhao, L., Chen, C., Yang, C., Zhang, H., Su, W., Cao, J., Shi, Q., & Tian, L. (2024). Structural and functional alterations of hippocampal subfields in patients with adult-onset primary hypothyroidism. *The Journal of Clinical Endocrinology and Metabolism, 109*(7), 1707–1717. https://doi.org/10.1210/clinem/dgae070

Zhu, D.-F., Wang, Z.-X., Zhang, D.-R., Pan, Z.-L., He, S., Hu, X.-P., Chen, X.-C., & Zhou, J.-N. (2006). fMRI revealed neural substrate for reversible working memory dysfunction in subclinical hypothyroidism. *Brain: A Journal of Neurology, 129*(Pt 11), 2923–2930. https://doi.org/10.1093/brain/awl215

Zhu, W., Wu, F., Li, J., Meng, L., Zhang, W., Zhang, H., Cha, S., Zhang, J., & Guo, G. (2022). Impaired learning and memory generated by hyperthyroidism is rescued by restoration of AMPA and NMDA receptors function. *Neurobiology of Disease, 171*, Article 105807.

Chapter 9

Thyroid hormones and Alzheimer's disease

Rakshit R. Sangolli, Ruturaj A. Kadu and Girdhari Lal Gupta
School of Pharmacy & Technology Management, SVKM's NMIMS Deemed-to-be University, Shirpur, Maharashtra, India

List of abbreviations

3xTg-AD Triple transgenic Alzheimer's Disease mouse model
AD Alzheimer's Disease
APP Amyloid Precursor Protein
APPswe Amyloid Precursor Protein—Swedish mutation
Aβ Amyloid-beta
CSF Cerebrospinal Fluid
FT_3 Free Triiodothyronine
FT_4 Free Thyroxine
GR Glucocorticoid Receptor
MRI Magnetic Resonance Imaging
NFTs Neurofibrillary Tangles
OA Okadaic Acid
PDAPP Transgenic mouse model expressing human APP
PS1 Presenilin one
PS1m146v Presenilin 1, mutant variant M146V
SERT Serotonin Transporter
SOD2 Superoxide Dismutase 2
T_3 Triiodothyronine
T_4 Thyroxine
TH Thyroid Hormone
THRs Thyroid Hormone Receptors
TSH Thyroid Stimulating Hormone
TT_4 Total Thyroxine

Introduction

Overview of Alzheimer's disease

Alzheimer's disease (AD) is a progressive neurodegenerative disorder marked by the buildup of abnormal protein aggregates, such as intracellular neurofibrillary tangles made of hyperphosphorylated tau and extracellular β-amyloid (Aβ) plaques. The most prevalent clinical manifestation is amnestic cognitive impairment, usually characterized by short-term memory loss. Nonetheless, additional cognitive domains are similarly impacted, including executive function (cognitive flexibility), expressive language, and visuospatial processing. Non-amnestic variants, while less prevalent, are acknowledged.

Most instances of Alzheimer's Disease (AD) are sporadic, rather than directly inherited, indicating a complex interaction of genetic and environmental influences. Pathologically, amyloid plaques build up outside cells, and tau tangles build up inside cells. These protein clumps mess up how neurons work, which causes cells to die and the brain to shrink, especially in areas related to memory like the hippocampus. The pathological process commences years prior to the manifestation of clinical symptoms and is characterized by significant decreases in essential neurotransmitters, particularly acetylcholine, which is vital for neuronal communication (Knopman et al., 2021).

The precise cause of AD remains elusive. Among the leading hypotheses, the cholinergic hypothesis posits that deficits in cholinergic neurotransmission are a major driver of disease progression, while the amyloid cascade hypothesis attributes disease onset to alterations in amyloid precursor protein metabolism and Aβ accumulation, triggering subsequent neurodegenerative changes (Breijyeh & Karaman, 2020).

Overview of thyroid hormones

Thyroid hormones (THs) are essential regulators of various physiological processes, encompassing growth, differentiation, and development in nearly all tissues. They affect how cells invade, survive, differentiate, and grow in neoplasms through complicated mechanisms specific to each cell type. These effects encompass angiogenesis, invasiveness, and the modulation of transcription factors, facilitated through genomic and nongenomic pathways. The correlation between TH status and the etiology and prognosis of human neoplasia highlights their essential role in tumor biology (Goemann et al., 2017).

Hormones in Neuroscience. https://doi.org/10.1016/B978-0-443-30261-9.00009-7

The main thyroid hormones, 3,5,3′,5′-tetraiodo-L -thyronine (T_4) and its active form, 3,5,3′-triiodo-L-thyronine (T_3), are important for controlling metabolism. Derivatives like 3,5-diiodo-L-thyronine (T_2) show similar benefits for metabolic health and lipid metabolism, usually with few negative effects. The thyroid gland makes and releases most of the T_4 hormone, which is mostly a prohormone. T_3, on the other hand, has most of the effects on cells. This chapter examines the synthesis, transport, and cellular uptake of thyroid hormones (THs), their metabolic pathways, and the receptor-mediated mechanisms that regulate gene expression. It also looks into TH-related disorders and drug treatments that try to change how the thyroid works (Petito et al., 2023).

Rationale for linking thyroid dysfunction with AD

Hypothyroidism and hyperthyroidism have been associated with the pathophysiology of Alzheimer's disease (AD) via various mechanisms, including neuronal apoptosis, compromised synaptic plasticity, oxidative stress, and neurochemical imbalances. The precise cause-and-effect relationship is still uncertain, but evidence indicates that thyroid dysfunction may be a notable modifiable risk factor for cognitive decline (Fig. 9.1).

AD, or late-onset dementia, is a progressive neurodegenerative disorder of the central nervous system (CNS). Its multifactorial etiology encompasses a complex interplay of physiological and genetic factors. Possible contributing factors encompass thyroid hormone imbalance, vitamin D deficiency, modifications in sex hormones, mitochondrial dysfunction, oxidative stress, and genetic predispositions. These factors can operate independently or synergistically to affect the initiation and advancement of disease.

The neuropathological characteristics of Alzheimer's Disease (AD)—intracellular neurofibrillary tangles consisting of hyperphosphorylated tau and extracellular senile plaques resulting from β-amyloid (Aβ) deposition—are extensively documented in the literature. Moreover, glial cell activation and enlarged endosomes are commonly noted, indicating extensive neuroinflammation and impaired intracellular transport. These mechanisms collectively justify the investigation of thyroid dysfunction as both a contributing factor and a potential therapeutic target in Alzheimer's disease (Li & Liu, 2024).

Thyroid hormones: Biosynthesis, regulation, and function

Thyroid hormones, namely thyroxine (T_4) and triiodothyronine (T_3), play a fundamental role in regulating various physiological systems' metabolic, growth, and developmental aspects. Thyroid-stimulating hormone (TSH) is synthesized and secreted from the anterior lobe of the pituitary gland. TSH stimulates the thyroid gland to produce T_4 and T_3, which feedback to regulate the production of TSH. The complex regulation of the thyroid axis is important for achieving metabolic homeostasis.

Synthesis and secretion (T_3, T_4)

Iodide uptake and transport

Iodide (I^-) is crucial for producing thyroid hormones and is actively accumulated from the blood serum by thyrocytes:

- **Iodide Basolateral Transport by NIS**

The sodium/iodide symporter (NIS) is a basolateral membrane protein that transports I^- against its electrochemical gradient by coupling with Na^+ (specifically, $2\,Na^+ + I^-$). The Na^+ gradient, essential for I^- transport by the NIS protein, is maintained by Na^+/K^+-ATPase. TSH can strongly induce NIS gene expression and trafficking to the membrane utilizing the cAMP/PKA pathway to increase NIS function. Voltage-gated K^+ channels (KCNQ1/KCNE2) operate to stabilize the membrane potential, allowing I^- to be passively taken up. Mutations in NIS can also result in congenital hypothyroidism and dyshormonogenic goiter.

- **Iodide Apical Transport to the Follicular Lumen**

Iodide is transported from the thyrocytes into the colloid via apical transport proteins. Pendrin is a pendrin (SLC26A4) Cl^-/I^- exchanger mutated in Pendred syndrome (a goiter with sensorineural hearing loss), and probably participates in I^- release. Anoctamin-1 (TMEM16A) is a Ca^{2+}-activated chloride channel, and CLC-5 is a voltage-gated chloride channel important for the I^- efflux from the cells. Gene knockout experiments have shown that CLC-5 deficient mice develop goiter, while the anoctamin-1 is responsible for ATP-stimulated I^- efflux.

Thyroglobulin: The hormone scaffold

Thyroglobulin (TG) is a relatively large glycoprotein dimer (660 kDa), the thyroid gland's scaffold for hormone production and iodine storage:

- Synthesis and processing: TG is synthesized in the endoplasmic reticulum (ER) of thyrocytes. Gene expression of the TG molecule is regulated by several transcription factors (PAX8, NKX2-1, and FOXE1), which are regulated by thyroid-stimulating hormone (TSH) through the cAMP pathway. Cysteine residues, through covalent disulfide bonds, also maintain proper

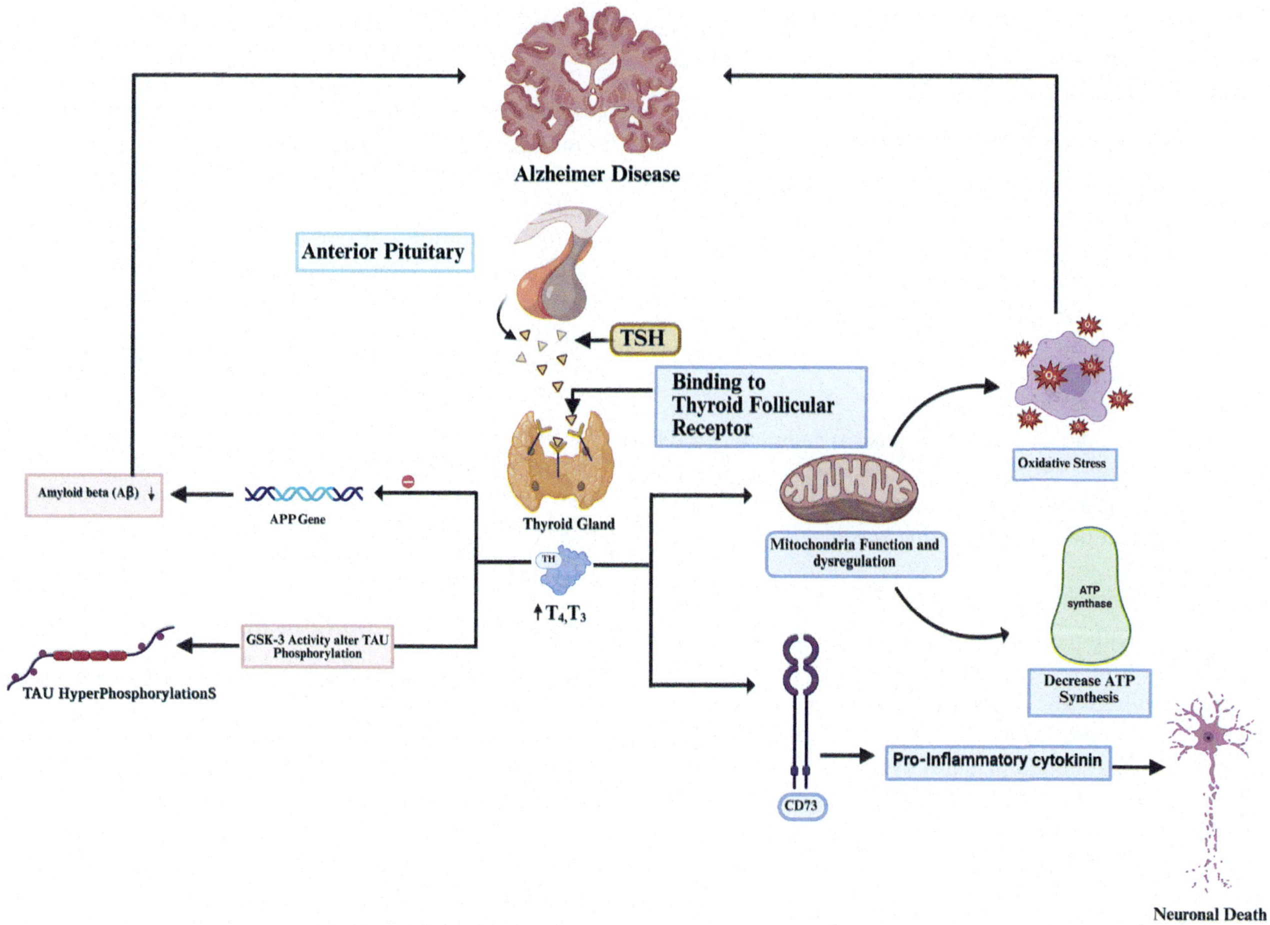

FIGURE 9.1 Role of thyroid dysfunction in Alzheimer's disease pathogenesis. The diagram shows how thyroid problems might be linked to Alzheimer's disease (AD). Changes in the anterior pituitary in AD cause changes in the secretion of thyroid-stimulating hormone (TSH), which binds to thyroid follicular receptors and changes the levels of thyroid hormones (T_4, T_3). When mitochondrial function is messed up and oxidative stress is high, ATP production decreases. When pro-inflammatory cytokines go up (through CD73 activation), they kill neurons. Also, abnormal signaling of thyroid hormones affects the expression of the amyloid precursor protein (APP) gene, which leads to the buildup of amyloid-beta (Aβ). When GSK-3 activity increases, it causes TAU to become hyperphosphorylated, making AD worse. These processes work together to create a feedback loop that speeds up the loss of nerve cells in Alzheimer's disease.

folding. Mutations in cysteine residues result in misfolded proteins, ER retention, and congenital goiter.

- Secretion and storage: Fully processed TG is secreted into the follicular lumen through exocytosis. TSH acutely stimulates exocytosis, resulting in an increase in the amounts of TG in the colloid. The colloid is a depot for iodinated TG; when TG is iodinated, it can contain two to eight T_4/T_3 residues per molecule of TG (Figueroa et al., 2021).

Thyroid peroxidase (TPO) and H_2O_2 generation

The biosynthesis of hormones occurs at the apical membrane-colloid layer and requires TPO and H_2O_2:

- TPO Catalysis: TPO is a membrane-bound hemoprotein that catalyzes the following two reactions.
- Iodination: H_2O_2 oxidizes I^- to generate reactive iodine species (I^+ or IOH) that iodinate tyrosine residues on TG to generate monoiodotyrosine (MIT) and diiodotyrosine (DIT).
- Coupling: The oxidative coupling of MIT + DIT yields T_3, while DIT + DIT yields T_4. Mutations in TPO are the most common cause of congenital hypothyroidism with organification defects.

H_2O_2 production by DUOX2

H_2O_2 is generated in the apical membrane by the NADPH oxidase, DUOX2. It requires maturation via DUOXA2 (which forms disulfide bonds to stabilize DUOX2) and is activated by Ca^{2+} influx and PKC phosphorylation. TSH increases DUOX2 expression, while excess iodide inhibits H_2O_2 generation, which helps understand the Wolff-Chaikoff effect.

Hormone output into the circulation

The thyroid hormones are released through a multi-step proteolytic process from stored TG:

- **TG Internalization and Degradation**

Iodinated TG is incorporated into the cell through fluid-phase endocytosis or megalin-mediated uptake. Upon incorporation, lysosomes will fuse with endocytotic vesicles and expose TG to acidic pHs. During this process, proteases like cathepsins B, K, and L will break down TG and release T_4 (~80%), T_3 (~20%), MIT, and DIT.

- **Iodine Recycling**

DEHAL1 (iodotyrosine dehalogenase) deiodinates MIT and DIT in the vesicles or at the apical membrane, allowing the iodide to be used. When DEHAL1 is mutated, hypothyroidism occurs with iodine wasting.

Hormone secretion

T_4 and T_3 will passively diffuse across the basolateral membrane (or in a carrier-dependent manner). The primary T_3 transporter is MCT8 (SLC16A2). When mutations in MCT8, there are severe neurodevelopmental defects known as Allan-Herndon-Dudley syndrome. There are also other avenues whereby intracellular T_4 may be converted into T_3 by deiodinases before being released into circulation (Carvalho & Dupuy, 2017).

Hypothalamic-pituitary-thyroid axis

The hypothalamic pPituitary–thyroid (HPT) axis is a central neuroendocrine feedback system regulating thyroid hormone synthesis, secretion, and overall metabolic homeostasis. It integrates signals from the hypothalamus, pituitary, and thyroid gland to maintain circulating levels of thyroid hormones. The thyrotropin-releasing hormone (TRH), a tripeptide, is synthesized and secreted at the hypothalamic level, stimulating thyrotrophs in the anterior pituitary. TRH secretion occurs in a pulsatile manner and is modulated by circadian rhythms, stress, temperature, and metabolic demands. Upon binding to its Gq-coupled receptor on anterior pituitary thyrotrophs, TRH activates membrane-bound phospholipase C, triggering calcium release from intracellular stores. This cascade promotes the synthesis and secretion of thyroid-stimulating hormone (TSH), a glycoprotein that acts directly on the thyroid gland. TSH binds to its receptor (TSHR) on thyroid follicular cells, initiating a sequence of events that includes the uptake of dietary iodide via the sodium–iodide symporter (NIS), the synthesis of thyroglobulin (TG), and the enzymatic action of thyroid peroxidase (TPO) to mediate iodination and coupling reactions. The proteolysis of TG subsequently releases thyroid hormones, predominantly thyroxine (T_4, approximately 80%) and triiodothyronine (T_3, approximately 20%) (Carvalho & Dupuy, 2017).

The HPT axis is tightly regulated by a negative feedback mechanism, primarily mediated by circulating thyroid hormones, with triiodothyronine (T_3) playing the most significant role. At the pituitary level, T_3 enters thyrotrophs and binds to nuclear thyroid hormone receptors, particularly the TRβ isoform, where it suppresses the transcription of the thyroid-stimulating hormone (TSH) gene, thereby reducing TSH synthesis and release. Concurrently, at the hypothalamic level, elevated T_3 concentrations inhibit the transcription of the TRH gene and interfere with TRH peptide processing, leading to reduced TRH availability. This dual-site feedback control ensures that thyroid hormone concentrations remain within a narrow physiological range, maintaining metabolic stability and preventing overproduction or deficiency of thyroid hormones (Giannocco et al., 2021).

Roles in CNS development and function

In the CNS, D2 and D3 are the principal deiodinases in the central nervous system (CNS). D2, predominantly observed in astrocytes and oligodendrocyte progenitor cells, converts T_4 to active T_3, providing the CNS with most of its T_3. T_3 responds to maintain appropriate steady state levels of T_3 by increasing when systemic thyroid hormone (TH) is low. D3, which is mainly expressed in neurons, application of D3 deactivates both T_4 and T_3 by producing rT_3 and T2, respectively. The development of T_4 leads to overproduction of TH, thereby decreasing the availability of T_3 by suppressing D2 and elevating D3. T_3 also regulates D3 expression via thyroid hormone receptor α1 (Thrα1) (Giannocco et al., 2021).

The cerebellum in the central nervous system governs motor coordination, involving movement coordination and balance maintenance through sensory information entered from the body's periphery. Also, TH governs the development of Bergmann's glia (BG), a particular type of glial cell connected to the proliferation of cells, differentiation at the popular embryonic brain, and cerebellum development (Hage & Azar, 2012).

Thyroid hormones (THs), with T_3 and T_4 being the most recognized, are critically important to brain health through their effects on neurotransmission, cellular metabolism, and gene expression via receptor-mediated pathways. At the molecular level, thyroid hormone receptors (TRs) act as ligand-dependent transcription factors. In the absence of TH, liganded TRs recruit corepressors and repress target genes, which partially explains the pathological consequences of hypothyroidism despite the presence of TRs. Upon binding of T_3, corepressor-release and coactivator recruitment begin transcription of genes important for synaptic plasticity and energy metabolism, showing the tightly regulated and multifaceted influence of THs on cognitive performance and brain homeostasis (Schroeder & Privalsky, 2014).

Pathophysiology of Alzheimer's disease

Alzheimer's disease (AD) is a neurodegenerative disorder defined by progressive cognitive decline and specific neuropathological changes. The pathophysiology of the disease involves interconnected mechanisms, such as amyloid-β plaque formation, tau hyperphosphorylation, neuroinflammation, oxidative stress, and neurodegeneration.

Amyloid-β plaque formation

Both β-secretase and γ-secretase break down amyloid precursor protein (APP) to make amyloid-β (Aβ) peptides. The imbalance between the production and removal of Aβ leads to the buildup and clumping of Aβ, especially Aβ42, which dissolves in water and can easily form insoluble fibrils. The accumulation of these fibrils in the brain leads to amyloid plaques, characteristic of Alzheimer's Disease (AD). For instance, familial Alzheimer's disease results from genetic mutations in APP, presenilin 1 (PSEN1), and presenilin 2 (PSEN2) that increase the production and aggregation of Aβ. Aβ plaques are toxic to neurons, cause inflammation in the area, and mess up how synapses work (Kai, 2023).

Tau hyperphosphorylation and neurofibrillary tangles

Tau is a protein that helps keep the cytoskeleton of neurons stable by attaching to microtubules. In Alzheimer's disease (AD), tau is hyperphosphorylated because kinases (like GSK3β, JNK, and CDK5) are activated too much, and phosphatases (like PP1 and PP2A) are not active enough. When tau gets too many phosphates, paired helical filaments build up inside cells and form neurofibrillary tangles (NFTs). These NFTs cannot stabilize microtubules anymore. NFTs interfere with synaptic function, hinder neuronal transport, and are strongly associated with neuronal death and the cognitive decline typically linked to Alzheimer's disease (AD) (Zhang et al., 2024).

Neuroinflammation and oxidative stress

Aβ plaques and tau aggregates cause astrocytes and microglia to stay active for a long time. This is known to cause more damage to neurons by releasing chemokines and pro-inflammatory cytokines, like IL-1β and TNF-α. Immune cells on the outside make inflammation even worse. Oxidative Stress: If the mitochondria are not working right, they make too many reactive oxygen species (ROS). Aβ and tau disease often cause this. Reactive oxygen species (ROS) harm lipids, proteins, and DNA, which speeds up neuronal dysfunction and cell death. Oxidative stress and neuroinflammation work together to make neurodegeneration happen over and over again (Kocahan & Doğan, 2017).

Neurodegeneration and cognitive decline

The clinical manifestations of Alzheimer's disease result from the gradual degeneration of neurons and synapses, particularly cholinergic neurons in the neocortex and basal forebrain. Synaptic Dysfunction: Both tau and Aβ pathological processes, which interfere with neurotransmitter systems and synaptic plasticity, make learning and remembering harder. Brain Shrinkage: The hallmark cognitive decline of Alzheimer's disease (AD) results from the cumulative impacts of amyloid plaques, neurofibrillary tangles, inflammation, and oxidative stress, resulting in extensive brain atrophy, especially in the cortex and hippocampus (Samuels, n.d.).

Thyroid dysfunction in aging and neurodegeneration

Thyroid hormones influence synaptic plasticity, neuronal development, and cognitive function over a lifetime and thus are very important for brain health. Aging dysfunction of the brain, more specifically due to hypothyroidism and subclinical hypothyroidism, can affect cognitive decline and neurodegenerative potential.

Hypothyroidism and subclinical hypothyroidism

In the elderly, overt hypothyroidism is often associated with major mood and cognitive dysfunctions. Patients tend to experience slow speech and thought, impaired attention, apathy, and sometimes depression or insanity ("myxedema madness"). Objective testing reveals impairments in general intelligence, attention, memory (especially verbal memory), perceptual skills, and executive functions. Levothyroxine (L-T_4) treatment generally resolves these symptoms (Eslami-Amirabadi & Sajjadi, 2021). Based on epidemiological studies, overt hypothyroidism is associated with an increased risk of developing Alzheimer's type dementia, which is likely more than double the risk compared to euthyroid individuals. The pathophysiology of AD has some hallmarks associated with a decreased phosphorylation of the tau protein, deleterious effects on cholinergic neurons, and increased expression of amyloid precursor protein, which enhances and promotes amyloid plaque formation (Kim et al., 2022).

Abnormal TSH (thyroid-stimulating hormone) levels with normal circulating thyroid hormones characterize SCH (subclinical hypothyroidism While some mild executive functioning impairment has been suggested, concrete evidence to establish a relationship with substantial neurodegenerative outcomes is lacking, while even minor alterations in thyroid hormone may be implicit in subtle alterations in brain structure and functioning, they are likely to be of lesser magnitude than overt hypothyroidism (Kim et al., 2022).

Hyperthyroidism and subclinical hyperthyroidism

Overt hyperthyroidism is associated with psychiatric symptoms like anxiety, irritability, and even psychosis, but also with overt cognitive deficits, such as memory, attention, and executive function. Some of the less obvious, subtle, and atypical symptoms in older individuals might sometimes be misattributed to just aging. An overabundance of thyroid hormone enhances sympathetic activity, can stimulate brain atrophy, and is often associated with vascular changes that may raise the incidence of stroke. Hyperthyroidism also leads to inflammatory responses in the brain, which may hasten the decline of neurons.

Risk of Dementia: Many studies and meta-analyses show that overt hyperthyroidism is related to increased risk of dementia and cognitive decline; however, there have been many methodological issues recently concerning studying cognitive impairment with hyperthyroidism, including new large cohort studies that found no statistically significant findings, perhaps due to small sample sizes (Ritchie & Yeap, 2015).

Epidemiological evidence linking thyroid disorders with AD

Increased Prevalence Among AD Patients: Multiple studies and meta-analyses show hypothyroidism is significantly more prevalent in AD patients than in control patients (6.4% vs. 2.4%, $P = .01$). According to very large population-based studies, individuals with hypothyroidism have an increased risk for AD (OR = 1.14, 95% CI = 1.00–1.30) (Kim et al., 2022).

Hyperthyroidism is similarly more common in this AD population than controls with an adjusted odds ratio of 1.13 (95% CI = 1.01–1.28). Epidemiological studies have linked higher circulating thyroid hormone levels, especially FT_4, to increased risk of AD. Both high and low TSH levels outside the normal range have been associated with increased dementia risk in population-based studies (van der Cammen et al., 2003).

Mechanistic interplay between thyroid hormones and AD pathogenesis

Impact on amyloid-β precursor protein (APP) processing

The APP gene codes for the amyloid-β precursor protein, the expression of which is directly controlled by thyroid hormone (TH), particularly T_3. TH suppresses expression of the APP gene by influencing histone acetylation and methylation, which decreases production of amyloid-β (Aβ).TH also plays a role in maintaining the brain's capacity to clear toxic Aβ aggregates by modifying the expression of Aβ clearance genes such as APOE, LRP1, TREM2, AQP4, and ABCB1. Transthyretin (TTR), important for transporting T_4 in the brain, regulates metabolism by inhibiting Aβ synthesis, hastening its clearance, and reducing its toxic effects. The disrupted function of TTR might alter AD, TH, and Aβ homeostasis (Li & Liu, 2024).

Influence on tau phosphorylation

An important step in neurofibrillary tangles development, which is important to AD pathology, is tau phosphorylation, which is elevated by abnormal thyroid hormone signaling. In AD models with altered TH levels, there is increased GSK-3 activity, a kinase that phosphorylates tau, suggesting TH imbalances may drive tau pathology through dysregulated kinases (Li & Liu, 2024).

Thyroid hormones and mitochondrial function in neurons

THs are important for neuronal energy metabolism, mitochondrial synthesis, and function. Dysregulation of THs has been shown to have increased oxidative stress, decreased ATP synthesis, and increased synaptic vulnerability, and will be linked to AD pathology. Increased oxidative stress has been noted in hypothyroidism and TH deficiency, which adds to mitochondrial dysfunction and worsens the development of tau and Aβ pathology (Li & Liu, 2024).

Hormonal regulation of neuroinflammation and synaptic plasticity

THs also directly act on the immunosuppressor gene CD73 and modify the immune response in the central nervous system. Pro-inflammatory cytokines such as IL-1β, IL-6, and TNF-α can induce autophagy and neuronal death, and are expressed more in the hippocampus in hypothyroid states, which may contribute to cognitive impairment. THs regulate autophagy, which is necessary to remove protein aggregates and cellular waste. Autophagy and apoptosis are increased in a state of hypothyroidism; however, excess or dysregulated autophagy may lead to Aβ accumulation and neurodegeneration (Li & Liu, 2024).

Clinical and biomarker evidence

The brain's growth, development, and function depend on thyroid hormones, T_3 and T_4, which affect neurogenesis, myelination, and repair. Age-related changes in TH levels, such as the decline in TSH and T_3, have been linked to an increased risk of dementia and AD. Based on various experimental studies, thyroid abnormalities could potentially lead to an increase in tau phosphorylation and Aβ accumulation and thus be involved in the pathogenesis of AD. Some supporting evidence supports this association, as brain atrophy and risk of dementia have been associated with high to normal FT_4 and low TSH. In advanced AD stages, reduced CSF T_3 and TT_4 levels indicate central thyroid dysregulation (Alzheimer's Disease Facts and Figures, 2024).

Thyroid hormone levels in AD patients

There is growing evidence that dysregulated TH signaling participates in AD at different stages of the disorder. Clinical image study reveals elevated thyroid-stimulating hormones correlate with reduced glucose metabolism in Alzheimer's affected brain while lowering serum-free T_4 (fT_4). These responses are supported by experimental animal models showing regulated thyroid hormone receptors (THRα, THRβ) (Choi et al., 2017). Increased FT_3-FT_4, TSH, and TRH. Early AD-like signs are possibly indicative of neuroprotection. The CSF of non-AD patients in advanced AD despite normal levels of serum thyroid hormones. The CSF shows an increase in the active product of the 3-Deiodinase pathway, reserve T_3 (rT_3), and a decrease in the active circulation. Suggesting brain Hypothyroidism due to expression of active T_3 in the brain. This monitoring may drive personalized therapy for AD (Sampaolo et al., 2005).

TSH, T_3, T_4 as potential biomarkers

Triiodothyronine (T_3) and Thyroxine (T_4) have a key role in Brain development, energy metabolism, and cognitive function, and increasing evidence from epidemiological and clinical trials supports the idea that changes in thyroid hormone could contribute to the onset and progression of Alzheimer's disease. Moreover, these could be used as early biomarkers (ADM van Osch et al., 2004). A large Mendelian randomized study found that genetically determined TSH levels in the normal range were associated with a lower risk of AD. Korean nationwide cohort studies also reported that individuals with thyroid disorders and low-normal TSH levels had an increased risk of AD (Kim et al., 2022). These findings suggest that normal TSH may be protective, while suppressed TSH and elevated fT_4 could signal increased Alzheimer's disease risk. Highlighting their potential risk as early biomarkers and therapeutic targets (Marouli et al., 2021a).

Thyroid autoimmunity and Alzheimer's disease risk

Recent studies have shown that thyroid functions show a strong association between the possible risk of AD and Thyroid disorders, especially autoimmune thyroid disorders. In some studies, thyroid function was even considered as a factor that can be modified for the risk of AD. In a large national cohort study in the Korean population with individuals having hyperthyroidism (OR = 1.13), Hypothyroidism (OR = 1.14), and Thyroiditis (OR = 1.12) as a condition that results from large immune-mediated responses in the thyroid and were more likely to develop AD (Kim et al., 2022). Mendelian randomization analyses in a genetic study revealed that higher TSH levels within the normal range significantly decreased AD risk (OR = 0.98). There is no direct evidence that Eddie changed thyroid function, implicating a possible causal relationship. Autoimmune thyroid diseases might predispose AD through systemic inflammation, damaged BBB, and disturbed hormone signaling. Neuroprotection by prevention or early treatment may be beneficial (Marouli et al., 2021a).

Experimental and animal models

Animal models used in experiments are an important part of biomedical research. Very important for learning about long-term illnesses like Alzheimer's disease. Researchers can study pathophysiology and try out treatments in controlled settings before doing studies on people. Transgenic mice are the most commonly used models in research; they have been genetically modified to have 36 genes linked to producing Aβ and tau proteins (Miklos et al., 2011). These models include single transgenic lines with mutant APP and presenilin genes and triple transgenic mice that express mutant APP, tau, and PS1. They copy important signs of Alzheimer's disease, such as the formation of tau tangles and plaques. They also show signs of neuroinflammation, neuronal death, and cognitive decline. Chemical lesions and viral vector models can also cause changes that are similar to Alzheimer's by making neurotoxins or amyloid peptides. No models completely replicate human Alzheimer's disease. These elucidate molecular pathways, identify early biomarkers and therapeutic strategies, and facilitate the transition from laboratory research to clinical application (Spires & Hyman, n.d.).

TABLE 9.1 Transgenic AD models with altered thyroid function.

Model name	Type	Key features	Thyroid relevance	References
Okadaic acid (OA)-mouse model	Chemically induced model	Aβ, tau pathology, and behavioral decline	Elevated levels of THs and THRs in the brain	Ren et al. (2023)
3xTg-AD mouse model	Triple transgenic model	Progressive AD features and memory deficits	T_3 improves behavior and alters gene regulation	Maglione et al. (2022)
Transgenic rat model (App-Swedish)	Viral-mediated transgenic rat model	Aβ plaques, cognitive impairment	Potential for thyroid-focused Alzheimer's studies	Maglione et al. (2022)
The PDAPP mouse model	APP transgenic mouse model	Hippocampal plaque formation and synaptic loss	Suitable for thyroid-based therapeutic strategies	Maglione et al. (2022)
Thyroid-disrupted AD mouse model	Endocrine manipulation model	AD features under hypo/hyperthyroid conditions	Shows a direct link between thyroid hormones and AD	Ren et al. (2023)

Transgenic AD models with altered thyroid function

Recent studies have established the relationship between thyroid hormone signaling and AD pathology. There are few experimental models to support this relationship (Table 9.1):

a. Okadaic Acid (OA) Mouse Model – Injecting OA into the hippocampus causes Aβ to build up, tau to become pathological, and cognitive problems. This model shows higher levels of TSH, TRH, FT_3, and FT_4 in the brain and blood, as well as higher levels of THRα and THRβ, which suggests that thyroid signaling is starting to work in AD (Ren et al, 2023).

b. 3xTg-AD Mouse Model –This transgenic strain has the human APPswe, tau P301L, and PS1m146v genes, which cause the same problems seen in Alzheimer's disease. Giving T_3 to these mice makes them less depressed, lowers the levels of AD-related genes (APP, ADAM10, SERT), and boosts the expression of antioxidant genes (Maglione et al., 2022; Nakai et al., 2021).

c. Transgenic rat via viral vector: This approach involves rats expressing APP 695 with the Swedish mutation through a vector. These exhibits accumulate Aβ42 protein with cognitive dysfunction. Despite having limited direct data, their human-like brain structure renders them suitable for studying thyroid connections (Maglione et al., 2022).

d. Thyroid-disrupted AD mouse model: The experimental system merges Alzheimer's disease genetics with methods such as iodine restrictions, medication administration, or Thyroidectomy to produce states of hypothyroidism (Ren et al, 2023).

e. The PDAPP mouse features mice that produce human APP with the V717F mutation and develop AD pathology, including Aβ plaques and neuronal degeneration. This study has not been designed to assist thyroid involvement, yet it offers a foundation to test thyroid-based treatments (Ren et al., 2023) (Table 9.2).

Behavioral and cognitive assessments

Thyroid hormones are crucial for neuroplasticity, brain development, and cognitive functioning. People with Alzheimer's disease often have emotional problems, and problems with the thyroid are strongly linked to cognitive decline (Warren et al., 2024). In thyroid-altered 3xTg-AD mice, spatial learning in the Morris Water Maze is largely preserved; however, deficits are observed in the Y-maze and T-maze tasks, as well as in the Novel Object Recognition (NOR) test, signifying impairments in working and recognition memory (Reul et al., 2017). These deficits are associated with dysfunction in the hippocampus and prefrontal cortex. The Mini-Mental State Examination (MMSE) and the Montreal Cognitive Assessment (MoCA) are used to find cognitive decline in people with thyroid disorders. MoCA is better at finding mild impairments. The Rey Auditory Verbal Learning Test (RAVLT) and Benton Visual Retention Test (BVRT) show that people with hypothyroidism have very bad memory problems. The Everyday Cognition (Ecog) scale can be used to measure functional impairment, and it is linked to the progression of Alzheimer's disease in people with thyroid problems (Webster et al., 2014).

Molecular insights from Preclinical studies

Rodents with hypothyroidism exhibit depression-like behavior in the Forced Swim Test and Tail Suspension Test. T_3 treatment produces antidepressant effects, partly by modulating the serotonergic system. At the molecular level, tau pathology in hypothyroid models is linked to reduced activity of protein phosphatase 2A (PP2A), promoting tau accumulation (Salehipour et al., 2023). In

TABLE 9.2 Potential therapeutic targets within the thyroid axis for Alzheimer's disease.

Target	Role in Alzheimer's disease	Therapeutic approach	Key challenges
Hypothalamic–Pituitary–Thyroid (HPT) axis	Dysregulation contributes to neuroinflammation and amyloid plaque formation.	Modulation through thyroid hormone receptor (TR) agonists	Achieving optimal Blood-Brain Barrier (BBB) penetration and timing of intervention
Thyroid hormone receptors (TRα, TRβ)	Crucial for cognitive function and neuroprotection	Development of brain-selective TR agonists	Efficient and targeted CNS delivery
Deiodinase enzymes	Catalyze conversion of T_4 to active T_3; activity impaired in AD	Pharmacological modulation to restore T_3 levels	Maintaining precise hormonal balance
TSH receptors in the brain	Disruption affects neurotransmission and neuronal growth	Targeting TSH signaling to reduce AD progression risk	Complexity of systemic and local hormonal crosstalk

elderly and thyroidectomized mice, hypothyroidism increases ApoE4 entry into the brain by upregulating ATP6AP1, facilitating exosomal ApoE4 trafficking. ApoE4 subsequently activates the canonical NLRP3 inflammasome, triggering neuroinflammation, oxidative stress, and cognitive as well as motor impairments (Ren et al., 2023).

These findings indicate that thyroid dysfunction contributes to AD pathogenesis through multiple mechanisms, including altered neurotransmitter regulation, exacerbation of tau pathology, and inflammasome-mediated neurodegeneration. Consequently, maintaining thyroid hormone homeostasis and targeting the ApoE4–NLRP3 axis may hold therapeutic promise for AD (Zhang et al., 2022).

Therapeutic and preventive perspectives

Therapeutic and preventative approaches to Alzheimer's disease (AD) in the context of thyroid dysfunction are centered on restoring thyroid hormone homeostasis and modifying pathways related to Thyroid hormones, notably T_3 and T_4, which are required for proper neuronal activity and healthy cognition (Accorroni et al., 2014). Both hypothyroidism and hyperthyroidism are associated with Alzheimer's disease pathology, including β-amyloid, tau phosphorylation, and oxidative stress. Early AD is characterized by local thyroid hormone dysregulation in the brain. T_3 supplementation and thyroid receptor targeting are expected to be potential strategies. Routine thyroid screening and care of autoimmune thyroid problems in at-risk persons may provide neuroprotective advantages in slowing the onset of Alzheimer's disease (Marouli et al., 2021a).

Thyroid hormone replacement in AD

The in vivo Preclinical Studies strongly support the probability of triiodothyronine (T_3) in Alzheimer's disease treatment. According to the evaluation in 3xTG-AD mice, 21 days of systemic D3 administration showed reduced depression-like behaviors, as shown by decreased immobility in forced swim and tail suspension tests (Maglione et al., 2022). T_3-treated groups also showed altered AD-related gene expression with decreased GSK3 and increased ADAM10 and antioxidant gene SOD2, reflecting the dual impact on Neurodegeneration. In an STZ-induced sporadic AD rat model, 30 days of T_3 (1.5 μg/100 g/day) showed improved short and long-term memory, enhanced insulin signaling (GSK3β), reduced neuroinflammation (IL-1β, IL-6, TNF-α), glial activation, and apoptosis (Bax) while increasing anti-apoptotic bcl-2 (Accorroni et al., 2014).

Challenges in hormonal therapy for neurodegeneration

Thyroid hormones (T_3 and T_4) and thyroid-stimulating hormone (TSH) play a vital role in brain development, metabolism, and cognition (Przybylak et al., 2021). Their dysregulation is associated with Alzheimer's disease (AD) pathogenesis. Despite having various benefits in preclinical models, several challenges limit therapy in AD (Liu & Brent, 2018). Brain thyroid hormone activity is tissue-specific and is regulated by enzymes such as DIO2 in activating T_4 &T_3 and DIO3 for THs. Reduced DIO2 activity in Alzheimer's disease lowers brain T_3 despite having normal serum levels. Transporters like MCT8 and OATP1C1 mediate thyroid hormone entry across the Blood-Brain Barrier (BBB), but various factors impair CNS TH supply (Liu & Brent, 2018). Accordingly, the interaction of the thyroid hormone and Alzheimer's disease varies by stage. Early hyperthyroidism can shift to central hypothyroidism, complicating treatment timing. Individual strategies are required to enhance thyroid hormone-based therapy for treating AD (Vancamp et al., 2020).

Potential of modulating the HPT axis

HPT axis, which controls the regulation of thyroid-stimulating hormone (TSH) and thyrotropin-releasing hormone (TRH), plays a key role in thyroid hormone synthesis (Canet et al., 2019). These are essential for synaptic plasticity, neuroprotection, mitochondrial functions, and cognition. In Alzheimer's disease, HPT axis disruption promotes neuroinflammation, amyloid deposition, and oxidative stress. T_3, acting via nuclear receptors TRα and TRβ, is a critical signaling pathway for memory and cognition, yet patients often exhibit low T_3 (Sharan & Vellapandian, 2024). Abnormal TSH regulation in clinical thyroid disorder further increases AD risk as TSH receptors in the brain affect neurotransmission and neuronal growth. Chronic stress and interactions with the HPA and HPG axes worsen the dysfunction. Therapeutic strategies include 11-HSD1 inhibitors, brain-selective TR agonist (e.g., sobetirome, eprotirome), and deiodinase modulation. Precision therapy targeting HPT axis function may slow or reverse AD progression (Canet et al., 2019; Sharan & Vellapandian, 2024).

Other hormones, endocrine tissues, and pathways: Translational perspectives

Thyroid hormones (T_3, T_4) and their regulatory pathways are fundamental to the neuroendocrine dimensions of Alzheimer's disease (AD); however, the extensive endocrine network is also vital for cognitive function, neuronal viability, and neurodegeneration. The interaction among various hormonal systems indicates that Alzheimer's disease pathophysiology arises not from singular thyroid dysregulation, but from integrated endocrine-neurocircuitry pathways. Comprehending these translational dimensions facilitates the development of multi-target therapeutic approaches.

Hypothalamic–pituitary–adrenal (HPA) axis

Chronic activation of the HPA axis results in increased glucocorticoid levels, especially cortisol, which can hinder hippocampal neurogenesis, disturb synaptic plasticity, and worsen amyloid-β and tau pathology. Glucocorticoid receptor (GR) signaling interacts with thyroid hormone pathways, and excessive cortisol may inhibit T_3 action in the brain. In Alzheimer's disease, elevated cortisol levels are associated with expedited cognitive decline, indicating that simultaneous modulation of the thyroid and adrenal axes may be advantageous. Translationally, GR antagonists or 11β-HSD1 inhibitors—currently being studied for metabolic and cognitive disorders—might be repurposed in conjunction with thyroid-targeted therapies.

Gonadal hormones and neuroprotection

Estrogens and androgens regulate synaptic density, mitochondrial activity, and amyloid metabolism. Estrogen, specifically, increases the expression of thyroid hormone receptors and facilitates neurotrophic signaling pathways, including BDNF-TrkB. Decreasing estrogen levels in postmenopausal women, coupled with modified thyroid function, may collaboratively elevate the susceptibility to Alzheimer's disease. This has spurred investigation into selective estrogen receptor modulators (SERMs) and bioidentical hormone replacement therapies, possibly in conjunction with CNS-selective thyroid analogues. Likewise, testosterone deficiency in men may exacerbate metabolic and cognitive decline, in part due to thyroid-androgen interactions in neuronal gene regulation.

Pancreatic hormones and brain metabolism

Insulin and glucagon-like peptide-1 (GLP-1) affect how the brain uses glucose and how long neurons live. People with type 2 diabetes often have insulin resistance, which is linked to thyroid problems and a higher risk of Alzheimer's disease. This has led to the idea of "type 3 diabetes." Thyroid hormones augment insulin-mediated glucose uptake and mitochondrial efficiency; conversely, insulin influences deiodinase activity and thyroid hormone metabolism. To bring the brain's metabolism back to normal, translational strategies could include intranasal insulin, GLP-1 receptor agonists (like semaglutide or liraglutide), or dual thyroid-insulin pathway modulators.

Adipose-derived hormones and inflammatory modulation

Adipokines, including leptin and adiponectin, are essential for maintaining energy balance, regulating inflammation, and controlling synaptic activity. Leptin improves the plasticity of synapses in the hippocampus and works with neurons in the hypothalamus that make TRH. In obesity and metabolic syndrome—both of which increase the risk of AD—leptin resistance and changed adipokine profiles go along with low-grade inflammation and thyroid signaling problems. Translational strategies may integrate lifestyle modifications, metabolic regulators, and thyroid hormone treatments to reestablish the adipose-brain endocrine axis.

Gut–brain–endocrine axis

New evidence shows that gut microbiota may play a role in regulating hormones and causing neurodegeneration. Gut microbes affect hormones like ghrelin, peptide YY (PYY), and GLP-1, which then affect the HPT axis. Dysbiosis may make it harder for the body to absorb thyroid hormones, change inflammation throughout the body, and break down the gut-brain barrier, which could make AD worse. In terms of translation, probiotics, prebiotics, and dietary changes that affect both the composition of the microbiota and the signaling of the endocrine system could make thyroid therapy work better for people with AD.

Integrated endocrine network in AD management

The translational challenge involves synchronizing interventions across these hormonal systems to tackle the multifactorial characteristics of Alzheimer's disease. Measuring TSH, FT_4, FT_3, cortisol, insulin, sex hormones, and adipokines for each person could help with precision medicine. Such strategies would customize treatment to each patient's endocrine-neurocognitive profile, integrating thyroid modulation with specific therapies for additional hormonal imbalances.

The interrelation of endocrine axes in AD indicates that interventions targeting a single hormone may be inadequate. Biomarker-guided patient selection should support future clinical trials that look into combination regimens like CNS-selective thyroid analogues with GLP-1 receptor agonists, SERMs, or GR modulators. A systems endocrinology approach may ultimately present the most promising strategy for decelerating or preventing the progression of Alzheimer's disease.

Current challenges and future directions

The connection between thyroid disorders and cognitive decline, particularly dementia, remains very complex. Meta-analyses highlight that hypothyroidism and subclinical hypothyroidism are associated with increased risk of dementia, suggesting that these could be modifiable risk factors. The evidence that hypothyroidism and subclinical hypothyroidism contribute to cognitive decline is inconsistent (Xu et al., n.d.). A longitudinal study has found no meaningful links between thyroid function and cognitive domains, such as memory and cognitive function or dementia risk. These may support the Methodological variations, including differences in how thyroid disease is defined and heterogeneity in demography, highlighting the need for a standardized approach (Van Vliet et al., 2021).

Gaps in knowledge

Despite extensive research, the relationship between thyroid dysfunction and cognitive decline remains elusive. Study findings are inconsistent, creating a major challenge to decode the relation (Ebbert et al., 2015). Observational and meta-analytic studies highlight the relation between hypothyroidism, especially its subclinical form, linked to an increased risk of dementia. At the same time, Hypothyroidism and subclinical hypothyroidism show very minimal or no association (Ebbert et al., 2015). Data on specific subgroups like age, sex, genetic, and risk remain limited and contradictory. The efficacy of therapies like levothyroxine and the long-term impact of subclinical thyroid disease on cognition also remain unclear. This requires standardized research.

Future research strategies

The association between thyroid hormones and dementia remains unclear, demanding well-focused longitudinal human studies and well-controlled animal models to determine and authenticate the relationship. Future research should focus on identifying whether thyroid abnormalities result in Alzheimer's disease (AD) by measuring early thyroid hormone levels and their regulation and receptor activity. Cohort studies stratified with cardiovascular health, age, sex, and genetic risks like APOE ε4Dose response meta-analysis could help refine TSH reference ranges and protective action against cognitive decline (Tang et al., 2021). Finally, well-defined interventional studies showing biomarkers, neuroimaging, and cognitive tests are needed to clarify that thyroid hormone therapies can reduce the risk of dementia and become a therapeutic strategy for cognitive improvement.

Integrative approaches in neuroendocrinology

Holistic neuroendocrinology explores the connection between thyroid regulation and cognition in AD. Evidences suggest that thyroid dysfunction may contribute to the development of early AD, especially when brain hormone levels and receptors are altered. Experimental AD models revealed significantly elevated TSH, free triiodothyronine (FT_3), and free thyroxine (FT_4) in brain tissue, which is believed to be a compensatory response to early neurological changes (Younesi & Hofmann-Apitius, 2013). Thyroid receptors are also upregulated, particularly in the hippocampal region, known for memory and learning. These findings highlighted the therapeutic potential of targeting thyroid hormone activity. Reduced Cognition is further influenced by a broader hormonal network involving melatonin, insulin, cortisol, and Oestrogen, integrating a theory suggesting the hippocampus and prefrontal cortex are key sites for these pathways.

Damaged glucocorticoid receptor (GR) signaling, often linked to increased cortisol levels and chronic stress, may interact with thyroid pathways, showing cognitive deficits and Alzheimer's disease progression (Younesi & Hofmann-Apitius, 2013).

Conclusion

Thyroid hormones (T_3, T_4) are vital for brain development, neuroplasticity, and cognitive function. There is a two-way link between thyroid problems and Alzheimer's disease (AD), and having too much or too little thyroid hormone can raise the risk of dementia. Changes in TSH and free T_4 levels are linked to the buildup of amyloid-β, the hyperphosphorylation of Tau, oxidative stress, and neuroinflammation. Late-stage Alzheimer's disease frequently demonstrates "brain hypothyroidism" despite normal serum hormone concentrations. Clinical and preclinical evidence indicates that thyroid modulation may provide neuroprotection, especially through T_3 therapy; however, obstacles persist in CNS delivery, treatment timing, and patient selection. Regular thyroid screening in older adults exhibiting cognitive symptoms, incorporating hormonal biomarkers with cognitive assessments, and interventions tailored to specific stages may improve early detection and targeted treatment. Future translational research should concentrate on CNS-selective thyroid analogues, deiodinase modulators, and integrated neuroendocrine-neuroimaging methodologies to formulate preventive and disease-modifying strategies for Alzheimer's Disease.

Applications of the material

The complex interplay between thyroid hormones (THs) and Alzheimer's disease (AD) delineated in this chapter has significant ramifications for clinical practice, diagnostics, therapeutics, and research methodologies. Comprehending the mechanistic connections between the hypothalamic–pituitary–thyroid (HPT) axis and neurodegeneration facilitates the application of this knowledge into concrete interventions focused on the prevention, early diagnosis, and management of Alzheimer's disease (AD).

Clinical screening and risk stratification

Because there is evidence that both hypo- and hyperthyroidism increase the risk of dementia, it makes sense to test older adults' thyroid function (TSH, FT_4, FT_3) on a regular basis, especially those who have cognitive symptoms. Early detection of subclinical thyroid disorders may facilitate prompt intervention, potentially postponing or alleviating neurodegenerative processes. Adding thyroid biomarker panels to known risk factors for Alzheimer's disease (like APOE ε4 status and cardiovascular health) could help doctors better group patients in primary care and neurology clinics.

The process of creating biomarkers

Thyroid-related biomarkers, including altered CSF T_3/T_4 ratios, elevated reverse T_3, or abnormal TSH levels within the high-normal or low-normal ranges, may function as preliminary indicators of Alzheimer's disease risk or progression. Combining these with neuroimaging (MRI, PET) and fluid biomarkers of amyloid-β and tau may produce multimodal diagnostic algorithms. These methods could help find problems earlier, before neurons start to die off permanently.

Strategies for therapy

Understanding TH–AD interactions facilitates various therapeutic pathways:

Hormone Replacement Therapy – Controlled T_3 or thyromimetic analogues for individuals with central hypothyroidism in Alzheimer's Disease, customized according to disease stage and unique endocrine profile.

HPT Axis Modulation – Pharmacological agents that target TRH, TSH, or thyroid receptor activity, especially CNS-selective TR agonists, to improve neuroprotection and synaptic function.

Combination Endocrine Therapies – Combining thyroid modulation with estrogen therapy, GLP-1 receptor agonists, or cortisol-lowering agents to fix problems with multiple hormones.

Interventions for prevention and lifestyle

Thyroid health can be a part of dementia prevention programs that are run by the government. Adequate nutrition in iodine and selenium, weight management to enhance leptin signaling, and interventions that mitigate chronic stress (thus stabilizing both HPA and HPT axes) may diminish the combined endocrine–neurodegenerative risk.

Research that moves from the lab to the clinic and clinical trials

Preclinical evidence indicating that T_3 therapy mitigates amyloid and tau pathology substantiates human trials focused on patients exhibiting early cognitive impairment and thyroid dysfunction. Longitudinal cohort studies can elucidate whether the preservation of optimal thyroid function attenuates the progression of Alzheimer's disease. Furthermore, animal models exhibiting engineered thyroid dysfunction in Alzheimer's disease contexts serve as platforms for the evaluation of novel agents, deiodinase modulators, and blood-brain barrier-penetrating thyromimetics.

Personalized medicine strategies

A personalized endocrine-neurocognitive profile could facilitate precision treatment by optimizing thyroid hormone levels for maximal cognitive enhancement while mitigating systemic side effects. This corresponds with developing "systems endocrinology" frameworks that examine the interactions among the thyroid, adrenal, gonadal, pancreatic, adipose, and gut–brain axes in neurodegenerative diseases.

Uses in education and training

This material emphasizes the significance of integrated endocrine-neurological assessment in geriatric, psychiatric, and neurological curricula. Using case scenarios of Alzheimer's disease (AD) with thyroid comorbidities in simulation-based learning can help improve diagnostic reasoning and skills for working with people from different fields.

The insights from this chapter go beyond just academic knowledge. They provide a guide for incorporating thyroid assessment into AD risk management, creating biomarker-guided interventions, designing targeted therapies, and training healthcare professionals. These applications facilitate a proactive, personalized, and multi-faceted strategy for the prevention and management of Alzheimer's disease by integrating mechanistic insights with clinical application.

Mini-dictionary

1. Amyloid-β (Aβ) – Peptide that aggregates into plaques in the brain, a hallmark of Alzheimer's disease.
2. Tau Protein – Microtubule-associated protein that, when hyperphosphorylated, forms neurofibrillary tangles in AD.
3. Thyroid Hormone Receptors (TRα, TRβ) – Nuclear receptors that mediate genomic effects of T_3, regulating gene transcription in neurons.
4. Deiodinases (D1, D2, D3) – Enzymes that activate or deactivate thyroid hormones by removing iodine atoms.
5. Hypothalamic–Pituitary–Thyroid (HPT) Axis – Endocrine system regulating thyroid hormone production via TRH, TSH, and feedback mechanisms.
6. Transthyretin (TTR) – Thyroid hormone transport protein in cerebrospinal fluid that also binds amyloid-β, reducing its toxicity.
7. Blood–Brain Barrier (BBB) – Selective barrier regulating passage of substances, including thyroid hormones, into the brain.
8. Glycogen Synthase Kinase-3 (GSK-3) – Kinase involved in tau phosphorylation and neurodegeneration in AD.
9. Subclinical Hypothyroidism – Elevated TSH with normal circulating thyroid hormone levels, sometimes linked to cognitive changes.
10. Morris Water Maze – Behavioral test used in rodents to assess spatial learning and memory.

Summary points

1. Thyroid hormones (T_3, T_4) are essential for neurodevelopment, synaptic plasticity, cognition, and metabolic regulation in the brain.
2. Hypothyroidism and hyperthyroidism both raise the risk of Alzheimer's disease (AD) through processes like amyloid-β buildup, tau hyperphosphorylation, oxidative stress, and neuroinflammation.
3. Even though serum thyroid hormone levels are normal, late-stage AD often shows signs of "brain hypothyroidism."
4. When thyroid hormone signaling isn't working right, it affects APP processing, tau phosphorylation, mitochondrial function, immune responses, and synaptic plasticity.
5. Epidemiological studies show that changes in TSH, FT_4, and FT_3 levels are linked to a higher risk of dementia. On the other hand, genetically higher-normal TSH may protect against dementia.
6. Animal models with altered thyroid function show changes in cognition, behavior, and AD pathology, which supports the idea that there is a causal link.
7. T_3 therapy in preclinical models makes cognition better, lowers the pathology of AD, and changes neuroinflammation and oxidative stress.
8. Some of the problems with thyroid-based AD treatment are getting through the blood-brain barrier, controlling hormones exactly, timing the intervention, and choosing the right patients.
9. Some possible therapeutic targets are the HPT axis, thyroid hormone receptors, deiodinase enzymes, and TSH receptors in the brain.
10. Future research should look into thyroid analogues that only affect the CNS, precise changes to the HPT axis, and combining hormonal biomarkers with cognitive tests.

References

Adm van Osch, L., Hogervorst, E., Combrinck, M., & David Smith, A. (2004). Low thyroid-stimulating hormone as an independent risk factor for Alzheimer's disease. www.neurology.org.

Alzheimer's disease facts and figures. *Alzheimer's and Dementia, 20*(5), (2024), 3708–3821. https://doi.org/10.1002/alz.13809

Breijyeh, Z., & Karaman, R. (2020). Comprehensive review on Alzheimer's disease: Causes and treatment. *Molecules, 25*(24). https://doi.org/10.3390/MOLECULES25245789

Canet, G., Hernandez, C., Zussy, C., Chevallier, N., Desrumaux, C., & Givalois, L. (2019). Is AD a stress-related disorder? focus on the HPA axis and its promising therapeutic targets. *Frontiers in Aging Neuroscience, 11*(SEP). https://doi.org/10.3389/fnagi.2019.00269

Carvalho, D. P., & Dupuy, C. (2017). Thyroid hormone biosynthesis and release. *Molecular and Cellular Endocrinology, 458*, 6–15. https://doi.org/10.1016/j.mce.2017.01.038

Choi, H. J., Byun, M. S., Yi, D., Sohn, B. K., Lee, J. H., Lee, J. Y., Kim, Y. K., & Lee, D. Y. (2017). Associations of thyroid hormone serum levels with in-vivo Alzheimer's disease pathologies. *Alzheimer's Research & Therapy, 9*(1). https://doi.org/10.1186/s13195-017-0291-5

Ebbert, M. T. W., Ridge, P. G., & Kauwe, J. S. K. (2015). Bridging the gap between statistical and biological epistasis in Alzheimer's disease. *BioMed research international, 2015*. https://doi.org/10.1155/2015/870123

Eslami-Amirabadi, M., & Sajjadi, S. A. (2021). The relation between thyroid dysregulation and impaired cognition/behavior: An integrative review. *Journal of neuroendocrinology, 33*(3). https://doi.org/10.1111/jne.12948

Figueroa, P. B. S., Ferreira, A. F. F., Britto, L. R., Doussoulin, A. P., & Torrão, A. da S. (2021). Association between thyroid function and Alzheimer's disease: A systematic review. *Metabolic brain disease* (Vol36,(7), 1523–1543. https://doi.org/10.1007/s11011-021-00760-1

Giannocco, G., Kizys, M. M. L., Maciel, R. M., & de Souza, J. S. (2021). Thyroid hormone, gene expression, and central nervous system: Where we are. *Seminars in cell and developmental biology, 114*, 47–56. https://doi.org/10.1016/j.semcdb.2020.09.007

Goemann, I. M., Romitti, M., Meyer, E. L. S., Wajner, S. M., & Maia, A. L. (2017). Role of thyroid hormones in the neoplastic process: An overview. *Endocrine-Related cancer, 24*(11), R367–R385. https://doi.org/10.1530/ERC-17-0192

Hage, M. P., & Azar, S. T. (2012). The link between thyroid function and depression. *Journal of thyroid research, 2012*. https://doi.org/10.1155/2012/590648

Kai, S. C. (2023). Pathophysiology of dementia. *Australian Journal of General Practice, 52*.

Kim, J. H., Lee, H. S., Kim, Y. H., Kwon, M. J., Kim, J. H., Min, C. Y., Yoo, D. M., & Choi, H. G. (2022). The association between thyroid diseases and Alzheimer's disease in a national health screening cohort in Korea. *Frontiers in Endocrinology, 13*. https://doi.org/10.3389/fendo.2022.815063

Knopman, D. S., Amieva, H., Petersen, R. C., Chételat, G., Holtzman, D. M., Hyman, B. T., Nixon, R. A., & Jones, D. T. (2021). Alzheimer disease. *Nature Reviews Disease Primers, 7*(1). https://doi.org/10.1038/s41572-021-00269-y

Kocahan, S., & Doğan, Z. (2017). Mechanisms of Alzheimer's disease pathogenesis and prevention: The brain, neural pathology, N-methyl-D-Aspartate receptors, tau protein, and other risk factors. *Clinical psychopharmacology and neuroscience, 15*(1), 1–8. https://doi.org/10.9758/cpn.2017.15.1.1

Li, Z., & Liu, J. (2024a). Thyroid dysfunction and Alzheimer's disease: A vicious circle. *Frontiers in endocrinology, 15*. https://doi.org/10.3389/fendo.2024.1354372

Liu, Y. Y., & Brent, G. A. (2018). Thyroid hormone and the brain: Mechanisms of action in development and role in protecting and promoting recovery after brain injury. *Pharmacology and therapeutics, 186*, 176–185. https://doi.org/10.1016/j.pharmthera.2018.01.007

Accorroni, A., Chiellini, G., & Origlia, N. (2014a). Send Orders for Reprints to reprints@benthamscience.ae. Neuroprotective effects of thyroid hormones and their metabolites. In Journal name, year (Vol. Vol. 1) ..

Maglione, A., Nascimento, B., Ribeiro, M., Souza, T., Silva, R., Sato, M., Penatti, C., Britto, L., Souza, J., Maciel, R., Conceição, R. R. da, Melo, R. L.-, & Giannocco, G. (2022). Triiodothyronine treatment reverses depression-like behavior in a triple-transgenic animal model of Alzheimer's disease. https://doi.org/10.21203/rs.3.rs-1277920/v1..

Marouli, E., Yusuf, L., Kjaergaard, A. D., Omar, R., Kuś, A., Babajide, O., Sterenborg, R., Åsvold, B. O., Burgess, S., Ellervik, C., Teumer, A., Medici, M., & Deloukas, P. (2021a). Thyroid function and the risk of Alzheimer's disease. *A Mendelian Randomisation study*. https://doi.org/10.1101/2021.06.20.21259140

Miklos, A. C., Li, C., Sorrell, C. D., Lyon, L. A., & Pielak, G. J. (2011). An upper limit for macromolecular crowding effects. *BMC Biophysics, 4*(1). https://doi.org/10.1186/2046-1682-4-13

Nakai, T., Yamada, K., & Mizoguchi, H. (2021). Alzheimer's disease animal models: Elucidation of biomarkers and therapeutic approaches for cognitive impairment. *International journal of molecular sciences, 22*(11). https://doi.org/10.3390/ijms22115549

Sharan, P., & Vellapandian, C. (2024). Hypothalamic-pituitary-adrenal (HPA) axis: Unveiling the potential mechanisms involved in stress-induced Alzheimer's disease and depression. *Cureus*. https://doi.org/10.7759/cureus.67595

Petito, G., Cioffi, F., Magnacca, N., de Lange, P., Senese, R., & Lanni, A. (2023). Adipose tissue remodeling in obesity: An overview of the actions of thyroid hormones and their derivatives. *Pharmaceuticals, 16*(4). https://doi.org/10.3390/ph16040572

Przybylak, M., Grabowski, J., & Bidzan, L. (2021). Cognitive functions and thyroid hormone secretion disorders. *Psychiatria Polska, 55*(2), 309–321. https://doi.org/10.12740/PP/112470

Ren, B., Ma, J., Tao, M., Jing, G., Han, S., Zhou, C., Wang, X., & Wang, J. (2023). The disturbance of thyroid-associated hormone and its receptors in the brain and blood circulation existed in the early stage of the mouse model of Alzheimer's disease. *Aging, 15*.

Reul, S., Lohmann, H., Wiendl, H., Duning, T., & Johnen, A. (2017). Can cognitive assessment discriminate early stages of Alzheimer's and behavioral variant frontotemporal dementia at initial clinical presentation? *Alzheimer's Research & Therapy, 9*(1). https://doi.org/10.1186/s13195-017-0287-1

Ritchie, M., & Yeap, B. B. (2015). Thyroid hormone: Influences on mood and cognition in adults. *Maturitas, 81*(2), 266–275. https://doi.org/10.1016/j.maturitas.2015.03.016

Salehipour, A., Dolatshahi, M., Haghshomar, M., & Amin, J. (2023). The role of thyroid dysfunction in Alzheimer's disease: A systematic review and meta-analysis. *Journal of prevention of Alzheimer's disease, 10*(2), 276–286. https://doi.org/10.14283/jpad.2023.20

Sampaolo, S., Campos-Barros, A., Mazziotti, G., Carlomagno, S., Sannino, V., Amato, G., Carella, C., & Di Iorio, G. (2005). Increased cerebrospinal fluid levels of 3,3′,5′- triiodothyronine in patients with Alzheimer's disease. *Journal of Clinical Endocrinology and Metabolism, 90*(1), 198–202. https://doi.org/10.1210/jc.2004-1083

Samuels, M. H. (n.d.). Psychiatric and cognitive manifestations of hypothyroidism. https://doi.org/10.1097/MED..

Schroeder, A. C., & Privalsky, M. L. (2014). Thyroid hormones, T_3 and T_4, in the brain. *Frontiers in endocrinology, 5*. https://doi.org/10.3389/fendo.2014.00040

Spires, T. L., & Hyman, B. T. (n.d.). Transgenic models of Alzheimer's disease: Learning from animals..

Tang, X., Song, Z. H., Wang, D., Yang, J., Cardoso, M. A., Zhou, J. B., & Simó, R. (2021). Spectrum of thyroid dysfunction and dementia: A dose–response meta-analysis of 344,248 individuals from cohort studies. *Endocrine Connections, 10*(4), 410–421. https://doi.org/10.1530/EC-21-0047

van der Cammen, T. J. M., Roos, M. P. F. M., & de Jong, M.-C. (2003). Lack of association between thyroid disorders and Alzheimer's disease in older persons: A cross-sectional observational study in a geriatric outpatient population. *Journal of the American Geriatrics Society, 51*.

Van Vliet, N. A., Van Heemst, D., Almeida, O. P., Åsvold, B. O., Aubert, C. E., Bae, J. B., Barnes, L. E., Bauer, D. C., Blauw, G. J., Brayne, C., Cappola, A. R., Ceresini, G., Comijs, H. C., Dartigues, J. F., Degryse, J. M., Dullaart, R. P. F., Van Eersel, M. E. A., Den Elzen, W. P. J., Ferrucci, L., … Trompet, S. (2021). Association of thyroid dysfunction with cognitive function: An individual participant data analysis. *JAMA internal medicine, 181* (11), 1440–1450. https://doi.org/10.1001/jamainternmed.2021.5078

Vancamp, P., Butruille, L., Demeneix, B. A., & Remaud, S. (2020). Thyroid hormone and neural stem cells: Repair potential following brain and spinal cord injury. *Frontiers in neuroscience, 14*. https://doi.org/10.3389/fnins.2020.00875

Warren, S. L., Reid, E., Whitfield, P., Helal, A. M., Abo Hamza, E. G., Tindle, R., Moustafa, A. A., & Hamid, M. S. (2024). Cognitive and behavioral abnormalities in individuals with Alzheimer's disease, mild cognitive impairment, and subjective memory complaints. *Current Psychology, 43*(1), 800–810. https://doi.org/10.1007/s12144-023-04281-1

Webster, S. J., Bachstetter, A. D., Nelson, P. T., Schmitt, F. A., & Van Eldik, L. J. (2014). Using mice to model Alzheimer's dementia: An overview of the clinical disease and the preclinical behavioral changes in 10 mouse models. *Frontiers in Genetics, 5*. https://doi.org/10.3389/fgene.2014.00088

Xu, J., Tiantan Hospital, B., Liu, G.-P., Yoshimura, R., Japan, H., Tan, L., L-y, M., Y-n, O., D-d, Z., Q-y, L., Ma, L.-Y., Zhao, B., Ou, Y.-N., Zhang, D.-D., & Li, Q.-Y. (n.d.). Open access edited by association of thyroid disease with risks of dementia and cognitive impairment: A meta-analysis and systematic review. www.R-project.org/..

Younesi, E., & Hofmann-Apitius, M. (2013). A network model of genomic hormone interactions underlying dementia and its translational validation through serendipitous off-target effect. http://bishop.scai..

Zhang, M., Gong, W., Zhang, D., Ji, M., Chen, B., Chen, B., Li, X., Zhou, Y., Dong, C., Wen, G., Zhan, X., Wu, X., Cui, L., Feng, Y., Wang, S., Yuan, H., Xu, E., Xia, M., Verkhratsky, A., & Li, B. (2022). Ageing related thyroid deficiency increases brain-targeted transport of liver-derived ApoE4-laden exosomes leading to cognitive impairment. *Cell Death and Disease, 13*(4). https://doi.org/10.1038/s41419-022-04858-x

Zhang, J., Zhang, Y., Wang, J., Xia, Y., Zhang, J., & Chen, L. (2024). Recent advances in Alzheimer's disease: Mechanisms, clinical trials and new drug development strategies. *Signal Transduction and Targeted Therapy, 9*(1). https://doi.org/10.1038/s41392-024-01911-3

Chapter 10

Diseases of the thyroid gland – an overview

Stanisław Surma[1] and Kazimierz Ciechanowski[2]
[1]*Department of Internal Medicine and Clinical Pharmacology, Medical University of Silesia, Katowice, Poland;* [2]*Department of Nephrology, Transplantology and Internal Medicine, Pomeranian Medical University of Szczecin, Szczecin, Poland*

List of abbreviations

^{131}I Radioactive Iodine-131
^{99m}Tc Technetium-99m
AITD Autoimmune Thyroid Disease
ALT Alanine Aminotransferase
AST Aspartate Aminotransferase
CAS Clinical Activity Score
CBC Complete Blood Count
CPK Creatine Phosphokinase
CRP C-Reactive Protein
CT Computed Tomography
ESR Erythrocyte Sedimentation Rate
FNAB Fine-Needle Aspiration Biopsy
fT3 Free Triiodothyronine
fT4 Free Thyroxine
GO Graves' Orbitopathy
L-T4 Levothyroxine
LDH Lactate Dehydrogenase
LDL Low-Density Lipoprotein
MRI Magnetic Resonance Imaging
NSAIDs Non-Steroidal Anti-Inflammatory Drugs
rT3 Reverse Triiodothyronine
SHBG Sex Hormone-Binding Globulin
T3 Triiodothyronine
T4 Thyroxine
Tg Thyroglobulin
TPO Thyroid Peroxidase
TRAb Thyrotropin Receptor Antibodies
TRH Thyrotropin-Releasing Hormone
TSH Thyroid-Stimulating Hormone
US Ultrasonography
WHO World Health Organization

Introduction

The thyroid is an unpaired, highly vascularized endocrine gland located in the anterior-lower part of the neck, anterior to the trachea. It consists of two lateral lobes connected by an isthmus, and in some individuals, an additional pyramidal lobe, a remnant of the thyroglossal duct, may be present. The thyroid gland has an average volume of 18–20 mL in women and 25 mL in men. The gland's primary function is the synthesis and secretion of thyroxine (T_4) and triiodothyronine (T_3), which play a key role in regulating metabolism, growth, and tissue differentiation. Additionally, the thyroid produces calcitonin, a hormone essential for calcium and phosphate metabolism (Hershman & Beck-Peccoz, 2023; Assi, 2023).

The basic structural unit of the thyroid gland is the thyroid follicle, composed of thyrocytes surrounding a colloid containing thyroglobulin. Iodine is an essential substrate for thyroid hormone synthesis, and its transport into thyrocytes is actively mediated by the sodium-iodine symporter. Subsequent stages of synthesis include iodine oxidation, iodination of thyroglobulin tyrosine residues, and the conjugation of iodotyrosines, leading to the formation of T_3 and T_4, with the participation of thyroid peroxidase (TPO). These hormones are then stored in the colloid and released into the circulation in response to body demand (Hershman & Beck-Peccoz, 2023; Assi, 2023).

In serum, over 99% of thyroid hormones are bound to transport proteins such as thyroxine-binding globulin, transthyretin, and albumin. Only a small fraction of free T_3

Hormones in Neuroscience. https://doi.org/10.1016/B978-0-443-30261-9.00010-3

and fT_4 exhibits biological activity and is used in laboratory diagnostics. Thyroid hormone metabolism occurs primarily through peripheral deiodination processes, catalyzed by deiodinases type 1, 2, and 3. The conversion of T_4 to the active form T_3 is a fundamental mechanism of hormonal regulation in target tissues, while the formation of inactive rT_3 serves as a protective mechanism against hormone excess (Hershman & Beck-Peccoz, 2023; Assi, 2023).

Thyroid hormone action is primarily mediated by nuclear receptors that modulate the expression of numerous genes, influencing metabolic processes, thermogenesis, and cardiovascular function. Regulation of the hypothalamic-pituitary-thyroid axis is based on a negative feedback mechanism, in which thyrotropin-releasing hormone (TRH) and thyroid-stimulating hormone (TSH) control thyroid hormone secretion, and their concentrations reciprocally inhibit the secretion of tropic hormones (Hershman & Beck-Peccoz, 2023; Assi, 2023) (Fig. 10.1).

Hypothyroidism: Definitions and causes

Hypothyroidism encompasses several key clinical and biochemical concepts. Overt hypothyroidism is defined as a set of clinical symptoms resulting from thyroid hormone deficiency (Zamwar & Muneshwar, 2023; Parretti et al., 2016; Bianco, 2024).

Subclinical hypothyroidism, also known as latent hypothyroidism, is diagnosed primarily based on laboratory tests. Characteristically, elevated serum TSH levels are present with normal fT_4 and fT_3 levels. Before making a diagnosis, it is necessary to exclude other potential causes of TSH elevation, such as the effects of certain medications (e.g., metoclopramide, haloperidol, or chlorpromazine), convalescence from serious illnesses, coexisting psychiatric disorders, or adrenal insufficiency (Zamwar & Muneshwar, 2023; Parretti et al., 2016; Bianco, 2024).

A specific, life-threatening form of extremely severe, untreated hypothyroidism is hypometabolic coma (myxoedema coma). This complication is associated with hypothermia, respiratory failure, hyponatremia, and hypoglycemia. It occurs rarely, primarily in the elderly. Triggering factors include infections, exposure to cold, heart attack, pulmonary embolism, circulatory failure, and the effects of certain medications, such as sedatives or general anesthesia (Zamwar & Muneshwar, 2023; Parretti et al., 2016; Bianco, 2024).

Hypothyroidism can be a consequence of primary dysfunction of the thyroid gland itself, which accounts for over 99% of cases, or result from disorders at the pituitary or hypothalamus level, leading to secondary (central) hypothyroidism (Zamwar & Muneshwar, 2023; Parretti et al., 2016; Bianco, 2024).

The most common cause of primary hypothyroidism in populations with adequate iodine intake remains chronic autoimmune thyroiditis, also known as Hashimoto's thyroiditis. Other causes of primary hypothyroidism include iatrogenic damage or removal of the thyroid gland, treatment with radioactive iodine, thyroidectomy, and neck radiation. Iodine deficiency also plays a significant role, particularly in developing countries. Hypothyroidism can also result from other forms of thyroiditis, such as

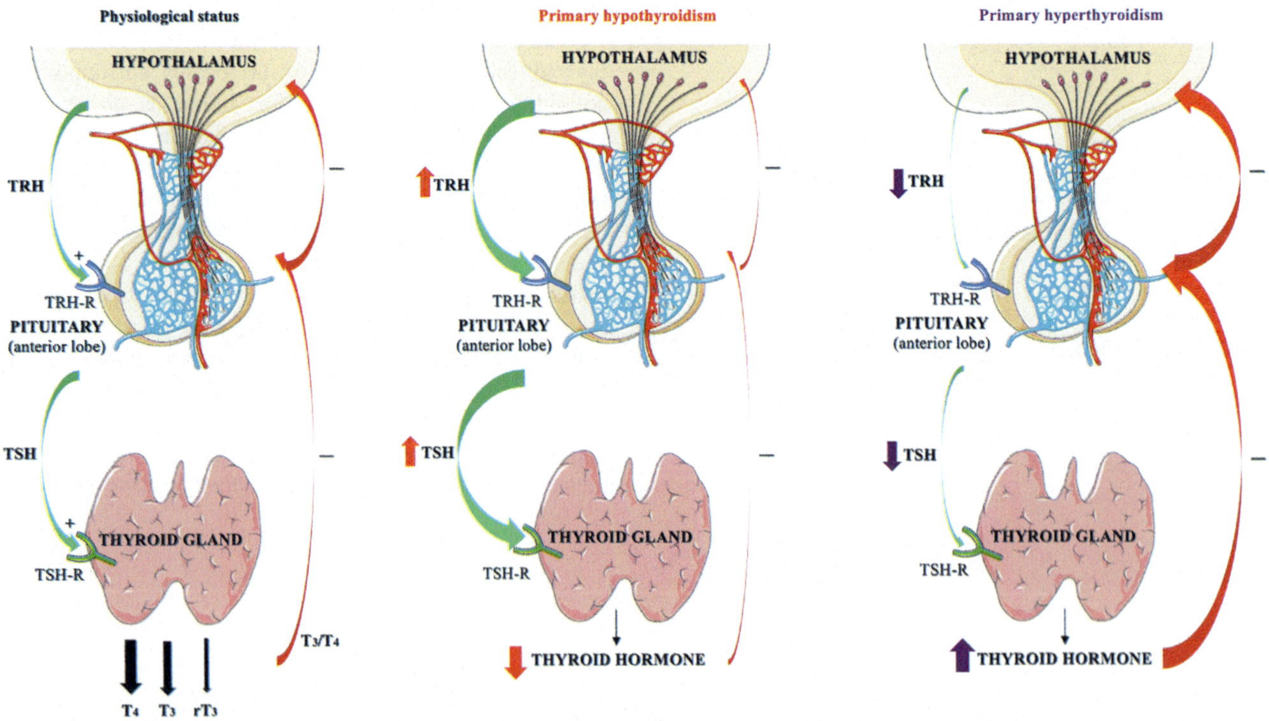

FIGURE 10.1 The hypothalamic-pituitary-thyroid axis in states of physiology and pathology.

subacute, postpartum, or silent thyroiditis, which can be transient or permanent (Zamwar & Muneshwa, 2023; Parretti et al., 2016; Bianco, 2024).

Furthermore, etiological factors include certain medications, including antithyroid agents, lithium salts, interferon α, amiodarone, and tyrosine kinase inhibitors. Congenital abnormalities of thyroid development, such as agenesis, hypoplasia, or ectopia, can also lead to hypothyroidism. Less commonly, it can be caused by thyroid destruction in conditions such as Riedel's goiter, hemochromatosis, sarcoidosis, or amyloidosis (Zamwar & Muneshwar, 2023; Parretti et al., 2016; Bianco, 2024).

Secondary hypothyroidism is most often associated with a pituitary tumor, skull trauma, pituitary apoplexy, neurosurgery, or radiation therapy to the hypothalamic-pituitary region. Isolated secondary hypothyroidism is exceptionally rare and usually coexists with multihormonal pituitary insufficiency, which also includes secondary adrenal and gonadal insufficiency (Zamwar & Muneshwar, 2023; Parretti et al., 2016; Bianco, 2024).

Hypothyroidism: Clinical symptoms

The clinical presentation of overt hypothyroidism in adults depends on the duration of the disease and the degree of thyroid hormone deficiency (Table 10.1).

Common symptoms of primary hypothyroidism include chronic fatigue, depressed mood, memory impairment, cold intolerance, weight gain, constipation, menstrual irregularities, and muscle cramps (Zamwar & Muneshwar, 2023; Parretti et al., 2016; Bianco, 2024).

Physical examination reveals psychomotor retardation, poor facial expression, as well as dry, rough, inelastic, and pale skin with a yellowish tint. Hyperkeratosis, particularly around the elbows, referred to as the "dirty elbow sign," may be characteristic. Facial swelling, especially around the eyelids, and swelling of the extremities that do not

TABLE 10.1 Clinical manifestations of hypothyroidism.

System/Organ	History	Physical examination
General symptoms	Fatigue, psychomotor slowing, somnolence, weakness	-
Metabolic disorders	Weight gain despite reduced appetite; decreased body temperature; cold intolerance; anemia; hypercholesterolemia; hyponatremia; elevated muscle and liver enzymes (AST, ALT, LDH, CPK); hyperprolactinemia	-
Nervous system	Cognitive impairment; memory disturbances; emotional instability; depression; in extreme cases dementia or coma; paresthesias	Psychomotor slowing; slowed speech; hearing impairment; mononeuropathies (e.g., carpal tunnel syndrome); decreased tendon reflexes
Skin changes	Hair loss; brittle nails	Pale, yellowish, dry, cold, thickened and scaling skin; non-pitting edema; dry, brittle, sparse hair
Musculoskeletal system	Muscle pain, cramps, weakness; stiffness, joint pain and swelling	Reduced muscle strength
Respiratory system	Exertional dyspnea	Hoarseness, muffled voice (vocal cord thickening, enlarged tongue); shallow breathing with reduced respiratory rate; upper airway obstruction symptoms; obstructive sleep apnea; in severe cases respiratory failure and pleural effusion
Cardiovascular system	Palpitations; reduced exercise tolerance	Bradycardia; ventricular arrhythmias; diastolic hypertension; in severe cases congestive heart failure with pericardial effusion
Gastrointestinal system	Abdominal bloating; chronic constipation; gallstones	In severe cases intestinal obstruction; ascites
Genitourinary system	Women: Menstrual disturbances (shortened cycle, heavy menses), infertility, miscarriages; men: Decreased libido and erectile dysfunction	

Based on information from Parretti H, Okosieme O, Vanderpump M. Current recommendations in the management of hypothyroidism: Developed from a statement by the British Thyroid Association Executive. British Journal of General Practice. October 2016;66(651):538-540. https://doi.org/10.3399/bjgp16X687493; Bianco AC. Emerging therapies in hypothyroidism. Annual Review of Medicine. January 29, 2024;75:307-319. https://doi.org/10.1146/annurev-med-060622-101007.

deform under pressure are also typical. Bradycardia, hoarseness of voice, and slowed reflexes are common (Zamwar & Muneshwar, 2023; Parretti et al., 2016; Bianco, 2024).

The thyroid gland may be normal in size, enlarged, or atrophic, sometimes with increased density, and sometimes with palpable nodules. In severe cases, fluid in the serous cavities, paralytic ileus, hypothermia, respiratory failure, impaired consciousness, and even coma may occur (Zamwar & Muneshwar, 2023; Parretti et al., 2016; Bianco, 2024).

Hypothyroidism in Hashimoto's thyroiditis usually develops gradually. Many symptoms are nonspecific and often appear in older individuals, which can lead to a delay in diagnosis of up to several years (Zamwar & Muneshwar, 2023; Parretti et al., 2016; Bianco, 2024).

It is also important to emphasize that patients with Hashimoto's thyroiditis have a higher incidence of coexisting autoimmune diseases than the general population. The most common include primary adrenal insufficiency, type 1 diabetes, Addison-Biermer anemia, myasthenia gravis, celiac disease, vitiligo, alopecia areata, and rheumatoid arthritis (Zamwar & Muneshwar, 2023; Parretti et al., 2016; Bianco, 2024).

Hypothyroidism: Diagnostic approach

Hypothyroidism is a relatively common condition, affecting approximately 1%–6% of people over 60. The risk of developing the disease increases with age, and women are 5–8 times more likely to develop the disease than men. Diagnosis can be delayed for several years. Diagnostic testing (Fig. 10.2) for hypothyroidism should be considered in patients:

- presenting clinical symptoms suggestive of hypothyroidism,
- with abnormalities detected on palpation of the thyroid gland,
- suffering from other autoimmune diseases (e.g., type 1 diabetes, vitiligo, pernicious anemia, celiac disease, Addison's disease),
- diagnosed with Down or Turner syndrome,
- having a family history of AITD,
- having undergone radiotherapy to the neck,
- having undergone radioiodine or surgical thyroid treatment,
- having psychiatric disorders,
- taking certain medications, such as lithium salts or interferon (Zamwar & Muneshwar, 2023; Parretti et al., 2016; Bianco, 2024; Qari, 2014).

If hypothyroidism is suspected, further diagnostic workup should focus on several key goals: laboratory confirmation of the disease and determination of its cause, especially in cases of primary hypothyroidism (Zamwar & Muneshwar, 2023; Parretti et al., 2016; Bianco, 2024).

The first step is to measure TSH and fT_4 levels. fT_3 levels may remain normal even in severe and long-standing hypothyroidism (Zamwar & Muneshwar, 2023).

The appearance of the thyroid on ultrasound depends on the cause of the hypothyroidism. In autoimmune inflammation, the gland may be heterogeneous and hypoechoic (Zamwar & Muneshwar, 2023).

In patients suspected of Hashimoto's thyroiditis, anti-TPO antibodies are tested, which are present in over 95% of patients. Anti-Tg antibodies are primarily tested when anti-TPO is negative (Zamwar & Muneshwar, 2023).

FNAB is performed in cases of focal thyroid lesions to rule out cancer. It should be noted that Hashimoto's

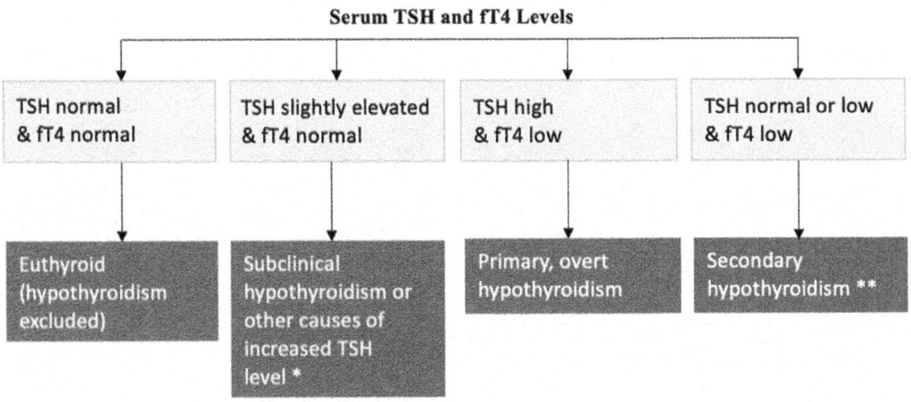

FIGURE 10.2 General principles of diagnosing hypothyroidism. *Based on information from Parretti H, Okosieme O, Vanderpump M. Current recommendations in the management of hypothyroidism: Developed from a statement by the British Thyroid Association Executive. British Journal of General Practice. October 2016;66 (651):538-540. https://doi.org/10.3399/bjgp16X687493; Bianco AC. Emerging therapies in hypothyroidism. Annual Review of Medicine. January 29, 2024;75:307-319. https://doi.org/10.1146/annurev-med-060622-101007. *Before diagnosing subclinical hypothyroidism, other causes of increased TSH level should be excluded, such as: medications (metoclopramide, haloperidol, chlorpromazine), recovery phase after severe illness, psychiatric disorders, adrenal insufficiency; ** If secondary hypothyroidism (damage of hypothalamic-pituitary axis) TSH may be decreased (most often), within normal level, or slightly elevated. Isolated secondary hypothyroidism is very rare - polyhypothyroidism is the most common cause.*

thyroiditis carries a small risk of developing thyroid lymphoma, and some cases may be seronegative (Zamwar & Muneshwar, 2023).

In patients with Hashimoto's thyroiditis, the possibility of coexisting autoimmune diseases should be considered (type 1 diabetes, celiac disease, etc.) and, if necessary, diagnostic workup should be expanded. In cases of concurrent hypothyroidism and adrenal insufficiency, treatment should begin with cortisol supplementation (Zamwar & Muneshwar, 2023).

Hypothyroidism: Treatment approach

Hypothyroidism treatment involves correcting thyroid hormone deficiency. The primary medication used in treatment is levothyroxine (L-T_4), but the indications for T_3 administration have not been clearly established (Zamwar & Muneshwar, 2023).

Indications for L-T_4 include:

- overt hypothyroidism,
- subclinical hypothyroidism in individuals under 70 years of age with a TSH >10 mIU/L,
- subclinical hypothyroidism in individuals under 70 years of age with a TSH between the upper limit of normal and 10 mIU/L, if symptoms, goiter, positive anti-TPO, or cardiovascular risk factors are present,
- subclinical hypothyroidism in women planning a pregnancy or who are pregnant,
- subclinical hypothyroidism in individuals over 70 years of age, with an individualized approach (Zamwar & Muneshwar, 2023).

The daily requirement depends on age, gender, body weight, and the etiology of the disease. In patients with overt hypothyroidism, it averages approximately 1.6 μg/kg body weight. Younger individuals can begin treatment with the full dose, while lower initial doses are recommended for patients over 50–60 years of age, particularly in patients with coronary artery disease. The drug should be taken on an empty stomach, approximately 30–60 minutes before a meal (Zamwar & Muneshwar, 2023).

Treatment monitoring is primarily performed by assessing TSH levels 4–8 weeks after initiating therapy or changing the dose. Target values depend on age. Once the optimal dose has been determined, TSH levels are monitored after 6–12 months. In central hypothyroidism, fT_4 levels are monitored (Zamwar & Muneshwar, 2023).

Adverse effects are rare, most often occurring at the beginning of treatment, such as palpitations, muscle pain, or headaches. If a change in medication is necessary, the dose should be adjusted again. Overdosing on L-T_4 can cause symptoms of hyperthyroidism (Zamwar & Muneshwar, 2023).

Some medications impair L-T4 absorption (e.g., iron, calcium, proton pump inhibitors, cholestyramine), others accelerate its metabolism (e.g., rifampicin, carbamazepine, phenytoin), and estrogens increase the concentration of thyroxine-binding proteins (Zamwar & Muneshwar, 2023).

Difficulties in treatment most often result from noncompliance with the medication or malabsorption (Zamwar & Muneshwar, 2023).

Patients with polyendocrine hypothyroidism, severe comorbidities, secondary hypothyroidism, as well as pregnant women, children, and individuals with treatment difficulties should be referred to an endocrinology clinic (Zamwar & Muneshwar, 2023).

Hyperthyroidism: Definitions and causes

Hyperthyroidism is a set of clinical symptoms resulting from increased synthesis and/or release of thyroid hormones, primarily T_4 and T_3, which leads to a generalized acceleration of metabolic processes in the body. Overt hyperthyroidism manifests clinical symptoms that are a consequence of the excessive action of thyroid hormones on numerous target tissues. It is important to emphasize that the term thyrotoxicosis refers to a state of excess thyroid hormones, regardless of their source, and therefore can encompass situations resulting from hormonal overproduction by the thyroid gland as well as cases of exogenous hormone origin, such as iatrogenic or drug-induced hyperthyroidism (Wiersinga et al., 2023).

A specific clinical form is thyroid storm, which poses a direct threat to life. It is a dramatic exacerbation of thyrotoxicosis, characterized by hyperpyrexia, severe cardiovascular disorders, and symptoms of central nervous system and gastrointestinal dysfunction, often in the context of multi-organ failure (Wiersinga et al., 2023).

The diagnosis of subclinical (latent) hyperthyroidism is based on the finding of a decreased serum TSH concentration with normal free thyroid hormone fractions (fT_4 and fT_3). This diagnosis requires the exclusion of other causes of TSH suppression, such as euthyroid sick syndrome, the effects of certain medications (e.g., dopamine, glucocorticoids, or somatostatin analogues), and physiological hormonal changes observed in the first trimester of pregnancy (Wiersinga et al., 2023).

The etiology of hyperthyroidism in the vast majority of cases is related to primary dysfunction of the thyroid gland itself, while secondary hyperthyroidism caused by a TSH-secreting pituitary adenoma is exceptionally rare. From a practical perspective, endogenous hyperthyroidism can be divided into two basic mechanistic groups: conditions associated with excessive thyroid hormone production and situations resulting from their excessive release during the destruction of the glandular parenchyma (Wiersinga et al., 2023).

The most common cause of hyperthyroidism overproduction is Graves' disease, in which autoantibodies that stimulate the TSH receptor on the surface of thyrocytes play a key role, leading to autonomous hypersecretion of hormones. This is the most common cause of thyrotoxicosis, with a peak incidence in the 3rd–5th decade of life. Excessive hormone synthesis can also occur in toxic nodular goiter and autonomous thyroid nodules, often due to mutations activating the TSH receptor or Gs protein, resulting in hormone secretion independent of pituitary control. Thyroid scintigraphy is helpful in diagnosis, demonstrating normal or increased iodine uptake, while treatment primarily relies on the use of antithyroid drugs (Chowdhury et al., 2024; Wiersinga et al., 2023).

A different mechanism is observed in cases where excess thyroid hormones result from their passive release following the destruction of glandular tissue, as occurs in thyroid inflammations, including subacute thyroiditis, postpartum thyroiditis, silent thyroiditis, and drug-induced thyroid damage, such as after amiodarone type 2 or interferon α. In these situations, scintigraphy reveals reduced iodine uptake, and treatment with antithyroid drugs is usually ineffective, as the process does not result from active hormone synthesis (Chowdhury et al., 2024; Wiersinga et al., 2023).

Hyperthyroidism: Clinical symptoms

The clinical presentation of hyperthyroidism is extremely heterogeneous (Table 10.2), which results from the presence of thyroid hormone receptors in almost all body cells.

TABLE 10.2 Clinical presentation of hyperthyroidism.

System/Organ	History	Physical examination
Metabolic disturbances	Heat intolerance due to increased metabolic rate; increased insulin resistance with impaired glucose tolerance; decreased total and LDL cholesterol; elevated serum calcium (enhanced bone resorption).	—
Nervous system	Behavioral and mood disturbances of varying severity (anxiety, irritability, insomnia, psychomotor agitation, depression, psychosis).	Fine tremor of the hands; hyperreflexia.
Ocular changes	Eye pain[a], photophobia[a], tearing[a], diplopia[a], visual impairment[a].	Eyelid retraction; graefe's sign; kocher's sign; möbius sign; stellwag's sign; exophthalmos[a], eyelid/conjunctival edema[a], eyelid/conjunctival erythema[a], inflammation of the caruncle and/or plica semilunaris.
Skin changes	Increased sweating, brittle nails, increased hair loss.	Warm, moist, smooth skin; pretibial myxedema[a].
Musculoskeletal system	Muscle weakness; osteoporosis and increased fracture risk in long-standing hyperthyroidism.	Reduced skeletal muscle mass; thyroid acropachy[a].
Neck changes	Increased neck circumference.	Thyroid of normal size or enlarged; thyroid bruit in vascular goiter[a]; thyroid nodules.
Respiratory system	Dyspnea caused by tracheal compression from goiter.	Hyperventilation due to increased oxygen consumption and carbon dioxide production.
Cardiovascular system	Palpitations, reduced exercise tolerance, worsening angina or exacerbation of heart failure.	Tachycardia; atrial fibrillation; systolic murmur; systolic hypertension with widened pulse pressure.
Gastrointestinal system	Weight loss despite preserved/increased appetite; malabsorption; increased intestinal motility; diarrhea.	Hepatomegaly and jaundice in severe hyperthyroidism.
Genitourinary system	Mild polyuria due to polydipsia and hypercalciuria. Women: Increased SHBG → menstrual disturbances. Men: Decreased libido, erectile dysfunction, gynecomastia.	Gynecomastia (in men).

LDL, low-density lipoprotein; *SHBG*, sex hormone-binding globulin.
[a]*Findings more typical of Graves' disease.*
Based on information from Chowdhury R, Turkdogan S, Silver JA, Hier J, Bursey S, Quttaineh D, Khoury M, Himdi L. Approach to hyperthyroidism. Journal of Otorhinolaryngology, Hearing and Balance Medicine. 2024; 5(2):20. https://doi.org/10.3390/ohbm5020020; Wiersinga WM, Poppe KG, Effraimidis G. Hyperthyroidism: Aetiology, pathogenesis, diagnosis, management, complications, and prognosis. Lancet Diabetes Endocrinol. April 2023;11(4):282-298. https://doi.org/10.1016/S2213-8587(23)00005-0.

Full-blown hyperthyroidism most often develops in young individuals, while in children and adolescents, symptoms can be nonspecific and mild, with psychomotor restlessness, tachycardia, hyperactivity, and thyroid enlargement predominating. In the elderly population, overt hyperthyroidism may present with minimal symptoms, taking the form of thyroid-cardiac syndrome, manifested by atrial fibrillation, exacerbation of coronary artery disease, or heart failure, or apathetic hyperthyroidism, characterized by depression and weight loss (Wiersinga et al., 2023).

Assessing the severity of hyperthyroidism and the risk of developing thyroid storm should be based primarily on the clinical picture, not solely on laboratory fT_4 and fT_3 values. In clinical practice, the Burch and Wartofsky scale is used for this purpose, which allows for the assessment of the risk of thyroid storm based on systemic symptoms (Wiersinga et al., 2023) (Table 10.3).

Hyperthyroidism: Diagnostic approach

In cases of suspected hyperthyroidism, proper diagnostic guidance is crucial, the scope of which depends on the severity of the patient's clinical condition. If a thyroid storm is suspected, posing an immediate threat to life, the

TABLE 10.3 **Risk of thyroid storm – Burch–Wartofsky clinical scoring system. Interpretation of total score: <25 points: Thyroid storm unlikely; 25–44 points: Impending thyroid storm; ≥45 points: Thyroid storm highly suggestive.**

Parameter	Finding	Points
Body temperature (°C)	37.2–37.7	5
	37.8–38.3	10
	38.3–38.8	15
	38.9–39.4	20
	39.4–39.9	25
	>40.0	30
Central nervous system effects	Absent	0
	Mild (agitation)	10
	Moderate (delirium, psychosis, extreme lethargy)	20
	Severe (seizures, coma)	30
Gastrointestinal/hepatic dysfunction	Absent	0
	Moderate (diarrhea, nausea/vomiting, abdominal pain)	10
	Severe (jaundice)	20
Cardiovascular dysfunction – tachycardia (beats/min)	100–109	5
	110–119	10
	120–129	15
	130–139	20
	>140	25
Cardiovascular dysfunction – heart failure	Absent	0
	Mild (pedal edema)	5
	Moderate (bibasilar rales)	10
	Severe (pulmonary edema)	15
Cardiovascular dysfunction – atrial fibrillation	Absent	0
	Present	10
Precipitating event[a]	Absent	0
	Present	10

[a]*Common precipitating factors include withdrawal of antithyroid drugs, radioiodine therapy (rare), iodinated contrast exposure, severe infection, diabetic ketoacidosis, myocardial infarction, acute heart failure, stroke, surgery, childbirth, and major trauma.*

patient requires immediate referral for emergency hospital treatment. In other cases, diagnostics should focus on two primary goals: biochemical confirmation of hyperthyroidism and determining its etiology based on the clinical picture and the results of additional tests (Wiersinga et al., 2023; Chowdhury et al., 2024; Lee & Pearce, 2023).

The first step is laboratory assessment of thyroid function. Typically, TSH and fT_4 levels are measured, while in selected clinical cases, fT_3 testing is also recommended, particularly in suspected T_3-dependent thyrotoxicosis. Interpretation of these parameters allows for the diagnosis of both overt and subclinical forms of hyperthyroidism (Wiersinga et al., 2023; Chowdhury et al., 2024; Lee & Pearce, 2023).

The next step is identifying the cause of the disorder, which requires integrating clinical data with immunological and imaging test results. The clinical picture itself can provide important diagnostic clues. The coexistence of a vascular goiter with a vascular murmur over the thyroid gland or typical features of Graves' orbitopathy (GO) often allows for the diagnosis of Graves' disease without the need for immediate further confirmatory testing (Wiersinga et al., 2023; Chowdhury et al., 2024; Lee & Pearce, 2023).

The assessment of antithyroid autoantibodies, particularly TSH receptor antibodies (TRAb), plays a significant role in differential diagnosis. Their presence confirms the autoimmune basis of hyperthyroidism and is found in the vast majority of patients with Graves' disease, accounting for approximately 95% of cases. TRAb testing also has prognostic significance and allows for monitoring disease activity (Wiersinga et al., 2023; Chowdhury et al., 2024; Lee & Pearce, 2023).

The primary imaging test used in the diagnosis of thyroid diseases remains ultrasonography (US), which allows for the assessment of the gland's size, echostructure, and the presence of focal lesions. This test is also a key tool in qualifying patients for fine-needle aspiration biopsy (FNAB) and assessing regional lymph nodes. Additionally, the use of Doppler techniques allows for the analysis of blood flow in the thyroid parenchyma, which can be helpful in differentiating the etiology of hyperthyroidism, especially between autoimmune and non-autoimmune processes (Wiersinga et al., 2023; Chowdhury et al., 2024; Lee & Pearce, 2023).

In cases requiring further assessment of thyroid function, isotope diagnostics are particularly important, primarily scintigraphy using radioiodine (most often ^{131}I) or technetium (usually ^{99m}Tc). This test allows for the simultaneous assessment of the morphology and metabolic activity of the gland. Scintigraphy is used in differentiating the causes of thyrotoxicosis, qualifying patients for radioiodine treatment by assessing iodine uptake, diagnosing retrosternal goiter, and in the evaluation of patients with thyroid cancer, particularly in the detection of iodine-sequestering metastases (Wiersinga et al., 2023; Chowdhury et al., 2024; Lee & Pearce, 2023).

Fine-needle aspiration biopsy (FNAB) is performed to diagnose focal thyroid lesions and to exclude cancer. It is the primary method for cytological evaluation of thyroid nodules. FNAB can also be helpful in diagnosing thyroiditis, including subacute thyroiditis, which may present with transient thyrotoxicosis resulting from destruction of the glandular parenchyma (Wiersinga et al., 2023; Chowdhury et al., 2024; Lee & Pearce, 2023).

The general principles of diagnosing hyperthyroidism are presented in Fig. 10.3.

The differential diagnosis of the most common causes of primary hyperthyroidism is presented in Table 10.4.

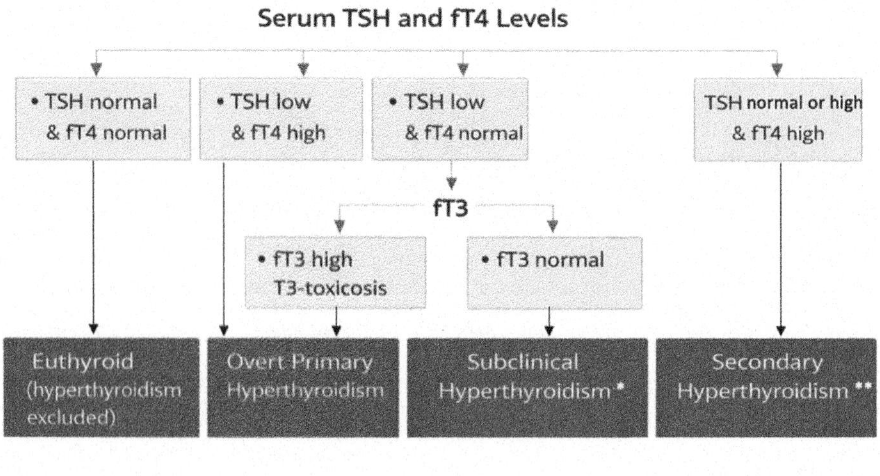

FIGURE 10.3 General principles of diagnosing hyperthyroidism. *Based on information from Wiersinga WM, Poppe KG, Effraimidis G. Hyperthyroidism: Aetiology, pathogenesis, diagnosis, management, complications, and prognosis. Lancet Diabetes Endocrinol. April 2023;11 (4):282-298. doi: 10.1016/S2213-8587(23) 00005-0; Chowdhury R, Turkdogan S, Silver JA, Hier J, Bursey S, Quttaineh D, Khoury M, Himdi L. Approach to hyperthyroidism. Journal of Otorhinolaryngology, Hearing and Balance Medicine. 2024; 5(2):20. https://doi.org/10.3390/ohbm5020020; Lee SY, Pearce EN. Hyperthyroidism: A review. Journal of the American Medical Association. October 17, 2023;330(15):1472-1483. https://doi.org/10.1001/jama.2023.19052. *Before diagnosing hyperthyroidism, other causes of decreased TSH should be excluded: effects of drugs such as glucocorticoids, dopamine, somatostatine analogs, non-thyroidal illness (e.g., acute illness, sepsis, starvation), first trimester of pregnancy; ** TSH-producing pituitary adenoma should be considered with a thyroid hormone resistancy.*

TABLE 10.4 Differential diagnosis of the most common causes of primary hyperthyroidism.

Parameter	Graves' disease	Toxic nodular goiter/Autonomous thyroid nodule	Subacute thyroiditis
Etiology	Autoimmune thyroid disease	TSHR or G protein mutation	Viral
Orbitopathy	~25%	Absent	Absent
Thyroid bruit	~15%	Absent	Absent
Inflammatory markers	Within normal range	Within normal range	Elevated ESR, leukocytosis, increased CRP
TRAb	~95% positive	Absent	Absent
Ultrasound (US)	Most often diffuse goiter, reduced echogenicity, increased vascularity[a]	Focal lesions present; normal echogenicity outside nodules	Heterogeneous reduced echogenicity, decreased vascularity
Thyroid scintigraphy/radioiodine uptake	Diffuse homogeneous tracer uptake; normal or increased uptake	Focal/heterogeneous tracer uptake; normal or increased uptake	Low radioiodine uptake in the initial phase of disease

CRP, C-reactive protein; *ESR*, erythrocyte sedimentation rate; *TRAb*, TSH receptor antibodies; *US*, ultrasonography.
[a]*Absence of goiter or presence of thyroid nodules does not exclude Graves' disease.*
Based on information from Wiersinga WM, Poppe KG, Effraimidis G. Hyperthyroidism: Aetiology, pathogenesis, diagnosis, management, complications, and prognosis. Lancet Diabetes Endocrinol. April 2023;11(4):282-298. https://doi.org/10.1016/S2213-8587(23)00005-0; Chowdhury R, Turkdogan S, Silver JA, Hier J, Bursey S, Quttaineh D, Khoury M, Himdi L. Approach to hyperthyroidism. Journal of Otorhinolaryngology, Hearing and Balance Medicine. 2024; 5(2):20. https://doi.org/10.3390/ohbm5020020; Lee SY, Pearce EN. Hyperthyroidism: A review. Journal of the American Medical Association. October 17, 2023;330(15):1472-1483. https://doi.org/10.1001/jama.2023.19052.

Hyperthyroidism: Treatment

The primary goal of hyperthyroidism treatment is to initially achieve euthyroidism, then maintain it long-term and minimize the risk of relapse. The vast majority of patients with overt hyperthyroidism require treatment and monitoring in a specialized endocrinology clinic.

Once thyroid hormone levels are balanced, further therapeutic management should be tailored to the individual patient. Clinical decisions should take into account the etiology and severity of hyperthyroidism, the presence of comorbidities, the patient's age, and their treatment preferences (Wiersinga et al., 2023; Chowdhury et al., 2024; Lee & Pearce, 2023).

Long-term pharmacological treatment is primarily used in patients with hyperthyroidism due to Graves' disease, in whom there is a realistic chance of achieving lasting clinical and biochemical remission. This applies particularly to patients with mild to moderate hyperthyroidism, moderately elevated anti-TSHR antibody titers, and without significant thyroid enlargement. This treatment is also preferred in patients with concomitant thyroid-associated orbitopathy. The adverse event profile of antithyroid medications is discussed in the previous sections of this chapter (Wiersinga et al., 2023; Chowdhury et al., 2024; Lee & Pearce, 2023).

Radioiodine therapy is the most commonly used treatment for hyperthyroidism in patients with preserved thyroid iodine uptake. Potential adverse effects include a transient worsening of hyperthyroidism symptoms and the possibility of exacerbation of active thyroid-associated orbitopathy (Wiersinga et al., 2023; Chowdhury et al., 2024; Lee & Pearce, 2023).

In patients with Graves' disease and concomitant mild, active orbitopathy, as well as in patients with risk factors for its development (such as smoking or high TRAb antibody levels), prophylactic glucocorticoid therapy is recommended. Prednisone is most commonly used at a dose of 0.4–0.5 mg/kg body weight daily for approximately 4 weeks, followed by a gradual dose reduction over the next 2 months. Before planned radioiodine treatment, thyroid ultrasound and fine-needle aspiration biopsy of focal lesions suspected of being malignant are recommended (Wiersinga et al., 2023; Chowdhury et al., 2024; Lee & Pearce, 2023).

The most common long-term consequence of radioiodine therapy is the development of hypothyroidism, with the highest risk occurring in patients with Graves' disease and the lowest in patients with a single autonomous thyroid nodule. Therefore, long-term monitoring of thyroid hormone function is necessary after radioiodine treatment (Wiersinga et al., 2023; Chowdhury et al., 2024; Lee & Pearce, 2023).

Surgical treatment is used in patients with known or suspected thyroid malignancy associated with hormonal hyperfunction, as well as in patients with a large goiter causing pressure symptoms such as shortness of breath, hoarseness, or dysphagia. The most serious complications of surgical treatment include hypoparathyroidism and damage to the recurrent laryngeal nerve (Wiersinga et al., 2023; Chowdhury et al., 2024; Lee & Pearce, 2023).

A permanent consequence of thyroidectomy is hypothyroidism, which requires lifelong levothyroxine (L-T4) replacement therapy (Wiersinga et al., 2023; Chowdhury et al., 2024; Lee & Pearce, 2023).

Management of thyroid storm

Thyroid storm is the most severe, life-threatening manifestation of hyperthyroidism, associated with secondary multi-organ failure. The clinical picture typically includes significant hyperthermia, severe cardiovascular disturbances, gastrointestinal symptoms, and neurological disorders of varying severity. It should be emphasized that the severity of clinical symptoms does not correlate closely with serum free thyroid hormone concentrations (De Almeida et al., 2022).

If thyroid storm is suspected, treatment should be immediate and include urgent hospitalization, preferably to an intensive care unit. Laboratory diagnostics include TSH, fT_3, and fT_4 levels, as well as basic tests to assess the patient's general condition, such as a complete blood count and liver function tests (aminotransferases) (De Almeida et al., 2022).

An important element of management is the exclusion of concomitant infections and the consideration of other conditions with a similar clinical picture in the differential diagnosis, including septic shock, pheochromocytoma, neuroinfection, stroke, withdrawal syndrome, acute liver failure, acute pancreatitis, gastrointestinal perforation, myocardial infarction, acute heart failure, or pulmonary embolism (De Almeida et al., 2022).

Treatment of thyroid storm includes intensive antithyroid therapy. High doses of antithyroid drugs are used, primarily thiamazole (80–240 mg/day in divided doses) or propylthiouracil (800–1600 mg/day), administered orally or by gavage. Beta-blockers are also administered, the dosage of which should be adjusted according to the heart rate, blood pressure, and symptoms of circulatory failure (e.g., propranolol 160–360 mg/day) (De Almeida et al., 2022).

Another component of therapy is glucocorticoids, used to compensate for relative adrenal insufficiency and inhibit the peripheral conversion of thyroxine (T_4) to triiodothyronine (T_3). Clinical practice utilizes hydrocortisone at a dose of 50–100 mg every 6–8 hours or dexamethasone 2 mg every 6 hours (De Almeida et al., 2022).

Inorganic iodine preparations, such as saturated potassium iodide solution or Lugol's iodine, should be administered at least 1 hour after the administration of the antithyroid agent to avoid iodine being used as a substrate for further thyroid hormone synthesis. In selected cases, iodine-containing contrast agents (e.g., iohexol) are also used, which further inhibit the conversion of T_4 to T_3 (De Almeida et al., 2022).

Supportive care includes symptomatic treatment (paracetamol), intensive intravenous hydration, correction of fluid and electrolyte imbalances, ensuring adequate caloric intake, and treatment of comorbidities. Sedatives, anticonvulsants, and thromboprophylaxis may also be necessary. A thyroid storm is an indication for further radical treatment of hyperthyroidism (De Almeida et al., 2022).

Management of thyroid-associated orbitopathy

Thyroid-associated orbitopathy (GO) is the most common extrathyroid manifestation of Graves' disease. Typical symptoms include widening of the palpebral fissure, eye pain, swelling and redness of the eyelids and conjunctiva, proptosis, diplopia, and in more advanced cases, decreased visual acuity and impaired color vision (Burch et al., 2022).

If GO is suspected, thyroid function should be assessed by measuring TSH, fT_4, and fT_3, as well as performing diagnostic tests for Graves' disease, including TRAb antibody testing and thyroid ultrasound. Orbital imaging (ultrasound, computed tomography, or magnetic resonance imaging) is recommended, particularly in equivocal situations or when complications are suspected (Burch et al., 2022).

After confirming the diagnosis of orbitopathy, its severity should be determined according to the EUGOGO classification and clinical activity assessed based on the CAS (Clinical Activity Score) (Burch et al., 2022).

Therapeutic management primarily involves eliminating risk factors, particularly complete smoking cessation, implementing topical therapy, and maintaining sustained euthyroidism. Patients with moderate to severe orbitopathy, especially those with vision-threatening symptoms, should be treated in specialized referral centers (Burch et al., 2022).

Glucocorticoids remain the first-line treatment for the active phase of moderate to severe orbitopathy and optic neuropathy. Intravenous pulses of high-dose methylprednisolone are preferred (Burch et al., 2022).

In cases where conservative treatment is ineffective or compression neuropathy is present, urgent surgical decompression of the orbits is indicated. Once the active phase subsides, planned, multi-stage treatment of the orbitopathy sequelae is possible, including decompression procedures, oculomotor muscle surgery, and eyelid surgery (Burch et al., 2022).

Orbital radiotherapy can be an adjunct to glucocorticoid therapy, and a combination of both methods can be more effective in achieving lasting clinical improvement (Burch et al., 2022).

Other thyroid diseases

Thyroid lesions

Thyroid lesions can include various pathologies that are diagnosed both during a clinical examination and through diagnostic imaging. The most common are thyroid nodules, which are palpable lesions within the gland, and focal lesions visualized on ultrasound. In clinical practice, an incidentaloma is also an important term, describing a lesion discovered incidentally during neck imaging studies performed for other reasons, such as carotid artery or salivary gland ultrasound, or chest computed tomography (CT) scans. Therefore, thyroid lesions can be detected both during a physical examination and during radiological diagnostics (Durante et al., 2023).

A goiter is any enlargement of the thyroid gland, with a cutoff value on ultrasound exceeding approximately 18–20 mL in women and 25 mL in men. The term also encompasses cases of ectopic thyroid tissue, such as a lingual or ovarian goiter. The World Health Organization classifies goiter size based on palpation: grade zero, in which the goiter is absent; grade one, when it is palpable but not visible with the neck in the correct position; and grade two, when neck enlargement is visible at rest and confirmed by palpation (Durante et al., 2023).

The assessment of a goiter considers not only its size but also its structure and hormonal activity. Goiters are classified as parenchymal or nodular, as well as neutral, hyperthyroid, and hypothyroid goiters, depending on their effect on thyroid function (Durante et al., 2023).

A goiter can develop asymptomatically for many years and be discovered incidentally during an imaging examination. In larger cases, patients may notice a gradual increase in neck circumference or the presence of a growing nodule in the thyroid area. It is especially important to first consider the possibility of malignancy in cases of rapid growth accompanied by pain (Durante et al., 2023).

A large goiter can lead to symptoms of compression resulting from pressure on adjacent structures. Compression on the trachea can cause shortness of breath, coughing, and stridor, while compression on the esophagus can result in difficulty swallowing. In some cases, compression also occurs on the jugular veins, leading to venous congestion and facial flushing, or on the recurrent laryngeal nerve, which can result in hoarseness or even loss of voice (Durante et al., 2023).

Neutral parenchymal goiter, also known as nontoxic goiter, is an enlargement of the thyroid gland without focal lesions or hormonal dysfunction. Historically, the main cause of its development was iodine deficiency, which led to decreased thyroid hormone synthesis and a secondary increase in TSH secretion. This hormone stimulates thyrocyte proliferation, which promotes gland enlargement (Durante et al., 2023).

In addition to iodine deficiency, causes of goiter include autoimmune processes, particularly chronic Hashimoto's thyroiditis, as well as the effects of goitrogens present in the environment and diet. Goiter can also develop following radiation to the head and neck area or in the setting of rare genetic disorders (Durante et al., 2023).

If goiter is suspected, diagnostic workup should include confirmation of thyroid enlargement by ultrasound, assessment of hormonal function based on TSH levels, and exclusion of an autoimmune basis by testing for antithyroid antibodies. In cases of pressure symptoms, it is also necessary to assess the indications for surgical treatment. Parenchymal goiter often requires no treatment, but in selected cases, a trial of L-thyroxine therapy may be considered (Durante et al., 2023).

A parenchymal goiter may develop into a nodular goiter over time. Nodular thyroid disease encompasses both solitary and multinodular forms, regardless of the gland's hormonal function. Thyroid nodules are palpable in approximately 5%–9% of the population, while ultrasound examinations reveal their presence in up to half of adults. The vast majority of nodules are benign, and only a small percentage, estimated at approximately 4%–6%, are malignant (Durante et al., 2023).

The diagnosis of a thyroid nodule requires further diagnostic workup aimed at assessing thyroid function, determining oncological risk, and assessing the local stage of the lesion.

Cancer risk assessment takes into account clinical factors such as rapid nodule growth, the firm, immobile nature of the lesion, lymph node metastases, and radiation exposure, as well as ultrasound features, including microcalcifications, hypoechogenicity, irregular borders, and chaotic vascular flow (Durante et al., 2023).

The most important additional test in the diagnosis of a thyroid nodule is fine-needle aspiration biopsy. It is particularly indicated for focal lesions larger than 1 cm, the presence of high-risk features, or suspicion of metastasis. Lesions smaller than 1 cm typically require ultrasound monitoring, and treatment decisions depend on the biopsy results and the patient's clinical presentation (Durante et al., 2023).

After performing a fine-needle aspiration biopsy (FNAB), the obtained result allows the lesion to be classified into one of several cytological diagnostic categories. Each category is associated with a specific risk of thyroid cancer and indicates the most recommended further management.

The first category includes non-diagnostic biopsy, in which the risk of thyroid cancer is approximately 5%–10%. This result can correspond to virtually any diagnosis, as the sample does not allow for a reliable assessment. In this situation, a second biopsy is necessary, and if there is clinical suspicion of cancer, diagnostics should be continued immediately. Further management depends primarily on the presence of clinical signs of malignancy (Durante et al., 2023).

The second category includes benign lesions, for which the risk of cancer is very low, less than 1%. These most commonly include nodular goiter, thyroiditis, or thyroid cysts. In such cases, repeat FNAB during follow-up is not required unless there are clinical concerns and the quality of the initial biopsy was good. Conservative management is recommended (Durante et al., 2023).

The third category is a follicular lesion of unknown origin, with a risk of malignancy of approximately 2%–5%. In this situation, a definitive cytological diagnosis cannot be established. A repeat biopsy is recommended, usually after 6–12 months, although in practice, follow-up may occur between 6 and 24 months. Conservative management is most often recommended, consisting of observation and ultrasound monitoring approximately every 6 months. For nodules larger than 4 cm or smaller lesions that grow rapidly, surgery is performed (Durante et al., 2023).

The fourth category includes suspected follicular neoplasm, with a risk of thyroid cancer of approximately 8%–20%. This finding may represent either a noncancerous lesion or a benign tumor, which cannot be distinguished cytologically from a malignant tumor. Indications for a second FNAB are not routine, but if surgery is planned, confirmation of the diagnosis by a second cytologist is necessary. In such situations, surgical treatment should be considered, especially if the focal lesion exceeds 2 cm (Durante et al., 2023).

The fifth category concerns suspected malignancy. The risk of thyroid cancer in this group is approximately 75%. This finding includes suspected thyroid cancer, including papillary carcinoma, medullary carcinoma, lymphoma, thyroid metastases, and anaplastic carcinoma. A second biopsy is not recommended, but confirmation by a second cytologist is necessary. In this situation, surgical treatment is recommended (Durante et al., 2023).

The sixth category indicates a diagnosis of malignancy, with a 95%–100% risk of cancer. This group includes papillary carcinoma, medullary carcinoma, anaplastic carcinoma, and other malignant thyroid tumors. As with the previous category, a second FNAB is not performed, but confirmation by a second cytologist is required. Surgery is the recommended treatment (Durante et al., 2023).

The treatment for nodular thyroid disease depends on the degree of toxicity of the nodular goiter. In the case of nontoxic nodular goiter, after clinical and ultrasound assessment of the focal lesions and consideration of the indications for biopsy, one of several management strategies can be chosen. Observation is most often used. In selected cases, levothyroxine treatment is attempted. Surgery is performed in patients with suspected thyroid cancer, a large goiter with compressive symptoms, or a retrosternal goiter. In selected cases, especially when surgery is contraindicated, radioactive iodine treatment is possible (Durante et al., 2023).

Toxic nodular goiter is diagnosed in a patient with nodular goiter and concomitant hyperthyroidism. In such a situation, in addition to clinical and ultrasound assessment of the focal lesions and consideration of the indications for FNAB, antithyroid drug treatment should be initiated. The nodules secrete thyroid hormones autonomously, regardless of TSH levels, therefore, discontinuation of antithyroid drug leads to a recurrence of hyperthyroidism. For this reason, after the patient has been euthyroid, radical treatment is necessary, which includes the use of radioactive iodine (^{131}I) or surgery. Surgery is always performed when malignant changes are suspected and in patients with a large goiter causing pressure on the trachea (Durante et al., 2023).

Thyroid cancers

Thyroid cancers are divided into malignant and benign types. Based on the tumor's origin, malignant thyroid cancers can be divided into several groups. These are most commonly cancers derived from follicular epithelial cells, or thyrocytes. These are differentiated cancers, which together constitute approximately 90% of cases. Papillary and follicular carcinomas are the most common, as are poorly differentiated or undifferentiated cancers, which occur approximately 2%–5% (Filetti et al., 2019).

Medullary carcinomas, which originate from parafollicular cells, or C cells, that produce calcitonin, are another group. They constitute approximately 5% of cases. Malignant thyroid cancers also include MALT lymphomas, sarcomas, and metastases from other organs to the thyroid gland (Filetti et al., 2019).

Diagnostic tests before treatment for thyroid cancer include a history and physical examination, as well as an ultrasound of the neck. Confirmation of the cancer with an ultrasound-guided FNAB is essential. TSH levels are also assessed to rule out thyroid dysfunction. Serum calcium levels should be measured before surgery to rule out calcium-phosphate metabolism disorders. Calcitonin, a marker of medullary carcinoma, is also measured (Filetti et al., 2019).

Additionally, a chest X-ray and an ear, nose, and throat examination are performed to assess vocal cord function. In selected cases, a CT scan of the neck and upper mediastinum is considered, especially if invasion of adjacent structures or distant metastases is suspected. However, it is important to remember that CT scans should be performed without contrast, as the use of iodinated contrast agents in patients with well-differentiated thyroid cancer is contraindicated, as it may delay adjuvant radioiodine therapy by several months (Filetti et al., 2019).

Treatment of thyroid cancer is multi-stage and depends on the type of cancer. In the case of differentiated thyroid cancers, surgery is the primary treatment. In all cases of cancer diagnosed preoperatively, total thyroidectomy, often with lymphadenectomy, is indicated. Surgery should be performed in specialized centers (Filetti et al., 2019).

Radioactive iodine therapy is used in patients with differentiated thyroid cancer as an adjuvant therapy after surgery, particularly in cases of distant metastases or palliative treatment. It is not used in medullary and anaplastic thyroid cancer (Filetti et al., 2019).

After surgery, levothyroxine must be administered in suppressive or semi-suppressive doses, depending on clinical risk. In the case of complete remission after 5 years, a dose change to replacement therapy may be considered (Filetti et al., 2019).

In medullary thyroid cancer, the treatment of choice is total thyroidectomy with lymphadenectomy. Calcitonin remains a sensitive marker of treatment efficacy. Molecular testing for the RET proto-oncogene mutation is performed in every patient, and if the result is positive, screening for family history is recommended (Filetti et al., 2019).

Anaplastic thyroid cancer is most often diagnosed at a stage where radical treatment is impossible. Radioactive iodine treatment is not recommended, and the effectiveness of radiotherapy and chemotherapy is low (Filetti et al., 2019).

The prognosis for differentiated thyroid cancers is excellent. Over 90% of patients achieve a 10-year survival rate with appropriate treatment. The prognosis for follicular thyroid cancer is slightly worse than for papillary thyroid cancer. Most patients with undifferentiated thyroid cancer die within 6 months of diagnosis (Filetti et al., 2019).

Thyroiditis

Thyroiditis is a heterogeneous group of diseases that differ in both etiopathogenesis, clinical presentation, and response to treatment. In clinical practice, the most common form is chronic autoimmune thyroiditis, which is considered the most common cause of this type of condition. Due to the diversity of pathogenetic mechanisms, numerous classifications of thyroiditis exist in the literature, but the most commonly used classifications take into account the clinical course of the disease, histological presentation, and the impact of the inflammatory process on the gland's hormonal function (Martinez Quintero et al., 2021).

Factors that can lead to the development of thyroiditis include primarily autoimmune processes, which are responsible for a large proportion of chronic forms of the disease. Infections of various etiologies, including viral, bacterial, parasitic, and fungal infections, also play a significant role. Thyroiditis can also result from mechanical trauma, such as aggressive palpation of the thyroid gland. Additionally, physical stimuli, such as radioiodine therapy, and chemical agents, particularly certain medications, are among the precipitating factors (Martinez Quintero et al., 2021).

Due to its diverse etiopathogenesis, the clinical presentation of thyroiditis is heterogeneous, and the type and severity of symptoms depend on both the precipitating factor and the duration of the disease process. Patients may present to a physician with a rapidly growing nodule or sudden enlargement of the thyroid gland, which is observed in acute bacterial and subacute thyroiditis, among other conditions. The presence of a goiter is also a common reason for consultation, particularly in autoimmune thyroiditis. Patients may also complain of neck pain, fever, and symptoms related to thyroid dysfunction (Martinez Quintero et al., 2021).

In some forms of thyroiditis, symptoms of hyperthyroidism predominate, as seen in postpartum thyroiditis, acute inflammation following radioiodine treatment, subacute inflammation, and postamiodarone thyroiditis. Some patients, especially in the initial phase of Hashimoto's thyroiditis, may experience transient thyrotoxicosis. In other cases, particularly chronic autoimmune thyroiditis and drug-induced thyroiditis, symptoms of thyroid hypofunction predominate (Martinez Quintero et al., 2021).

The differential diagnosis of thyroiditis relies largely on a detailed clinical interview, which is particularly important in diagnosing specific forms of the disease. Diagnostic procedures also include a range of laboratory tests, including TSH, fT_4, and free fT_3. Evaluation of antithyroid autoantibodies, such as anti-thyroid peroxidase (anti-TPO), anti-thyroglobulin (anti-Tg), and anti-TSH receptor (anti-TSHR), is also crucial. Additional tests include C-reactive protein (CRP), erythrocyte sedimentation rate (ESR), and a complete blood count (CBC) with a differential (Martinez Quintero et al., 2021).

Imaging methods are also used in diagnosis, primarily thyroid ultrasound with Doppler assessment of vascular flow, and in selected cases, scintigraphy and radioiodine uptake testing. In situations requiring further evaluation, FNAB is performed, while if an infectious etiology is suspected, bacteriological tests, including culture, are indicated (Martinez Quintero et al., 2021).

Treatment of thyroiditis depends on the etiological factor and the clinical presentation of the disease. Treatment can be divided into causative treatment, used in cases where the causative factor is known, for example, in bacterial or drug-induced thyroiditis, and symptomatic treatment, which is applicable to all types of thyroiditis. Antibiotic therapy is used for causal treatment of acute bacterial thyroiditis, and surgical drainage may be necessary if an abscess is present. In cases of drug-induced thyroiditis, discontinuation of the causative drug is considered, although this decision always requires individual clinical evaluation (Martinez Quintero et al., 2021).

Symptomatic treatment primarily includes analgesic therapy with nonsteroidal anti-inflammatory drugs (NSAIDs), such as ibuprofen or ketoprofen, and paracetamol. In more severe cases, glucocorticosteroids are used. In cases of thyrotoxicosis, β-blockers, such as propranolol or metoprolol, are recommended. However, thyrostatic drugs are ineffective because in thyroiditis, the excess hormone results from follicular destruction, not increased synthesis. If hypothyroidism develops, replacement therapy with levothyroxine is implemented.

Among the chronic forms of thyroiditis, the most common is chronic autoimmune (lymphocytic) thyroiditis, also known as AITD. It is considered the most common cause of both thyroiditis and hypothyroidism. The disease affects all age groups, but is more common in women. Anti-TPO and/or anti-Tg antibodies are present in over 90% of cases. Clinically, AITD can present with goiter, with preserved thyroid function, or in an atrophic form, leading to subclinical or overt hypothyroidism (Martinez Quintero et al., 2021).

Painless thyroiditis, also known as silent thyroiditis, is characterized by a four-phase course similar to subacute thyroiditis, but without the presence of a painful goiter. Postpartum thyroiditis, on the other hand, manifests in the first year after childbirth and manifests as transient or permanent thyroid dysfunction. It is most often self-limiting but can also lead to permanent hypothyroidism (Martinez Quintero et al., 2021).

Conclusions

Hypothyroidism, usually of autoimmune origin, leads to a decreased metabolic rate and numerous clinical symptoms. Hyperthyroidism, most often resulting from Graves' disease, results in thyrotoxicosis and cardiovascular complications. Diagnosis is based on TSH and free thyroid hormone levels, as well as imaging studies. Treatment includes replacement therapy, antithyroid drugs, radioiodine therapy, or surgical intervention.

Summary points

- The thyroid gland produces the hormones T_3 and T_4, which are essential for proper regulation of the body's metabolism.
- The most common cause of hypothyroidism is chronic autoimmune Hashimoto's thyroiditis.
- Hyperthyroidism most often develops in the course of Graves' disease or toxic nodular goiter.
- The basis of laboratory diagnosis is the measurement of TSH and free thyroxine (fT_4) levels.
- Treatment of hypothyroidism involves long-term levothyroxine replacement therapy.
- Hyperthyroidism requires treatment with antithyroid drugs, radioiodine, or surgery.

Mini-dictionary of terms

- Deiodinases – enzymes that catalyze the peripheral conversion of T4 to the active form T3 or the inactive rT3, regulating hormonal availability in tissues.
- TRAb – autoantibodies directed against the TSH receptor, characteristic of autoimmune hyperthyroidism in Graves' disease.
- Hypometabolic coma – a life-threatening complication of severe, untreated hypothyroidism, associated with hypothermia, hyponatremia, and multiorgan failure.
- Thyroid storm – an acute, critical form of thyrotoxicosis characterized by hyperthermia, circulatory disturbances, and multiorgan failure.
- Fine-needle aspiration biopsy (FNAB) – the primary method for the cytological assessment of thyroid nodules, used in the differential diagnosis of benign and malignant lesions.

Other hormones, endocrine tissues, and pathways

The thyroid gland functions in close integration with other structures of the endocrine system, and its activity is regulated by the hypothalamic-pituitary-thyroid axis. The hypothalamus secretes TRH, which stimulates the pituitary gland to produce TSH. TSH affects thyrocytes, increasing the synthesis and secretion of thyroid hormones. A negative feedback mechanism ensures homeostatic control of T_3 and T_4 concentrations in the circulation.

Thyroid hormones exert a multifaceted influence on other endocrine axes. Hyperthyroidism is associated with increased insulin resistance and impaired carbohydrate metabolism. Furthermore, the thyroid gland participates in the regulation of calcium and phosphate metabolism through the secretion of calcitonin by parafollicular cells (C cells), which inhibits bone resorption.

Autoimmune thyroid diseases often coexist with other immune-mediated conditions, such as type 1 diabetes or primary adrenal insufficiency, requiring a comprehensive assessment of multiple endocrine gland function. Thyroid hormones also affect the cardiovascular system by increasing the sensitivity of β-adrenergic receptors, which predisposes to tachyarrhythmias and heart failure. Understanding these relationships is fundamental to proper clinical management.

Applications of the material

Understanding the pathophysiology and clinical presentation of thyroid diseases is crucial in medical practice, as thyroid dysfunction is among the most frequently diagnosed endocrine disorders. Early identification of hypothyroidism allows for the implementation of thyroid replacement therapy and the prevention of metabolic complications such as dyslipidemia, anemia, and the development of cardiovascular disease. Diagnosis is particularly important in pregnant women and those planning a pregnancy, as even subclinical hormonal imbalances can affect fetal development.

Hyperthyroidism, however, requires prompt diagnosis and treatment due to the risk of serious complications, including atrial fibrillation and thyroid storm. In clinical practice, pharmacological treatment (thyroid agonists), radioiodine therapy, and surgical procedures are used, selected depending on the etiology and the patient's condition.

This material is also useful in the diagnosis of thyroid nodules, which are a common ultrasound finding. Fine-needle aspiration biopsy enables oncological risk assessment and surgical qualification. Furthermore, awareness of coexisting autoimmune diseases allows for a holistic approach to the patient and early detection of multiglandular disorders. Therefore, the presented issues constitute the basis for rational diagnostic and therapeutic procedures in endocrinology.

References

Assi, M. H. (2023). Thyroid gland basics: A comprehensive review. *Mustansiriya Medical Journal, 22*(2), 172–181. https://doi.org/10.4103/mj.mj_43_23

Bianco, A. C. (2024). Emerging therapies in hypothyroidism. *Annual Review of Medicine, 75*, 307–319. https://doi.org/10.1146/annurev-med-060622-101007

Burch, H. B., Perros, P., Bednarczuk, T., Cooper, D. S., Dolman, P. J., Leung, A. M., Mombaerts, I., Salvi, M., & Stan, M. N. (2022). Management of thyroid eye disease: A consensus statement by the American Thyroid Association and the European Thyroid Association. *European Thyroid Journal, 11*(6). https://doi.org/10.1530/ETJ-22-0189

Chowdhury, R., Turkdogan, S., Silver, J. A., Hier, J., Bursey, S., Quttaineh, D., Khoury, M., & Himdi, L. (2024). Approach to hyperthyroidism. *Journal of Otorhinolaryngology, Hearing and Balance Medicine, 5*(2), 20. https://doi.org/10.3390/ohbm5020020

De Almeida, R., McCalmon, S., & Cabandugama, P. K. (2022). Clinical review and update on the management of thyroid storm. *Missouri Medicine, 119*(4), 366–371.

Durante, C., Hegedüs, L., Czarniecka, A., Paschke, R., Russ, G., Schmitt, F., Soares, P., Solymosi, T., & Papini, E. (2023). European Thyroid Association Clinical Practice Guidelines for thyroid nodule management. *European Thyroid Journal, 12*(5). https://doi.org/10.1530/ETJ-23-0067

Filetti, S., Durante, C., Hartl, D., Leboulleux, S., Locati, L. D., Newbold, K., Papotti, M. G., Berruti, A., & ESMO Guidelines Committee. Electronic address: clinicalguidelines@esmo.org. (2019). Thyroid cancer: ESMO clinical practice guidelines for diagnosis, treatment and follow-up. *Annals of Oncology, 30*(12), 1856–1883. https://doi.org/10.1093/annonc/mdz400

Hershman, J. M., & Beck-Peccoz, P. (2023). Discoveries around the hypothalamic-pituitary-thyroid axis. *Thyroid, 33*(7), 785–790. https://doi.org/10.1089/thy.2022.0258

Martinez Quintero, B., Yazbeck, C., & Sweeney, L. B. (2021). Thyroiditis: Evaluation and treatment. *American Family Physician, 104*(6), 609–617.

Lee, S. Y., & Pearce, E. N. (2023). Hyperthyroidism: A review. *Journal of the American Medical Association, 330*(15), 1472–1483. https://doi.org/10.1001/jama.2023.19052

Parretti, H., Okosieme, O., & Vanderpump, M. (2016). Current recommendations in the management of hypothyroidism: Developed from a statement by the British Thyroid Association Executive. *British Journal of General Practice, 66*(651), 538–540. https://doi.org/10.3399/bjgp16X687493

Qari, F. (2014). Hypothyroidism in clinical practice. *Journal of Family Medicine and Primary Care, 3*(2), 98–101. https://doi.org/10.4103/2249-4863.137609

Wiersinga, W. M., Poppe, K. G., & Effraimidis, G. (2023). Hyperthyroidism: Aetiology, pathogenesis, diagnosis, management, complications, and prognosis. *Lancet Diabetes Endocrinol, 11*(4), 282–298. https://doi.org/10.1016/S2213-8587(23)00005-0

Zamwar, U. M., & Muneshwar, K. N. (2023). Epidemiology, types, causes, clinical presentation, diagnosis, and treatment of hypothyroidism. *Cureus, 15*(9). https://doi.org/10.7759/cureus.46241

Chapter 11

The thyroid hormone drives bidirectional communication along the brain-gut axis in vertebrate models

Dennis P. Jacob[1], Meenu Viswanath[2, 3] and M. C. Subhash Peter[2, 3, 4]

[1]*Department of Medicine, Saint Francis Hospital, Hartford, CT, United States;* [2]*Inter-University Centre for Evolutionary and Integrative Biology-iCEIB, School of Life Sciences, University of Kerala, Thiruvananthapuram, Kerala, India;* [3]*Department of Zoology, University of Kerala, Thiruvananthapuram, Kerala, India;* [4]*Sastrajeevan CIS-Life, F-17, Thiruvananthapuram, Kerala, India*

Introduction

The main forms of thyroid hormones (TH) are thyroxine (T_4), the primary circulating form, and triiodothyronine (T_3), the biologically active form (Gavrila & Hollenberg, 2019). They play a crucial role in regulating metabolism across all vertebrates. As products of the hypothalamo-pituitary-thyroid (HPT) axis, THs are essential for maintaining energy balance and cellular homeostasis (McAninch & Bianco, 2014; Peter et al., 2007). Any disruption of this balance can trigger compensatory cellular signals from the brain or peripheral organs that aim to modify TH release from the HPT axis or its availability in the periphery (Deal & Volkoff, 2020; Peter, 2007). In addition to energy redistribution, THs regulate various physiological processes, including ion and osmotic homeostasis (Peter, 2011; Peter et al., 2000) and developmental processes (Sudhakaran & Peter, 2024).

This chapter examines the role of TH in the brain-gut axis of vertebrate models, mainly in fish and mammals. Although TH is known for directly controlling the brain and gut tissues, its overall control of the functional axis remains limited. We thus explore how TH influences the brain-gut axis in these vertebrate models to better understand the integrative physiology of TH, considering its significant role in maintaining physiological balance and disease management.

Brain-gut axis

The brain-gut (B-G) axis is a bidirectional communication system that links the gastrointestinal and central nervous systems, primarily through the enteric nervous system. This physiological network further interacts through endocrine, humoral, metabolic, and immune pathways (Appleton, 2018). However, its comprehensive role in vertebrate models, including zebrafish (Sudhakaran & Peter, 2024) and mice (Nair et al., 2024), remains unclear. The bidirectional network includes the central nervous system (CNS), comprising the brain and spinal cord; the autonomic nervous system (ANS); and the enteric nervous system (ENS). The autonomic system, comprising the sympathetic and parasympathetic branches, transmits afferent signals from the lumen via enteric, spinal, and vagal pathways to the CNS and sends efferent signals from the CNS to the intestinal wall (Carabotti et al., 2015). The ANS and the neural network surrounding the gut connect it to the brain, enabling the brain to regulate intestinal functions, including immune cell activity, and allowing the gut to influence mood, cognition, and mental health (Appleton, 2018). The B-G axis integrates brain-gut functions, linking the emotional and cognitive centers of the brain with peripheral intestinal mechanisms such as immune activation, intestinal permeability, enteric reflexes, and enteroendocrine signaling (Mayer, 2011).

There are two main neuroanatomical pathways through which the gut communicates with the CNS, specifically the brain. One is a direct exchange of information between the gut and brain via the ANS and the vagus nerve (VN) within the spinal cord. Similarly, bidirectional communication occurs between the ENS in the gut and the ANS, with the VN serving as an additional pathway. These systems interact to form an integrated cascade network that regulates gut function (Mulak & Bonaz, 2015). The ENS, comprising the myenteric and submucosal ganglia and

Hormones in Neuroscience. https://doi.org/10.1016/B978-0-443-30261-9.00011-5

enteric glial cells (Schemann & Neunlist, 2004), integrates enteric signals with the CNS. The third level includes the spinal cord, brainstem centers, and the dorsal motor nucleus of the vagus nerve, along with afferent and efferent vagal fibers. Cholinergic neurons in the gut wall and the dorsal motor nucleus regulate vagal activity, influencing the upper gastrointestinal tract (Chang et al., 2003). The highest centers in the brain, including cortical and subcortical regions such as the basal ganglia, communicate downward to specific brainstem nuclei involved in gut regulation, enabling the final fourth-level integration (Fig. 11.1). The vagus nerve, originating from the brainstem, primarily mediates direct neural communication with the gut and also links the gut microbiota to the brain (Mulak & Bonaz, 2015). Disruptions or abnormalities at any of these levels can impair gut function by affecting both local reflexes and external neural regulation (Mulak & Bonaz, 2015). Microbial signals can activate ENS afferent neurons, and vagal signaling can, in turn, stimulate anti-inflammatory pathways that protect against microbial infections (Borre et al., 2014). However, brain function may occur independently of vagal nerve activation (Bauer et al., 2016). Moreover, resident gut bacteria are essential for postnatal development and the maturation of key physiological systems, including the immune, endocrine, and nervous systems (Borre et al., 2014). Despite pathways that regulate the B-G axis in vertebrate models, the role of TH in integrating this axis remains poorly understood. In this chapter, we examine the integrative mechanisms of TH that regulate the brain and gut in vertebrate models.

Like all other tissues that respond to TH, brain and gut tissues, which respond to THs, are regulated by converting T_4 into either more active or less active forms of TH in peripheral tissues through the activity of iodothyronine deiodinase (Dio) enzymes. Vertebrates have three types of iodothyronine deiodinase enzymes: Deo1, Deo2, and Deo3, which remove iodine atoms from either the outer or inner ring of THs, ultimately controlling TH action in target tissues (Orozco & Valverde, 2005). Deo1 catalyzes both outer- and inner-ring deiodination. Deo2 catalyzes

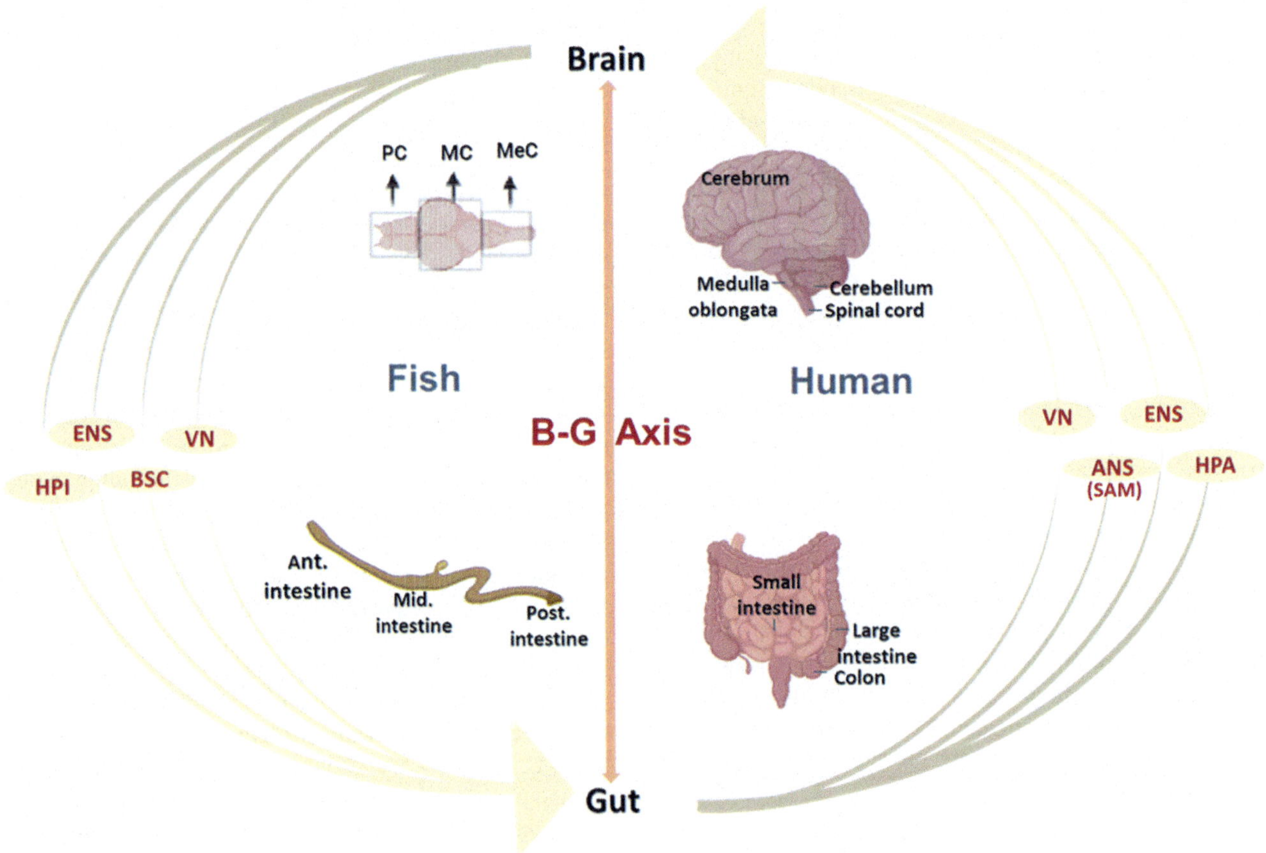

FIGURE 11.1 Diagram illustrating the current bidirectional communication between the brain-gut (B–G) axis in fish and humans. The primary neuroanatomical pathways that facilitate communication between the brain and the gut in humans include the vagus nerve (VN), the autonomic nervous system (ANS) (which comprises the sympathetic adrenomedullary system (SAM), the enteric nervous system (ENS), and the hypothalamic-pituitary-adrenal (HPA) axes). In the fish model, the corresponding mammalian communication centers are the vagus nerve (VN), the autonomic nervous system (ANS), the brain-sympathetic-chromaffin axis (BSC), the enteric nervous system (ENS), and the hypothalamic-pituitary-interrenal (HPI) axis.

outer-ring deiodination by removing iodine from the outer ring of T_4, thereby converting it to T_3. Meanwhile, Deo3 acts as an inner-ring deiodinase, removing iodine from the inner ring of T_4 and T_3, thereby converting them into the inactive forms, reverse triiodothyronine (rT_3) and diiodothyronine (T_2) (Darras & Van Herck, 2012).

THs in the plasma bind to TH-binding proteins such as albumin, transthyretin (TTR), and thyroxine-binding globulin (TBG) before entering the cell membrane (Power et al., 2000). At the membrane level, their transmembrane movement occurs via facilitated transport mediated by several ATP-dependent solute carrier (SLC) proteins, including monocarboxylate transporters (MCTs) and organic anion transporter polypeptides (OATPs) (Visser et al., 2011). Among the MCT family, only MCT8 (solute carrier family 16 member 2) and MCT10 (solute carrier family 16 member 10) are involved in TH transport (Visser et al., 2011). Studies in both humans and rats have shown that MCT8 proteins exhibit high specificity for THs and transport T_4, T_3, rT_3, and T_2 (Visser et al., 2011).

TH and brain-gut axis

Fish and mammals share the basic vertebrate brain plan, but they differ significantly in the development of their neural regions. In fish, the optic tectum (midbrain) is the main center for visual processing, displaying visual information and integrating it with sensorimotor systems (Butler & Hodos, 2005). In many teleosts, the cerebellum is well-developed, coordinating swimming and balance, whereas the telencephalon remains relatively small and primarily processes olfaction and basic sensory functions (Nieuwenhuys, 2011). In contrast, mammals show a significant expansion of the telencephalon, especially the cerebral cortex, which supports higher cognitive functions, complex sensory integration, memory, and voluntary motor control (Northcutt, 2002). The thalamus and hypothalamus regulate endocrine and autonomic functions and are crucial for sensory relay. In contrast, the cerebellum and brainstem continue to play roles in motor coordination and autonomic control (Kandel et al., 2013). Therefore, although both fish and mammalian brains share the same fundamental structure, evolutionary divergence emphasizes sensory-motor dominance in fish and cognitive specialization in mammals, reflecting differences in ecological demands and behavioral complexity (Butler & Hodos, 2005).

TH influences CNS development by mediating through thyroid hormone receptors (TRs), which belong to the nuclear hormone receptor family. TRs regulate gene expression by binding to DNA response elements and interacting with coregulators, thereby activating or repressing transcription. T_3 is a biologically active form of TH because it binds TRs with higher affinity than T_4. T_4 primarily acts as a prohormone, converted to T_3 by tissue-specific deiodinases. T_4 can cross the blood–brain barrier (BBB) to supply the brain for direct TR activation, although T_3 and T_4 Ind.uce similar conformational changes in TRs (Sandler et al., 2004). When bound to TH, the TRs undergo conformational changes that release corepressors and recruit coactivators, thus activating the transcription of positive-response target genes (Cheng et al., 2010). In mammals, two genes—thra and thrb—encode the TRα and TRβ isoforms (Lazar, 1993). In teleosts, additional paralogs are often present following whole-genome duplication (Nelson & Habibi, 2009). For example, two thra genes have been identified in zebrafish (*Danio rerio*) (Takayama et al., 2008) and in other teleost species (Bertrand et al., 2007), while two thrb genes have been reported in Japanese conger eel (*Conger myriaster*) (Kawakami et al., 2003). The maturation of several neural cell types during brain development may depend on TRα1 expression, which is mediated by complex interactions among TRs, THs, and coregulators. TRα1 also remains involved in neurodevelopment, including in adult neurons (Wallis et al., 2010).

In this review, we examine the multiple regulatory mechanisms that govern the action of THs across the cells. Firstly, regulation of free versus bound THs occurs through reversible binding to serum carrier proteins (Williams & Larsen, 2003). Then, cell-specific expression of TH membrane transporters drives TH entry into the target cell (Friesema et al., 2008). This facilitated transport process is mediated by several ATP-dependent transporters, including monocarboxylate transporters (MCTs; e.g., MCT8; Bernal et al., 2015). TH transport mechanisms are well characterized in mammals, but in fish, they remain limited, apart from studies on the role of MCT8 in zebrafish (*Danio rerio*) development. In zebrafish, MCT8 mRNA is expressed in the brain, spinal cord, and vascular system (Groeneweg et al., 2019). In zebrafish, MCT8 facilitates the transport of T3 across the blood–brain barrier (BBB) (Groeneweg et al., 2019), and MCT8 deficiency has been linked to alterations in nervous system development (Vatine et al., 2013).

Later, intracellular deiodination of TH, particularly the conversion of T_4 to T_3, determines which TH form is vital for that cell (Bianco & Kim, 2006). Only a small part of circulating T_4 is converted to T_3 through outer-ring 5′-monodeiodination (St Germain et al., 2009). Deiodinase 2 (Dio2) is the primary enzyme responsible for the T_4-to-T_3 conversion inside the brain, while deiodinase one is mainly found in the liver (Bianco & Kim, 2006). In the CNS, Dio2 is mainly expressed in astrocytes and tanycytes, glial cells located in the hypothalamus (Guadano-Ferraz et al., 1999). In neurons, deiodinase 3 (DIO3) inactivates T_4 and T_3 by converting them to rT_3 and T_4, thereby preventing TH overload. In this process, astrocytes supply active T_3 from circulating T_4, while neurons regulate local hormone levels by degrading excess THs (Figs. 11.2 and 11.3). Similarly, Dio2 expression increases in hypothyroid conditions, whereas Dio3 expression rises in hyperthyroid states, collectively protecting the brain from TH imbalance (Heuer, 2007).

The endothelial cells in brain capillaries, formed by astrocytic end-feet, restrict the entry of TH into the brain, requiring transporters for TH to cross the BBB (Dratman et al., 1991). Transport mainly occurs via the MCT8 and OATP1C1 transporters at the choroid plexus, with a preference for T_4 over T_3 due to these transporters' higher affinity (Chatonnet et al., 2011). OATP1C1 facilitates the uptake of T4 into astrocytes, where DIO2 converts it into T_3. The T_3 in astrocytes uses MCT8 to enter neurons, where it binds to TRs. A notable cross-talk exists between astrocytes and neurons that lack Dio2, the enzyme responsible for converting T_4 to T_3, even when TRs are present. This physiological mechanism maintains TH homeostasis in brain clusters, although THs can also reach the CNS via the cerebrospinal fluid (Mathiisen et al., 2010). These are key regulators of growth, and disruptions in thyroid function impair brain development (Molla et al., 2019). TH action in the brain appears to be region- and stage-dependent (Parhar et al., 2000).

The pattern of differential expression of transcripts of the thyroid receptor-regulator (TR-R) system (*thraa, thrb, deo1, deo2, deo3*) observed in various brain segments of zebrafish treated with L-NAME, which altered the status of NO and air exposure (Sudhakaran & Peter, 2024), revealed that the TR-R system in the MC segment of stressed zebrafish brains, following the altered NO status, showed a reversal of TR-R. This provides evidence for the TR-R system's involvement in the stress response. Likewise, under stress, the TR-R system in the final gut segment (posterior intestine) showed a reciprocal response to altered NO, further supporting the role of the TR-R system in the gut in TH-mediated stress acclimation (Fig. 11.2). Similarly, the pattern of differential expression of the TR-R system (trα, trβ, trαb, *deo1, deo2, deo3, mct8, mct10*) across different brain regions, especially in the MC segment of hypoxic *Anabas* brains, showed a reversal of TR-R expression with T_3, indicating the TR-R system's role in recovery from hypoxic stress. Likewise, upon stress, the TR-R system in the anterior intestine exhibited a recovery response with T_3 treatment, further emphasizing the role of the TR-R system in the gut in TH-mediated stress acclimation. As seen in zebrafish, this study identified a shift in the integration of the TR-R system across the B-G segment during stress acclimation in an air-breathing fish (Fig. 11.3).

As the primary organ for absorption and digestion in vertebrates, the intestine maintains water and electrolyte balance, regulates the endocrine functions of digestion and metabolism, and supports immunity (Ringo et al., 2003). The gastrointestinal tract (GIT) in mammals shows significant variation in its structural organization. In contrast, the fish gut is continuous with the external environment. It can relay signals to physiological targets through paracrine/autocrine and/or luminescent signaling. At the same time, the gut can send signals to distant targets via endocrine signaling. Additionally, neuropeptides and neurotransmitters released from peripheral nervous system axon terminals can contribute to regulatory signaling in the gut. The B-G axis can also operate in parallel with humoral factors to regulate digestive processes by modulating digestive enzyme secretion, motility, and other bodily functions, including those of the brain (Ringo et al., 2003). They influence growth and metabolism, including feeding, drinking, and behavior, by regulating nutrient and ion/water absorption and by modulating the secretion and functions of other hormones. Additionally, they act at distant sites in the brain to modulate these processes (Holzer & Farzi, 2014).

As a target organ of TH action, the gut undergoes significant remodeling, involving an initial phase of apoptosis followed by a burst of cell proliferation (Heimeier et al., 2010). Similarly, TH plays a crucial role in mammalian postnatal intestinal development, with TRs, particularly TRα1, also playing a critical role (Kress et al., 2009). The gut acts as an essential reservoir for thyroid hormones, especially T_3, and can influence the activity of other hormones. Additionally, in the gut, both T_4 and T_3 are absorbed, although the process is incomplete, and various factors influence this absorption (Hays, 1999). The intestinal wall may also contribute to systemic T_3 levels, which are deiodinase-active. A significant portion of circulating T_4 and T_3 is taken up by the liver, which then releases them back into the gut via bile secretion. Moreover, T_4 and T_3 can enter the gut directly through the mesenteric circulation. Furthermore, the composition of gut microbiota affects the bioavailability of key micronutrients necessary for TH metabolism. Nutrients such as iodine, iron, and copper are essential for hormone synthesis, whereas selenium and zinc are critical for the conversion of T_4 to T_3 (Knezevic et al., 2020). Similarly, gut bacteria can bind TH, even competing with albumin for this interaction (Virili, Centanni, et al., 2018; Virili, Fallahi, et al., 2018).

Other hormones, endocrine tissues, and pathways

This chapter also examines how TH affects other pathways that modulate neurotransmission dynamics in the brain and gut, which have receptors for various hormones, thereby coordinating central and enteric neural activity. In the

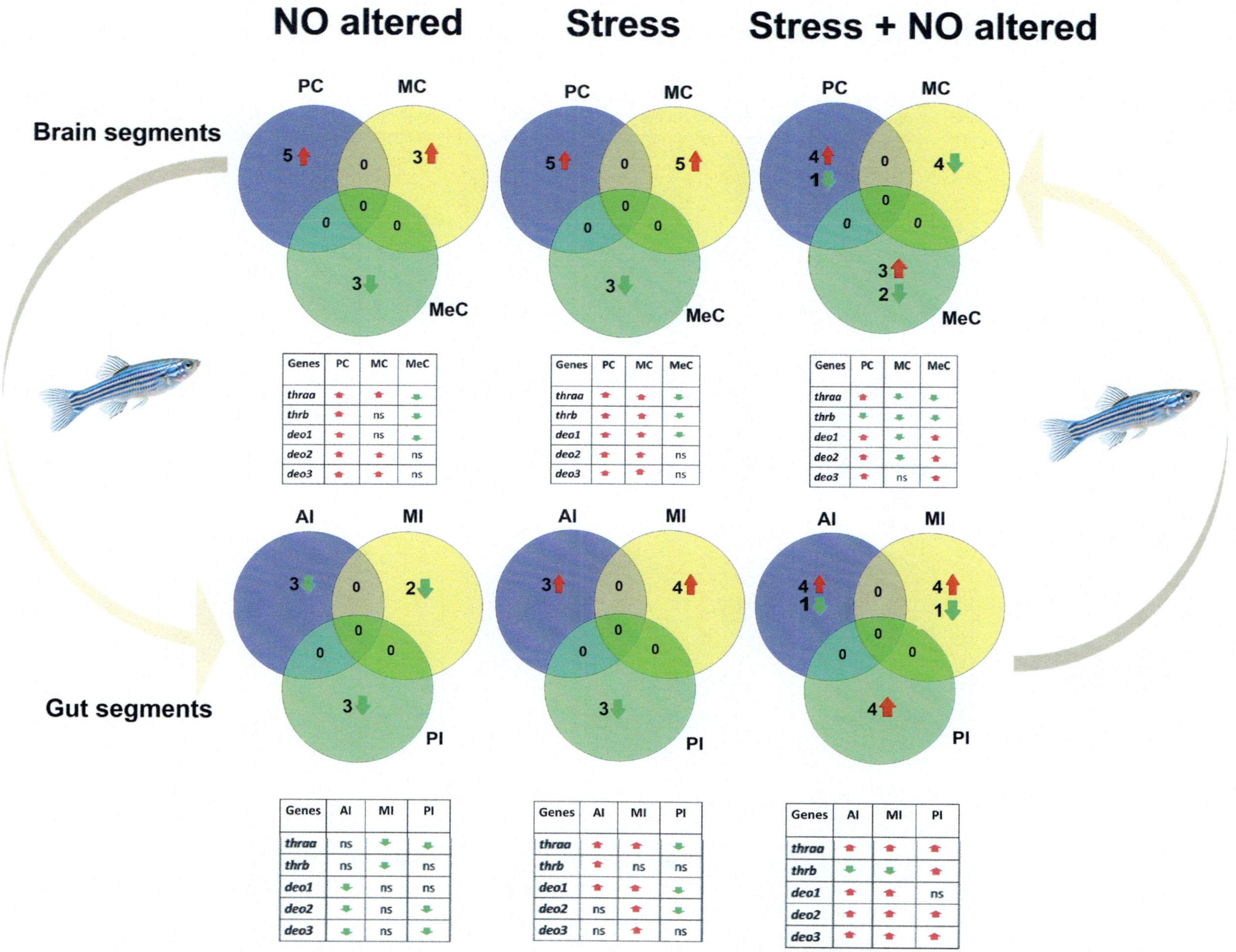

FIGURE 11.2 Venn diagram showing the pattern of differential expression (upregulated and downregulated) of transcripts of thyroid receptor-regulator (TR-R) system (*thraa, thrb, deo1, deo2, deo3*) observed in different brain segments such as prosencephalon (PC), mesencephalon (MC), and metencephalon (MeC) and gut segments such as anterior intestine (AI), middle intestine (MI), and posterior intestine (PI) in zebrafish treated with L-NAME that altered the status of NO. Here, a group of air-exposed zebrafish was held with or without L-NAME, as demonstrated by Sudhakaran and Peter (2024). The pattern analysis revealed that, following altered NO status, the TR-R system in the MC segment of stressed zebrafish brains reversed, providing evidence for the involvement of the TR-R system in recovery from air-exposed stress. Likewise, upon stress, the TR-R system in the final gut segment (posterior intestine) showed a reciprocal response to altered NO, further supporting the role of the TR-R system in the gut in TH-mediated stress acclimation. Here, a shift in the integration of the TR-R system across the B-G segments of zebrafish is evident.

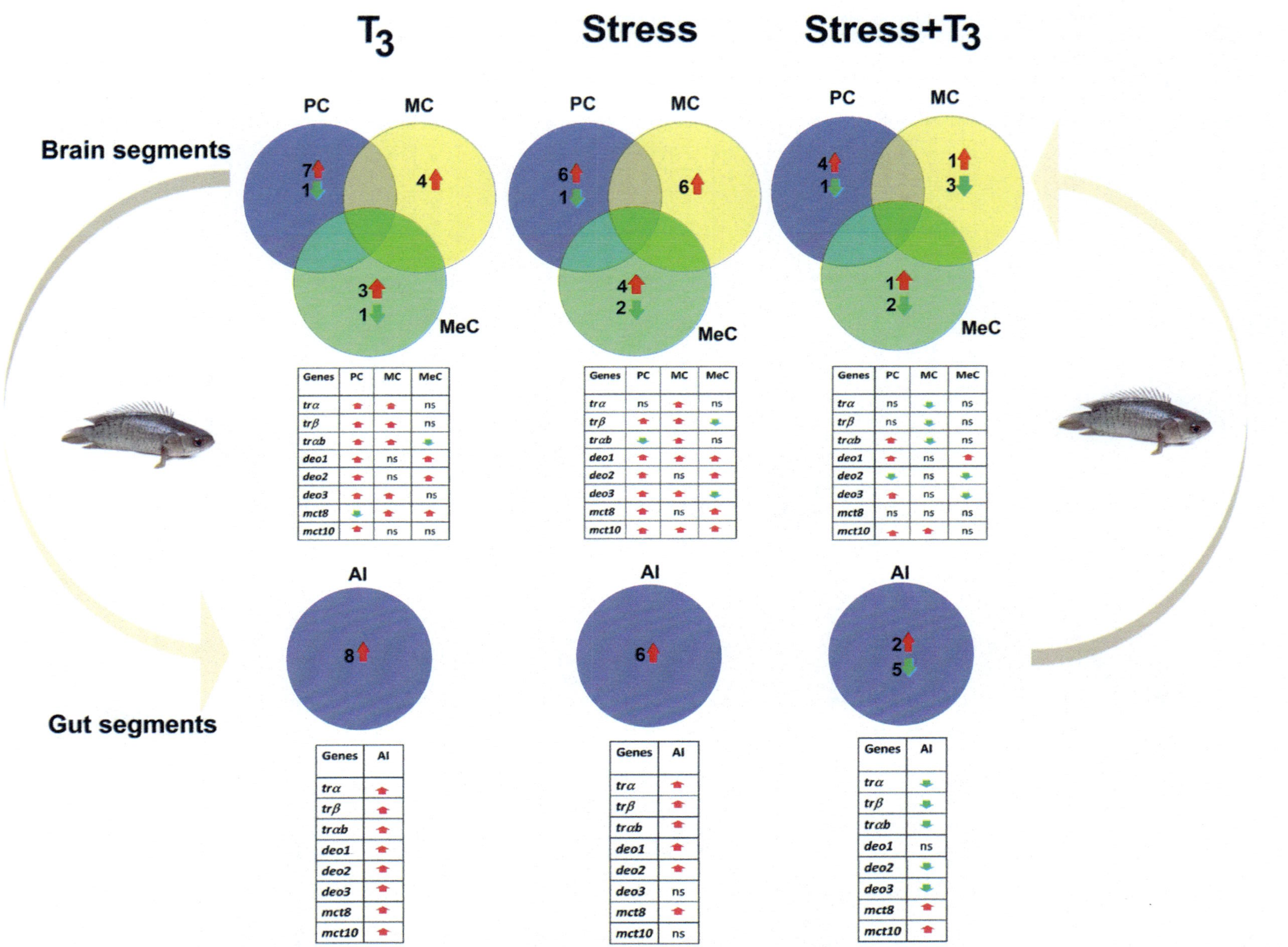

FIGURE 11.3 Venn diagram illustrating the pattern of differential expression (upregulated and downregulated) of transcripts in the thyroid receptor-regulator (TR-R) system (tr*α*, tr*β*, tr*ab*, *deo1*, *deo2*, *deo3*, *mct8*, *mct10*) across different brain regions such as prosencephalon (PC), mesencephalon (MC), and metencephalon (MeC), as well as the gut segment, specifically the anterior intestine (AI), in the air-breathing fish *Anabas* treated with T_3, hypoxia stress, or both. In this experiment, a group of hypoxic *Anabas* was maintained with or without T_3, as investigated by Peter et al. (Unpublished data). The pattern analysis revealed that the TR-R system in the MC segment of hypoxic *Anabas* brains showed a reversal of TR-R expression by T_3, providing evidence for the TR-R system's role in recovery from hypoxic stress. Similarly, under stress, the TR-R system in the anterior intestine exhibited a recovery response to T_3 treatment, further supporting the role of the gut TR-R system in TH-mediated stress acclimation. As observed in zebrafish, this study identified a shift in the integration of the TR-R system across the B-G segment in this air-breathing fish.

brain, TH enhances catecholaminergic signaling by upregulating tyrosine hydroxylase—the rate-limiting enzyme in dopamine and norepinephrine synthesis—and by increasing monoamine oxidase activity, thereby influencing neurotransmitter turnover and mood regulation (Bauer et al., 2002). Similarly, TH modulates serotonergic release by regulating tryptophan hydroxylase and the serotonin transporter; hypothyroidism can lower 5-HT levels and contribute to depressive symptoms, while T_3 supplementation enhances serotonergic transmission (Bauer et al., 2008). Additionally, TH maintain excitatory–inhibitory balance by regulating glutamate decarboxylase, NMDA, and AMPA receptor subunits, which influence cognitive and mood processes (Ahmed et al., 2008; Sui & Chen, 2012). Likewise, in the gut, TH regulates acetylcholinesterase activity and muscarinic receptor expression, thereby modulating cholinergic signaling and motility (Fregly, 1989). In brain regions, TH can alter serotonin transporter expression, thereby affecting intestinal 5-HT availability and peristalsis (de Vries et al., 2015). Furthermore, TH regulates nitric oxide synthase activity, thereby affecting smooth muscle relaxation and gastrointestinal motility (Moneva et al., 2011). Likewise, the coexistence of mood and gastrointestinal disturbances is commonly observed in thyroid dysfunction (de Vries et al., 2015). This clearly shows that TH can coordinate neurotransmission across the brain–gut axis (Fig. 11.4).

TH and stress-responsive endocrine signaling

This chapter explains how the mammalian brain, a target of other hormones, responds to stressful challenges and coordinates the body's activities through two main neuroendocrine axes. First, a rapid autonomic pathway sends efferent sympathetic nerves from the brain to the adrenal medulla, forming the sympatho-adrenomedullary system (SAM). This pathway releases adrenaline, a quick emergency stress hormone. In fish, as in mammals, adrenaline is also released in response to stress, triggered by the brain-sympathetic-chromaffin (BSC) axis. In both systems, activation of the adrenal medulla releases catecholamines, such as norepinephrine and epinephrine, into the bloodstream (Lyte et al., 2011). The second axis, which constitutes the hypothalamic-pituitary-adrenal (HPA) axis in mammals and the hypothalamic-pituitary-interrenal (HPI) axis in fish, secretes cortisol, a powerful corticosteroid with delayed effects (Martin et al., 2018).

The HPA axis and its corresponding HPI axis in fish serve as the central stress efferent pathways, coordinating the organism's adaptive responses to various stressors (Tsigos & Chrousos, 2002). This brain region, which is part of the limbic system, is involved in memory and emotional responses. It responds to both intrinsic and extrinsic stress stimuli, as well as increased pro-inflammatory cytokines, which can activate this system, leading to the release of corticotropin-releasing factor (CRF) from the hypothalamus (Mayer et al., 2014). CRF stimulates the release of adrenocorticotropic hormone (ACTH) from the pituitary gland, which in turn leads to cortisol production by the adrenal glands. As a neuroendocrine product, cortisol, the well-known stress hormone, affects nearly all tissues in the body and communicates with neural pathways, enabling the diverse intestinal tissue populations to regulate brain activity (Carabotti et al., 2015). Cortisol, through its network of actions, suppresses immune and digestive functions, modifying metabolic pathways to maintain an energy boost by shutting down non-essential systems, equipping the animal to respond quickly to potentially life-threatening stress stimuli (Tan, 2023).

During the stress response, the vagus nerve functions efferently to promote intestinal barrier health and reduce inflammation, thereby acting as a brake on sympathetic responses that affect gut function (Bonaz et al., 2018; Breit et al., 2018). Stress weakens these vagus nerve efferent fibers, allowing sympathetic effects on the gut to dominate (Bonaz et al., 2018; Breit et al., 2018), thereby compromising the integrity of the intestinal barrier. Chronic stress can alter intestinal permeability, enabling enteric bacteria and their ligands to enter the bloodstream from the gut and reach the brain, leading to systemic inflammation and brain dysfunction (Foster et al., 2017). Subsequently, the vagus nerve's afferent fibers in the gut send signals to the brain about the state of the intestine, which then regulates the HPA axis (Breit et al., 2018). This illustrates a classic bidirectional communication between the gut and brain that helps mitigate the effects of stress. Notably, extensive evidence from various fish models indicates that the HPT axis plays a key role in modulating the brain-gut axis. This is primarily due to interactions between these axes and elements of the TR-R system along the B-G axis; however, direct communication across the B-G axis also occurs via the TR-R system (Fig. 11.4).

Induction of stress alters neurotransmitter function in both the brain and the gut, disrupting the coordination of the brain–gut axis. Activation of the HPA axis, which increases glucocorticoid release, can modify the metabolism of catecholamines and serotonin in stress-responsive regions such as the hippocampus, amygdala, and prefrontal cortex (Joëls & Baram, 2009; McEwen, 2007). Additionally, higher glucocorticoid levels can boost tyrosine hydroxylase and monoamine oxidase activities, initially elevating catecholamine levels. However, prolonged stress may deplete dopamine and serotonin, potentially leading to

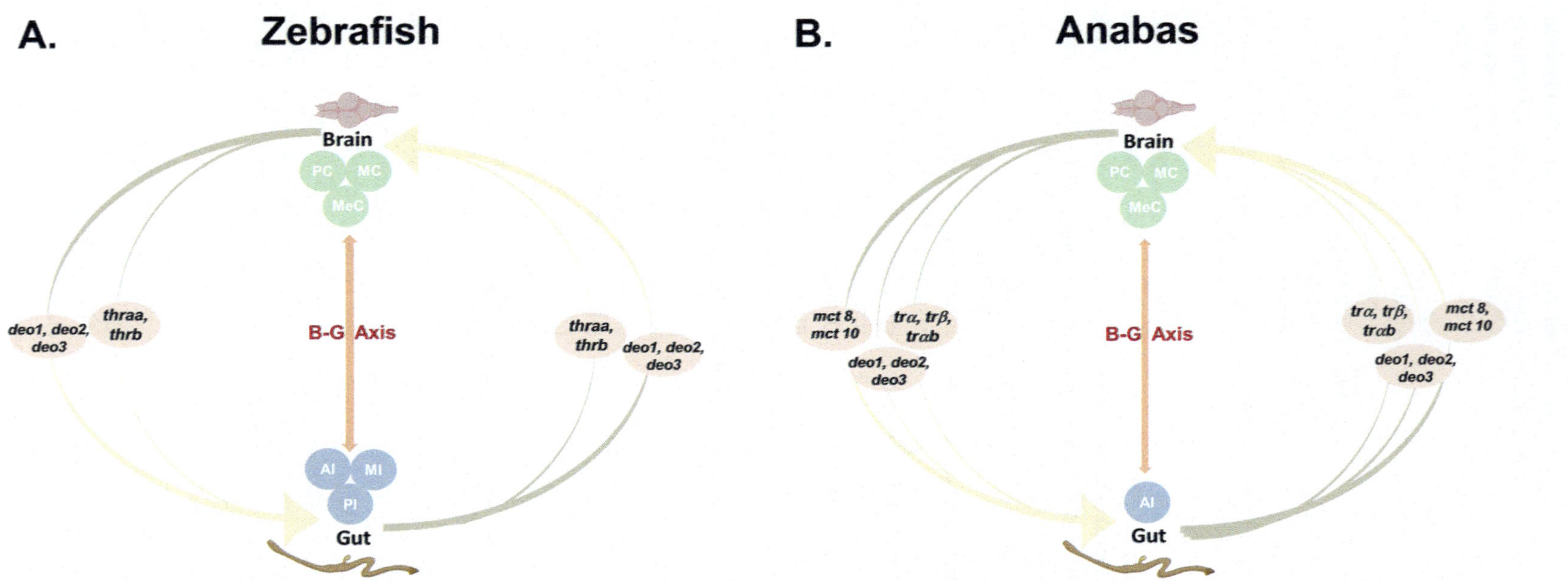

FIGURE 11.4 Diagram illustrating how the TR-R system mediates brain-gut communication through thyroid hormone receptors (TRs: Trα, Trβ, Trαb), TH regulators such as deiodinases (*Dio1, Dio2, Dio3*), and TH transporters, including MCT (monocarboxylate transporter), *Mct8, and Mct10*. By differentially expressing the TR-R components, each part of the B-G axis modulates the availability of TH and its binding to receptors, thereby influencing its function across the axis. This differential regulation of each component and their final interaction is evident in the TR-R expression pattern, confirming a direct role of TH in controlling the B-G axis in vertebrate models. TR: Thyroid hormone receptor; PC: prosencephalon, MC: mesencephalon, MeC: metencephalon, AI: anterior intestine; MI: middle intestine; PI: Posterior intestine.

anxiety and depression (Herman & Tasker, 2016; Rau et al., 2005). In mammals, stress can reduce thyroid hormone secretion and peripheral availability, and it also affects neurotransmitter production and receptor function (Fekete & Lechan, 2014).

Stress influences CRH-driven enteric neurotransmission by altering the release of acetylcholine and serotonin from enteric neurons and enterochromaffin cells (Tache et al., 2009). Acute stress may boost gut motility through increased release of 5-HT and acetylcholine, whereas chronic stress leads to desensitization of serotonin receptors and impaired cholinergic signaling, resulting in dysmotility and visceral hypersensitivity (Mayer, 2011; Mulak & Bonaz, 2015). Stress-related oxidative and inflammatory responses also disrupt nitric oxide signaling, affecting smooth muscle relaxation and gut barrier function (Konturek et al., 2011). Overall, stress affects neurotransmitter synthesis, release, and receptor sensitivity in both the brain and gut, with these effects exacerbated by stress-induced suppression of thyroid hormone activity—ultimately contributing to mood disorders and gastrointestinal issues.

TH is involved in nearly all physiological processes in fish that are crucial for growth and development. TH helps maintain Na^+ and water balance during osmotic challenges, and a healthy thyroid state is essential for an optimal stress response (Grau, 1988; Peter, 2011). Altered thyroid function exhibits distinct stress-response patterns when fish are exposed to stressors such as net confinement and air exposure (Peter, 2007; Peter & Wendelaar Bonga, 2006). In air-breathing fish, the combined function of the thyroid and interrenal glands is necessary for recovery from prolonged air-breathing (Peter, 2007). Fish regulate ionic and osmotic homeostasis under the influence of corticosteroids and THs (Peter, 2011), and cortisol can decrease plasma T_3 concentrations (Babitha & Peter, 2010).

Fish show changes in their stress responses, supporting the idea that a healthy thyroid axis is crucial for properly regulating the interrenal axis and stress response in fish (Peter, 2007). It has been shown that, in fish, THs can alter the pattern and strength of stress responses by either modulating their own actions or interacting with stress hormones (Peter, 2011). Additionally, several lines of evidence suggest that interactions between hypothalamic and pituitary hormones of the thyroid and interrenal axes may regulate the effects of THs and cortisol (Peter, 2007). Changes in TH levels have been shown to affect the severity of the stress response in fish, indicating a link with the endocrine stress axes (Fig. 11.5), especially the interrenal axis (Peter, 2011). An interaction between TH and cortisol during fish acclimation to net confinement reveals a vital role for TH in stress acclimation (Peter, 2009).

TH influences hypothalamic–pituitary signaling pathways that regulate feeding, reproduction, and metabolism. Several studies have documented the role of THs in regulating feeding behavior and appetite, where THs interact with neuropeptides such as neuropeptide Y (NPY), orexin, and CART in the hypothalamus to control food intake (Volkoff, 2016). Hypothyroidism can reduce feeding drive, while exogenous THs may boost appetite during growth phases. Additionally, stress influences TH levels, which, in turn, affect gut function and energy distribution. Cortisol–TH interactions modify gut permeability and nutrient absorption in fish (Vijayan et al., 2003). TH drives significant gut and brain remodeling during metamorphosis (e.g., in flatfish and amphibians). Changes in intestinal length, epithelial structure, and neural innervation of the gut align digestive functions with habitat shifts (McMenamin & Parichy, 2013; Power et al., 2001).

A functional thyroid status is essential for many physiological processes of the B-G axis. For example, when thyroid activity is suppressed, it affects hypothalamic GnRH neurons and digestive functions in species such as catfish and tilapia, indicating that THs regulate both central and peripheral processes (Deal & Volkoff, 2020). Studies emphasize THs' role in gut remodeling and nutrient absorption. During metamorphosis and growth, THs alter the intestinal tract, enhancing nutrient uptake and digestive enzyme activity (McMenamin & Parichy, 2013; Power et al., 2001). Furthermore, the brain–gut–thyroid axis integrates signals from feeding behavior and gut nutrient status, allowing fish to adapt to changing metabolic demands in aquatic environments (Volkoff, 2016). TH thus plays a crucial role in connecting neural and digestive functions through the B-G axis in both fish and mammals.

While reviewing the definitive role of TH in B-G axis, it's evident that in mammals, TH integrates the B-G axis by modifying the neural and endocrine pathways, affecting the enteric nervous system and vagal signaling (Fig. 11.5). Hyperthyroidism increases gut motility, while hypothyroidism slows it, demonstrating direct central–peripheral integration (Furness, 2012; Sawin & Becker, 1997). Furthermore, THs and gut microbiota interact bidirectionally. Microbiota modulate essential micronutrients for thyroid function (iodine, selenium, zinc), while THs influence microbial diversity and intestinal barrier function (Knezevic et al., 2020; Virili, Centanni, et al., 2018; Virili, Fallahi, et al., 2018). Intestinal deiodinases locally convert T_4 to T_3, linking peripheral thyroid hormone metabolism to systemic energy balance (Nguyen, Chapa, et al., 1993; Nguyen, Distefano, et al., 1993). The gut acts as a reservoir for THs, with enterohepatic circulation returning TH to the intestine and affecting their availability (Hay & Wass, 2009). Both fish and mammalian systems highlight THs as central regulators of neuroendocrine control, digestive capacity, and systemic energy homeostasis; however, mammals exhibit additional layers of complexity through their microbiota and higher cortical control. THs affect brain regions such as the hypothalamus, hippocampus, and

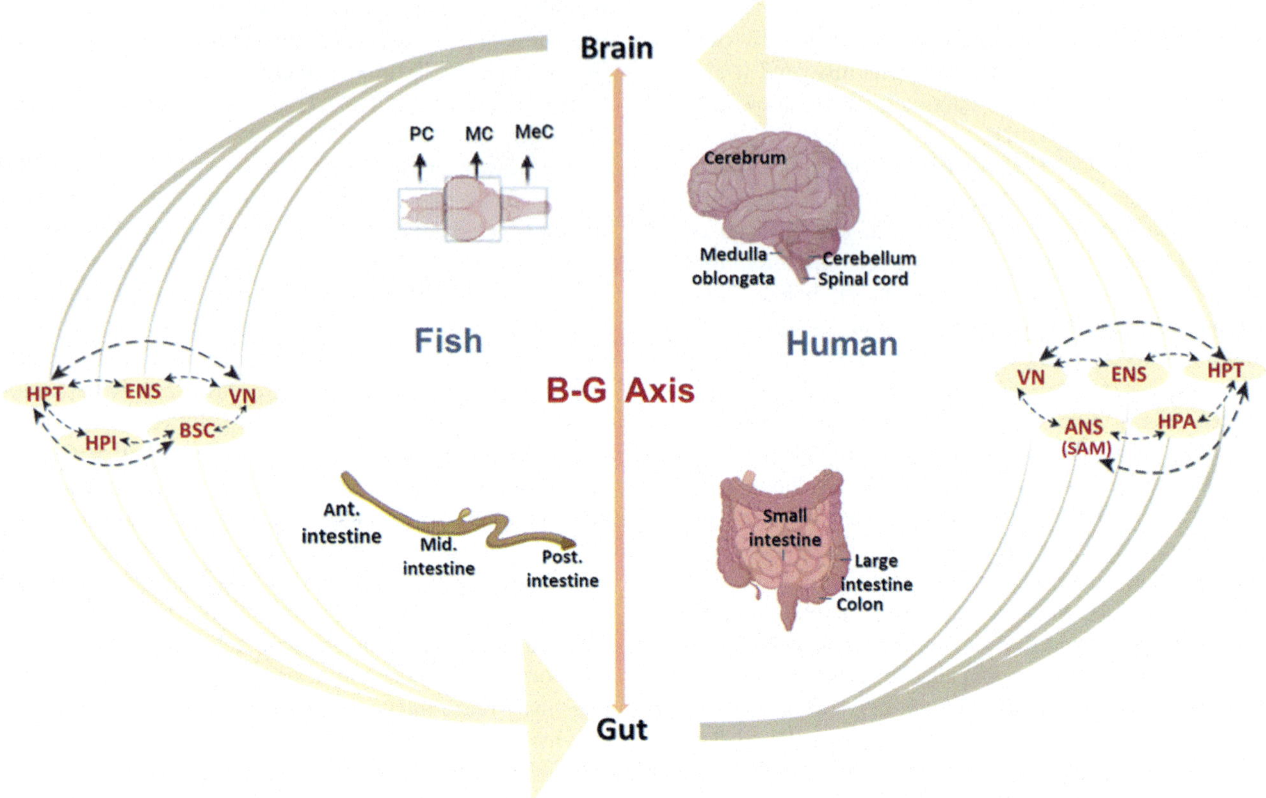

FIGURE 11.5 Diagram illustrating the novel bidirectional communication with a critical role for the HPT axis, which can drive the B-G axis in fish and mammalian models. The primary neuroanatomical pathways that facilitate communication between the brain and the gut in humans include the vagus nerve (VN), the autonomic nervous system (ANS), comprising the sympathetic-adrenomedullary (SAM) system, the enteric nervous system (ENS), and the hypothalamic-pituitary-adrenal (HPA) axis. In the fish model, the analogous mammalian communication centers are the vagus nerve (VN), the autonomic nervous system (ANS), the brain-sympathetic-chromaffin axis (BSC), the enteric nervous system (ENS), and the hypothalamic-pituitary-interrenal (HPI) axis. While emphasizing the vital role of the HPT axis, which releases T_3 and T_4 as primary hormones interacting with the aforementioned axes in both fish and mammalian models, it is clear that the HPT axis exerts significant control over the B-G axis. Similarly, considering the integration of TR-R systems across the B-G axis, this novel model further highlights TH's role in integrating the B-G axis.

cortex, which influence appetite, energy expenditure, and vagal control of gut motility (Furness, 2012; Koehler et al., 2005). Hypothyroidism has been linked to increased gut inflammation and permeability ("leaky gut") (Virili, Centanni, et al., 2018; Virili, Fallahi, et al., 2018). A portion of circulating THs is excreted in bile and reabsorbed in the intestine, creating a feedback loop among the liver, gut, and brain (Hay & Wass, 2009). Dysbiosis can impair TH metabolism, while probiotics enhance TH status and gut homeostasis (Knezevic et al., 2020). Some bacteria produce enzymes that bind THs and alter their local availability (Nguyen, Chapa, et al., 1993; Nguyen, Distefano, et al., 1993). Changes in TH levels can modify vagal tone, thereby affecting motility and secretion (Furness, 2012).

TR-R system and B-G axis

Given the compelling evidence for the role of the TR-R system in B-G communication, it is clear that TR-R regulates its function through TR subtypes and TH regulators, including deiodinases and TH transporters, as demonstrated in zebrafish (Sudhakaran & Peter, 2024). They found activation of stress- and ease-responsive receptors along the B-G axis in zebrafish under NO-altered conditions, demonstrating the clear role of the TR-R system in modulating stress and ease responses during stress acclimation. By differentially expressing components of the TR-R, each part of the B-G axis responds to TH availability and its binding to receptors, thereby affecting its function throughout the axis. This differential regulation of each component and their final interaction is evident in the TR-R expression pattern, confirming an integrative role for TH in B-G axis function in vertebrate models (Fig. 11.4).

This chapter, while reviewing the role of TH in B-G axes, shows that, from a neural perspective, the vagus nerve acts as a crucial link between TH-influenced brain centres and gut activity by connecting the ANS and ENS

(Fig. 11.5). Additionally, from a hormonal perspective, it is clear that TH plays a vital role in integrating the B-G axis via TR-R system in two main ways. First, TH connects to major neural components such as the ANS and ENS, which coordinate the system. Second, by interacting with inter-hormonal axes like BSC and HPI—similar to the SAM and HPA axes in mammals—the TH and TR-R system facilitates hormonal integration to coordinate B-G functions (Fig. 11.5). In this context, the TH's central role in integrating the B-G axis across all vertebrate models highlights its essential contribution to physical and mental health in vertebrates, including humans. The crucial role of TH and TR-R system in regulating mitochondrial function and oxidative metabolism in the brain and gut—supporting energy supply during high-demand conditions (Viswanath & Peter, 2025)—further supports this idea.

Applications of the material

Taken together, this chapter thoroughly models the vertebrate brain–gut axis, which functions bidirectionally through the HPT axis along with neural, immune, endocrine, and microbial signals. Any disruption or interference with the B-G axis, likely caused by stress-response effects, can compromise TH homeostasis and impair its metabolic and growth-promoting functions. Stress activates the SAM and HPA axes, influencing both B-G physiology—which is linked to thyroid activity—and impairing intestinal integrity and neuroendocrine balance. Additionally, TH interacts with many neuroendocrine factors and helps maintain homeostasis during stress by modulating physiological flexibility. This chapter highlights the vital role of TH in integrating the B-G axis along with other signals that regulate it. Recognizing the importance of thyroid physiology in disease management is crucial, as it can improve healthcare outcomes. It is also essential for managing diseases in wild and domestic animals, supporting animal welfare.

At the same time, further research into how TH interacts with other regulators of the B-G axis is crucial for understanding the importance of HPT-driven B-G axis regulation and integration in vertebrates, particularly in humans. Additionally, an integrative stress approach, along with associated methodologies, is essential for assessing the role of the HPT axis in stress adaptation. When exploring the potential impact of the HPT axis on brain and gut health, it would be helpful to identify HPT-driven mechanisms, particularly in diseases associated with the B-G axis.

Mini-dictionary terms

Brain-gut axis: The brain-gut axis is a bidirectional interorgan system connecting the central nervous system with the enteric nervous system. A functional B-G axis is essential for the optimal physiological functioning of the brain and gut. Thyroid hormone is now recognized as an integrator of the B-G axis.

Stress Response: The organism's reaction at all levels to a stressful challenge is known as the stress response. This response disrupts homeostasis and triggers allostasis. Stress hormones like adrenaline and cortisol initiate the stress response.

Ease Response: The process of restoring disrupted homeostasis in the stress response to its baseline state is known as the ease response. Melatonin and serotonin can facilitate this process by attenuating the stress response through their effects on stress hormones.

Thyroid Hormone Receptor and Regulator (TR-R) system: Thyroid hormone actions are mediated by thyroid hormone receptors (TRs). Similarly, TH availability to the TRs is regulated by TH regulators, such as deiodinases, through various deiodination processes. TH is transported across membranes by TH regulators, including monocarboxylate transporters (MCTs). The TRs and TH regulators, together with TH, form the TR-R system that controls TH action.

Hypothalamic-pituitary-thyroid (HPT) axis: The hypothalamic-pituitary-thyroid axis is an endocrine system that integrates stimuli and negative feedback on thyroid hormone synthesis, metabolism, and secretion. This axis is sensitive to both intrinsic and extrinsic stimuli and can detect signals from other endocrine and cellular sources.

Summary Points

- Little information is available on the integrative control of thyroid hormone (TH) on brain-gut (B-G) functions
- The thyroid hormone receptor regulator (TR-R) system operates across the B-G axis of vertebrate models
- The TR-R network in the B-G axis connects with the autonomic and enteric nervous systems
- The stress response modulates the TR-R system and the sympathetic-adrenomedullary (SAM) and hypothalamic-pituitary-adrenal (HPA) axes.
- This review highlights the novel insight into the bidirectional communication of TH across the B-G axis of vertebrates
- Recognizes the significance of TH in B-G axis physiology and its implications on healthcare outcomes.

Acknowledgments

MCS and MV acknowledge the facility of iCEIB of the University of Kerala for pursuing integrative stress physiology research. MCS values the UGC BSR Faculty Fellowship and acknowledges "Sastrajeevan Initiative" for their continuous support in exploring the "Stress and Ease" programme.

References

Ahmed, O. M., Abd El-Tawab, S. M., & Ahmed, R. G. (2008). Effects of experimentally induced hypothyroidism and hyperthyroidism on the development of the rat cerebellar cortex. *International Journal of Developmental Neuroscience, 26*(5), 437–446.

Appleton, J. (2018). The gut-brain axis: Influence of microbiota on mood and mental health. *Integrative Medicine: A Clinician's Journal, 17*(4), 28.

Babitha, K. S., & Peter, V. S. (2010). Cortisol inhibits thyroidal activity and modifies hydromineral balance in the freshwater fish *Clarias gariepinus*. *Fish Physiology and Biochemistry, 36*(3), 735–742. https://doi.org/10.1007/s10695-009-9343-7

Bauer, M., Goetz, T., Glenn, T., & Whybrow, P. C. (2008). The thyroid–brain interaction in thyroid disorders and mood disorders. *Journal of Neuroendocrinology, 20*(10), 1101–1114.

Bauer, P. V., Hamr, S. C., & Duca, F. A. (2016). Regulation of energy balance by a gut–brain axis and involvement of the gut microbiota. *Cell Metabolism, 23*(3), 421–428. https://doi.org/10.1016/j.cmet.2016.02.008

Bauer, M., Heinz, A., & Whybrow, P. C. (2002). Thyroid hormones, serotonin and mood: Of synergy and significance in the adult brain. *Molecular Psychiatry, 7*(2), 140–156. https://doi.org/10.1038/sj.mp.4000963

Bernal, J., Guadaño-Ferraz, A., & Morte, B. (2015). Thyroid hormone transporters—Functions and clinical implications. *Nature Reviews. Endocrinology, 11*(7), 406–417. https://doi.org/10.1038/nrendo.2015.66

Bertrand, S., Thisse, B., Tavares, R., Sachs, L., Chaumot, A., Bardet, P. L., et al. (2007). Unexpected novel relational links uncovered by extensive developmental profiling of nuclear receptor expression. *PLoS Genetics, 3*(11), 2085–2100. https://doi.org/10.1371/journal.pgen.0030188

Bianco, A. C., & Kim, B. W. (2006). Deiodinases: Implications of the local control of thyroid hormone action. *Journal of Clinical Investigation, 116*, 2571–2579. https://doi.org/10.1172/JCI29812

Bonaz, B., Bazin, T., & Pellissier, S. (2018). The vagus nerve at the interface of the microbiota-gut-brain axis. *Frontiers in Neuroscience, 12*, 49. https://doi.org/10.3389/fnins.2018.00049

Borre, Y. E., O'Keeffe, G. W., Clarke, G., Stanton, C., Dinan, T. G., & Cryan, J. F. (2014). Microbiota and neurodevelopmental windows: Implications for brain disorders. *Trends in Molecular Medicine, 20*(9), 509–518. https://doi.org/10.1016/j.molmed.2014.05.002

Breit, S., Kupferberg, A., Rogler, G., & Hasler, G. (2018). Vagus nerve as modulator of the brain–gut axis in psychiatric and inflammatory disorders. *Frontiers in Psychiatry, 9*, 44. https://doi.org/10.3389/fpsyt.2018.00044

Butler, A. B., & Hodos, W. (2005). *Comparative vertebrate neuroanatomy: Evolution and adaptation*. Hoboken, NJ: Wiley.

Carabotti, M., Scirocco, A., Maselli, M. A., & Severi, C. (2015). The gut–brain axis: Interactions between enteric microbiota, central and enteric nervous systems. *Annals of Gastroenterology, 28*(2), 203–209. PMID: 25830558.

Chang, H. Y., Mashimo, H., & Goyal, R. K. (2003). Musings on the wanderer: What's new in our understanding of vago-vagal reflex? IV. Current concepts of vagal efferent projections to the gut. *American Journal of Physiology - Gastrointestinal and Liver Physiology, 284*(3), G357–G366. https://doi.org/10.1152/ajpgi.00493.2002

Chatonnet, F., Picou, F., Fauquier, T., & Flamant, F. (2011). Thyroid hormone action in cerebellum and cerebral cortex development. *Journal of Thyroid Research.* , Article 145762. https://doi.org/10.4061/2011/145762

Cheng, S. Y., Leonard, J. L., & Davis, P. J. (2010). Molecular aspects of thyroid hormone actions. *Endocrine Reviews, 31*, 139–170. https://doi.org/10.1210/er.2009-0007

Darras, V. M., & Van Herck, S. L. J. (2012). Iodothyronine deiodinase structure and function: From ascidians to humans. *Journal of Endocrinology, 215*(2), 189–206. https://doi.org/10.1530/JOE-12-0204

de Vries, F., Lobatto, D. J., Bakker, S. J., Links, T. P., Kema, I. P., & Sluiter, W. J. (2015). Thyroid function and serotonin transporter binding in the human hypothalamus and midbrain: A PET study. *European Journal of Endocrinology, 172*(5), 553–560. https://doi.org/10.1530/EJE-15-0288

Deal, C., & Volkoff, H. (2020). The role of the thyroid axis in fish. *Frontiers in Endocrinology, 11*, Article 596585. https://doi.org/10.3389/fendo.2020.596585

Dratman, M. B., Crutchfield, F. L., & Schoenhoff, M. B. (1991). Transport of iodothyronines from bloodstream to brain: Contributions by blood-brain and choroid plexus–cerebrospinal fluid barriers. *Brain Research, 554*, 229–236. https://doi.org/10.1016/0006-8993(91)90194-Z

Fekete, C., & Lechan, R. M. (2014). Negative feedback regulation of hypophysiotropic thyrotropin-releasing hormone (TRH) neurons: Role of neuronal afferents and type 2 deiodinase. *Frontiers in Neuroendocrinology, 35*(4), 465–473.

Foster, J. A., Rinaman, L., & Cryan, J. F. (2017). Stress & the gut–brain axis: Regulation by the microbiome. *Neurobiology of Stress, 7*, 124–136. https://doi.org/10.1016/j.ynstr.2017.03.001

Fregly, M. J. (1989). Thyroid hormones and cholinergic function. *Pharmacology & Therapeutics, 41*(2), 277–304.

Friesema, E. C., Jansen, J., Jachtenberg, J. W., Visser, W. E., Kester, M. H., & Visser, T. J. (2008). Effective cellular uptake and efflux of thyroid hormone by human monocarboxylate transporter 10. *Molecular Endocrinology, 22*, 1357–1369. https://doi.org/10.1210/me.2007-0112

Furness, J. B. (2012). The enteric nervous system and neurogastroenterology. *Nature Reviews Gastroenterology & Hepatology, 9*(5), 286–294. https://doi.org/10.1038/nrgastro.2012.32

Gavrila, A., & Hollenberg, A. N. (2019). The hypothalamic-pituitary-thyroid axis. In S. Melmed, K. S. Polonsky, P. R. Larsen, & H. M. Kronenberg (Eds.), *Williams textbook of endocrinology* (14th ed., pp. 380–423). Elsevier.

Grau, E. G. (1988). Environmental and endocrinological control of osmoregulation in teleosts. *American Zoologist, 28*(1), 73–89. https://doi.org/10.1093/icb/28.1.73

Groeneweg, S., van Geest, F. S., Peeters, R. P., Heuer, H., & Visser, W. E. (2019). Thyroid hormone transporters. *Endocrine Reviews, 41*(2), 146–201. https://doi.org/10.1210/endrev/bnz008

Guadano-Ferraz, A., Escamez, M. J., Rausell, E., & Bernal, J. (1999). Expression of type 2 iodothyronine deiodinase in hypothyroid rat brain indicates an important role of thyroid hormone in the development of specific primary sensory systems. *Journal of Neuroscience, 19*(9), 3430–3439. https://doi.org/10.1523/JNEUROSCI.19-09-03430

Hay, I. D., & Wass, J. A. (Eds.). (2009). *Clinical endocrine oncology*. John Wiley & Sons.

Hays, M. T. (1999). Localization of human thyroxine absorption. *Thyroid, 9*(7), 623–629. https://doi.org/10.1089/thy.1999.9.623

Heimeier, R. A., Das, B., Buchholz, D. R., & Shi, Y. B. (2010). The xenoestrogen bisphenol a inhibits postembryonic development by antagonizing gene regulation by thyroid hormone. *Endocrinology, 150*(6), 2964–2973. https://doi.org/10.1210/en.2008-1418

Herman, J. P., & Tasker, J. G. (2016). Paraventricular hypothalamic mechanisms of chronic stress adaptation. *Frontiers in Endocrinology, 7*, 137. https://doi.org/10.3389/fendo.2016.00137

Heuer, H. (2007). The importance of thyroid hormone transporters for brain development and function. *Best Practice & Research Clinical Endocrinology & Metabolism, 21*(2), 265–276. https://doi.org/10.1016/j.beem.2007.03.003

Holzer, P., & Farzi, A. (2014). Neuropeptides and the microbiota–gut–-brain axis. *Advances in Experimental Medicine and Biology, 817*, 195–219. https://doi.org/10.1007/978-1-4939-0897-4_9

Joëls, M., & Baram, T. Z. (2009). The neuro-symphony of stress. *Nature Reviews Neuroscience, 10*(6), 459–466. https://doi.org/10.1038/nrn2632

Kandel, E. R., Schwartz, J. H., Jessell, T. M., Siegelbaum, S. A., & Hudspeth, A. J. (2013). *Principles of neural science* (5th ed.). New York: McGraw-Hill.

Kawakami, Y., Tanda, M., Adachi, S., & Yamauchi, K. (2003). cDNA cloning of thyroid hormone receptor betas from the conger eel, *Conger myriaster*. *General and Comparative Endocrinology, 131*, 232–240. https://doi.org/10.1016/S0016-6480(03)00006-8

Knezevic, J., Starchl, C., Tmava Berisha, A., & Amrein, K. (2020). Thyroid-gut-axis: How does the microbiota influence thyroid function? *Nutrients, 12*(6), 1769. https://doi.org/10.3390/nu12061769

Koehler, V. F., Lehmphul, I., & Hoefig, C. S. (2005). Thyroid hormone action in the central nervous system. *Current Opinion in Endocrinology Diabetes and Obesity, 12*(5), 390–395. https://doi.org/10.1097/01.med.0000186479.88353.0f

Konturek, P. C., Brzozowski, T., & Konturek, S. J. (2011). Stress and the gut: Pathophysiology, clinical consequences, diagnostic approach and treatment options. *Journal of Physiology & Pharmacology, 62*(6), 591–599.

Kress, E., Rezza, A., Nadjar, J., Samarut, J., & Plateroti, M. (2009). The thyroid hormone receptor α1 directly controls transcription of the β-catenin gene in intestinal epithelial cells. *Molecular and Cellular Biology, 29*(22), 5959–5970. https://doi.org/10.1128/MCB.00714-09

Lazar, M. A. (1993). Thyroid hormone receptors: Multiple forms, multiple possibilities. *Endocrine Reviews, 14*, 184–193. https://doi.org/10.1210/edrv-14-2-184

Lyte, M., Vulchanova, L., & Brown, D. R. (2011). Stress at the intestinal surface: Catecholamines and mucosa–bacteria interactions. *Cell and Tissue Research, 343*(1), 23–32. https://doi.org/10.1007/s00441-010-1050-0

Martin, E. I., Ressler, K. J., Binder, E., & Nemeroff, C. B. (2018). The neurobiology of anxiety disorders: Brain imaging, genetics, and psychoneuroendocrinology. *The Psychiatric Clinics of North America, 32*(3), 549–575. https://doi.org/10.1016/j.psc.2009.05.004

Mathiisen, T. M., Lehre, K. P., Danbolt, N. C., & Ottersen, O. P. (2010). The perivascular astroglial sheath provides a complete covering of the brain microvessels: An electron microscopic 3D reconstruction. *Glia, 58*, 1094–1103. https://doi.org/10.1002/glia.20990

Mayer, E. A. (2011). Gut feelings: The emerging biology of gut-brain communication. *Nature Reviews Neuroscience, 12*, 453–466.

Mayer, E. A., Knight, R., Mazmanian, S. K., Cryan, J. F., & Tillisch, K. (2014). Gut microbes and the brain: Paradigm shift in neuroscience. *Journal of Neuroscience, 34*(46), 15490–15496. https://doi.org/10.1523/JNEUROSCI.3299-14.2014

McAninch, E. A., & Bianco, A. C. (2014). Thyroid hormone signaling in energy homeostasis and energy metabolism. *Annals of the New York Academy of Sciences, 1311*, 77–87. https://doi.org/10.1111/nyas.12374

McEwen, B. S. (2007). Physiology and neurobiology of stress and adaptation: Central role of the brain. *Physiological Reviews, 87*(3), 873–904.

McMenamin, S. K., & Parichy, D. M. (2013). Metamorphosis in teleosts. *Current Topics in Developmental Biology, 103*, 127–165. https://doi.org/10.1016/B978-0-12-385979-2.00005-8

Molla, M. H. R., Hasan, M. T., Jang, W. J., Soria Diaz, C. D., Appenteng, P., Marufchoni, H., et al. (2019). Thyroid hormone–induced swim bladder and eye maturation are transduced by IGF-1 in zebrafish embryos. *Aquaculture Research., 50*(11), 3462–3470. https://doi.org/10.1111/are.14305

Moneva, M. H., Videla, L. A., & Fernández, V. (2011). Influence of thyroid status on nitric oxide synthase activity and protein nitration in rat gastric mucosa. *Thyroid, 21*(7), 739–744.

Mulak, A., & Bonaz, B. (2015). Brain–gut–microbiota axis in Parkinson's disease. *World Journal of Gastroenterology, 21*(37), 10609–10620. https://doi.org/10.3748/wjg.v21.i37.10609

Nair, I. B., Manish, K., & Peter, M. S. (2024). Intermittent restraint stress and recovery on GPCR expression in brain/gut axis of mice: Activation of adrenergic, dopamine, serotonin and melatonin receptors during stress and ease response. *Proceedings of the Zoological Society, 77*(1), 58–74. https://doi.org/10.1007/s12595-023-00508-y

Nelson, E. R., & Habibi, H. R. (2009). Thyroid receptor subtypes: Structure and function in fish. *General and Comparative Endocrinology, 161*(1), 90–96. https://doi.org/10.1016/j.ygcen.2008.09.014

Nguyen, T. T., Chapa, F., DiStefano, J. J., & Hays, M. T. (1993). Binding of thyroxine and triiodothyronine by intestinal bacteria. *American Journal of Physiology. Endocrinology and Metabolism, 265*(5), E924–E929. https://doi.org/10.1152/ajpendo.1993.265.5.E924

Nguyen, T. T., Distefano, J. J., Huang, L. M., Yamada, H., & Cahnmann, H. J. (1993). 5′- and 5-deiodinase activities in adult rat cecum and large bowel contents inhibited by intestinal microflora. *American Journal of Physiology. Endocrinology and Metabolism, 265*, E521–E524. https://doi.org/10.1152/ajpendo.1993.265.3.E521

Nieuwenhuys, R. (2011). The development and general morphology of the telencephalon of actinopterygian fishes: Synopsis, documentation and commentary. *Brain structure & Function, 215*(3–4), 141–157. https://doi.org/10.1007/s00429-010-0285-6

Northcutt, R.,G. (2002). Understanding vertebrate brain evolution. *Integrative and Comparative Biology, 42*(4), 743–756. https://doi.org/10.1093/icb/42.4.743

Orozco, A., & Valverde, R. C. (2005). Thyroid hormone deiodination in fish. *Thyroid, 15*(8), 799–813. https://doi.org/10.1089/thy.2005.15.799

Parhar, I. S., Soga, T., & Sakuma, Y. (2000). Thyroid hormone and estrogen regulate brain region–specific messenger ribonucleic acids encoding three gonadotropin–releasing hormone genes in sexually immature male fish, *Oreochromis niloticus*. *Endocrinology, 141*(5), 1618–1626. https://doi.org/10.1210/endo.141.5.7460

Peter, M. C. S. (2007). *reportThyroid hormones and hydromineral regulation during stress response in fish*. DSc Dissertation. The Netherlands: Radboud University Nijmegen https://respository.ubn.ru.nl.

Peter, V. S. (2009). Stress response in Mozambique tilapia (*Oreochromis mossambicus*): Temporal and inverse interaction of cortisol and thyroid hormone when confined to net. *Journal of Endocrinology and Reproduction, 13*(2), 87–96.

Peter, M. C. S. (2011). The role of thyroid hormones in stress response of fish. *General and Comparative Endocrinology, 172*(2), 198–210. https://doi.org/10.1016/j.ygcen.2011.02.023

Peter, V. S., Joshua, E. K., Bonga, S. E. W., & Peter, M. S. (2007). Metabolic and thyroidal response in air-breathing perch (*Anabas testudineus*) to water-borne kerosene. *General and Comparative Endocrinology, 152*(2–3), 198–205. https://doi.org/10.1016/j.ygcen.2007.05.015

Peter, M. C. S., Lock, R. A., & Bonga, S. E. W. (2000). Evidence for an osmoregulatory role of thyroid hormones in the freshwater mozambique tilapia *Oreochromis mossambicus*. *General and Comparative Endocrinology, 120*(2), 157–167. https://doi.org/10.1006/gcen.2000.7542

Peter, M. C. S., & Wendelaar Bonga, S. E. (2006). Ionoregulatory and interrenal responses in hyper- and hypothyroid tilapia (*Oreochromis mossambicus*) to thyroid hormones. *Comparative Biochemistry and Physiology A, 143*, 56–57.

Power, D. M., Llewellyn, L., Faustino, M., Nowell, M. A., Björnsson, B. T., Einarsdottir, I. E., Canario, A. V. M., & Sweeney, G. E. (2000). Thyroid hormones in growth and development of fish. *Comparative Biochemistry and Physiology - Part C: Toxicology & Pharmacology, 125*(4), 441–453. https://doi.org/10.1016/S0742-8413(00)00128-4

Power, D. M., Llewellyn, L., Faustino, M., Nowell, M. A., Björnsson, B. T., Einarsdottir, I. E., Canario, A. V. M., & Sweeney, G. E. (2001). Thyroid hormones in growth and development of fish. *Comparative Biochemistry and Physiology - Part C: Toxicology & Pharmacology, 130*(4), 447–459. https://doi.org/10.1016/S1532-0456(01)00271-X

Rau, V., DeCola, J. P., & Fanselow, M. S. (2005). Stress-induced enhancement of fear learning: An animal model of posttraumatic stress disorder. *Neuroscience & Biobehavioral Reviews, 29*(8), 1207–1223.

Ringo, E., Lødemel, J. B., Myklebust, R., Jensen, L., Lund, V., & Mayhew, T. M. (2003). Epithelium-associated bacteria in the gastrointestinal tract of arctic charr (*Salvelinus alpinus* L.). an electron microscopical study. *Journal of Applied Microbiology, 95*(1), 109–118. https://doi.org/10.1046/j.1365-2672.2003.01956.x

Sandler, B., Webb, P., Apriletti, J. W., Huber, B. R., Togashi, M., Cunha Lima, S. T., et al. (2004). Thyroxine-thyroid hormone receptor interactions. *Journal of Biological Chemistry, 279*, 55801–55808. https://doi.org/10.1074/jbc.M410124200

Sawin, C. T., & Becker, D. V. (1997). Radioiodine and the treatment of hyperthyroidism: The early history. *Thyroid, 7*(2), 163–176. https://doi.org/10.1089/thy.1997.7.163

Schemann, M., & Neunlist, M. (2004). The human enteric nervous system. *Neuro-Gastroenterology and Motility, 16*(Suppl. 1), 55–59. https://doi.org/10.1111/j.1743-3150.2004.00476.x

St Germain, D. L., Galton, V. A., & Hernandez, A. (2009). Minireview: Defining the roles of the iodothyronine deiodinases: Current concepts and challenges. *Endocrinology, 150*, 1097–1107. https://doi.org/10.1210/en.2008-1588

Sudhakaran, A., & Peter, M. C. S. (2024). Effects of L-NAME and air exposure on mitochondrial energetic markers, thyroid hormone receptor/regulator system and stress/ease-responsive receptor expression in the brain/gut axis of zebrafish. *Comparative Biochemistry and Physiology - Part C: Toxicology & Pharmacology, 287*, Article 110043. https://doi.org/10.1016/j.cbpc.2024.110043

Sui, L., & Chen, M. (2012). Prenatal hypothyroidism alters expression of NMDA receptor subunits and impairs long-term potentiation in rat hippocampus. *Neuroendocrinology, 96*(4), 272–281.

Tache, Y., Million, M., Nelson, A. G., Lamy, C., & Wang, L. (2009). Role of corticotropin-releasing factor pathways in stress-related alterations of colonic function. *Journal of Animal Science, 87*(14 Suppl. l.), E101–E108.

Takayama, S., Hostick, U., Haendel, M., Eisen, J., & Darimont, B. (2008). An F-domain introduced by alternative splicing regulates activity of the zebrafish thyroid hormone receptor α. *General and Comparative Endocrinology, 155*(1), 176–189. https://doi.org/10.1016/j.ygcen.2007.03.013

Tan, C. C. (2023). Stress responses: From physiology to disease. *Frontiers in Endocrinology, 14*, Article 1123456. https://doi.org/10.3389/fendo.2023.1123456

Tsigos, C., & Chrousos, G. P. (2002). Hypothalamic–pituitary–adrenal axis, neuroendocrine factors and stress. *Journal of Psychosomatic Research, 53*(4), 865–871. https://doi.org/10.1016/S0022-3999(02)00429-4

Vatine, G. D., Zada, D., Lerer-Goldshtein, T., Tovin, A., Malkinson, G., Yaniv, K., et al. (2013). Zebrafish as a model for monocarboxyl transporter 8–deficiency. *Journal of Biological Chemistry, 288*(1), 169–180. https://doi.org/10.1074/jbc.M112.413831

Vijayan, M. M., Raptis, S., & Sathiyaa, R. (2003). Cortisol treatment affects glucocorticoid receptor and glucocorticoid-responsive genes in the liver of rainbow trout. *General and Comparative Endocrinology, 132*(2), 256–263. https://doi.org/10.1016/S0016-6480(03)00092-3

Virili, C., Centanni, M., Maccora, C., Santaguida, M. G., Brusca, N., Capriello, S., & Gargano, L. (2018). Gut microbiota and thyroid

interaction in health and disease. *Reviews in Endocrine & Metabolic Disorders, 19*(3), 293–300. https://doi.org/10.1007/s11154-018-9467-3
Virili, C., Fallahi, P., Antonelli, A., Benvenga, S., & Centanni, M. (2018). Gut microbiota and hashimoto's thyroiditis. *Reviews in Endocrine & Metabolic Disorders, 19*, 293–300. https://doi.org/10.1007/s11154-018-9452-9
Visser, W. E., Friesema, E. C. H., & Visser, T. J. (2011). Minireview: Thyroid hormone transporters: The knowns and the unknowns. *Molecular Endocrinology, 25*(1), 1–14. https://doi.org/10.1210/me.2010-0095
Viswanath, M., & Peter, M. C. S. (2025). Thyroid hormones activate TH/E_2 receptor/regulator system and drive Na^+/K^+-ATPase in the ovarian wall of hypothyroid air-breathing fish (*Anabas testudineus*). *General and Comparative Endocrinology, 360*, Article 114640. https://doi.org/10.1016/j.ygcen.2024.114640
Volkoff, H. (2016). The neuroendocrine regulation of food intake in fish: A review of current knowledge. *Frontiers in Neuroscience, 10*, 540. https://doi.org/10.3389/fnins.2016.00540
Wallis, K., Dudazy, S., Van Hoger linden, M., Nordstrom, K., Mittag, J., & Vennstrom, B. (2010). The thyroid hormone receptor alpha1protein is expressed in embryonic post-mitotic neurons and persists in most adult neurons. *Molecular Endocrinology, 24*, 1904–1916. https://doi.org/10.1210/me.2010-0175
Williams, R. H., & Larsen, P. R. (2003). *Williams textbook of endocrinology*. Philadelphia, PA: Saunders.

Chapter 12

Scenarios of astrocyte dysfunction linked to thyroid hormone availability

Marina Guillén-Yunta[1,3], Víctor Valcárcel-Hernández[2] and Ana Montero-Pedrazuela[1]

[1]Department of Neurological Diseases, Instituto de Investigaciones Biomédicas Sols-Morreale, Consejo Superior de Investigaciones Científicas-Universidad Autónoma de Madrid, Madrid, Spain; [2]Laboratory Molecular Physiology and Adaptation, UMR 7221, Department Adaptations of Life, Muséum National d'Histoire Naturelle, Paris, France; [3]Department of Biosciences, Faculty of Biomedical and Health Sciences, Universidad Europea de Madrid, Madrid, Spain

List of abbreviations

AHDS Allan-Herndon-Dudley syndrome (aka MCT8 deficiency)
ALDH1L1 Aldehyde dehydrogenase 1 family member L1
AQP4 Aquaporin 4
BBB Blood-brain barrier
CNS Central nervous system
DIO Deiodinase
EDC Endocrine-disrupting chemical
GABA Gamma aminobutyric acid
GFAP Glial fibrillary acidic protein
GLAST Glutamate aspartate transporter (aka EAAT1)
GLT1 Glutamate transporter 1 (aka EAAT2)
GLUT1 Glucose transporter 1
KO Knock-out
LAT Large neutral amino acid transporters
LPS Lipopolysaccharide (bacterial endotoxin)
MCT Monocarboxylate transporter
NPC Neural Progenitor Cell
NSC Neural stem cell
OATP Organic anion transporter polypeptides
OL Oligodendrocyte
PPII Pyroglutamyl peptidase II
T3 3-5-3′-triiodothyronine
T4 Thyroxine or 3-5-3′-5′-tetraiodothyronine
TH Thyroid hormones
TR Thyroid hormone receptor
TRH Thyrotropin-releasing hormone
TSH Thyroid-stimulating hormone or thyrotropin

Introduction

Astrocytes, named after their star-like morphology, were firstly described by Rud olf (Virchow, 1846) and represent the major glial cell type in the central nervous system (CNS), constituting 20%–40% of brain parenchymal cells. Astrocytes are generated from neural progenitor cells (NPC) derived from neural stem cells (NSC) in the neuroectoderm. As the brain develops, neuroepithelial cells transform into radial glia, which, during late embryonic development, give rise to the vast majority of neurons in the brain parenchyma before acquiring the capacity to generate glial cells, a process known as the gliogenic switch (Akdemir et al., 2020; Lattke & Guillemot, 2022). These gliogenic NPC can produce both astrocytes and oligodendrocytes, which will differentiate through a complex process of functional and morphological maturation specific to each cell type.

Following their maturational program, these cells sequentially express (from immature to mature state) the microfilaments nestin, vimentin, and glial fibrillary acidic protein (GFAP). Thus, fully mature astrocytes are usually identified by GFAP expression (although radial glia and NSC also express this protein). Astrocytes can also be specifically detected by the expression of other markers such the aldehyde dehydrogenase 1 family member L1 (ALDH1L1), the glutamine synthetase, and several transporter proteins as glutamate-aspartate transporter (GLAST), glutamate transporter 1 (GLT1), glucose transporter 1 (GLUT1) and aquaporin 4 (AQP4) (Jurga et al., 2021).

Astrocytes comprise a very heterogeneous population of cells, showing a high morphological and functional diversity, as well as important inter-species differences (Oberheim et al., 2012). Both the time course of astrocytic lineage development and the morphological and molecular regional heterogeneities are relatively well-characterized in the mouse cortex, but not in other brain regions or species. Several brain regions display specialized forms of

Hormones in Neuroscience. https://doi.org/10.1016/B978-0-443-30261-9.00012-7

astrocytes with specific functions and molecular features, such as the Bergmann glia in the cerebellar cortex, or the tanycytes that line the walls of the third ventricle (Oberheim et al., 2012). In the cerebral cortex, diverse subtypes of astrocytes have been described based on morphological and localization criteria (Tabata, 2015) (Fig. 12.1). In rodents, two main astrocytic subtypes are recognized: protoplasmic astrocytes (located in the gray matter, showing low GFAP expression and highly branched bushy processes) and fibrous astrocytes (mainly found in the white matter, expressing high levels of GFAP with long straight processes). Besides, other subtype are Layer I specialized astrocytes, with a bushy morphology and high GFAP expression covering the brain outer surface beneath the pia matter. Primates present two additional specific subtypes: interlaminar astrocytes (restricted to Layer I, with long processes reaching Layers II-IV) and varicose projection astrocytes (with long varicose processes, coexisting with protoplasmic astrocytes in Layers V-VI).

Astrocytes functions in health and disease

Astrocytes have emerged as key regulators of the CNS beyond their assigned classical roles in ionic buffering,

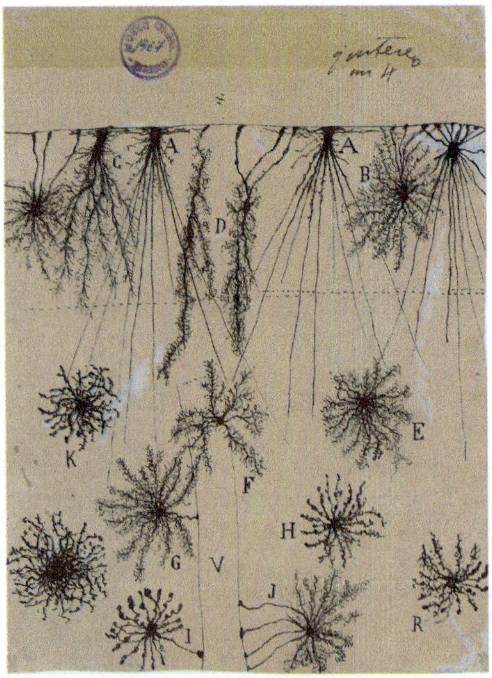

FIGURE 12.1 Astroglia of the superficial layers of the human cerebral cortex. Original drawing by Santiago Ramóny Cajal (1904) . The image shows the distribution and morphology of astrocytes within the upper cortical layers, emphasizing their structural complexity and spatial organization in early neuroanatomical descriptions. *Reproduced with permission of the Legado Cajal, CSIC.*

metabolic support, neurotransmitter clearance, and blood-brain barrier (BBB) maintenance (Yuan et al., 2023). Although the identity and specific functions of each astrocytic differentiation stage are not well-characterized to date, it is known that immature astrocytes support neuronal maturation during embryonic development by promoting neurite outgrowth and synaptic formation, synapse-specific pruning, and secretion of extracellular proteins to establish guidelines for axonal growth. Astrocytes also participate in the process of brain angiogenesis and in the BBB ontogeny, and contribute to developmental myelination (Lattke & Guillemot, 2022).

Mature astrocytes acquire new functions to ensure adult brain homeostasis, providing trophic and metabolic support for neurons, and modulating neuronal activity through the local control of extracellular ions and neurotransmitters, the release of gliotransmitters, and the modulation of synaptic plasticity (Fig. 12.2). In addition, mature astrocytes are also involved in the maintenance and regulation of the BBB as essential components of the neurovascular unit (NVU), controlling ion and water homeostasis between the blood and the CNS, and conducting immune functions as chemokines producers, affecting other cells' functions, such as microglia (Fig. 12.2) (Lattke & Guillemot, 2022; Yuan et al., 2023).

Under pathological conditions, astrocytes undergo profound morphological and functional alterations in which they acquire inflammatory properties that compromise their normal homeostatic roles (Tiwari et al., 2025). These astrocytes, widely known as reactive astrocytes, display increased number and length of processes together with an elevated GFAP expression, becoming hypertrophic compared to their physiological state. Beyond this structural remodeling, they also fail to maintain glutamate and mitochondrial metabolic balance, contributing to excitotoxicity, oxidative stress, impaired neuronal energy supply, cytokine and chemokine release, and a bidirectional crosstalk with microglia that amplifies neuroinflammation. In parallel, protein aggregation, acute injury, or demyelinating processes may further promote reactive gliosis, extracellular matrix deposition, and glial scar formation. Importantly, these alterations also affect the NVU, increasing BBB permeability and disrupting water homeostasis through changes in AQP4 expression (Fig. 12.2). In further sections of this chapter, we will deepen into the astrocyte pathophysiology regarding the BBB and neuroinflammation in thyroid hormones (TH)-related disorders.

Astrocytes as mediators for thyroid hormone action in the CNS

Astrocytes are the most abundant CNS cells that express the enzyme type 2 deiodinase or DIO2 (Fig. 12.3), which

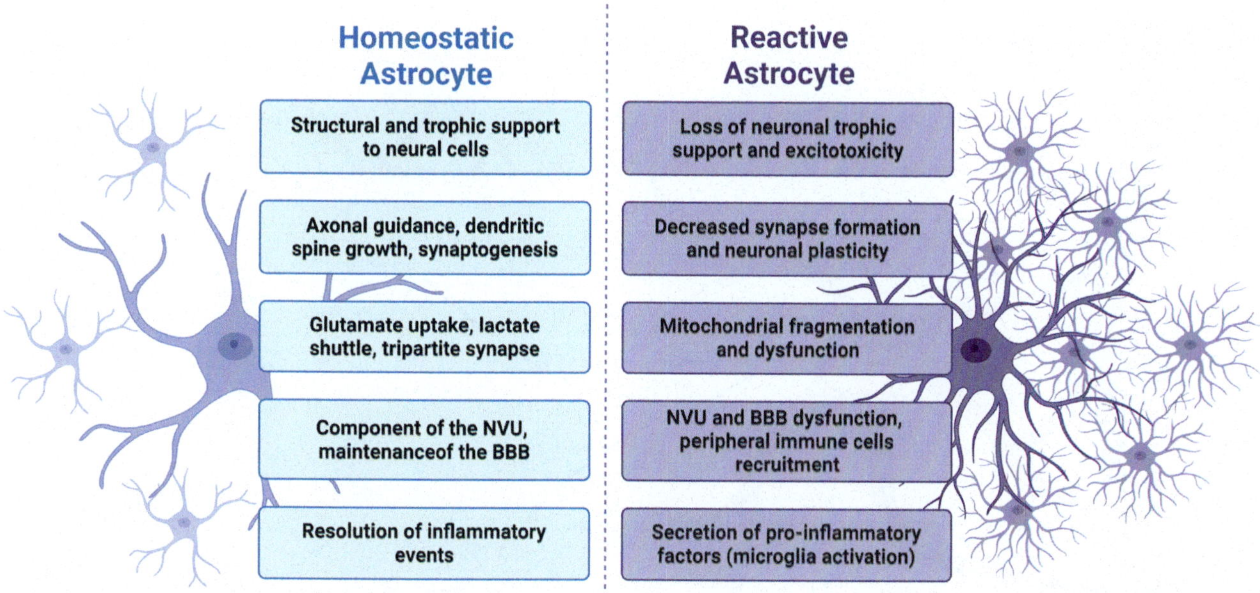

FIGURE 12.2 Astrocyte functions in health and disease. Overview of mature astrocyte functions under physiological (left) and pathological (right) conditions. *BBB*, blood-brain barrier; *NVU*, neurovascular unit. *Figure created in BioRender.*

catalyzes the conversion of T4 to T3, ensuring a local and steady supply of the active hormone to neurons and other glial cells (Guadaño-Ferraz et al., 1997). Most T3 receptor occupancy in the cerebral cortex derive from this conversion, as demonstrated by a seminal study (Crantz et al., 1982), underscoring the relevance of local T3 production. Although tanycytes also express DIO2, astrocytic DIO2 activity is prevalent as demonstrated by the 90% reduction in DIO2 activity after astrocyte-specific *Dio2* inactivation (Fonseca et al., 2014).

Due to their location at the BBB and DIO2 expression, astrocytes are uniquely positioned to modulate TH availability and signaling in the CNS, by controlling both hormone transport and metabolism. Thus, astrocytes are crucial mediators of TH bioavailability, shaping neuronal and glial development and function.

This key role of astrocytes has been highlighted by the identification of the underlying cause of the Allan-Herndon-Dudley syndrome (AHDS) (Allan et al., 1944) and the differences between humans and mouse models. AHDS, also known as MCT8 deficiency, is a rare X-linked disease caused by inactivating mutations of *SLC16A2*, the gene coding for the TH-specific transporter MCT8 (Dumitrescu et al., 2004; Friesema et al., 2004). This severe neurodevelopmental disorder is characterized by motor dysfunction, intellectual disability, and altered serum TH profiles (Sarret et al., 1993). In humans, MCT8 is the main TH transporter at the BBB and its mutation impedes TH entry to brain parenchyma (López-Espíndola et al., 2019; Roberts et al., 2008). This way, the brain renders hypothyroid, suffering severe brain impairments.

By contrast, mice present a compensatory mechanism that is barely expressed in humans: a high expression of the T4 transporter OATP1C1 at the BBB. Consequently, MCT8 knock-out (KO) mice do not exhibit significant neurological alterations, as astrocytic DIO2 activity supplies T3 to the brain parenchyma from the OATP1C1-derived T4 (Ceballos et al., 2009). These differences emphasize the importance of astrocytes as essential providers for TH availability and signaling in the CNS and underscore the significance of their dysfunction.

Astrocytes are thyroid hormone target cells

In addition to DIO2, astrocytes also express the main TH effectors: TH transporters and TH receptors (TR) (Fig. 12.3). It has been shown that astrocytes express high levels of OATP1C1, lower levels of LAT1 and minimum levels of MCT8 in mice (Bernal et al., 2015), based on previously published single cell RNAseq data (Zhang et al., 2014) (Fig. 12.4). In humans, our current analysis of published single cell RNAseq data (Zhang et al., 2016) reveal similar results (Fig. 12.4). These TH transporters, along with the ones expressed at the BBB, allow the uptake of circulating T4 and T3 into the astrocytic cytoplasm, where T4 will be converted into T3 by DIO2. Single cell RNAseq data also demonstrate the expression of the two main TR isoforms, THRA and THRB, in both mouse and human astrocytes (Fig. 12.4), to which T3 will bind to exert its genomic actions. Rapid non-genomic TH actions have also been described in astrocytes; however, TH effects are

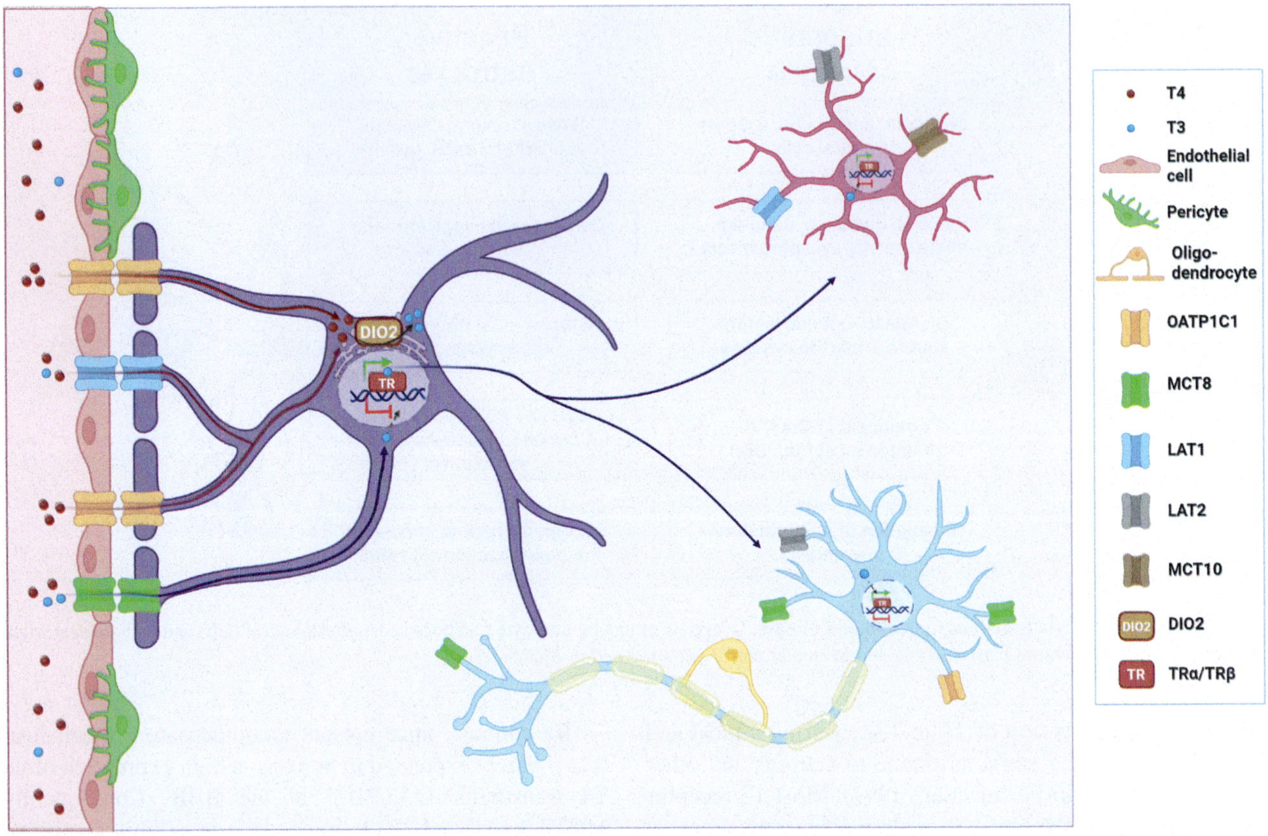

FIGURE 12.3 Scheme of the main effectors for thyroid hormones in neural cells. The figure illustrates an astrocyte (purple), neuron (blue) and microglia (pink) with the main thyroid hormone effectors: Transmembrane transporters (OATP1C1, MCT8, LAT1, LAT2, MCT10), type 2 deiodinase (DIO2), and thyroid hormone receptors (TRs). The blood-brain barrier, composed of endothelial cells, pericytes, and astrocytic end-feet, is also shown. Refer to the included legend at the right-side of the figure to identify each element. *Figure created in BioRender.*

predominantly mediated by T3 binding to the TR, which act as transcriptional factors regulating the expression of specific TH-responsive genes. On the other hand, astrocytes do not express the TH inactivating enzyme DIO3, which is mainly expressed by neurons (Tu et al., 1999).

The expression of these TH effectors points to astrocytes not only as mediators, but also as direct targets of TH action. Indeed, TH play a fundamental role in the process of morphogenesis and homeostatic development of different astroglial populations, suggesting that TH-related disorders can actually affect astrocyte development and function.

Astrocyte defects associated to thyroid hormone alterations

Several studies have explored the effects of TH on astrocyte development and function. Most evidence has been reported using primary cultures, which display a particular differentiation process, different from their in vivo maturation, probably due to the lack of most of the surrounding key signaling factors, essential for proper astrocyte development. Besides, cultured astrocytes show high variability on TH response depending on their origin (age and region) (Lima et al., 1998). Thus, in this chapter we will focus mainly on in vivo studies using experimental models of TH-related disorders.

Perinatal TH deficiency induced an increase in astrocytes and Bergmann glia cell number, which also presented altered maturation during rat cerebellum development (Clos & Legrand, 1973; Clos et al., 1980; Faivre-Sarrailh et al., 1991). Delayed astrocyte development and accumulation in the cerebellar white matter was also described in hypothyroid mice (Morte et al., 2004). Moreover, this study indicates that a balanced TRβ1 and TRα1 activity is needed for proper astrocyte differentiation in the cerebellum. Another evidence of TH-dependent astrocyte dysfunction is the altered nestin-vimentin-GFAP transition in the hypothyroid rat cerebellum (Manzano et al., 2007). The transition from radial glia to mature astrocytes is also impaired by fetal TH deficiency in the rat hippocampus (Faivre-Sarrailh et al., 1991; Martinez-Galan et al., 1997).

On the other hand, neonatal hyperthyroidism accelerates astroglial maturation and induces premature disappearance of radial glia in the rat cerebral cortex (Gould

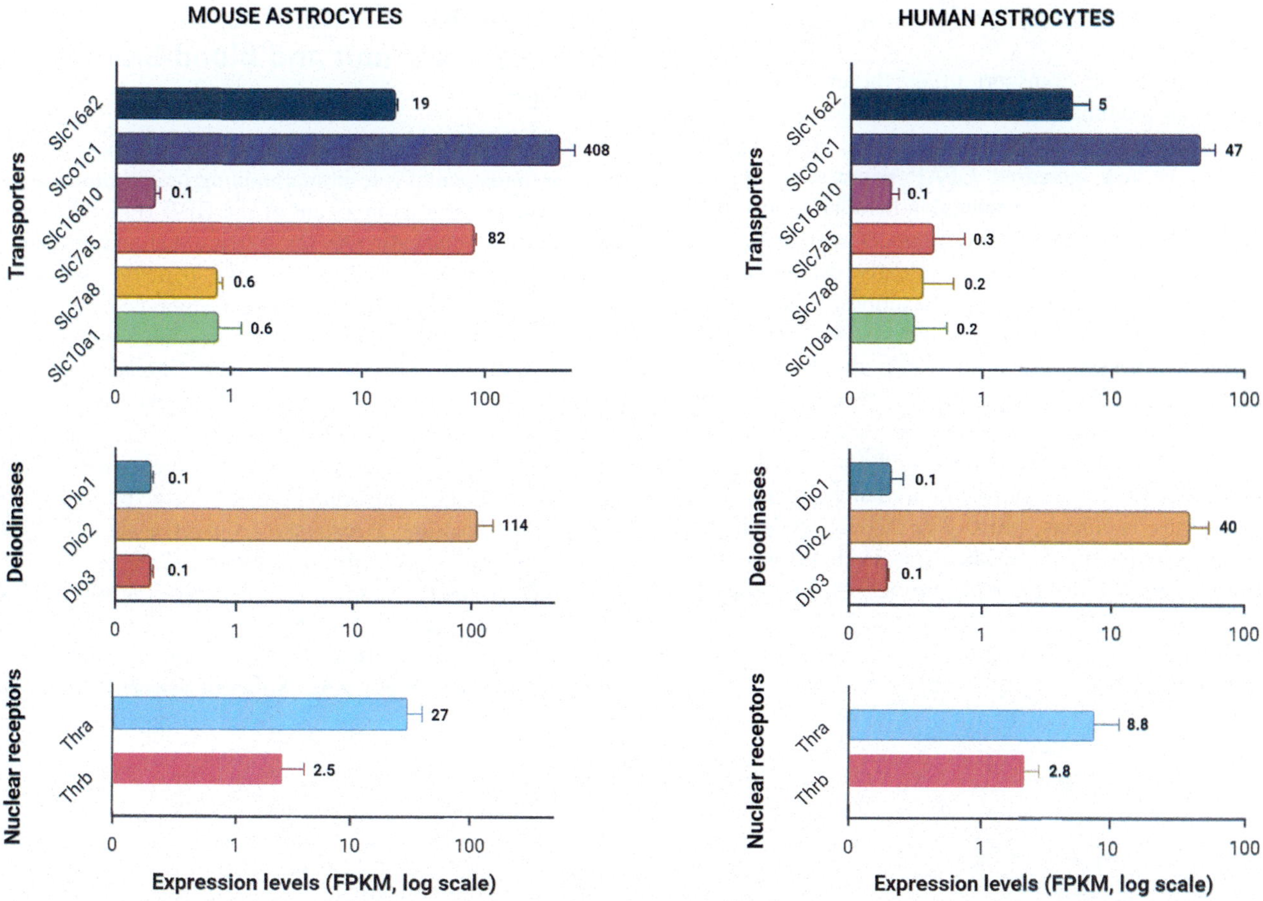

FIGURE 12.4 Expression levels of thyroid hormone effectors in mouse and human astrocytes. Graphics depicting RNA expression levels of thyroid hormone transporters (*Scl16a2* for MCT8, *Slco1c1* for OATP1C1, *Slc16a10* for MCT10, *Slc7a5* for LAT1, *Slc7a8* for LAT2, *Slc10a1* for NTCP), deiodinases (*Dio1*, *Dio2*, *Dio3*), and nuclear receptors (*Thra* for TRα, *Thrb* for TRβ) in mouse (left) and human (right) astrocytes. Note the logarithmic scale at the x axis to account for large expression range differences (expressed as fragments per kilobase of transcript sequence per million mapped fragments, FPKM). Mean expression values are displayed adjacent to each bar. RNAseq mouse data were obtained from Zhang et al., 2014, and human data from Zhang et al., 2016. *Graphs created in BioRender. from Zhang, Y., Chen, K., Sloan, S. A., Bennett, M. L., Scholze, A. R., O'Keeffe, S., . . . Wu, J. Q. (2014). An RNA-sequencing transcriptome and splicing database of glia, neurons, and vascular cells of the cerebral cortex. Journal of Neuroscience, 34(36), 11929-11947. Data from Zhang, Y., Sloan, S. A., Clarke, L. E., Caneda, C., Plaza, C. A., Blumenthal, P. D., . . . Barres, B. A. (2016). Purification and characterization of progenitor and mature human astrocytes reveals transcriptional and functional differences with mouse. Neuron, 89(1), 37-53.*

et al., 1990). Similar results have been found for the Bergmann glia in the cerebellum (Clos et al., 1980).

Adult-onset hypothyroidism has shown to modulate astrocytic GFAP expression although, controversially, different studies present opposed results. For example, in the rat hippocampus one study reveals an important increase in both apoptosis of astrocytic cells and GFAP expression, suggestive of astrogliosis (Cortes et al., 2012), whereas other shows a reduction in GFAP expression, indicating astroglial atrophy (Nam et al., 2013) due to TH deficiency. These differences could be due to different rat strains, or to the different anti-thyroid drug, dose and/or length of treatment used. However, it is clear that adult-onset disruption of TH homeostasis induces astrocyte alterations.

Thus, although the effect of TH deficit or excess have been widely explored, few studies have assessed the contribution of TH transporters to astrocyte function. Regarding the main TH transporter expressed by astrocytes, OATP1C1 deficiency in humans presents with severe neurological symptoms and cognitive decline, and the only patient studied to date showed elevated GFAP levels in the cerebrospinal fluid, suggesting alterations in the astrocytic population (Strømme et al., 2018). However, the mouse model of the disease had no apparent neurological phenotype (Mayerl et al., 2012), with no study of the population of astrocytes in this model to date. Nevertheless, studies in *oatp1c1*-deficient zebrafish showed morphological alterations of astrocytes and increased Gfap expression (Admati et al., 2020), suggesting the relevance of the OAPT1C1 transporter for astrocyte morphology and function.

Deficient TH transport to the brain (due to MCT8 mutations in humans and in *Mct8/Dio2* deficient mice) also induces alterations in cortical astrocytes: higher number of astrocytes and increased GFAP expression, suggesting astrogliosis, as well as a reduction in radial glia cells fibers during development (Guillén-Yunta et al., 2024).

Altogether, this evidence clearly establishes astrocytes as essential targets of TH action in the developing and adult brain. Their unique role as the main source of local T3 production in the CNS, together with their expression of TH transporters and receptors, makes astrocytes extremely sensitive to fluctuations in TH availability, inevitably impacting the neuronal networks and other glial cells given the pivotal crosstalk between astrocytes, neurons, microglia and oligodendrocytes. Understanding these mechanisms is therefore essential for elucidating the pathophysiology of TH-related disorders and for designing therapeutic strategies aimed at restoring proper TH availability and astrocyte function within the CNS.

Astrocyte functions in the neurovascular unit and blood-brain barrier

Astrocytes are the most abundant glial cells in the CNS and play a fundamental role in maintaining brain homeostasis. They are a central component of the BBB (Fig. 12.5) and the neurovascular unit (NVU), a functional and structural complex comprising endothelial cells, pericytes, astrocyte end-feet, the endothelial basement membrane, tight junctions and adherent junctions, and ultimately, neurons and microglial cells (Friedman et al., 2025). The NVU ensures a tight regulation of the molecular and water traffic between the blood and the brain parenchyma (Fig. 12.6). The dynamic changes in permeability and functionality of the BBB allow for adaptability to different physiological states, coordinating a response to diverse signals (from inside the brain and from the periphery).

In physiological conditions, astrocytes contribute to the NVU formation and maintenance by extending specialized perivascular end-feet that almost completely enwrap the brain capillaries (Fig. 12.5). Adjacent astrocytic end-feet

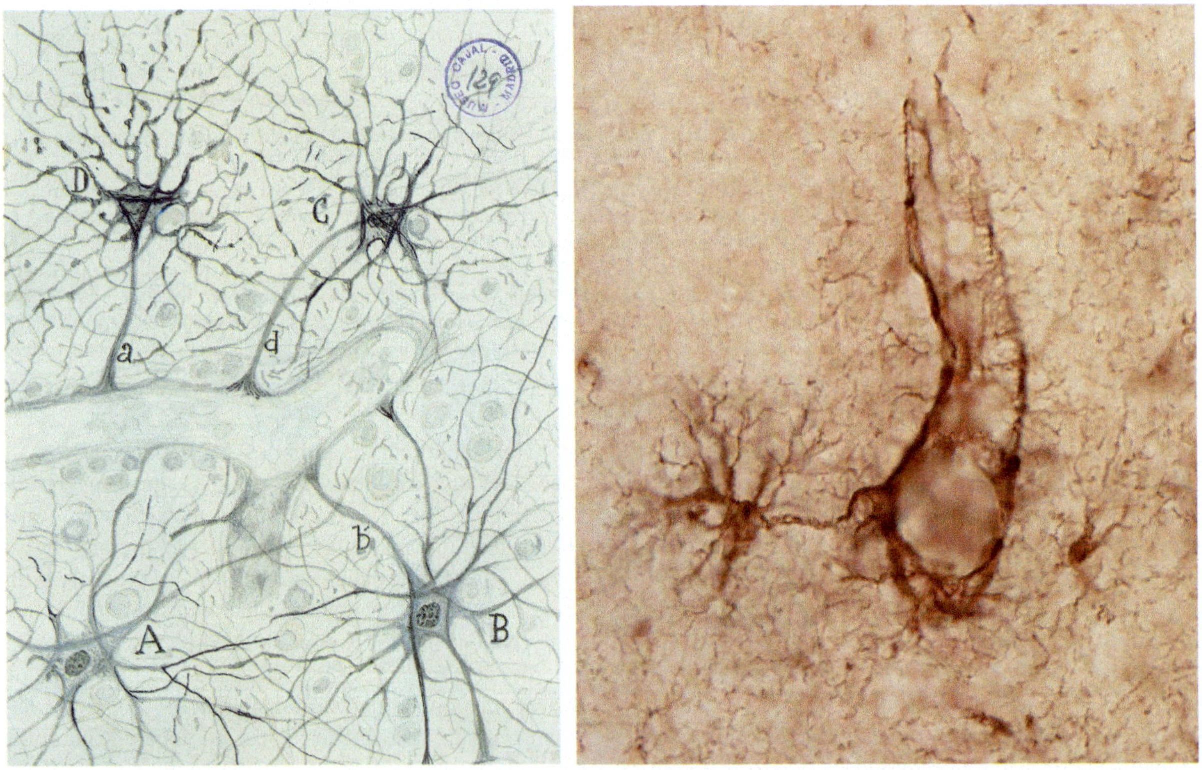

FIGURE 12.5 Astrocyte end-feet contact blood vessels at the blood-brain barrier interface. Astrocytes of the white matter of the adult human brain, some of them showing perivascular feet contacting a blood vessel, as originally drawn by Ramón y Cajal, 1913 (reproduced with permission of Legado Cajal, CSIC). The drawing highlights the early morphological descriptions of astrocyte-vascular interactions. A corresponding GFAP-stained mouse brain astrocyte contacting a blood vessel is shown at the right side to illustrate the same structural organization using modern histological methods.

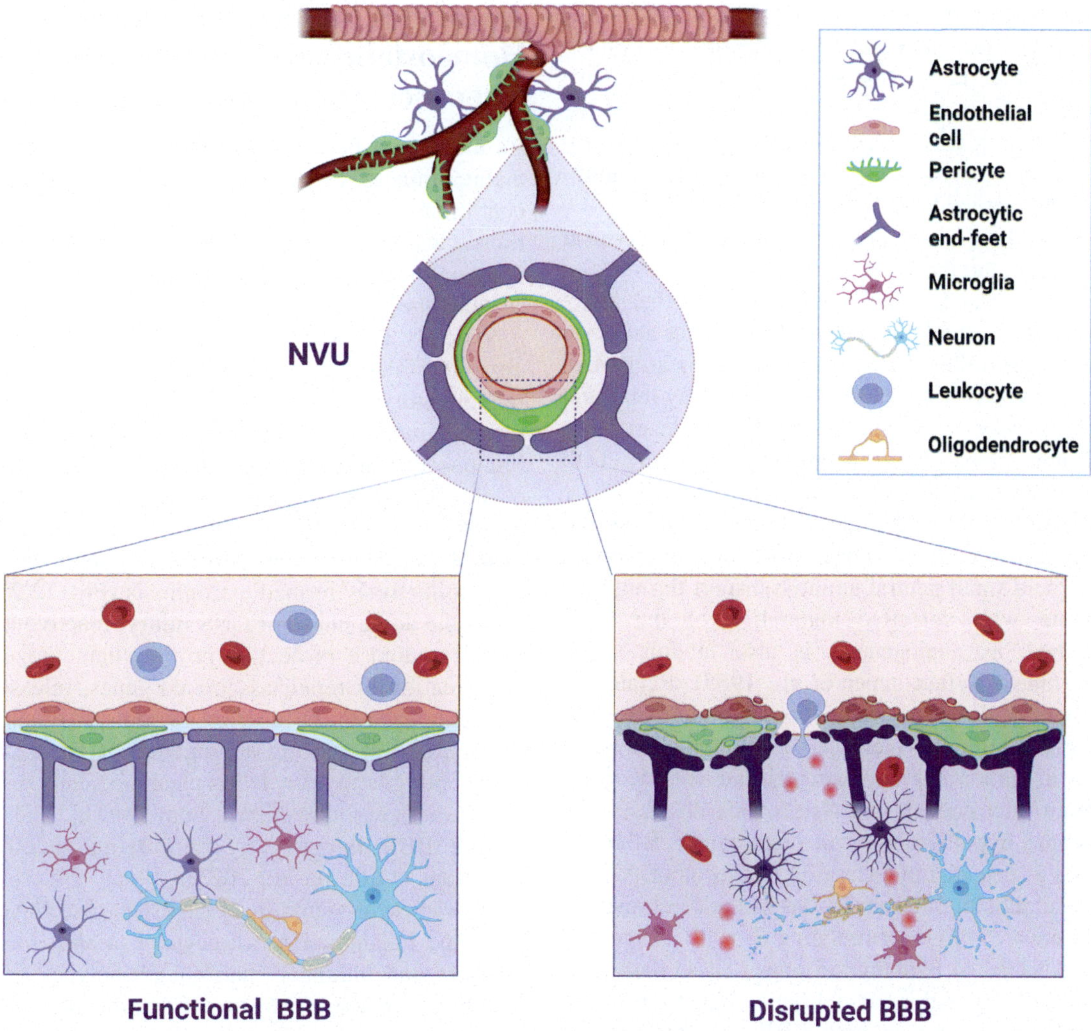

FIGURE 12.6 Neurovascular unit in health and disease and astrocyte association to BBB dysfunction. Scheme of the main components of the NVU and their status under physiological, homeostatic conditions (left, intact and functional BBB) versus pathological conditions (right, compromised and disrupted BBB), as states of thyroid hormone deficiency. *BBB*, blood-brain barrier; *NVU*, neurovascular unit. *Figure created in BioRender.*

form a functional syncytium and also interact with endothelial cells and pericytes. Through these end-feet, astrocytes secrete molecules such as sonic hedgehog, angiopoietin-1, apolipoprotein E and different growth factors that modulate tight-junction protein expression, thereby regulating barrier integrity and permeability (Friedman et al., 2025). Astrocytes also regulate neurovascular coupling, the mechanism that links neuronal activity to local blood flow. Ca^{+2} waves within astrocytic end-feet trigger the release of vasoactive mediators that signal pericytes and endothelial cells to dilate or constrict the vessel, ensuring neuronal metabolic supply and the secretion of toxic and metabolic residues to the peripheral blood flow through the BBB (Manu et al., 2023). They also express AQP4 water channels, Kir4.1 inward-rectifying potassium channels, and glutamate and GABA transporters responsible of maintaining extracellular osmolarity, ionic homeostasis and buffering neurotransmitters (Mack & Wolburg, 2013). Thus, astrocytes maintain the ionic and metabolic microenvironment of the CNS protecting neurons from osmotic and excitotoxic stress.

Neurovascular unit and blood-brain barrier alterations associated to astrocyte dysfunction

Under pathological conditions, astrocytes undergo morphological, distributional, transcriptional, and functional changes, acquiring inflammatory properties and impairing their ability to perform normal homeostatic functions that can profoundly affect BBB and NVU integrity (McConnell et al., 2019) (Fig. 12.6). This state is known as reactive astrogliosis, and can contribute to the progression of the pathology. While reactive astrocytes can initially play protective roles, chronic or excessive activation often contributes to barrier disruption and neuronal injury.

On one hand, astrocyte functions at the BBB are critical for TH passage into the brain.

On the other hand, pathological conditions that affect TH availability in the brain can severely affect astrocyte function and capability to respond to triggering stimuli.

Thus, astrocytic dysfunction at the NVU may represent a critical pathway by which defects in TH availability translate into global brain pathology, impacting both neuronal connectivity and vascular integrity.

It has been demonstrated that perinatally hypothyroid rats showed reduced brain vascular volume, cerebral blood flow and small neutral amino transport through the BBB, along with defective maturation of the brain capillary bed, as angiogenesis is also modulated by astrocytic clues (Lefauconnier et al., 1985). Related to this pro-angiogenic astrocyte function, hypothyroidism in aged mice in addition to increase GFAP immunopositive astrocytes in cortex and hippocampus, induce a decrease in the brain microvessel area (Niedowicz et al., 2023). Also, chronic hypothyroidism in dogs causes BBB abnormalities as detected by increased total protein, VEGF, S–100B, and albumin concentrations in the cerebrospinal fluid. Moreover, a higher circulating S–100B concentration was detected in hypothyroid dogs. These data indicate a BBB disruption in chronic hypothyroidism, pointing to dysfunction in both BBB endothelium and astrocytes (Pancotto et al., 2010).

Deficits in TH transport, such as those caused by MCT8 mutations, not only disrupt neuronal TH signaling but also impair NVU maturation and vascular development. Recent work identified neurovascular and BBB abnormalities as novel pathophysiological features of MCT8 deficiency, including disruption of endothelial tight junctions, increased BBB permeability evidenced by IgG and Evans-Blue leakage, reduced microvessel density, dysregulated angiogenic factors and NVU ultrastructural defects in human MCT8-deficient brain tissue and *Mct8/Dio2*-KO mice as a model of the disease (Guillén-Yunta et al., 2023). Interestingly, the ultrastructural defects included swollen astrocytic end-feet, partially detached from the endothelial basal lamina, and increased endothelial vesicular transport, compromising NVU integrity. These findings suggest that deficient TH signaling in astrocytes likely impairs the transcription of genes required for vascular maintenance (e.g., AQP4, VEGFA, or tight-junction modulators), thereby weakening astrocyte-endothelial signaling.

Astrocytes in neuroinflammation: Homeostatic role and reactive responses

As mentioned before, astrocytes play essential roles in maintaining CNS homeostasis. Under physiological conditions, they regulate extracellular ion concentrations, clear neurotransmitters such as glutamate, support neuronal metabolism through the lactate shuttle, and sustain the integrity of the BBB. These astrocytes (classically referred to as "resting" or "homeostatic") also help dampen inflammation by expressing anti-inflammatory mediators, modulating microglial responses, maintaining synaptic homeostasis, and limiting immune cell infiltration by supporting the BBB (Tiwari et al., 2025). Nevertheless, upon CNS insult -whether infectious, traumatic, ischemic, or neurodegenerative- astrocytes transition into a reactive phenotype in a process collectively termed astrogliosis (Fig. 12.2).

In the acute phase of CNS injury, reactive astrocytes initially adopt a protective profile characterized by the upregulation of repair-associated genes, release of neurotrophic factors, production of anti-inflammatory mediators, and promoting angiogenesis (Fig. 12.7). Thus, contributing to myelin debris clearance, help re-establish ionic and neurotransmitter homeostasis, and work to restore BBB integrity by reinforcing endothelial interactions (Valles et al., 2023).

However, in scenarios of persistent, excessive, or dysregulated inflammation, astrocytes can shift toward maladaptive and ultimately harmful phenotypes (Fig. 12.7). Chronically reactive astrocytes often downregulate GLT1, promoting excitotoxicity, and lose potassium buffering capacity alongside depolarization of AQP4, disrupting osmotic homeostasis. Sustained secretion of pro-inflammatory cytokines and reactive oxygen species contributes directly to neuronal injury, and at the vascular level, astrocyte end-feet retract from the capillaries, alter the composition of the basement membrane and weaken BBB properties, leading to NVU breakdown. Impaired phagocytic capacity can also exacerbate the accumulation of myelin debris and misfolded proteins, while extracellular matrix remodeling impairs axonal regrowth and synaptic plasticity (Liddelow & Barres, 2017; Valles et al., 2023). Opposed to this neuroprotective-neurotoxic dichotomy, recent transcriptomic and single-cell studies support the view that astrocyte reactivity lies on a multidimensional continuum influenced by injury type, anatomical region, and molecular environment (Escartin et al., 2021).

A central aspect of astrocyte function in neuroinflammation involves their bidirectional crosstalk with microglia (Fig. 12.7). Microglia release cytokines that can

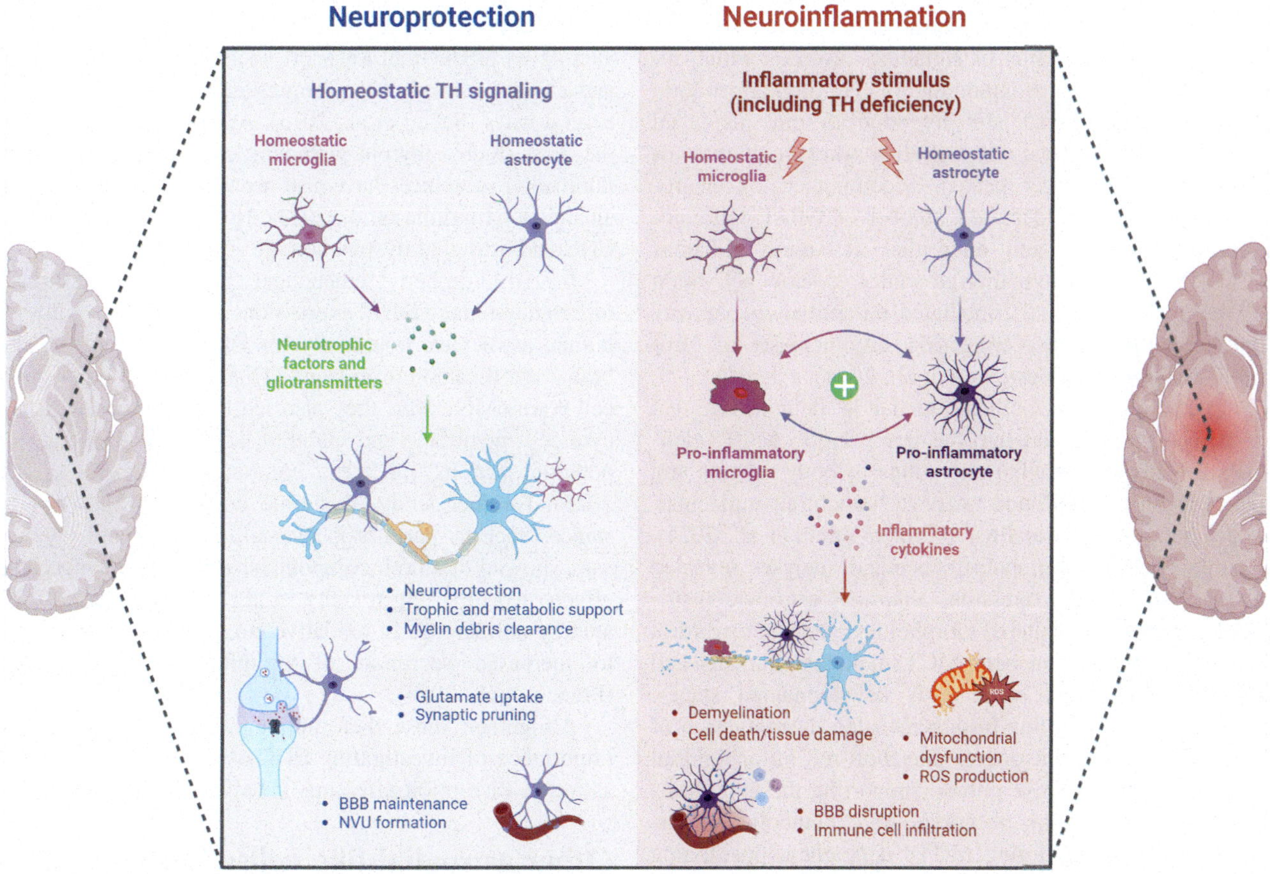

FIGURE 12.7 **Main roles of astrocytes in physiological and inflammatory conditions.** Scheme illustrating the main roles of astrocytes during physiological (left) and neuroinflammatory conditions (right), as occurs in thyroid hormone deficiency, and the associated processes. *Figure created with BioRender.*

transform astrocytes into neurotoxic phenotypes, reducing their ability to maintain synaptic and metabolic support, and release ATP, nitric oxide and prostaglandins, further modulating astrocytic calcium dynamics and inflammatory gene expression. Conversely, astrocytes regulate microglial activation, either by suppressing excessive microglial activation, facilitating inflammation resolution (neuroprotective role) or by recruiting microglial cells and promoting pro-inflammatory polarization, for the neurotoxic phenotype (Matejuk & Ransohoff, 2020). This reciprocal communication positions astrocytes as key modulators of the magnitude and trajectory of microglial responses during both acute and chronic neuroinflammation.

Altogether, the transition from adaptive reactivity to chronic dysfunction underscores the dualistic nature of astrocytes in neuroinflammation, as well as the importance of their dynamic interplay with microglia and the vascular compartment in determining neurodegenerative trajectories.

Astrocyte dysfunction during neuroinflammation and thyroid hormone-related disorders

On one hand, systemic inflammation is associated with profound changes in TH metabolism and secretion, a condition known as nonthyroidal illness syndrome, which is characterized by low serum TH levels (Wiersinga, 2000). As a model for this condition, systemic inflammation induced by the administration of bacterial lipopolysaccharide (LPS) increases DIO2 activity in mouse and rat hypothalamus, leading to an increase in local T3 availability, while the T3 inactivating enzyme DIO3 activity decreases (Boelen et al., 2011). In accordance, in vitro studies showed that LPS treatment of cultured rat glioma cells and cerebellar primary astrocytes increased DIO2 activity (Xu et al., 2013).

Conversely, different studies have demonstrated that TH alterations induce neuroinflammation. Adult-onset hypothyroidism in mice alters the neuroinflammation signaling pathway at the hippocampus and prefrontal cortex as revealed by the activation of the inflammasome, leading to the activation of signaling cascades which increase the levels of pro-inflammatory cytokines (Bortolotto et al., 2025). Likewise, adult hypothyroid mice displayed disturbed hippocampal TH signaling (downregulation of DIO3 and TH target genes), accompanied by neuroinflammation, with increased number of GFAP immunopositive astrocytes and microglial activation (Chamas et al., 2022). Also, hypothyroid adult rats show activation of the cyclooxygenase 2-mediated inflammation pathway at the hippocampus, triggering an increase of pro-inflammatory cytokines (Nam et al., 2018).

In a recent work, Guillén-Yunta et al. demonstrate that defective TH transport into the brain due to MCT8 deficiency leads to astroglial alterations in both humans and mice, including profound astrocyte structural, molecular, and functional abnormalities (Guillén-Yunta et al., 2024). Histological and immunohistochemical analyses revealed changes in GFAP expression, abnormal astrocyte distribution, density, and altered morphology, shifting towards a reactive phenotype in both MCT8-deficient humans and *Mct8/Dio2*-KO mice since early developmental stages. Transcriptomic profiling further revealed dysregulation of genes involved in astrocyte metabolism, mitochondrial respiration, and reactive gliosis, supporting the hypothesis of an astrogliosis-like phenotype and metabolic impairment in astrocytes during MCT8 deficiency, positioning astrocytes as novel therapeutic targets of the disease.

All these findings suggest that astrocytic TH deprivation (resulting from impaired cellular T3 availability) leads to a loss of normal homeostatic and neuroprotective functions, predisposing the brain to neuroinflammatory and vascular vulnerability. Likely, astrocytes, normally key regulators of the NVU and inflammatory signaling, require sufficient TH signaling to maintain their normal metabolic and structural integrity. Without proper TH levels, astrocytes shift toward a maladaptive phenotype (less able to respond to inflammatory cues, more prone to pro-inflammatory mediator release and less capable of vascular support and barrier repair) exacerbating neuroinflammation, disrupting glial-vascular coupling, and impairing BBB support. These astroglial changes may underlie part of the neurological and vascular dysfunctions observed in TH-related disorders, including AHDS.

Other endocrine alterations

Aside from the TH-availability alterations described in this chapter, it is interesting to consider the actual context where the amount and diversity of chemicals present in our direct environment have exponentially increased in the past decades.

Among these chemicals, some of the so-called endocrine-disrupting chemicals (EDC) can interfere with TH signaling, altering its production, transport, metabolism and reception; and a lot have already been demonstrated to affect the CNS in several ways (Schug et al., 2015). At the interface, some of the most studied thyroid disrupting chemicals such as perfluoroalkyl substances have also been linked to neurological alterations (Butruille et al., 2023). However, studies linking EDC and astroglial dysfunction are scarce.

In 2015, diphenyl ethers and their metabolites were reported to reduce DIO2 expression and activity in cultured human astrocytes, potentially hindering T3 supply to the brain, and thus compromising TH signaling in other brain cell types. Moreover, they also led to alterations in astrocyte cell membrane and viability, adding another layer of potential actions for these compounds (Roberts et al., 2015). Another study, this time using well-known substances such as perfluorooctanesulfonic acid and glyphosate, showed that both compounds, with potent synergistic effects, led to strong toxicity in astrocyte cell lines mediated by an increase in oxidative stress, that eventually led to increased secretion of proinflammatory cytokines (Franco et al., 2024).

Altogether, these data, although scarce, highlight the importance of investigating EDC as a potential source of astrocyte disruption affecting TH signaling.

Other astroglial-like cells: Tanycytes and bergmann glia

Tanycytes are GFAP-positive specialized ependymoglial cells lining the wall of the third ventricle and projecting into the median eminence, where they regulate TH availability and central feedback of the hypothalamic-pituitary-thyroid axis. Their strategic position enables them to sense circulating TH, control hypophysiotropic TRH access to the portal vasculature, and locally adjust T3 supply to hypothalamic neurons. A hallmark of tanycytes is the expression of a highly enriched TH "gatekeeper module" consisting of MCT8, OATP1C1, DIO2, and occasionally DIO3, which together govern TH uptake, conversion, and clearance, adapting to hormonal fluctuations.

Tanycyte-mediated feedback also relies on pyroglutamyl peptidase II (PPII), a tanycyte-specific enzyme that degrades TRH in the median eminence. Its activity rises in hyperthyroidism, reducing TSH release, while hypothyroidism lowers PPII and upregulates DIO2 to enhance TRH availability (Sanchez et al., 2009). Moreover, TSH directly modulates tanycytes via functional TSH receptors, increasing *Slco1c1*, *Dio2*, and *Dio3* expression and thus reshaping their gatekeeping capacity (Chandrasekar et al., 2024). Through these mechanisms, tanycytes integrate TH, TRH, and TSH signaling, linking endocrine feedback to local morphological and metabolic plasticity and shaping central thyroid homeostasis.

Similarly, Bergmann glia are specialized radial astroglia of the cerebellar cortex that contact Purkinje cells and regulate neuronal migration, synaptic refinement, glutamate clearance, and ion homeostasis. Increasing evidence indicates that Bergmann glia are also responsive to TH and significantly contribute to TH-dependent cerebellar development and function. Bergmann glia express TH transporters, TR, and deiodinases (especially DIO2), facilitating local T3 production required for Purkinje cell dendritogenesis, granule cell migration, and proper synaptic organization (Fauquier et al., 2014). Hypothyroidism is known to delay Bergmann glia maturation, reducing radial fiber formation and impairing glutamate clearance, which increases Purkinje cell vulnerability to excitotoxicity and disrupts granule cell migration and dendritic arborization (Faustino & Ortiga-Carvalho, 2014), whereas hyperthyroidism accelerates their morphological development (Clos et al., 1980). As for tanycytes, altered TH availability for Bergmann glia leads to significant neurodevelopmental defects, underscoring their shared role as TH-responsive glial regulators essential for brain structure, function, and the neurological outcomes of thyroid dysfunction.

Applications of the material: *Therapeutic perspectives*

Given the centrality of astrocytes in mediating TH effects in the brain, various therapeutic strategies to improve astrocytic dysfunctions in different pathologies can be designed to target several nodes:

- ***Replacement therapy****: Systemic TH replacement (mostly T4, as astrocytes generate local T3). Challenges include crossing the BBB, ensuring proper transporter and deiodinase functions, or avoiding peripheral side effects. This approach may also be useful to tackle non-thyroidal astroglial alterations, using TH signaling to ameliorate astroglial homeostasis.*
- ***Use of TH analogs****: Molecules mimicking TH actions and able to enter cells through alternative pathways or bind different transporters are promising, especially for transporter mutation disorders like AHDS and for avoiding some of the aforementioned side-effects, as some TH analogs exhibit preferential affinity for some TR isoforms.*
- ***Modulating extracellular matrix and/or neurotrophic factors****: Since astrocytes influence neuronal growth via extracellular matrix and growth factor secretion, treatments that mimic or enhance these astrocyte-derived signals might compensate partially for upstream defects and may also serve to alleviate some of the alterations derived from TH availability alterations.*
- ***Gene therapy/molecular correction****: In cases of monogenic defects, gene therapy targeting astrocytes or endothelial cells could restore its function. Similarly, strategies to locally enhance expression of TH transporters or DIO2 in astrocytes may improve local TH availability and homeostasis, both in TH-related and unrelated conditions.*

The election of these therapeutic strategies will depend on the nature of each pathology. This highlights the importance of the variety of model systems, such as in vitro human astrocyte cultures, organoids, induced pluripotent stem cell models, and humanized animal models.

Mini-dictionary of terms

Astrocyte Glial cell type ubiquitously distributed throughout the brain that provides metabolic, structural, and signaling support to neurons and other glial cells, regulates synaptic activity, maintains extracellular homeostasis, and contributes to the BBB structure and function.

Astrogliosis Response of astrocytes to injury or disease, acquiring a reactive phenotype characterized by cellular hypertrophy, increased GFAP expression, and remodeling of the local microenvironment, which can be either protective or detrimental depending on context.

Blood-brain barrier Specialized interface formed by endothelial cells, astrocytic end-feet, and pericytes that tightly regulates the exchange of molecules between the bloodstream and the CNS, ensuring brain homeostasis.

Deiodinases Enzymes responsible for thyroid hormone metabolism by iodide removal. This metabolism can either promote activation (main function of DIO2) or inactivation (DIO3).

Neuroinflammation Immune response within the CNS involving microglia, astrocytes, and infiltrating immune cells, aimed at restoring tissue integrity but potentially harmful when chronically activated.

Neurovascular unit Structural and functional complex composed of endothelial cells, astrocytes, pericytes, neurons, and extracellular matrix components that coordinates cerebral blood flow, BBB permeability, and metabolic coupling in the brain.

Thyroid hormone transporter Membrane protein that facilitates the translocation of thyroid hormones across cellular membranes, essential for proper cellular T3 and T4 availability and signaling, such as MCT8 and OATP1C1.

Summary points

- Astrocytes are both mediators of TH availability and target cells for TH action.
- Astrocyte-dependent TH signaling regulates key neural processes including neurogenesis, neurite outgrowth,

synaptogenesis, and metabolic support during development and adulthood.
- Pathological conditions that alter TH transport, metabolism, or receptor signaling disrupt astrocyte homeostasis and impair their neurotrophic and metabolic functions.
- BBB dysfunction alters the regulated entry of TH into the CNS, disrupting astrocyte-mediated T4-to-T3 conversion and impairing local TH supply to neurons.
- Neuroinflammatory states induce astrocyte reactivity and modify their ability to buffer excitotoxicity, maintain redox balance, and sustain local TH activation through DIO2.
- Disruption of astrocyte function in TH-related disorders compromises TH-dependent neuronal support and contributes to cognitive, motor, and sensory deficits.

References

Admati, I., Wasserman-Bartov, T., Tovin, A., Rozenblat, R., Blitz, E., Zada, D., ... Appelbaum, L. (2020). Neural alterations and hyperactivity of the hypothalamic-pituitary-thyroid axis in Oatp1c1 deficiency. *Thyroid, 30*(1), 161–174.

Akdemir, E. S., Huang, A. Y., & Deneen, B. (2020). Astrocytogenesis: Where, when, and how. *F1000Res, 9*.

Allan, W., Herndon, C. N., & Dudley, F. C. (1944). Some examples of the inheritance of mental deficiency: Apparently sex-linked idiocy and microcephaly. *American Journal of Mental Deficiency, 48*, 325–334.

Bernal, J., Guadaño-Ferraz, A., & Morte, B. (2015). Thyroid hormone transporters–functions and clinical implications. *Nature Reviews. Endocrinology, 11*(7), 406–417.

Boelen, A., Kwakkel, J., & Fliers, E. (2011). Beyond low plasma T3: Local thyroid hormone metabolism during inflammation and infection. *Endocrine Reviews, 32*(5), 670–693.

Bortolotto, V. C., Zborowski, V. A., Dahleh, M. M. M., Borstmann, S. M. A., Viana, C. E., Pinheiro, F. C., ... Prigol, M. (2025). Neuroinflammation in hypothyroidism? Chrysin treatment as a strategy to reduce neuroinflammation induced by hypothyroidism. *Neuroscience, 584*, 395–403.

Butruille, L., Jubin, P., Martin, E., Aigrot, M. S., Lhomme, M., Fini, J. B., ... Remaud, S. (2023). Deleterious functional consequences of perfluoroalkyl substances accumulation into the myelin sheath. *Environment International, 180*, Article 108211.

Ceballos, A., Belinchon, M. M., Sanchez-Mendoza, E., Grijota-Martinez, C., Dumitrescu, A. M., Refetoff, S., ... Bernal, J. (2009). Importance of monocarboxylate transporter 8 for the blood-brain barrier-dependent availability of 3,5,3'-triiodo-L-thyronine. *Endocrinology, 150*(5), 2491–2496.

Chamas, L., Seugnet, I., Poirier, R., Clerget-Froidevaux, M. S., & Enderlin, V. (2022). A fine regulation of the hippocampal thyroid signalling protects hypothyroid mice against glial cell activation. *International Journal of Molecular Sciences, 23*(19).

Chandrasekar, A., Schmidtlein, P. M., Neve, V., Rivagorda, M., Spiecker, F., Gauthier, K., ... Muller-Fielitz, H. (2024). Regulation of thyroid hormone gatekeepers by thyrotropin in tanycytes. *Thyroid, 34*(2), 261–273.

Clos, J., & Legrand, J. (1973). Effects of thyroid deficiency on the different cell populations of the cerebellum in the young rat. *Brain Research, 63*, 450–455.

Clos, J., Legrand, C., & Legrand, J. (1980). Effects of thyroid state on the formation and early morphological development of Bergmann glia in the developing rat cerebellum. *Developmental Neuroscience, 3*(4–6), 199–208.

Cortes, C., Eugenin, E., Aliaga, E., Carreno, L. J., Bueno, S. M., Gonzalez, P. A., ... Riedel, C. A. (2012). Hypothyroidism in the adult rat causes incremental changes in brain-derived neurotrophic factor, neuronal and astrocyte apoptosis, gliosis, and deterioration of postsynaptic density. *Thyroid, 22*(9), 951–963.

Crantz, F. R., Silva, J. E., & Larsen, P. R. (1982). An analysis of the sources and quantity of 3,5,3'-triiodothyronine specifically bound to nuclear receptors in rat cerebral cortex and cerebellum. *Endocrinology, 110*(2), 367–375.

Dumitrescu, A. M., Liao, X. H., Best, T. B., Brockmann, K., & Refetoff, S. (2004). A novel syndrome combining thyroid and neurological abnormalities is associated with mutations in a monocarboxylate transporter gene. *The American Journal of Human Genetics, 74*(1), 168–175.

Escartin, C., Galea, E., Lakatos, A., O'Callaghan, J. P., Petzold, G. C., Serrano-Pozo, A., ... Verkhratsky, A. (2021). Reactive astrocyte nomenclature, definitions, and future directions. *Nature Neuroscience, 24*(3), 312–325.

Faivre-Sarrailh, C., Rami, A., Fages, C., & Tardy, M. (1991). Effect of thyroid deficiency on glial fibrillary acidic protein (GFAP) and GFAP-mRNA in the cerebellum and hippocampal formation of the developing rat. *Glia, 4*(3), 276–284.

Fauquier, T., Chatonnet, F., Picou, F., Richard, S., Fossat, N., Aguilera, N., ... Flamant, F. (2014). Purkinje cells and Bergmann glia are primary targets of the TRalpha1 thyroid hormone receptor during mouse cerebellum postnatal development. *Development, 141*(1), 166–175.

Faustino, L. C., & Ortiga-Carvalho, T. M. (2014). Thyroid hormone role on cerebellar development and maintenance: A perspective based on transgenic mouse models. *Frontiers in Endocrinology, 5*, 75.

Fonseca, T. L., Werneck-De-Castro, J. P., Castillo, M., Bocco, B. M., Fernandes, G. W., McAninch, E. A., ... Bianco, A. C. (2014). Tissue-specific inactivation of type 2 deiodinase reveals multilevel control of fatty acid oxidation by thyroid hormone in the mouse. *Diabetes, 63*(5), 1594–1604.

Franco, G. A., Molinari, F., Marino, Y., Tranchida, N., Inferrera, F., Fusco, R., ... Britti, D. (2024). Enviromental endocrine disruptor risks in the central nervous system: Neurotoxic effects of PFOS and glyphosate. *Environmental Toxicology and Pharmacology, 109*, Article 104496.

Friedman, A., Prager, O., Serlin, Y., & Kaufer, D. (2025). Dynamic modulation of the blood-brain barrier in the healthy brain. *Nature Reviews Neuroscience, 26*, 749–764.

Friesema, E. C. H., Grueters, A., Biebermann, H., Krude, H., von Moers, A., Reeser, M., ... Visser, T. J. (2004). Association between mutations in a thyroid hormone transporter and severe X-linked psychomotor retardation. *The Lancet, 364*(9443), 1435–1437.

Gould, E., Frankfurt, M., Westlind-Danielsson, A., & McEwen, B. S. (1990). Developing forebrain astrocytes are sensitive to thyroid hormone. *Glia, 3*(4), 283–292.

Guadaño-Ferraz, A., Obregon, M. J., St Germain, D. L., & Bernal, J. (1997). The type 2 iodothyronine deiodinase is expressed primarily in glial cells in the neonatal rat brain. *Proceedings of the National Academy of Sciences of the United States of America, 94*(19), 10391–10396.

Guillén-Yunta, M., García-Aldea, A., Valcárcel-Hernández, V., Sanz-Bógalo, A., Munoz-Moreno, E., Matheus, M. G., … Bárez-López, S. (2024). Defective thyroid hormone transport to the brain leads to astroglial alterations. *Neurobiology of Disease, 200*, Article 106621.

Guillén-Yunta, M., Valcárcel-Hernández, V., García-Aldea, A., Soria, G., García-Verdugo, J. M., Montero-Pedrazuela, A., & Guadaño-Ferraz, A. (2023). Neurovascular unit disruption and blood-brain barrier leakage in MCT8 deficiency. *Fluids and Barriers of the CNS, 20*(1), 79.

Jurga, A. M., Paleczna, M., Kadluczka, J., & Kuter, K. Z. (2021). Beyond the GFAP-astrocyte protein markers in the brain. *Biomolecules, 11*(9).

Lattke, M., & Guillemot, F. (2022). Understanding astrocyte differentiation: Clinical relevance, technical challenges, and new opportunities in the omics era. *WIREs Mechanisms of Disease, 14*(5), Article e1557.

Lefauconnier, J. M., Lacombe, P., & Bernard, G. (1985). Cerebral blood flow and blood-brain influx of some neutral amino acids in control and hypothyroid 16-day-old rats. *Journal of Cerebral Blood Flow & Metabolism, 5*(2), 318–326.

Liddelow, S. A., & Barres, B. A. (2017). Reactive astrocytes: Production, function, and therapeutic potential. *Immunity, 46*(6), 957–967.

Lima, F. R., Goncalves, N., Gomes, F. C., de Freitas, M. S., & Moura Neto, V. (1998). Thyroid hormone action on astroglial cells from distinct brain regions during development. *International Journal of Developmental Neuroscience, 16*(1), 19–27.

López-Espíndola, D., Garcia-Aldea, A., Gomez de la Riva, I., Rodriguez-Garcia, A. M., Salvatore, D., Visser, T. J., … Guadaño-Ferraz, A. (2019). Thyroid hormone availability in the human fetal brain: Novel entry pathways and role of radial glia. *Brain Structure & Function, 224*(6), 2103–2119.

Mack, A. F., & Wolburg, H. (2013). A novel look at astrocytes: Aquaporins, ionic homeostasis, and the role of the microenvironment for regeneration in the CNS. *The Neuroscientist, 19*(2), 195–207.

Manu, D. R., Slevin, M., Barcutean, L., Forro, T., Boghitoiu, T., & Balasa, R. (2023). Astrocyte involvement in blood-brain barrier function: A critical update highlighting novel, complex, neurovascular interactions. *International Journal of Molecular Sciences, 24*(24).

Manzano, J., Bernal, J., & Morte, B. (2007). Influence of thyroid hormones on maturation of rat cerebellar astrocytes. *International Journal of Developmental Neuroscience, 25*(3), 171–179.

Martinez-Galan, J. R., Pedraza, P., Santacana, M., Escobar del Ray, F., Morreale de Escobar, G., & Ruiz-Marcos, A. (1997). Early effects of iodine deficiency on radial glial cells of the hippocampus of the rat fetus. A model of neurological cretinism. *Journal of Clinical Investigation, 99*(11), 2701–2709.

Matejuk, A., & Ransohoff, R. M. (2020). Crosstalk between astrocytes and Microglia: An overview. *Frontiers in Immunology, 11*, 1416.

Mayerl, S., Visser, T. J., Darras, V. M., Horn, S., & Heuer, H. (2012). Impact of Oatp1c1 deficiency on thyroid hormone metabolism and action in the mouse brain. *Endocrinology, 153*(3), 1528–1537.

McConnell, H. L., Li, Z., Woltjer, R. L., & Mishra, A. (2019). Astrocyte dysfunction and neurovascular impairment in neurological disorders: Correlation or causation? *Neurochemistry International, 128*, 70–84.

Morte, B., Manzano, J., Scanlan, T. S., Vennstrom, B., & Bernal, J. (2004). Aberrant maturation of astrocytes in thyroid hormone receptor alpha 1 knockout mice reveals an interplay between thyroid hormone receptor isoforms. *Endocrinology, 145*(3), 1386–1391.

Nam, S. M., Kim, J. W., Yoo, D. Y., Jung, H. Y., Chung, J. Y., Kim, D. W., … Yoon, Y. S. (2018). Hypothyroidism increases cyclooxygenase-2 levels and pro-inflammatory response and decreases cell proliferation and neuroblast differentiation in the hippocampus. *Molecular Medicine Reports, 17*(4), 5782–5788.

Nam, S. M., Kim, Y. N., Yoo, D. Y., Yi, S. S., Choi, J. H., Hwang, I. K., … Yoon, Y. S. (2013). Hypothyroidism affects astrocyte and microglial morphology in type 2 diabetes. *Neural Regeneration Research, 8*(26), 2458–2467.

Niedowicz, D. M., Wang, W. X., Price, D. A., Xie, K., Patel, E., & Nelson, P. T. (2023). Impact of thyroid hormone perturbations in adult mice: Brain weight and blood vessel changes, gene expression variation, and neurobehavioral outcomes. *Neurobiology of Aging, 128*, 74–84.

Oberheim, N. A., Goldman, S. A., & Nedergaard, M. (2012). Heterogeneity of astrocytic form and function. *Methods in Molecular Biology, 814*, 23–45.

Pancotto, T., Rossmeisl, J. H., Jr., Panciera, D. L., & Zimmerman, K. L. (2010). Blood-brain-barrier disruption in chronic canine hypothyroidism. *Veterinary Clinical Pathology, 39*(4), 485–493.

Roberts, S. C., Bianco, A. C., & Stapleton, H. M. (2015). Disruption of type 2 iodothyronine deiodinase activity in cultured human glial cells by polybrominated diphenyl ethers. *Chemical Research in Toxicology, 28*(6), 1265–1274.

Roberts, L. M., Woodford, K., Zhou, M., Black, D. S., Haggerty, J. E., Tate, E. H., … Zerangue, N. (2008). Expression of the thyroid hormone transporters monocarboxylate transporter-8 (SLC16A2) and organic ion transporter-14 (SLCO1C1) at the blood-brain barrier. *Endocrinology, 149*(12), 6251–6261.

Sanchez, E., Vargas, M. A., Singru, P. S., Pascual, I., Romero, F., Fekete, C., … Lechan, R. M. (2009). Tanycyte pyroglutamyl peptidase II contributes to regulation of the hypothalamic-pituitary-thyroid axis through glial-axonal associations in the median eminence. *Endocrinology, 150*(5), 2283–2291.

Sarret, C., Petit, I. O., & Tonduti, D. (1993). Allan-Herndon-Dudley syndrome. In M. P. Adam, H. H. Ardinger, R. A. Pagon, S. E. Wallace, L. J. H. Bean, K. Stephens, & A. Amemiya (Eds.), *GeneReviews*. https://www.ncbi.nlm.nih.gov/books/NBK26373/. Seattle (WA).

Schug, T. T., Blawas, A. M., Gray, K., Heindel, J. J., & Lawler, C. P. (2015). Elucidating the links between endocrine disruptors and neurodevelopment. *Endocrinology, 156*(6), 1941–1951.

Strømme, P., Groeneweg, S., Lima de Souza, E. C., Zevenbergen, C., Torgersbraten, A., Holmgren, A., … Visser, T. J. (2018). Mutated thyroid hormone transporter OATP1C1 associates with severe brain hypometabolism and juvenile neurodegeneration. *Thyroid, 28*(11), 1406–1415.

Tabata, H. (2015). Diverse subtypes of astrocytes and their development during corticogenesis. *Frontiers in Neuroscience, 9*, 114.

Tiwari, A., Rout, S., Deep, P., Sahu, C., & Samal, P. K. (2025). The dual role of astrocytes in CNS homeostasis and dysfunction. *Neuroglia, 6* (4), 38.

Tu, H. M., Legradi, G., Bartha, T., Salvatore, D., Lechan, R. M., & Larsen, P. R. (1999). Regional expression of the type 3 iodothyronine deiodinase messenger ribonucleic acid in the rat central nervous system and its regulation by thyroid hormone. *Endocrinology, 140*(2), 784–790.

Valles, S. L., Singh, S. K., Campos-Campos, J., Colmena, C., Campo-Palacio, I., Alvarez-Gamez, K., & Jorda, A. (2023). Functions of astrocytes under normal conditions and after a brain disease. *International Journal of Molecular Sciences, 24*(9).

Virchow, R. (1846). Über das granulierte Ansehen der Wandungen der Gehirnventrikel. Allgemeine Zeitschrift für Psychiatrie, 3, 242–250.

Wiersinga, W. M. (2000). Nonthyroidal illness. In L. E. Braverman, & R. D. Utiger (Eds.), *The thyroid* (7 ed., pp. 281–295). Philadelphia: Lippincott.

Xu, M., Iwasaki, T., Shimokawa, N., Sajdel-Sulkowska, E. M., & Koibuchi, N. (2013). The effect of low dose lipopolysaccharide on thyroid hormone-regulated actin cytoskeleton modulation and type 2 iodothyronine deiodinase activity in astrocytes. *Endocrine Journal, 60*(11), 1221–1230.

Yuan, W. Q., Huang, W. P., Jiang, Y. C., Xu, H., Duan, C. S., Chen, N. H., ... Fu, X. M. (2023). The function of astrocytes and their role in neurological diseases. *European Journal of Neuroscience, 58*(9), 3932–3961.

Zhang, Y., Chen, K., Sloan, S. A., Bennett, M. L., Scholze, A. R., O'Keeffe, S., ... Wu, J. Q. (2014). An RNA-sequencing transcriptome and splicing database of glia, neurons, and vascular cells of the cerebral cortex. *Journal of Neuroscience, 34*(36), 11929–11947.

Zhang, Y., Sloan, S. A., Clarke, L. E., Caneda, C., Plaza, C. A., Blumenthal, P. D., ... Barres, B. A. (2016). Purification and characterization of progenitor and mature human astrocytes reveals transcriptional and functional differences with mouse. *Neuron, 89*(1), 37–53.

Section 3

Psychology and behavior as a consequence of changes in the endocrine status

Chapter 13

Thyroid function and psychosis

Eloi Giné-Servén[1, 2] and Javier Labad[3, 4]

[1]Department of Psychiatry, Hospital Universitario de Navarra, Pamplona, Spain; [2]IdiSNA, Navarra Institute for Health Research, Pamplona, Spain; [3]Department of Mental Health and Addictions, Hospital Universitari de Mataró, Consorci Sanitari del Maresme, Universitat Autònoma de Barcelona, Mataró, Spain; [4]Centro de Investigación biomédica en Red de Salud Mental (CIBERSAM), Instituto de Salud Carlos III, Madrid, Spain

List of abbreviations

Anti-TG Anti–thyroglobulin
Anti-TPO Anti-thyroid peroxidase
BDNF Brain-derived neurotrophic factor
BNSS Brief Negative Symptom Scale
BPRS Brief Psychiatric Rating Scale
CAR Cortisol awakening response
CRT Cognitive remediation therapy
DUP Duration of untreated psychosis
EEG Electroencephalography
FEP First-episode psychosis
FT3 Free triiodothyronine
FT4 Free thyroxine
HE Hashimoto's encephalopathy
HPA Hypothalamic–pituitary–adrenal
HPT Hypothalamic–pituitary–thyroid
IGT Impaired glucose tolerance
IR Insulin resistance
MP Myxedema psychosis
MRI Magnetic resonance imaging
NSS Neurological soft signs
PANSS Positive and Negative Syndrome Scale
SPECT Single-photon emission computed tomography
T3 Triiodothyronine
T4 Thyroxine
TRH Thyrotropin-releasing hormone
TSH Thyroid-stimulating hormone

Introduction

Thyroid hormones are crucial regulators of brain development and adult brain function, with established roles in neuronal migration, synaptogenesis, myelination, and neurotransmitter activity (Bernal, 2007). Through triiodothyronine (T3) and thyroxine (T4), they influence neuronal excitability, serotonergic and dopaminergic transmission, and cerebral energy metabolism. These broad neurobiological actions provide a strong rationale for the long-recognized association between hypothalamic– pituitary–thyroid (HPT) axis disturbances and psychiatric syndromes. Among them, psychosis represents one of the most a clinically significant, albeit infrequent, manifestations.

The relationship between thyroid dysfunction and psychosis is bidirectional. On the one hand, overt thyroid diseases, including hypothyroidism, hyperthyroidism, and autoimmune thyroiditis, can present with psychotic syndromes. The classic entity of "myxedema madness" described by Asher, (1949) remains a paradigmatic example of how untreated hypothyroidism can cause delusions and hallucinations that remit with thyroid hormone replacement therapy. More recent systematic reviews have confirmed that myxedema psychosis may occur even in the absence of somatic hypothyroid symptoms and is highly responsive to levothyroxine (Mohamed et al., 2021). Hyperthyroidism has also been linked to agitation, irritability, insomnia, and paranoid delusions, while Hashimoto's encephalopathy illustrates how autoimmune thyroid disease can present with steroid-responsive psychosis (Menon et al., 2017; Shaw et al., 1991). Although rare, these conditions highlight the need to evaluate thyroid function in patients presenting with acute psychosis.

On the other hand, subtle alterations of thyroid function appear to modulate the vulnerability and clinical phenotype of primary psychotic disorders. Systematic review and meta-analytic studies have shown that drug-naïve patients with first-episode psychosis (FEP) exhibit significantly lower thyroid-stimulating hormone (TSH) compared with controls, suggesting HPT axis dysregulation at illness onset (Cavaleri et al., 2023; Misiak et al., 2021). Prospective studies have indicated that free thyroxine (FT4) levels, even within the reference range, are associated with specific phenotypes: higher FT4 has been linked to acute onset and subsequent affective psychosis, whereas lower FT4 correlates with prodromal symptoms, insidious onset, and

Hormones in Neuroscience. https://doi.org/10.1016/B978-0-443-30261-9.00013-9

poorer attentional performance (Barbero et al., 2015; Gine-Serven et al., 2023; Labad et al., 2016). In addition, the presence of thyroid autoantibodies has been associated with more severe negative symptoms and worse functioning in early psychosis cohorts (Barbero et al., 2020). These findings suggest that thyroid function influences the heterogeneity of schizophrenia spectrum disorders.

Clinically, this evidence underscores the importance of routine thyroid assessment in patients with psychosis. While the primary aim is to exclude secondary psychoses, subtle deviations in FT4 or TSH levels may also have prognostic value. Moreover, autoimmune thyroiditis should be considered in atypical or treatment-resistant presentations, given the therapeutic implications of corticosteroid responsiveness. From a research perspective, thyroid hormones and antibodies have emerged as potential biomarkers for the stratification of psychosis and as targets for novel therapeutic approaches. Evidence from mood disorders, where triiodothyronine (T3) augmentation improves antidepressant response and cognition (Bauer et al., 2008), suggests transdiagnostic relevance and raises the question of whether similar strategies could be beneficial in schizophrenia.

This chapter reviews the current knowledge on the relationship between thyroid function and psychosis. We first analyze psychosis associated with overt thyroid diseases and then examine thyroid abnormalities in schizophrenia and FEP with attention to clinical phenotypes and longitudinal outcomes. We next discuss the role of thyroid hormones in cognition in psychosis before considering therapeutic implications and future directions. Our aim is to provide an integrated perspective that bridges endocrinology and psychiatry, emphasizing both the clinical importance of recognizing secondary psychoses and the potential of thyroid hormones as modulators of primary psychotic disorders.

Thyroid disease and psychosis

Hypothyroidism and myxedema psychosis

The association between hypothyroidism and psychosis has been known for more than a century. Early case reports emphasized that untreated hypothyroidism could cause prominent delusions, hallucinations, and thought disorder. Asher, (1949) coined the term myxedema madness to describe these presentations, which was later reframed as myxedema psychosis (MP). Since then, MP has been documented repeatedly in the literature.

A recent systematic review of 75 published cases provides the most comprehensive characterization of this condition (Mohamed et al., 2021). The median age at presentation was 42 years, with a slight female predominance, and more than half of the patients had no previous diagnosis of hypothyroidism. Delusions were present in 91% of cases, most often persecutory in nature (84%), and hallucinations occurred in 78% of cases, mainly auditory. Strikingly, overt physical signs of hypothyroidism were absent in 26% of patients, and one-third presented without clear somatic symptoms. These findings emphasize that psychosis can be the first and sometimes the only manifestation of hypothyroidism. Laboratory evaluation revealed markedly elevated TSH levels (median 93 mIU/L) and low FT4 levels, frequently accompanied by thyroid autoantibodies (75% of cases). Neuroimaging and electroencephalography (EEG) were usually normal. Treatment with levothyroxine led to remission in >90% of cases, often within 2 weeks, with short-term antipsychotics used for symptomatic control but rarely required long-term. Complete recovery was achieved in the majority of patients.

The biological mechanisms underlying MP remain incompletely understood. Hypothyroidism has been associated with altered monoaminergic neurotransmission (Bauer et al., 2002) as well as with reduced cerebral blood flow and glucose metabolism (Constant et al., 2001). Preclinical evidence from animal models further supports that thyroid hormone deficiency disrupts hippocampal and amygdala function, regions central to psychosis (Bauer et al., 2002). Overall, MP demonstrates that hormone deficiency alone can precipitate a reversible psychosis.

Hyperthyroidism and psychosis

Hyperthyroidism, particularly in Graves' disease, has also been linked to psychosis, although it is less common than hypothyroidism. Case reports have described patients presenting with agitation, irritability, insomnia, emotional lability, and psychotic features such as persecutory delusions or hallucinations (Fukao et al., 2020). In many cases, psychiatric symptoms dominate the clinical picture, leading to misdiagnosis of primary psychosis. As in hypothyroidism, somatic signs such as weight loss, tremors, or tachycardia may be subtle or absent.

Pathophysiologically, thyroid hormone excess enhances monoaminergic activity and adrenergic receptor responsiveness, leading to hyperexcitability within cortical and limbic circuits (Hassan et al., 2013). Psychotic symptoms typically remit once euthyroidism is restored, whether through antithyroid drugs, radioactive iodine, or surgery (Whybrow, 1969). Antipsychotics may provide short-term symptom relief, but endocrine treatment is definitive. These cases underscore the necessity of excluding hyperthyroidism in acute psychosis, especially when accompanied by anxiety or mood lability.

Hashimoto's encephalopathy

Hashimoto's encephalopathy (HE), also known as steroid-responsive encephalopathy associated with autoimmune thyroiditis, is another thyroid-related cause of psychosis. First clearly delineated by Shaw et al., (1991), HE is characterized by subacute onset of psychiatric and neurological symptoms, usually in middle-aged women. Presentations include psychosis, mood disorders, seizures, confusion and cognitive decline. Elevated anti-thyroid peroxidase (anti-TPO) antibodies are a hallmark of the disease, although titers do not correlate with disease severity. Thyroid function may be normal, hypothyroid, or hyperthyroid, indicating that autoimmunity, rather than hormonal imbalance, is key.

A systematic review of 46 cases confirmed psychosis as the most frequent psychiatric manifestation, followed by depressive and bipolar syndromes (Menon et al., 2017). EEG abnormalities, usually diffuse slowing, and nonspecific magnetic resonance imaging (MRI) changes are common but not diagnostic. Corticosteroid therapy, typically high-dose intravenous methylprednisolone followed by tapering, produces dramatic improvement in most patients, supporting the autoimmune nature of the disorder. Immunosuppressants are used in refractory cases. The pathogenesis likely involves autoimmune-mediated neuroinflammation, blood–brain barrier dysfunction, and intrathecal synthesis of antibodies cross-reacting with neuronal or thyroid antigens (Ferracci et al., 2004). Cerebral hypoperfusion in single-photon emission computed tomography (SPECT) has also been reported, suggesting impaired cerebral perfusion related to autoimmune vasculitic mechanisms (Payer et al., 2012).

Clinically, HE is important because it can mimic schizophrenia or affective psychosis but responds to immunotherapy rather than long-term antipsychotics. Suspicion should be raised in patients with acute or atypical psychosis, neurological signs, or poor response to antipsychotics, especially if thyroid antibodies levels are elevated.

Clinical implications and differential diagnosis

Taken together, these conditions illustrate how thyroid dysfunction can precipitate psychotic syndromes through diverse mechanisms (Table 13.1): hormone deficiency in hypothyroidism, hormone excess in hyperthyroidism, and autoimmunity in HE. Psychiatric manifestations often resemble primary psychotic disorders and may lack overt somatic signs, complicating diagnosis. Therefore, routine thyroid testing, including TSH, free T4, and thyroid antibodies, is essential for all patients presenting with FEP or atypical psychiatric syndromes.

Treatment implications are important. While antipsychotics may provide short-term relief, definitive management requires addressing underlying thyroid disorders. Levothyroxine replacement reverses MP, antithyroid drugs resolve psychosis in Graves' disease, and corticosteroids are life-saving in HE. Failure to recognize these conditions increases the risk of chronic disability and unnecessary long-term exposure to antipsychotics. Conversely, timely diagnosis can improve prognosis, as most patients achieve complete recovery with appropriate therapy (Payer et al., 2012).

Thyroid abnormalities in schizophrenia and psychosis

Beyond secondary psychoses due to overt thyroid disease, converging evidence indicates that thyroid hormones and the HPT axis shape the onset, clinical phenotype, and prognosis of primary psychotic disorders, particularly in the schizophrenia spectrum. Psychopathology in these disorders spans positive (delusions, hallucinations, disorganization), negative (blunted affect, alogia, asociality, anhedonia, avolition), and affective symptoms (depression, mania), alongside cognitive impairment—domains tightly

TABLE 13.1 Psychotic symptoms in thyroid disease (secondary psychoses).

Condition	Main psychiatric features	Thyroid findings	Treatment	Outcome
Myxedema psychosis	Delusions (persecutory), hallucinations (auditory), thought disorder	Markedly ↑TSH, ↓FT4; may lack overt somatic symptoms	Levothyroxine ± short-term antipsychotics	>90% remission, usually within 2–3 weeks
Hyperthyroidism (Graves' disease, toxic goiter)	Agitation, irritability, insomnia, mood lability, sometimes delusions/hallucinations	↑FT4/FT3, suppressed TSH	Antithyroid drugs, radioiodine, surgery ± short-term antipsychotics	Symptoms remit with euthyroidism
Hashimoto's encephalopathy	Psychosis, mood disorder, confusion, seizures, cognitive decline	↑ anti-TPO, anti-TG antibodies; thyroid status variable	Corticosteroids (IV methylprednisolone), immunosuppressants if refractory	Dramatic response in most; relapse possible

linked to functional outcome (McCutcheon et al., 2020). The etiology is multifactorial, with neurodevelopmental, genetic, epigenetic, and environmental components (Smeland et al., 2020). Given the roles of thyroid hormones in neurodevelopment and adult brain homeostasis, hormone dysregulation is a biologically plausible contributor to psychosis (Hayes et al., 2012).

Biological plausibility: Neurodevelopment and glia–neuron crosstalk

Thyroid hormones are essential for neuronal proliferation, migration, and synaptogenesis and remain active modulators of adult brain function through their effects on astroglia and microglia (Dezonne et al., 2015; Noda, 2015). Glia–neuron crosstalk, regulated by thyroid hormones, influences immune signaling, neurotransmitter release, and cellular metabolism—processes repeatedly implicated in schizophrenia (Misiak et al., 2021). Mechanistically, thyroid hormones modulate dopaminergic, serotonergic, and glutamatergic transmission and support myelination, hippocampal plasticity, and mitochondrial energetics, offering multiple pathways through which subtle alterations could bias psychosis phenotypes (Bauer et al., 2002; Bernal, 2007).

Biologically anchored correlates further support the relevance of thyroid status. In antipsychotic-naïve FEP, approximately one-third of patients exhibited elevated TSH, which was inversely associated with peripheral brain-derived neurotrophic factor (BDNF) levels and hippocampal volume, suggesting that thyroid dysfunction may contribute to hippocampal vulnerability early in the illness (Toll et al., 2023). This accords with the role of thyroid hormones in neuronal plasticity and with the hippocampus as a structural locus for cognitive and affective dysfunction in psychosis.

From an integrative standpoint, phase-dependent TSH changes can be conceptualized as the net of pituitary gating under dopaminergic drive in acute psychosis (suppressing TSH) and homeostatic or iatrogenic shifts later (Misiak et al., 2021). Peripherally, inflammatory and autoimmune signals (Barbero et al., 2020; Jiang & Gong, 2025) may modulate the availability and transport of thyroid hormones. Centrally, thyroid hormones influence myelination and oligodendrocyte biology (affecting processing speed), hippocampal plasticity (memory/learning), and prefrontal dopaminergic and glutamatergic balance (executive control) (Bernal, 2007; Noda, 2015; Tost et al., 2019). The association between the severity of neurological soft signs and lower FT4 levels in males (Treiber et al., 2025) suggests potential sex-specific developmental interactions. Finally, the TSH–BDNF–hippocampus triad in FEP (Toll et al., 2023) embeds thyroid markers within broader neurotrophic frameworks relevant to the trajectory and outcome.

Cross-sectional and meta-analytic evidence on HPT-axis dysregulation

Evidence from meta-analytic studies indicates altered HPT axis activity in schizophrenia, particularly in early stages of the illness. In drug-naïve first-episode psychosis (FEP), TSH levels tend to be lower than those in healthy controls, indicating downregulation of the HPT axis at onset (Cavaleri et al., 2023). Another meta-analysis reported a phase-dependent endocrine pattern, with reduced TSH and elevated FT4 in FEP, but increased TSH and unchanged FT4 in multi-episode schizophrenia (Misiak et al., 2021). The same study also reported lower total T3 levels in FEP. The authors noted that this pattern may reflect mechanisms resembling the non-thyroidal illness syndrome, including altered feedback between TSH and thyroid hormones, and stress-induced suppression of TSH secretion and T4-to-T3 conversion by glucocorticoids. Differences between FEP and multi-episode schizophrenia were attributed to greater cumulative exposure to antipsychotic treatment in the latter group (Misiak et al., 2021). This biphasic pattern reinforces the importance of considering illness stage when interpreting thyroid markers.

A key limitation of many cross-sectional studies is medication exposure. Antipsychotics—and lithium in schizoaffective/bipolar contexts—can influence thyroid indices (Czarnywojtek et al., 2020; Vedal et al., 2018). Consequently, drug-naïve or minimally exposed cohorts are critical to disentangle primary illness effects from treatment-related changes.

Prospective evidence and phenotype associations in early psychosis

Longitudinal designs clarify how the baseline thyroid status maps onto clinical trajectories. In prospective FEP cohorts with minimal antipsychotic exposure, higher FT4 levels within the reference range have been linked to a phenotype of fewer prodromal symptoms, shorter duration of untreated psychosis (DUP), and abrupt onset, alongside a greater likelihood of an affective psychosis diagnosis at follow-up (Barbero et al., 2015; Gine-Serven et al., 2023). Conversely, lower FT4 levels are related to insidious onset, prodromal features, and poorer attention/vigilance, even after adjusting for demographic and treatment covariates (Labad et al., 2016). These findings suggest that FT4 operates as a dimensional modulator of presentation: relatively higher values track with an "affective/acute" pole, and lower values with a "deficit/cognitive–negative" pole.

This polarity aligns with clinical heuristics— hyperactivation versus hypoactivation—and echoes the neuropsychiatric phenotypes of thyrotoxicosis (agitation, affective lability) and hypothyroidism (psychomotor slowing, apathy). Importantly, the effects are detectable within the normal laboratory range, underscoring that interindividual variation in euthyroid indices carries a clinical signal in psychosis (Gine-Serven et al., 2023; Labad et al., 2016).

Comparative studies in early phase samples have shown higher FT4 levels in affective psychoses (bipolar disorder with psychosis, psychotic depression) than in non-affective psychoses (schizophrenia spectrum), a pattern replicated across cohorts (Labad et al., 2016). Given that FEP diagnoses often evolve, these endocrine differences may assist in early stratification, complementing clinical and neurobiological markers to anticipate the diagnostic trajectory.

Severity of psychopathology and dimensional correlates

The associations between thyroid hormones and global symptom severity are heterogeneous across studies, likely reflecting differences in illness stage, exposure, and instruments. Some reports found no significant associations between FT4 and global scales (Barbero et al., 2020; Gine-Serven et al., 2023; Jose et al., 2015), whereas others described divergent trends: higher total T4 correlating with greater severity (Brief Psychiatric Rating Scale [BPRS]) (Baumgartner et al., 2000), lower FT3 associated with higher Positive and Negative Syndrome Scale (PANSS) total (Ichioka et al., 2012), and higher TSH correlating positively with PANSS, while T3/T4 correlated inversely with severity in a recent clinical series (Jiang & Gong, 2025). While inconsistent, these patterns are broadly compatible with a model in which relative thyroid hypoactivity (lower T3/T4, higher TSH) is accompanied by more severe psychopathology in the chronic phases.

Emerging data support nuanced associations at the level of specific dimensions. In schizophrenia-spectrum disorders, lower FT4 levels are related to greater severity of neurological soft signs (NSS)—putative neurodevelopmental markers—in men (Treiber et al., 2025), hinting at sex-specific interactions between thyroid status and neurodevelopmental burden. Regarding negative and depressive symptoms (key drivers of functional disability), mild negative correlations between TSH and negative symptoms have been described (Telo et al., 2016), higher TSH levels are linked to more severe Brief Negative Symptom Scale (BNSS) "lack-of-distress" scores (Rodrigues et al., 2024), and lower free triiodothyronine (FT3) is associated with more pronounced negative symptoms (Treiber et al., 2025). For suicidal ideation, studies have reported mixed results; one small study linked higher FT4 levels to a greater suicide risk (Jose et al., 2015), and an observational study suggested a link between thyroid autoimmunity and suicidality in first-episode major depressive disorder with psychotic symptoms (Feng et al., 2023). Taken together, these findings suggest that thyroid status maps onto clinically meaningful dimensions (positive, negative, affective, neurodevelopmental, suicidal) rather than merely tracking categorical diagnoses.

Thyroid autoimmunity as an immunoendocrine marker

Autoimmune thyroid disease appears to index particularly disabling phenotypes in early psychosis. In an outpatient cohort with illness <3 years, thyroid antibody–positive patients showed more severe negative and depressive symptoms and poorer functioning; anti-TPO titers correlated with PANSS negative/depressive domains and disability, while anti–thyroglobulin (anti-TG) related to functioning (Barbero et al., 2020). Population data link autoimmune thyroid disease to an increased psychosis risk (Eaton et al., 2010), and a meta-analysis described increased mortality due to suicide in patients with Hashimoto's thyroiditis (Heiberg Brix et al., 2019). Mechanistic bridges likely include cytokine signaling, blood–brain barrier alterations, and molecular mimicry, consistent with an immunoendocrine model of psychosis in which thyroid autoimmunity demarcates a high-impairment subgroup.

Longitudinal change in thyroid indices across the course of illness

Evidence points to dynamic shifts in thyroid hormone levels across acute episodes and recovery. Naturalistic inpatient studies have reported decreases in FT3 and increases in TSH after the resolution of acute psychosis; importantly, less symptomatic improvement was observed among those who also showed a decline in FT4 (Bunevicius et al., 2014). In early non-affective psychosis, FT4 levels were lower at 1 year than at baseline, while TSH remained unchanged, consistent with the attenuation of FT4 over continued treatment (Obdržálková et al., 2023). Pediatric and adolescent cohorts reveal declines in total T4 or FT4 after 3–6 months, echoing the progressive thyroid involvement during early treatment (Fraguas et al., 2008; Schimmelmann et al., 2007). These longitudinal patterns dovetail with the meta-analytic observation of stage-dependent HPT-axis profiles (Misiak et al., 2021) and underscore the need to consider time, treatment, and development when interpreting thyroid laboratory results in psychosis.

Medication effects and other confounders

Lithium—a mainstay treatment for affective and schizoaffective disorders—induces hypothyroidism in a substantial subset of patients (Czarnywojtek et al., 2020).

Several antipsychotics (e.g., quetiapine and clozapine) have been associated with reduced thyroid hormone levels or altered TSH levels (Bou Khalil & Richa, 2011; Kontaxakis et al., 2009). Systematic studies on severe mental disorders have confirmed that antipsychotic exposure can influence FT4/TSH (Vedal et al., 2018). Given these effects, drug-naïve or minimally exposed FEP cohorts are preferable for establishing primary associations; in clinical practice, serial monitoring helps disentangle treatment effects from disease-related endocrine changes.

Clinical synthesis and forward look

Clinically, the clearest and most replicated signals are (Table 13.2): (i) lower TSH at FEP (Cavaleri et al., 2023; Misiak et al., 2021); (ii) FT4 as a dimensional marker of presentation and short-term course—higher FT4 aligning with acute/affective features and lower FT4 with deficit/cognitive—negative phenotypes (Barbero et al., 2015; Gine-Serven et al., 2023; Labad et al., 2016); (iii) autoimmune thyroid markers indexing worse negative/depressive symptoms and functioning (Barbero et al., 2020), with possible links to suicidality (Feng et al., 2023); and (iv) longitudinal drift in thyroid indices across treatment and development (Bunevicius et al., 2014; Fraguas et al., 2008; Obdržálková et al., 2023; Schimmelmann et al., 2007). These threads converge on an immunoendocrine–neurodevelopmental model, wherein euthyroid-range variation and autoimmunity contribute to psychosis heterogeneity.

Two practical implications follow. First, baseline thyroid assessment (TSH, FT4, ± antibodies) in FEP provides diagnostic and prognostic signals beyond ruling out secondary psychoses. Second, thyroid measures may help explain the variance in treatment response, as discussed in Section Therapeutic implications, given early suggestions that thyroid hypofunction (or its drift over time) is related to poorer outcomes. Future studies should integrate serial endocrine, immune, and neurotrophic profiling with dimensional psychopathology and cognition to define actionable subgroups for tailored intervention.

Thyroid and cognition in psychosis

Cognitive impairment is a defining and disabling feature of schizophrenia spectrum disorders. Deficits are consistently observed across domains, such as processing speed, attention, working memory, verbal learning, visual memory, and executive function (Fioravanti et al., 2012). They often predate the FEP, remain relatively stable over the course of the illness, and represent a major determinant of functional outcome. Therefore, understanding the biological contributors to cognitive heterogeneity is critical. While neurodevelopmental, genetic, and environmental factors are central, endocrine influences, particularly thyroid hormones, are increasingly recognized as modulators of cognitive performance in psychosis.

Cognitive dysfunction in thyroid disease

The relevance of thyroid hormones to cognition is well illustrated by thyroid disorders. Adult hypothyroidism is associated with slowed processing speed, attentional deficits, impaired memory, and executive dysfunction (Samuels, 2014). Even subclinical hypothyroidism has been linked to poorer cognitive performance, especially in the working memory and executive domains (Jorde et al., 2006). In contrast, hyperthyroidism may manifest as distractibility, poor concentration, and agitation.

TABLE 13.2 Thyroid hormones and clinical features in psychosis (primary disorders).

Study	Sample	Thyroid marker	Clinical finding
Misiak et al. (2021) (meta-analysis)	FEP (drug-naïve)	TSH ↓	Reduced pituitary response at onset
Cavaleri et al. (2023) (meta-analysis)	FEP (drug-naïve)	TSH ↓	Abnormal anterior pituitary hormone secretion
Giné-Servén et al. (2023)	FEP, minimal exposure	FT4 ↑	Acute onset, short DUP, affective trajectory
Barbero et al. (2015)	Early psychosis (<3 years of illness)	FT4 ↑	Affective psychosis; better cognition (attention)
Labad et al. (2016)	FEP, 1y follow-up	FT4 ↓	Poorer attentional performance (U relationship with extreme quartiles)
Jiang and Gong (2025)	Schizophrenia	↑TSH, ↓T3/T4	Greater PANSS severity
Treiber et al. (2025)	Schizophrenia	FT4 ↓ (men)	More neurological soft signs

Importantly, restoration of euthyroidism often improves cognition, highlighting the causal role of thyroid hormones (Dugbartey, 1998). These observations provide a foundation for examining thyroid function and cognition in patients with psychosis.

Thyroid hormones and cognition in psychosis

Thyroid hormones are associated with specific cognitive domains in patients with early psychosis (Table 13.3). In a cross-sectional study, higher free thyroxine (FT4) levels were associated with better cognitive performance, particularly in attention and vigilance, as measured by the MATRICS Consensus Cognitive Battery (Barbero et al., 2015). These associations remained significant after adjusting for demographic and clinical covariates, indicating their robustness.

Longitudinally, thyroid hormones may also shape the cognitive trajectory. In a one-year follow-up of the same cohort, a U-shaped pattern was observed: patients with intermediate FT4 levels at baseline showed improvement or stability in attention and vigilance, whereas those in the lowest and highest FT4 quartiles exhibited cognitive deterioration (Labad et al., 2016). These findings suggest that both low and high FT4 levels, even within the reference range, may have detrimental effects on cognition in psychosis, highlighting the importance of optimal thyroid homeostasis for cognitive functioning.

In chronic schizophrenia, the results are more heterogeneous, partly due to medication confounders. In a cohort of 33 patients, FT4 levels were associated with executive function, problem-solving abilities, and verbal fluency, whereas FT3 levels were exclusively correlated with reasoning and problem-solving skills (Kaya & Ayik, 2022). Similarly, elevated FT3 levels were linked to improved memory performance, as assessed by the Mini-Mental State Examination (Ichioka et al., 2012). Although less consistent than early phase studies, these data support the hypothesis that thyroid hormones contribute to cognitive heterogeneity in psychosis.

Another prospective study extended these findings by examining the response to cognitive remediation therapy (CRT) in patients with schizophrenia. Estrada et al., (2020) found that FT4 moderated treatment outcomes: patients with higher FT4 worsened executive functions and verbal memory but improved attention following CRT. This positions thyroid hormones as potential biomarkers of treatment response, paving the way for personalized cognitive rehabilitation strategies for psychosis.

Potential mechanisms

Multiple neurobiological mechanisms may explain the influence of thyroid hormones on cognition in psychosis. At the structural level, thyroid hormones regulate myelination and oligodendrocyte function, both of which are critical for processing speed. They also influence hippocampal neurogenesis and synaptic plasticity, which are directly relevant to learning and memory (Bernal, 2007). At the metabolic level, thyroid hormones enhance mitochondrial activity and cerebral glucose utilization, which aligns with the bioenergetic models of schizophrenia-related cognitive deficits.

Neurotransmitter modulation offers an alternative pathway. Thyroid hormones regulate dopaminergic receptor sensitivity and turnover (Bauer et al., 2002), which may affect the prefrontal circuits governing executive function. They also enhance serotonergic and glutamatergic signaling, processes implicated in working memory and attentional control (Noda, 2015). These diverse effects suggest that thyroid hormones fine-tune the influence of multiple neurotransmitter systems on cognition.

TABLE 13.3 Thyroid hormones and cognition in psychosis.

Study	Sample	Marker	Cognitive domain	Main result
Barbero et al. (2015)	Early psychosis (<3years of illness)	FT4	Vigilance/Attention	Higher FT4 → better performance
Labad et al. (2016)	FEP, 1y follow-up	FT4	Vigilance/Attention	U Relationship: High and low FT4 → decline; Mid FT4 → stable
Estrada et al. (2020)	FEP, CRT trial	FT4	Executive function, attention, verbal memory	Higher FT4 → better response to CRT in sustained attention; poorer response in executive functions and verbal memory
Kaya and Ayik (2022)	Chronic schizophrenia	FT3, FT4	Verbal memory, executive	Low FT3/Low FT4 linked to deficits
Barbero et al. (2020)	Early psychosis	Anti-TPO, anti-TG	Executive, fluency	Antibody+ → poorer cognition, more negative symptoms

Autoimmunity and cognition

Although both thyroid hormone levels and autoimmunity were examined in the same cohort of patients with early psychosis, their associations with clinical outcomes were distinct. Whereas higher FT4 concentrations correlated with better cognitive performance in attention and vigilance domains (Barbero et al., 2015), the presence of thyroid autoantibodies was unrelated to cognitive scores but was associated with greater severity of negative and depressive symptoms and poorer overall functioning (Barbero et al., 2020). These findings suggest that hormonal and autoimmune thyroid alterations may represent complementary but independent pathways contributing to different dimensions of impairment in psychosis.

Clinical implications

These findings have important clinical implications. Routine thyroid testing in psychosis could help identify patients at risk of more severe cognitive impairment. FT4 levels may serve as prognostic biomarkers of cognitive performance and as moderators of response to CRT. Incorporating thyroid markers into clinical decision-making may enable the early identification of patients who require more intensive cognitive interventions. Moreover, thyroid hormones could be investigated as adjunctive pro-cognitive treatments for psychosis, particularly in patients with low-normal FT4 levels or antibody positivity. Although preliminary, this approach is consistent with the broader move toward biomarker-informed precision psychiatry.

Despite the promising evidence, significant gaps remain. Most studies are cross-sectional or involve small cohorts of patients. Large-scale longitudinal studies are required to establish whether thyroid hormone alterations precede cognitive decline or simply reflect illness severity. Randomized controlled trials of thyroid hormone supplementation in psychosis are scarce and warrant further investigation. Mechanistic studies integrating endocrine, immune, and neuroimaging approaches could clarify the pathways linking thyroid hormones to cognition. Ultimately, predictive models incorporating thyroid markers may help stratify cognitive outcomes and guide individualized interventions.

Therapeutic implications

The evidence linking thyroid dysfunction and psychosis has major therapeutic relevance. The clinical implications can be divided into the management of secondary psychoses due to overt thyroid disease and potential interventions in primary psychotic disorders, particularly schizophrenia.

Secondary psychoses

The most robust evidence concerns psychosis secondary to thyroid disease. Thyroid hormone replacement with levothyroxine is the definitive therapy for MP. Antipsychotics may provide short-term relief, especially for agitation or insomnia, but are rarely needed long term. Most patients achieve complete remission within weeks of hormone replacement therapy (Mohamed et al., 2021). Recognizing this syndrome prevents unnecessary chronic exposure to antipsychotics and ensures rapid recovery.

In HE, corticosteroids are the cornerstone of treatment, producing dramatic improvements in most cases (Menon et al., 2017). Immunosuppressants, such as azathioprine or intravenous immunoglobulins, may be used in refractory cases. Early identification of this autoimmune entity is crucial because untreated patients can develop chronic cognitive impairment or seizures. Therefore, thyroid antibody testing should be considered in cases of atypical psychosis or poor antipsychotic response.

In hyperthyroidism-related psychosis, antithyroid drugs, radioactive iodine, or surgery are required to restore euthyroidism. Psychiatric symptoms typically resolve once hormonal excess is corrected (Bunevicius & Prange, 2006; Golub & Rodack, 2018). Antipsychotics can be used transiently, but endocrine treatment is essential.

Adjunctive thyroid hormone therapy in primary psychoses

Interest has grown in the possibility of thyroid hormone supplementation as an adjunctive treatment for psychiatric disorders in recent years. Most evidence comes from major depression, where T3 augmentation of antidepressants accelerates the response and improves remission rates (Bauer et al., 2008). T3 has also been shown to have pro-cognitive effects in patients undergoing electroconvulsive therapy (Tremont & Stern, 1997). These findings highlight the potential of thyroid hormones as neuropsychiatric modulators.

In schizophrenia, the evidence is limited and mostly observational. FT4 has been shown to moderate the response to cognitive remediation therapy, suggesting that thyroid status influences treatment efficacy (Estrada et al., 2020). Some studies have highlighted preliminary associations between thyroid function and antipsychotic response, with low thyroid hormone levels linked to poorer treatment outcomes and a higher risk of treatment resistance (Bunevicius et al., 2014; Kelly & Conley, 2005). These observations reinforce the hypothesis that thyroid status may partly explain the heterogeneity in treatment response in psychosis.

However, no large randomized controlled trials have evaluated thyroid hormone supplementation as an adjunctive treatment in psychosis. Although small exploratory studies and case reports have suggested potential benefits, rigorous evidence is lacking. Given the strong rationale, future clinical trials are warranted, particularly those targeting cognitive deficits, negative symptoms, and treatment-resistant schizophrenia.

Risks and cautions

Thyroid hormone augmentation is not without risk. Excessive supplementation can induce anxiety, insomnia, or manic symptoms and carries cardiovascular risks such as arrhythmias. Therefore, any use of T3 or T4 in patients with psychosis should be carefully monitored, including regular thyroid function testing and cardiovascular assessment. Moreover, there is concern that increased thyroid activity could exacerbate affective instability in vulnerable individuals, potentially precipitating manic or mixed episodes. These considerations argue for cautious, individualized approaches if thyroid hormone augmentation is pursued in schizophrenia. Future studies should delineate optimal dosing strategies and identify clinical or biological predictors of benefit versus risk.

Future perspectives

Several directions emerge for future research in this area. First, large-scale prospective studies should clarify whether thyroid hormone alterations can predict treatment response or relapse in schizophrenia. Second, randomized trials should evaluate the efficacy and safety of thyroid hormones as adjunctive therapies, focusing on cognitive and negative symptoms of schizophrenia. Third, integrative biomarker approaches combining thyroid measures with immune and HPA axis markers could identify patient subgroups that are most likely to benefit. Finally, mechanistic studies are needed to understand how thyroid modulation affects brain plasticity and neurotransmission in patients with psychosis.

Conclusions

The relationship between thyroid function and psychosis is complex and clinically relevant. Overt thyroid disorders—such as hypothyroidism, hyperthyroidism, and autoimmune thyroiditis—can directly present with psychotic syndromes that typically remit once euthyroidism is restored. Conversely, subtle variations in thyroid hormones and autoimmunity appear to modulate the expression and prognosis of primary psychotic disorders. Meta-analytic and prospective evidence indicates that lower TSH and altered FT4 levels are linked to distinct symptom dimensions, cognitive profiles, and treatment responses, underscoring the role of thyroid status as a biological modifier within schizophrenia spectrum disorders.

Recognizing thyroid-related psychoses ensures effective, often curative, endocrine treatment, while systematic thyroid screening in psychosis provides valuable diagnostic and prognostic information. Although adjunctive thyroid hormone therapy shows promise in depression, its application in psychosis remains exploratory. Future research should determine whether thyroid hormones can improve cognition or negative symptoms and clarify the mechanisms linking endocrine and psychiatric dysfunction. Bridging endocrinology and psychiatry may ultimately lead to more precise and effective care for patients with psychosis.

Other hormones, endocrine tissues, and pathways

While thyroid function represents a paradigmatic example of endocrine influence on psychosis, other hormonal systems play equally significant roles in brain regulation and psychiatric vulnerability. The hypothalamic–pituitary–adrenal (HPA) axis, gonadal steroids, metabolic hormones, and peripheral endocrine organs, such as adipose tissue and the gut, constitute an integrated network that interacts with the neural circuits, immune processes, and energy metabolism. Exploring these translational aspects broadens our understanding of psychosis as a systemic disorder.

Hypothalamic–pituitary–adrenal (HPA) axis dysregulation

Alterations in the HPA axis function are among the most replicated biological findings in psychotic disorders. Elevated baseline cortisol levels have been consistently reported in FEP, supporting early HPA hyperactivity and the "neural diathesis–stress" model (Hubbard & Miller, 2019). A flattened cortisol awakening response (CAR) is observed in FEP and schizophrenia, suggesting blunted stress adaptation (Berger et al., 2016), whereas in at-risk individuals, cortisol alterations correlate with subthreshold symptoms (Labad et al., 2018). These findings identify HPA dysregulation as a potential biomarker of stress-system imbalance in psychosis.

Gonadal steroids and sex differences

Sex hormones have profound effects on brain development, synaptic pruning, and neurotransmission. Estrogens have neuroprotective and anti-inflammatory effects and modulate dopamine receptor sensitivity, contributing to the later onset and better prognosis of schizophrenia in females. Selective estrogen receptor modulators, such as raloxifene and pregnenolone, have shown pro-cognitive effects in patients with schizophrenia (Soria et al., 2018). Higher testosterone levels in older men with schizophrenia have been associated with the severity of negative symptoms (Li et al., 2024).

Prolactin, which is regulated by dopamine and estrogen, is often elevated in patients with psychosis, even before antipsychotic exposure, indicating intrinsic hypothalamic–pituitary dysregulation (Aymerich et al., 2023). In treated patients, chronic hyperprolactinemia, often linked to D2 receptor blockade, which is most pronounced with amisulpride, risperidone, and paliperidone, can exacerbate mood, cognition, and metabolic disturbances (Peuskens et al., 2014).

These findings highlight the importance of the hormonal milieu in modulating psychosis expression and treatment response, supporting the development of sex-informed and endocrine-targeted interventions for treating psychosis.

Metabolic hormones

Insulin, leptin, and adiponectin act as endocrine bridges between metabolism and brain function. Insulin resistance (IR) and impaired glucose tolerance (IGT) are highly prevalent in drug-naïve FEP, indicating intrinsic metabolic dysregulation (Perry et al., 2016). IGT is associated with higher negative and total symptom severity in patients in the early stage (Chen et al., 2016).

Glycated hemoglobin has been associated with poorer cognitive (Montalvo et al., 2020), and increased CSF glucose relates to disorganized symptoms in FEP (Giné-Servén et al., 2022). IR has been associated with a reduced response to antipsychotic treatment (Tomasik et al., 2019). These findings suggest that a more severe FEP phenotype is related to glucose abnormalities.

Interventions that improve insulin sensitivity, including metformin (Qin et al., 2025), GLP-1 receptor agonists (Flintoff et al., 2021), and lifestyle changes, have demonstrated benefits for both metabolic parameters and cognitive function, highlighting shared neuroendocrine pathways.

Pineal hormones

Melatonin, secreted by the pineal gland, regulates circadian rhythms, oxidative balance, and neuronal repair. In schizophrenia, reduced melatonin secretion and disrupted circadian rhythmicity contribute to sleep–wake disturbances, cognitive impairment, and negative symptoms (Anderson & Maes, 2012). Beyond its chronobiological role, melatonin exerts neuroprotective, anti-inflammatory, and antioxidant effects. Adjunctive melatonin therapy has shown beneficial effects on sleep quality, metabolic parameters, and tardive dyskinesia in patients with schizophrenia (Duan et al., 2021). Restoration of circadian homeostasis and melatonin function could contribute to improving sleep and overall quality of life in these patients.

Adipose and metabolic signaling

Adipose tissue acts as an active endocrine organ that releases adipokines to regulate energy balance and neuroendocrine pathways. Dysregulation of this system appears early in the course of psychosis, before exposure to antipsychotic medications. In FEP, altered circulating levels of leptin, adiponectin, resistin, and chemerin suggest intrinsic adipose dysfunction independent of confounding factors such as body composition, childhood trauma, or minority status (Veru-Lesmes et al., 2021). In chronic schizophrenia, higher leptin levels and leptin/adiponectin ratios are strongly associated with the presence and severity of metabolic syndrome, as well as with anthropometric measures, such as body fat percentage and BMI (Mednova et al., 2020). These findings indicate that adipose-derived hormonal alterations are an early and stable feature of psychosis, contributing to metabolic vulnerability and potentially interacting with inflammatory and neuroendocrine mechanisms over the course of the illness.

Integrative translational perspective

The endocrine axes described above—HPA, gonadal, metabolic, pineal, and adipose—form an interconnected network linking peripheral hormonal regulation with brain function and behavior. Alterations in these systems contribute to stress sensitivity, sex-related differences, metabolic vulnerability, and circadian disruption, all of which shape psychosis clinical heterogeneity. Integrating hormonal and metabolic profiling with clinical assessment can inform personalized interventions aimed at restoring endocrine balance, improving outcomes, and enhancing quality of life.

Applications of the material (final version)

This chapter highlights how thyroid function interacts with brain and systemic processes relevant to psychosis. Clinically, the evidence reviewed underscores that both overt and subtle thyroid alterations can influence symptom expression, disease trajectory, and treatment outcomes.

In secondary psychosis, overt thyroid dysfunction—either hypo- or hyperthyroidism—can precipitate affective or psychotic symptoms that often remit after normalization of thyroid status, emphasizing the importance of routine endocrine evaluation in acute psychiatric presentations.

In primary psychotic disorders, particularly first-episode psychosis (FEP), thyroid hormone variations within the reference range also appear clinically significant. Elevated free thyroxine (FT4) levels have been associated with acute and more severe psychotic states, whereas lower FT4 or higher thyroid-stimulating hormone (TSH) levels tend to correlate with negative symptom predominance and poorer functional outcomes. These findings suggest that thyroid hormones contribute to the phenotypic variability observed in psychosis.

The presence of thyroid autoantibodies has also been linked to specific symptom patterns and poorer functioning, indicating that immune–endocrine interactions may modulate clinical presentation even in the absence of overt thyroid disease.

Taken together, these findings have direct translational implications. Routine assessment of thyroid parameters and antibodies should be incorporated into the diagnostic work-up of new-onset psychosis to identify secondary cases and refine prognosis in primary disorders. Moreover, longitudinal monitoring may help anticipate treatment responses and functional recovery, particularly in patients with metabolic or hormonal vulnerability.

Finally, as discussed in Section Other hormones, endocrine tissues, and pathways, these thyroid-related mechanisms are integrated within broader endocrine networks, including the adrenal, gonadal, and metabolic axes, which collectively shape stress sensitivity, energy balance, and circadian regulation. Understanding psychosis as a disorder of multisystem endocrine imbalance opens new avenues for precision approaches combining hormonal, metabolic, and neuropsychiatric assessments to improve clinical outcomes and quality of life.

Mini-dictionary of terms

Adipokines. Bioactive hormones secreted by adipose tissue (e.g., leptin and adiponectin) regulate energy balance and interact with neuroendocrine and immune pathways.

Cortisol Awakening Response (CAR). The physiological rise in cortisol levels occurring within the first hour after waking serves as an indicator of hypothalamic–pituitary–adrenal (HPA) axis reactivity.

First-Episode Psychosis (FEP). The initial clinical manifestations of psychotic symptoms typically precede long-term treatment exposure and provide a key window for studying intrinsic biological alterations.

Free Thyroxine (FT4). The unbound fraction of thyroxine circulating in the plasma represents the biologically active component that enters target tissues, including the brain.

Hypothalamic–Pituitary–Adrenal (HPA) Axis. The HPA axis is a major neuroendocrine system that coordinates the stress response through hormonal cascades involving corticotropin-releasing hormone, adrenocorticotropic hormone, and cortisol.

Prolactin. A pituitary hormone regulated by dopamine and estrogen, frequently elevated in psychosis, and modulated by antipsychotic treatment.

Thyroid Autoantibodies. Immunoglobulins directed against thyroid antigens (e.g., thyroid peroxidase or thyroglobulin), whose presence in psychosis correlates with specific symptom patterns and functional impairment, even in the absence of overt thyroid disease.

Thyroid-Stimulating Hormone (TSH). It is a pituitary hormone that regulates thyroid gland activity and serves as a sensitive indicator of central thyroid axis regulation.

Translational Biomarkers. A measurable biological indicator that links mechanistic research to clinical outcomes facilitates the early identification of at-risk or treatment-responsive patients.

Summary points

- Both overt and subclinical thyroid dysfunction can modulate the onset and trajectory of psychotic disorders.
- In first-episode psychosis, higher FT4 levels are associated with acute symptomatology, whereas lower FT4 or elevated TSH levels are associated with predominant negative symptoms and poorer functional outcomes.
- The presence of thyroid autoantibodies correlates with specific symptom dimensions and reduced psychosocial functioning, indicating immune–endocrine contributions to clinical variability.
- Comprehensive thyroid assessment facilitates the differentiation between primary and secondary psychoses, thereby improving diagnostic precision.
- Longitudinal monitoring of thyroid parameters may serve as a prognostic marker and guide individualized treatment.
- Interactions between thyroid regulation and other endocrine axes (HPA, gonadal, metabolic, and circadian) contribute to the multisystem nature of psychosis.
- Integrating endocrine and metabolic profiling into psychiatric care enhances precision diagnostics and supports holistic intervention strategies aimed at improving long-term outcomes.

Acknowledgments

During the preparation of this work, the authors used ChatGPT-3.5 to assist with grammar and style corrections. All content was reviewed and edited by the authors, who take full responsibility for the publication's content.

References

Anderson, G., & Maes, M. (2012). Melatonin: An overlooked factor in schizophrenia and in the inhibition of anti-psychotic side effects. *Metabolic Brain Disease, 27*(2), 113–119. https://doi.org/10.1007/s11011-012-9307-9

Asher, R. (1949). Myxoedematous madness. *British Medical Journal, 2* (4627), 555–562. https://doi.org/10.1136/bmj.2.4627.555

Aymerich, C., Pedruzo, B., Pacho, M., Laborda, M., Herrero, J., Pillinger, T., McCutcheon, R. A., Alonso-Alconada, D., Bordenave, M., Martínez-Querol, M., Arnaiz, A., Labad, J., Fusar-Poli, P., González-Torres, M.Á., & Catalan, A. (2023). Prolactin and morning cortisol concentrations in antipsychotic naïve first episode psychosis: A systematic review and meta-analysis. *Psychoneuroendocrinology, 150*, Article 106049. https://doi.org/10.1016/j.psyneuen.2023.106049

Barbero, J. D., Gutiérrez-Zotes, A., Montalvo, I., Creus, M., Cabezas, Á., Solé, M., Algora, M. J., Garcia-Parés, G., Vilella, E., & Labad, J. (2015). Free thyroxine levels are associated with cognitive abilities in subjects with early psychosis. *Schizophrenia Research, 166*(1–3), 37–42. https://doi.org/10.1016/j.schres.2015.04.030

Barbero, J. D., Palacín, A., Serra, P., Solé, M., Ortega, L., Cabezas, Á., Montalvo, I., Algora, M. J., Martorell, L., Vilella, E., Sánchez-Gistau, V., & Labad, J. (2020). Association between anti-thyroid antibodies and negative symptoms in early psychosis. *Early Intervention in Psychiatry, 14*(4), 470–475. https://doi.org/10.1111/eip.12873

Bauer, M., Goetz, T., Glenn, T., & Whybrow, P. C. (2008). The thyroid-brain interaction in thyroid disorders and mood disorders. *Journal of Neuroendocrinology, 20*(10), 1101–1114. https://doi.org/10.1111/j.1365-2826.2008.01774.x

Bauer, M., Heinz, A., & Whybrow, P. C. (2002). Thyroid hormones, serotonin and mood: Of synergy and significance in the adult brain. *Molecular Psychiatry, 7*(2), 140–156. https://doi.org/10.1038/sj.mp.4000963

Baumgartner, A., Pietzcker, A., & Gaebel, W. (2000). The hypothalamic–pituitary–thyroid axis in patients with schizophrenia. *Schizophrenia Research, 44*(3), 233–243. https://doi.org/10.1016/S0920-9964(99)00187-5

Berger, M., Kraeuter, A. K., Romanik, D., Malouf, P., Amminger, G. P., & Sarnyai, Z. (2016). Cortisol awakening response in patients with psychosis: Systematic review and meta-analysis. *Neuroscience & Biobehavioral Reviews, 68*, 157–166. https://doi.org/10.1016/j.neubiorev.2016.05.027

Bernal, J. (2007). Thyroid hormone receptors in brain development and function. *Nature Clinical Practice Endocrinology & Metabolism, 3* (3), 249–259. https://doi.org/10.1038/ncpendmet0424

Bou Khalil, R., & Richa, S. (2011). Thyroid adverse effects of psychotropic drugs. *Clinical Neuropharmacology, 34*(6), 248–255. https://doi.org/10.1097/WNF.0b013e31823429a7

Bunevicius, R., & Prange, A. J., Jr. (2006). Psychiatric manifestations of Graves' hyperthyroidism: Pathophysiology and treatment options. *CNS Drugs, 20*(11), 897–909. https://doi.org/10.2165/00023210-200620110-00003

Bunevicius, R., Steibliene, V., & Prange, A. J. (2014). Thyroid axis function after in-patient treatment of acute psychosis with antipsychotics: A naturalistic study. *BMC Psychiatry, 14*, 279. https://doi.org/10.1186/s12888-014-0279-7

Cavaleri, D., Capogrosso, C. A., Guzzi, P., Bernasconi, G., Re, M., Misiak, B., Crocamo, C., Bartoli, F., & Carrà, G. (2023). Blood concentrations of anterior pituitary hormones in drug-naïve people with first-episode psychosis: A systematic review and meta-analysis. *Psychoneuroendocrinology, 158*, Article 106392. https://doi.org/10.1016/j.psyneuen.2023.106392

Chen, D. C., Du, X. D., Yin, G. Z., Yang, K. B., Nie, Y., Wang, N., Li, Y. L., Xiu, M. H., He, S. C., Yang, F. D., Cho, R. Y., Kosten, T. R., Soares, J. C., Zhao, J. P., & Zhang, X. Y. (2016). Impaired glucose tolerance in first-episode drug-naïve patients with schizophrenia: Relationships with clinical phenotypes and cognitive deficits. *Psychological Medicine, 46*(15), 3219–3230. https://doi.org/10.1017/S0033291716001902

Constant, E. L., de Volder, A. G., Ivanoiu, A., Bol, A., Labar, D., Seghers, A., Cosnard, G., Melin, J., & Daumerie, C. (2001). Cerebral blood flow and glucose metabolism in hypothyroidism: A positron emission tomography study. *The Journal of Clinical Endocrinology & Metabolism, 86*(8), 3864–3870. https://doi.org/10.1210/jcem.86.8.7749

Czarnywojtek, A., Zgorzalewicz-Stachowiak, M., Czarnocka, B., Sawicka-Gutaj, N., Gut, P., Krela-Kazmierczak, I., & Ruchala, M. (2020). Effect of lithium carbonate on the function of the thyroid gland: Mechanism of action and clinical implications. *Journal of Physiology & Pharmacology: An Official Journal of the Polish Physiological Society, 71*(2). https://doi.org/10.26402/jpp.2020.2.03

Dezonne, R. S., Lima, F. R. S., Trentin, A. G., & Gomes, F. C. (2015). Thyroid hormone and astroglia: Endocrine control of the neural environment. *Journal of Neuroendocrinology, 27*(6), 435–445. https://doi.org/10.1111/jne.12283

Duan, C., Jenkins, Z. M., & Castle, D. (2021). Therapeutic use of melatonin in schizophrenia: A systematic review. *World Journal of Psychiatry, 11*(8), 463–476. https://doi.org/10.5498/wjp.v11.i8.463

Dugbartey, A. T. (1998). Neurocognitive aspects of hypothyroidism. *Archives of Internal Medicine, 158*(13), 1413. https://doi.org/10.1001/archinte.158.13.1413

Eaton, W. W., Pedersen, M. G., Nielsen, P. R., & Mortensen, P. B. (2010). Autoimmune diseases, bipolar disorder, and non-affective psychosis. *Bipolar Disorders, 12*(6), 638–646. https://doi.org/10.1111/j.1399-5618.2010.00853.x

Estrada, F., Crosas, J. M., Ahuir, M., Pérez-Muñoz, S., Zabala, W., Aguayo, R., Barbero, J. D., Montalvo, I., Tost, M., Llauradó, L., Guardia, A., Palao, D., Monreal, J. A., & Labad, J. (2020). Free thyroxine concentrations moderate the response to a cognitive remediation therapy in people with early psychosis: A pilot randomized clinical trial. *Frontiers in Psychiatry, 11*. https://doi.org/10.3389/fpsyt.2020.00636

Feng, X.-Z., Wang, K., Li, Z., Wang, F., Li, Z.-Y., Tan, X., Zhu, W.-L., Zhang, X.-Y., & Li, S.-X. (2023). Association between thyroid autoimmunity and clinical characteristics in first-episode and drug-naive depressed patients with suicide attempts. *General Hospital Psychiatry, 83*, 156–163. https://doi.org/10.1016/j.genhosppsych.2023.05.008

Ferracci, F., Bertiato, G., & Moretto, G. (2004). Hashimoto's encephalopathy: Epidemiologic data and pathogenetic considerations. *Journal of the Neurological Sciences, 217*(2), 165–168. https://doi.org/10.1016/j.jns.2003.09.007

Fioravanti, M., Bianchi, V., & Cinti, M. E. (2012). Cognitive deficits in schizophrenia: An updated metanalysis of the scientific evidence. *BMC Psychiatry, 12*(1), 64. https://doi.org/10.1186/1471-244X-12-64

Flintoff, J., Kesby, J. P., Siskind, D., & Burne, T. H. (2021). Treating cognitive impairment in schizophrenia with GLP-1RAs: An overview of their therapeutic potential. *Expert Opinion on Investigational Drugs, 30*(8), 877–891. https://doi.org/10.1080/13543784.2021.1951702

Fraguas, D., Merchan-Naranjo, J., Laita, P., Paralleda, M., Moreno, D., Ruiz-Sancho, A., Cifuentes, A., Giraldez, M., & Arango, C. (2008). Metabolic and hormonal side effects in children and adolescents treated with second-generation antipsychotics. *The Journal of Clinical Psychiatry, 69*(7), 1166–1175. https://doi.org/10.4088/JCP.v69n0717

Fukao, A., Takamatsu, J., Arishima, T., Tanaka, M., Kawai, T., Okamoto, Y., Miyauchi, A., & Imagawa, A. (2020). Graves' disease and mental disorders. *Journal of Clinical & Translational Endocrinology, 19*, Article 100207. https://doi.org/10.1016/j.jcte.2019.100207

Giné-Servén, E., Martinez-Ramirez, M., Boix-Quintana, E., Davi-Loscos, E., Guanyabens, N., Casado, V., Muriana, D., Torres-Rivas, C., Crespo-Facorro, B., & Labad, J. (2022). Routine cerebrospinal fluid parameters as biomarkers in first-episode psychosis: A prospective observational study. *Progress in Neuro-Psychopharmacology and Biological Psychiatry, 112*, Article 110424. https://doi.org/10.1016/j.pnpbp.2021.110424

Gine-Serven, E., Martinez-Ramirez, M., Boix-Quintana, E., Davi-Loscos, E., Guanyabens, N., Casado, V., Muriana, D., Torres-Rivas, C., Cuesta, M. J., & Labad, J. (2023). Association between free thyroxine levels and clinical phenotype in first-episode psychosis: A prospective observational study. *PeerJ, 11*, Article e15347. https://doi.org/10.7717/peerj.15347

Golub, D., & Rodack, V. (2018). Antipsychotics in hyperthyroid-related psychosis: Case report and systematic review. *Neuro Endocrinology Letters, 39*(1), 65–74.

Hassan, W. A., Abdel Rahman, T., Aly, M. S., & Shahat, A. S. (2013). Alterations in monoamines level in discrete brain regions and other peripheral tissues in young and adult male rats during experimental hyperthyroidism. *International Journal of Developmental Neuroscience, 31*(4), 311–319.

Hayes, E., Gavrilidis, E., & Kulkarni, J. (2012). The role of oestrogen and other hormones in the pathophysiology and treatment of schizophrenia. *Schizophrenia Research and Treatment, 2012*, 1–8. https://doi.org/10.1155/2012/540273

Heiberg Brix, T., Ferløv-Schwensen, C., Thvilum, M., & Hegedüs, L. (2019). Death by unnatural causes, mainly suicide, is increased in patients with Hashimoto's thyroiditis. A nationwide Danish register study. *Endocrine, 65*(3), 616–622. https://doi.org/10.1007/s12020-019-01946-5

Hubbard, D. B., & Miller, B. J. (2019). Meta-analysis of blood cortisol levels in individuals with first-episode psychosis. *Psychoneuroendocrinology, 104*, 269–275. https://doi.org/10.1016/j.psyneuen.2019.03.014

Ichioka, S., Terao, T., Hoaki, N., Matsushita, T., & Hoaki, T. (2012). Triiodothyronine may be possibly associated with better cognitive function and less extrapyramidal symptoms in chronic schizophrenia. *Progress in Neuro-Psychopharmacology and Biological Psychiatry, 39*(1), 170–174. https://doi.org/10.1016/j.pnpbp.2012.06.008

Jiang, Q.-H., & Gong, W.-D. (2025). Correlation analyse between thyroid hormone levels and severity of schizophrenia symptoms. *World Journal of Psychiatry, 15*(1). https://doi.org/10.5498/wjp.v15.i1.100880

Jorde, R., Waterloo, K., Storhaug, H., Nyrnes, A., Sundsfjord, J., & Jenssen, T. G. (2006). Neuropsychological function and symptoms in subjects with subclinical hypothyroidism and the effect of thyroxine treatment. *The Journal of Clinical Endocrinology & Metabolism, 91*(1), 145–153. https://doi.org/10.1210/jc.2005-1775

Jose, J., Nandeesha, H., Kattimani, S., Meiyappan, K., Sarkar, S., & Sivasankar, D. (2015). Association between prolactin and thyroid hormones with severity of psychopathology and suicide risk in drug free male schizophrenia. *Clinica Chimica Acta, 444*, 78–80. https://doi.org/10.1016/j.cca.2015.02.003

Kaya, H., & Ayik, B. (2022). Effects of TSH, fT3 and fT4 levels on neurocognitive symptoms in patients with schizophrenia. *Cukurova Medical Journal, 47*(3), 1263–1272. https://doi.org/10.17826/cumj.1135374

Kelly, D. L., & Conley, R. R. (2005). Thyroid function in treatment-resistant schizophrenia patients treated with quetiapine, risperidone, or fluphenazine. *The Journal of Clinical Psychiatry, 66*(01), 80–84. https://doi.org/10.4088/JCP.v66n0111

Kontaxakis, V. P., Karaiskos, D., Havaki-Kontaxaki, B. J., Ferentinos, P., & Papadimitriou, G. N. (2009). Can quetiapine-induced hypothyroidism be reversible without quetiapine discontinuation? *Clinical Neuropharmacology, 32*(5), 295–296. https://doi.org/10.1097/WNF.0b013e3181a8cbcc

Labad, J., Armario, A., Nadal, R., Solé, M., Gutiérrez-Zotes, A., Montalvo, I., Moreno-Samaniego, L., Martorell, L., Sánchez-Gistau, V., & Vilella, E. (2018). Clinical correlates of hypothalamic-pituitary-adrenal axis measures in individuals at risk for psychosis and with first-episode psychosis. *Psychiatry Research, 265*, 284–291. https://doi.org/10.1016/j.psychres.2018.05.018

Labad, J., Barbero, J. D., Gutiérrez-Zotes, A., Montalvo, I., Creus, M., Cabezas, Á., Solé, M., Algora, M. J., Garcia-Parés, G., & Vilella, E. (2016). Free thyroxine levels are associated with cognitive changes in individuals with a first episode of psychosis: A prospective 1-year follow-up study. *Schizophrenia Research, 171*(1–3), 182–186. https://doi.org/10.1016/j.schres.2016.01.036

Li, S., Liu, W., Huang, Z., Lin, H., Ning, Y., & Li, Z. (2024). Sex-specific associations between sex hormones and clinical symptoms in late-life schizophrenia. *Schizophrenia Research, 267*, 1–7. https://doi.org/10.1016/j.schres.2024.03.002

McCutcheon, R. A., Reis Marques, T., & Howes, O. D. (2020). Schizophrenia—An overview. *JAMA Psychiatry, 77*(2), 201. https://doi.org/10.1001/jamapsychiatry.2019.3360

Mednova, I. A., Boiko, A. S., Kornetova, E. G., Parshukova, D. A., Semke, A. V., Bokhan, N. A., Loonen, A. J. M., & Ivanova, S. A. (2020). Adipocytokines and metabolic syndrome in patients with schizophrenia. *Metabolites, 10*(10), 410. https://doi.org/10.3390/metabo10100410

Menon, V., Subramanian, K., & Thamizh, J. S. (2017). Psychiatric presentations heralding hashimoto's encephalopathy: A systematic review and analysis of cases reported in literature. *Journal of Neurosciences in Rural Practice, 08*(02), 261–267. https://doi.org/10.4103/jnrp.jnrp_440_16

Misiak, B., Stańczykiewicz, B., Wiśniewski, M., Bartoli, F., Carra, G., Cavaleri, D., Samochowiec, J., Jarosz, K., Rosińczuk, J., & Frydecka, D. (2021). Thyroid hormones in persons with schizophrenia: A systematic review and meta-analysis. *Progress in Neuro-Psychopharmacology and Biological Psychiatry, 111*, Article 110402. https://doi.org/10.1016/j.pnpbp.2021.110402

Mohamed, M. F. H., Danjuma, M., Mohammed, M., Mohamed, S., Siepmann, M., Barlinn, K., Suwileh, S., Abdalla, L., Al-Mohanadi, D., Silva Godínez, J. C., Elzouki, A.-N., & Siepmann, T. (2021). Myxedema psychosis: Systematic review and pooled analysis. *Neuropsychiatric Disease and Treatment, 17*, 2713–2728. https://doi.org/10.2147/NDT.S318651

Montalvo, I., González-Rodríguez, A., Cabezas, Á., Gutiérrez-Zotes, A., Solé, M., Algora, M. J., Ortega, L., Martorell, L., Sánchez-Gistau, V., Vilella, E., & Labad, J. (2020). Glycated haemoglobin is associated with poorer cognitive performance in patients with recent-onset psychosis. *Frontiers in Psychiatry, 11*. https://doi.org/10.3389/fpsyt.2020.00455

Noda, M. (2015). Possible role of glial cells in the relationship between thyroid dysfunction and mental disorders. *Frontiers in Cellular Neuroscience, 9*. https://doi.org/10.3389/fncel.2015.00194

Obdržálková, M., Ustohal, L., Hlaváčová, N., Mayerová, M., Češková, E., Kašpárek, T., & Ježová, D. (2023). Selected neuroendocrine factors as potential molecular biomarkers of early non-affective psychosis course in relation to treatment outcome: A pilot study. *Heliyon, 9*(10), Article e21173. https://doi.org/10.1016/j.heliyon.2023.e21173

Payer, J., Petrovic, T., Lisy, L., Langer, P., & Ďurovcová, D. (2012). Hashimoto's encephalopathy: A rare intricate syndrome. *International Journal of Endocrinology and Metabolism, 10*(2), 506–514.

Perry, B. I., McIntosh, G., Weich, S., Singh, S., & Rees, K. (2016). The association between first-episode psychosis and abnormal glycaemic control: Systematic review and meta-analysis. *The Lancet Psychiatry, 3*(11), 1049–1058. https://doi.org/10.1016/S2215-0366(16)30262-0

Peuskens, J., Pani, L., Detraux, J., & De Hert, M. (2014). The effects of novel and newly approved antipsychotics on serum prolactin levels: A comprehensive review. *CNS Drugs, 28*(5), 421–453. https://doi.org/10.1007/s40263-014-0157-3

Qin, Z.-J., Shi, Z.-M., Li, L.-J., Wei, X., Hu, H.-L., Wei, W., Xie, Z.-Y., Ji, H.-X., Wei, Y.-H., & Zheng, W. (2025). Metformin for neurocognitive dysfunction in schizophrenia: A systematic review. *Frontiers in Psychiatry, 15*. https://doi.org/10.3389/fpsyt.2024.1540153

Rodrigues, A., Santos, H. C., Ferreira, S., Diogo, V., Costa, M., Brissos, S., Marques, J. G., & Prata, D. (2024). An exploration of blood-based biomarkers of negative symptoms of psychosis in men. *Journal of Psychiatric Research, 177*, 256–263. https://doi.org/10.1016/j.jpsychires.2024.06.050

Samuels, M. H. (2014). Thyroid disease and cognition. *Endocrinology and Metabolism Clinics of North America, 43*(2), 529–543. https://doi.org/10.1016/j.ecl.2014.02.006

Schimmelmann, B. G., Mehler-Wex, C., Lambert, M., Schulze-zur-Wiesch, C., Koch, E., Flechtner, H. H., Gierow, B., Maier, J., Meyer, E., & Schulte-Markwort, M. (2007). A prospective 12-Week study of quetiapine in adolescents with schizophrenia spectrum disorders. *Journal of Child and Adolescent Psychopharmacology, 17*(6), 768–778. https://doi.org/10.1089/cap.2007.0048

Shaw, P. J., Walls, T. J., Newman, P. K., Cleland, P. G., & Cartlidge, N. E. F. (1991). Hashirnoto's encephalopathy. *Neurology, 41*(2_part_1). https://doi.org/10.1212/WNL.41.2_Part_1.228, 228–228.

Smeland, O. B., Frei, O., Dale, A. M., & Andreassen, O. A. (2020). The polygenic architecture of schizophrenia — Rethinking pathogenesis and nosology. *Nature Reviews Neurology, 16*(7), 366–379. https://doi.org/10.1038/s41582-020-0364-0

Soria, V., González-Rodríguez, A., Huerta-Ramos, E., Usall, J., Cobo, J., Bioque, M., Barbero, J. D., García-Rizo, C., Tost, M., Monreal, J. A., & Labad, J. (2018). Targeting hypothalamic-pituitary-adrenal axis hormones and sex steroids for improving cognition in major mood disorders and schizophrenia: A systematic review and narrative synthesis. *Psychoneuroendocrinology, 93*, 8–19. https://doi.org/10.1016/j.psyneuen.2018.04.012

Telo, S., Bilgic, S., & Karabulut, N. (2016). Thyroid hormone levels in chronic schizophrenic patients: Association with psychopathology. *West Indian Medical Journal, 65*(2), 312–315. https://doi.org/10.7727/wimj.2015.186

Toll, A., Blanco-Hinojo, L., Berge, D., Manzano, A., El Abidi, K., Perez-Solà, V., & Mané, A. (2023). Relationship between thyroid-stimulating hormone, BDNF levels, and hippocampal volume in antipsychotic-naïve first-episode psychosis patients. *Frontiers in Psychiatry, 14*. https://doi.org/10.3389/fpsyt.2023.1301714

Tomasik, J., Lago, S. G., Vázquez-Bourgon, J., Papiol, S., Suárez-Pinilla, P., Crespo-Facorro, B., & Bahn, S. (2019). Association of insulin resistance with schizophrenia polygenic risk score and response to antipsychotic treatment. *JAMA Psychiatry, 76*(8), 864. https://doi.org/10.1001/jamapsychiatry.2019.0304

Tost, M., Monreal, J. A., Armario, A., Barbero, J. D., Cobo, J., García-Rizo, C., Bioque, M., Usall, J., Huerta-Ramos, E., Soria, V., Pnecat, G., & Labad, J. (2019). Targeting hormones for improving cognition in major mood disorders and schizophrenia: Thyroid hormones and prolactin. *Clinical Drug Investigation, 39*(12), 1215–1229. https://doi.org/10.1007/s40261-019-00854-w

Treiber, M. C., Tsapakis, E.-M., Athanasiou, S., Chovardas, K., Kyziridis, T., & Fountoulakis, K. N. (2025). Neurological soft signs and thyroid hormones in schizophrenia spectrum disorders. *Progress in Neuro-Psychopharmacology and Biological Psychiatry, 138*, Article 111338. https://doi.org/10.1016/j.pnpbp.2025.111338

Tremont, G., & Stern, R. A. (1997). Use of thyroid hormone to diminish the cognitive side effects of psychiatric treatment. *Psychopharmacology Bulletin, 33*(2), 273–280. http://www.ncbi.nlm.nih.gov/pubmed/9230642.

Vedal, T. S. J., Steen, N. E., Birkeland, K. I., Dieset, I., Reponen, E. J., Laskemoen, J. F., Rødevand, L., Melle, I., Andreassen, O. A., Molden, E., & Jönsson, E. G. (2018). Free thyroxine and thyroid-stimulating hormone in severe mental disorders: A naturalistic study with focus on antipsychotic medication. *Journal of Psychiatric Research, 106*, 74–81. https://doi.org/10.1016/j.jpsychires.2018.09.014

Veru-Lesmes, F., Guay, S., Shah, J. L., Schmitz, N., Giguère, C.-É., Joober, R., Iyer, S. N., & Malla, A. K. (2021). Adipose tissue dysregulation at the onset of psychosis: Adipokines and social determinants of health. *Psychoneuroendocrinology, 123*, Article 104915. https://doi.org/10.1016/j.psyneuen.2020.104915

Whybrow, P. C. (1969). Mental changes accompanying thyroid gland dysfunction. *Archives of General Psychiatry, 20*(1), 48. https://doi.org/10.1001/archpsyc.1969.01740130050004

Chapter 14

Thyroid cancers and cognition

Nazan Dolu
Department of Physiology, School of Medicine, University of Istanbul Medipol, İstanbul, Türkiye

Abbreviations

CNS Central nervous system
DTC Differentiated thyroid cancer
ESS Euthyroid sick syndrome
RAI Radioactive Iodine
ST Suppression therapy

Introduction

Thyroid hormones influence the physiological processes of many organ systems, including basal metabolic rate, cardiac output, respiratory rate and depth, bone growth, and central nervous system maturation. Thyroid hormones also play important roles in regulating cognitive functions. Cognitive functions include mental functions such as attention, memory, perception, and intelligence (Jung et al., 2022). Impairments in working memory and executive functions have been demonstrated in thyroid dysfunctions such as Graves' thyrotoxicosis (Riguetto et al., 2019), experimentally induced subclinical hypothyroidism (Correia et al., 2009), overt hyperthyroidism (George et al., 2019), and primary hypothyroidism treated with L-T4 replacement therapy (Kramer et al., 2009). Depression observed in thyroid dysfunction has also been shown to be related to deficits in working memory and executive functions (Vogel et al., 2007).

Cancer-related cognitive impairments are a complex neurocognitive condition. The methods employed during cancer diagnosis and treatment cause impairments in many cognitive functions (Zheng et al., 2024). Factors such as fluctuations in thyroid hormone levels, fear of recurrence, and social isolation during treatment can place a significant burden on cognitive performance.

This chapter focuses on changes in cognitive function in thyroid cancer from a pretreatment, posttreatment (thyroidectomy, iodine, and medication), and survival perspective.

Types of thyroid carcinoma

The incidence of thyroid cancer has also increased with the use of imaging studies and fine-needle aspiration biopsies. The 5-year relative survival rate, based on the weighted average of all disease stages, is 98.5%.

Three types of thyroid carcinoma are recognized: differentiated, undifferentiated, and medullary thyroid carcinoma.

Differentiated thyroid carcinoma (DTC, originating from epithelial follicular cells) is further classified as papillary (84%), follicular (4%), and oncocytic (2%). Approximately 5% of papillary carcinomas are aggressive. Follicular carcinomas are highly differentiated, while oncocytic carcinomas are poorly differentiated.

Anaplastic thyroid carcinoma, which originates from epithelial follicular cells (1%), is a subtype of undifferentiated thyroid carcinoma and is more aggressive.

Medullary thyroid carcinoma (4%) originates from parafollicular C cells (Boucai et al., 2024).

Methods used in the diagnosis of thyroid cancers

Thyroid nodules can be diagnosed incidentally during physical examination or on diagnostic imaging studies. Patients with thyroid cancer are generally asymptomatic. Recent guidelines indicate that 7%–15% of thyroid nodules may be malignant. Malignancy is determined based on ultrasound and cytology findings. American Thyroid Association guidelines recommend not biopsying nodules smaller than 1 cm (Boucai et al., 2024; Haugen et al., 2016). Thyroid ultrasound helps determine when fine-needle aspiration is necessary. Sonographic features such as nodule size, hypoechogenicity, irregular margins, increased vascularity, and the presence of microcalcifications are specific in detecting malignancy (Kim et al., 2023).

Hormones in Neuroscience. https://doi.org/10.1016/B978-0-443-30261-9.00014-0

According to the Bethesda classification, cytology for benign (Bethesda II) or malignant (Bethesda V or VI) does not require molecular testing. In cases of indeterminate cytology, molecular testing can help with diagnosis (Ali et al., 2023).

Methods used in thyroid cancer treatment

Active Surveillance: Ultrasound monitoring is recommended instead of biopsy for thyroid nodules smaller than 1.5 cm and thyroid nodules smaller than 1 cm that are not suspicious for malignancy based on ultrasound findings (Haugen et al., 2016).

Surgery: Generally, the risk of recurrence is low in DTCs, and surgical intervention is sufficient (Haugen et al., 2016).

A neck ultrasound should be performed before surgery to evaluate the thyroid and adjacent neck regions. Thyroid lobectomy is generally performed for unilateral, smaller than 4 cm thyroid cancers (Boucai et al., 2024). Total thyroidectomy is recommended for bilateral DTCs larger than 4 cm (Haugen et al., 2016).

Clinical Staging and Risk Classification: The American Joint Committee on Cancer (AJCC) Tumor, Lymph Node, and Metastasis (TNM) staging system assesses disease according to patient age, tumor size, the presence of lymph node metastases, and distant metastases.

Immediately after surgery and approximately 6 months later, staging is performed according to the American Thyroid Association (ATA) (Haugen et al., 2016) and the European Thyroid Association (ETA) risk classification systems (Pacini et al., 2006). Treatment and surveillance frequency are determined based on histopathological features and certain molecular markers (Boucai et al., 2024; Haugen et al., 2016).

Adjuvant Therapy in Thyroid Cancer: In patients with DTC, postoperative radioactive iodine (RAI) is administered after total thyroidectomy. RAI aims to reduce recurrence, or the treatment of patients at high risk of progression (Haugen et al., 2016).

RAI therapy has been shown not to reduce the 3-year recurrence rate in low-risk thyroid cancer cases (Leboulleux et al., 2022). Therefore, it is recommended for high-risk patients. It can also be used in some patients at risk of recurrence (Haugen et al., 2016; Lamartina et al., 2015). The follicular papillary thyroid carcinoma responds to RAI doses. Thyroid cancers with BRAF variants develop resistance to RAI therapy (Chakravarty et al., 2011).

TSH suppression therapy (ST) with supraphysiological doses of levothyroxine is administered to prevent the growth of thyroid cancer cells and thyroglobulin production. In patients with high-risk thyroid cancer, suppressing TSH below 0.1 mU/L improves overall survival, but it is not beneficial in patients with intermediate or low risk of recurrence (Gubbi et al., 2024). ST is not recommended in medullary thyroid cancer because parafollicular C cells do not express the TSH receptor (Boucai et al., 2024).

Advanced Cancer Treatment: In the treatment of advanced thyroid cancer, active treatment options were not available when the patient did not respond to RAI. The US Food and Drug Administration (FDA) has approved systemic treatment options for advanced thyroid cancer patients. Lenvatinib and sorafenib (a tyrosine kinase inhibitor) are traditional first-line agents for the systemic treatment of metastatic DTC resistant to RAI therapy. Cabozantinib (a multitargeted tyrosine kinase inhibitor) may be an alternative or advanced treatment option. Targeted therapies such as selpercatinib and pralsetinib are effective FDA-approved options, particularly in patients with RET mutations/fusions (Chen et al., 2023).

Thyroid cancers and cognition

Numerous studies have demonstrated that thyroid gland dysfunction leads to cognitive deficits by affecting attention, memory, and executive functions (Gan & Pearce, 2012; Samuels, 2014). Cognitive functions in hypothyroid, hyperthyroid, and euthyroid thyroid patients were compared (Botella-Carretero et al., 2003; Cordes et al., 2020; Jaracz et al., 2012). Cognitive dysfunction has also been demonstrated in thyroid cancer patients (Botella-Carretero et al., 2003; Cordes et al., 2020; Jaracz et al., 2012; Moon et al., 2014; Samuels et al., 2014). Research is ongoing to determine whether this impairment stems from hormonal imbalances, the psychological stress of cancer, fear of recurrence, social isolation during treatment, or the difficulties or side effects of the treatment process.

Cognitive functions between thyroid cancer patients and healthy individuals have often been examined by comparing neuropsychological tests and self-report questionnaire scores. These studies have demonstrated changes in general cognition (Moon et al., 2014), attention and executive functions (Botella-Carretero et al., 2003; Jaracz et al., 2012; Jung & Visovatti, 2017), working memory (Jung & Visovatti, 2017; Samuels et al., 2014), and visual memory (Cordes et al., 2020).

Cognition before thyroid cancer treatment

The prevalence of cognitive deficits in patients newly diagnosed with breast, colorectal, and testicular cancer ranges from 13% to 46% (Wefel et al., 2010; Vardy et al.,

2015; Wefel et al., 2011). In patients diagnosed with non-CNS (Central nervous system) cancer, cognitive functions have often been investigated postoperatively and before postoperative adjuvant therapy (Vardy et al., 2015; Wefel et al., 2010). Neuroimaging data suggest structural changes in brain regions that regulate cognitive functions, particularly the temporal and frontal lobes, in patients diagnosed with non-CNS cancer before chemotherapy (Shiroishi et al., 2017). Studies investigating cognitive functions before thyroid cancer treatment are quite limited.

78.2% of women newly diagnosed with thyroid cancer were found to have low-level cognitive impairment before and after any treatment, according to both the Global Deficit Score (GDS) and the International Cognition and Cancer Task Force (ICCTF) criteria (Jung et al., 2022).

Some factors, mediators, or mechanisms related to pretreatment cognitive dysfunction in patients newly diagnosed with thyroid cancer have been suggested. Advanced age and fewer years of education have been shown to reduce attention and cognitive control functions (Jung et al., 2022). The type of cancer, tumor diameter, the presence of extraglandular infiltration (Jia et al., 2025), length of hospital stay (Larrabee et al., 2023), and thyroid hormone levels will affect prognosis, as well as psychological status and cognitive functions. It has been shown that peripheral inflammatory mediators (cytokines, immune cells) originating from the tumor microenvironment can affect cognitive processes in the brain (Low et al., 2015; Ni et al., 2025).

Cognitive dysfunction in patients with thyroid cancer may result from the effects of cancer itself or from preexisting psychological stress. With a new cancer diagnosis, patients may experience mental fatigue due to the challenges of understanding the cancer diagnosis, managing the emotional distress associated with it, making treatment decisions, keeping up with medical appointments, figuring out how to adjust their work/life schedules for treatment, learning new information about their illness, and coping with their illness themselves (Ferrante et al., 2022; Jung et al., 2022).

While awaiting their first surgical treatment, 26%–34% of cancer patients experienced depressive mood (Foster et al., 2016; Jung et al., 2022). Studies have shown a relationship between depressive mood and cognitive impairment in thyroid cancer patients (Jung et al., 2022). The immune system of individuals reporting depressed mood is affected by its response to cancer and the psychological stress associated with the diagnosis. Similar to changes in cognitive function, depressed mood can also result from cytokine-mediated symptoms and behavioral changes (Low et al., 2015). The impact of the tumor and tumor burden on the inflammatory response and cytokine-mediated symptoms influences the degree to which depression or cognitive dysfunction develops (Jung et al., 2022).

The relationship between thyroid dysfunction, such as hypothyroidism, hyperthyroidism, or thyroiditis, and thyroid cancer is controversial. Hypothyroidism (Kitahara et al., 2020), hyperthyroidism (Kitahara et al., 2020), and autoimmune thyroiditis (Song et al., 2021) may also increase the risk of developing thyroid cancer. Thyroid hormone and TSH can stimulate the proliferation and growth of thyroid cancer cells (Kimura et al., 2001; Song et al., 2021). Frequent thyroid ultrasound examinations may increase the risk of proliferation of even benign thyroid cancers. Based on all these findings, cognitive impairments that occur before thyroid cancer treatment may be due to thyroid dysfunction (Yang et al., 2022).

The thyroidectomy effect on cognitive functions

The mechanisms underlying cognitive impairments associated with thyroidectomy treatment can be examined as prethyroidectomy, perithyroidectomy, and postthyroidectomy.

Prethyroidectomy (Preoperative) Period: Before surgery, cognitive disorders may be observed due to fear of surgery (Yang et al., 2022).

Perithyroidectomy (Perioperative) Period: Thyroid hormone dysregulation can disrupt cognitive circuits and contribute to postoperative cognitive changes. Thyroid hormone dysregulation may result from direct thyroid hormone changes due to surgery or from the physiological stress associated with surgery. It is hypothesized that thyroid hormones may influence central cholinergic function and amyloid precursor regulation, and that perioperative hormonal changes may affect postoperative cognition (Mafrica & Fodale, 2008).

Postthyroidectomy (Postoperative) Period: When cognitive performance was examined in women with papillary thyroid cancer after thyroidectomy, they performed worse on tests of attention and working memory than individuals without thyroid cancer (Jung & Visovatti, 2017). Early postoperative neurocognitive changes, detected both clinically and by neuroimaging, have been observed in the postoperative period following thyroidectomy. In patients undergoing thyroidectomy under general anesthesia, the observed decrease in functional networks in the supplementary motor area on fMRI 7 days postoperatively was found to be consistent with errors in the Warrington Recognition Memory Test (WRMT) (Yang et al., 2022).

After surgery for solid tumors, anesthesia-induced neurocognitive deficits are common (regardless of tumor location) (Stern et al., 2023). These cognitive changes are often transient and reversible. They may be related to the synergistic effect or neurotoxicity caused by general anesthetic drugs (Mafrica & Fodale, 2008).

Surgical stress and postoperative pain elicit neuroendocrine and tissue responses aimed at defending against injury. These responses are described as a "hormonal storm" caused by the release of inflammatory mediators from the sympathetic nervous system, sympathetic-adrenal system, hypothalamic-pituitary-adrenal system, and the surgical site (Mafrica & Fodale, 2008). Postoperatively, low triiodothyronine, total thyroxine, and TSH concentrations are observed not only in thyroid surgeries but also in many nonthyroidal surgeries. Furthermore, IL-6 levels increase postoperatively (Neff et al., 2022). After major surgery, inflammatory stress (especially increased IL-6) occurs as a result of neuroendocrine system activation. This, in turn, reduces thyroid hormone concentrations, creating surgical stress. Euthyroid sick syndrome (ESS), characterized by suppressed plasma T3, increased rT3, and decreased T4 without increasing TSH levels, is also common after surgery. ESS can occur a few hours after major surgery and then recover. Cognitive impairments can be observed during this period. Furthermore, the conversion of T4 to T3 in the liver, which metabolizes anesthetic drugs, may be reduced, leading to ESS and impairing cognitive function (Mafrica & Fodale, 2008).

Postoperative decreases in thyroid hormone levels may reduce the effect of thyroid hormone in inhibiting the transcription of the gene encoding AβPP. Increased AβPP expression leads to cognitive impairment. Thyroid hormone or its analogs have been shown to reduce AβPP gene expression and amyloid-β peptide accumulation (O'Barr et al., 2006).

Persistent or transient hypoparathyroidism and associated hypocalcemia after thyroid surgery impact quality of life and are a possible contributor to cognitive symptoms; guideline-based prevention and management are recommended to reduce morbidity (Reinke et al., 2023).

RAI treatment effect on cognitive functions

RAI therapy is frequently administered to DTC after thyroidectomy to reduce the risk of cancer recurrence. A low-iodine diet is typically implemented before RAI treatment, and levothyroxine therapy is discontinued before RAI treatment to achieve a TSH level of 30 μIU/mL or higher. RAI treatment is generally safe. However, RAI treatment can cause side effects such as depression and anxiety, nausea and vomiting, transient neck pain and edema, and nasolacrimal obstruction, as well as changes in cognitive function (Ming et al., 2022). Clinical symptoms of hypothyroidism may also accompany these effects.

The findings of studies on various cognitive domains, such as memory, attention, and executive functions, following radioactive iodine (RAI) treatment in thyroid cancer patients, are complex and present a multifaceted picture. Differences in patient populations, methodological variations in cognitive function assessment methods, and follow-up durations make it difficult to synthesize findings. Conclusions are limited by the limited number of studies conducted on specific types of thyroid cancer and the paucity of long-term follow-up. Thus, the specific effects of RAI therapy on various cognitive domains are not yet fully understood. RAI therapy may contribute to cognitive deterioration through mechanisms involving neurotoxic effects, interacting with subclinical hyperthyroidism, TSH suppression, hypothyroidism, and other systemic factors (Jaracz et al., 2012; Ming et al., 2022).

The effect of RAI alone on cognitive outcomes in RAI-treated patients could not be assessed because other treatments were also received concomitantly, and relative cognitive risks were reported (Saeed et al., 2019). This limits conclusions about the specificity of RAI-associated cognitive impairments (Ming et al., 2022; Saeed et al., 2019).

Neuropsychological assessment utilizes a variety of standardized objective tests targeting attention, working memory, and executive functions, and is often supplemented by mood and quality-of-life questionnaires. This multifaceted assessment approach highlights the complex interaction between cognitive symptoms and psychological factors such as anxiety, depression, and fatigue, which themselves contribute to or influence cognitive outcomes. However, separating the specific contributions of RAI from associated treatment modalities and psychological distress remains challenging because mood disorders and subclinical hyperthyroidism frequently co-occur and impact cognitive function (Changchien et al., 2023).

TSH suppression therapy (ST) effect on cognitive functions

Thyroid hormone replacement therapy aims to alleviate the symptoms of hypothyroidism and normalize circulating TSH. Supraphysiological thyroid hormone replacement induces hypothalamic and pituitary feedback inhibition, suppressing TSH and, consequently, thyroid cancer proliferation (Gigliotti & Jasim, 2024).

TSH suppression may be beneficial in tumors at high risk of recurrence or distant metastasis. However, it has been reported to be ineffective in low- and intermediate-risk tumors (Leboulleux et al., 2022). International guidelines, based on the risk of recurrence of DTC, recommend a TSH suppression level of 0.1–0.5 mIU/L for intermediate-risk DTC and <0.1 mIU/L for high-risk DTC (Haugen et al., 2016).

The long-term risks associated with TSH suppression in DTC are similar to those in patients with subclinical or overt hyperthyroidism. These include cardiac arrhythmias (atrial tachycardia and atrial fibrillation), bone density loss, increased fracture risk, impaired cognitive function and quality of life, and increased risk of other cancers (Biondi & Cooper, 2019; Gigliotti & Jasim, 2024; Samuels et al., 2014). These risks may vary depending on age, the dose of TSH suppression, and the duration of suppressive therapy.

Findings regarding the effects of TSH-ST on cognitive function are conflicting. These discrepancies may arise from its use in patients with thyroidectomy and/or RAI therapy, whether the patient is in remission, the type of cancer, differences in cognitive tests administered, the thyroxine dose, the duration of TSH-ST, and age and gender.

TSH Suppression Therapy and Its Beneficial Effect on Cognitive Functions: The duration of hypothyroidism and the patient's age are important factors in the impact of TSH-ST on cognitive function. In the effect of TSH suppression or LT4 therapy on cognitive function in thyroid cancer, short-term hypothyroidism may not be sufficient to cause irreversible damage to the brain's memory region, and LT4 therapy may improve cognitive function. Furthermore, thyroxine's increased cholinergic activity in the frontal cortex and hippocampus on cognitive function may also potentiate this effect (Smith et al., 2002).

In DTC patients, cognitive function was assessed with the Montreal Cognitive Assessment-Philippines (MoCA-P) before levothyroxine (LT4) discontinuation, during hypothyroidism, and after LT4 therapy. A decrease in MoCA-P score from baseline to hypothyroidism was observed, with improvement observed after LT4 therapy. Visuospatial and attention functions, which decline with hypothyroidism, improved after LT4 treatment. It has been shown that exposure to hypothyroidism for as short as 4 weeks can lead to a significant decrease in general cognitive functions (Israel et al., 2023). It has been reported that cognitive decline in attention, memory, and visuospatial domains in hypothyroidism can be improved after 6 months of LT4 treatment (Capet et al., 2000).

In elderly DTC patients, long-term TSH suppression (at least 5 years) was associated with higher serum free T4 (fT4) levels compared to the control group. These patients also performed better on the Mini-Mental State Examination (MMSE) and Trail Making Test A, which examines cognitive functions. It has been reported that mild thyroid hormone excess produced by exogenous levothyroxine administration may be beneficial in some cognitive domains if patients have endogenous thyroid hormone deficiency, such as severe overt hypothyroidism or thyroidectomy (Moon et al., 2014). It has been reported that administration of L-T4 replacement therapy together with T3 hormone after total thyroidectomy may be more effective than administration alone in improving cognitive functions (Bononi et al., 2010).

TSH Suppression Therapy and No Effect on Cognitive Functions: Studies report that TSH-ST does not negatively impact cognitive function. In women (mean age 35.8) with low-risk DTC who were in remission and underwent total thyroidectomy and received RAI therapy and levothyroxine TSH-ST, no impairments were observed in memory and executive functions compared to healthy controls when performed on tests such as declarative memory, paragraph recall, working memory, letter cancellation test, trail making test, and Iowa Gambling Test (Samuels et al., 2014). In patients with DTC who underwent postoperative TSH suppression, no impairments were observed in general cognitive function (Mini-Mental State test), visuospatial function, long-term memory (Wechsler Memory Scale), information processing speed (Digit Symbol Substitution Test), or working memory at 6 and 12 months. However, short-term memory impairment (Wechsler Memory Scale), decreased attention, and word selection anomia (using the Aphasia Battery of Chinese Test) have been reported in these patients (Jin et al., 2019). Furthermore, long-term TSH suppression is reported to have a safe effect on cognitive functions in elderly DTC patients (aged 65 and over) (Moon et al., 2014).

TSH Suppression Therapy and Its Reducing Effect on Cognitive Functions: The majority of studies report the deterioration of cognitive functions with TSH-ST. A study of 495 elderly patients aged 65 and older found a significant association between decreased serum TSH levels and cognitive impairment. Inhibition of serum TSH levels can also lead to cognitive impairment and emotional distress in normal individuals (Kim et al., 2010).

It has been suggested that the decline in cognitive functions due to low TSH levels may be due to various mechanisms. Iatrogenic subclinical hyperthyroidism is generally blamed for cognitive impairment. Neurodegenerative changes in the brain resulting from the toxic effects of excess thyroid hormone on the CNS have been reported to affect TRH and TSH secretion. It has been suggested that a decrease in TRH, which also acts as a neurotransmitter in the CNS, may be responsible for cognitive impairment (Gan & Pearce, 2012). Furthermore, oxidative stress in the brain due to thromboembolism resulting from excess thyroid hormone or the cardiac effects of mild hyperthyroidism can affect cognitive centers (Moon et al., 2014). In addition to changes in thyroid hormone and other hormones, psychological disorders, increased disease symptoms, low education level, or other health problems may contribute to cognitive dysfunction (Jung & Visovatti, 2017; Ni et al., 2025).

When neurocognitive tests were performed in patients with thyroid cancer during levothyroxine (LT4) treatment and 5–6 weeks after LT4 treatment was discontinued, a decrease in attentional perception and memory was found after LT4 treatment was discontinued (Burmeister et al.,

2001). When cognitive functions were examined with the Wais Digit Span test in 18 female patients with thyroid carcinoma the day before levothyroxine discontinuation (subclinical hyperthyroidism), 4–7 days after (normal serum T3 and T4), and approximately 30 days after LT4 discontinuation (in case of hypothyroidism), Wais Digit Span test results were impaired on the 4–7 day measurements. Memory, attention, and problem-solving worsened further in the hypothyroid state. It was concluded that both excess and deficiency of thyroid hormones may negatively affect cognitive performance in patients with DTC (Botella-Carretero et al., 2003). In DTC patients treated with RAI and TSH suppression after thyroid surgery, executive functions and attention levels were found to be reduced when the Wisconsin Card Sorting Test, Trail Making Test, Stroop, and Digit Span Test were applied (Jaracz et al., 2012). During LT4 replacement in DTC patients, decreased attention and working memory were found according to the Digit Span Test, Controlled Oral Word Association Test, and Attentional Function Index (AFI) test results (Jung & Visovatti, 2017). Short-term memory impairment (Wechsler Memory Scale), decreased attention, and word selection anomia (using the Aphasia Battery of Chinese Test) were detected at 6 and 12 months during TSH suppression in DTC thyroidectomy patients (Jin et al., 2019). In addition to TSH-ST, acute hypothyroidism following 1 month of L-thyroxine withdrawal in patients with DTC has been shown to result in decreased cognitive function, with a prolongation of auditory evoked potential P300 wave latency (Ünlühizarcı et al., 2024).

TSH-ST and LT4 administration not only lead to a decrease in cognitive function but also lead to high medical costs, poor quality of life (QoL), and depression. Therefore, improving cognitive function, depression, and anxiety is crucial for improving patients' quality of life (Hedman et al., 2018).

Systemic chemotherapy

Because chemotherapy drugs are cytotoxic drugs that affect cancer cells and even normal cells, a decline in various neuropsychological measurements is observed in cancer patients after chemotherapy treatment. Decreases in attention and memory negatively impact patients' quality of life and work skills.

Cognitive impairment can begin as early as 1 week after chemotherapy (Megari, 2021). This impairment is generally mild and transient, gradually decreasing within 6–12 months after treatment (Khan et al., 2019). Cognitive impairment following chemotherapy may be due to alterations in the normal course of cognitive aging, acceleration of the aging process, increased oxidative stress and inflammation, decreased cerebral vascularity, changes in brain structural networks and neurogenesis, and increased catecholamine production (Fardel et al., 2011). The duration and severity of postchemotherapy cognitive impairment may vary depending on the patient's advanced age, previous cognitive reserve, the dose and duration of chemotherapy, and the presence of anxiety, fatigue, and depression. Thyroid cancer patients who have not received any treatment other than chemotherapy have been reported to perform worse immediately after chemotherapy compared to prechemotherapy, with performance stabilizing 6 months after chemotherapy (Megari, 2021).

In conclusion, this chapter evaluates cognitive functions in patients with thyroid cancer before, during, and after treatment. Cognitive effects can be influenced by many factors, including the patient's age, gender, level of education, thyroid hormone and neurotransmitter changes, and the dose and duration of treatment. However, T4 replacement therapy, in particular, has been reported to support cognitive recovery. As treatment duration increases, patients' understanding of the disease and psychological adjustment improve, and depression and anxiety levels decrease. This demonstrates that patients adhere to treatment in the long term. The treatment process in patients with thyroid cancer should be monitored not only from biological but also from neuropsychological perspectives.

Other hormones and pathways affecting cognition in thyroid cancer

Other hormones and pathways that affect cognitive functions in patients with thyroid cancer include the hypothalamic-pituitary-adrenal axis, cortisol, and estrogen/testosterone. During the thyroid cancer diagnosis and treatment process (surgical stress, postoperative pain), the sympathetic nervous, sympatho-adrenomedullary, and hypothalamic-pituitary-adrenal systems are activated. Increased release of corticotropin-releasing hormone and ACTH from the pituitary gland increases the release of cortisol from the adrenal cortex (Mafrica & Fodale, 2008). Cortisol and catecholamines are important in stress adaptation. However, high and prolonged stress levels negatively impact mental health. Chronic stress appears to increase the risk of developing dementia, and individuals experiencing cognitive decline in older age are exposed to chronic stressful events (Stuart & Padgett, 2020).

Estrogen plays a crucial role in maintaining cognitive function by enhancing neuronal connections and promoting synaptic plasticity. Postmenopausal women with papillary thyroid carcinoma may experience declines in cognitive functions such as learning and memory. Postmenopausal women with papillary thyroid carcinoma exhibited delayed recall knowledge and decreased general cognition compared to premenopausal women. This finding further demonstrates the role of estrogen in supporting cognitive function and the decrease in this protective effect with menopause (Zheng et al., 2024). Furthermore, estrogen reduces circulating free thyroid hormone levels by increasing thyroxine-binding globulin (TBG) (Brown et al., 2023). This, in turn, may affect thyroid function, predisposing to cognitive disorders. Thus, it appears that estrogen may exert cognitive effects directly or through thyroid hormones. Testosterone also affects cognitive functions; cognitive performance decreases when testosterone levels are low (Jevtovic-Todorovic, 2022).

In translational terms, these findings suggest that cognitive dysfunction in thyroid cancer arises from a multihormonal imbalance rather than isolated thyroid deficiency. Integrative therapeutic strategies may offer novel avenues for preserving cognitive health during and after thyroid cancer treatment.

Applications of the material

This section reviews potential cognitive impairments in thyroid cancer, both pretreatment and posttreatment. The information presented here can be used by clinicians and applied in a variety of research settings. Clinicians can understand the importance of diagnosing cognitive impairments before treatment or early in the treatment process. Preoperative psychological assessment and counseling can help reduce anxiety-related cognitive decline before thyroidectomy (Yang et al., 2022). Preventing hormonal fluctuations and minimizing oxidative stress during surgery may also reduce the risk of postoperative cognitive dysfunction (Neff et al., 2022). During radioactive iodine (RAI) treatment, clinicians should consider the potential neurotoxic and hormonal effects on cognitive function, especially in patients with depressive symptoms and low thyroid hormone levels. In these patients, neurocognitive testing can facilitate early detection of treatment-related cognitive changes (Jung & Visovatti, 2017; Ming et al., 2022). Including patient monitoring parameters such as cytokine profiles, oxidative stress markers, and thyroid hormone levels can aid in the diagnosis of cognitive changes (Jia et al., 2025; Larrabee et al., 2023). The combined application of electrophysiological techniques such as electroencephalography (EEG), event-related potentials (ERPs), and functional near-infrared spectroscopy (fNIRS) can provide new insights into thyroid-related cognitive impairment and serve as biomarkers (Ünlühizarcı et al., 2024). Integrating thyroid cancer treatment methods with behavioral and cognitive rehabilitation programs can enhance recovery of attention, memory, and executive functions during thyroid cancer treatment and after treatment. These rehabilitation programs will also be beneficial in enhancing cognitive function and alleviating fatigue in cancer survivors (Megari, 2021; Fardel et al., 2011). In this context, not ignoring cognitive dysfunction in thyroid cancer is important for both clinical treatment and translational neuroscience research.

Mini-dictionary of terms

- **Aphasia.** It is a language disorder.
- **Recovery:** Improvement
- **Amyloid.** It is an insoluble protein that is not normally found in the body. It causes Alzheimer's.
- **Working memory.** The part of short-term memory.
- **Atrial fibrillation.** It is an irregular heart rhythm that begins in the atria.

Summary points

- Cognitive dysfunction may occur before treatment for thyroid cancer, after thyroidectomy, RAI therapy, TSH suppression therapy, or systemic chemotherapy.
- Pretreatment cognitive dysfunction may vary depending on the type and extent of cancer spread, immune system activity, and the patient's age.
- Cognitive functions are impaired due to fear of surgery before thyroidectomy and anesthesia afterward.
- RAI treatment is not used as a stand-alone; its effects on cognitive functions have not been fully measured.
- TSH suppression therapy impairs, has no effect, or improves cognitive functions.
- Systemic chemotherapy impairs cognitive functions through oxidative stress or neurotoxic effects.

Acknowledgement

During the preparation of this work, the author used ChatGPT 5 from OpenAI to improve readability. Under no circumstances has it been used to draw insights from data, conclude, or produce recommendations. After using this service, the authors reviewed and edited the content as needed and take full responsibility for the content of the publication.

References

Ali, S. Z., Baloch, Z. W., Cochand-Priollet, B., Schmitt, F. C., Vielh, P., & VanderLaan, P. A. (2023). The 2023 bethesda system for reporting thyroid cytopathology. *Thyroid, 33*(9), 1039–1044.

Biondi, B., & Cooper, D. S. (2019). Thyroid hormone suppression therapy. *Endocrinology and Metabolism Clinics of North America, 48*(1), 227–237.

Bononi, M., De Toma, G., Scarpini, M., Miccini, M., De Cesare, A., Meucci, M., et al. (2010). Hormone replacement therapy after total thyroidectomy. Can the combined treatment be considered effective to get metabolic adequacy? Preliminary results. *Il Giornale di Chirurgia, 31*(6–7), 303–307.

Botella-Carretero, J. I., Gal, J. M., Caballero, C., Sancho, J., & Escobar-Morreale, H. F. (2003). Quality of life and psychometric functionality in patients with differentiated thyroid carcinoma. *Endocrine-Related Cancer, 10*(4), 601–610.

Boucai, L., Zafereo, M., & Cabanillas, M. E. (2024). Thyroid cancer: A review. *JAMA, 331*(5), 425–435.

Brown, E. D., Obeng-Gyasi, B., Hall, J. E., & Shekhar, S. (2023). The thyroid hormone axis and female reproduction. *International Journal of Molecular Sciences, 24*(12), 9815.

Burmeister, L. A., Ganguli, M., Dodge, H. H., Toczek, T., Dekosky, S. T., & Nebes, R. D. (2001). Hypothyroidism and cognition: Preliminary evidence for a specific defect in memory. *Thyroid, 11*(12), 1177–1185.

Capet, C., Jego, A., Denis, P., Noel, D., Clerc, I., Cornier, A. C., et al. (2000). Is cognitive change related to hypothyroidism reversible with replacement therapy? *La Revue de Médecine Interne, 21*(8), 672–678.

Chakravarty, D., Santos, E., Ryder, M., Knauf, J. A., Liao, X., West, B. L., et al. (2011). Small-molecule MAPK inhibitors restore radioiodine incorporation in mouse thyroid cancers with conditional BRAF activation. *The Journal of Clinical Investigation, 121*(12), 4700–4711.

Changchien, T. C., Yen, Y. C., & Lu, Y. C. (2023). Mental health and quality of life of patients with differentiated thyroid cancer prior to and after radioactive iodine treatment: A prospective study. https://doi.org/10.21203/rs.3.rs-3095990/v1.

Chen, D. W., Lang, B. H., McLeod, D. S., Newbold, K., & Haymart, M. R. (2023). Thyroid cancer. *The Lancet, 401*(10387), 1531–1544.

Cordes, J., Woite, M., Engelke, C., Regenbrecht, G., Kahl, K. G., Schmidt-Kraepelin, C., et al. (2020). Hormone replacement therapy with L-thyroxine promotes working memory and concentration in thyroidectomized female patients after differentiated thyroid carcinoma. *The International Journal of Psychiatry in Medicine, 55*(2), 114–122.

Correia, N., Mullally, S., Cooke, G., Tun, T. K., Phelan, N., Feeney, J., et al. (2009). Evidence for a specific defect in hippocampal memory in overt and subclinical hypothyroidism. *The Journal of Clinical Endocrinology & Metabolism, 94*(10), 3789–3797.

Fardell, J. E., Vardy, J., Johnston, I. N., & Winocur, G. (2011). Chemotherapy and cognitive impairment: Treatment options. *Clinical Pharmacology and Therapeutics, 9*, 366–376.

Ferrante, M., Distefano, G., Distefano, C., Copat, C., Grasso, A., Oliveri Conti, G., et al. (2022). Benefits of physical activity during and after thyroid cancer treatment on fatigue and quality of life: A systematic review. *Cancers, 14*(15), 3657.

Foster, C., Haviland, J., Winter, J., Grimmett, C., Chivers Seymour, K., Batehup, L., et al. (2016). Pre-surgery depression and confidence to manage problems predict recovery trajectories of health and well-being in the first two years following colorectal cancer: Results from the CREW cohort study. *PLoS One, 11*(5), Article e0155434.

Gan, E. H., & Pearce, S. H. (2012). Clinical review: The thyroid in mind: Cognitive function and low thyrotropin in older people. *The Journal of Clinical Endocrinology & Metabolism, 97*(10), 3438–3449.

George, K. M., Lutsey, P. L., Selvin, E., Palta, P., Windham, B. G., & Folsom, A. R. (2019). Association between thyroid dysfunction and incident dementia in the atherosclerosis risk in communities neurocognitive study. *Journal of Endocrinology and Metabolism, 9*(4), 82.

Gigliotti, B. J., & Jasim, S. (2024). Differentiated thyroid cancer: A focus on post-operative thyroid hormone replacement and thyrotropin suppression therapy. *Endocrine, 83*(2), 251–258.

Gubbi, S., Al-Jundi, M., Foerster, P., Cardenas, S., Butera, G., Auh, S., Wright, E. C., & Klubo-Gwiezdzinska, J. (2024). The effect of thyrotropin suppression on survival outcomes in patients with differentiated thyroid cancer: A systematic review and meta-analysis. *Thyroid, 34*(6), 674–686.

Haugen, B. R., Alexander, E. K., Bible, K. C., Doherty, G. M., Mandel, S. J., Nikiforov, Y. E., et al. (2016). American Thyroid Association management guidelines for adult patients with thyroid nodules and differentiated thyroid cancer: The American Thyroid Association guidelines task force on thyroid nodules and differentiated thyroid cancer. *Thyroid, 2*, 1–133.

Hedman, C., Djärv, T., Strang, P., & Lundgren, C. I. (2018). Fear of recurrence and view of life affect health-related quality of life in patients with differentiated thyroid carcinoma: A prospective Swedish population-based study. *Thyroid, 28*(12), 1609–1617.

Israel, S. C., Cunanan, E. C., Galan, J. L. A., & Ampil, E. R. (2023). Evaluating cognitive function in patients with differentiated thyroid cancer who experienced shortterm hypothyroidism by levothyroxine withdrawal using Montreal cognitive assessment-Philippines (MoCA-P) tool. *Neurology Asia, 28*(1).

Jaracz, J., Kucharska, A., Rajewska-Rager, A., & Łaącka, K. (2012). Cognitive functions and mood during chronic thyrotropin-suppressive therapy with L-thyroxine in patients with differentiated thyroid carcinoma. *Journal of endocrinological investigation, 35*(8), 760–765.

Jevtovic-Todorovic, V. (2022). Testosterone: Much more for the brain than a sex hormone. *British Journal of Anaesthesia, 128*(3), 411–413.

Jia, J. L., Han, J. H., Pang, R., Bi, W., Liu, B., & Yang, K. (2025). Predictors of poor prognosis in long-term survivors of differentiated thyroid cancer with psychiatric disorders. *World Journal of Psychiatry, 15*(5), Article 103628.

Jin, S., Yang, Y. T., Bao, W., Bai, Y., Ai, J. W., Liu, Y., et al. (2019). Naming difficulties after thyroid-stimulating hormone suppression therapy in patients with differentiated thyroid carcinoma: A prospective cohort study. *Endocrine, 65*(2), 327–337.

Jung, M. S., & Visovatti, M. (2017). Post-treatment cognitive dysfunction in women treated with thyroidectomy for papillary thyroid carcinoma. *Supportive Care in Cancer, 25*(3), 915–923.

Jung, M. S., Visovatti, M., Kim, M., Cha, K., Dlamini, N., & Cui, X. (2022). Cognitive impairment in women newly diagnosed with thyroid cancer before treatment. *Supportive Care in Cancer, 30*(11), 8959–8967.

Khan, O. F., Cusano, E., Raissouni, S., Pabia, M., Haeseker, J., Bosma, N., et al. (2019). Immediate-term cognitive impairment following intravenous (IV) chemotherapy: A prospective pre-post design study. *BMC Cancer, 19*(1), 150.

Kim, D. H., Kim, S. W., Basurrah, M. A., Lee, J., & Hwang, S. H. (2023). Diagnostic performance of six ultrasound risk stratification systems for thyroid nodules: A systematic review and network meta-analysis. *American Journal of Roentgenology, 220*(6), 791–803.

Kim, J. M., Stewart, R., Kim, S. Y., Bae, K. Y., Yang, S. J., Kim, S. W., et al. (2010). Thyroid stimulating hormone, cognitive impairment and depression in an older Korean population. *Psychiatry investigation, 7*(4), 264.

Kimura, T., Van Keymeulen, A., Golstein, J., Fusco, A., Dumont, J. E., & Roger, P. P. (2001). Regulation of thyroid cell proliferation by TSH and other factors: A critical evaluation of in vitro models. *Endocrine Reviews, 22*, 631–656.

Kitahara, C. M., de Vathaire, F., Boutron-Ruault, M. C., & Journy, N. (2020). Thyroid dysfunction and cancer incidence: A systematic review and meta-analysis. *Endocrine-Related Cancer, 27*(4), 245–259.

Kramer, C. K., von Mühlen, D., Kritz-Silverstein, D., & Barrett-Connor, E. (2009). Treated hypothyroidism, cognitive function, and depressed mood in old age: The rancho bernardo study. *European Journal of Endocrinology, 161*(6), 917–921.

Lamartina, L., Durante, C., Filetti, S., & Cooper, D. S. (2015). Low-risk differentiated thyroid cancer and radioiodine remnant ablation: A systematic review of the literature. *The Journal of Clinical Endocrinology & Metabolism, 100*(5), 1748–1761.

Larrabee, K., Meeks, N., Williams, A. M., Springer, K., Siddiqui, F., Chang, S. S., et al. (2023). Cognitive function and postoperative outcomes in patients with head and neck cancer. *The Laryngoscope, 133*(11), 2999–3005.

Leboulleux, S., Bournaud, C., Chougnet, C. N., Ghuzlan, A., Catargi, B., Cao, C. D., et al. (2022). Thyroidectomy without radioiodine in patients with low-risk thyroid cancer. *New England Journal of Medicine, 386*(10), 923–932.

Low, C. A., Kalinski, P., & Bovbjerg, D. H. (2015). Neurocognitive impairment as one facet of cancer-related sickness behavior symptoms. *Journal of the National Cancer Institute, 107*(8), Article djv176.

Mafrica, F., & Fodale, V. (2008). Thyroid function, alzheimer's disease and postoperative cognitive dysfunction: A tale of dangerous liaisons. *Journal of Alzheimer's Disease, 14*(1), 95–105.

Megari, K. (2021). Neuropsychological functioning among patients with different types of cancer: Postchemotherapy cognitive impairment and implications for rehabilitation. *Neuropsychiatrie, 35*(2), 92–97.

Ming, H., Yu, H., Liu, Y., Yang, L., & Chen, Y. (2022). Effect of radioiodine therapy under thyroid hormone withdrawal on health-related quality of life in patients with differentiated thyroid cancer. *Japanese Journal of Clinical Oncology, 52*(10), 1159–1166.

Moon, J. H., Ahn, S., Seo, J., Han, J. W., Kim, K. M., Choi, S. H., et al. (2014). The effect of long-term thyroid-stimulating hormone suppressive therapy on the cognitive function of elderly patients with differentiated thyroid carcinoma. *The Journal of Clinical Endocrinology & Metabolism, 99*(10), 3782–3789.

Neff, T. A., Braun, J., Rana, D., Puhan, M., Filipovic, M., Seeberger, et al. (2022). Interleukin-6 is an early plasma marker of severe postoperative complications in thoracic surgery: Exploratory results from a substudy of a randomized controlled multicenter trial. *Anesthesia & Analgesia, 134*(1), 123–132.

Ni, T., Sun, J., He, Q., Dai, Y., Wang, X., Yu, E., & Shen, G. (2025). Risk factors and prediction model for cancer-related cognitive impairment in thyroid cancer patients. *American Journal of Cancer Research, 15*(1), 153.

O'Barr, S. A., Oh, J. S., Ma, C., Brent, G. A., & Schultz, J. J. (2006). Thyroid hormone regulates endogenous amyloid-beta precursor protein gene expression and processing in both in vitro and in vivo models. *Thyroid, 16*, 1207–1213.

Pacini, F., Schlumberger, M., Dralle, H., Elisei, R., Smit, J. W., Wiersinga, W., et al. (2006). European consensus for the management of patients with differentiated thyroid carcinoma of the follicular epithelium. *European Journal of Endocrinology, 154*(6), 787–803.

Reinke, R., Londero, S. C., Almquist, M., Rejnmark, L., & Rolighed, L. (2023). Total thyroidectomy: Reduction in postoperative hypoparathyroidism. *Endocrine Connections, 12*(9), Article e230198.

Riguetto, C. M., Neto, A. M., Tambascia, M. A., & Zantut-Wittmann, D. E. (2019). The relationship between quality of life, cognition, and thyroid status in graves' disease. *Endocrine, 63*(1), 87–93.

Saeed, O., Bernstein, L. J., Fazelzad, R., Samuels, M. H., Burmeister, L. A., Thabane, L., et al. (2019). Cognitive functioning in thyroid cancer survivors: A systematic review and meta-analysis. *Journal of Cancer Survivorship, 13*(2), 231–243.

Samuels, M. H. (2014). Psychiatric and cognitive manifestations of hypothyroidism. *Current Opinion in Endocrinology Diabetes and Obesity, 21*(5), 377–383.

Samuels, M. H., Kolobova, I., Smeraglio, A., Peters, D., Janowsky, J. S., & Schuff, K. G. (2014). The effects of levothyroxine replacement or suppressive therapy on health status, mood, and cognition. *The Journal of clinical endocrinology and metabolism, 99*(3), 843.

Shiroishi, M. S., Gupta, V., Bigjahan, B., Cen, S. Y., Rashid, F., Hwang, D. H., et al. (2017). Brain cortical structural differences between non-central nervous system cancer patients treated with and without chemotherapy compared to non-cancer controls: A cross-sectional pilot MRI study using clinically-indicated scans. *Proceedings of SPIE-The International Society for Optical Engineering, 10572*, Article 105720G.

Smith, J. W., Evans, A. T., Costall, B., & Smythe, J. W. (2002). Thyroid hormones, brain function and cognition: A brief review. *Neuroscience & Biobehavioral Reviews, 26*(1), 45–60.

Song, Y. S., Kim, K. S., Kim, S. K., Cho, Y. W., & Choi, H. G. (2021). Screening leads to overestimated associations of thyroid dysfunction and thyroiditis with thyroid cancer risk. *Cancers, 13*(21), 5385.

Stern, M., Nieuwenhuijs-Moeke, G. J., Absalom, A., van Leeuwen, B., van der Wal-Huisman, H., Plas, M., & Bosch, D. J. (2023). Association between anaesthesia-related factors and postoperative neurocognitive disorder: A post-hoc analysis. *BMC Anesthesiology, 23*(1), 368.

Stuart, K. E., & Padgett, C. (2020). A systematic review of the association between psychological stress and dementia risk in humans. *Journal of Alzheimer's Disease, 78*(1), 335–352.

Ünlühizarcı, K., Cengiz, B., Dolu, N., Hacıoğlu, A., Karaca, Z., Tanrıverdi, F., et al. (2024). Changes in cognitive functions caused by acute hypothyroidism during withdrawal of l-thyroxine in patients with differentiated thyroid carcinoma. *Endocrinology Research and Practice, 28*(3), 164–168.

Vardy, J. L., Dhillon, H. M., Pond, G. R., Rourke, S. B., Bekele, T., Renton, C., et al. (2015). Cognitive function in patients with colorectal cancer who do and do not receive chemotherapy: A prospective, longitudinal, controlled study. *Journal of Clinical Oncology, 33*(34), 4085–4092.

Vogel, A., Elberling, T. V., Hørding, M., Dock, J., Rasmussen, Å. K., Feldt-Rasmussen, U., Perrild, H., & Waldemar, G. (2007). Affective symptoms and cognitive functions in the acute phase of graves' thyrotoxicosis. *Psychoneuroendocrinology, 32*(1), 36–43.

Wefel, J. S., Saleeba, A. K., Buzdar, A. U., & Meyers, C. A. (2010). Acute and late onset cognitive dysfunction associated with chemotherapy in women with breast cancer. *Cancer, 116*(14), 3348–3356.

Wefel, J. S., Vidrine, D. J., Veramonti, T. L., Meyers, C. A., Marani, S. K., Hoekstra, H. J., et al. (2011a). Cognitive impairment in men with testicular cancer prior to adjuvant therapy. *Cancer, 117*(1), 190–196.

Yang, X., Yu, B., & Ma, L. (2022). Functional MRI changes in patients after thyroidectomy under general anesthesia. *BioMed Research International, 2022*(1), Article 1935125.

Zheng, Y., Zhao, J., Shi, Y., Gui, Z., Xu, C., Wu, et al. (2024). Impact of menopausal status on cognitive function in female papillary thyroid carcinoma patients: A longitudinal propensity score matched study. *BMC Women's Health, 24*(1), 642.

Chapter 15

The impact of thyroid hormones on learning and memory

Mohammad Rahmanian[1, 2], Roshanak Ghorbani[3], Amirmohammad Jabbarpour[4], Mona Minadi[5], Mahdiyeh Nozad Varjovi[6] and Andisheh Iranfar[7]

[1]*Gastroenterology and Liver Diseases Research Center, Research Institute for Gastroenterology and Liver Diseases, Shahid Beheshti University of Medical Sciences, Tehran, Iran;* [2]*Student Research Committee, School of Medicine, Shahid Beheshti University of Medical Sciences, Tehran, Iran;* [3]*Universal Scientific Education and Research Network (USERN), Tehran, Iran;* [4]*Student Research Committee, Faculty of Medicine, Zanjan University of Medical Sciences, Zanjan, Iran;* [5]*Student Research Committee, Kermanshah University of Medical Sciences, Kermanshah, Iran;* [6]*Student Research Committee, Faculty of Pharmacy, Tabriz University of Medical Sciences, Tabriz, Iran;* [7]*School of Pharmacy, Tehran University of Medical Sciences, Tehran, Iran*

Abbreviations

CNS Central Nervous System
DIO1 Deiodinase type 1
DIO2 Deiodinase type 2
DIO3 Deiodinase type 3
DIT Diiodotyrosine
IQ Intelligence Quotient
L-T4 Levothyroxine
LAT1 L-type amino acid transporter 1
LAT2 L-type amino acid transporter 2
MAPK Mitogen-Activated Protein Kinase
MBP Myelin Basic Protein
MCT10 Monocarboxylate transporter 10
MCT8 Monocarboxylate transporter 8
MIT Monoiodotyrosine
OATP1C1 Organic anion-transporting polypeptide 1C1
PI3K Phosphatidylinositol 3-kinase
PKA Protein Kinase A
PKC Protein Kinase C
Reln Reelin
T3 Triiodothyronine
T4 Thyroxine
Tg Thyroglobulin
TH(s) Thyroid Hormone(s)
TPO Thyroid Peroxidase
TR Thyroid Hormone Receptor
TRE Thyroid Hormone Response Element
TSH Thyroid-Stimulating Hormone

Introduction

T3 and T4 are important for optimal performance of diverse mammals, including the central nervous system (CNS). They also regulate metabolism, support development, and contribute to tissue development (Accorroni et al., 2017). Additionally, these hormones affect the production of new neurons and neuroglia from neural stem cells (NSCs), facilitating their differentiation and maturation into fully functional brain cells (Valcárcel-Hernández et al., 2024). T3, the biologically active form of thyroid hormone, influences both neuronal maturation and the development and organization of various glial cell types, including astrocytes, oligodendrocytes, microglia, and radial glial cells (Noda, 2018). Thyroid hormones are crucial for brain development and cognitive function. A deficiency during pregnancy or early childhood, due to maternal iodine deficiency or congenital hypothyroidism, can cause serious and irreversible brain damage that might result in intellectual disabilities and slow growth. In adults, thyroid dysfunction is linked to neurological issues, mood disorders, and memory problems (Bernal). These effects are mainly mediated by alterations in the hippocampus, where disrupted synaptic activity results in persistent impairments in learning and memory. Moreover, inadequate thyroid hormone levels during brain development can

Hormones in Neuroscience. https://doi.org/10.1016/B978-0-443-30261-9.00015-2

impair hippocampal synaptic function, further reducing performance on learning and memory tasks dependent on the hippocampus (Rivas & Naranjo, 2007, pp. 1601–1848). Changes in thyroid hormone levels are associated with deficits in hippocampus-dependent learning and memory, as well as impairments in synaptic plasticity and neurogenesis (Cooke et al., 2014, pp. 1557–9077).

Thyroid hormones T3 and T4 significantly affect the nervous system from fetal development through adulthood and are recognized as key contributors to Alzheimer's disease (AD). Recent studies have demonstrated a link between both hypothyroidism and hyperthyroidism and AD (Bavarsad et al., pp. 1097–4652). The effects of thyroid hormones are not limited to the brain and the central nervous system, but also encompass their peripheral metabolic derivatives. These metabolites, which were previously regarded as inactive products, have now been demonstrated to exert significant biological effects (Accorroni et al., 2017). In a study conducted on rats, both short-term and long-term administration of L-T4 improved spatial memory performance and reduced the deleterious effects of scopolamine, which was accompanied by increased cholinergic activity in the frontal cortex and hippocampus (Smith et al.). Adolescents with congenital hypothyroidism (CH), who lack thyroid hormones during late gestation and early life, exhibit impaired verbal recall abilities, reduced hippocampal volume, and abnormal hippocampal functioning in tasks involving visual association (Bernal, 2017). As previously noted, the thyroid gland plays a crucial role in memory and learning, since the hippocampus—the brain's center for these processes—is regulated in its growth and function by thyroid hormones (Espina & Bagamasbad). Even mild changes in thyroid function, even if within the normal range, can have a significant impact on the cognitive abilities of the elderly. Various cognitive impairments that may be associated with thyroid failure do not always follow a specific pattern, and treatment with L-thyroxine may not fully restore cognitive abilities to normal in all patients with hypothyroidism (Ettleson & Papaleontiou, 2022).

Physiology and molecular basis

The thyroid hormones (THs) are biosynthesized by thyroid gland, thyroxine (T4) and triiodothyronine (T3) (Accorroni et al., 2017). This process involves of four key stages, iodide absorption and oxidation, binding iodine to tyrosine residues, combining components to produce hormones, and eventually being released through protein breakdown (Barbaro et al., 2019). The synthesis is initiated with absorbing iodide through its sodium-iodide symporter, then being oxidized by enzyme thyroid peroxidase. Once iodide has been oxidized, it becomes integrated into the thyroglobulin (Tg) protein by binding to specific tyrosine residues, consequently forming diiodotyrosine (DIT) or monoiodotyrosine (MIT). The combination of MIT with DIT is facilitated by thyroid peroxidase to produce T3, or joining two DIT molecules to form thyroxine T4. THs are released upon breakdown of Tg molecule through proteolysis remaining embedded (Accorroni et al., 2017; Barbaro et al., 2019). Producing THs in the thyroid gland depends on the signals from thyroid-stimulating hormone (TSH), produced by the pituitary gland and the harmonized expression of specific thyroid proteins (Accorroni et al., 2017).

Fig. 15.1 presents an overview of thyroid hormone transport and its impact on the brain. As displayed in this figure, once THs are released by the thyroid gland, they transfer via the bloodstream to target tissues, being converted into active forms there that control biological functions. THs enter cells by passing through the membrane to mediate physiological functions via specialized transport proteins. This uptake is mediated by various transporter families, such as monocarboxylate transporters (MCT8 and MCT10), organic anion-transporting polypeptides (especially OATP1C1), and L-type amino acid transporters like LAT1 and LAT2 (Lasa & Contreras-Jurado, 2022). Meanwhile, MCT8 is a highly specialized transporter that allows THs to enter cells in many different tissues throughout the body (Chen et al., 2018). After entering cells via dedicated transport proteins, T4 is converted into the active hormone T3 (Huang et al., 2019). The conversion of THs is mediated by three selenoprotein deiodinase enzymes (including DIO1, DIO2, and DIO3), each performing specific roles in different tissues (Sabatino & Vassalle, 2025).

Following the entry of T3 into target cells and its binding to nuclear thyroid hormone receptors (TRs), the genomic signaling pathway is initiated (Huang et al., 2019). Thyroid hormones primarily act through genomic signaling, which is a classic and the most widely studied mechanism of their function, controlling the transcription of numerous genes involved in growth, metabolism, and cell differentiation (Huang et al., 2019). When T3 binds to TRs, the receptor changes shape in its ligand-binding domain and this change causes corepressors to detach and allows coactivator proteins to bind. The switch from corepressor to coactivator binding stimulates histone acetylation, which relaxes chromatin structure and activates gene transcription (Schoenmakers et al., 2015). TRs selectively bind to thyroid hormone response elements (TREs), which are specific DNA sequences located in the promoter and enhancer regions of target genes (Huang et al., 2019). TREs are composed of distinctive half-site sequences, which are organized in direct repeat or inverted repeat configurations to facilitate receptor binding and transcriptional regulation (Huang et al., 2019). In addition to their genomic effects, THs can initiate rapid biological

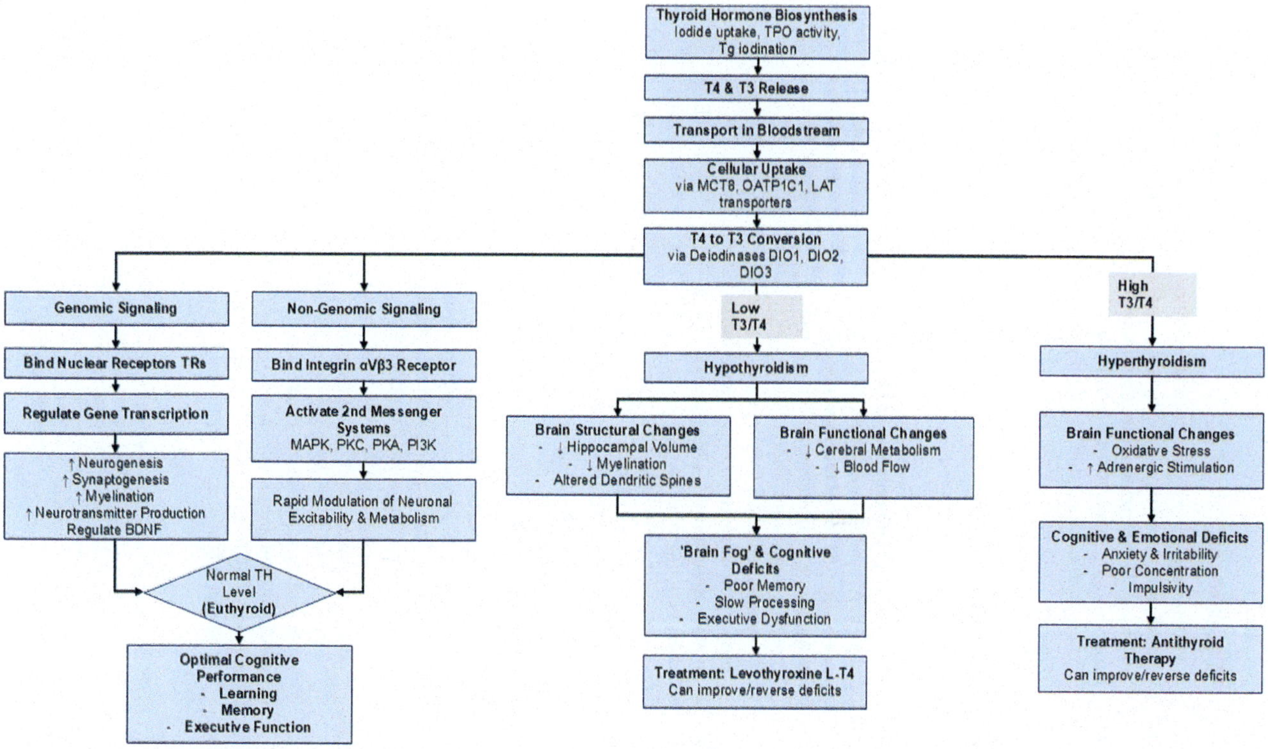

FIGURE 15.1 A summary of thyroid hormone's journey and impact on the brain. Thyroid hormones (THs) are synthesized in the thyroid gland and released into the bloodstream. Upon reaching target brain cells, T4 is converted to the active T3 hormone. T3 exerts its effects through two primary pathways: (1) Genomic signaling: Slow-acting regulation of gene expression critical for brain development and plasticity. (2) Non-genomic signaling: Rapid activation of second messenger systems to modulate neuronal function. A euthyroid state (normal TH levels) supports optimal cognition. Hypothyroidism leads to structural and metabolic deficits in the brain, causing symptoms of "brain fog." hyperthyroidism causes functional deficits through excitotoxicity and oxidative stress, leading to anxiety and attention problems. Fortunately, appropriate treatment can often reverse these cognitive impairments. *BDNF*, Brain-derived neurotrophic Factor; *Tg*, Thyroglobulin; *TPO*, Thyroid peroxidase.

responses by non-genomic pathways that independent function of direct interactions with DNA or TRs.

When THs binds to integrin αvβ3 on the cell surface, a receptor that is not structurally similar to nuclear TRs, activates key non-genomic signaling pathway (Liu et al., 2019) such as protein kinase C (PKC), protein kinase A (PKA), phosphatidylinositol 3-kinase pathways (PI3K), and mitogen-activated protein kinase (MAPK) (Galow et al., 2023; Incerpi et al., 2022).

Overall, THs exhibit their effects through both of genomic and non-genomic mechanisms. Among all hormones produced in the human body, THs exert the broadest range of physiological effects. They act on nearly all cells and subcellular structures playing a key role in keeping cells and the body in balance.

The sources of THs, their mechanism of action and role in cognitive performance are summarized in Table 15.1.

Developmental role in cognition

THs are essential for brain development during fetal and postnatal stages and play a critical role in maintaining healthy brain function in adulthood (Sönmez et al., 2025). Experimental studies in rodent models have shown that THs control the activity of many genes in the central nervous system (CNS), affecting many processes important for brain growth and function, including neurogenesis, neuronal migration, synaptogenesis, and myelination (Godbole et al.; Hernandez & Stohn, 2018). Insufficient thyroid hormone levels during fetal and neonatal development can disrupt important neurodevelopmental processes, including myelination and production of essential neurotransmitters (Godbole et al.; Hernandez & Stohn, 2018).

Proper neuronal migration is regulated by THs. Hypothyroidism lowers brain weight and causes tightly packed neurons in the cerebellum, showing problems with migration (Velasco et al., 2018). Disrupted neuronal migration has profound and irreversible consequences. For instance, in maternal thyroid hormone deficiency, neurons fail to reach the upper cortical layers, leading to abnormal positioning and disruption of the laminar architecture of the cerebral cortex (Velasco et al., 2018).

TABLE 15.1 Overview of THs, sources, mechanism of action and their role in cognitive performance.

Hormone	Source	Cellular mechanism	Cognitive role	Reference
Thyroid hormones (THs: T4, T3)	From maternal supply (during fetal period) and endogenous thyroid gland; transported into CNS through specific transporters; controlled by deiodinases (activating DIO2, inactivating DIO3) in neurons/glia.	They organize multiple processes: neurogenesis, neuronal/glial differentiation and migration; synaptogenesis; myelination; regulation of gene networks (adhesion molecules, axon growth factors); timing of developmental events; also regulate adhesion, cytoskeleton, etc. **They are key regulators of adult neurogliogenesis** by controlling **NSC proliferation, lineage determination** (neuron vs. oligodendrocyte), **differentiation/maturation**, and **apoptosis** across niches (SVZ, SGZ; hypothalamus; emerging in striatum/cortex). **TRα1** predominates in neuronal lineage; **TRβ** contributes to SGZ regulation	TH deficiency/dysfunction lead to intellectual disability, learning deficits, Alzheimer's disease risk; normal function contributes to cognition, memory, brain structure; both developmental and adult TH dysregulation contributes to mood disorders and neurological impairment.	(Cooke et al., 2014; Noda, 2018; Rivas & Naranjo, 2007; Valcárcel-Hernández et al., 2024)
T4 (thyroxine)	**Main hormone secreted by the thyroid gland**; contributes to brain TH supply (CSF **TTR**; transfer via **MCT8/OATP1C1**) and serves as **precursor for local T3**	Genomic via nuclear TH receptors (TRs); also described non-nuclear (cytoplasm, cytoskeleton, membranes, organelles). **Non-genomic: reversible, non-competitive** inhibition of nAChR currents; **not via αVβ3** (Tetrac-insensitive); **increases agonist cooperativity** (n_H ↑) and **enhances desensitization**; potency similar to T3. Supports niche TH availability that guides lineage decisions; in demyelination paradigms, T4/T3 pulses during recovery are used after a transient hypothyroid phase to aid remyelination by newly generated OPCs.	As for T3, **fast modulation of cholinergic signaling** provides a mechanistic link between **thyroid imbalance and cognitive disturbance** (learning/memory), complementary to genomic effects. **T4 supplementation** can improve **hippocampus-related cognitive perturbations in subclinical** hypothyroidism (not in overt hypothyroidism). Essential for brain maturation; maternal T4 during gestation critical for fetal neurodevelopment; insufficiency → cretinism, mental retardation, cognitive impairment, poorer intellectual outcomes; it contributes indirectly via T3 to cognitive functions.	(Bernal, 2024; Moreno et al., 2008; Noda, 2018; Puia & Ravazzini, 2020; Valcárcel-Hernández et al., 2024)
T3 (3,5,3′-triiodothyronine)	**Produced locally in** brain astrocytes and peripheral tissues through 5′-deiodination of T4 **mediatited by type-2 deiodinase** in glial cells. Present in adult neurogliogenic niches (SVZ, SGZ, hypothalamus; emerging in striatum/cortex).	Genomic actions via binding to ligand-dependent TRα/TRβ, thus regulating transcription of target genes involved in synaptic plasticity (e.g., GAP-43, RC3, CAMKII). The expression of genes is associated with myelin proteins, adhesion molecules, Reelin, etc. that control neuronal and glial differentiation, cell migration, terminal differentiation, and synaptogenesis. Nongenomic actions on ion channels & signaling cascades: rapidly reduces nAChR-mediated currents (nicotine/ACh) in SH-SY5Y cells; **reversible, non-competitive antagonism; not blocked by Tetrac (so not via integrin αVβ3); enhances receptor desensitization.**	Normal T3 is needed for regulating correct myelination, neuronal development, synaptic plasticity, learning and memory; insufficiency impairs hippocampal-dependent tasks (Morris water maze, fear conditioning, radial arm maze, object recognition). Proposed contributor to intellectual deficits, in thyroid dysfunction via **cholinergic (nAChR) modulation**; aligns with reports of memory/working-memory changes in hypo/hyperthyroidism and TH treatment effects.	(Bernal, 2024; Noda, 2018; Puia & Ravazzini, 2020; Valcárcel-Hernández et al., 2024)

Abbreviations: *CNS*, central nervous system; *DIO2*, type-2 iodothyronine deiodinase; *DIO3*, type-3 iodothyronine deiodinase; *TRα1 / TRβ*, thyroid hormone nuclear receptors; *nAChR*, nicotinic acetylcholine receptor; *αVβ3*, integrin *αVβ3*, a cell surface integrin receptor that some thyroid-hormone (or analog) non-genomic effects may involve; *OPCs*, oligodendrocyte precursor cells that differentiate into oligodendrocytes, which myelinate neurons in CNS; *SVZ*, subventricular zone, a neurogenic niche in the brain where neural stem cells persist and generate new neurons/glia; *SGZ*, subgranular zone, a region in the dentate gyrus of the hippocampus that is neurogenic in adults; *GAP-43*, growth Associated Protein 43 (also known as neuromodulin, B-50), a protein enriched in growth cones during axonal growth, involved in regulation of actin cytoskeleton and neurite outgrowth; *RC3/Neurogranin*, RC3 is another name for neurogranin, a small postsynaptic protein that binds calmodulin (CaM), enriched in neurons, involved in modulating calcium/calmodulin signaling, thus affecting synaptic plasticity; *CAMKII*, calcium/calmodulin-dependent Protein Kinase II, a serine/threonine kinase critical for synaptic plasticity, long-term potentiation (LTP), learning & memory. Has several isoforms: α, β, γ, δ; *Reelin*, a large extracellular matrix glycoprotein (encoded by RELN) that regulates neuronal migration during development, dendrite and spine growth, and synaptic plasticity in the adult brain.

THs serve as critical modulators of synaptogenesis, and their deficiency disrupts this process, resulting in impaired synaptic formation and compromised neural connectivity (Velasco et al., 2018). THs modulate synaptogenesis via both genomic and non-genomic mechanisms. By binding to nuclear receptors in neurons and glial cells, they influence synapse formation and play a pivotal role in nervous system development. The critical importance of THs in synaptogenesis is highlighted by congenital hypothyroidism, which causes disrupted neuronal migration, defective axonal development, and impaired synapse formation (Dezonne et al., 2013). Such disturbances may cause permanent neurological issues, including deficits in learning and memory, changes in motor activity and emotional behaviors (Dezonne et al., 2013).

Myelination is strongly regulated by THs, as demonstrated in the 1960s when researchers observed that neonatal thyroidectomy led to reduced myelination, a deficit that could be reversed through thyroid hormone replacement therapy (Lariosa-Willingham et al., 2022). In demyelination models, T3 and T4 promote remyelination by restoring myelin basic protein (MBP) expression and activating oligodendrocyte precursor cells (Lariosa-Willingham et al., 2022). The precise timing of thyroid hormone signaling is essential for proper myelination. Inadequate levels of T3 during early development result in diminished myelin density throughout the brain and spinal cord (Farías-Serratos et al., 2021).

T3 and T4 are crucial for CNS development during fetal development, infancy, and early childhood, periods of rapid brain growth. Maternal THs are crucial for fetal brain development during prenatal life, acting from the very beginning of pregnancy. The fetal brain starts responding to THs around the 10th week of pregnancy, with T3 receptors appearing, type 2 deiodinase activity rising, and T3 levels increasing in the cortex, peaking by weeks 18–20 (Bernal, 2017). Even without obvious clinical symptoms, moderate maternal thyroid dysfunction during pregnancy can significantly affect neurodevelopment. Children of hypothyroid mothers exhibit an increased risk of learning difficulties, neuropsychological deficits, and lower IQ scores (Fotakis et al., 2022).

After birth, neonates must produce their own THs. Preterm infants may not have a fully developed thyroid system, putting them at higher risk for thyroid problems and neurodevelopmental issues (Rajan & Fame, 2024). Studies show that insufficient levels of THs during fetal and newborn stages can lead to serious, long-term neurological problems, including difficulties with movement coordination, intellectual disability, and loss of hearing or speech (Bagamasbad et al., 2019). The importance of THs for brain development is evident in cretinism, a severe disorder caused by low maternal thyroid function and iodine deficiency during pregnancy (Bagamasbad et al., 2019).

In adults, hypothyroidism can cause memory problems and depression, which usually improve upon restoration of normal thyroid function (Omar, 2023). The hippocampus (central to memory formation) is very sensitive to THs and vulnerable to their deficiency, which can disrupt its structure and function and reduce learning and cognition (Omar, 2023).

Rodent studies have been key in demonstrating how thyroid hormone deficiency affects the brain. Genetic investigations using thyroid hormone receptor mutant mice have demonstrated that impaired thyroid signaling causes behavioral and cognitive problems, including memory deficits, locomotor problems, and anxiety (Sabatino et al., 2024). Human neuroimaging studies support animal studies, showing that patients with hypothyroidism show smaller hippocampi and memory deficits (Uma et al., 2021).

In summary, both animal and human studies highlight that proper thyroid hormone signaling is crucial for neural development and function, especially in the hippocampus, which controls learning and memory. Also, thyroid hormone deficiency during sensitive developmental stages can cause lasting structural and functional brain abnormalities and cognitive impairments. Therefore, accurate thyroid hormone regulation is crucial to avoid neurodevelopmental issues. The evidence regarding role of THs in neurodevelopmental processes, molecular mechanisms, and promising interventions are summarized in Table 15.2.

Adult brain: Learning, memory and neuroplasticity

THs impact brain development, particularly in the hippocampus, an important area for memory, learning, and spatial navigation that is highly sensitive to the effects of these hormones. T3 and T4 play pivotal role in the function of adult brain, with increasing the survival of neuronal progenitor cells after division, affecting their migration and dendritic branching, and supporting myelin generation (Anderson et al., 2003).

The molecular signaling pathways are regulated by THs, underscoring their role in hippocampal neurogenesis. For instance, T3 controls the RC3/Neurogranin gene, which produces a calmodulin-binding protein that boosts synaptic plasticity by positioning calmodulin near dendritic spine membranes (Talhada et al., 2019). Moreover, THs reinforce brain development by controlling Reelin (Reln), a protein organizing neural movement and synapse (Talhada et al., 2019). Overall, THs control neurogenesis in adult

TABLE 15.2 Thyroid hormones and neurodevelopmental processes.

Process	Role of T3/T4	Mechanism	Clinical significance	Experimental evidence	Potential interventions
Neurogenesis (Bernal, 2017; Yajima et al., 2021)	Promotes differentiation of neural stem cells into neurons and glia, particularly in the hippocampus. TH deficiency reduces new neuron production and impairs memory and learning.	Regulation of BDNF and RC3/Neurogranin genes via nuclear receptors (TRs) and MAPK/PI3K pathways. TH deficiency disrupts hippocampal neurogenesis.	Impaired neurogenesis causes spatial memory and learning deficits, as seen in cretinism. Early TH treatment mitigates deficits, but delays cause permanent damage.	Rodent models show reduced neurogenesis in TH deficiency, improved with T3. Human studies confirm adult neurogenesis links.	Levothyroxine therapy in fetal or adult periods; iodine supplements for pregnant mothers.
Neuronal Migration (Talhada et al., 2019; Velasco et al., 2018)	Regulates neuronal migration in cortex and cerebellum for proper positioning. TH deficiency disrupts brain layering and neural connections.	Modulation of Reelin via genomic (TREs) and nongenomic (integrin $\alpha v\beta 3$, PKC/PKA) pathways. TH deficiency causes incomplete migration and reduced brain weight.	Disrupted migration causes permanent neurological deficits (reduced IQ, motor issues). Maternal preventive treatment reduces risk.	Neonatal thyroidectomy in rodents shows cerebellar migration issues, reversible with TH. Human studies confirm cortical layering disruptions.	Thyroid screening in pregnancy; TH treatment for mothers.
Synaptogenesis (Yajima et al., 2021)	Enhances synapse formation and synaptic plasticity. TH deficiency reduces synapse density and impairs learning.	Upregulation of synaptic proteins (synaptophysin) and BDNF via TREs and PI3K/MAPK. TH deficiency reduces BDNF and disrupts LTP.	Synaptogenesis deficits cause attention and memory issues. TH treatment improves synaptic function, but early damage may be permanent.	Neuronal cultures show reduced synapses in TH deficiency, improved with T3. Hypothyroid rodents confirm LTP deficits.	TH therapy; combination with neurotrophics like BDNF for plasticity
Myelination (Lariosa-Willingham et al., 2022)	Stimulates myelin formation by oligodendrocytes and remyelination. TH deficiency reduces myelin density, slowing neural signaling.	Increased MBP expression via T3 binding to TRs. TH deficiency reduces MBP, causing demyelination.	Inadequate myelination causes slowed thinking and motor deficits. TH treatment improves cognitive function.	Thyroidectomy studies show reduced myelin, reversible with TH. Modern models confirm T4's role in remyelination.	TH therapy for thyroid-related demyelination; monitoring TH in neurodegenerative diseases.

BDNF, Brain-Derived Neurotrophic Factor; *LTP*, Long-Term Potentiation; *MAPK*, Mitogen-Activated Protein Kinase; *MBP*, Myelin Basic Protein; *PI3K*, Phosphatidylinositol 3-Kinase.; *PKA*, Protein Kinase A; *PKC*, Protein Kinase C; *TREs*, Thyroid Hormone Response Elements; *TRs*, Thyroid Hormone Receptors.

brain by regulating brain-derived neurotrophic factor (BDNF) and neurotransmitter system components preserving cognitive funstion (Yu et al., 2017).

Dendritic architecture and synaptic development in the hippocampus are regulated by THs. How thyroid hormones shape dendrites and synapses depends on where in the brain, the stage of development, and the kind of neuron (Alcaide Martin & Mayerl, 2023). Despite in direct formation of synapse by THs, they intensely regulate synaptic proteins, particularly at the presynaptic level (Yajima et al., 2021).

THs influence hippocampal development through a hierarchical signaling pathway in which T3 regulates BDNF expression, subsequently promoting dendritic growth and synaptic formation. Therefore, T3 deficiency in early stages of human development noticeably lowers BDNF mRNA in hippocampal neurons (Yajima et al., 2021).

By adjusting gene expression, THs control neurotransmitter production and intercellular signaling (Yajima et al., 2021). For example, thyroid hormones attach to synaptic membranes, modifying enzyme activity and ion flow, and triggering calcium and nitric oxide based signals. In adults, they behave like neurotransmitters, particularly within adrenergic systems (Chakrabarti et al., 2023). THs are essential neuromodulators, controlling neural development, neurotransmitter activity, and overall brain metabolism (Li & Jia Liu, 2024).

Generally, THs are key players in preserving either the functional performance or structural integrity of the hippocampus by the mediating neurogenesis, improving dendritic architecture, enhancing synaptic plasticity, and the equalizing of neurotransmitter activity.

Experimental models consistently indicate memory deficits linked to thyroid hormone deficiency, highlighting their role in cognition and neural circuitry. Studies on radial arm water maze tasks have revealed that rats with lack of thyroid glands have poor performance in memory tests and show both short and long-term memory deficits (Alzoubi & Alkadhi, 2023). Additionally, in adult patients, both inadequate and excessive thyroid activity lead to distinct neuropsychiatric symptoms.

Hypothyroidism can cause memory problems and difficulties with attention, focus, and handling mentally challenging tasks, as well as psychological and cognitive issues, such as depression and anxiety (Remaud et al., 2014). Neuroimaging studies have clearly documented structural alterations underlying cognitive deficits in hypothyroidism. Neuroimaging of people with hyperthyroidism shows changes in areas important for cognition, especially lower choline levels in the posterior parietal and dorsolateral prefrontal cortices, which are essential for focus and handling complex information (Kumar et al., 2021).

In hyperthyroidism, animal studies show that hyperthyroidism induced by T3/T4 supplementation impairs cognition, causing deficits in learning, spatial navigation, and recognition memory across both short- and long-term domains (Agnieszka Piekiełko-Witkowska et al., 2025). In hyperthyroid animal models, learning and memory are markedly impaired, as shown by poor performance in Y-maze, novel object recognition, and Morris water maze tasks. These deficits are associated with lower mature dendritic spine density in the hippocampus, representing disrupted synaptic structure in important cognitive regions (Martin & Sarkar, 2023). Furthermore, neuroimaging data displays reduced hippocampal volume in hypothyroidism, significantly linked to cognitive impairments in children and adolescents (Madhusudhan et al., 2022).

The association between thyroid hormone levels and intellectual abilities demonstrates a reversed U-shaped curve, signifying the harmful effects of insufficient and extra levels of TH on memory and learning performance. Therefore, normal thyroid hormone levels lead to optimal cognitive performance (Martin & Sarkar, 2023).

In behavioral tasks, the noticeable impact of hypothyroidism on performance is manifested, affecting multiple cognitive domains. Hypothyroidism in adults can cause difficulties in psychomotor function, language memory, and executive functioning (Ettleson & Papaleontiou, 2022). Similar results in cognitive dysfunction are indicated in animal and humans studies on hypothyroidism. Studies in rodents confirm that hypothyroidism leads to poor performance in tests such as impaired motor coordination, weakened spatial memory, and novel object recognition (Sabatino et al., 2024), slower information processing, motor learning deficits, and pseudo-dementia (Uma et al., 2021). Thyroid hormone replacement therapy, specifically with levothyroxine, has the promise to effectively reverse cognitive deficits, by enhancing spatial consciousness and reaction speed (Kale et al., 2022). In contrast, hyperthyroid patients may have and difficulties in visual-motor functioning and attention, despite being less serious in hypothyroidism (Bégin et al., 2008).

Behavioral research in animal models demonstrates that hyperthyroidism impairs learning and memory. Rodents with high thyroid hormone levels perform poorly in spatial memory tasks (including the Y-maze, object recognition tests, and Morris water maze), which is linked to fewer mature dendritic spines in the hippocampus (Martin & Sarkar, 2023). Treating hyperthyroidism usually helps restore attention and processing speed, with many cognitive problems improving once thyroid hormone levels return to normal (Ekinci et al., 2021).

It is shown that L-thyroxine (L-T4) therapy efficiently preserve mental function andimprove intellectual disability resulted by hypothyroidism. Experimental animal studies strongly support the memory-enhancing effects of L-T4 therapy. In hypothyroid mice, L-T4 therapy significantly improved performance in object recognition tests, reaching normal levels (Rutigliano et al., 2023). Furthermore, case studies have shown that L-T4 treatment improves cognitive function and memory. In one report, an 81-year-old patient demonstrated improved Mini-Mental Scores following L-T4 therapy, along with restored memory and orientation (Uma et al., 2021).Generally, supplementation with T3 and T4 may provide therapeutic benefits. Specifically, T3 treatment has been shown to restore both short- and long-term memory in object recognition tasks, an effect associated with increased hippocampal BDNF expression and neuroprotective actions (Sepúlveda et al., 2024). The association between thyroid diseases and cognitive performance, neuronal mechanisms, and transcriptional modulations are summarized in Table 15.3.

Clinical implications

Thyroid disorders in adults have significant clinical implications for cognitive health, and their effects extend beyond simple metabolic disturbances (Alšauskė et al., 2024). Thyroid hormones, particularly triiodothyronine and thyroxine, are major regulators of brain metabolism, synaptic plasticity, and neurotransmitter balance (Li & Liu, 2024). Even mild changes in the levels of these hormones can affect memory, attention, and executive function; therefore, physicians should be aware of neuropsychological symptoms in patients with thyroid diseases (Piekiełko-Witkowska et al., 2025).

In clinical practice, hypothyroidism often manifests with a phenomenon known as "brain fog", a condition that patients describe as mental slowness, poor concentration, and forgetfulness (Ettleson et al., 2022). Objective assessments indicate deficits in short-term memory, reduced psychomotor speed, and impairment in executive functions such as problem-solving and planning. These changes are primarily attributed to reduced glucose metabolism and decreased blood flow in sensitive brain regions such as the hippocampus and prefrontal cortex. In more severe cases, these manifestations can appear similar to dementia unless a thyroid function assessment is specifically performed (Rasmussen et al., 2023).

In contrast, hyperthyroidism is associated with a different pattern of cognitive impairment, often accompanied by psychomotor and emotional hyperactivity (Eslami-Amirabadi & Sajjadi, 2021; Zhu et al., 2022). Patients typically complain of anxiety, irritability, reduced concentration, and impulsive decision-making. Neuropsychological tests also often reveal deficits in sustained attention and working memory (Eslami-Amirabadi & Sajjadi, 2021). Underlying mechanisms include increased adrenergic stimulation, increased excitatory neurotransmitters, and oxidative stress, which lead to pressure on neural networks and reduced synaptic stability. If left untreated, hyperthyroidism can severely disrupt patients' work performance and quality of life (Zhu et al., 2022).

Over a long period, thyroid disorders have been proposed as a potential risk factor for dementia (Kim et al., 2022; Li & Liu, 2024). Epidemiological studies have shown that hypothyroidism is associated with an increased incidence of Alzheimer's disease and other dementias (Kim et al., 2022). Possible mechanisms include chronic brain hypometabolism, neuroinflammation, and increased beta-amyloid deposition (Li & Liu, 2024). These factors can interact with other risk factors and accelerate cognitive decline in susceptible individuals (Kim et al., 2022). Although the evidence regarding hyperthyroidism is more limited, the data suggest that neuronal damage caused by oxidative stress and excitotoxicity can contribute to future cognitive decline (Yashkin et al., 2024). These findings highlight the importance of evaluating thyroid function in older adults with cognitive complaints (Elbadawy et al., 2020).

One important clinical question is whether these cognitive symptoms are reversible with treatment. In hypothyroidism, levothyroxine replacement therapy usually leads to significant improvements in attention, memory, and processing speed, especially if treatment is started promptly (Uma et al., 2021). Functional imaging studies have shown that after returning to a euthyroid state, cerebral blood flow and metabolism also return to normal (Uma et al., 2021). However, delays in treatment can lead to permanent changes in brain structure or function that are not fully reversible even with hormone replacement (Li & Liu, 2024).

In hyperthyroidism, treatment with antithyroid medications or radioactive iodine also reduces anxiety, improves concentration, and restores working memory function (Holmberg et al., 2019). Nevertheless, treatment outcomes are variable and depend heavily on the severity and duration of the illness prior to intervention (Holmberg et al., 2019). In some patients, persistent damage from oxidative stress prevents complete cognitive recovery. Therefore, early diagnosis and treatment are of vital importance (Li & Liu, 2024).

This evidence has clear clinical implications. First, cognitive symptoms in adults with thyroid disorders should not be dismissed as non-specific complaints or primary psychiatric disorders. Rather, they should be considered potential indicators of glandular disorders and hormonal evaluation should be performed (National Institute for Health and Care Excellence: Clinical Guidelines, 2023). Second, the link between thyroid diseases and the risk of dementia highlights the importance of including thyroid testing in the assessment of older adults with cognitive decline (Sharma et al., 2020). Third, the partial reversibility of symptoms with treatment indicates that timely

TABLE 15.3 The link between thyroid disease and cognitive performance, associated neuronal mechanisms, and gene expressions.

Condition	Brain Region	Cognitive/Functional Impact	Mechanism	Transcriptional Modulation	Reference
TH deficiency during late brain development	Rodent whole brain, specially hippocampus, cerebellum, cerebral cortex, optic nerve	Permanent deficits in learning, intellectual impairment; dendritic arborization; axonal and dendritic growth; ↓ myelination; ↓white matter tract size; permanent reduction of brain weight and function.	↓TH (maternal or fetal), delayed TH replacement; asynchronous development of neurons, glia; impaired differentiation, maturation, synapse formation, migration; ↓ survival of maturing oligodendrocytes; apoptosis of oligodendrocyte precursors	Genes involved include MBP, MAG, PLP (myelin genes), RC3 (neurogranin), Pcp-2, transcription factors like BTEB; many of these genes show **transient TH-responsivity** during a sensitive postnatal period; magnitude of change often modest (2-3 fold) for brain genes versus much larger in periphery	(Anderson et al., 2003)
Acute ischemic stroke / after stroke (rodent & human studies)	Brain in general; specific regions including peri-infarct area, infarct core; hippocampus (CA1)	Poor functional outcomes in patients with low T_3; greater infarct volume; worse neuronal damage; slower recovery; neurological impairment correlated with low TH levels.	↓ active T_3 (due to altered deiodinase activity, transport, thyroid dysfunction); disrupted thyroid hormone homeostasis (T_3/T_4/TSH changes) after stroke; non-thyroidal illness syndrome; age dependency. TH actions include influence over neurogenesis, angiogenesis, cytoskeletal dynamics, intracellular transport, excitotoxicity.	Several genes are modulated by TH in the context of recovery: e.g. Bcl2 (neuronal survival), Vegfa (angiogenesis), Sox2 (stem cell proliferation), neurotrophins, NRGN / Nrgn (synaptic plasticity), transporters (MCT8), deiodinases (DIO2, DIO3).	(Talhada et al., 2019)
T3-free culture medium (−T3) vs +T3 at 10 DIV (primary hippocampal neurons)	Hippocampus (primary cultured neurons)	↓ dendritic arborization in −T3 group at **10 DIV**; fewer secondary & tertiary dendrites; growth retardation; no difference at 7 DIV or by 14 DIV.	Lack of T3 delays dendrite growth during a specific time window (10 DIV); effect is transient ("developmental delay")	↓ **BDNF** mRNA at 10 DIV in −T3 vs +T3; ↓ Syn1 mRNA also at that time; at 7 DIV, Nrgn and Hr mRNAs are lower; at 14 DIV, Nrgn, Hr, and Thrb mRNA levels ↓; **Thra** not significantly changed.	(矢島弘之, 2021)
Hyperthyroidism	Hippocampus; temporal lobe; limbic system; ↓ hippocampus–cortex functional connectivity	Impaired memory; ↓ LTP / ↑ LTD; faster forgetting	Oxidative stress; neuroinflammation with necroptosis (RIPK3/MLKL); ↓ dendritic spine density and impaired AMPAR/NMDAR function; ↓glutamate–glutamine cycle efficiency; ↓glucose metabolism in limbic system; cholinergic disturbance (↑acetylcholinesterase; TRH depletion); ↑neuroserpin → promotes β-amyloid (Aβ) deposition; cardiovascular "two-hit" contribution to Aβ accumulation	Downregulation of **GRIN2B** (NMDAR subunit) in hippocampus; TRH–MAPK–GSK3 signaling favors tau hyperphosphorylation (non-genomic signaling noted)	(Zhaoqing Li & Jia Liu, 2024)
Hypothyroidism	Hippocampus (dentate gyrus, perforant path); memory-related regions	Broad cognitive deficits (esp. memory); impaired LTP and input–output function	↓neuromodulation; ↓adult neurogenesis; ↓MAPK/ERK-CREB and Ca^{2+}/calmodulin pathway components (rescued by TH); ↓BDNF (rescued by BDNF); ↓glutamate release and ↓NR1 expression; heightened glutamate neurotoxicity (astrocytic transporter effects); cholinergic deficits (↓ACh, ↓ChAT; altered excitability);	TH represses **APP** transcription via ↓ H3 acetylation and H3K4 methylation; TH regulates **TREM2** expression in microglia (linking immune response to Aβ); transcriptomic note: **DIO2** ↓ in late-onset AD (context of AD but tied to TH activation)	(Zhaoqing Li & Jia Liu, 2024)

Continued

TABLE 15.3 The link between thyroid disease and cognitive performance, associated neuronal mechanisms, and gene expressions.—cont'd

Condition	Brain Region	Cognitive/Functional Impact	Mechanism	Transcriptional Modulation	Reference
			↑IL-1β/IL-6/TNF; autophagy upregulation → neuronal death and facilitates Aβ accumulation; ER stress and mitochondrial dysfunction; impaired insulin signaling (↓GSK3 activation, tau changes with TH); ↑APOE4-exosome brain delivery; ↓ cerebral blood flow to memory regions		
Adult hypothyroidism (or TRα2-/- overexpression)	Hippocampus (SGZ, dentate gyrus)	↓ hippocampal neurogenesis; memory impairment; anxiety/depression-like symptoms; ↓ dendritic arborization.	↓ progenitor proliferation; impaired survival/differentiation of neuroblasts; structural alterations in hippocampus. TH supplementation rescues these modifications.	T3 via TRα1 represses pluripotency genes like **Sox2**; also repression of cell cycle regulators (CyclinD1, c-Myc) in SVZ; expression of TR isoforms (TRα1, TRβ1, β2) in SGZ; adult-onset hypothyroidism changes expression of these regulators.	(Remaud et al., 2014)
Autoimmune hyperthyroidism (Graves' disease)	Hippocampus	Exacerbated cognitive impairment; ↑ Aβ plaques in AD model	T3 drives pro-inflammatory microglial shift (↑TNFα, NOS, IL-1β; ↓Arg1, IL-4, CD206), reduces Aβ phagocytosis; RIPK3/MLKL-mediated necroptosis		(Piekiełko-Witkowska et al., 2025)
Adult hypothyroidism	Hippocampus, parahippocampus, amygdala (plus white/gray-matter changes reported)	↓ cognition and memory; anxiety/depression reported in some adults	↓ BDNF expression in hippocampus and amygdala; ↓ hippocampal volume, fewer neurons in CA1/CA3, dendritic loss; thyroid hormone modulates 5-HT metabolism/receptors which in turn influences BDNF levels	TH modulates 5-HT signaling; 5-HT modulates hippocampal BDNF **mRNA** (down-regulation under stress); overall, TH is "necessary for BDNF expression" in hippocampus/parahippocampus/cortex (as stated in the text)	(Madhusudhan et al., 2022)
L-T4 + T1AM treatment (hypothyroid mice)	Hippocampus	Memory rescued	↑ DCX⁺ cells vs hypothyroid	Significant upregulation of Mapk3 and Neurog2 vs L-T4; treatment effects also noted for Kdr, L1cam (and trends for Ngf, Kit, Ntf3)	(Rutigliano et al., 2023)
STZ-induced sporadic Alzheimer's-like model (icv STZ)	Hippocampus (CA1, CA3, dentate gyrus)	Impaired novel object recognition (short- and long-term memory)	↓ p-GSK3α/β (less inhibitory phosphorylation), neuroinflammation ↑ (Iba-1 microglia, GFAP astrocytes), cytokines ↑ (TNF-α, IL-6, IL-1β)	BDNF mRNA ↓ in hippocampus	(Sepúlveda et al., 2024)

Ach: Acetylcholine; Akt (Protein Kinase B, PKB): A serine/threonine kinase involved in cell survival, growth, proliferation, metabolism, and inhibition of apoptosis. Activated (phosphorylated) downstream of PI3K signaling; AMPAR/NMDAR: AMPA/NMDA glutamate receptors; Aβ: Amyloid-beta; Bcl2: B-cell lymphoma 2; BTEB: Basic Transcription Element-Binding protein; ChAT: Choline Acetyltransferase; CREB: cAMP response element-binding protein, a transcription factor that is activated by phosphorylation (e.g. by kinases like Akt, PKA), binds to cAMP response elements in DNA to regulate transcription of target genes important for synaptic plasticity, learning/memory; DM20: A splice isoform of PLP (Proteolipid Protein); ER: Endoplasmic Reticulum; GSK-3 (Glycogen Synthase Kinase-3): A serine/threonine kinase (two main isoforms GSK-3α and GSK-3β) that is active in resting cells, when phosphorylated at certain serine residues (e.g. Ser9 in GSK-3β) becomes inhibited. It regulates many downstream targets related to metabolism, cytoskeleton, transcription factors (e.g. tau in Alzheimer's disease); IL-1β: Interleukin-1 beta; IL-6: Interleukin-6; LTD: Long-Term Depression; LTP: Long-Term Potentiation; MAG: Myelin-Associated Glycoprotein; MAPK (Mitogen-Activated Protein Kinase)/ERK (ERK Extracellular signal-Regulated Kinase): These are part of signaling cascades that transmit extracellular signals (growth factors, hormones) into cellular responses including gene transcription, cell growth, differentiation.; MBP: Myelin Basic Protein; MCT8: Monocarboxylate Transporter 8; MLKL: Mixed Lineage Kinase Domain-Like protein; NR1: NMDA receptor subunit 1; Nrgn: Neurogranin (NRGN); Pcp-2: Purkinje cell protein 2; PLP: Proteolipid Protein; RIPK3: Receptor-Interacting Protein Kinase 3; Sox2: SRY-box transcription factor 2; TNF-α: Tumor Necrosis Factor alpha; VEGFA: Vascular Endothelial Growth Factor A.

intervention can significantly improve patients' quality of life (van Vliet et al., 2021).

Optimal management requires a multidisciplinary approach involving collaboration between endocrinologists, neurologists, and neuropsychologists to ensure accurate diagnosis, timely treatment, and targeted cognitive rehabilitation (Sharma et al., 2020). Additionally, given the heterogeneity of treatment response, long-term monitoring of glandular and cognitive function is of particular importance (Thiruvengadam & Luthra, 2021). Ultimately, future research should elucidate the linking mechanisms between thyroid disorders and neurodegeneration, identify predictive biomarkers for cognitive improvement, and develop more effective therapeutic interventions to protect brain health (Li & Liu, 2024).

Overall, thyroid disorders have widespread effects on cognition, ranging from brain fog and memory problems in hypothyroidism to attention deficits and impulsivity in hyperthyroidism (Piekiełko-Witkowska et al., 2025). The risk of dementia and the relative reversibility of symptoms with treatment double the importance of early diagnosis and active management (Kim et al., 2022). A comprehensive and multidisciplinary approach can lead to improved cognitive outcomes and enhance patients' quality of life (Leso et al., 2020). Clinical Implications of thyroid dysfunction on cognitive and neurological function, diagnostic approaches and treatment outcomes are summarized in Table 15.4.

Other hormones, endocrine tissues, and pathways

Hypothalamus exerts critical control over systemic homeostasis via regulation of pituitary output and orchestrated neural signaling, despite constituting under 2% of brain volume. It is categorized into different nuclei, caudal nuclei allocate metabolic resources to defensive, exploratory, and adaptive responses, while rostral nuclei modulate energy balance and reproductive functions. The neurohypophysial and tuberoinfundibular systems link hypothalamic output to anterior and posterior pituitary secretion, prompting not only thyroid regulation but also circadian rhythms, nutrient and fluid intake, lactation, and thermoregulation. During serious infection or illness, a syndrome known as nonthyroidal illness (NTIS) occurs, in which serum T3 reduces, reverse T3 (rT3) increases, T4 remains measurable, and TSH may not properly increase. Cytokines including IL-1, IL-6, and TNF-α inhibit thyroid hormone synthesis, while inflammation alters peripheral deiodinase expression (e.g., in liver and muscle), disrupting T4 to T3 conversion resulting in inactivation. In acute inflammation, hypothalamic tanycytes activate DIO2 locally, increasing T3 concentration at the cost of suppressing TRH mRNA and dampening hypothalamic drive. Clinically, lower serum T4 in NTIS correlates with worse prognosis and higher mortality (Feldt-Rasmussen et al., 2021).

Local deiodinase activity in median basal hypothalamus (MBH) and the pituitary links circulating T4 to the regulation of TRH/TSH feedback. In the brain, most T3 arises from local conversion rather than direct plasma influx. In iodine deficiency, cortex DIO2 induction helps maintain local T3 despite low plasma T4. In cold exposure, DIO2 induction boosts local TH signaling and thermogenesis via upregulation of UCP1 (Bianco et al., 2019). Glial cells convert T4 to T3 via DIO2, which acts on neurons and oligodendrocytes to regulate genes controlling myelination and connectivity. DIO2 knockout mice exhibit preserved serum T3 but sharply reduced brain T3, demonstrating reliance on local synthesis, though residual T3 from circulation partially mitigates damage. Under hypothyroid or iodine-deficient states, upregulation of DIO2 and suppression of DIO3 act to preserve local TH signaling (Sabatino et al., 2021). DIO3 opposes local T3 action, increased DIO3 attenuates thyroid signaling and under some conditions can precipitate systemic hypothyroidism. During development, reciprocal expression of DIO2 and DIO3 tailors local thyroid action in tissues (Bianco et al., 2019).

Steroid-mediated suppression of the HPA axis may constitute a protective adaptation to metabolic stress; chronically, this may reduce energy expenditure and predispose to obesity. It remains unclear whether glucocorticoids entirely block cold responses, or instead impose a ceiling effect that overshadows further TH induction. TRH neurons in the rostral PVN increase pro-TRH mRNA in response to both glucocorticoids and cold, suggesting a shared integration of stress and thermal inputs distinct from classical hypophysiotropic neurons. Yet both stimuli also activate the HPA axis; over prolonged exposure, steroids suppress HPA output via negative feedback. The interplay of low temperature, stress, and glucocorticoids thus modulates TRH suppression and HPA axis activity in both basal and cold states (Petrowski & Kahaly, 2025).

Thyroid hormone also modulates cholesterol homeostasis directly and via interactions with FXR, LXR, PPAR, and coactivator PGC-1α. It stimulates both lipogenesis and lipolysis contextually, and through bile acid pathways enhances D2 activity and TH-driven energy expenditure. Because lipid and carbohydrate metabolism intersect, TH also enhances hepatic insulin sensitivity and suppresses gluconeogenesis. D2 gene polymorphisms (e.g., Thr92Ala) are unpredictably related to insulin resistance, type 2 diabetes, and obesity, particularly when concomitant with variants in interacting pathways like PPARγ, highlighting the integrated role of TH signaling in systemic metabolism (Deng et al., 2023; Mullur et al., 2014).

TABLE 15.4 Clinical implications of thyroid dysfunction on cognitive and neurological function.

Thyroid Condition	Cognitive and neurological effects	Environmental/Genetic risk factors	Diagnostic approaches	Association with neuro-degenerative diseases	Treatment outcomes
Hypothyroidism (Eslami-Amirabadi and Sajjadi, 2021; Kim et al., 2022; Li & Liu, 2024; Uma et al., 2021)	Short- and long-term memory deficits, brain fog(mental slowness, poor focus), slowed processing, executive dysfunction (e.g., planning, problem-solving), depression, anxiety, and pseudo-dementia symptoms. Reduced psychomotor speed and speech issues are also reported.	Iodine deficiency, autoimmune diseases (Hashimoto's), family history of hypothyroidism, and genetic mutations in TH receptors (TRα/β). Chronic stress and aging increase risk.	Regular TSH and free T4 screening in older adults and high-risk groups (pregnant women, autoimmune patients). MRI imaging to assess hippocampal atrophy in chronic cases.	81% increased risk of Alzheimer's and dementia in those over 65 due to neuroinflammation, beta-amyloid deposition, and reduced cerebral blood flow. Consistent treatment reduces risk, but prolonged hypothyroidism elevates it.	Levothyroxine (L-T4) improves memory, processing speed, and psychiatric symptoms. Early treatment is typically reversible, but delays may cause hippocampal atrophy and permanent deficits. Combined T3/T4 therapy may be more effective in some patients.
Hyperthyroidism (Eslami-Amirabadi and Sajjadi, 2021; Holmberg et al., 2019; Yashkin et al., 2024)	Severe anxiety, irritability, reduced sustained attention, working memory and object recognition deficits, impulsive decision-making, and occasional psychiatric symptoms like mania or paranoia. Visuomotor impairments and reduced performance in complex cognitive tasks are reported.	Graves' disease, thyroid nodules, excessive iodine intake, and genetic mutations in TSHR pathways. Acute stress and environmental toxins increase risk.	TSH, free T3, and free T4 testing; thyroid imaging (ultrasound or scintigraphy) for nodules or Graves'. Neuropsychological tests to assess cognitive deficits.	65% increased dementia risk in older adults due to oxidative stress, excitotoxicity, and reduced dendritic spine density in the hippocampus. Early treatment mitigates risk, but hyperthyroidism history elevates it.	Antithyroid drugs (methimazole) or radioactive iodine improve anxiety, attention, and memory. Outcomes depend on disease severity and duration; oxidative damage may not fully reverse.
Congenital Hypothyroidism (Bernal, 2017)	Reduced IQ (15–20 points), impaired verbal recall, learning and spatial memory issues, reduced hippocampal volume, motor and speech problems, and deficits in attention and visuospatial processing. Symptoms are more severe in untreated children, potentially leading to cretinism.	Maternal iodine deficiency, genetic mutations in TH synthesis (TPO, DUOX2), and pituitary-thyroid axis defects. Environmental iodine deficiency in endemic areas is a key factor.	Neonatal TSH screening; free T4 and iodine testing in pregnant mothers. Periodic neuropsychological assessments for at-risk children.	Long-term risk of cognitive deficits and increased dementia in adulthood due to early brain damage (disrupted neurogenesis, synaptogenesis). Preventive treatment reduces risk.	Early levothyroxine (within 2 weeks of birth) reduces cognitive and neurological deficits, but structural damage (hippocampal volume reduction) may persist. Neonatal screening is critical.
Euthyroid (Normal) (Martin & Sarkar, 2023)	Optimal cognitive and neurological function with normal memory, attention, and learning. No psychiatric or neurological symptoms, serving as a reference for comparison	Balanced neuronal metabolism, high BDNF expression, and normal synaptic function. Normal TH levels maintain neural plasticity and prevent neuroinflammation.	Periodic TSH and free T4 monitoring in older adults or high-risk groups (autoimmune patients). Balanced diet with sufficient iodine	Lower risk of dementia and neurodegenerative diseases in euthyroid state. Deviation to hypo- or hyperthyroidism increases risk.	Not applicable; maintaining euthyroid state via diet (adequate iodine) and healthy lifestyle preserves cognitive function.

Endocrine dysfunction is progressively associated with Alzheimer's disease (AD) and other types of dementias. The thyroid gland, directed by the HPT axis, produces T4 and some T3, with TH guiding myelination, neuronal differentiation, and synaptic formation. With AD progression and β-amyloid accumulation, hypothyroid states often emerge. In early studies, patients with AD exhibited lower TRH, TSH, and TH levels, and feedback correlations among these axes were lost, indicating breakdown of regulatory control. Moreover, in AD patients the level of TSH was reduced compared to controls with lack of circadian rhythm, leading to pituitary or hypothalamic degeneration that weakened secretion of TRH/TSH and feedback regulation (Li & Jia Liu, 2024).

Thyroid hormones participate in intricate crosstalk with other endocrine axes, including the pancreatic/metabolic systems and HPA axis that modulate metabolism, systemic homeostasis, and stress responses. The hypothalamus-pituitary-thyroid (HPT) axis is central to sustaining proper TH output. The deiodinase system is essential for activating or inactivating THs at the tissue level. Disruption of these endocrine pathways is increasingly implicated in neurodegenerative processes and intellectual disability.

Applications of the material

Levothyroxine sodium is recognized as a leading therapy for hypothyroidism (Chiovato et al., 2019). In addition, hyperthyroidism is usually treated with antithyroid drugs, including methimazole, carbimazole and propylthiouracil (PTU) (Joshi et al., 2024). Research indicates that treating hypothyroidism with levothyroxine can reversed cognitive and memory issues (Uma et al., 2021). Levothyroxine treatment cannot always restore cognitive function completely. Although therapy can help repair brain connections in hippocampus, this regeneration may be inadequate for complete recovery in all patients (Li & Jia Liu, 2024). Accurate Levothyroxine dose is important to avoid cognitive issues. Excessive drugs can cause hyperthyroidism, which can increase metabolism, anxiety and attention problems. Conversely, very few drugs can cause frequent hypothyroidism, which reduces the energy required for appropriate memory and decisions (Ouled Toumi et al., 2025).

The primary method to prevent intellectual disability from congenital hypothyroidism is to detect newborn screening and high TSH levels - an indication of low thyroid hormones - current screening protocols successfully identify 90% of congenital hypothyroidism cases (Guerri et al., 2019). Early levothyroxine treatment is important for children with congenital hypothyroidism, but also requires extensive follow-up care and developmental programs to ensure optimal cognitive development (Ontiveros-Mendoza et al., 2025). Additionally, supportive treatment such as T3 supplementation and BDNF enhancers, shows promise in reducing neurodevelopmental loss related to congenital hypothyroidism. For example, a study on hypothyroid mice has shown that treatment with T3 restore the necessary developmental processes, such as reeline MRNA expression in both in vitro and vivo settings, indicating that the installation of radial neurogenic gradiants is dependent on the thyroid hormone (Uchida & Suzuki, 2021). In addition, in another study, administering hypothyroid mice to administer BDNF or NT -3 prevents neurons' death, suggesting that these neurotrophin can directly compensate for developmental problems caused by lack of thyroid (Uchida & Suzuki, 2021).

Because iodine deficiency is the primary cause of global thyroid disorders, comprehensive iodine supplementation is an important preventive measure (Chiovato et al., 2019). Iodine supplementation should be provided universally, with special attention to weak groups such as pregnant and breastfeeding women, which require additional iodine. Overall, comprehensive health management, from prevention to long-term monitoring, require coordination between many healthcare professionals, including endocrinologists and neurologists (Chiovato et al., 2019).

Mini-dictionary of terms

- **Hippocampus.** A nervous structure in the brain that plays an important role in learning and memory, located in the temporal lobe.
- **Neurogenesis.** The formation of new neurons from stem cells in areas including the hippocampus contributes to learning and memory.
- **Neuroplasticity.** A feature required for memory and learning processes, which helps them and improves cognitive functions.
- **Synaptic Transmission.** A procedure involved in memories or learning is that the signals are transmitted from one neuron to another via synapses.
- **Synapse.** The site where two neurons communicate with each other.
- **Cerebral Cortex.** The cortical layer of the brain, responsible for processing information, also plays a role in learning and memory formation.
- **Neuroendocrine System.** Refers to endocrine glands that, with the help of hormones and the nervous system, regulate body activities, including effects on memory, learning, and the hippocampus.
- **Epigenetics.** Changes in gene expression that occur without altering the DNA sequence.
- **Synaptogenesis.** In this process, new synapses are formed between neurons, which increases neuronal connections and helps in the transmission of nerve signals.

- **Myelination.** This refers to the process in which a myelin sheath is formed around the axon of a neuron by cells called oligodendrocytes. Myelination of the axon increases the speed of nerve signal transmission and contributes to the formation of memory and learning.

Summary points

- The central nervous system depends on the thyroid hormones T3 and T4, which also control brain growth and cognitive function.
- Especially during the prenatal and early postnatal stages, T3 and T4 have an impact on neurogenesis, myelination, synaptogenesis, neuronal migration, and synaptic transmission.
- Irreversible neurological abnormalities, including intellectual deficits, can result from thyroid hormone deficiencies throughout critical developmental periods.
- Thyroid hormones in adults support neuroplasticity, neurotransmitter balance, and hippocampus activity, all of which contribute to cognitive performance.
- Cognitive deficiencies like brain fog, memory problems, and reduced processing speed are linked to hypothyroidism.
- Hyperthyroidism can affect cognitive function by causing anxiety, impatience, and attentional problems.
- There is a U-shaped curve in the association between thyroid hormone levels and cognitive function, with both hypo- and hyperthyroidism affecting brain function. Early identification and intervention are crucial since thyroid abnormalities are associated with a higher risk of dementia. Prompt hormone replacement or antithyroid medication can frequently reverse cognitive symptoms, improving patient outcomes.
- Thyroid-related cognitive deficits require a multidisciplinary approach from endocrinologists, neurologists, and neuropsychologists.

Declaration of AI assistance

ChatGPT and QuillBot were employed to support final language refinement in preparing this manuscript. All content was subsequently reviewed and revised by the authors, who accept full responsibility for the final published work.

Conflicts of interest

The authors confirm that they have no conflicts of interest to disclose.

References

Accorroni, A., Chiellini, G., & Origlia, N. (2017). Effects of thyroid hormones and their metabolites on learning and memory in normal and pathological conditions. *Current Drug Metabolism, 18*(3), 225–236.

Alcaide Martin, A., & Mayerl, S. (2023). Local thyroid hormone action in brain development. *International Journal of Molecular Sciences, 24* (15), Article 12352.

Alšauskė, S. V., Liseckienė, I., & Verkauskienė, R. (2024). The association of thyroid disease with risk of dementia and cognitive impairment: A systematic review. *Medicina (Kaunas), 60*(12). https://doi.org/10.3390/medicina60121917

Alzoubi, K., & Alkadhi, K. (2023). Thyroidectomy and PTU-induced hypothyroidism: Effect of L-thyroxine on suppression of spatial and non-spatial memory related signaling molecules. *Current Molecular Pharmacology, 16*(6), 654–663.

Anderson, G. W., Schoonover, C. M., & Jones, S. A. (2003). Control of thyroid hormone action in the developing rat brain. *Thyroid, 13*(11), 1039–1056.

Bagamasbad, P. D., Espina, J. E. C., Knoedler, J. R., Subramani, A., Harden, A. J., & Denver, R. J. (2019). Coordinated transcriptional regulation by thyroid hormone and glucocorticoid interaction in adult mouse hippocampus-derived neuronal cells. *PLoS One, 14*(7), Article e0220378.

Barbaro, D., Orrù, B., & Unfer, V. (2019). Iodine and myo-inositol: A novel promising combination for iodine deficiency. *Frontiers in Endocrinology, 10*, 457.

Bavarsad, K., Hosseini, M., Hadjzadeh, M. A., & Sahebkar, A. A.-O. *The effects of thyroid hormones on memory impairment and Alzheimer's disease.* (1097-4652 ((Electronic))).

Bégin, M., Langlois, M., Lorrain, D., & Cunnane, S. (2008). Thyroid function and cognition during aging. *Current gerontology and geriatrics research, 2008*(1), Article 474868.

Bernal, J. *Thyroid hormones in brain development and function*. BTI - Endotext.

Bernal, J. (2017). Thyroid hormone regulated genes in cerebral cortex development. *Journal of Endocrinology, 232*(2), R83–R97.

Bernal, J. (2024). *Thyroid and Brain: Understanding the Actions of Thyroid Hormones in Brain Development and Function*. Bentham Science Publishers.

Bianco, A. C., Dumitrescu, A., Gereben, B., Ribeiro, M. O., Fonseca, T. L., Fernandes, G. W., & Bocco, B. M. (2019). Paradigms of dynamic control of thyroid hormone signaling. *Endocrine Reviews, 40*(4), 1000–1047.

Chakrabarti, N., Sarkar, P. K., Ray, A. K., & Martin, J. V. (2023). Unveiling the nongenomic actions of thyroid hormones in adult mammalian brain: The legacy of Mary B. Dratman. *Frontiers in Endocrinology, 14*, Article 1240265.

Chen, C., Xie, Z., Shen, Y., & Xia, S. F. (2018). The roles of thyroid and thyroid hormone in pancreas: Physiology and pathology. *International Journal of Endocrinology, 2018*(1), Article 2861034.

Chiovato, L., Magri, F., & Carlé, A. (2019). Hypothyroidism in context: Where we've been and where we're going. *Advances in Therapy, 36* (Suppl. 2), 47–58.

Cooke, G. E., Mullally, S., Correia, N., O'Mara, S. M., & Gibney, J. (2014). Hippocampal volume is decreased in adults with hypothyroidism. *Thyroid, 24*(3), 433–440. https://doi.org/10.1089/thy.2013.0058

Deng, Y., Han, Y., Gao, S., Dong, W., & Yu, Y. (2023). The physiological functions and polymorphisms of type II deiodinase. *Endocrinology and Metabolism, 38*(2), 190–202.

Dezonne, R. S., Stipursky, J., Araujo, A. P., Nones, J., Pavão, M. S., Porcionatto, M., & Gomes, F. C. (2013). Thyroid hormone treated astrocytes induce maturation of cerebral cortical neurons through modulation of proteoglycan levels. *Frontiers in Cellular Neuroscience, 7*, 125.

Ekinci, İ., Ekinci, E., Buyukkaba, M., Cinar, A., Utku, İ. K., Ozkan, H., Tunc, M., Kumbasar, A., & Tabak, O. (2021). Selective attention and information processing speed in graves' disease: Stroop interference effect. *Cureus, 13*(3).

Elbadawy, A. M., Mansour, A. E., Abdelrassoul, I. A., & Abdelmoneim, R. O. (2020). Relationship between thyroid dysfunction and dementia. *The Egyptian Journal of Internal Medicine, 32*(1), 9. https://doi.org/10.1186/s43162-020-00003-2

Eslami-Amirabadi, M., & Sajjadi, S. A. (2021). The relation between thyroid dysregulation and impaired cognition/behaviour: An integrative review. *Journal of Neuroendocrinology, 33*(3), Article e12948. https://doi.org/10.1111/jne.12948

Espina, J. E. C., & Bagamasbad, P. D. *Synergistic gene regulation by thyroid hormone and glucocorticoid in the hippocampus*. (0083–6729 (Print)).

Ettleson, M. D., & Papaleontiou, M. (2022). Evaluating health outcomes in the treatment of hypothyroidism. *Frontiers in Endocrinology, 13*, Article 1026262.

Ettleson, M. D., Raine, A., Batistuzzo, A., Batista, S. P., McAninch, E., Teixeira, M., Jonklaas, J., Laiteerapong, N., Ribeiro, M. O., & Bianco, A. C. (2022). Brain fog in hypothyroidism: Understanding the patient's perspective. *Endocrine Practice, 28*(3), 257–264. https://doi.org/10.1016/j.eprac.2021.12.003

Farías-Serratos, B. M., Lazcano, I., Villalobos, P., Darras, V. M., & Orozco, A. (2021). Thyroid hormone deficiency during zebrafish development impairs central nervous system myelination. *PLoS One, 16*(8), Article e0256207.

Feldt-Rasmussen, U., Effraimidis, G., & Klose, M. (2021). The hypothalamus-pituitary-thyroid (HPT)-axis and its role in physiology and pathophysiology of other hypothalamus-pituitary functions. *Molecular and Cellular Endocrinology, 525*, Article 111173.

Fotakis, C., Moros, G., Kontogeorgou, A., Iacovidou, N., Boutsikou, T., & Zoumpoulakis, P. (2022). Uncontrolled thyroid during pregnancy alters the circulative and exerted metabolome. *International Journal of Molecular Sciences, 23*(8), 4248.

Galow, A.-M., Brenmoehl, J., & Hoeflich, A. (2023). Synergistic effects of hormones on structural and functional maturation of cardiomyocytes and implications for heart regeneration. *Cellular and Molecular Life Sciences, 80*(8), 240.

Godbole, M. M., Pal, A., Pathak, A., Kumar, P., Sinha, R. A., & Mohan, V. Spatio-temporal effects of iodine/thyroid hormone deficiency on neurodevelopment and plausible amelioration approaches beyond the critical window. *Annals of the National Academy of Medical Sciences*, 1–7.

Guerri, G., Bressan, S., Sartori, M., Costantini, A., Benedetti, S., Agostini, F., Tezzele, S., Cecchin, S., Scaramuzza, A., & Bertelli, M. (2019). Hypothyroidism and hyperthyroidism. *Acta BioMedica: Atenei Parmensis, 90*(Suppl. 10), 83.

Hernandez, A., & Stohn, J. P. (2018). The type 3 deiodinase: Epigenetic control of brain thyroid hormone action and neurological function. *International Journal of Molecular Sciences, 19*(6), 1804.

Holmberg, M. O., Malmgren, H., Berglund, P., Bunketorp-Käll, L., Heckemann, R. A., Johansson, B., Klasson, N., Olsson, E., Skau, S., & Nystrom Filipsson, H. (2019). Structural brain changes in hyperthyroid Graves' disease: Protocol for an ongoing longitudinal, case-controlled study in Göteborg, Sweden-the CogThy project. *BMJ Open, 9*(11), Article e031168. https://doi.org/10.1136/bmjopen-2019-031168

Huang, P.-S., Wang, C.-S., Yeh, C.-T., & Lin, K.-H. (2019). Roles of thyroid hormone-associated microRNAs affecting oxidative stress in human hepatocellular carcinoma. *International Journal of Molecular Sciences, 20*(20), 5220.

Incerpi, S., Gionfra, F., De Luca, R., Candelotti, E., De Vito, P., Percario, Z. A., Leone, S., Gnocchi, D., Rossi, M., & Caruso, F. (2022). Extranuclear effects of thyroid hormones and analogs during development: An old mechanism with emerging roles. *Frontiers in Endocrinology, 13*, Article 961744.

Joshi, J. S., Shanoo, A., Patel, N., Gupta, A., & Patel, J.,N. (2024). From conception to delivery: A comprehensive review of thyroid disorders and their far-reaching impact on feto-maternal health. *Cureus, 16*(2).

Kale, J. S., Katole, N. T., Rode, S. B., Gade, S. A., & Gade, S. (2022). Effect on reaction time in primary hypothyroid patients before and after thyroxin treatment. *Cureus, 14*(6).

Kim, J. H., Lee, H. S., Kim, Y. H., Kwon, M. J., Kim, J. H., Min, C. Y., Yoo, D. M., & Choi, H. G. (2022). The association between thyroid diseases and Alzheimer's disease in a national health screening cohort in Korea. *Frontiers in Endocrinology, 13*, Article 815063. https://doi.org/10.3389/fendo.2022.815063

Kumar, M., Singh, S., Rana, P., Kumar, P., Sekhri, T., Kanwar, R., D'Souza, M., & Khushu, S. (2021). Neurometabolite changes in hyperthyroid patients before and after antithyroid treatment: An in vivo 1H MRS study. *Frontiers in Human Neuroscience, 15*, Article 739917.

Lariosa-Willingham, K., Leonoudakis, D., Bragge, T., Tolppanen, L., Nurmi, A., Flanagan, M., Gibson, J., Wilson, D., Stratton, J., & Lehtimäki, K. K. (2022). An in vivo accelerated developmental myelination model for testing promyelinating therapeutics. *BMC Neuroscience, 23*(1), 30.

Lasa, M., & Contreras-Jurado, C. (2022). Thyroid hormones act as modulators of inflammation through their nuclear receptors. *Frontiers in Endocrinology, 13*, Article 937099.

Leso, V., Vetrani, I., De Cicco, L., Cardelia, A., Fontana, L., Buonocore, G., & Iavicoli, I. (2020). The impact of thyroid diseases on the working life of patients: A systematic review. *International Journal of Environmental Research and Public Health, 17*(12), 4295.

Li, Z., & Liu, J. (2024). Thyroid dysfunction and Alzheimer's disease, a vicious circle. *Frontiers in Endocrinology, 15*, Article 1354372. https://doi.org/10.3389/fendo.2024.1354372

Liu, Y.-C., Yeh, C.-T., & Lin, K.-H. (2019). Molecular functions of thyroid hormone signaling in regulation of cancer progression and anti-apoptosis. *International Journal of Molecular Sciences, 20*(20), 4986.

Madhusudhan, U., Kalpana, M., Singaravelu, V., Ganji, V., John, N., Gaur, A., Madhusudhan, U., & Singaravelu, V. (2022). Brain-derived neurotrophic factor-mediated cognitive impairment in hypothyroidism. *Cureus, 14*(4).

Martin, J. V., & Sarkar, P. K. (2023). Nongenomic roles of thyroid hormones and their derivatives in adult brain: Are these compounds putative neurotransmitters? *Frontiers in Endocrinology, 14*, Article 1210540.

Moreno, M., de Lange, P., Lombardi, A., Silvestri, E., Lanni, A., & Goglia, F. (2008). Metabolic effects of thyroid hormone derivatives. *Thyroid, 18*(2), 239–253.

Mullur, R., Liu, Y.-Y., & Brent, G. A. (2014). Thyroid hormone regulation of metabolism. *Physiological Reviews, 94*(2), 355–382.

National Institute for Health and Care Excellence: Clinical Guidelines. (2023). In *Thyroid disease: Assessment and management*. National Institute for Health and Care Excellence (NICE). © nice 2024.

Noda, M. (2018). Chapter thirteen - Thyroid hormone in the CNS: Contribution of neuron–glia interaction. In G. Litwack (Ed.), *Vitamins and hormones* (Vol 106, pp. 313–331). Academic Press. https://doi.org/10.1016/bs.vh.2017.05.005

Omar, S. H. (2023). Unravelling the threads: A brief insight into vascular dementia. *Journal of Vascular Diseases, 2*(4), 419–437.

Ontiveros-Mendoza, E., González-Medrano, J. A., Rivera-González, R., Sánchez-Huerta, K., Sánchez, C., & Barragán-Mejía, G. (2025). Congenital hypothyroidism and associated visual-motor and intellectual development. *Pediatric Research*, 1–10.

Ouled Toumi, R., Kassaw, C., & Demareva, V. (2025). Gender disparities in cognitive impairment across neurological autoimmune disorders: A systematic review. *Frontiers in Neurology, 16*, Article 1555407.

Petrowski, K., & Kahaly, G. J. (2025). Stress & thyroid function–from bench to bedside. *Endocrine Reviews*, Article bnaf015.

Piekiełko-Witkowska, A., Duda, M. K., Bogusławska, J., & Mackiewicz, U. (2025). The impact of autoimmune thyroid disease on cognitive and psychiatric disorders: Focus on clinical, pre-clinical and molecular studies. *European Thyroid Journal, 14*(3). https://doi.org/10.1530/etj-24-0406

Puia, G., & Ravazzini, F. (2020). Thyroid hormones reduce nicotinic receptor mediated currents in SH-SY5Y neuroblastoma cells. *Pharmacological Reports, 72*(6), 1766–1771.

Rajan, A., & Fame, R. M. (2024). Brain development and bioenergetic changes. *Neurobiology of Disease, 199*, Article 106550.

Rasmussen, M. J., Robbins, H., Fipps, D. C., Rustad, J. K., Pallais, J. C., & Stern, T. A. (2023). Neuropsychiatric sequelae of thyroid dysfunction: Evaluation and management. *Primary Care Companion for CNS Disorders, 25*(6). https://doi.org/10.4088/PCC.23f03570

Remaud, S., Gothié, J.-D., Morvan-Dubois, G., & Demeneix, B. A. (2014). Thyroid hormone signaling and adult neurogenesis in mammals. *Frontiers in Endocrinology, 5*, 62.

Rivas, M., & Naranjo, J. R. (2007). Thyroid hormones, learning and memory. *Genes, Brain, and Behavior, 6*(Suppl 1), 40–44. https://doi.org/10.1111/j.1601-183X.2007.00321

Rutigliano, G., Bertolini, A., Grittani, N., Frascarelli, S., Carnicelli, V., Ippolito, C., Moscato, S., Mattii, L., Kusmic, C., & Saba, A. (2023). Effect of combined levothyroxine (L-T4) and 3-iodothyronamine (T1am) supplementation on memory and adult hippocampal neurogenesis in a mouse model of hypothyroidism. *International Journal of Molecular Sciences, 24*(18), Article 13845.

Sabatino, L., Lapi, D., & Del Seppia, C. (2024). Factors and mechanisms of thyroid hormone activity in the brain: Possible role in recovery and protection. *Biomolecules, 14*(2), 198.

Sabatino, L., & Vassalle, C. (2025). Thyroid hormones and metabolism regulation: Which role on brown adipose tissue and browning process? *Biomolecules, 15*(3), 361.

Sabatino, L., Vassalle, C., Del Seppia, C., & Iervasi, G. (2021). Deiodinases and the three types of thyroid hormone deiodination reactions. *Endocrinology and Metabolism, 36*(5), 952–964.

Schoenmakers, N., Alatzoglou, K. S., Chatterjee, V. K., & Dattani, M. T. (2015). Recent advances in central congenital hypothyroidism. *Journal of Endocrinology, 227*(3), R51–R71.

Sepúlveda, P., Ferreira, A. F. F., Sandoval, C., Bergoc, G., Moreno, A. C. R., Nunes, M. T., & Torrão, A.d. S. (2024). Thyroid hormone supplementation restores cognitive deficit, insulin signaling, and neuroinflammation in the hippocampus of a sporadic Alzheimer's-like disease rat model. *Cells, 13*(21), 1793.

Sharma, S. K., Mudgal, S. K., & Mandal, A. (2020). Thyroid disease in older people: Nursing perspectives. *Thyroid Research and Practice, 17*(3), 110–117. https://doi.org/10.4103/trp.trp_25_20

Smith, J. W., Evans, A.T. At Fau - Costall, B., Costall B Fau - Smythe, J. W., & Smythe, J. W. *Thyroid hormones, brain function and cognition: A brief review*. (0149–7634 (Print)).

Sönmez, G., Yazarkan, Y., Aki, Ö. E., & Bodur, E. (2025). Unraveling the enigma: Exploring the periphery's influence in Alzheimer's pathophysiology—cause or consequence? *The Eurasian Journal of Medicine, 57*(1), Article e24592.

Talhada, D., Santos, C. R. A., Gonçalves, I., & Ruscher, K. (2019). Thyroid hormones in the brain and their impact in recovery mechanisms after stroke. *Frontiers in Neurology, 10*, 1103.

Thiruvengadam, S., & Luthra, P. (2021). Thyroid disorders in elderly: A comprehensive review. *Disease-a-Month, 67*(11), Article 101223. https://doi.org/10.1016/j.disamonth.2021.101223

Uchida, K., & Suzuki, M. (2021). Congenital hypothyroidism and brain development: Association with other psychiatric disorders. *Frontiers in Neuroscience, 15*, Article 772382.

Uma, D., Rabbani, R., Lee, J. H., Gavini, D. R., Shah, P. H., & Hamid, P. (2021). Does hormone supplementation with levothyroxine improve hypothyroid impaired cognitive dysfunction? *Cureus, 13*(9), Article e17885. https://doi.org/10.7759/cureus.17885

Valcárcel-Hernández, V., Mayerl, S., Guadaño-Ferraz, A., & Remaud, S. (2024). Thyroid hormone action in adult neurogliogenic niches: The known and unknown. *Frontiers in Endocrinology, 15*, Article 1347802. https://doi.org/10.3389/fendo.2024.1347802

van Vliet, N. A., van Heemst, D., Almeida, O. P., Åsvold, B. O., Aubert, C. E., Bae, J. B., Barnes, L. E., Bauer, D. C., Blauw, G. J., Brayne, C., Cappola, A. R., Ceresini, G., Comijs, H. C., Dartigues, J.-F., Degryse, J.-M., Dullaart, R. P. F., van Eersel, M. E. A., den Elzen, W. P. J., Ferrucci, L., … Collaboration, T. S. (2021). Association of thyroid dysfunction with cognitive function: An individual participant data analysis. *JAMA Internal Medicine, 181*(11), 1440–1450. https://doi.org/10.1001/jamainternmed.2021.5078

Velasco, I., Bath, S. C., & Rayman, M. P. (2018). Iodine as essential nutrient during the first 1000 days of life. *Nutrients, 10*(3), 290.

Yajima, H., Amano, I., Ishii, S., Sadakata, T., Miyazaki, W., Takatsuru, Y., & Koibuchi, N. (2021). Absence of thyroid hormone induced delayed dendritic arborization in mouse primary hippocampal neurons through insufficient expression of brain-derived neurotrophic factor. *Frontiers in Endocrinology, 12*, Article 629100.

Yashkin, A. P., Kolpakov, S., Ukraintseva, S., Yashin, A., & Akushevich, I. (2024). Graves disease is associated with increased risk of clinical Alzheimer's disease: Evidence from the medicare system. *Clinical Diabetes and Endocrinology, 10*(1), 11. https://doi.org/10.1186/s40842-024-00170-z

Yu, D., Zhou, H., Zou, L., Jiang, Y., Wu, X., Jiang, L., Zhou, Q., Yang, Y., Xu, L., & Mao, R. (2017). Hippocampal administration of levothyroxine impairs contextual fear memory consolidation in rats. *Frontiers in Cellular Neuroscience, 11*, 223.

Zhu, W., Wu, F., Li, J., Meng, L., Zhang, W., Zhang, H., Cha, S., Zhang, J., & Guo, G. (2022). Impaired learning and memory generated by hyperthyroidism is rescued by restoration of AMPA and NMDA receptors function. *Neurobiology of Disease, 171*, Article 105807. https://doi.org/10.1016/j.nbd.2022.105807

矢島弘之 (2021). Absence of thyroid hormone induced delayed dendritic arborization in mouse primary hippocampal neurons through insufficient expression of brain-derived neurotrophic factor. Gunma University.

Chapter 16

Thyroid hormone signaling shaping fear memory: Implications for PTSD and other stress-related disorders

Olga Ponomareva[1], Radha Shah[2] and Kerry J. Ressler[1]
[1]*McLean Hospital, Belmont, MA, United States;* [2]*University of Toledo, Toledo, OH, United States*

List of abbreviations

D2 Type 2 deiodinase
D3 Type 3 deiodinase
GABA Gamma-aminobutyric acid
PFC Prefrontal Cortex
PTSD Posttraumatic Stress Disorder
SSRIs Selective serotonin reuptake inhibitors
TH Thyroid hormone
TR Thyroid hormone receptor
TRH Thyrotropin-releasing hormone
TRHR Thyrotropin-releasing hormone receptor

Introduction

Thyroid testing to identify potential underlying medical causes of depression, and use of thyroid hormone as an augmentation to antidepressants, are established as standards of care for treatment-resistant mood disorders with defined guidelines for dosing, monitoring, and duration of treatment (Rosenthal et al., 2011; Touma et al., 2017). Notably, trauma and other stress-related disorders, characterized by dysregulated expression of fear and threat memories, hypervigilance, hyperarousal, avoidance, and negative mood are also frequently treated with the same pharmacologic agents as those used for mood disorders, with emerging clinical evidence suggesting thyroid dysfunction in Posttraumatic Stress Disorder (PTSD) (Toloza et al., 2020). To date, however, no studies have examined thyroid supplementation for disorders of fear learning; yet preclinical animal studies are starting to mechanistically define thyroid pathway signaling during formation and consolidation of fear, threat, and trauma-related memories which underlie these disorders.

This chapter examines insights from clinical studies linking thyroid signaling to mental disorders with dysregulated fear and trauma learning, as well as preclinical animal models offering mechanistic insights into the role of thyroid signaling during distinct phases of fear learning and expression of fear memories. Of note, although both PTSD and thyroid related disorders are more common in females, mechanistic animal models have predominantly focused on male animals. Thus, mechanisms of active thyroid signaling during fear memory formation in females remain less explored and remains an important area for future research.

Fear memory: Definitions and implications in disease

Auditory fear conditioning, a form of Pavlovian conditioning, is a frequently used behavioral paradigm to study how animals acquire and express fear memories (Trent et al., 2025). In this task, an animal learns to associate a neutral tone with an aversive foot shock, typically measured by an increased freezing response when the tone is played later. Immediately after fear memory learning, the memory undergoes a consolidation phase – a protein synthesis-dependent process during which short-term memories are stabilized into long-term memories (Schafe & LeDoux, 2000). In the animal is later transferred to a different context and presented with the same tones without receiving a shock, its freezing response gradually declines. This reduction in fear behavior is known as extinction, demonstrating the animal learning that the tone no longer predicts danger in the new context (Fig. 16.1).

Hormones in Neuroscience. https://doi.org/10.1016/B978-0-443-30261-9.00016-4

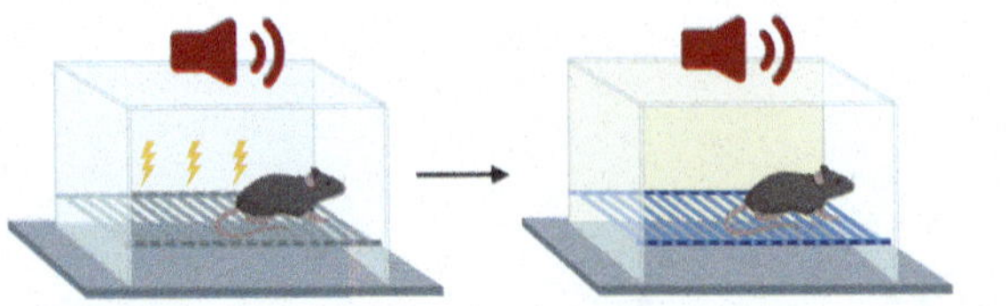

FIGURE 16.1 Acoustic fear conditioning and extinction. Fear conditioning is performed in context a with paired presentations of tones with brief foot shocks. Following fear learning, the memory of this paired association can be extinguished with repeated presentations of the tone in a new context.

Decades of clinical research, human neuroimaging, and preclinical circuit-mapping studies have identified key brain regions that work together to support fear learning and extinction (Fenster et al., 2018; Giustino & Maren, 2015; Greco & Liberzon, 2016). These regions include the Prefrontal Cortex (PFC), hippocampus, amygdala, and the brainstem (Fig. 16.2).

The PFC integrates sensory signals (such as contextually relevant sights, sounds, smells, pain), with emotional signals to help regulate expression of learned fear. The hippocampus contributes by encoding and retrieving contextual information about the environment both during the initial fear learning and during fear extinction. Finally, the amygdala is a central hub for fear memory formation by associating the conditioned stimulus (tone) with a natural, unconditioned response (freezing). Downstream structures, including the hypothalamus and brainstem, then translate amygdala output into the behavioral expression of fear.

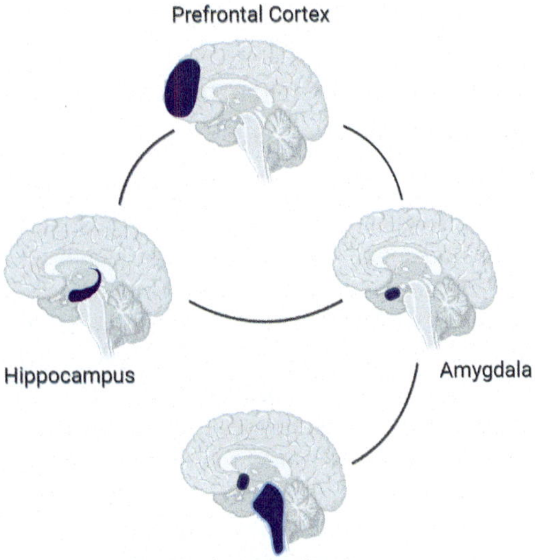

FIGURE 16.2 Brain regions involved in fear learning circuitry. Integration of sensory and emotional information in the prefrontal cortex (also known as the PFC) coordinates with the hippocampus, which is vital for processing cues, or context relevant information. The amygdala integrates this information and consolidates the fear memory, behavioral manifestation of which (such as freezing) is mediated by the hypothalamus and the brainstem.

Disorders of fear learning are characterized by maladaptive fear learning, in which individuals experience a deficit in fear learning and continue to generalize the learned fear to new (or previously safe) contexts (Kausche et al., 2025). The following section will examine evidence of thyroid signaling role during fear learning, extinction, and related disorders in a clinical setting. Although disordered fear learning has been observed in a variety of other clinical diagnoses marked by dysregulation of negative emotion, including a range of anxiety and mood disorders, for brevity this chapter will limit the scope to clinical examples related to PTSD (Fig. 16.3) (Trent et al., 2025).

Thyroid signaling in fear learning: Insights from clinical studies

PTSD introduction and initial studies linking thyroid disorders

The role of the thyroid and related hormones has been established in multiple neuropsychiatric disorders, most notably in Major Depressive Disorder (MDD). Depressive symptoms are often linked to hypothyroidism or subclinical hypothyroidism and the central nervous system is highly sensitive to changes in thyroid hormone levels (Toloza et al., 2021). As a result, clinical guidelines now recommend the use of triiodothyronine (T3) augmentation, especially when selective reuptake inhibitors (SSRIs) provide insufficient relief of symptoms (Dwyer et al., 2020; Rosenthal et al., 2011; Touma et al., 2017). This strong connection between thyroid function and mood regulation has broadened interest in the endocrine system's role in mental health. However, thyroid involvement in fear learning and in PTSD is less well understood, and its role in neurocircuit modulation of negative emotion is lacking, despite shared features such as cognitive difficulties and heightened stress responses.

Although most PTSD research has traditionally focused on the hypothalamic–pituitary–adrenal (HPA) axis, recent work has broadened attention to other neuroendocrine systems, including the hypothalamic–pituitary–thyroid (HPT) axis. Early studies in combat-exposed veterans have shown thyroid hormone profiles were often altered in individuals with PTSD. In one 1994 study, circulating thyroid hormones were measured in 96 combat veterans with chronic PTSD. Even though thyroid-stimulating hormone (TSH) levels were normal, the researchers observed higher T3/T4 ratios, increased total and free T3, and elevated thyroxine-binding globulin levels (Mason, 1994). This pattern pointed to an unusual "high T-3" state that did not match typical primary thyroid disease. These findings were

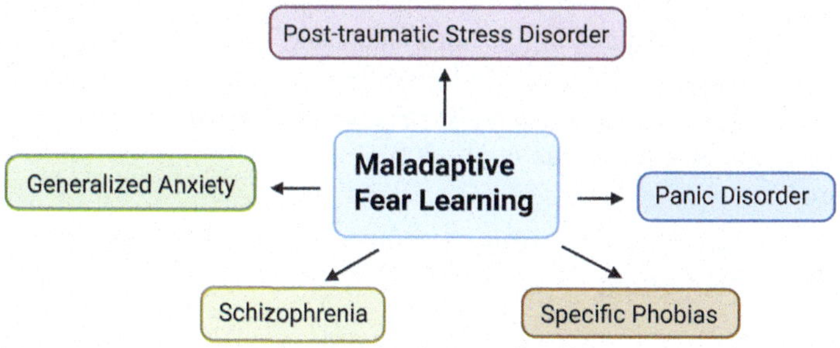

FIGURE 16.3 Clinical disorders associated with disrupted fear learning. Although most frequently associated PTSD, fear learning abnormalities are observed in a variety of psychiatric disorders.

interpreted as evidence of elevated deiodinase activity converting T4 to T3, along with sympathetic nervous system activation, which may reflect an adaptive response to chronic stress exposure. A follow-up study further showed that higher serum T3 and T4 levels were associated with more severe hyperarousal symptoms, including insomnia and exaggerated startle responses (Wang et al., 1995). Together, these studies suggested that thyroid hormones may amplify noradrenergic signaling and contribute to heightened vigilance in PTSD, providing early evidence that thyroid function is altered in trauma-related disorders and highlighting the importance of examining the HPT axis.

Findings from multicenter clinical studies

Large-scale and multicenter studies have expanded upon early findings to determine whether thyroid dysregulation is a consistent biomarker of PTSD across diverse demographics. A key study of 65 Vietnam veterans reported elevated levels of total and free T3 levels and a positive correlation between T3/T4 ratios and hyperarousal scores, reinforcing that this patterns was not unique to a single cohort (Wang & Mason, 1999). Of note, these early studies were conducted in men, despite the recognition that both thyroid disorders and PTSD are more common in women. Later research broadened participation to include women and civilians exposed to a wide range of traumas. Meta-analysis suggests that individuals with PTSD have significantly higher levels of T3 compared to controls, even after adjusting for factors such as age, sex, body mass index, and psychiatric medication use (Mandagere et al., 2025). One of the most comprehensive studies, using data from over 45,000 participants in the Nurses' Health Study II, found that women with increased reported PTSD symptoms were significantly more likely to develop hypothyroidism at follow-up (Jung et al., 2019). Notably, those who experienced PTSD earlier in life had an even stronger association, suggesting that long-term stress-related neuroendocrine disruption may increase risk for future thyroid dysfunction.

Taken together, multicenter results show that increased T3 levels and altered feedback regulation may represent physiological signatures of PTSD across different populations and trauma exposures (Fig. 16.4).

Emerging work: Thyroid signaling as a predictive PTSD biomarker

A recent study in n = 417 trauma survivors used machine learning of biomarkers obtained within 48 hours of trauma to predict development of PTSD. They found that levels of TSH, free T4 and T3 are among 15 of the most relevant predictors of PTSD 12 months after trauma exposure (Schultebraucks et al., 2021). Although this work was limited by its observational nature and somewhat limited number of subjects, it suggests that altered thyroid pathway signaling may precede and be involved in the pathogenesis of trauma and stressor-related disorders. Such prospective and mechanistic studies may be better explored in animal models, which are described in the following section.

Thyroid signaling in fear learning: Insights from animal studies

Animal models of stress: Fear learning

Despite recent advances in clinical studies, the underlying mechanisms linking trauma-related disorders and thyroid abnormalities remain unknown. Animal models have

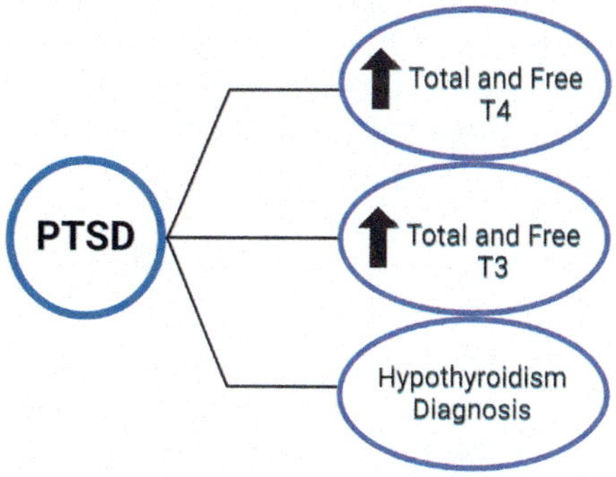

FIGURE 16.4 Clinical associations of thyroid status with PTSD. Increased T3, T4, and diagnosis of hypothyroidism are associated with PTSD.

begun to define thyroid signaling mechanisms during initial fear memory formation, a critical first step in the development of trauma and stressor-related disorders in humans. Auditory Pavlovian fear conditioning, an amygdala-dependent memory system, has emerged as an excellent animal model with which to examine the neurobiological mechanisms, at the cellular, molecular and physiological levels, which underlie the formation of traumatic fear memories (Blair et al., 2001; Johansen et al., 2011; Maren & Quirk, 2004). The strength of this paradigm is a robust behavioral response (increased freezing) in response to a conditioned stimulus (tone), along with the ability to reliably measure extinction of fear behaviors, which is frequently disrupted in patients with trauma-related disorders (Garfinkel et al., 2014; Norrholm et al., 2011).

Thyroid signaling in animal models of stress: Hippocampus

Much of the work in rodent models has focused on the role of thyroid signaling in the hippocampus, a structure critical for processing of contextual information during fear memory formation (Eichenbaum, 2000). In the hippocampus, free T3 and T4 levels are elevated after systemic administration of levothyroxine (synthetic form of T4) in adult male rats (Yu et al., 2017). Moreover, when levothyroxine is given immediately after fear conditioning, rats show impaired fear memory. In contrast, administering levothyroxine directly into the hippocampus before fear learning does not affect either the acquisition or retrieval of the fear memory. However, when levothyroxine is delivered during the memory consolidation phase, the period during which short-term memories are stabilized through molecular and synaptic changes, rats display deficits in later fear memory retrieval. Taken together, these findings suggest that thyroid signaling in the hippocampus is critical during the stage of fear memory consolidation, but does not significantly influence the initial learning of the fear association (Yu et al., 2017).

In the brain, the transcriptionally active thyroid hormone T3 is produced from T4 through the action of two enzymes, the type 2 (D2) and type 3 (D3) deiodinases (Courtin et al., 2005). Mice that lack D2 have decreased T3 in the brain, increased anxiety-like behaviors, and increased fear memory, which are likely due to altered gene expression downstream of T3 signaling (Bárez-López et al., 2017). Interestingly, local infusion of T3 directly into the rat hippocampus (before or after fear learning), also enhances fear memory, suggesting that timing and dynamic nature of T3 signaling, rather than absolute amounts, are vital for downstream gene signaling during fear memory formation (Sui et al., 2006). Such downstream gene expression and intracellular changes downstream of T3 are mediated by thyroid receptors alpha and beta (TRα and TRβ). Interestingly, mice that lack TRα1 have increased freezing 1 week after initial fear conditioning, suggesting that this receptor plays a critical role in fear memory (Guadaño-Ferraz et al., 2003).

Thyroid signaling in animal models of stress: Amygdala

Recent work using unbiased RNA sequencing approaches as well as local, mechanistic testing of thyroid role in fear learning, have begun to elucidate the role of thyroid-mediated transcriptional changes during fear learning in the amygdala (Lori et al., 2018; Maddox et al., 2025). Interestingly, local activation of thyroid signaling in the amygdala with an infusion of active T3 hormone prior to fear conditioning, decreases subsequent fear extinction. Conversely, local antagonism of thyroid signaling with an injection of a small molecule thyroid hormone receptor antagonist, has the opposite effect of increasing fear extinction. These effects appear to be moderated by transcriptional changes downstream of T3 hormone (Maddox et al., 2025). Interestingly, mRNA levels of known TR and other TH pathway regulatory genes, including thyrotropin-releasing hormone (*Trh*), transthyretin (*Ttr*), thyrotropin-releasing hormone receptor (*Trhr*), type 2 iodothyronine deiodinase (*Dio2*, also known as D2), mediator complex subunit 12 (*Med12/Trap230*) and retinoid X receptor gamma (*Rxrg*) are dynamically altered in the amygdala after fear conditioning, suggesting dynamic transcriptional regulation by TR (Table 16.1).

Conversely, inhibiting thyroid signaling in the amygdala with a TR antagonist decreases fear memory consolidation; while activation of TR with T3 increases fear memory formation. Interestingly, infusing T3 directly into the amygdala rescues fear memory deficits observed in a hypothyroid mouse model. Collectively, these results suggest that dynamic thyroid hormone signaling and TR modulation in the amygdala are critical regulators of fear memory plasticity (Maddox et al., 2025).

Other hormones and pathways

Recognition of increased prevalence of PTSD, anxiety disorders, and thyroid abnormalities in females has raised the question of potential interaction of thyroid signaling with sex hormones and their derivatives. Estrogen and androgens modulate thyroid signaling by changing levels of the thyroid binding globulin (TBG), while progesterone can directly regulate gene expression in thyroid cells (Santin & Furlanetto, 2011; Tahboub & Arafah, 2009).

Estrogen increases synthesis and decreases clearance of TBG which leads to an increase in total T3/T4 concentrations (Ain et al., 1988). This is frequently observed

TABLE 16.1 Transcriptional thyroid changes during fear learning. Components of the thyroid pathway are differentially regulated after fear learning in the mouse amygdala.

	Amygdala mRNA after fear learning
Thyroid hormone receptor alpha (Thra)	Increased
Thyroid hormone receptor beta (Thrb)	Increased
Thyrotropin-releasing hormone receptor (Trhr)	Decreased
Thyrotropin releasing hormone (Trh)	Decreased
Type 2 iodothyronine deiodinase (Dio2)	Increased
Transthyretin (Ttr)	Decreased

during periods of high estrogen signaling, such as in pregnancy or with estrogen replacement therapy. In contrast to estrogens, androgens, including testosterone, down-regulate TBG.

Multiple studies have shown that progesterone and its derivatives can directly regulate gene expression in thyroid cells (Benvenga et al., 2018; Bertoni et al., 2015). Of particular interest in the recent years is the recognition of GABA-A modulation by allopregnanolone, a steroid hormone derivative of progesterone. Disrupted GABAergic signaling is recognized as a hallmark of PTSD and allopregnanolone has become a PTSD therapeutic target under active investigation since the recognition that allopregnanolone levels are correlated with response to trauma (Amir Hamzah et al., 2025; Huang et al., 2023).

Like allopregnanolone, thyroid signaling is also a modulator of GABA-A signaling in the brain (Puia & Losi, 2011; Wiens & Trudeau, 2006). Hypothyroidism is associated with a reduction in GABAergic signaling in key brain regions mediating fear learning, including mPFC; furthermore, restoration of hypothyroid patients to euthyroid status normalizes GABA levels and psychiatric symptoms (Liu et al., 2017, 2020). In particular, T3 can inhibit the GABA-A receptor, and studies have demonstrated that T3 and allopregnanolone compete for binding on GABA-A receptor (Westergard et al., 2015). Intriguingly, allopregnanolone is increased in thyroidectomized patients, although direct causation is not established (Bicikova et al., 1998). Fiinally, the gene encoding the rate limiting enzyme for allopregnanolone synthesis, *5-alpha reductase,* contains predicted thyroid response elements, suggesting that thyroid signaling may directly influence transcriptional levels of allopregnanolone (Chiao et al., 1999; Kala et al., 2002; Ram & Waxman, 1990) (Fig. 16.5).

Taken together, these findings suggest that thyroid hormone levels may interact with allopregnanolone signaling to influence fear learning.

Applications of the material

The thyroid hormone pathway has long been generally known to be important for brain function, and to potentially be important in mood-, stress-, and trauma-related disorders. Thyroid testing and supplementation of antidepressants is established as standard of care for treatment-resistant mood disorders with defined guidelines for dosing, monitoring, and duration of treatment (Bauer & Whybrow, 2021). The same pharmacologic agents are used for the treatment of PTSD, with emerging clinical evidence suggesting thyroid dysfunction in this disorder, however, no recommendations regarding thyroid testing or supplementation are in place for trauma-related disorders to date.

Recent studies described here have begun to define thyroid pathway signaling during formation and consolidation of trauma- or stress-related memories in clinical studies as well as in animal models. Notably, although both PTSD and thyroid related disorders are more common in females, to date mechanistic animal work has been performed in male animals, and studies of thyroid signaling in females are lacking. Following rigorous demonstration of location, timing, and behavioral sequelae of thyroid hormone pathway activity in the context of fear learning in animals of both sexes, pharmacological targets for thyroid supplementation should be tested in fear- and stress-related mental disorders. Although in the early stages of development, predictive models of PTSD in individuals who undergo trauma may use thyroid levels as biomarkers for symptom development and, possibly, response to treatment.

In conclusion, thyroid signaling and downstream transcriptional changes during fear learning remains an active target of investigation, with promising insights into the development, progression, and predicted development of fear and stress-related disorders.

Mini-dictionary of terms

Allopregnanolone – A neurosteroid that modulates GABAergic signaling and is implicated in PTSD.

Auditory fear conditioning – A behavioral paradigm in which an animal learns to associate a neutral sound (tone) with an aversive stimulus (shock), often measured by freezing behavior.

Deiodinases (D2 and D3) – Enzymes that convert T4 into the active T3 or inactive forms, regulating local thyroid hormone activity.

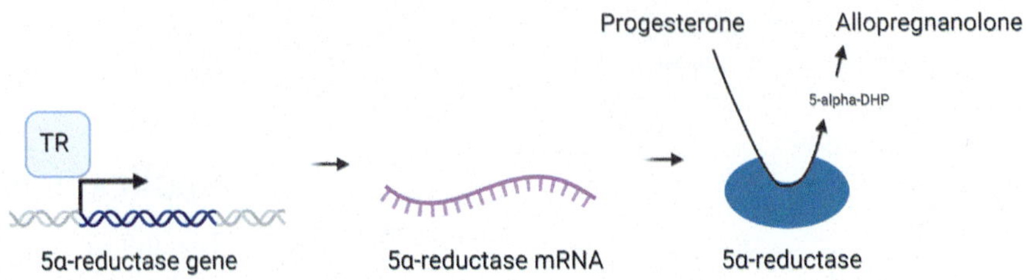

FIGURE 16.5 Hypothesized pathway by which a putative thyroid response element in the 5-alpha reductase gene mediates allopregnanolone levels during fear learning. Thyroid hormone receptor (TR) binding to the thyroid response element moderates mRNA and, subsequently, levels of functional 5-alpha reductase protein, which then converts progesterone into allopregnanolone.

Fear Memory Consolidation - The process by which short-term memories are stabilized and stored as long-term memories, often requiring protein synthesis.

Fear Memory Extinction – The gradual decrease of a learned fear response when the conditioned stimulus is repeatedly presented without the aversive event.

Freezing behavior – A common defensive response in rodents, used as a measure of fear.

Levothyroxine – A synthetic form of T4 used experimentally or clinically to increase thyroid hormone levels.

T3 downstream transcription - changes in gene expression as a result of activated thyroid signaling.

Summary points

- Early clinical studies observed higher T3/T4 ratios, increased total and free T3, and elevated thyroxine-binding globulin levels in male veterans with PTSD.
- Subsequent studies suggest that individuals with increased number of PTSD symptoms are significantly more likely to develop a diagnosis of hypothyroidism.
- Hippocampal thyroid signaling is critical during fear memory consolidation but likely does not significantly influence the initial learning of the fear association.
- Transcriptional changes downstream of TH receptor mediate initial formation of fear memories in the amygdala.
- Further work on modulation of GABA-A signaling by allopregnanolone and thyroid pathways during fear learning may be an intriguing subject of future investigation.

References

Ain, K. B., Refetoff, S., Sarne, D. H., & Murata, Y. (1988). Effect of estrogen on the synthesis and secretion of thyroxine-binding globulin by a human hepatoma cell line, HEP G2. *Molecular Endocrinology, 2*, 313–323. https://doi.org/10.1210/mend-2-4-313

Amir Hamzah, K., Lipp, O. V., & Ney, L. J. (2025). Allopregnanolone and intrusive memories: A potential therapeutic target for PTSD treatment? *Progress in Neuro-Psychopharmacology and Biological Psychiatry, 136*, 111168. https://doi.org/10.1016/j.pnpbp.2024.111168

Bárez-López, S., Montero-Pedrazuela, A., Bosch-García, D., Venero, C., & Guadaño-Ferraz, A. (2017). Increased anxiety and fear memory in adult mice lacking type 2 deiodinase. *Psychoneuroendocrinology, 84*, 51–60. https://doi.org/10.1016/j.psyneuen.2017.06.013

Bauer, M., & Whybrow, P. C. (2021). Role of thyroid hormone therapy in depressive disorders. *Journal of Endocrinological Investigation, 44*, 2341–2347. https://doi.org/10.1007/s40618-021-01600-w

Benvenga, S., Capodicasa, G., Perelli, S., Ferrari, S. M., Fallahi, P., & Antonelli, A. (2018). Increased requirement of replacement doses of levothyroxine caused by liver cirrhosis. *Frontiers in Endocrinology, 9*, 150. https://doi.org/10.3389/fendo.2018.00150

Bertoni, A. P. S., Brum, I. S., Hillebrand, A. C., & Furlanetto, T. W. (2015). Progesterone upregulates gene expression in normal human thyroid follicular cells. *International Journal of Endocrinology, 2015*, 1–6. https://doi.org/10.1155/2015/864852

Bicikova, M., Putz, Z., Hill, M., Hampl, R., Diebbelt, L., Tallova, J., & Starka, L. (1998). Serum levels of neurosteroid allopregnanolone in patients with premenstrual syndrome and patients after thyroidectomy. *Endocrine Regulations, 32*, 87–92.

Blair, H. T., Schafe, G. E., Bauer, E. P., Rodrigues, S. M., & LeDoux, J. E. (2001). Synaptic plasticity in the lateral amygdala: A cellular hypothesis of fear conditioning. *Learning & Memory, 8*, 229–242. https://doi.org/10.1101/lm.30901

Chiao, Y. C., Lee, H. Y., Wang, S. W., Hwang, J. J., Chien, C. H., Huang, S. W., Lu, C. C., Chen, J. J., Tsai, S. C., & Wang, P. S. (1999). Regulation of thyroid hormones on the production of testosterone in rats. *Journal of Cellular Biochemistry, 73*, 554–562.

Courtin, F., Zrouri, H., Lamirand, A., Li, W. W., Mercier, G., Schumacher, M., Goascogne, C. L., & Pierre, M. (2005). Thyroid hormone deiodinases in the central and peripheral nervous system. *Thyroid, 15*, 931–942. https://doi.org/10.1089/thy.2005.15.931

Dwyer, J. B., Aftab, A., Radhakrishnan, R., Widge, A., Rodriguez, C. I., Carpenter, L. L., Nemeroff, C. B., McDonald, W. M., Kalin, N. H., & APA Council of Research Task Force on Novel Biomarkers and Treatments. (2020). Hormonal treatments for major depressive disorder: State of the art. *The Australian Journal of Pharmacy, 177*, 686–705. https://doi.org/10.1176/appi.ajp.2020.19080848

Eichenbaum, H. (2000). A cortical–hippocampal system for declarative memory. *Nature Reviews Neuroscience, 1*, 41–50. https://doi.org/10.1038/35036213

Fenster, R. J., Lebois, L. A. M., Ressler, K. J., & Suh, J. (2018). Brain circuit dysfunction in post-traumatic stress disorder: From mouse to man. *Nature Reviews Neuroscience, 19*, 535–551. https://doi.org/10.1038/s41583-018-0039-7

Garfinkel, S. N., Abelson, J. L., King, A. P., Sripada, R. K., Wang, X., Gaines, L. M., & Liberzon, I. (2014). Impaired contextual modulation of memories in PTSD: An fMRI and psychophysiological study of extinction retention and fear renewal. *Journal of Neuroscience, 34*, 13435–13443. https://doi.org/10.1523/JNEUROSCI.4287-13.2014

Giustino, T. F., & Maren, S. (2015). The role of the medial prefrontal cortex in the conditioning and extinction of fear. *Frontiers in Behavioral Neuroscience, 9*, 298. https://doi.org/10.3389/fnbeh.2015.00298

Greco, J. A., & Liberzon, I. (2016). Neuroimaging of fear-associated learning. *Neuropsychopharmacology, 41*, 320–334. https://doi.org/10.1038/npp.2015.255

Guadaño-Ferraz, A., Benavides-Piccione, R., Venero, C., Lancha, C., Vennström, B., Sandi, C., DeFelipe, J., & Bernal, J. (2003). Lack of thyroid hormone receptor α1 is associated with selective alterations in behavior and hippocampal circuits. *Molecular Psychiatry, 8*, 30–38. https://doi.org/10.1038/sj.mp.4001196

Huang, J., Xu, F., Yang, L., Tuolihong, L., Wang, X., Du, Z., Zhang, Y., Yin, X., Li, Y., Lu, K., & Wang, W. (2023). Involvement of the GABAergic system in PTSD and its therapeutic significance. *Frontiers in Molecular Neuroscience, 16*, 1052288. https://doi.org/10.3389/fnmol.2023.1052288

Johansen, J. P., Cain, C. K., Ostroff, L. E., & LeDoux, J. E. (2011). Molecular mechanisms of fear learning and memory. *Cell, 147*, 509–524. https://doi.org/10.1016/j.cell.2011.10.009

Jung, S. J., Kang, J. H., Roberts, A. L., Nishimi, K., Chen, Q., Sumner, J. A., Kubzansky, L., & Koenen, K. C. (2019). Posttraumatic stress disorder and incidence of thyroid dysfunction in women. *Psychological Medicine, 49*, 2551–2560. https://doi.org/10.1017/S0033291718003495

Kala, N., Ravisankar, B., Govindarajulu, P., & Aruldhas, M. M. (2002). Impact of foetal-onset hypothyroidism on the epididymis of mature rats. *International Journal of Andrology, 25*, 139–148. https://doi.org/10.1046/j.1365-2605.2002.00338.x

Kausche, F. M., Carsten, H. P., Sobania, K. M., & Riesel, A. (2025). Fear and safety learning in anxiety - and stress-related disorders: An updated meta-analysis. *Neuroscience & Biobehavioral Reviews, 169*, 105983. https://doi.org/10.1016/j.neubiorev.2024.105983

Liu, B., Wang, Z., Lin, L., Yang, H., Gao, F., Gong, T., Edden, R. A. E., & Wang, G. (2020). Brain GABA + changes in primary hypothyroidism patients before and after levothyroxine treatment: A longitudinal magnetic resonance spectroscopy study. *NeuroImage. Clinical, 28*, 102473. https://doi.org/10.1016/j.nicl.2020.102473

Liu, B., Yang, H., Gao, F., Wang, Q., Zhao, B., Gong, T., Wang, Z., Chen, W., Wang, G., & Edden, R. A. E. (2017). Investigation of brain GABA + in primary hypothyroidism using edited proton MR spectroscopy. *Clinical Endocrinology, 86*, 256–262. https://doi.org/10.1111/cen.13177

Lori, A., Maddox, S. A., Sharma, S., Andero, R., Ressler, K. J., & Smith, A. K. (2018). Dynamic patterns of threat-associated gene expression in the amygdala and blood. *Front Psychiatry, 9*, 778. https://doi.org/10.3389/fpsyt.2018.00778

Maddox, S. A., Ponomareva, O. Y., Zaleski, C. E., Chen, M. X., Vella, K. R., Hollenberg, A. N., Klengel, C., & Ressler, K. J. (2025). Evidence for thyroid hormone regulation of amygdala-dependent fear-relevant memory and plasticity. *Molecular Psychiatry, 30*, 201–212. https://doi.org/10.1038/s41380-024-02679-2

Mandagere, K., Stoy, S., Hammerle, N., Zapata, I., & Brooks, B. (2025). Systematic review and meta-analysis of post-traumatic stress disorder as a risk factor for multiple autoimmune diseases. *Front. Psychiatry, 16*, 1523994. https://doi.org/10.3389/fpsyt.2025.1523994

Maren, S., & Quirk, G. J. (2004). Neuronal signalling of fear memory. *Nature Reviews Neuroscience, 5*, 844–852. https://doi.org/10.1038/nrn1535

Mason, J. (1994). Elevation of serum free triiodothyronine, total triiodothyronine, thyroxine-binding globulin, and total thyroxine levels in combat-related posttraumatic stress disorder. *Archives of General Psychiatry, 51*, 629. https://doi.org/10.1001/archpsyc.1994.03950080041006

Norrholm, S. D., Jovanovic, T., Olin, I. W., Sands, L. A., Karapanou, I., Bradley, B., & Ressler, K. J. (2011). Fear extinction in traumatized civilians with posttraumatic stress disorder: Relation to symptom severity. *Biological Psychiatry, 69*, 556–563. https://doi.org/10.1016/j.biopsych.2010.09.013

Puia, G., & Losi, G. (2011). Thyroid hormones modulate GABAA receptor-mediated currents in hippocampal neurons. *Neuropharmacology, 60*, 1254–1261. https://doi.org/10.1016/j.neuropharm.2010.12.013

Ram, P. A., & Waxman, D. J. (1990). Pretranslational control by thyroid hormone of rat liver steroid 5 alpha-reductase and comparison to the thyroid dependence of two growth hormone-regulated CYP2C mRNAs. *Journal of Biological Chemistry, 265*(31), 19223–19229. PMID: 2172247.

Rosenthal, L. J., Goldner, W. S., & O'Reardon, J. P. (2011). T_3 augmentation in major depressive disorder: Safety considerations. *The Australian Journal of Pharmacy, 168*, 1035–1040. https://doi.org/10.1176/appi.ajp.2011.10030402

Santin, A. P., & Furlanetto, T. W. (2011). Role of estrogen in thyroid function and growth regulation. *Journal of Thyroid Research, 2011*, 875125. https://doi.org/10.4061/2011/875125

Schafe, G. E., & LeDoux, J. E. (2000). Memory consolidation of auditory pavlovian fear conditioning requires protein synthesis and protein kinase a in the amygdala. *Journal of Neuroscience, 20*. https://doi.org/10.1523/JNEUROSCI.20-18-j0003.2000. RC96–RC96.

Schultebraucks, K., Sijbrandij, M., Galatzer-Levy, I., Mouthaan, J., Olff, M., & Van Zuiden, M. (2021). Forecasting individual risk for long-term posttraumatic stress disorder in emergency medical settings using biomedical data: A machine learning multicenter cohort study. *Neurobiology of Stress, 14*, 100297. https://doi.org/10.1016/j.ynstr.2021.100297

Sui, L., Wang, F., Liu, F., Wang, J., & Li, B. M. (2006). Dorsal hippocampal administration of triiodothyronine enhances long-term memory for trace cued and delay contextual fear conditioning in rats. *Journal of Neuroendocrinology, 18*, 811–819. https://doi.org/10.1111/j.1365-2826.2006.01480.x

Tahboub, R., & Arafah, B. M. (2009). Sex steroids and the thyroid. *Best Practice & Research Clinical Endocrinology & Metabolism, 23*, 769–780. https://doi.org/10.1016/j.beem.2009.06.005

Toloza, F. J. K., Mao, Y., Menon, L. P., George, G., Borikar, M., Erwin, P. J., Owen, R. R., & Maraka, S. (2020). Association of thyroid function with posttraumatic stress disorder: A systematic review and meta-analysis. *Endocrine Practice, 26*, 1173–1185.

Toloza, F. J. K., Mao, Y., Menon, L., George, G., Borikar, M., Thumma, S., Motahari, H., Erwin, P., Owen, R., & Maraka, S. (2021). Association of thyroid function with suicidal behavior: A systematic review and meta-Analysis. *Medicina (Kaunas), 57*, 714.

Touma, K. T. B., Zoucha, A. M., & Scarff, J. R. (2017). Liothyronine for depression: A review and guidance for safety monitoring. *Innovations in Clinical Neuroscience, 14*, 24–29.

Trent, S., Abdullah, M. H., Parwana, K., Valdivieso, M. A., Hassan, Z., & Müller, C. P. (2025). Fear conditioning: Insights into learning, memory and extinction and its relevance to clinical disorders. *Progress in Neuro-Psychopharmacology and Biological Psychiatry, 138*, 111310. https://doi.org/10.1016/j.pnpbp.2025.111310

Wang, S., & Mason, J. (1999). Elevations of serum T3 levels and their association with symptoms in world war II veterans with combat-Related posttraumatic stress disorder: Replication of findings in Vietnam combat veterans. *Psychosomatic Medicine, 61*, 131–138. https://doi.org/10.1097/00006842-199903000-00001

Wang, S., Mason, J., Southwick, S., Johnson, D., Lubin, H., & Charney, D. (1995). Relationships between thyroid hormones and symptoms in combat-Related posttraumatic stress disorder. *Psychosomatic Medicine, 57*, 398–402. https://doi.org/10.1097/00006842-199507000-00012

Westergard, T., Salari, R., Martin, J. V., & Brannigan, G. (2015). Interactions of L-3,5,3'-Triiodothyronine, allopregnanolone, and ivermectin with the GABAA receptor: Evidence for overlapping intersubunit binding modes. *Public Library of Science One, 10*, e0139072. https://doi.org/10.1371/journal.pone.0139072

Wiens, S. C., & Trudeau, V. L. (2006). Thyroid hormone and γ-aminobutyric acid (GABA) interactions in neuroendocrine systems. *Comparative Biochemistry and Physiology Part A: Molecular & Integrative Physiology, 144*, 332–344. https://doi.org/10.1016/j.cbpa.2006.01.033

Yu, D., Zhou, H., Zou, L., Jiang, Y., Wu, X., Jiang, L., Zhou, Q., Yang, Y., Xu, L., & Mao, R. (2017). Hippocampal administration of levothyroxine impairs contextual fear memory consolidation in rats. *Frontiers in Cellular Neuroscience, 11*, 223. https://doi.org/10.3389/fncel.2017.00223

Further reading

Ayers, S., Switnicki, M. P., Angajala, A., Lammel, J., Arumanayagam, A. S., & Webb, P. (2014). Genome-wide binding patterns of thyroid hormone receptor beta. *Public Library of Science One, 9*, e81186. https://doi.org/10.1371/journal.pone.0081186

Chatonnet, F., Guyot, R., Benoît, G., & Flamant, F. (2013). Genome-wide analysis of thyroid hormone receptors shared and specific functions in neural cells. *Proceedings of the National Academy of Sciences of the United States of America, 110*. https://doi.org/10.1073/pnas.1210626110

Section 4

Treatment regimens modulating the thyroid status and effects on neurological tissues

Chapter 17

Levothyroxine treatments in pregnancy and impact on cognition in offspring

Alice Batistuzzo[1] and Miriam Oliveira Ribeiro[2]

[1]*Institute for Mind and Biology, Neuroscience Institute, Department of Psychology, University of Chicago, Chicago, IL, United States;* [2]*Human Developmental Sciences Graduate Program, Center for Biological Sciences and Health, Mackenzie Presbyterian University, Sao Paulo, Brazil*

Abbreviations

Arc Activity-regulated cytoskeleton-associated protein
BDNF Brain-derived neurotrophic factor
CH Clinical hypothyroidism
D2 Type 2 deiodinase
D3 Type 3 deiodinase
Egr-1 Early growth response protein 1
ERK Extracellular signal-regulated
fT4 Free T4
hCG Human chorionic gonadotropin
L-T4 Levothyroxine sodium
LAT System-L amino acid transporters
LTP Long-term potentiation
MAPK Mitogen-activated protein kinases
MCT Monocarboxylate transporter
MEK MAPK/ERK kinase
NMDA N-mety-D-aspartate
OATP Organic anion transporting polypeptides
PSD-95 Postsynaptic density protein 95
Rap-1 Ras-related protein 1
SCH Subclinical hypothyroidism
SHH Sonic Hedgehog
T3 Thiiodothyronine
T4 Thyroxine
TBG Thyroxine-binding globulin
TG Thyroglobulin
TH Thyroid hormone
TPO Thyroid peroxidase
TR Thyroid hormone receptor
TSH Thyroid-stimulating hormone
TSHR Thyroid-stimulating hormone receptor

Introduction

Thyroid hormones (THs) are essential regulators of development, growth, and metabolism across vertebrates, exerting critical effects from the earliest stages of embryogenesis through adulthood. Over the past decades, advances in our understanding of thyroid physiology have underscored the pivotal role of THs during fetal development, particularly the necessity an adequate maternal supply from the onset of pregnancy.

In humans, the fetal thyroid gland begins to produce hormones only in the second trimester, well after early developmental events have taken place. However, thyroid hormone receptor (TR) transcripts are already detectable in fetal tissues by the ninth week of gestation, demonstrating that thyroid signaling is active before the fetus is capable of its own hormone synthesis. Thus, maternal thyroid hormones are the exclusive source during the first trimester, a period that coincides with organogenesis and the establishment of critical neurodevelopmental processes. Consequently, insufficient maternal thyroid function during this window has been associated with adverse obstetric outcomes and long-term effects on offspring cognition and behavior.

These findings reinforce the importance of early diagnosis and appropriate management of thyroid dysfunction during pregnancy. In this chapter, we review the etiology, diagnostic criteria, and treatment strategies for clinical and subclinical hypothyroidism in pregnant woman, as well as the implications for fetal development.

Thyroid organogenesis

The thyroid gland is the first endocrine organ to be formed in the embryo, initiating its development early in gestation. In humans, thyroid organogenesis begins in the third week of gestation (embryonic day 20), when endodermal cells in the floor of the primitive pharynx proliferate to form the thyroid primordium. In mice and rats, this event occurs

Hormones in Neuroscience. https://doi.org/10.1016/B978-0-443-30261-9.00017-6

earlier, at embryonic days 8.5 and 9–10, respectively (De Felice & Di Lauro, 2015) (Fig. 17.1).

Following its formation, the primordium invaginates into a thyroid bud that migrates caudally, ultimately reaching its definitive cervical position anterior to the trachea (De Felice & Di Lauro, 2015). By the seventh week in humans (E44–48) and the end of the second week in rodents (E13.5 in mice, E15 in rats), this migration is complete, and follicular precursors begin to differentiate (Fig. 17.1). During this stage, differentiation is marked by the progressive activation of thyroid-specific genes, including thyroglobulin (TG), thyroid peroxidase (TPO), the TSH receptor (TSHR), the sodium/iodide symporter (NIS), and pendrin, which together establish the machinery required for hormone biosynthesis. Organogenesis and differentiation are completed between the 10th and twelfth weeks in humans and by approximately day 16.5 in mice. Hormone secretion begins shortly thereafter—between weeks 12 and 14 in humans and at E17–18 in rats—and increases progressively, reaching maximal production capacity around the 28th week of human gestation (Fig. 17.1) (Brown, 2004; De Felice & Di Lauro, 2015). However, thyroid hormone receptors expression in fetal tissues were identified in the ninth week in humans, at embryonic day 11.5 in mice, and at day 13 in rats, suggesting that tissues are responsive to thyroid signaling before the fetal gland is fully functional (Morvan-Dubois et al., 2013).

Hormonal and metabolic adaptations affecting thyroid function during pregnancy

Since the fetal thyroid it is fully functional around week 12–14, reaching full functional capacity much later in gestation (Brown, 2004; De Felice & Di Lauro, 2015), maternal THs remains the primary source of these hormones during early development. However, pregnancy induces several hormonal and metabolic changes that alter thyroid economy such as increased renal clearance of iodide, high placental expression of type 3 deiodinase (D3), estrogen-driven increases in thyroxine-binding globulin (TBG), and stimulation of maternal TSH receptors by human chorionic gonadotropin (hCG) (Glinoer et al., 1990; Huang et al., 2005; Medici et al., 2015; Organization, 1994). All of them will be discussed below.

- **Reduction in Plasma Iodide**
  In the early stages of pregnancy, renal plasma flow and glomerular filtration rate increase, leading to enhanced plasma iodide clearance. As a result, plasma iodide

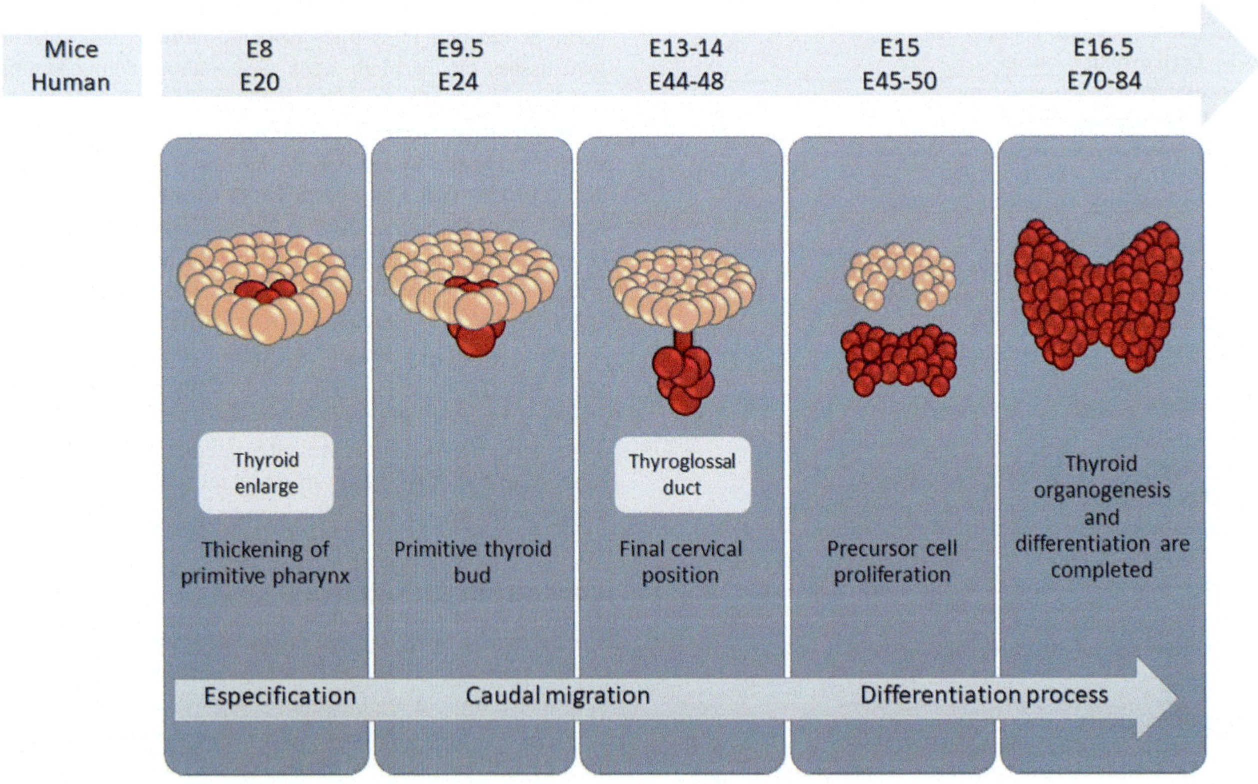

FIGURE 17.1 Stages of thyroid gland development. *Adapted from Batistuzzo, A., Ribeiro, M.O. (2020) Archives of Endocrinology and metabolism 64 (1):89–95, CC BY 4.0. (https://creativecommons.org/licenses/by/4.0/). Changes made: recolored.*

concentrations decrease, thereby raising the requirement for dietary iodine intake. For non-pregnant women, the recommended daily iodine intake is 150 μg. During pregnancy, this requirement increases to approximately 220–250 μg per day to support maternal thyroid hormone synthesis and fetal neurodevelopment. International guidelines, including those from the World Health Organization and the American Thyroid Association, recommend that pregnant women receive adequate iodine through diet and, when necessary, supplementation, as iodine deficiency during gestation is associated with adverse outcomes such as maternal hypothyroidism, impaired fetal brain development, and increased risk of cretinism. In Japan, a large nationwide birth cohort study identified that maternal iodine intake outside the adequate range was linked to developmental risks in children. Women with insufficient intake during pregnancy had offspring who were more likely to experience delays in communication, fine motor skills, and problem-solving abilities at ages one and three. In contrast, mothers with higher iodine intake, mainly from seaweed, did not show these risks, and in some domains, such as fine motor and problem-solving, their children had a lower risk of developmental delay (Hisada et al., 2022). However, a randomized trial in Thailand found that daily iodine supplementation (200 mcg) in mildly deficient pregnant women did not improve maternal thyroid function or offspring neurodevelopment up to 6 years of age. Unexpectedly, supplementation was associated with slight reductions in maternal thyroxine levels and borderline lower motor scores in children (Verhagen et al., 2020). These results emphasize that ensuring sufficient iodine intake during pregnancy is essential for optimal neurodevelopment, but routine supplementation in populations with only mild iodine deficiency may not provide clear benefits and could have unintended effects. Therefore, dietary iodine supplementation is recommended primarily in iodine-deficient regions only. Treating iodine deficiency is straightforward and should ensure a daily dietary intake of at least 250 mcg of iodine (Organization, 1994).

- **Presence of D3 in the Placenta**
  The placenta expresses type 2 deiodinase (D2), which converts T4 to the active T3, and D3, which inactivates T4 to reverse T3 (rT3) and T3 to T2. In the placenta, D3 is expressed at levels approximately 200 times higher than D2, as estrogen upregulates its gene expression. Despite D3's role in metabolizing T4 to T3r in the placenta, significant amounts of maternal T4 are still transported to the fetus through thyroid hormone transporters, including monocarboxylate transporters (MCT8 and MCT10), system-L amino acid transporters (LAT1 and LAT2), and organic anion-transporting polypeptides (OATPs) (Huang et al., 2005).
- **Action of hCG on TSH receptors**
  During the early weeks of gestation, the placenta synthesizes human chorionic gonadotropin (hCG), which reaches peak concentrations between the 9th and 11th weeks. Due to its structural homology with TSH, hCG acts as an agonist at the TSH receptor, thereby increasing thyroid hormone production. As a result, maternal serum concentrations of total T3 and total T4 increase during the first trimester, with a subsequent reduction in serum TSH levels. In the second trimester, hCG levels gradually decline, reaching a stable concentration after the 20th week, followed by a progressive rise in serum TSH levels (Carvalho et al., 2013; Medici et al., 2015).
- **Estrogen Effects on TBG**
  The hyperestrogenism observed during pregnancy stimulates hepatic synthesis and glycosylation of plasma transport proteins, particularly thyroxine-binding globulin (TBG), thereby increasing its half-life. As a result, although a slight elevation in serum free T4 levels may be observed during the first trimester, in the second and third trimesters serum free T3 and free T4 concentrations decrease to approximately 30% below those in non-pregnant women. It is important to note that only free T4 is available for tissue uptake, while the remainder of T4 is bound to serum proteins, primarily TBG (Glinoer et al., 1990).
  All these adaptations are summarized in Table 17.1 (Alexander et al., 2017; Sgarbi et al., 2013).

Considering all the physiological changes described above, the reference ranges for thyroid function tests in pregnant women differ from non-pregnant women and also along trimesters of pregnancy (Table 17.2) (Alexander et al., 2017; Sgarbi et al., 2013).

Hypothyroidism

According to Brenta et al., 2013 and summarized in Table 17.3, gestational subclinical hypothyroidism is defined by a normal free T4 with elevated TSH—specifically, 2.5–10.0 mIU/L (μIU/mL) in the first trimester and 3.5–10.0 mIU/L in the second and third trimesters. In contrast, gestational clinical (overt) hypothyroidism is diagnosed when TSH >10 mIU/L (μIU/mL) regardless of free T4, or when TSH exceeds the trimester-specific upper reference limit in conjunction with a subnormal free T4.

Assays for free T4 that are albumin-dependent may be biased in pregnancy, often underestimating concentrations by up to 50% owing to the decline in maternal serum albumin; as a result, reference intervals vary substantially across assay platforms. For the evaluation of free T4 in

TABLE 17.1 Hormonal and metabolic adaptations affecting thyroid function during pregnancy.

Mechanism	Process	Effect on thyroid physiology	Consequence for pregnancy
Reduction of plasma iodide	Increased renal plasma flow and glomerular filtration enhance clearance of iodide.	Decreased plasma iodide concentrations.	Increased dietary iodine requirement (≥250 μg/day.
Placental deiodinase activity (D3 > D2)	D3 highly expressed (≈200× D2), metabolizes T4 → rT3 and T3 → T2.	Reduced maternal TH activity within placenta.	Despite D3 activity, significant T4 still reaches fetus via TH transporters (MCT8/10, LAT1/2, OATPs).
hCG stimulation of TSH receptors	hCG peaks at 9–11 weeks, acts as TSH receptor agonist.	Increases maternal thyroid hormone production (↑ total T3/T4).	Suppression of maternal serum TSH in 1st trimester, gradual normalization by 2nd trimester.
Estrogen effect on TBG	Hyperestrogenism increases hepatic synthesis and glycosylation of TBG.	Increased half-life of TBG; higher total THs but ↓ free THs (~30% below baseline in 2nd/3rd trimesters).	Only free T4 available to tissues; reference ranges differ for pregnancy versus non-pregnant state.

TABLE 17.2 Reference values for TSH during pregnancy.

Trimester	TSH (μIU/mL)
1st trimester	0.1–2.5
2nd trimester	0.2–3.0/3.5
3rd trimester	0.3–3.0/3.5

From Alexander, E. K., Pearce, E. N., Brent, G. A., Brown, R. S., Chen, H., Dosiou, C., Grobman, W. A., Laurberg, P., Lazarus, J. H., Mandel, S. J., Peeters, R. P., & Sullivan, S. (2017). 2017 guidelines of the American thyroid association for the diagnosis and management of thyroid disease during pregnancy and the postpartum. Thyroid, 27(3), 315–389. https://doi.org/10.1089/thy.2016.0457; From Sgarbi, J. A., Teixeira, P. F., Maciel, L. M., Mazeto, G. M., Vaisman, M., Montenegro Junior, R. M., & Ward, L. S. (2013). The Brazilian consensus for the clinical approach and treatment of subclinical hypothyroidism in adults: Recommendations of the thyroid department of the Brazilian society of endocrinology and metabolism. Arquivos Brasileiros de Endocrinologia and Metabologia, 57, 166–183.

pregnant women, method-specific, pregnancy- and trimester-adjusted reference intervals provided by the manufacturer should be applied. Total T4 is a reliable alternative if the kit's reference limits explicitly account for the ~50% increase in thyroxine-binding globulin (TBG) observed during gestation (Alexander et al., 2017). Therefore, serial serum TSH remains the primary laboratory parameter for assessing thyroid function in pregnancy.

Prevalence and clinical manifestations

The prevalence of hypothyroidism varies widely across geographic regions, influenced by differences in diagnostic criteria, population characteristics, the sensitivity of assays historically employed to assess thyroid function, and disparities in iodine intake (Chiovato et al., 2019). In Europe, the prevalence of overt (symptomatic) hypothyroidism in the general population ranges from 0.2% to 5.3% (Chaker et al., 2017; Garmendia Madariaga et al., 2014). A meta-

TABLE 17.3 Reference values of TSH and free T4 for the diagnosis of clinical and subclinical hypothyroidism.

Condition	TSH	fT4
Clinical hypothyroidism	Above 10 μIU/mL	Regardless of free T4
Clinical hypothyroidism	1st trimester: >2.5–10.0 μIU/mL 2nd & 3rd trimester: >3.5–10.0 μIU/mL	Low
Subclinical hypothyroidism	1st trimester: >2.5–10.0 μIU/mL 2nd & 3rd trimester: >3.5–10.0 μIU/mL	Normal

Brenta, G., Vaisman, M., Sgarbi, J. A., Bergoglio, L. M., Andrada, N. C. d., Bravo, P. P., Orlandi, A. M., & Graf, H. (2013). Clinical practice guidelines for the management of hypothyroidism. Arquivos Brasileiros de Endocrinologia and Metabologia, 57, 265–291.

analysis of seven studies conducted across nine European countries estimated the prevalence of undiagnosed hypothyroidism—including both overt and subclinical forms—at approximately 5%, with an incidence rate of 226.2 (222.26–230.17) per 100,000 person-years (Garmendia Madariaga et al., 2014). In the United States, overt hypothyroidism has been reported in 0.3% of the population, while subclinical hypothyroidism affects 4.3% (Hollowell et al., 2002).

The prevalence of hypothyroidism is consistently higher in women than in men. A recent global analysis estimated that up to 17.5% of adult women are affected (Hu et al., 2022). Population-based studies demonstrate a 3- to 7-fold higher frequency of overt or subclinical hypothyroidism in women compared with men (Chiovato et al., 2019). For example, a Croatian population-based study reported a combined prevalence of 10.5% for overt and subclinical hypothyroidism, with women showing approximately 2.4 greater odds of clinical hypothyroidism than men (Strikic Dula et al., 2022).

Hypothyroidism is also a frequent endocrine disorder in pregnancy and is the most common thyroid disease among pregnant women (Lazarus, 2000; Teng et al., 2013). In Brazil, prevalence among women has been reported around 12% (Sgarbi et al., 2013). In other countries, clinical hypothyroidism during pregnancy occurs in 0.3%–0.5% of women, while subclinical hypothyroidism affects 2%–5% (Lazarus, 2000; Teng et al., 2013).

In iodine-sufficient areas, autoimmune thyroiditis is the predominant cause of hypothyroidism. Other etiologies include genetic disorders, radioactive iodine ablation therapy, surgical thyroidectomy, and hypothalamic-pituitary dysfunction.

Laboratory parameters of hypothyroidism in pregnancy

Because the symptoms of hypothyroidism (e.g., fatigue, weight gain, constipation) overlap with physiologic changes of euthyroid pregnancy, clinical assessment alone is insufficient to evaluate thyroid function (2). Accordingly, biochemical testing is essential for diagnosis and longitudinal monitoring—principally measurement of serum TSH and free thyroxine (free T4) (Brenta et al., 2013; Carvalho et al., 2013).

Currently, the reference values for TSH (4) are 0.4–4.5 mUI/L and free T4 are 0.7–1.8 ng/dL in non-pregnant adults up to 60 years of age. Based on these thresholds, overt (clinical) hypothyroidism is defined by a subnormal free T4 in the setting of a markedly elevated TSH (>10 mIU/L [μIU/mL]). In contrast, subclinical hypothyroidism is defined by an elevated TSH (>4.5 mIU/L) with a normal free T4 concentration. In routine practice, once an elevated TSH is detected, free T4 must be measured to distinguish overt from subclinical hypothyroidism (Brenta et al., 2013).

Treatment

Levothyroxine sodium (L-T4) is the treatment of choice for hypothyroidism. Following oral administration, L-T4 is absorbed predominantly in the small intestine (jejunum and ileum), with peak serum concentrations occurring approximately 2–4 h post-dose. Its elimination half-life is about 8 days, enabling stable serum levels at steady state and permitting once-daily dosing (Brenta et al., 2013).

In non-pregnant adults younger than 60 years, the typical starting dose of levothyroxine for overt hypothyroidism is approximately 1.6–1.8 μg per kilogram of body weight per day (Brenta et al., 2013). During gestation, however, maternal thyroid hormone synthesis must expand by roughly 20%–50% to preserve euthyroidism (Alexander et al., 2004). Insufficient T4 availability at this stage can adversely affect both maternal health and fetal development; therefore, prompt initiation of therapy following diagnosis is essential. The physiologic demand for T4 begins to rise between the fourth and sixth weeks of pregnancy, continues to increase until approximately weeks 16–20, and remains relatively stable thereafter until delivery (Alexander et al., 2004).

Treatment with levothyroxine has demonstrated significant benefits during pregnancy. In euthyroid pregnant women who are positive for thyroid peroxidase (TPO) antibodies, levothyroxine therapy can reduce the risk of premature birth up to 72%. Women with subclinical hypothyroidism and positive anti-TPO antibodies who become pregnant may also benefit from levothyroxine treatment. However, there is no consensus regarding the management of women with subclinical hypothyroidism who are negative for anti-TPO antibodies. In these cases, current recommendations are to monitor TSH and free T4 every 4 weeks until the 16th–20th week of gestation, and at least once between the 26th and 43rd weeks, in order to detect progression to clinical hypothyroidism.

Levothyroxine treatment during pregnancy has been shown to provide significant clinical benefits. Among euthyroid women who test positive for thyroid peroxidase (TPO) antibodies, supplementation with levothyroxine can lower the risk of preterm delivery by as much as 72%. Pregnant women with subclinical hypothyroidism who are also TPO-antibody positive appear to derive similar benefit from treatment. In contrast, management of subclinical hypothyroidism in antibody-negative women remains uncertain, as evidence supporting universal therapy is lacking. For these patients, current guidelines recommend periodic surveillance of thyroid function—measuring TSH and free T4 every 4 weeks through the 16th–20th week of gestation, and at least once more between weeks 26 and 43—to identify potential progression to overt hypothyroidism.

When pregnancy is anticipated, the levothyroxine dose should be optimized in advance to maintain serum TSH concentrations below 2.5 mU/L. Women with preexisting hypothyroidism who conceive before seeing a physician are advised to increase their levothyroxine dose empirically by approximately 25%–30%, equivalent to two additional tablets per week, as soon as pregnancy is confirmed. If, at any point, TSH exceeds 2.5 mU/L—irrespective of free T4 concentrations—levothyroxine therapy should be adjusted to restore TSH values within the trimester-specific reference range (Brenta et al., 2013).

According to the most recent clinical guidelines, levothyroxine requirements typically return to pre-pregnancy levels following delivery. The American Thyroid Association (ATA) specifically recommends reducing the dose to the preconception dose immediately postpartum and reassessing thyroid function at approximately 6 weeks (Alexander et al., 2017). The American College of Obstetricians and Gynecologists (ACOG) likewise emphasizes individualized postpartum follow-up and monitoring of thyroid function, particularly in women with autoimmune thyroid disease or those requiring substantial dose adjustments during pregnancy (ACOG, 2020).

During pregnancy, thyroid hormone replacement should consist of levothyroxine (L-T4) monotherapy (6,13). The use of liothyronine (T3), combined L-T4/L-T3 therapy, or desiccated thyroid is **not** recommended, as T3 does not cross the fetal blood–brain barrier to a degree sufficient to support the developing brain (Calvo et al., 1990). Consequently, T3-containing regimens may normalize maternal TSH yet leave the fetus relatively hypothyroid, with inadequate thyroid hormone availability for neurodevelopment.

Complications during pregnancy and effects on the offspring

Gestational hypothyroidism is linked to multiple adverse outcomes for both mother and child. Thyroid hormone deficiency during the first trimester can impair neuronal differentiation and disrupt central nervous system development, increasing the risk of congenital anomalies, low birth weight, and long-term deficits in motor and cognitive performance (Negro & Stagnaro-Green, 2014; Springer et al., 2017; Zhou et al., 2019).

Maternal psychological stress plays a significant role in modulating the association between thyroid hormone levels and pregnancy outcomes. For example, sustained low concentrations of free T4 in conjunction with elevated maternal anxiety have been linked to a higher probability of delivering infants who are larger than expected for their gestational age. In contrast, women with higher free T4 levels and low anxiety appear to have a reduced risk of this outcome. Collectively, these observations underscore that maternal thyroid function during pregnancy is shaped not only by endocrine mechanisms but also by psychological factors, emphasizing the complex interplay between physiological and emotional regulation in maternal–fetal health (Wu et al., 2025).

Role of thyroid hormones in neurodevelopment

Thyroid hormones are essential for fetal brain development from the earliest stages of gestation. During the first trimester, maternal hormones regulate neuronal proliferation and the initiation of neuronal migration within the developing cerebral cortex, hippocampus, and medial ganglionic eminence (Williams, 2008).

As gestation progresses into the second trimester, neurogenesis intensifies alongside axonal growth, dendritic arborization, synapse formation, and the differentiation of glial cells, marking the onset of myelination. The proper development of structures such as the cortex, hippocampus, cochlea, and cerebellum continues to depend on adequate thyroid hormone availability. At this stage, fetal thyroid hormone production also begins to increase progressively (Williams, 2008).

From approximately the 28th week onward, brain maturation accelerates as maternal and fetal thyroid hormone contributions become roughly equivalent. After birth, the newborn relies entirely on its own thyroid function. Ongoing postnatal processes include continued myelination; migration of granule cells in the dentate gyrus of the hippocampus and cerebellum; and migration of pyramidal and Purkinje cells within cortical and cerebellar layers (Williams, 2008) (Fig. 17.2).

Disruption of maternal fetal thyroid hormone supply during these critical stages can therefore lead to long-lasting neurological and developmental complications, as discussed in the following sections.

Complications in offspring resulting from gestational hypothyroidism in humans

Both clinical hypothyroidism (CH) and subclinical hypothyroidism (SCH) during pregnancy have been linked to reductions in offspring intelligence quotient (IQ). Evidence from longitudinal observations indicates that children born to mothers diagnosed with hypothyroidism as early as the 12th week of gestation exhibit delays in motor and cognitive development during the first 2 years of life, when compared to children of euthyroid mothers. These findings reinforce the importance of adequate maternal thyroid

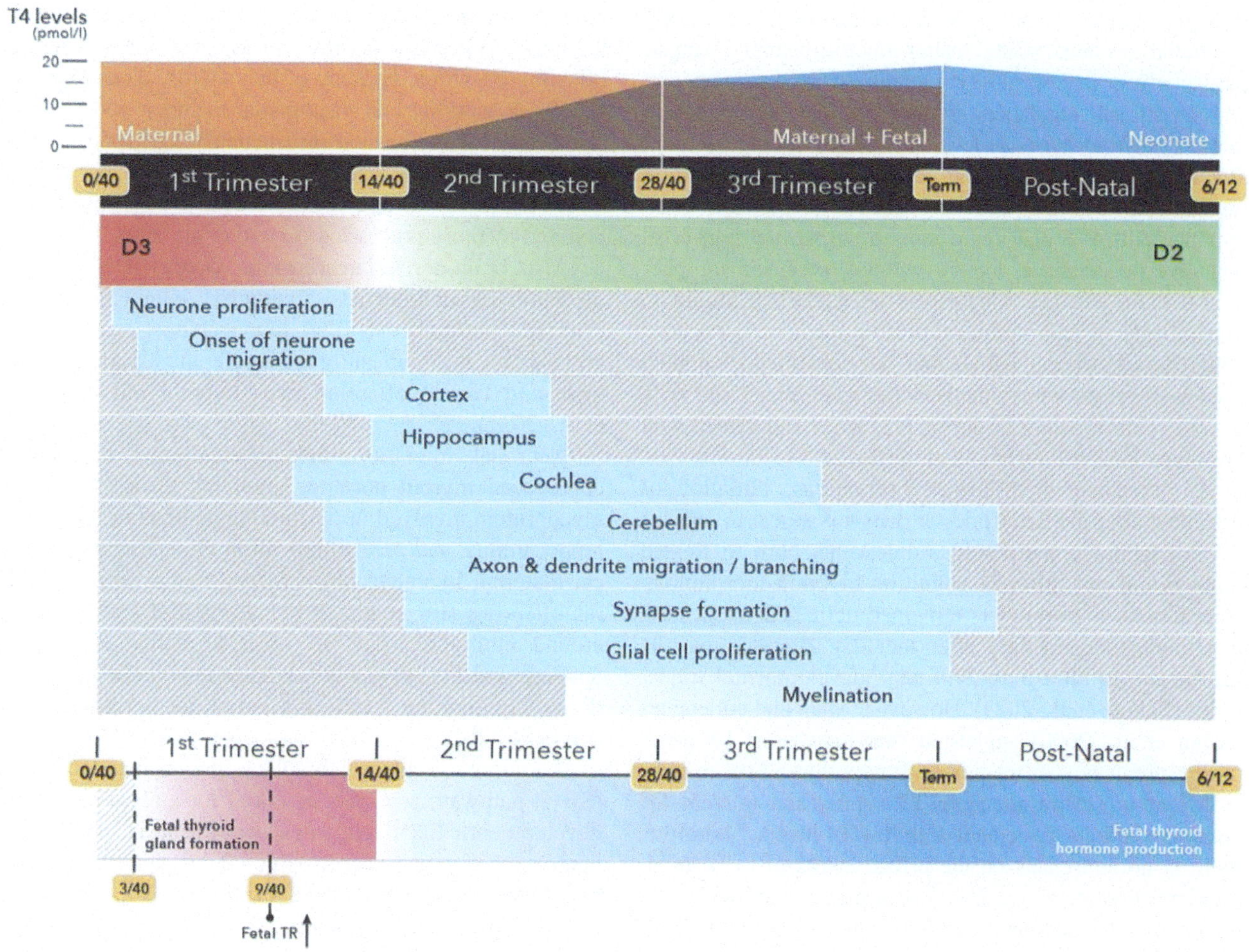

FIGURE 17.2 Central nervous system development during the gestational period in humans. *Reproduced from Batistuzzo, A., Ribeiro, M.O. (2020) Archives of Endocrinology and metabolism 64(1):89–95, CC BY 4.0 (https://creativecommons.org/licenses/by/4.0/). No changes were made.*

function for optimal neurodevelopmental outcomes in early childhood (Pop et al., 2003). In line with previous evidence, maternal thyroid dysfunction—encompassing overt and subclinical hypothyroidism as well as isolated hypothyroxinemia—has been associated with poorer neuropsychological performance in offspring. Developmental deficits can be identified as early as 6 months of age and, in some instances, persist through the second year of life. These observations emphasize the critical influence of maternal thyroid status on early cognitive and behavioral development (Chen et al., 2023).

Extending into early school years, evidence from a large cohort study involving approximately 4000 mother–child pairs demonstrated that children born to mothers with either hypothyroidism or hyperthyroidism during pregnancy exhibited significantly lower IQ scores compared to those of euthyroid mothers. Neuroimaging assessments at 6 years of age further revealed reductions in gray matter and overall cortical volume in these children, suggesting lasting structural and functional consequences of maternal thyroid imbalance on brain development (Korevaar et al., 2018).

Beyond hormone levels alone, maternal thyroglobulin concentrations also appear to influence offspring outcomes: elevated levels have been associated with lower IQ and alterations in cortical morphology in children, independent of other thyroid parameters. These findings suggest that thyroglobulin may serve as an integrated marker of iodine and thyroid hormone sufficiency during pregnancy (Mulder et al., 2025).

However, not all neurodevelopmental outcomes appear to be influenced by maternal thyroid function. In a prospective cohort of nearly 3000 pregnancies, first-trimester maternal TSH, free T4, and thyroid peroxidase antibody concentrations—whether elevated or low within the normal range—were not associated with attention-deficit/hyperactivity disorder in children followed at 7.5 and 15 years of age (Fetene et al., 2018).

Long-term findings further indicate that both low and high maternal free T4 concentrations in early gestation can alter offspring brain structure. A follow-up study into early adulthood revealed white matter microstructural changes, particularly in frontotemporal regions involved in cognition and emotional regulation, suggesting that deviations in maternal thyroid hormone levels may leave lasting neuroanatomical signatures (Bjornholm et al., 2023). Additional evidence shows that children born to mothers with hypothyroidism display reductions in verbal and nonverbal cognitive performance, though without detectable morphological brain changes (Ghassabian et al., 2014). Similarly, impairments in short-term declarative memory, associated with reduced hippocampal volume, were observed in children aged 9–12 years whose mothers had been diagnosed with hypothyroidism before or during pregnancy and treated with levothyroxine (Willoughby et al., 2014).

In agreement with these observations, initiation of treatment after the first trimester does not appear to prevent developmental impairments. Adolescents born to treated mothers did not differ in cognition or brain morphology from those whose mothers remained untreated, suggesting that therapy begun only after the first trimester may be insufficient to alter long-term neurodevelopmental trajectories (Scholz et al., 2025). However, Allan and colleagues (Allan et al., 2000) suggested that increasing T4 doses during pregnancy in women with hypothyroidism may be sufficient to ensure adequate T3 supply and prevent the cognitive deficits associated with low T4 levels. Therefore, there is no consensus on the effectiveness of L-T4 treatment in preventing cognitive impairments in offspring.

Reinforcing the importance of timing, a focused review on gestational hypothyroxinemia underscored that both the degree and onset of maternal free T4 deficiency are critical determinants of outcome. Low free T4 in the first trimester was most strongly associated with impaired cognitive and motor development, whereas hypothyroxinemia emerging later in pregnancy showed weaker effects (Chen & Xue, 2020).

Overall, the major impacts of gestational hypothyroidism on offspring development can be summarized as stemming from inadequate maternal thyroid hormone supply during early neurogenesis. This deficiency disrupts neuronal differentiation and central nervous system formation, thereby increasing the risk of congenital anomalies, low birth weight, and persistent impairments in motor and cognitive performance (Negro & Stagnaro-Green, 2014; Springer et al., 2017; Zhou et al., 2019).

Complications in offspring resulting from experimental gestational hypothyroidism in rodents

Animal models have been instrumental in elucidating the mechanisms by which maternal hypothyroidism disrupts fetal brain development. In both rats and mice, maternal thyroid hormone deficiency produces measurable impairments in offspring cognition. Progeny of hypothyroid dams exhibited deficits in working and visuospatial memory, accompanied by altered expression and distribution of synaptic proteins such as postsynaptic density protein 95 (PSD-95) and the NMDA receptor subunit 1, both critical for synapse formation and memory processing (Opazo et al., 2008).

Also, brain-derived neurotrophic factor (BDNF) - a key mediator of neurogenesis, neuronal maintenance, and synaptic plasticity in the hippocampus, cortex, and brainstem-was significantly reduced in the hippocampus of male and female offspring from hypothyroid dams (Lu et al., 2012).

Neuronal migration and differentiation depend on appropriate thyroid hormone signaling through reelin, a glycoprotein involved in positioning neurons in the cortex, hippocampus, and cerebellum. Both clinical and subclinical maternal hypothyroidism reduce reelin expression, in rat offspring (Lu et al., 2012). Early L-T4 replacement around embryonic day 13 reversed deficits in spatial learning and memory, whereas later intervention failed to do so, highlighting a critical window for hormone action (Lu et al., 2012).

Additional evidence involves the Sonic Hedgehog (SHH) pathway, a protein essential for embryonic neurodevelopment. SHH regulates neurogenesis and exerts neuroprotective and neurotrophic effects. In rats, maternal hypothyroidism reduces SHH expression and causes corresponding cerebellar abnormalities (Wang et al., 2014).

Several other T3-regulated proteins also play important roles in learning and memory. These include EGR-1, a transcription factor upregulated during LTP; ARC, which mediates post-learning plasticity through dendritic signaling; and RAP-1, which participates in neuronal proliferation, adhesion, and differentiation. Mitogen-activated protein kinases (MAPKs), such as MAPK kinase (MEK) and extracellular signal-regulated kinase (ERK), further integrate thyroid-dependent growth signaling. Decreased expression of these proteins in the hippocampus of offspring from dams with gestational hypothyroidism has been correlated with deficits in spatial learning and exploratory behavior (Ge et al., 2012).

Interestingly, both deficiency and excess of thyroid hormones during gestation can influence neurobehavioral outcomes. In mice, maternal thyroxine overexposure produced male offspring with long-term reductions in anxiety-like behavior (Laureano-Melo et al., 2019). Together with evidence from models of reduced hormone availability, these findings underscore that deviations in thyroid hormone levels—whether insufficient or excessive—can program persistent structural and behavioral changes in the offspring brain.

Possible strategies for stimulating development and cognition

Evidence that aspects of intellectual disability in congenital or neonatal hypothyroidism can be improved, dates to 1848. The Swiss physician Johann J. Guggenbühl described notable gains in physical and cognitive function in children with cretinism following multisensory stimulation that included warm aromatic baths, exercise and sports, affective socialization, and stimulation through sound, color, images, textures, and diverse foods (Crétinisme, 1848).

It is now well established that the central nervous system retains structural and functional adaptability throughout life. This neuroplasticity—the capacity of the brain to produce durable changes in synaptic and cellular architecture in response to experience—manifests via axonal and dendritic remodeling, synaptogenesis, adult neurogenesis, and activity-dependent modulation of signaling pathways. Neurotrophic factors and related molecular systems mediate these processes.

Understanding the mechanisms that regulate neural plasticity provides the basis for strategies aimed at preventing, delaying, or mitigating cognitive and motor impairments. Several studies have explored how multisensory stimulation promotes neuroplasticity with the goal of preserving or enhancing cognitive function (Inguaggiato et al., 2013; Paraskevopoulos et al., 2012; Shams & Seitz, 2008).

To date, only one experimental study has evaluated whether postnatal multisensory stimulation can mitigate the effects of developmental thyroid hormone insufficiency. In a rat model of moderate gestational hypothyroidism, 8 weeks of multisensory stimulation administered in adulthood restored short-term memory performance and reduced anxiety-like and depressive-like behaviors (Batistuzzo et al., 2022).

Complementary human evidence, though available only for congenital hypothyroidism, also supports the concept that experience-dependent interventions can promote structural brain recovery. Children with congenital hypothyroidism who received long-term music lessons showed hippocampal volumes comparable to those of healthy controls, whereas untreated peers exhibited reduced hippocampal size, suggesting that sensory–cognitive engagement can promote structural compensation (Zendel et al., 2013).

At present, no randomized or cohort studies have examined similar interventions in children born to mothers with gestational hypothyroidism. Nonetheless, findings from animal models showing partial behavioral recovery following postnatal stimulation, together with human data linking experiential training to hippocampal normalization, point to the potential of targeted cognitive–sensory programs to support neurodevelopment in this population. Continued longitudinal work is needed to determine which intervention strategies are most effective and which aspects of cognition and behavior are most responsive to such approaches.

Applications of the material

The material summarized in this chapter has several translational implications for clinical care, public health, and research.

Clinical applications

The recognition that maternal thyroid hormones are indispensable during the first trimester provides a strong rationale for early monitoring. Management of hypothyroidism in pregnancy relies on levothyroxine therapy, which has proven effective in reducing complications such as premature birth, miscarriage, postpartum hemorrhage, and anemia. For women already on treatment before conception, a pre-emptive approach is especially valuable in settings where immediate medical follow-up is not accessible. However, data also indicate that therapy initiated only after the first trimester does not prevent offspring cognitive impairments, highlighting the limitations of treatment once critical windows of neurodevelopment have passed. These findings emphasize the need for both rapid initiation and ongoing monitoring of therapy to maximize maternal and child benefits.

Public health applications

The association between iodine intake and offspring outcomes demonstrates the continued need for nutritional interventions. Inadequate iodine intake is associated with delayed communication, motor, and problem-solving skills, whereas excessive supplementation in mildly deficient populations may even produce adverse effects. These findings inform public health policies, showing that universal supplementation is not always advantageous. Instead, regional monitoring of iodine sufficiency should guide whether supplementation, dietary recommendations, or food fortification are indicated.

Research applications

In humans, an important research application involved long-term developmental follow-up. Human studies have shown that early maternal thyroid dysfunction is associated not only with reduced IQ and gray matter volume in childhood but also with white matter microstructural changes persisting into adulthood. Insights into neuroplasticity provide a basis for non-pharmacological interventions. Structured enrichment programs that engage sensory, cognitive, and motor systems may help improve developmental outcomes in children at risk, even when early maternal therapy is insufficient.

Conclusion

The insights reviewed in this chapter have direct implications for clinical care, public health, and future research. In clinical practice, they underscore the urgency of early screening, accurate diagnosis, and timely treatment. In public health, they highlight the importance of ensuring universal access to thyroid function screening during pregnancy, alongside iodine supplementation strategies that must be tailored to regional nutritional needs. In research, they reveal molecular and developmental pathways that could serve as targets for future therapeutic or preventive interventions. Together, these applications translate fundamental insights about thyroid hormone biology into actionable steps to safeguard maternal health and optimize offspring neurodevelopment.

Mini-dictionary of terms

- ARC: Activity-regulated cytoskeleton-associated protein involved in synaptic plasticity and dendritic remodeling, which is regulated by thyroid hormone availability.
- BDNF: Brain-derived neurotrophic factor that supports neuronal survival, differentiation, and synaptic plasticity; levels are reduced in offspring of hypothyroid dams.
- Iodothyronine Deiodinases (D2/D3): Enzymes that convert T4 to active T3 (D2) or inactivate thyroid hormones (D3).
- Hippocampus: A brain structure involved in memory and learning; sensitive to maternal thyroid hormone levels during gestation.
- Neurodevelopment: The process of brain growth and maturation, including neuronal proliferation, migration, and synapse formation; highly dependent on maternal endocrine status.

Summary Points

- Maternal thyroid hormones are essential for fetal development during the first trimester, when the fetal thyroid is not yet functional.
- Gestational hypothyroidism is associated with miscarriage, preterm birth, placental abruption, intrauterine growth restriction, and low birth weight.
- Inadequate maternal thyroid hormone supply is linked to impaired motor and cognitive development, reduced IQ, altered cortical morphology, decreased gray matter volume, and long-lasting changes in white matter microstructure.
- Levothyroxine therapy reduces complications during pregnancy, but initiation after the first trimester may not prevent offspring cognitive impairments.
- Early detection, trimester specific screening, and integrative prenatal care are essential applications of thyroid hormone research for improving maternal and offspring health.

Use of AI acknowledgment

The authors acknowledge the use of ChatGPT-5 (OpenAI) solely for grammar correction and language refinement during the preparation of this chapter. The tool was not involved in generating or interpreting the intellectual content. All authors reviewed and edited the text after its use and accept full responsibility for the final content of publication.

References

ACOG. (2020). Thyroid disease in pregnancy: ACOG practice bulletin, number 223. *Obstetrics & Gynecology, 135*(6), e261–e274. https://doi.org/10.1097/AOG.0000000000003893

Alexander, E. K., Marqusee, E., Lawrence, J., Jarolim, P., Fischer, G. A., & Larsen, P. R. (2004). Timing and magnitude of increases in levothyroxine requirements during pregnancy in women with hypothyroidism. *New England Journal of Medicine, 351*(3), 241–249. https://doi.org/10.1056/NEJMoa040079

Alexander, E. K., Pearce, E. N., Brent, G. A., Brown, R. S., Chen, H., Dosiou, C., Grobman, W. A., Laurberg, P., Lazarus, J. H., Mandel, S. J., Peeters, R. P., & Sullivan, S. (2017). 2017 guidelines of the American thyroid association for the diagnosis and management of thyroid disease during pregnancy and the postpartum. *Thyroid, 27*(3), 315–389. https://doi.org/10.1089/thy.2016.0457

Allan, W., Haddow, J., Palomaki, G., Williams, J., Mitchell, M., Hermos, R., Faix, J., & Klein, R. (2000). Maternal thyroid deficiency and pregnancy complications: Implications for population screening. *Journal of Medical Screening, 7*(3), 127–130.

Batistuzzo, A., De Almeida, G. G., Brás, T. S., Zucato, V. P., Arnold, A. J., Giannocco, G., ... Ribeiro, M. O. (2022). Multisensory stimulation improves cognition and behavior in adult male rats born to LT4-treated thyroidectomized dams. *Endocrinology, 163*(9), bqac105.

Bjornholm, L., Orell, O., Kerkela, M., Honka, U., Laasonen, S., Riekki, T., Surcel, H. M., Suvanto, E., & Veijola, J. (2023). Maternal thyroid function during pregnancy and offspring white matter microstructure in early adulthood: A prospective birth cohort study. *Thyroid, 33*(10), 1245–1254. https://doi.org/10.1089/thy.2022.0699

Brenta, G., Vaisman, M., Sgarbi, J. A., Bergoglio, L. M., Andrada, N. C. d., Bravo, P. P., Orlandi, A. M., & Graf, H. (2013). Clinical practice guidelines for the management of hypothyroidism. *Arquivos Brasileiros de Endocrinologia and Metabologia, 57*, 265–291.

Brown, R. S. (2004). Minireview: Developmental regulation of thyrotropin receptor gene expression in the fetal and newborn thyroid. *Endocrinology, 145*(9), 4058–4061.

Calvo, R., Obregon, M. J., Ruiz de Ona, C., Escobar del Rey, F., & Morreale de Escobar, G. (1990). Congenital hypothyroidism, as studied in rats. Crucial role of maternal thyroxine but not of 3,5,3'-triiodothyronine in the protection of the fetal brain. *Journal of Clinical Investigation, 86*(3), 889–899. https://doi.org/10.1172/JCI114790

Carvalho, G. A.d., Perez, C. L. S., & Ward, L. S. (2013). Utilização dos testes de função tireoidiana na prática clínica. *Arquivos Brasileiros de Endocrinologia and Metabologia, 57*(3), 193–204.

Chaker, L., Bianco, A. C., Jonklaas, J., & Peeters, R. P. (2017). Hypothyroidism. *Lancet, 390*(10101), 1550–1562. https://doi.org/10.1016/S0140-6736(17)30703-1

Chen, Y., Luo, Z. C., Zhang, T., Fan, P., Ma, R., Zhang, J., & Ouyang, F. (2023). Maternal thyroid dysfunction and neuropsychological development in children. *The Journal of Cinical Endocrinology and Metabolism, 108*(2), 339–350. https://doi.org/10.1210/clinem/dgac577

Chen, Y., & Xue, F. (2020). The impact of gestational hypothyroxinemia on the cognitive and motor development of offspring. *Journal of Maternal-Fetal and Neonatal Medicine, 33*(11), 1940–1945. https://doi.org/10.1080/14767058.2018.1529749

Chiovato, L., Magri, F., & Carle, A. (2019). Hypothyroidism in context: Where we've been and where we're going. *Advances in Therapy, 36* (Suppl. 2), 47–58. https://doi.org/10.1007/s12325-019-01080-8

Crétinisme, C. p. l. (1848). Rapport de la commission créée par SM le roi de Sardaigne pour étudier le crétinisme. Impr. royale.

De Felice, M., & Di Lauro, R. (2015). Anatomy and development of the thyroid. In L. J. De Groot, G. Chrousos, K. Dungan, K. R. Feingold, A. Grossman, J. M. Hershman, C. Koch, M. Korbonits, R. McLachlan, M. New, R. Purnell, R. Rebar, F. Singer, & A. Vinik (Eds.), *Endotext*. MDText.com.Inc.

Fetene, D. M., Betts, K. S., & Alati, R. (2018). Maternal prenatal thyroid function and offspring ADHD: Findings from the ALSPAC cohort. *The Journal of Nervous and Mental Disease, 206*(11), 859–864. https://doi.org/10.1097/NMD.0000000000000891

Garmendia Madariaga, A., Santos Palacios, S., Guillen-Grima, F., & Galofre, J. C. (2014). The incidence and prevalence of thyroid dysfunction in Europe: A meta-analysis. *The Journal of Cinical Endocrinology and Metabolism, 99*(3), 923–931. https://doi.org/10.1210/jc.2013-2409

Ge, J. F., Peng, L., Hu, C. M., & Wu, T. N. (2012). Impaired learning and memory performance in a subclinical hypothyroidism rat model induced by hemi-thyroid electrocauterisation. *Journal of Neuroendocrinology, 24*(6), 953–961.

Ghassabian, A., El Marroun, H., Peeters, R. P., Jaddoe, V. W., Hofman, A., Verhulst, F. C., Tiemeier, H., & White, T. (2014). Downstream effects of maternal hypothyroxinemia in early pregnancy: Nonverbal IQ and brain morphology in school-age children. *The Journal of Cinical Endocrinology and Metabolism, 99*(7), 2383–2390. https://doi.org/10.1210/jc.2013-4281

Glinoer, D., Nayer, P.d., Bourdoux, P., Lemone, M., Robyn, C., Stertghem, A. V., Kinthaert, J., & Lejeune, B. (1990). Regulation of maternal thyroid during pregnancy. *Journal of Clinical Endocrinology and Metabolism, 71*(2), 276–287.

Hisada, A., Takatani, R., Yamamoto, M., Nakaoka, H., Sakurai, K., Mori, C., The Japan, E., & Children's Study Jecs, G. (2022). Maternal iodine intake and neurodevelopment of offspring: The Japan environment and children's study. *Nutrients, 14*(9). https://doi.org/10.3390/nu14091826

Hollowell, J. G., Staehling, N. W., Flanders, W. D., Hannon, W. H., Gunter, E. W., Spencer, C. A., & Braverman, L. E. (2002). Serum TSH, T(4), and thyroid antibodies in the United States population (1988 to 1994): National health and nutrition examination survey (NHANES III). *The Journal of Cinical Endocrinology and Metabolism, 87*(2), 489–499. https://doi.org/10.1210/jcem.87.2.8182

Hu, X., Chen, Y., Shen, Y., Tian, R., Sheng, Y., & Que, H. (2022). Global prevalence and epidemiological trends of Hashimoto's thyroiditis in adults: A systematic review and meta-analysis. *Frontiers in Public Health, 10*, Article 1020709. https://doi.org/10.3389/fpubh.2022.1020709

Huang, S. A., Mulcahey, M. A., Crescenzi, A., Chung, M., Kim, B. W., Barnes, C., Kuijt, W., Turano, H., Harney, J., & Larsen, P. R. (2005). Transforming growth factor-β promotes inactivation of extracellular thyroid hormones via transcriptional stimulation of type 3 iodothyronine deiodinase. *Molecular Endocrinology, 19*(12), 3126–3136.

Inguaggiato, E., Sgandurra, G., Perazza, S., Guzzetta, A., & Cioni, G. (2013). Brain reorganization following intervention in children with congenital hemiplegia: A systematic review. *Neural Plasticity, 2013*, Article 356275. https://doi.org/10.1155/2013/356275

Korevaar, T. I., Tiemeier, H., & Peeters, R. P. (2018). Clinical associations of maternal thyroid function with foetal brain development: Epidemiological interpretation and overview of available evidence. *Clinical Endocrinology, 89*(2), 129–138.

Laureano-Melo, R., Souza, J. S., da Conceicao, R. R., Albuquerque, J. M. L., Rodrigues, N. C., Marinho, B. G., Olivares, E. L., Giannocco, G., & Cortes, W. D. S. (2019). Prenatal thyroxine treatment promotes anxiolysis in male Swiss mice offspring. *Hormones and Behavior, 108*, 10–19. https://doi.org/10.1016/j.yhbeh.2018.12.008

Lazarus, J. (2000). Thyroid regulation and dysfunction in the pregnant patient. In K. R. Feingold, B. Anawalt, A. Boyce, G. Chrousos, W. W. de Herder, K. Dhatariya, K. Dungan, J. M. Hershman, J. Hofland, S. Kalra, G. Kaltsas, C. Koch, P. Kopp, M. Korbonits, C. S. Kovacs, W. Kuohung, B. Laferrere, M. Levy, E. A. McGee, et al.D. P. Wilson (Eds.), *Endotext*.

Lu, L., Yu, X., Teng, W., & Shan, Z. (2012). Treatment with levothyroxine in pregnant rats with subclinical hypothyroidism improves cell migration in the developing brain of the progeny. *Journal of Endocrinological Investigation, 35*(5), 490–496.

Medici, M., Korevaar, T. I., Visser, W. E., Visser, T. J., & Peeters, R. P. (2015). Thyroid function in pregnancy: What is normal? *Clinical Chemistry, 61*(5), 704–713.

Morvan-Dubois, G., Fini, J. B., & Demeneix, B. A. (2013). Is thyroid hormone signaling relevant for vertebrate embryogenesis? *Curr. Topic. Dev. Biol., 103*, 365–396.

Mulder, T. A., Guxens, M., Rebagliato, M. L., Dineva, M., Bath, S. C., Hunziker, S., Sunyer, J., Delgado-Saborit, J. M., Irizar Loibide, A., Lertxundi, N., Muetzel, R. L., Tiemeier, H., Peeters, R. P., & Korevaar, T. I. M. (2025). Association of maternal thyroglobulin with gestational thyroid function and offspring IQ and brain morphology. *The Journal of Cinical Endocrinology and Metabolism, 110*(7), 2007–2015. https://doi.org/10.1210/clinem/dgae679

Negro, R., & Stagnaro-Green, A. (2014). Clinical aspects of hyperthyroidism, hypothyroidism, and thyroid screening in pregnancy.

Endocrine Practice, 20(6), 597–607. https://doi.org/10.4158/EP13350.RA

Opazo, M. C., Gianini, A., Pancetti, F., Azkcona, G., Alarcon, L., Lizana, R., Noches, V., Gonzalez, P. A., Marassi, M. P., Mora, S., Rosenthal, D., Eugenin, E., Naranjo, D., Bueno, S. M., Kalergis, A. M., & Riedel, C. A. (2008). Maternal hypothyroxinemia impairs spatial learning and synaptic nature and function in the offspring. *Endocrinology, 149*(10), 5097–5106. https://doi.org/10.1210/en.2008-0560

Organization, W. H. (1994). *United nations children's fund: International council for control of iodine deficiency disorders and their control through salt iodization* (Vol 94, pp. 1–55). WHO NUT/.

Paraskevopoulos, E., Kuchenbuch, A., Herholz, S. C., & Pantev, C. (2012). Evidence for training-induced plasticity in multisensory brain structures: An MEG study. *PLoS One, 7*(5), Article e36534. https://doi.org/10.1371/journal.pone.0036534

Pop, V. J., Brouwers, E. P., Vader, H. L., Vulsma, T., van Baar, A. L., & de Vijlder, J. J. (2003). Maternal hypothyroxinaemia during early pregnancy and subsequent child development: A 3-year follow-up study. *Clinical Endocrinology, 59*(3), 282–288. https://doi.org/10.1046/j.1365-2265.2003.01822.x

Scholz, A., McNabb, C. B., Bloomfield, L., Bhargava, R., Hales, C., Dayan, C. M., Taylor, P. N., Lazarus, J. H., Okosieme, O., Ludgate, M., Jones, D. K., & Rees, D. A. (2025). Controlled antenatal thyroid screening study III: Effects of gestational thyroid status on adolescent brain morphology. *The Journal of Cinical Endocrinology and Metabolism, 110*(4), e1094–e1102. https://doi.org/10.1210/clinem/dgae338

Sgarbi, J. A., Teixeira, P. F., Maciel, L. M., Mazeto, G. M., Vaisman, M., Montenegro Junior, R. M., & Ward, L. S. (2013). The Brazilian consensus for the clinical approach and treatment of subclinical hypothyroidism in adults: Recommendations of the thyroid department of the Brazilian society of endocrinology and metabolism. *Arquivos Brasileiros de Endocrinologia and Metabologia, 57*, 166–183.

Shams, L., & Seitz, A. R. (2008). Benefits of multisensory learning. *Trends in Cognitive Sciences, 12*(11), 411–417. https://doi.org/10.1016/j.tics.2008.07.006

Springer, D., Jiskra, J., Limanova, Z., Zima, T., & Potlukova, E. (2017). Thyroid in pregnancy: From physiology to screening. *Critical Reviews in Clinical Laboratory Sciences, 54*(2), 102–116. https://doi.org/10.1080/10408363.2016.1269309

Strikic Dula, I., Pleic, N., Babic Leko, M., Gunjaca, I., Torlak, V., Brdar, D., Punda, A., Polasek, O., Hayward, C., & Zemunik, T. (2022). Epidemiology of hypothyroidism, hyperthyroidism and positive thyroid antibodies in the Croatian population. *Biology, 11*(3). https://doi.org/10.3390/biology11030394

Teng, W., Shan, Z., Patil-Sisodia, K., & Cooper, D. S. (2013). Hypothyroidism in pregnancy. *Lancet Diabetes and Endocrinology, 1*(3), 228–237. https://doi.org/10.1016/S2213-8587(13)70109-8

Verhagen, N. J. E., Gowachirapant, S., Winichagoon, P., Andersson, M., Melse-Boonstra, A., & Zimmermann, M. B. (2020). Iodine supplementation in mildly iodine-deficient pregnant women does not improve maternal thyroid function or child development: A secondary analysis of a randomized controlled trial. *Frontiers in Endocrinology, 11*, Article 572984. https://doi.org/10.3389/fendo.2020.572984

Wang, Y., Wang, Y., Dong, J., Wei, W., Song, B., Min, H., Yu, Y., Lei, X., Zhao, M., Teng, W., & Chen, J. (2014). Developmental hypothyroxinemia and hypothyroidism reduce proliferation of cerebellar granule neuron precursors in rat offspring by downregulation of the sonic hedgehog signaling pathway. *Molecular Neurobiology, 49*(3), 1143–1152. https://doi.org/10.1007/s12035-013-8587-3

Williams, G. R. (2008). Neurodevelopmental and neurophysiological actions of thyroid hormone. *Journal of Neuroendocrinology, 20*(6), 784–794. https://doi.org/10.1111/j.1365-2826.2008.01733.x

Willoughby, K. A., McAndrews, M. P., & Rovet, J. F. (2014). Effects of maternal hypothyroidism on offspring hippocampus and memory. *Thyroid, 24*(3), 576–584. https://doi.org/10.1089/thy.2013.0215

Wu, P., Ouyang, J., Cai, W., Chen, L., Tong, J., Gao, G., Wu, X., Han, Y., Yan, S., Tao, F., & Huang, K. (2025). Interaction between maternal thyroid hormone levels and pregnancy-related anxiety on birth outcomes-A birth cohort study. *Journal of Affective Disorders, 389*, Article 119646. https://doi.org/10.1016/j.jad.2025.119646

Zendel, B. R., Willoughby, K. A., & Rovet, J. F. (2013). Neuroplastic effects of music lessons on hippocampal volume in children with congenital hypothyroidism. *NeuroReport, 24*(17), 947–950. https://doi.org/10.1097/WNR.0000000000000031

Zhou, M., Wang, M., Li, J., Luo, X., & Lei, M. (2019). Effects of thyroid diseases on pregnancy outcomes. *Experimental and Therapeutic Medicine, 18*(3), 1807–1815. https://doi.org/10.3892/etm.2019.7739

Chapter 18

Nanomedicines with triiodothyronine and their effects on neurological tissues

Tanu Dixit, Sohini Chattopadhyay, Tan Thakur, Mansi Sawardekar and Selvan Ravindran
Symbiosis School of Biological Sciences, Symbiosis International (Deemed University), Pune, Maharashtra, India

Introduction

The formation, maintenance, and functional integrity of the central nervous system (CNS) are all considerably influenced by the thyroid hormones, which play a vital role in sustaining physiological homeostasis. Triiodothyronine (T3) is the metabolically active hormone among them, according to Sabatino et al., (2024). Both genetic and non-genomic effects are exerted by it, and it plays a crucial role in neural growth, synaptogenesis, myelination, and neurotransmitter modulation. The existence of thyroid hormone receptors (TRalpha, TRbeta) in glia and neurons signals the imminent importance of T3 signaling in adult brain plasticity and neurodevelopment (Mokle & Sanap, 2023). It is significant to the study of neurology because fluctuations in thyroid hormone levels (largely T3) have been associated with neurocognitive anomalies of depressed mood and cortical activity change. Both the clinical and experimental sectors have shown a growing amount of theoretical interest in the neurological symptoms of thyroid disease. It is stated that hypothyroidism intensifies the process of neuroinflammation and contributes to the onset of vascular and neurodegenerative conditions (Churilov et al., 2019). The studies state that even minor variations in the level of T3 could modify the outcomes of stroke and cognitive resilience and presuppose the fragile neuroendocrine axis (Sidorov et al., 2023). Experimental results have reinforced the therapeutic prospect of T3 in such acute neurological injuries and found that after ischemia, treatment led to stabilization of the "blood-brain barrier (BBB) and enhanced neurovascular restoration" (Ullrich et al., 2023). However, it remains the problem of precise and effective delivery of T3 to the tissues of the CNS due to its ineffective pharmacokinetics and poor permeability through the BBB.

The BBB is a highly discerning endothelial blockade that controls the flow of chemicals between the central nervous system and the circulatory system. Even though such a barrier provides homeostatic preservation of the brain structures, it is considered to be a great challenge against the delivery of pharmacological compounds to the brain, like T3 (Andersen & Andersen, 2024). The regular delivery mode also tends to cause systemic side effects and subtherapeutic CNS levels due to rapid deiodination and washout of T3. Besides, the endogenous thyroid hormone is transported by a special type of transmembrane protein, among which there are MCT8 and OATP1C1, the violation of the functioning of which also limits the availability of T3 in the brain, in both normal and pathological conditions (Liu et al., 2025).

There is a paradigm shift in terms of the new mechanisms of delivering T3 to the neural targets that are based on nanotechnology. The physicochemical properties of nanoparticles (NPs) exploited by nanomedicine allow one to avoid crossing the BBB and contribute to controlled release, increase in bioavailability, and decreased peripheral toxicity (Kaushik et al., 2018). Some of the platforms that have undergone modification on their surface are liposomes, polymeric nanoparticles, and inorganic nanocarriers, which have been modified to enhance penetration of the CNS through either receptor-mediated endocytosis or transcytosis (Tan et al., 2023). By means of individualized nanomedicine approaches, it is also possible to target some receptors in neurons or inflamed tissue and enhance the accuracy of the process (Sim et al., 2020).

The neuroprotective medicine and thyroid analogs reached the CNS compartments successfully due to recent advancement in transnasal delivery, ligand-conjugated NPs, and biodegradable carriers (Mohapatra et al., 2024; Zhang et al., 2023). It has also been established that naturally occurring bioenhancers or anti-inflammatories, on co-delivery, could cross the BBB's tiny junctions

Hormones in Neuroscience. https://doi.org/10.1016/B978-0-443-30261-9.00018-8

without harming its anatomical integrity (Liu et al., 2025). This implies that T3-based nanomedicines are being used more often for neurological conditions, which will change how hypothyroidism causes stroke, neurodegeneration, and cognitive impairment.

Therapeutic potential and delivery of triiodothyronine in neurological applications

Triiodothyronine (T3), which is an active thyroid hormone, is implicated in central nervous system (CNS) function and structure. It regulates neuroplasticity, metabolism, development, and repair through both non-genomic and genomic pathways. Due to so many effects of the hormone, any alteration of T3 availability or signal has profound neurological consequences from mild impairments of cognition to debilitating developmental disorders. There has been research in the recent past in which T3 improvement or reconstitution has been demonstrated to have beneficial therapeutic effects on most CNS disorders. For example, Ullrich et al., (2023) demonstrated that T3 therapy significantly improved ischemic stroke outcomes by reducing infarct volume, enhancing BBB integrity, and reducing neurological deficits. T3 is a potential therapeutic drug for acute CNS injury, chronic neurodegenerative disease, and mental illness due to its neuroprotective potential.

Although T3 possesses potential therapeutic applications, some of the limitations lie in preventing it from becoming popular on a large scale. Its systemic bioavailability is limited by its quick metabolism, short half-life, extensive binding to plasma proteins, and low ability to cross the BBB. To avoid these pharmacokinetic issues, the medication has to be administered in a standard dose, which does not make therapy less complicated but rather enhances the possibility of side effects like cardiac arrhythmias. In addition, serious neurological disease can follow abnormalities of the specialized transporters that tightly control permeability across the blood-brain barrier, even when systemic T3 is normal. After efforts have been made to overcome these problems, new transport systems that are skilled of stabilizing T3, avoiding transporter dependency, and ensuring direct transport to the CNS have been the subject of current research. One of the effective strategies that can augment neuronal recovery and minimize systemic toxicity is the use of exosome-delivery devices, which transport T3 transversely the BBB in a highly targeted fashion. Another effective strategy is the use of ligand-conjugated nanocarriers for site-specific targeting in the CNS (Goyal et al., 2014). T3 therapy may potentially benefit from other, more general nanotechnological developments that improve the stability and water solubility of hydrophobic compounds (Dabiri et al., 2024). These results collectively show how nanomedicine and T3 biology work together to enable novel therapeutic possibilities.

Role of triiodothyronine in brain development and function

Triiodothyronine regulates brain activity by both genetic and non-genomic mechanisms, fostering neurodevelopment and plasticity of the adult brain. The genomic actions are regulated by thyroid hormone receptors TRα1 and TRβ1. Following binding to T3, the receptors dimerize with retinoid X receptors and bind to thyroid hormone response elements on DNA to regulate the transcription of genes (Bernal, 2007; Gil-Ibañez et al., 2017). These gene programs play key roles in brain differentiation, cytoskeletal arrangement, mitochondrial metabolism, and myelin synthesis. For example, T3-regulated genes influence microtubule-associated proteins critical for axonal growth and dendritic branching, as well as myelin basic and proteolipid proteins critical for myelination (Remaud et al., 2014). Interruptions during the prenatal and neonatal periods might lead to irreversible consequences, including intellectual incapacitation, motor impairments, and behavioral abnormalities.

Apart from these transcriptional functions, T3 also causes rapid non-genomic responses that do not involve gene transcription. These include the stimulation of intracellular signal transduction cascades like *mitogen-activated protein kinase* and *phosphatidylinositol-3-kinase*, modulation of ion channels, and mitochondrial function (Gothié et al., 2017; Morte & Bernal, 2014). T3 modulates calcium influx, neuronal excitability, synaptic transmission, and metabolic responses in real time through several mechanisms. Such rapid responses are particularly important in acute injury or metabolic insult, when the immediate activation of protective mechanisms may be the difference between neuron survival and death. Thyroid hormone regulation in the CNS is also augmented by astrocytes. Through type 2 deiodinase-catalyzed conversion of thyroxine to T3, these glial cells ensure that neurons receive a localized and stable availability of the active hormone. Astrocytes also express thyroid hormone receptors that respond to T3 by modulating the release of growth factors, cytokines, and metabolic substrates. This intercellular communication system maintains CNS homeostasis (Gil-Ibañez et al., 2017; Morte & Bernal, 2014). This local signaling and conversion network provides important buffering in states of systemic hormone instability.

During the progress and role of the brain, the two receptor isoforms play distinct but complementary roles. TRα1, with its strong presence in the hippocampus, cortex, and cerebellum during early brain development, plays a supportive role in axonal growth, myelination, and structural maturation. Especially vital for the energy metabolism, maintenance of synapses, and modulation of neurotransmitters, TRβ1 becomes more evident postnatally (Gil-Ibañez et al., 2017). When either of these receptors is interrupted, neural plasticity and connections are impaired, resulting in measurable abnormalities of cognition and behavior. T3 can control the vital processes of neurogenesis, synaptogenesis, and myelination through the coordination of multiple genetic, non-genomic, and astrocyte-based mechanisms. T3 regulates cytoskeletal and chromatin remodeling to ensure proper maturation while stimulating stem cell proliferation, enabling cell cycle exit, and leading progenitors to neuronal differentiation in hippocampus neurogenic niches (Remaud et al., 2014). It enhances plasticity and connection at synapses through upregulation of neurotransmitter receptors and structural proteins. It ensures differentiation and the formation of myelin proteins in oligodendrocytes, which are necessary for rapid signal conduction (Gothié et al., 2017). Cognitive impairment, learning deficit, and motor dysfunction are all clinical manifestations of insufficient T3 signaling, which may be brought about by mutation of the receptor, failure of the transporter, or overall deficiency.

Neurological disorders linked to T3 dysregulation

Dysregulation of T3 has neurological manifestations that appear in all phases of life, ranging from emotional, cognitive, and degenerative diseases in adults to abnormalities in child development. Memory, executive function, attention, and learning deficits are strongly associated with hypothyroidism (Bernal, 2007; Remaud et al., 2014). These deficiencies are mechanistically linked with reduced glucose metabolism and perfusion in thyroid-sensitive regions such as the prefrontal cortex and hippocampal regions (Gil-Ibañez et al., 2017; Gothié et al., 2017). Thyroid hormone replacement therapy can treat hypothyroid individuals for their hippocampus reduction and compromised frontotemporal networks, as established by neuroimaging tests (Morte & Bernal, 2014; Ullrich et al., 2023). Depression and other mood disorders also correlate with low T3. The hormone regulates monoaminergic pathways, which are critical to mood regulation and comprise the serotonergic, dopaminergic, and noradrenergic systems. Similar imbalances of neurotransmitters to major depressive illness are induced by deficiencies that disrupt such pathways (Goldsmith et al., 2014; Kumar et al., 2022; Mdzinarishvili et al., 2013). These findings explain why antidepressant response in treatment-resistant depression is enhanced by adjunctive T3 therapy (Patel et al., 2024). T3 deficiency accelerates neuronal damage in Alzheimer's disease by inducing mitochondrial dysfunction, oxidative stress, and disruption of amyloid precursor protein (Bernal, 2007; Remaud et al., 2014). T3 enhances the production of dopamine and synaptic preservation in Parkinson's disease, while a deficit aggravates motor dysfunctions and dopaminergic degeneration (Gothié et al., 2017; Goyal et al., 2014; Morte & Bernal, 2014). The severity of the altered T3 and TSH of PD patients is associated with their hormonal imbalance, as per clinical studies. Even if systemic levels are normal, transporter defects can also cause localized CNS hypothyroidism (Gil-Ibañez et al., 2017). T3 plays a very critical role during development. Congenital hypothyroidism results in developmental delay, motor dysfunction, and profound intellectual disability if untreated. While the worst outcomes are prevented with early hormone replacement, less severe impairments such as memory and concentration difficulties often persist (Remaud et al., 2014). Poor transplacental delivery of T3 is likely to be responsible for maternal thyroid dysfunction, which increases the risk of autistic spectrum disorder, ADHD, and language delay in children (Gil-Ibañez et al., 2017). Mutations within the MCT8 transporter led to Allan-Herndon-Dudley syndrome, which has been described as marked intellectual impairment and motor dysfunction without normal or elevated systemic T3. These figures reveal that both systemic and local T3 deficiency led to severe and irreversible neurological consequences (Sahoo & Labhasetwar, 2003).

Pharmacokinetics and challenges in T3 delivery

There are major drawbacks to the therapeutic application of T3 secondary to its pharmacokinetic properties. Hepatic and renal deiodination and sulfation rapidly metabolize T3, leading to unpredictable plasma concentrations. Its half-life is only approximately 24 hours, compared with some 7 days for thyroxine (Anderson, 2001; Oppenheimer et al., 1994). To ensure systemic levels, there should be frequent dosing, but this increases the risk of cardiovascular side effects such as arrhythmias and tachycardia (Peeters et al., 2006). T3 is also tightly bound to plasma proteins such as transthyretin and albumin, decreasing the free fraction that could be absorbed by the brain. By activating or inactivating thyroid hormone through deiodinase enzymes, neural tissues also regulate local T3 levels (Heuer & Visser, 2009; Visser et al., 2011). The greatest hurdle is penetrating the BBB. Even though T3 is lipophilic, it needs some transporters to penetrate the brain, such as MCT8 and OATP1C1 (Friesema et al., 2003; Mayerl et al., 2014). Disabling diseases such as Allan-Herndon-Dudley syndrome are brought about by mutations in transporters (Visser et al., 2011). Although efflux pumps like *P-glycoprotein* and *breast cancer resistance protein* have the active function of pumping out substances, structural components of the BBB, comprising endothelial tight junctions, pericytes, and astrocytic end-feet, impose additional restriction (Pardridge, 2012). T3 transport remains strictly regulated even under conditions such as ischemia or neuroinflammation that make the BBB more permeable (Saraiva et al., 2016). The necessity for delivery systems capable of stabilizing T3, escaping transporter dependency, and achieving customized CNS distribution is emphasized by these pharmacokinetic and structural limitations (Esposito et al., 1997) (Table 18.1).

Nanomedicine platforms for T3 delivery

Through the stabilization of the hormone, prolongation of circulation, and enhancement of BBB penetration, nanomedicine provides a range of strategies for the optimization of T3 administration. T3 is entrapped in liposomes and nanostructured lipid carriers, which stabilize it and inhibit degradation (Beloqui et al., 2016; Masserini, 2013). Surface functionalization with ligands such as lactoferrin or transferrin enhances CNS selectivity and allows for receptor-mediated endocytosis (Ceña & Játiva, 2018). Fine regulation of particle size, charge, and release kinetics can be achieved with polymers such as PLGA and PEGylated chitosan (Kreuter, 2014). Selective release is enabled under pathological CNS conditions by stimuli-responsive concepts, including pH-sensitive formulations (Gao, 2016). Compared with free T3, polymeric nano-T3 facilitates neurological recovery and reduces systemic side effects in preclinical stroke models (Mdzinarishvili et al., 2013). Receptor-mediated BBB penetration is taken advantage of by the conjugation of ligands to lactoferrin, insulin, or transferrin (Dong, 2018; Pardridge, 2012). PEGylation makes it possible to functionalize with targeted ligands while increasing circulation time and reducing immune clearance (Gao, 2016; Tosi et al., 2013). Sophisticated methods that enhance uptake and selectivity but must be properly balanced to avoid cytotoxicity are cell-penetrating peptides, aptamers, and regulated surface charge modulation (Karakus et al., 2021; Kreuter, 2001). Together, these nanocarriers offer unprecedented control over the pharmacokinetics and biodistribution of T3 and represent promising delivery systems for CNS-targeted therapy (Bauer et al., 2008) (Table 18.2).

Preclinical and clinical evidence

Most preclinical studies show that T3 nanoformulated is superior to free hormone administration. In comparing nanoparticle-encapsulated T3 with free hormone in ischemic stroke models, the latter significantly reduced infarct size, enhanced neurological recovery, and compact cerebral edema (Mdzinarishvili et al., 2013; Patel et al., 2024). Nano-T3 suppressed systemic toxicity, increased synaptic plasticity, initiated hippocampus neurogenesis, and corrected hormone levels better in hypothyroid rats (Karakus et al., 2021; Kreuter, 2001, 2014). While free T3 treatment requires high doses to deliver CNS effects due to its short half-life, fast metabolism, and systemic toxicity (Anderson, 2001; Heuer & Visser, 2009; Oppenheimer et al., 1994; Visser et al., 2011). Nanoformulated T3 circumvents these limitations by employing ligand-mediated transcytosis, PEGylation, and controlled release to generate consistent CNS levels (Masserini, 2013; Saraiva et al., 2016). In contrast to free T3, nano-T3 performs superiorly on memory, synaptic density, neuroprotection, and anti-inflammation (Patel et al., 2024). These results are strong evidence supporting additional translational research and clinical trials of nanoformulated T3 as a therapeutic intervention for illnesses of the CNS, such as ischemic stroke, neurodegenerative disease, and developmental disorders.

TABLE 18.1 Pharmacokinetic limitations of free T3 and advantages of nanoformulations.

Parameter	Free triiodothyronine (T3)	Nanoformulated T3	References
Half-life	Short (approximately 1 day)	Prolonged circulation through controlled or sustained release	Ullrich et al., 2023; Mdzinarishvili et al., 2013
Blood–Brain barrier (BBB)	Limited permeability	Enhanced BBB penetration via surface modification and ligand targeting	Goyal et al., 2014; Abdelsalam et al., 2023
Systemic toxicity	High at therapeutic doses	Reduced due to localized delivery and controlled release	Patel et al., 2024; Kaur et al., 2016
Bioavailability	Variable and often low	Increased through improved solubility and absorption	Kumar et al., 2022; Goldsmith et al., 2014
Metabolic degradation	Rapid enzymatic breakdown	Protected within a nanocarrier shell, enabling slower degradation	Kreuter, 2014; Kaushik et al., 2018

TABLE 18.2 Nanocarrier types and T3 targeting strategies.

Nanocarrier type	Material	Targeting strategy	Advantages	References
Liposomes	Phospholipids	Surface PEGylation, antibody-conjugation	Biocompatible, easily modifiable	Goyal et al., 2014; Lammers et al., 2012
Polymeric nanoparticles	PLGA, chitosan	Ligand–receptor (e.g., transferrin)	Sustained release, stable in circulation	Kaur et al., 2016; Tosi et al., 2013
Solid lipid nanoparticles	Compritol, stearic acid	ApoE coating, lactoferrin conjugates	BBB crossing, low toxicity	Kumar et al., 2022; Goldsmith et al., 2014
Dendrimers	PAMAM	Surface functionalization with peptides	Precise size control, multivalent functionality	Abdelsalam et al., 2023

Applications of the material

Clinical potential of T3-loaded nanomedicines

Nanotechnology has made a breakthrough in transport of therapeutic drugs to CNS and the difficult process of crossing BBB (Fig. 18.1). Triiodothyronine (T3) is one of the most essential thyroid hormones of neurodevelopment and neurorepair, and is linked to issues with systemic breakdown and poor BBB permeability. Nanocarriers provide a new method to entrap and deliver T3 to the neural tissues in an efficient way. As another example, Mdzinarishvili et al., (2013) exhibited the possibility of T3-loaded nanoparticles in an ischemic brain model with high neuroprotection and low infarct volume. Such nanocarriers mitigate systemic toxicity, maintain T3 bioactivity, and deliver it to the neural injury site properly.

Precision nanomedicine has since then more perfected this, allowing targeting of a certain cell, releasing at a

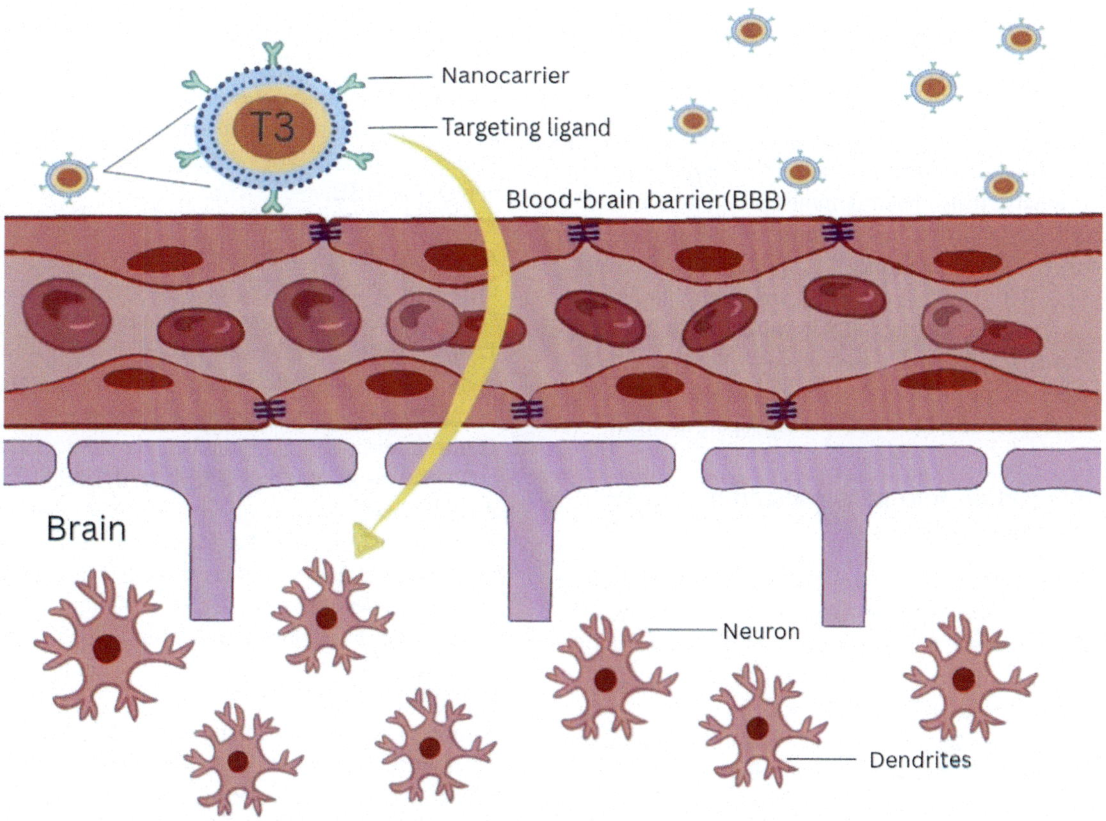

FIGURE 18.1 Mechanistic illustration of T3-Loaded nanocarrier crossing the blood–brain barrier.

desired point, and even reducing the dose. How nanocarrier systems may be selectively modified with targeting ligands (e.g., transferrin or apolipoprotein E) that mediate receptor-mediated cellular endocytosis into neurones is also highlighted by Goldsmith et al., (2014) and Lammers et al., (2012). Such targeted systems may be especially beneficial in reducing off-target activities of the systems as well as improving the therapeutic benefit of T3 in neurodegenerative illnesses.

Treatment of CNS hypothyroidism, brain injury, and demyelinating disorders

Treatment of CNS hypothyroidism, which presents itself in the form of cognitive deterioration, depressive issues, and developmental delays, is one of the most promising uses of nanoformulated T3. Goyal et al., (2014) glance through some nano-based approaches to overcome the low permeability of the BBB in the case of such endocrine-related neurological disorders. Nanoparticles loaded with T3 can provide a solution for restoring the local hormone levels in the brain without any chances of systemic overdoses.

Besides, new preclinical studies indicate that T3 can promote the regeneration of nerves after injury. According to a research by Ullrich et al., (2023), post-ischemic use of T3 showed positive results by stabilizing the BBB and lowering the secondary inflammation. Such an observation has great relevance to stroke survivors, whereby secondary brain damage caused by breaching the BBB can prove to be as damaging as the initial one.

T3 also acts in remyelination and oligodendrocyte differentiation, which are imperative in treating diseases linked with demyelinating illness such as *multiple sclerosis* (MS). Previously, it has been demonstrated that chitosan-engineered nanoparticles, as outlined by Kaur et al., (2016), are effective in neuropeptide and hormone delivery through the BBB, which is why this approach would also be effective in delivering nano-T3 in patients with MS and similar disorders. T3 levels could be controlled accurately locally at the site of the lesion, and this would control the neurodegeneration and facilitate axonal repair as well.

Future integration into personalized medicine and nanopharmacology

Nanomedicine platforms are well-positioned with industry advancement in the field of personalized medicine to incorporate hormonal therapies, such as T3, in an individualized therapy plan. The article by Kumar et al., (2022) defines the intersection of the nano-biotechnology and artificial intelligence (AI) fields to allow implementation in providing smart drug delivery in the context of neurodegenerative disease management. In that regard, the patient-specific variation in T3 levels according to genomic and endocrine patterns is achievable with responsive nanocarriers.

T3 delivery through an exosomal-based delivery system, as examined by Abdelsalam et al., (2023), develops another arena of T3 delivery. Such endogenous vesicle T3-deposits are extremely selective and can produce insignificant immunological reaction upon passage athwart the BBB. Rack and Makela (2000) consider the exosomes as possessing a heterogeneous repertoire of molecular cues that may, potentially, cooperate with T3 to promote neuroplasticity and increase the effectiveness of therapy.

In general, the integration of endocrinology and nanotechnology can expand the horizon of therapeutic opportunities of using T3 in CNS disorders. Nano-T3 preclinical research is poised to open up a new dimension in neurotherapeutics with hormone therapy driven by optimal effectiveness, safety, and personalization instead of pharmacokinetics or delivery efficiency.

Mini-dictionary of terms

Astrocytes: Star-shaped glial cells in the spinal cord and brain are responsible for maintaining the BBB and neuronal support.

The blood–brain barrier (BBB): A selective barrier to prevent destructive substances in the blood from incoming the brain while permitting necessary molecules to pass; a significant hindrance to CNS drug delivery.

Chitosan Nanoparticles: Biodegradable and biocompatible polymeric nanocarriers are often used to improve the transport of neuropeptides and hormones across the BBB.

CNS Hypothyroidism: A neurological condition resulting from reduced thyroid hormone availability or action within the central nervous system, affecting cognitive and physiological functions.

Demyelinating Diseases: Disorders characterized by damage to the myelin sheath of neurons, often treated by hormone or regenerative therapies.

Dendrimers: Highly branched synthetic macromolecules are used as precision drug delivery systems in nanomedicine.

Exosomes: Nano-sized extracellular vesicles that facilitate intercellular communication and have potential as natural carriers for CNS-targeted therapy.

Lipid Nanoparticles: Nano-formulations composed of lipids that enhance drug solubility, stability, and delivery to the brain.

Monocarboxylate Transporters (MCT8): Specialized proteins that facilitate the transport of thyroid hormones across cellular membranes, particularly in neural tissues.

Nanocarriers: Engineered nano-scale carriers are employed to deliver drugs or biomolecules to targeted cells or tissues, enhancing therapeutic accuracy and efficiency.

Nanomedicine: The utilization of nanotechnology in medicine for drug delivery, diagnostic monitoring, and therapeutic monitoring.

Neurogenesis: The mechanism of producing new brain neurons is controlled by hormones such as triiodothyronine (T3).

Polymeric Nanoparticles: Nanocarriers derived from natural or synthetic polymers are engineered for targeted and controlled delivery of drugs in neurological therapy.

Thyroid Hormones (THs): Endocrine hormones such as T4 and T3 control the metabolic processes and neural growth.

Triiodothyronine (T3): A functioning thyroid hormone is vital for brain evolution, neuroplasticity, and metabolism, which is also finding novel applications in nanomedicine.

Summary points

- **Triiodothyronine (T3)** performs a crucial role in neural progress, synaptogenesis, and myelination of the central nervous system.
- **Blood–brain barrier (BBB)** confines the effective delivery of T3 to brain tissues, which presents difficulties in the treatment of CNS hypothyroidism and associated disorders.
- **Nanocarrier systems** such as polymeric, lipid-based, and chitosan nanoparticles significantly enhance the targeted delivery and stability of T3.
- **Engineered T3 nanoparticles** have demonstrated improved neuroprotective effects in ischemic stroke models and other neurodegenerative conditions.
- **Exosome-based delivery platforms** offer promising, biocompatible alternatives for transporting T3 across the BBB with minimal immune activation.
- **T3-loaded nanomedicines** show potential for reversing demyelination and promoting neuronal regeneration in experimental studies.
- **Personalized nanomedicine strategies** are being explored to tailor T3 dosing and delivery based on individual genetic and physiological profiles.
- **Future integration** of nanotechnology and endocrinology may enable real-time monitoring and precision targeting of thyroid hormone therapies in the CNS.
- **Clinical translation** of T3 nanotherapeutics is underway but requires further validation through long-term safety and efficacy trials.

References

Abdelsalam, M., Ahmed, M., Osaid, Z., Hamoudi, R., & Harati, R. (2023). Insights into exosome transport through the blood–brain barrier and the potential therapeutic applications in brain diseases. *Pharmaceuticals, 16*(4), 571.

Andersen, S., & Andersen, S. L. (2024). Response to fitzgerald et al. re: "Thyroid stimulating hormone and thyroid hormones (Triiodothyronine and Thyroxine): An American thyroid association-commissioned review of current clinical and laboratory status". *Thyroid, 34*(5).

Anderson, G. W. (2001). Thyroid hormones and the brain. *Frontiers in Neuroendocrinology, 22*(1), 1–17.

Bauer, M., Goetz, T., Glenn, T., & Whybrow, P. C. (2008). The thyroid-brain interaction in thyroid disorders and mood disorders. *Journal of Neuroendocrinology, 20*(10), 1101–1114.

Beloqui, A., Solinís, M.Á., Rodríguez-Gascón, A., Almeida, A. J., & Préat, V. (2016). Nanostructured lipid carriers: Promising drug delivery systems for future clinics. *Nanomedicine: Nanotechnology, Biology and Medicine, 12*(1), 143–161.

Bernal, J. (2007). Thyroid hormone receptors in brain development and function. *Nature Clinical Practice Endocrinology & Metabolism, 3*(3), 249–259.

Ceña, V., & Játiva, P. (2018). Nanoparticle crossing of blood–blood-brain barrier: A road to new therapeutic approaches to central nervous system diseases. *Nanomedicine, 13*(13), 1513–1516.

Churilov, L. P., Sobolevskaia, P. A., & Stroev, Y. I. (2019). Thyroid gland and brain: Enigma of hashimoto's encephalopathy. *Best Practice & Research Clinical Endocrinology & Metabolism, 33*(6), Article 101364.

Dabiri, S., Jafari, S., & Molavi, O. (2024). Advances in nanocarrier-mediated delivery of chrysin: Enhancing solubility, bioavailability, and anticancer efficacy. *BioImpacts: BI, 15*, Article 30269.

Dong, X. (2018). Current strategies for brain drug delivery. *Theranostics, 8*(6), 1481.

Esposito, S., Prange Jr, A. J., & Golden, R. N. (1997). The thyroid axis and mood disorders: Overview and future prospects. *Psychopharmacology Bulletin, 33*(2), 205.

Friesema, E. C., Ganguly, S., Abdalla, A., Fox, J. E. M., Halestrap, A. P., & Visser, T. J. (2003). Identification of monocarboxylate transporter 8 as a specific thyroid hormone transporter. *Journal of Biological Chemistry, 278*(41), 40128–40135.

Gao, H. (2016). Progress and perspectives on targeting nanoparticles for brain drug delivery. *Acta Pharmaceutica Sinica B, 6*(4), 268–286.

Gil-Ibañez, P., García-García, F., Dopazo, J., Bernal, J., & Morte, B. (2017). Global transcriptome analysis of primary cerebrocortical cells: Identification of genes regulated by triiodothyronine in specific cell types. *Cerebral Cortex, 27*(1), 706–717.

Goldsmith, M., Abramovitz, L., & Peer, D. (2014). Precision nanomedicine in neurodegenerative diseases. *ACS Nano, 8*(3), 1958–1965.

Gothié, J. D., Demeneix, B., & Remaud, S. (2017). Comparative approaches to understanding thyroid hormone regulation of neurogenesis. *Molecular and Cellular Endocrinology, 459*, 104–115.

Goyal, K., Koul, V., Singh, Y., & Anand, A. (2014). Targeted drug delivery to central nervous system (CNS) for the treatment of neurodegenerative disorders: Trends and advances. *Central Nervous System Agents in Medicinal Chemistry, 14*(1), 43–59.

Heuer, H., & Visser, T. J. (2009). Pathophysiological importance of thyroid hormone transporters. *Endocrinology, 150*(3), 1078–1083.

Karakus, O. O., Darwish, N. H., Sudha, T., Salaheldin, T. A., Fujioka, K., Dickinson, P. C. T., … Mousa, S. A. (2021). Development of triiodothyronine polymeric nanoparticles for targeted delivery in the cardioprotection against ischemic insult. *Biomedicines, 9*(11), 1713.

Kaur, S., Bhararia, A., Sharma, K. K., Mittal, S., Jain, R., Wangoo, N., & Sharma, R. K. (2016). Thyrotropin-releasing hormone-loaded and chitosan-engineered polymeric nanoparticles: Towards effective delivery of neuropeptides. *Journal of nanoscience and nanotechnology, 16*(5), 5324–5332.

Kaushik, A., Jayant, R. D., Bhardwaj, V., & Nair, M. (2018). Personalized nanomedicine for CNS diseases. *Drug Discovery Today, 23*(5), 1007–1015.

Kreuter, J. (2001). Nanoparticulate systems for brain delivery of drugs. *Advanced Drug Delivery Reviews, 47*(1), 65–81.

Kreuter, J. (2014). Drug delivery to the central nervous system by polymeric nanoparticles: What do we know? *Advanced Drug Delivery Reviews, 71*, 2–14.

Kumar, R., Aadil, K. R., Mondal, K., Mishra, Y. K., Oupicky, D., Ramakrishna, S., & Kaushik, A. (2022). Neurodegenerative disorders management: State-of-the-art and prospects of nano-biotechnology. *Critical Reviews in Biotechnology, 42*(8), 1180–1212.

Lammers, T., Rizzo, L. Y., Storm, G., & Kiessling, F. (2012). Personalized nanomedicine. *Clinical Cancer Research, 18*(18), 4889–4894.

Liu, S., Jin, X., Ge, Y., Dong, J., Liu, X., Pei, X., & Yu, X. A. (2025). Advances in brain-targeted delivery strategies and natural product-mediated enhancement of blood–brain barrier permeability. *Journal of Nanobiotechnology, 23*(1), 1–32.

Masserini, M. (2013). Nanoparticles for brain drug delivery. *International Scholarly Research Notices, 2013*(1), Article 238428.

Mayerl, S., Müller, J., Bauer, R., Richert, S., Kassmann, C. M., Darras, V. M., & Heuer, H. (2014). Transporters MCT8 and OATP1C1 maintain murine brain thyroid hormone homeostasis. *The Journal of Clinical Investigation, 124*(5), 1987–1999.

Mdzinarishvili, A., Sutariya, V., Talasila, P. K., Geldenhuys, W. J., & Sadana, P. (2013). Engineering triiodothyronine (T3) nanoparticle for use in ischemic brain stroke. *Drug Delivery and Translational Research, 3*, 309–317.

Mohapatra, P., Gopikrishnan, M., Doss C, G. P., & Chandrasekaran, N. (2024). How precise are nanomedicines in overcoming the blood--brain barrier? A comprehensive review of the literature. *International Journal of Nanomedicine*, 2441–2467.

Mokle, K., & Sanap, G. (2023). Review on thyroid disease. *International Journal of Pharmaceutical Sciences, 1*(12), 1-1.

Morte, B., & Bernal, J. (2014). Thyroid hormone action: Astrocyte–neuron communication. *Frontiers in Endocrinology, 5*, 82.

Oppenheimer, J. H., Schwartz, H. L., & Strait, K. A. (1994). Thyroid hormone action 1994: The plot thickens. *European Journal of Endocrinology, 130*(1), 15–24.

Pardridge, W. M. (2012). Drug transport across the blood–brain barrier. *Journal of cerebral blood flow & metabolism, 32*(11), 1959–1972.

Patel, H., Patel, J., & Patel, A. (2024). Enhancement of neuroprotective and anti-edema action in a mouse ischemic stroke model using T3 loaded nanoparticles. *Medical and Pharmaceutical Journal, 3*(1), 1–12.

Peeters, R. P., van der Deure, W. M., & Visser, T. J. (2006). Genetic variation in thyroid hormone pathway genes: Polymorphisms in the TSH receptor and the iodothyronine deiodinases. *European Journal of Endocrinology, 155*(5), 655–662.

Rack, S. K., & Makela, E. H. (2000). Hypothyroidism and depression: A therapeutic challenge. *Annals of Pharmacotherapy, 34*(10), 1142–1145.

Remaud, S., Gothié, J. D., Morvan-Dubois, G., & Demeneix, B. A. (2014). Thyroid hormone signaling and adult neurogenesis in mammals. *Frontiers in Endocrinology, 5*, 62.

Sabatino, L., Lapi, D., & Del Seppia, C. (2024). Factors and mechanisms of thyroid hormone activity in the brain: Possible role in recovery and protection. *Biomolecules, 14*(2), 198.

Sahoo, S. K., & Labhasetwar, V. (2003). Nanotech approaches to drug delivery and imaging. *Drug Discovery Today, 8*(24), 1112–1120.

Saraiva, C., Praça, C., Ferreira, R., Santos, T., Ferreira, L., & Bernardino, L. (2016). Nanoparticle-mediated brain drug delivery: Overcoming–brain barrier to treat neurodegenerative diseases. *Journal of Controlled Release, 235*, 34–47.

Sidorov, E., Paul, A., Xu, C., Nouh, C. D., Chen, A., Gosmanova, A., & Mdzinarishvili, A. (2023). A decrease in thyroid function after ischemic stroke is related to stroke severity. *Thyroid Research, 16*(1), 28.

Sim, T. M., Tarini, D., Dheen, S. T., Bay, B. H., & Srinivasan, D. K. (2020). Nanoparticle-based technology approaches to the management of neurological disorders. *International Journal of Molecular Sciences, 21*(17), 6070.

Tan, Q., Zhao, S., Xu, T., Wang, Q., Zhang, M., Yan, L., & Lan, M. (2023). Inorganic nano-drug delivery systems for crossing the blood–brain barrier: Advances and challenges. *Coordination Chemistry Reviews, 494*, Article 215344.

Tosi, G., Bortot, B., Ruozi, B., Dolcetta, D., Vandelli, M. A., Forni, F., & Severini, G. M. (2013). Potential use of polymeric nanoparticles for drug delivery across the blood-brain barrier. *Current Medicinal Chemistry, 20*(17), 2212–2225.

Ullrich, D., Führer-Sakel, D., Heuer, H., Mayerl, S., Haupeltshofer, S., Schmitt, L. I., & Langhauser, F. (2023). Post-ischemic triiodothyronine treatment improves stroke outcome by stabilizing the blood-brain barrier. *bioRxiv*. https://doi.org/10.1101/2023.11.21.568025, 2023-11.

Visser, W. E., Friesema, E. C., & Visser, T. J. (2011). Minireview: Thyroid hormone transporters: The knowns and the unknowns. *Molecular Endocrinology, 25*(1), 1–14.

Zhang, X., Wang, M., Liu, Z., Wang, Y., Chen, L., Guo, J., & Guan, B. (2023). Transnasal-brain delivery of nanomedicines for neurodegenerative diseases. *Frontiers in Drug Delivery, 3*, Article 1247162.

Further reading

Nouri, Z., Barfar, A., Perseh, S., Motasadizadeh, H., Maghsoudian, S., Fatahi, Y., … Atyabi, F. (2024). Exosomes as therapeutic and drug delivery vehicle for neurodegenerative diseases. *Journal of Nanobiotechnology, 22*(1), 463.

Tosi, G., Costantino, L., Ruozi, B., Forni, F., & Vandelli, M. A. (2008). Polymeric nanoparticles for the drug delivery to the central nervous system. *Expert Opinion on Drug Delivery, 5*(2), 155–174.

Chapter 19

Hypothyroidism and improved spatial working memory

Satkarjit Kaur Jhandi[1] **and Ajaypal Singh**[2]
[1]Department of Physiotherapy, Khalsa College, Amritsar, Punjab, India; [2]Fortis Escort Hospital, Amritsar, Punjab, India

List of abbreviations

5 HTP 5-Hydroxytryptophan
CANTAB Cambridge Neuropsychological Testing Automated Battery
CNS Central nervous system
LT-4 levothyroxine
RELN reelin
TSH thyroid stimulating hormone

Introduction

The incidence of hypothyroidism varies depending on the population investigated and the diagnostic criteria employed, but it is believed to affect roughly 1%–5% of the worldwide population. It is with higher frequency among women (with a female-to-male ratio of 5:1) and elderly population (Chiovato et al., 2019). Common reasons of hypothyroidism include autoimmune thyroiditis (Hashimoto's disease), surgical thyroidectomy, radioactive iodine therapy, and drugs including lithium, amiodarone, and several cancer treatments (Gaitonde et al., 2012). Thyroid hormone insufficiency affects both thyroxine and triiodothyronine, with triiodothyronine being the active form that exerts metabolic effects at the cellular level.

Symptoms of hypothyroidism may include tiredness, weight gain, cold sensitivity, constipation, dry skin, hair loss, or depression (Bianco, 2025). The diagnosis is often verified by evaluating blood thyroid stimulating hormone (TSH) and free thyroxine levels, with excessive TSH (AlAnazi et al., 2023).

Thyroid hormones and hypothyroidism

In the brain, thyroid hormones, including TSH, play a vital role in its development and function. These hormones influence neural differentiation, synaptic plasticity, and cognitive functions (Amano et al., 2025).

Hormones secreted by the thyroid gland, including thyroxine and triiodothyronine are important from prenatal development, such as growth and differentiation of the nervous system, to adult functions, such as maintenance of metabolism and nutrition, thermogenesis (Cheng et al., 2010). Additionally, they are believed to be allied to cognitive processes (Loosen, 1992; Timiras & Nzekwe, 1989; Zhu et al., 2006). There are different target genes for thyroid hormones; many of them are critical for brain functioning, and more than 1100 are important for brain development (Bernal, 2017). Among these genes several genes perform important functions in the nervous system; for example, RELN, by producing Reelin, an extracellular matrix protein, is tangled in neural migration in brain (Alvarez-Dolado et al., 1999); the next is the protein kinase C substrate NRGN (RC3/neurogranin) (Iñiguez et al., 1992), which is tangled in neural plasticity procedures such as long-term potentiation (Kaleka & Gerges, 2016). Thyroid hormones additionally modulate the appearance of neuromodulin/GAP-43, thereby regulating synaptic malleability and processes related to memory.

Thyroid dysfunction is recognized as a possibly reversible contributor to cognitive decline. For many years, psychological and cognitive alterations have been observed in adults exhibiting deviations in thyroid function and TSH is a sensitive indicator for thyroid dysfunction (Booth et al., 2013). Dating back to the late 19th century, the initial reports established an association between hypothyroidism, thyrotoxic diseases, and mental disorders (Khaleghzadeh-Ahangar et al., 2022). Animal models have illustrated that adult hypothyroidism induces alterations in brain function and morphology within regions responsible for cognitive functions (Fernández-Lamo et al., 2009).

Hormones in Neuroscience. https://doi.org/10.1016/B978-0-443-30261-9.00019-X

Hypothyroidism induces a reduction in brain serotonin levels along with its precursor, tryptophan (5-HTP) (Bauer et al., 2002). Given the essential role of the brain serotonin system in enhancing memory functions, this alteration may represent a plausible mechanism underlying hypothyroidism-induced cognitive impairment (Schmitt et al., 2006). Bortolotto et al. (2020) in their study revealed that hippocampal Na+, K + -ATPase commotion and glutamate levels are diminished in hypothyroidism conditions, potentially contributing to memory deficits observed in mice. With decreasing thyroid hormone levels, multiple genes are downregulated, including laminin, a critical guidance protein in neural cell migration, and brain-derived neurotrophic factor, which is linked to synapse structural alterations. As a result, hypothyroidism may affect cognition by disrupting processes such as neuronal migration and synapse architecture (Koromilas et al., 2010). Furthermore, hypothyroidism exacerbates cognitive impairment in all age groups by limiting energy-consuming tasks necessary for crucial central nervous system activities, particularly neurotransmission, memory, and higher-order cognitive functions (Schraml, 2011). Did a pilot research to look at the relationship between TSH levels in hypothyroidism and cognitive symptoms. Eleven individuals who had thyroidectomy were evaluated for cognitive performance and thyroid function while in overt hypothyroidism and then after undergoing thyroid hormone replacement treatment. Performance on the only two cognitive tests in our cohort that showed a detrimental influence from hypothyroidism was inversely related to TSH levels during hypothyroidism. This suggests that during the acute stage of the disease, the severity of "brain hypothyroidism" may be reflected in the blood TSH level. Another longitudinal research was undertaken by Beydoun et al. (2015) to determine the relationship between thyroid hormones and long-term cognitive changes. Cognitive test findings were used to assess learning, memory, language or verbal skills, sustained attention, psychomotor speed, cognitive flexibility, and visuospatial abilities. They concluded that among urban US adults, greater baseline TSH was associated with a faster rate of cognitive decline over time, particularly in the domains of working memory and visuospatial ability.

Hypothyroidism and working memory and neuropsychological symptoms

Working memory is described as the "sketchbook of sensible thoughts" and serves as a platform for manipulating and organizing thoughts (Diamond, 2013). Spatial working memory is a type of working memory, which is considered as the ability to sustain and manipulate info in the mind over a period of seconds is a crucial executive function that impacts one's ability to learn, make decisions in daily life. It is an essential element of several cognitive functions (learning, comprehension, language, decision making, and reasoning).

Elevated TSH levels can affect domains of cognitive function. Working memory is the most consistently found affected domain in patients with hypothyroidism. Patients with hypothyroidism showed hippocampal atrophy, leading to memory loss, cerebral blood flow impairment, and a reduction in working memory. These patients may experience cognitive impairments, some of which can be improved with LT-4. Deficits in cognitive processes, such as the capability to inhibit info that is not relevant, to shift attention between various tasks, and to process info hastily have been shown to be responsible for the specific memory deficits that occur in normal aging. In animal models of hypothyroidism, deficits in memory tasks that are reversible with thyroid hormone therapy have been seen. In rats with hypothyroidism decreased hippocampal neurogenesis, provided a potential mechanistic explanation for memory deficits (Desouza et al., 2005). Thyroid hormones exert a broad spectrum of effects on the central nervous system (CNS). Hypothyroidism, especially during CNS development, can lead to structural and functional changes (mostly resulting in mental retardation). The hippocampus is considered as one of the most important CNS structures, while the investigation and understanding of its direct and indirect interactions with the thyroid hormones could provide crucial information on the neurobiological basis of the (frequently-faced in clinical practice) hypothyroidism-induced mental retardation and neurobehavioral dysfunction. THs-deficiency during the fetal and/or the neonatal period produces deleterious effects for neural growth and development (such as reduced synaptic connectivity, delayed myelination, disturbed neuronal migration, deranged axonal projections, decreased synaptogenesis and alterations in neurotransmitters' levels). On the other hand, the adult-onset thyroid dysfunction is usually associated with neurological and behavioral abnormalities. In both cases, genomic and proteomic changes seem to occur.

A positron emission tomography study revealed that in subclinical hypothyroidism population had aberrations in glucose metabolism in the limbic system. Specifically, in amygdala (responsible for emotions), hippocampal region (responsible for learning and memory), and also in cingulate gyrus (memory and decision making or planning) (Bauer et al., 2002; Koromilas et al., 2010) revealed that these had significant decrease in volume of gray matter numerous brain regions, including frontal gyrus (responsible for executive functions), right occipital gyrus (responsible for visual processing), left precentral and postcentral gyrus (responsible for voluntary movements), and right upper temporal pole (responsible for visual spatial information). Moreover, Kumar et al. (2018) additionally revealed that in hypothyroidism population there were variations in intrinsic resting-state functional connectivity within the somato-motor and right frontoparietal attention network (responsible for such as attention and working memory).

Clinically, hypothyroidism population frequently express complaints of cognitive and psychiatric issues, common complaints comprise lack of memory and neuropsychological symptoms, especially depressive symptoms. 63.5% of depressive symptoms has been found on Hamilton Depression Rating Scale (Demartini et al., 2014). Also, some studies (Galkin et al., 2020; Li et al., 2018) explored cognitive deficits in patients with depression demonstrated changes in executive functions, specifically working memory. Extensive empirical evidence from multiple studies (Li et al., 2018; Weiland-Fiedler et al., 2004) indicates that depression is specifically allied with notable functional deficits in WM. Patients in spite of being in euthyroid state after LT-4 treatment, present with distressing neuropsychological symptoms. Also, despite achieving normal TSH levels, a considerable number of patients undergoing LT-4 treatment frequently report persisting complaints related to lack of memory or forgetfulness, apathy and mental slowness. A TSH value of 2.5 mU/L as the optimal threshold for predicting the presence of depression symptoms in patients with hypothyroidism undergoing L-T4 treatment (Talaei et al., 2017). It is significant that concomitant depression would be present in patients with TSH $\geq$ 2.5 mU/L who are experiencing neuropsychological symptoms.

Levothyroxine (LT4) and spatial working memory

The administration of thyroid hormones in thyroid disorders can substantially enhance cognitive and emotional indices (Bunevicius, 2009). The central role of frontal executive functions in various cognitive processes, providing appropriate treatment to address cognitive deficit would be highly beneficial for hypothyroidism population. Consequently, the question arises regarding whether treatment of hypothyroidism is warranted to potentially mitigate further cognitive dysfunction.

In clinicals, hypothyroidism population demonstrated deficits in memory functions due to reduced cerebral blood flow to the hippocampus area of the brain. Functional neuroimaging study has been reported that impaired working memory was improved with levothyroxine (LT-4) treatment, including changed task induced deactivation in neural activity within the defaulting mode network were found before and recovered after LT-4 treatment. LT4 is the traditional medicine prescribed to manage hypothyroidism. This therapy has greatly improved the lives of millions of patients with hypothyroidism since its primer in 1949.

L-T4 treatment is usually effective in treating cognitive impairments, although there may not be complete reversal. The impairment of frontal executive function in hypothyroidism is reversible with LT-4 treatment (Bono et al., 2004; Zhu et al., 2006). A review study conducted by (Koromilas et al., 2010) documented that animal studies have demonstrated the significant impact of L-T4 treatment on brain regions responsible for cognitive functioning. However, a recent cohort study by Beltrão et al. (2026) reported that LT-4 treatment can mitigate certain cognitive dysfunctions, it may not fully restore all aspects of cognitive impairment. Although controversies exist in the literature but left untreated, longstanding elevation in the level of TSH can lead to irretrievable cognitive loss.

Aghili et al. (2012) evaluated whether memory quotient improved after LT-4 treatment in hypothyroidism. Authors recruited 60 patients with TSH range of 4.5–10 mIU/L, with mean age of 32.32 years, and LT-4 treatment was given for 3 months. There was significant improvement in logical memory and memory quotient. Jorde et al. (2006) examined the relation between neuropsychological function and subclinical hypothyroidism. Participants were older than 29 years, and the TSH range set was between 3.5 and 10.0 mIU/L (absence of hypothyroidism symptoms), and for healthy controls, 0.50–3.49 mIU/L. LT-4 treatment was given for 12 months to the patients. However, no significant differences were seen between the patients with SCH and the healthy controls in working memory functions. An observational trial conducted by Kalra et al. (2020) determined the impairments of neuropsychological functions in hypothyroidism population when compared to euthyroid population. Thirty-nine patients with mean age of 30.28 years and TSH levels 6.36 $\pm$ 1.3 mIU/L; 23 euthyroid controls with TSH range 0.45–4.5 mIU/L were recruited. Authors determined that the patients have deficits in memory recall (delayed recall), impairment of memory may warrant the use of hormonal replacement therapy.

The discrepancies among the studies may be accredited to the alterations in TSH range of the participants and may be the different tools or tests used to measure cognitive functions. More importantly, none of these studies clearly define and assess spatial working memory with tools which are specifically utilized for it. Whereas, in a recent study, Jhandi et al. (2025) used a specific tool, the Cambridge Neuropsychological Test Automated Battery (CANTAB), to assess spatial working memory function in a subclinical hypothyroidism population. Authors reported that there was impaired spatial working memory in this population. Newly diagnosed patients demonstrated poorer performance on the CANTAB task than those who were on LT-4 treatment. Additionally, authors also reported more improvement in the patients who have been on LT-4 for a decade than patients who have been medicated for a year.

The study above clearly indicates that long-term LT-4 medication results in a drop in TSH and improved spatial working memory in hypothyroidism.

Screening of spatial working memory

Spatial working memory should be tested in hypothyroidism patients. Assessment can be performed using a specialized and trustworthy tool. CANTAB is a recognized and reliable neuropsychological instrument that uses computerized test batteries to measure cognitive performance. It consists comprehensive subsets allow for the measurement of multiple cognitive subdomains within executive functions, including spatial working memory, using a single battery, as opposed to traditional neuropsychological tests. The automated nature of tests like the CANTAB allows for the research of executive function across the lifetime. With limited reliance on language and the capacity to continually and instantaneously adjust to the participant's performance level, the CANTAB is especially well-suited. This automated battery illustrates the effective combination of traditional cognitive test paradigms with novel forms. Furthermore, its administrative procedures are more simplified and less laborious. It takes less training to administer and is less vulnerable to practice effects than typical neuropsychological examinations.

The move from paper-and-pencil to computer-based evaluation is challenging. Each method of assessment has unique benefits and hitches. Computerized tests have several advantages, including the ability to assess a broader range of abilities while reducing floor and ceiling effects, more standardized administration, the availability of multiple versions for repeated testing, and the ability to precisely record response accuracy and speed with a sensitivity that traditional methods cannot. Another possible advantage of computerized test batteries over traditional paper-and-pencil examinations is their ability to rapidly adjust to an individual's performance levels. The foregoing research clearly shows that the computerized test is more reliable for assessing executive functions. As a result, using CANTAB to examine spatial working memory allows for the documentation of correct modifications.

Conclusion

The neural basis of spatial working memory involves sustained neural activity in the prefrontal cortex, parietal cortex, and occipital area. This network maintains spatial information over short periods through the sustained firing of neurons. Additionally, this network is closely linked with spatial attention and eye movements. Recent research (Abubaker et al., 2024; Benchenane et al., 2011) has shown that brain oscillations, specifically theta gamma coupling play a crucial role in organizing and processing spatial information within the prefrontal cortex. This suggests that the neural activity observed in spatial working memory tasks may reflect a prospective motor code for upcoming eye movements, rather than just memory storage itself.

FMRI studies (Zhu et al., 2006) have shown that untreated individuals may have abnormal brain activity in frontal areas, specifically the right dorsolateral PFC and parietal lobe (right posterior), which are critical for spatial working memory. PET investigations revealed that untreated patients with hypothyroidism had reduced regional glucose metabolism in brain regions crucial for cognition, which was reversed following 3 months of LT4.

It is imperative to evaluate the spatial working memory function in patients with hypothyroidism and assessed it in follow up after treating them with LT-4.

Mini-dictionary of terms

Executive functions: These are higher mental functions or mental skills which are crucial for everyday mental decisions, planning or reasoning.
Spatial working memory: The capability to recall the incidence of an event or the site of any object, which is likely to be affected utmost.
Neuropsychological symptoms: Patients with hypothyroidism commonly present with neuropsychological symptoms such as depressive symptoms, fatigue, forgetfulness, insomnia, and mental fogginess.
Levothyroxine: Levothyroxine (LT-4) is hormonal replacement therapy which is traditionally recommended to patients with elevated TSH levels.
CANTAB: It is an automated computerized battery, which is widely used for the examination of executive functions. It is reliably documenting the changes in these functions.

Summary

- Hypothyroidism (i.e., raised TSH levels) had a direct impact on higher cognitive processes. Spatial working memory is most impaired cognitive function in hypothyroidism.
- Hypothyroidism is directly linked with neuropsychological symptoms, which may also lead to memory deficits.
- Spatial working memory function recovers as a result of normalization in TSH levels following LT-4 treatment.
- LT-4 treatment may be beneficial in patients with mild increase TSH levels or it can be initiate when patients present with neuropsychological symptoms.
- Evaluation of impairment of spatial working memory in hypothyroidism is imperative with a specific tool.

References

Abubaker, M., Al Qasem, W., Pilátová, K., Ježdík, P., & Kvašňák, E. (2024). Theta-gamma-coupling as predictor of working memory performance in young and elderly healthy people. *Molecular Brain, 17*(1), 74. https://doi.org/10.1186/s13041-024-01149-8

Aghili, R., Khamseh, M. E., Malek, M., Hadian, A., Baradaran, H. R., Najafi, L., & Emami, Z. (2012). Changes of subtests of Wechsler memory scale and cognitive function in subjects with subclinical hypothyroidism following treatment with levothyroxine. *Archives of Medical Science: AMS, 8*(6), 1096–1101. https://doi.org/10.5114/aoms.2012.32423

AlAnazi, F. H., Al-kuraishy, H. M., Alexiou, A., Papadakis, M., Ashour, M. H. M., Alnaaim, S. A., Elhussieny, O., Saad, H. M., & Batiha, G. E.-S. (2023). Primary hypothyroidism and alzheimer's disease: A tale of two. *Cellular and Molecular Neurobiology, 43*(7), 3405–3416. https://doi.org/10.1007/s10571-023-01392-y

Alvarez-Dolado, M., Ruiz, M., Del Río, J. A., Alcántara, S., Burgaya, F., Sheldon, M., Nakajima, K., Bernal, J., Howell, B. W., Curran, T., Soriano, E., & Muñoz, A. (1999). Thyroid hormone regulates reelin and dab1 expression during brain development. *Journal of Neuroscience: The Official Journal of the Society for Neuroscience, 19*(16), 6979–6993. https://doi.org/10.1523/JNEUROSCI.19-16-06979.1999

Amano, I., Ninomiya, A., & Koibuchi, N. (2025). Neurological consequences of adult-onset hypothyroidism. *Endocrine Journal, 72*(11), 1175–1187. https://doi.org/10.1507/endocrj.EJ25-0163

Bauer, M., Heinz, A., & Whybrow, P. C. (2002). Thyroid hormones, serotonin and mood: of synergy and significance in the adult brain. *Molecular Psychiatry, 7*(2), 140–156. https://doi.org/10.1038/sj.mp.4000963

Beltrão, F. E. de L., Carvalhal, G., Meneghini, V., Matos, D. A. R., Beltrão, D. C. de A., Andrade, B. A. R. M., Beltrão, F. N. de L., Ramos, H. E., Ribeiro, M. O., Golovko, G., Ettleson, M. D., & Bianco, A. C. (2026). Treatment of hypothyroidism that contains liothyronine is associated with reduced risk of dementia and mortality. *The Journal of Clinical Endocrinology & Metabolism, 111*(2), 561–571. https://doi.org/10.1210/clinem/dgaf367. dgaf367.

Benchenane, K., Tiesinga, P. H., & Battaglia, F. P. (2011). Oscillations in the prefrontal cortex: A gateway to memory and attention. *Current Opinion in Neurobiology, 21*(3), 475–485. https://doi.org/10.1016/j.conb.2011.01.004

Bernal, J. (2017). Thyroid hormone regulated genes in cerebral cortex development. *Journal of Endocrinology, 232*(2), R83–R97. https://doi.org/10.1530/JOE-16-0424

Beydoun, M. A., Beydoun, H. A., Rostant, O. S., Dore, G. A., Fanelli-Kuczmarski, M. T., Evans, M. K., & Zonderman, A. B. (2015). Thyroid hormones are associated with longitudinal cognitive change in an urban adult population. *Neurobiology of Aging, 36*(11), 3056–3066. https://doi.org/10.1016/j.neurobiolaging.2015.08.002

Bianco, A. C. (2025). Shared decisionmaking in the treatment of hypothyroidism. *Clinical Endocrinology, 103*(1), 106–112. https://doi.org/10.1111/cen.15228

Bono, G., Fancellu, R., Blandini, F., Santoro, G., & Mauri, M. (2004). Cognitive and affective status in mild hypothyroidism and interactions with L-thyroxine treatment. *Acta Neurologica Scandinavica, 110*(1), 59–66. https://doi.org/10.1111/j.1600-0404.2004.00262.x

Booth, T., Deary, I. J., & Starr, J. M. (2013). Thyroid stimulating hormone, free thyroxine and cognitive ability in old age: The Lothian birth cohort study 1936. *Psychoneuroendocrinology, 38*(4), 597–601. https://doi.org/10.1016/j.psyneuen.2012.07.018

Bortolotto, V. C., Araujo, S. M., Pinheiro, F. C., Poetini, M. R., de Paula, M. T., Meichtry, L. B., de Almeida, F. P., Musachio, E. A. S., Guerra, G. P., & Prigol, M. (2020). Modulation of glutamate levels and Na+,K+-ATPase activity contributes to the chrysin memory recovery in hypothyroidism mice. *Physiology & Behavior, 222*, Article 112892. https://doi.org/10.1016/j.physbeh.2020.112892

Bunevicius, R. (2009). Thyroid disorders in mental patients. *Current Opinion in Psychiatry, 22*(4), 391–395. https://doi.org/10.1097/YCO.0b013e328329e1ae

Cheng, S.-Y., Leonard, J. L., & Davis, P. J. (2010). Molecular aspects of thyroid hormone actions. *Endocrine Reviews, 31*(2), 139–170. https://doi.org/10.1210/er.2009-0007

Chiovato, L., Magri, F., & Carlé, A. (2019). Hypothyroidism in context: Where we've been and where we're going. *Advances in Therapy, 36*(S2), 47–58. https://doi.org/10.1007/s12325-019-01080-8

Demartini, B., Ranieri, R., Masu, A., Selle, V., Scarone, S., & Gambini, O. (2014). Depressive symptoms and major depressive disorder in patients affected by subclinical hypothyroidism: A cross-sectional study. *The Journal of Nervous and Mental Disease, 202*(8), 603–607. https://doi.org/10.1097/NMD.0000000000000168

Desouza, L. A., Ladiwala, U., Daniel, S. M., Agashe, S., Vaidya, R. A., & Vaidya, V. A. (2005). Thyroid hormone regulates hippocampal neurogenesis in the adult rat brain. *Molecular and Cellular Neurosciences, 29*(3), 414–426. https://doi.org/10.1016/j.mcn.2005.03.010

Diamond, A. (2013). Executive functions. *Annual Review of Psychology, 64*, 135–168. https://doi.org/10.1146/annurev-psych-113011-143750

Fernández-Lamo, I., Montero-Pedrazuela, A., Delgado-García, J. M., Guadaño-Ferraz, A., & Gruart, A. (2009). Effects of thyroid hormone replacement on associative learning and hippocampal synaptic plasticity in adult hypothyroid rats. *European Journal of Neuroscience, 30*(4), 679–692. https://doi.org/10.1111/j.1460-9568.2009.06862.x

Gaitonde, D. Y., Rowley, K. D., & Sweeney, L. B. (2012). Hypothyroidism: An update. *American Family Physician, 86*(3), 244–251.

Galkin, S. A., Peshkovskaya, A. G., Simutkin, G. G., Vasil'eva, S. N., Roshchina, O. V., Ivanova, S. A., & Bokhan, N. A. (2020). Impairments to the functions of spatial working memory in mild depression and their neurophysiological correlates. *Neuroscience and Behavioral Physiology, 50*(7), 825–829. https://doi.org/10.1007/s11055-020-00973-4

Iñiguez, M. A., Rodriguez-Peña, A., Ibarrola, N., Morreale de Escobar, G., & Bernal, J. (1992). Adult rat brain is sensitive to thyroid hormone. Regulation of RC3/neurogranin mRNA. *The Journal of Clinical Investigation, 90*(2), 554–558. https://doi.org/10.1172/JCI115894

Jhandi, S. K., Shenoy, S., Singh, A. P., & Singh, A. (2025). Spatial working memory in hypothyroidism – an observational study on different ranges of thyroid stimulating hormone. https://doi.org/10.15584/ejcem.2025.1.25.

Jorde, R., Waterloo, K., Storhaug, H., Nyrnes, A., Sundsfjord, J., & Jenssen, T. G. (2006). Neuropsychological function and symptoms in subjects with subclinical hypothyroidism and the effect of thyroxine treatment. *The Journal of Clinical Endocrinology & Metabolism, 91*(1), 145–153. https://doi.org/10.1210/jc.2005-1775

Kaleka, K. S., & Gerges, N. Z. (2016). Neurogranin restores amyloid β-mediated synaptic transmission and long-term potentiation deficits. *Experimental Neurology, 277*, 115–123. https://doi.org/10.1016/j.expneurol.2015.12.013

Kalra, P., Kumaraswamy, D. R., Dharmalingam, M., Saini, J., & Yadav, R. (2020). Neuropsychological impairments in young patients with subclinical hypothyroidism: A case control study. *Annals of Neurosciences, 27*(3–4), 169–174. https://doi.org/10.1177/0972753121990177

Khaleghzadeh-Ahangar, H., Talebi, A., & Mohseni-Moghaddam, P. (2022). Thyroid disorders and development of cognitive impairment: A review study. *Neuroendocrinology, 112*(9), 835–844. https://doi.org/10.1159/000521650

Koromilas, C., Liapi, C., Schulpis, K. H., Kalafatakis, K., Zarros, A., & Tsakiris, S. (2010). Structural and functional alterations in the hippocampus due to hypothyroidism. *Metabolic Brain Disease, 25*(3), 339–354. https://doi.org/10.1007/s11011-010-9208-8

Kumar, M., Modi, S., Rana, P., Kumar, P., Kanwar, R., Sekhri, T., D'souza, M., & Khushu, S. (2018). Alteration in intrinsic and extrinsic functional connectivity of resting state networks associated with subclinical hypothyroid. *Journal of Neuroendocrinology*. https://doi.org/10.1111/jne.12587

Li, M., Feng, L., Liu, X., Zhang, M., Fu, B., Wang, G., Lu, S., Zhong, N., & Hu, B. (2018). Emotional working memory in patients with major depressive disorder. *Journal of International Medical Research, 46*(5), 1734–1746. https://doi.org/10.1177/0300060518758225

Loosen, P. T. (1992). Effects of thyroid hormones on central nervous system in aging. *Psychoneuroendocrinology, 17*(4), 355–374. https://doi.org/10.1016/0306-4530(92)90041-5

Schmitt, J.a. J., Wingen, M., Ramaekers, J. G., Evers, E.a. T., & Riedel, W. J. (2006). Serotonin and human cognitive performance. *Current Pharmaceutical Design, 12*(20), 2473–2486. https://doi.org/10.2174/138161206777698909

Schraml, F. V. (2011). Thyroid stimulating hormone and cognition during severe transient hypothyroidism. *Journal of Neuroendocrinology*.

Talaei, A., Rafee, N., Rafei, F., & Chehrei, A. (2017). TSH cut off point based on depression in hypothyroid patients. *BMC Psychiatry*, 1–5. https://doi.org/10.1186/s12888-017-1478-9

Timiras, P. S., & Nzekwe, E. U. (1989). Thyroid hormones and nervous system development. *Biology of the Neonate, 55*(6), 376–385. https://doi.org/10.1159/000242941

Weiland-Fiedler, P., Erickson, K., Waldeck, T., Luckenbaugh, D. A., Pike, D., Bonne, O., Charney, D. S., & Neumeister, A. (2004). Evidence for continuing neuropsychological impairments in depression. *Journal of Affective Disorders, 82*(2), 253–258. https://doi.org/10.1016/j.jad.2003.10.009

Zhu, D.-F., Wang, Z.-X., Zhang, D.-R., Pan, Z.-L., He, S., Hu, X.-P., Chen, X.-C., & Zhou, J.-N. (2006). fMRI revealed neural substrate for reversible working memory dysfunction in subclinical hypothyroidism. *Brain: A Journal of Neurology, 129*(Pt 11), 2923–2930. https://doi.org/10.1093/brain/awl215

Chapter 20

Levothyroxine and effects on the brain: MRI and beyond

Sweta Sahu[1], Salma Younas[2], Muhammad Nafees[3,4], Soha Farooq[5], Rajkumar Rajendram[6,7,8] and Victor R. Preedy[9]

[1]Department of Medicine, JJM Medical College, Davanagere, Karnataka, India; [2]Department of Pharmacy, University of the Punjab, Lahore, Pakistan; [3]Department of Pharmacy, University of Faisalabad, Faisalabad, Pakistan; [4]Department of Pharmacy, Bahauddin Zakariya University, Multan, Pakistan; [5]Department of Pharmacy, Shifa Tameer e Millat University, Islamabad, Pakistan; [6]Research Unit, Department of Medical Education, College of Medicine, King Saud bin Abdulaziz University for Health Sciences, Riyadh, Saudi Arabia; [7]Department of Medicine, King Abdulaziz Medical City, Ministry of National Guard Health Affairs, Riyadh, Saudi Arabia; [8]Research Offices, King Abdullah International Medical Research Center, Riyadh, Saudi Arabia; [9]School of Life Course and Population Sciences, Faculty of Life Science and Medicine, King's College London, London, United Kingdom

List of abbreviations

AD Alzheimer's disease
ALFF Amplitude of low-frequency fluctuations
ASL Arterial spin labeling
BBB Blood–brain barrier
BDNF Brain-derived neurotrophic factor
CBF Cerebral blood flow
CNS Central nervous system
Cr Creatine (magnetic resonance spectroscopy metabolite)
CSF Cerebrospinal fluid
D1/D2/D3 Type 1/Type 2/Type 3 iodothyronine deiodinase
DMN Default mode network
DTI Diffusion tensor imaging
EEG Electroencephalography
FA Fractional anisotropy
fALFF Fractional amplitude of low-frequency fluctuations
FDG Fluorodeoxyglucose
FDG-PET Fluorodeoxyglucose positron emission tomography
fMRI Functional magnetic resonance imaging
GMV Gray matter volume
HPA Hypothalamic–pituitary–adrenal axis
HPG Hypothalamic–pituitary–gonadal axis
HPT Hypothalamic–pituitary–thyroid axis
IQ Intelligence quotient
L-T4 Levothyroxine (synthetic thyroxine)
MCT8 Monocarboxylate transporter 8
MD Mean diffusivity
MEG Magnetoencephalography
mI Myo-inositol (magnetic resonance spectroscopy metabolite)
MRI Magnetic resonance imaging
MRS Magnetic resonance spectroscopy
NAA N-acetylaspartate (neuronal marker)
OATP1C1 Organic anion transporting polypeptide 1C1
OGTT Oral glucose tolerance test
P300 P300 event-related potential
PET Positron emission tomography
rCBF Regional cerebral blood flow
rs-fMRI Resting-state functional magnetic resonance imaging
SPECT Single-photon emission computed tomography
T3 Triiodothyronine
T4 Thyroxine
TBG Thyroxine-binding globulin
TRα/TRβ Thyroid hormone receptor alpha/beta
TRH Thyrotropin-releasing hormone
TSH Thyroid-stimulating hormone

Introduction

Thyroid hormones—primarily thyroxine (T4) and the more active triiodothyronine (T3)—have far-reaching systemic roles. They regulate basal metabolism, impact cardiovascular function, and modulate growth and development across nearly all organ systems. Critically, thyroid hormones also exert profound effects on the central nervous system (CNS) (Uchida & Suzuki, 2021). This thyroid–brain axis is vital for normal neurodevelopment and adult brain function. Disruptions of thyroid homeostasis can produce cognitive deficits and psychiatric disturbances: overt hypothyroidism often leads to slowed cognition, memory impairment, and depression, whereas hyperthyroidism may cause agitation, anxiety, and even psychosis in severe cases (Lekurwale et al., 2023).

Hormones in Neuroscience. https://doi.org/10.1016/B978-0-443-30261-9.00020-6

Levothyroxine (L-T4) is the standard therapy for hypothyroidism and related thyroid conditions. It is a synthetic form of T4 that restores circulating hormone levels to the euthyroid range, with the goal of reversing the systemic and neurological symptoms of thyroid deficiency (Eghtedari & Correa, 2023). Studying levothyroxine's neurological effects is important for several reasons. Hypothyroidism is common worldwide, and cognitive or mood symptoms are frequent complaints in these patients. Effective thyroid replacement should ideally normalize brain function, yet a subset of patients report persistent issues like "brain fog," fatigue, or depression despite biochemically normal thyroid levels (Taylor et al., 2024). Approximately 10% of patients on levothyroxine continue to experience ill-health symptoms even after thyroid function tests normalize (Taylor et al., 2024). This has led to heightened interest in how well levothyroxine truly restores euthyroid brain state and whether standard T4 therapy adequately addresses all neural needs. Second, levothyroxine dosing must be carefully balanced—overtreatment can push patients into subclinical hyperthyroidism (low TSH levels) with potential adverse neurological and cardiovascular consequences, while undertreatment leaves residual hypothyroid symptoms (Alaeddin et al., 2023).

Given the thyroid hormones' influence on the brain, advanced neuroimaging has emerged as a valuable approach to study the brain–thyroid relationship. Techniques such as magnetic resonance imaging (MRI), functional MRI, diffusion tensor imaging, and others allow researchers to visualize structural and functional brain changes associated with thyroid dysfunction and its treatment.

Physiology of thyroid hormones and the brain

Normal thyroid hormone synthesis and regulation

Thyroid hormones are synthesized in the thyroid gland via the iodination of tyrosine residues on thyroglobulin, producing primarily T4 and a smaller amount of T3. The hypothalamic–pituitary–thyroid (HPT) axis tightly regulates this process: Thyrotropin-releasing hormone (TRH) from the hypothalamus stimulates pituitary release of thyroid-stimulating hormone (TSH), which in turn drives the thyroid gland to secrete T4 (about 80% of output) and T3 about 20%) (Freitas et al., 2010). T4 serves largely as a prohormone—in peripheral tissues and the brain, it is converted to the active T3 by deiodinase enzymes. This local conversion is crucial: roughly 50% of circulating T4 is deiodinated to T3, and within the CNS the type 2 deiodinase in astrocytes generates T3 from T4 (Nillni, 2010). Thyroid hormones enter the brain via specific transporters on the blood–brain barrier, notably monocarboxylate transporter 8 (MCT8) and organic anion transporting polypeptide 1C1 (OATP1C1). MCT8 is especially critical for T3 uptake into neurons; genetic deficiency of MCT8 (Allan–Herndon–Dudley syndrome) results in severely impaired brain–thyroid hormone signaling with profound intellectual and motor disability (van Geest et al., 2021), highlighting the importance of thyroid hormone transport to the developing brain. Once inside cells, T3 binds to nuclear thyroid hormone receptors (TRα and TRβ), which function as transcription factors to regulate gene expression (Uchida & Suzuki, 2021).

This overall thyroid–brain axis, from hormone synthesis to transport across the blood–brain barrier and neural targets, is summarized schematically in Fig. 20.1.

Mechanisms of thyroid hormone action in the CNS

T3 is the biologically active form that binds thyroid receptors in the nucleus of neurons and glial cells, directly influencing DNA transcription of target genes (Rousset et al., 2000). This genomic action modulates proteins involved in neuronal growth, differentiation, and metabolism. During brain development, thyroid hormone signaling is indispensable for processes such as neuronal migration, synaptogenesis, myelination of white matter, and neurogenesis (Uchida et al., 2005). T3-responsive genes drive the maturation of oligodendrocytes (for myelin production) and the proper layering of the cortex, among other effects. Thyroid hormone deficiency in utero or early life leads to abnormal cortical layering, impaired synaptic development, and deficits in myelin formation (Uchida & Suzuki, 2021). These changes underlie the cognitive and motor delays seen in untreated congenital hypothyroidism. Notably, the actions of thyroid hormone are spatially and temporally specific: the developing fetal brain is highly sensitive to maternal thyroid hormone in early gestation, before the fetal thyroid is fully functional (Remaud et al., 2014). Certain brain regions (hippocampus, cortex, cerebellum) have high expression of thyroid receptors and deiodinases, reflecting a high demand for local T3. There are also nongenomic actions of thyroid hormones (at the cell membrane or mitochondria) that can rapidly influence neural excitability and energy utilization, though these are less well understood than the genomic effects (Ishii et al., 2021).

Within the adult brain, a unique paracrine system operates: astrocytes take up circulating T4 and convert it to T3 via type 2 deiodinase (D2), then transport this T3 to adjacent neurons. Neurons themselves express thyroid hormone receptors and are the main cellular targets of T3, but notably lack D2—instead, they express type 3

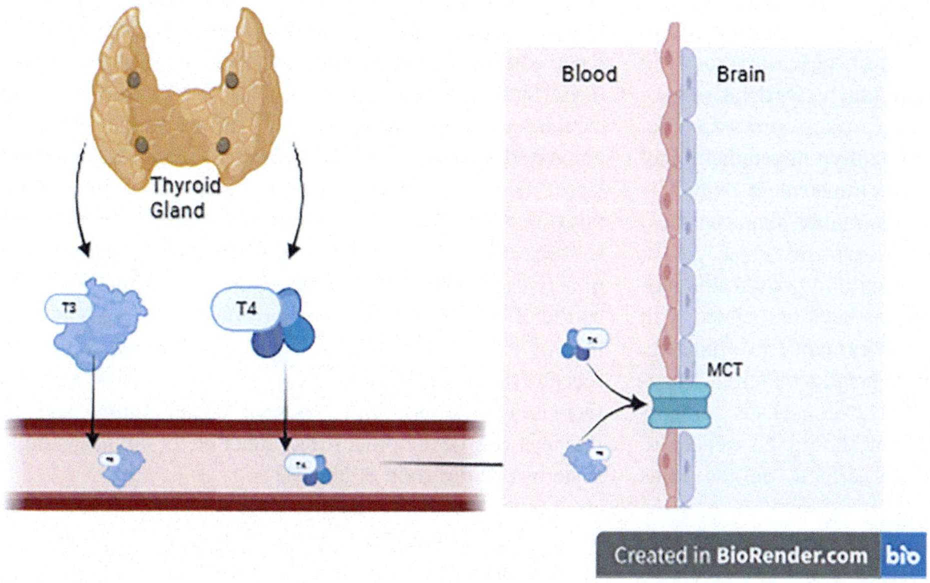

FIGURE 20.1 The thyroid–brain axis.

deiodinase (D3), which inactivates thyroid hormones (Freitas et al., 2010). This arrangement means neurons rely on glial cells for their T3 supply. Indeed, studies have shown that astrocyte-derived T3 can act in a paracrine fashion to activate gene expression in nearby neurons (Bernal et al., 2003). This astrocyte–neuron coupling for thyroid hormone metabolism ensures that brain T3 levels are buffered against fluctuations in serum T3, and that even if circulating T3is low, the brain can locally convert T4 to T3 to meet its needs. In conditions of systemic hypothyroidism (low T4), the brain compensates by upregulating deiodinase activity (D2) to preserve intracellular T3 levels (Guadaño-Ferraz et al., 1999).

Key regional roles and mechanisms are summarized in Table 20.1.

Key developmental and neurophysiological roles

Adequate thyroid hormone levels are especially important during prenatal and early postnatal brain development. Thyroid hormones drive the timely progression of neurodevelopmental milestones—they promote neural stem cell proliferation and differentiation, neuronal migration to form cortical layers, extension of axons and dendrites, synapse formation, and the maturation of glial cells (Oppenheimer & Schwartz, 1997; Uchida & Suzuki, 2021). If maternal thyroid hormone is insufficient in early pregnancy, the fetus (which depends on maternal T4 in the first trimester) can suffer irreversible brain damage. Severe congenital hypothyroidism (cretinism) historically led to

TABLE 20.1 Physiological roles of thyroid hormones in the brain.

Brain region	Role of thyroid hormones	Key mechanisms
Hippocampus (Uchida & Suzuki, 2021)	Memory formation; adult neurogenesis	Neuronal TRα/β–mediated transcription; astrocyte D2–derived T3 supply to neurons; D3 in neurons limits excess T3
Cerebral cortex (prefrontal) (Bernal, 2022)	Executive function; synaptogenesis	T3-regulated cortical gene programs controlling neuronal maturation and synaptic development
Cerebellum (Uchida & Suzuki, 2021)	Motor coordination; developmental myelination	TH drives oligodendrocyte maturation and myelin; region highly TH-sensitive in development
Astrocyte–neuron unit (Freitas et al., 2010)	Local T4→T3 conversion and paracrine delivery	D2 in astrocytes generates T3 that activates neuronal genes; neurons express TRs and D3
BBB transport and availability (Sabatino et al., 2021)	Central TH availability	Transporters + local deiodinases govern brain T3 levels; most brain T3 is produced locally from T4

profound intellectual disability, spasticity, and deafness. Fortunately, neonatal screening and prompt treatment with levothyroxine at birth prevent most of these outcomes (de Escobar et al., 1987; Rastogi & LaFranchi, 2010). Even so, subtle deficits can persist in some cases—for example, children with early-treated congenital hypothyroidism may still show mild neuropsychological impairment or differences in brain structure (subtle white matter abnormalities) later in life(Cooper et al., 2019; Lischinsky et al., 2016; Uchida & Suzuki, 2021). Maternal mild hypothyroidism or isolated low T4during pregnancy has been associated with slightly lower IQ and academic performance in offspring, though clinical trial data on treating mild maternal hypothyroxinemia have been mixed. The consensus remains that overt maternal hypothyroidism should be treated to avoid adverse neurodevelopmental effects on the fetus (Wang et al., 2022).

In the adult brain, thyroid hormones continue to play significant roles in neurophysiology. They contribute to adult neurogenesis in areas like the hippocampus and subventricular zone, supporting cognitive flexibility and mood regulation (Remaud et al., 2014). Even after the developmental period, thyroid hormone influences synaptic plasticity—for instance, hypothyroid adult rats exhibit impaired long-term potentiation in the hippocampus, correlating with memory deficits. Adult-onset hypothyroidism is well documented to cause symptoms such as poor memory, reduced attention, slowed motor function, and depressive or anxious mood (Remaud et al., 2014; Samuels, 2014). Many of these issues stem from thyroid hormone's effects on neurotransmitter systems (serotonin and norepinephrine pathways are blunted in hypothyroid states) and on structural aspects (adult hypothyroidism is associated with smaller hippocampal volume, which is thought to recover after treatment) (Martin & Sarkar, 2023; Remaud et al., 2014; Uchida & Suzuki, 2021).

Levothyroxine therapy: Pharmacology and clinical use

Pharmacokinetics and pharmacodynamics

Levothyroxine (L-T4) is a synthetic form of thyroxine that is bioidentical to endogenous T4. After oral administration, it is absorbed mainly in the small intestine (jejunum and ileum) with an absorption rate of about 60%–80% under fasting conditions. Absorption is decreased by food, fiber, and certain medications, so patients are advised to take L-T4 on an empty stomach (typically 30–60 minutes before breakfast). Once absorbed, levothyroxine is highly protein-bound in serum (over 99% bound to thyroxine-binding globulin, transthyretin, and albumin) (Eghtedari & Correa, 2023). It has a long elimination half-life of approximately 7 days in euthyroid adults, which allows for once-daily dosing and leads to a steady state being achieved in 6–8 weeks. In hypothyroid patients, the half-life may extend to ~9–10 days, whereas in hyperthyroid conditions it shortens to ~3–4 days (Drugbank, 2023; Garber et al., 2012), due to altered clearance. Levothyroxine itself is a prohormone—it must be converted to T3 by deiodinases in the liver, kidneys, and target tissues (including the brain) to exert most of its physiological effects (Bruinstroop et al., 2023; Eghtedari & Correa, 2023). This conversion yields about equal amounts of active T3 and inactive reverse T3. The pharmacodynamic action of levothyroxine, once converted to T3, is identical to that of endogenous thyroid hormone: it binds to nuclear receptors in cells to regulate gene transcription and thus increases metabolic rate, oxygen consumption, and influences the development and function of multiple organ systems (Bruinstroop et al., 2023).

At the neuronal level, these steps involve uptake of T4, its local conversion to T3, binding of T3 to nuclear receptors, and regulation of gene transcription, as illustrated in Fig. 20.2.

Indications for Therapy: Levothyroxine monotherapy is the treatment of choice for virtually all forms of hypothyroidism (Taylor et al., 2024). Indications include:

- **Primary hypothyroidism:** Dysfunction of the thyroid gland itself, most commonly due to Hashimoto's thyroiditis (autoimmune thyroid destruction) or iatrogenic causes (postthyroidectomy or after radioactive iodine ablation). Primary hypothyroidism accounts for the majority of cases, and L-T4 replacement is typically lifelong in these patients.
- **Central (secondary/tertiary) hypothyroidism:** Pituitary or hypothalamic failure leading to inadequate TSH or TRH production. Although the pathophysiology differs (the thyroid gland is normal but understimulated), treatment is still levothyroxine to provide the lacking T4. In central hypothyroidism, dose adjustments are guided by free T4 levels rather than TSH (since TSH is not a reliable marker in these patients).
- **Congenital hypothyroidism:** Infants born with an absent or defective thyroid gland (incidence ~1:2000–1:4000 births) are started on levothyroxine immediately upon diagnosis to ensure normal growth and neurodevelopment (Uchida & Suzuki, 2021).
- **Pregnancy-related hypothyroidism:** Women with hypothyroidism who become pregnant require levothyroxine dose adjustments (typically a 20%–30% increase in dose by the end of the first trimester) to meet the increased thyroid hormone demands of pregnancy (Gaitonde et al., 2012). Adequate maternal thyroid hormone is crucial to prevent obstetric complications and support fetal brain development.
- **Post-thyroidectomy or radiation therapy:** Patients who have had thyroid surgery (for benign goiter,

Graves' disease, or thyroid cancer) or radioiodine treatment will often require levothyroxine to maintain normal hormone levels. In thyroid cancer patients, higher-dose levothyroxine is sometimes given to intentionally suppress TSH (since TSH can stimulate cancer cell growth) (Eghtedari & Correa, 2023); this is known as TSH-suppressive therapy.

- **Subclinical hypothyroidism:** This refers to elevated TSH with normal T4 levels and minimal symptoms. Treatment with levothyroxine in subclinical cases is somewhat controversial and usually considered if TSH is > 10 mIU/L or if the patient has symptoms, positive antithyroid antibodies, or is pregnant (Eghtedari & Correa, 2023).

Common clinical scenarios, initial dosing, and monitoring targets are outlined in Table 20.2.

Dosing Considerations and Variability in Response: The initial dose of levothyroxine is typically calculated based on body weight and clinical context. For a young, otherwise healthy adult with overt hypothyroidism, a full replacement dose of about 1.6 μg per kg of body weight per day is often recommended (Liu et al., 2022). A 70 kg individual might start around 100–125 μg daily. Elderly patients or those with coronary artery disease are usually started at a lower dose (25–50 μg/day) to avoid overstressing the heart (Wikipedia Contributors, 2025).

Effects of levothyroxine on brain function

Once hypothyroid patients are on levothyroxine therapy and their thyroid levels normalize, one expects improvement in the neurologic and psychiatric symptoms that accompany thyroid dysfunction.

Cognitive outcomes

Thyroid dysfunction, particularly hypothyroidism, is well known to impair cognitive performance. Patients often report "brain fog," memory lapses, slowed thinking, and difficulty concentrating when hypothyroid. The encouraging news is that many of these cognitive deficits are reversible with adequate levothyroxine treatment. In cases of overt hypothyroidism, restoring euthyroid levels typically leads to noticeable improvement in mental clarity and memory. Indeed, longstanding severe hypothyroidism can cause a reversible dementia syndrome—sometimes called "myxedema dementia"—which improves once thyroid hormone is replaced (Lekurwale et al., 2023). Clinical studies and patient reports indicate better performance on attention, executive function, and memory tasks after a period of L-T4 therapy, compared to the hypothyroid state (Kaur Jhandi et al., 2025; Samuels, 2014).

However, the extent of recovery may depend on factors like patient age, duration of hypothyroidism, and whether there was developmental impact. In younger and middle-aged adults, cognitive function usually normalizes with treatment. In older adults, especially those with only mild thyroid insufficiency, the benefits of levothyroxine on cognition have been harder to demonstrate. Notably, two randomized placebo-controlled trials in older patients with subclinical hypothyroidism (elevated TSH ~6–12 mIU/L, normal T4 found that levothyroxine therapy did not significantly improve cognitive scores compared to placebo (Effraimidis et al., 2021; Zhao et al., 2022).

For patients with overt hypothyroidism, there is little doubt that untreated thyroid hormone deficiency can lead to significant cognitive impairment, and thus treatment is necessary. Imaging studies provide additional insight: hypothyroid patients have been found to have reduced hippocampal volume and decreased cerebral blood flow/metabolism in certain regions responsible for information processing (Remaud et al., 2014). After achieving a euthyroid state on L-T4, follow-up scans show partial or complete reversal of these changes in many cases, correlating with cognitive improvement. For example, one MRI volumetric study reported that adult patients with untreated hypothyroidism had smaller hippocampal volumes than controls, but volume increased after months of levothyroxine therapy alongside memory improvement (Cooke et al., 2014). MR spectroscopy has similarly shown that cerebral metabolic abnormalities (altered choline/creatine ratios) in hypothyroid brains tend to normalize on therapy, although some changes may lag behind clinical recovery (Kumar et al., 2025).

Mood and affective disorders

Thyroid dysfunction can present with a variety of mood disturbances. Hypothyroidism is classically associated with low mood, depression, and blunted affect, while hyperthyroidism can cause nervousness, anxiety, and irritability. Patients with hypothyroidism often experience fatigue, loss of interest, and even clinically significant depression (Lekurwale et al., 2023). Encouragingly, many mood symptoms abate once the patient is on adequate levothyroxine replacement. In fact, one of the key reasons to treat even mild hypothyroidism is the potential impact on quality of life and mood—several studies have noted improvements in depression scores and overall well-being after L-T4 therapy in those who were symptomatic before (AlAwaji & Alhamwy, 2023; Nuguru et al., 2022).

Most affective symptoms reverse with T4 supplementation, as thyroid levels normalize in the brain (Remaud et al., 2014). A hypothyroid patient with depressed mood and apathy will often report feeling more energetic, mentally "brighter," and less depressed once on the right

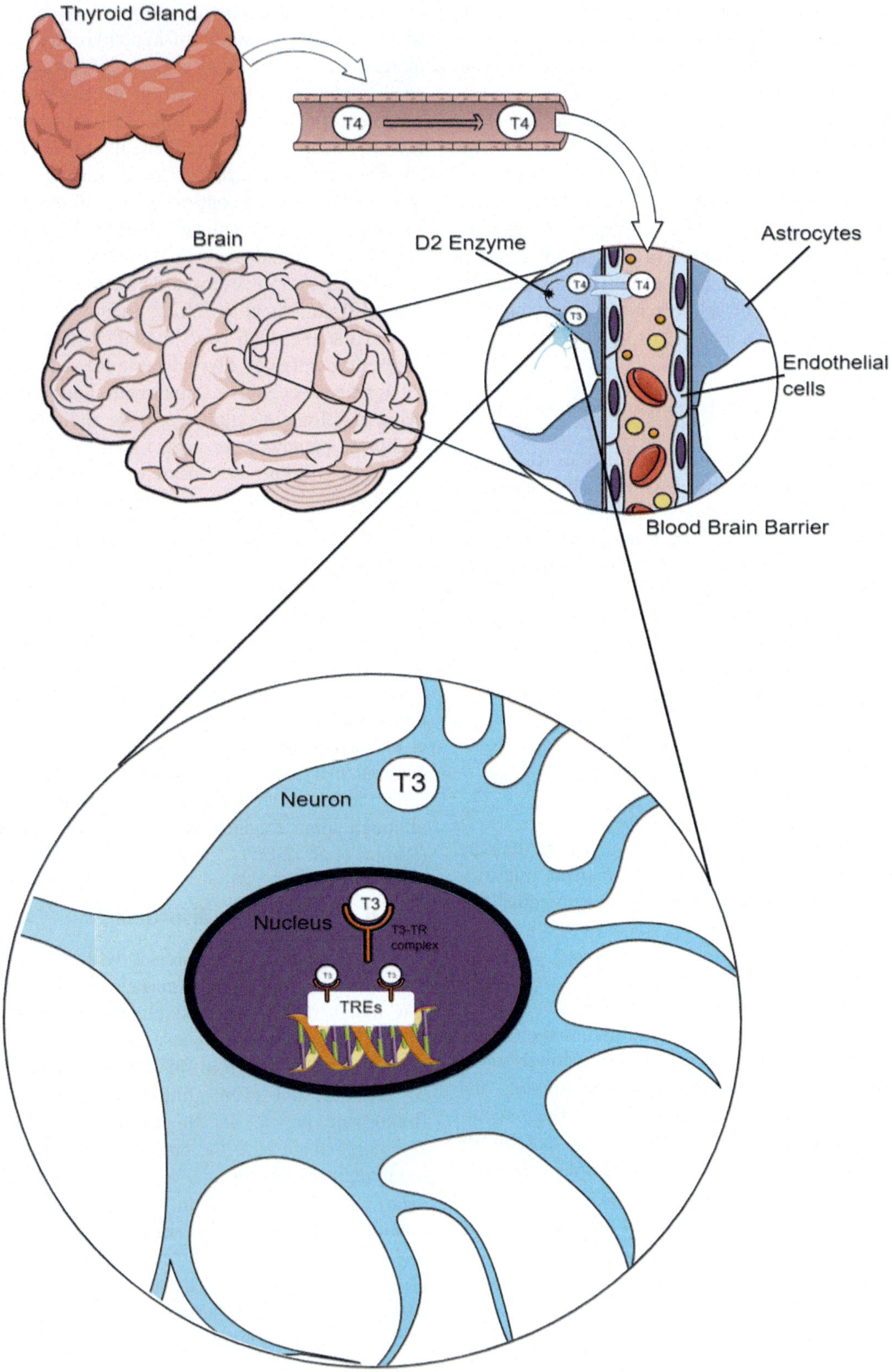

FIGURE 20.2 Mechanism of action of levothyroxine in neurons.

TABLE 20.2 Clinical indications and dosing of levothyroxine (L-T4).

Condition	Typical starting dose	Adjustment considerations	Monitoring parameters
Primary (overt) hypothyroidism	~1.6–1.8 μg/kg/day (younger, otherwise healthy adults)	Dose needs vary with BMI/lean mass, residual thyroid function, drug/food interactions; avoid brand switching when possible (Garber et al., 2012).	TSH every 6–8 weeks after changes until in range, then every 6–12 months (Eghtedari & Correa, 2023).
Central (secondary/tertiary) hypothyroidism (Jonklaas et al., 2019).	Individualize (often 1.2–1.6 μg/kg/day)	Titrate to keep free T4 in upper-normal range (TSH not reliable)	Free T4 (not TSH) every 6–8 weeks during titration
Post-thyroidectomy replacement (Garber et al., 2012; Pearce, 2019).	~1.6 μg/kg/day, then titrate	Older adults often need lower per-kg doses; consider lean mass/BSA-based dosing	TSH (or free T4 if central)
Pregnancy (preexisting hypothyroidism) (Jonklaas et al., 2019).	Increase total daily dose by ~20%–30% as soon as pregnant	Higher TH requirements in pregnancy; reassess frequently each trimester	TSH/Free T4 q4–6 weeks during pregnancy
Older adults (Eghtedari & Correa, 2023; Gavigan et al., 2023).	Start lower (e.g., 12.5–50 μg/day) or ~1.1 μg/kg when rapid correction needed	Avoid overreplacement (AFib, bone risk); "start low, go slow" unless clinical need	TSH q6–8 weeks until stable

dose of levothyroxine. Biologically, thyroid hormone interacts with neurotransmitter systems involved in mood regulation: T3 can increase serotonergic activity and enhance the action of antidepressants, which is why in resistant depression, sometimes liothyronine (T3) is added as an augmenting agent. In hypothyroidism, deficient thyroid hormone likely leads to low serotonin and dopamine activity, contributing to depression-like symptoms (Remaud et al., 2014).

However, just as with cognition, there are cases where mood symptoms persist despite apparently adequate thyroid hormone replacement. This could be due to permanent changes that occurred during a period of untreated hypothyroidism, or simply reflect the multifactorial nature of mood disorders (i.e., not all depression in a patient with hypothyroidism is because of the thyroid alone) (Ritchie & Yeap, 2015). A notable example of persistent effects is in Hashimoto's thyroiditis, which can sometimes be associated with neuropsychiatric symptoms even before overt hypothyroidism develops (Şorodoc et al., 2023).

Representative cognitive and affective outcomes reported with levothyroxine are summarized in Table 20.3.

Neurodevelopmental implications (perinatal and pediatric)

Levothyroxine's impact on the developing brain is arguably the most critical aspect of thyroid replacement therapy. Thyroid hormone is absolutely required for normal brain maturation, as detailed in section Physiology of thyroid hormones and the brain, so any deficiency during pregnancy or early childhood can have devastating consequences if not promptly corrected. The introduction of routine neonatal screening for congenital hypothyroidism (measuring TSH on heel-prick blood samples) and immediate levothyroxine treatment is a triumph of preventive medicine. Without levothyroxine, babies with congenital hypothyroidism would develop severe intellectual disability and neurological impairment (Uchida & Suzuki, 2021).

Levothyroxine is also important for maternal hypothyroidism during pregnancy. If a pregnant woman has untreated hypothyroidism, the fetus may be exposed to insufficient thyroid hormone in utero, particularly in the first trimester when it cannot produce its own. This has been linked to impaired cognitive development and lower neuropsychological scores in the child. Therefore, guidelines strongly recommend that pregnant women with known hypothyroidism increase their levothyroxine dose (often by ~25%–30% as soon as pregnancy is confirmed) (Wikipedia Contributors, 2025), and that women with even subclinical hypothyroidism (TSH >4 mIU/L) be treated during pregnancy.

Neurodegenerative disorders and the aging brain

The relationship between thyroid hormone and neurodegenerative disease (such as Alzheimer's disease and other dementias) is complex and still being unraveled. It is

TABLE 20.3 Cognitive and neuropsychiatric outcomes reported with levothyroxine.

Domain	Study findings	Population
Global cognition (SCH, older adults) (Stott et al., 2017)	No improvement in cognitive scores versus placebo with L-T4	RCT, ≥65 y with persistent SCH (TRUST).
Working memory/executive function (Samuels, 2014)	Frontal hypoactivation and performance deficits in (sub)clinical hypothyroidism improve after achieving euthyroidism	fMRI/task studies; adults with (sub) clinical hypothyroidism.
Memory (overt hypothyroidism) (Alzoubi et al., 2009)	Reversal of hippocampus-dependent cognitive impairment with thyroxine; human imaging shows smaller hippocampal volume that improves after treatment	Animal and human translational evidence; adult-onset hypothyroidism
Quality of life/symptoms (SCH, older adults) (Zhao et al., 2022)	Minimal to no benefit on hypothyroid symptom/fatigue scales in pooled analyses	Older SCH cohorts; systematic review/meta-analytic summary
Processing speed/attention in autoimmune thyroiditis (Waliszewska-Prosół et al., 2021)	Prolonged P300 latency and altered MRS metabolites even when euthyroid; supports residual CNS effects beyond serum levels	Euthyroid Hashimoto's thyroiditis versus controls

known that profound hypothyroidism can cause cognitive slowing that resembles dementia, and as noted, this is reversible with treatment. The question is whether milder thyroid dysfunction or variations in thyroid levels within the normal range influence the risk of true neurodegenerative conditions like Alzheimer's disease (AD). Some studies have suggested a link: for example, excess thyroid hormone over time (even subclinical hyperthyroidism) has been associated with an increased risk of dementia in older adults (Lekurwale et al., 2023).

When it comes to levothyroxine, the relevant consideration is avoiding iatrogenic subclinical hyperthyroidism in the elderly. If an older patient is slightly over-replaced (TSH suppressed), their risk for atrial fibrillation and bone loss is elevated, and some studies indicate their cognitive function could suffer as well (Apostu et al., 2020).

Neuroimaging insights: MRI and advanced techniques

An overview of imaging modalities, key findings, and clinical implications is provided in Table 20.4.

Structural MRI findings

Gray and White Matter Changes: In subclinical hypothyroidism, one study found global and regional GMV reductions, notably in the hippocampus and prefrontal cortex. These volumetric deficits are thought to reflect thyroid hormone's role in maintaining neuronal size, synaptic density, and possibly glial support. White matter may also be affected; thyroid hormone is crucial for oligodendrocyte maturation and myelination, so prolonged deficiency can lead to subtle white matter loss or delayed myelination visible on MRI (Zhang et al., 2021). Some cerebellar volume reduction has been noted as well (consistent with the cerebellum's sensitivity to thyroid hormone during development) (Chambers et al., 2020; Zhang et al., 2021).

Hippocampal and Cortical Alterations: Among specific brain regions, the hippocampus—critical for memory and mood regulation—appears particularly vulnerable to thyroid dysfunction. Hypothyroid patients (including those with subclinical disease) have exhibited smaller hippocampal volumes compared to euthyroid controls (Zhang et al., 2021). Given that thyroid receptors (TRα, TRβ) are densely expressed in the limbic system (including the hippocampus), thyroid hormone deficiency likely disrupts adult neurogenesis and dendritic branching in this region. Cortical thickness analyses also reveal changes: overt hyperthyroidism, conversely, can cause cortical thickening (Kumar et al., 2024), whereas hypothyroidism tends toward slight cortical thinning or volume loss in frontal and parietal areas that subserve executive function (Samuels, 2014; Zhang et al., 2021).

Functional MRI (fMRI) studies

Resting-State Connectivity: Beyond static anatomy, hypothyroidism produces discernible changes in brain functional networks. Resting-state fMRI, which measures spontaneous neural activity and connectivity, has shown that thyroid dysfunction can alter the brain's intrinsic connectivity patterns. In hypothyroid (and subclinical hypothyroid) patients, functional connectivity within the default mode network (DMN)—a network involved in introspective and memory-related processes—appears

TABLE 20.4 MRI and advanced imaging findings relevant to levothyroxine.

Imaging modality	Key finding(s)	Clinical implication
Structural MRI (VBM/volumetry) (Zhang et al., 2021)	Reduced GMV (prefrontal/temporal) and smaller hippocampi in untreated/subclinical states; partial reversal after L-T4	Structural deficits are largely state-dependent and can improve with treatment
Resting-state fMRI (Zhang et al., 2021)	Altered DMN/visual–cerebellar connectivity; aberrations partly normalize after euthyroidism	Functional network disruption tracks thyroid status; supports objective "brain fog" correlates
Task-based fMRI (Samuels, 2014)	Reduced prefrontal activation during working-memory tasks improves after L-T4	Objective correlate of cognitive recovery with therapy
Diffusion tensor imaging (DTI)	Lower FA/higher MD across associative tracts; memory inversely correlates with FA; abnormalities can improve after euthyroidism (limited longitudinal data)	White-matter microstructure is TH-sensitive; links to cognitive complaints
MR spectroscopy (^{1}H-MRS) (Singh et al., 2014)	Reduced NAA (neuronal marker) and elevated myoinositol in hypothyroid/HT brains; hippocampal metabolites improve postthyroxine	Neurochemical dysregulation is reversible with treatment in many patients
PET/SPECT (Bauer et al., 2009)	Reduced CMRglc and rCBF in hypothyroid states; several studies show increases after TH replacement	Functional hypometabolism/hypoperfusion provides treatment-responsive biomarkers

disrupted. For example, one study found increased low-frequency fluctuation amplitude (fALFF) in frontal regions (middle and superior frontal gyrus) of subclinical hypothyroid patients at rest (Zhang et al., 2021). This suggests a compensatory hyperactivation of certain networks, possibly reflecting the brain's attempt to maintain cognitive function amid hormone deficiency. Concurrently, reduced connectivity has been noted between other region pairs: Salas-Lucia et al. summarize two fMRI studies that detected decreased functional connectivity between the cuneus (visual cortex) and cerebellum, as well as between the cuneus and medial prefrontal cortex, in hypothyroid patients, correlating with worse cognitive and emotional scores (Salas-Lucia, 2024; Zhang et al., 2021).

Task-Based fMRI and Cognitive Correlates: Complementing resting-state studies, task-based fMRI provides insight into how hypothyroidism affects brain activation during specific cognitive challenges. Research consistently demonstrates that untreated hypothyroidism blunts activation of key cortical regions during cognitive tasks, aligning with patients' subjective cognitive slowdowns. In a landmark fMRI study, subclinical hypothyroid individuals performing a working memory task (n-back test) showed significantly less activation in the dorsolateral prefrontal cortex (middle frontal gyrus) and supplementary motor area compared to healthy controls (Zhang et al., 2021).

Diffusion tensor imaging (DTI)

White Matter Integrity: DTI is an MRI technique that maps white matter microstructure by measuring water diffusion along axons. Hypothyroidism—especially if prolonged—can disrupt white matter integrity due to thyroid hormone's role in myelin production. Indeed, DTI studies reveal microstructural abnormalities in white matter tracts of hypothyroid patients. A notable study by Singh et al. compared adults with untreated hypothyroidism to healthy controls and found significant changes in DTI metrics across multiple major white matter tracts (Singh et al., 2014). Specifically, fractional anisotropy (FA) (an indicator of fiber coherence and myelination) was lower in hypothyroid patients in certain tracts, and mean diffusivity (MD) (an indicator of water mobility in tissue) was higher, suggesting microstructural disorganization (Singh et al., 2014). These changes were not localized to one area; hypothyroid subjects showed alterations in long associative tracts like the inferior frontooccipital fasciculus and anterior thalamic radiation (Singh et al., 2014). Such diffuse involvement aligns with the global role of thyroid hormone in brain development and maintenance (e.g., supporting oligodendrocytes and preventing demyelination). Crucially, DTI findings have been linked to cognitive deficits: in Singh's study, memory scores inversely correlated with FA in frontooccipital pathways, meaning patients with more white matter disruption had worse memory performance. This implies that slowed cognition in hypothyroidism may stem partly from suboptimal network communication due to myelin loss or edema. There is evidence that appropriate thyroid hormone replacement can improve white matter metrics, though DTI data posttreatment are limited. Some pediatric research and case reports in congenital hypothyroidism have shown myelination catches up after thyroid therapy, reflected in DTI by increasing FA values toward normal. Adult neuroimaging review also notes that DTI abnormalities (reduced interhemispheric connectivity) tend to resolve once euthyroidism is restored (Salas-Lucia, 2024).

MR spectroscopy

Neurochemical Alterations: Proton MR spectroscopy (MRS) offers a noninvasive window into the brain's biochemical status by measuring concentrations of metabolites such as N-acetylaspartate (NAA), choline (Cho), creatine (Cr), and myoinositol (mI). Thyroid dysfunction has been associated with characteristic neurochemical shifts on MRS, reflecting neuronal and glial changes during hypothyroidism. For instance, studies of patients with hypothyroid symptoms have reported reduced NAA/Cr ratios in brain regions such as the hippocampus and cortex (Salas-Lucia, 2024). NAA is a marker of neuronal viability, so a lower NAA suggests reversible neuronal dysfunction or loss of dendritic density when thyroid hormone is lacking. Concurrently, myoinositol (mI), a glial marker that often increases with gliosis or astrocytic activation, can be elevated in hypothyroid states (Waliszewska-Prosół et al., 2021). In a spectroscopy study of euthyroid Hashimoto's thyroiditis patients (who often have had hypothyroid episodes), researchers found an increased mI/Cr ratio in parietal white matter alongside prolonged cognitive processing (longer P300 latency) (Waliszewska-Prosół et al., 2021). This points to subtle astrocytic changes and slower neurotransmission even when systemic levels are normalized, possibly due to past or fluctuating thyroid hormone exposure. Encouragingly, several MRS-detected abnormalities improve with levothyroxine therapy. A longitudinal MRS study noted that decreased NAA levels in the hippocampus and cerebellum of hypothyroid patients rose back to normal after thyroid hormone replacement, accompanied by improvements in symptoms (Salas-Lucia, 2024). Likewise, elevated myoinositol tends to decrease toward normal with treatment as brain metabolism normalizes (Salas-Lucia, 2024; Waliszewska-Prosół et al., 2021).

Beyond MRI: Multimodal and emerging techniques

Hypothyroidism's impact on the brain can be examined through other imaging modalities and investigative techniques that complement MRI. These emerging or adjunct approaches provide additional perspectives on thyroid hormone's role in brain function:

Positron emission tomography (PET) and SPECT

PET and SPECT scans measure brain metabolism and perfusion, respectively, using radiotracers. In hypothyroid patients, FDG-PET studies demonstrate globally reduced cerebral glucose metabolism, reflecting the brain's slowed metabolic state when thyroid hormone is low. A meta-analysis of PET/SPECT findings identified consistent hypometabolism in regions like the anterior cingulate, parietal cortex (including precuneus), and occipital cortex in hypothyroidism, compared to euthyroid controls (Waliszewska-Prosół et al., 2021). SPECT scans similarly show decreased cerebral blood flow (CBF) in hypothyroid states—for example, reduced perfusion in the postcentral gyrus and diffuse cortical areas. These changes likely underpin some of the neuropsychiatric symptoms (lethargy and slowed mentation). Interestingly, hyperthyroidism can produce the opposite pattern (focal hypermetabolism), underlining thyroid hormones' bidirectional influence on brain activity. The advantage of PET/SPECT is their high sensitivity to functional changes; they can quantify how restoring euthyroidism increases brain glucose utilization and blood flow back to normal levels (indeed, many deficits are reversible with treatment (Pak et al., 2020)). Limitations include exposure to radiation, high cost, and limited availability. Additionally, PET/SPECT findings have been somewhat variable across studies (Pak et al., 2020), partly due to small cohorts and different analytic methods.

Electroencephalography (EEG) and magnetoencephalography (MEG)

EEG measures the brain's electrical activity, while MEG records magnetic fields from neuronal firing—both offer millisecond-level temporal resolution of brain function. Thyroid dysfunction can lead to characteristic EEG changes. Classic EEG findings in severe hypothyroidism include generalized slowing of the background rhythm (e.g., a shift toward more theta/delta waves), consistent with a generalized slowing of cerebral activity. Even in milder cases, more subtle changes can be captured with cognitive evoked potentials: for instance, the P300 wave (an EEG marker of attention and working memory) is often delayed and reduced in amplitude in hypothyroid patients (Waliszewska-Prosół et al., 2021). One study on patients with Hashimoto's thyroiditis (many of whom were biochemically euthyroid on treatment) showed significantly prolonged P300 latencies and lower P300 amplitudes compared to controls (Waliszewska-Prosół et al., 2021), indicating slowed information processing despite "normal" hormone levels—possibly due to residual CNS effects or autoimmunity. These EEG alterations correlated with neuropsychological complaints and even with MRS metabolites (e.g., higher myo-inositol levels were associated with longer P300 latency (Waliszewska-Prosół et al., 2021)), reinforcing a link between thyroid-related metabolic changes and electrophysiological function.

Biomarker studies (CSF, serum, and neurotrophic factors)

Investigating body fluids for biomarkers can shed light on thyroid hormone's neurochemical influence. For instance, cerebrospinal fluid (CSF) analyses in hypothyroid patients allow measurement of intrathecal thyroid hormone levels and neurotransmitter metabolites. One could measure CSF T4/T3 to see if central hormone availability lags behind serum levels in treated patients. Serum biomarkers have been extensively studied: for example, brain-derived neurotrophic factor (BDNF) is a key protein for neuronal health that may be low in hypothyroidism. Some studies suggest reduced BDNF levels accompany cognitive impairment in hypothyroid patients, and levels may rise after treatment (Toll et al., 2023).

Comparative strengths and limitations of adjunct and emerging methods are summarized in Table 20.5.

Clinical implications and case studies

Diagnosis and Treatment Monitoring: The neuropsychiatric and neuroimaging findings in hypothyroidism underscore the need for clinicians to maintain a high index of suspicion for thyroid dysfunction in patients with unexplained cognitive or mood changes. While diagnosis of hypothyroidism is ultimately biochemical (TSH and free T4 levels), recognizing the brain-related symptoms is crucial. A patient presenting with depression, fatigue, and mild cognitive impairment should prompt a TSH test, as overt hypothyroidism can mimic a depressive disorder or early dementia ("myxedema madness" in extreme cases) (Samuels, 2014). Once hypothyroidism is diagnosed, levothyroxine remains the standard of care and typically results in improvement of neurocognitive symptoms over weeks to months. Monitoring treatment involves regular TSH measurements (in primary hypothyroidism) and clinical assessment of symptom resolution. Neuroimaging is usually not needed in routine management, but research suggests it could play a role in special situations. For example, if a patient's cognitive symptoms persist despite normalized TSH, an MRI might reveal another pathology—or conversely, a normal MRI could reassure that no structural brain issue exists. In severe cases like myxedema coma, imaging (CT/MRI) is sometimes done to rule out other causes of unconsciousness, although diffuse slowing on EEG and global hypoperfusion on SPECT would be expected findings in myxedema (Samuels, 2014).

Personalized Medicine and Tailoring Therapy: One emerging insight is that one size may not fit all in levothyroxine therapy when it comes to neurocognitive outcomes. Many patients do fine on standard T4 monotherapy and regain full cognitive function. However, a subset of patients continue to experience fatigue, "brain fog," or depressive symptoms despite having normal TSH on levothyroxine (Samuels, 2014). This raises the question of whether certain individuals might benefit from personalized approaches—for example, combination therapy with liothyronine (T3) or targeting a different TSH goal. Research into genetic polymorphisms provides some clues: a common variant in the type 2 deiodinase gene (Thr92Ala DIO2 polymorphism) has been associated with poorer psychological well-being on T4 alone, possibly because of less efficient conversion of T4 to T3 in the brain (Samuels, 2014; Zinck et al., 2023). Patients with this variant or other

TABLE 20.5 Emerging and adjunct methods beyond MRI.

Method	Applications in thyroid–brain research	Advantages	Limitations
FDG-PET (Bauer et al., 2009)	Maps regional cerebral glucose metabolism in hypo/hyperthyroid states; shows reversibility with therapy.	Sensitive to functional change; complements MRI	Cost, radiation, smaller cohorts
SPECT (perfusion) (Krausz et al., 2004)	Quantifies rCBF deficits in hypothyroidism; treatment-related rCBF increases reported	Widely available perfusion metric; treatment monitoring	Lower spatial resolution than MRI/PET
EEG/ERPs (P300) (Waliszewska-Prosół et al., 2021)	Detects slowed processing and attentional deficits in HT (even when euthyroid)	Direct millisecond-scale neural readout; repeatable	Limited spatial localization; nonspecific findings
MRS (advanced metabolites incl. GABA+) (Li et al., 2022)	Tracks neuronal/glial metabolites and inhibitory tone changes before/after L-T4	Noninvasive neurochemistry; longitudinal tracking	Region-of-interest constraints; technical expertise
Multimodal integration (imaging + genetics/biomarkers) (Hernandez, 2018)	Tests whether deiodinase/transporter variants predict residual CNS deficits on L-T4	Toward precision endocrinology	Requires large, interdisciplinary datasets

polymorphisms (e.g., MCT10 transporter) might have "tissue-level" hypothyroidism in the brain even if serum levels are normalized. This concept of central hypothyroidism (tissue specific) despite normal blood tests is still being explored, but it dovetails with the imaging evidence: for example, if a patient's fMRI still shows reduced prefrontal activation and PET shows low metabolism after standard treatment, that might indicate a need for a tailored approach. In practice, some endocrinologists already adopt personalized strategies: starting doses are adjusted for age and comorbidities (elderly or cardiac patients get lower initial doses to avoid overshoot (Eghtedari & Correa, 2023), and target TSH ranges might be tighter (e.g., TSH in the lower half of normal) for patients with persistent symptoms. Combination T4+T3 therapy is another personalized approach; while large trials have not conclusively shown its superiority, certain patients report better cognitive/mood outcomes on it, and ongoing studies are trying to identify if genetic markers predict this response (Ettleson et al., 2024).

Case Illustrations (Pre- and Posttreatment): Real-world cases vividly demonstrate levothyroxine's impact on the brain. Case 1: A 50-year-old woman with long-standing untreated hypothyroidism presented with apathy, forgetfulness, and slowed speech. MRI showed diffuse mild cerebral atrophy and white matter hyperintensities. After 1 year of levothyroxine replacement (titrated to a TSH of 1.5 μIU/mL), her symptoms resolved—interestingly, a follow-up MRI noted reduction in the white matter lesions and improved brain volume, consistent with reversal of myxedema-associated changes (similar reversible MRI abnormalities have been documented in case reports of severe hypothyroidism) (Samuels, 2014). Case 2: A 34-year-old man with subclinical hypothyroidism (TSH 8 mIU/L) complained of "fuzzy thinking" and memory lapses, though standard cognitive tests were largely normal. He enrolled in a study and underwent fMRI during a memory task, which revealed reduced activation in his dorsolateral prefrontal cortex. After 4 months of levothyroxine (bringing TSH to 2 mIU/L), he reported subjective improvement in concentration. Repeat fMRI showed restoration of frontal cortex activation to near-normal patterns, paralleling his clinical improvement (Samuels, 2014).

Other hormones, endocrine tissues, and pathways

Although levothyroxine (L-T4) targets thyroid hormone deficiency, brain outcomes in hypothyroidism are shaped by wider endocrine networks that influence neurovascular function, neurotransmission, inflammation, and energy metabolism. The hypothalamic–pituitary–thyroid (HPT) axis is embedded within the hypothalamus, which also coordinates the hypothalamic–pituitary–adrenal (HPA) and hypothalamic–pituitary–gonadal (HPG) axes. Consequently, hypothyroidism and its treatment may have downstream effects on stress reactivity, mood, and cognition through endocrine cross-talk (Bernal, 2022).

The HPA axis is especially relevant for neuropsychiatric symptoms. Hypothyroidism can alter cortisol dynamics indirectly via changes in metabolism, sleep architecture, and inflammatory tone. Cortisol excess or dysregulation can impair hippocampal-dependent memory and contribute to anxiety/depression phenotypes—symptoms that overlap with "brain fog" complaints in treated hypothyroid patients. Clinically, this reinforces the importance of considering comorbid adrenal dysfunction or chronic stress when patients remain symptomatic despite normalized TSH (Sharan & Vellapandian, 2024).

Insulin and metabolic pathways also interact with thyroid hormone signaling in the brain. Thyroid hormones modulate basal metabolic rate, lipid handling, and glucose utilization; hypothyroidism is associated with insulin resistance in some patients, while overtreatment may increase catabolism. Since the brain depends on tightly regulated glucose delivery, peripheral insulin resistance and vascular dysfunction may compound neurocognitive symptoms and imaging findings. Translationally, evaluating cardiometabolic risk (BMI, lipids, glucose tolerance) alongside thyroid tests can help interpret persistent cognitive complaints (El-Akad et al., 2025).

Sex steroids (estrogens and androgens) influence cognition, mood, and cerebral perfusion and can modify thyroid hormone economy through effects on thyroxine-binding globulin (TBG) and hepatic metabolism. Estrogen exposure (pregnancy, oral contraceptives, hormone therapy) can increase TBG and change total T4 levels, sometimes complicating interpretation of thyroid status and dose titration. In perinatal contexts, placental endocrine function and maternal thyroid status jointly shape fetal neurodevelopment, making pregnancy a key translational setting where thyroid replacement strategy directly impacts brain outcomes (Fröhlich & Wahl, 2022).

Prolactin and growth hormone/IGF-1 pathways can be indirectly affected by hypothalamic tone and pituitary function. Central hypothyroidism, pituitary lesions, or prior pituitary therapy may require free T4-guided levothyroxine dosing and simultaneous evaluation of other pituitary axes. This is clinically important because untreated pituitary hormone deficits can mimic or worsen cognitive and mood symptoms and may confound neuroimaging interpretation (Yuen et al., 2025).

Immune–endocrine interactions are particularly relevant in Hashimoto's thyroiditis, where neuropsychiatric symptoms may not fully track serum thyroid indices. Cytokine-driven neuroinflammation can influence white matter integrity, neurotransmitter function, and fatigue. In translational terms, persistent symptoms in biochemically euthyroid patients warrant a broader differential (sleep disorders, anemia, depression, autoimmunity, metabolic syndrome, medication interactions affecting L-T4 absorption) rather than reflexive dose escalation alone (Zhang et al., 2025).

Applications of the material

One application is patient stratification: neurocognitive complaints (attention slowing, memory lapses, reduced processing speed) can be contextualized using evidence that hypothyroidism affects hippocampal volume, cortical network activation, and white matter microstructure. Clinicians can use this framework to counsel patients that many brain-related changes improve after restoration of euthyroidism, while also acknowledging that recovery may lag behind serum normalization in a subset.

A second application is treatment optimization and safety. Because overtreatment may push patients toward subclinical hyperthyroidism (with potential cognitive, cardiovascular, and skeletal risks), the neuroimaging perspective reinforces careful titration rather than symptom-driven dose escalation. Conversely, undertreatment may perpetuate perfusion/metabolic deficits that contribute to fatigue and cognitive slowing. The chapter's emphasis on imaging and functional measures supports a balanced approach: target biochemical euthyroidism while actively evaluating nonthyroid contributors to persistent symptoms (sleep, depression/anxiety, cardiometabolic factors, and medication-related malabsorption).

A third application lies in special populations, particularly pregnancy and congenital hypothyroidism. The chapter's neurodevelopmental discussion translates into a clear clinical message: early identification and appropriate levothyroxine dosing are time-sensitive interventions that prevent or reduce long-term neurocognitive impairment. In pregnancy, the application is proactive management, anticipating increased levothyroxine requirements and monitoring to maintain adequate maternal free T4 for fetal brain development.

In research and advanced clinical settings, multimodal techniques provide a platform for biomarker development. Structural MRI, fMRI, DTI, and MRS can be integrated with PET/SPECT or EEG/evoked potentials to create composite endpoints for clinical trials (measuring changes in DMN connectivity or white matter integrity after therapy). Such endpoints are valuable for studying why a subgroup remains symptomatic despite normal thyroid function tests, and for testing personalized approaches (investigating whether deiodinase polymorphisms or transporter variability corresponds to persistent functional deficits).

A further application is differential diagnosis. Cognitive impairment in hypothyroidism may resemble mild cognitive impairment or depressive pseudodementia. The chapter's synthesis supports routine thyroid testing in neuropsychiatric presentations and helps interpret neuroimaging when performed for other indications. Importantly, normalizing thyroid function can reverse a potentially reversible contributor to cognitive symptoms, reducing unnecessary escalation to dementia workups in appropriate contexts.

Finally, the framework enables patient communication: explaining that thyroid hormone effects on the brain involve metabolism, perfusion, myelination, and neurotransmission can improve adherence to levothyroxine administration instructions (fasting dosing, avoidance of interfering agents) and can set realistic expectations for symptom improvement over weeks to months.

Future directions

Future research will focus on advanced imaging, larger clinical trials, and personalized treatment approaches to better understand and optimize how thyroid hormones support brain health. Improved MRI, PET, and AI techniques may allow earlier detection and monitoring of brain changes. Long-term studies will clarify which patients benefit most from therapy and how brain function recovers over time. Translational work in neurology and psychiatry may expand the use of thyroid hormones in conditions like depression and demyelinating disorders. Precision endocrinology will help tailor treatment to genetics, symptoms, and individual needs, aiming to protect cognitive function and improve overall neurological outcomes.

Conclusion

Hypothyroidism can negatively affect brain structure and function, but levothyroxine treatment can reverse many of these changes. Maintaining normal thyroid levels is essential for cognitive and emotional health. Ongoing research will continue to improve how we protect brain health in patients with thyroid disorders.

Mini-dictionary of terms

1. **Levothyroxine (L-T4):** Synthetic thyroxine used as standard replacement therapy for hypothyroidism; converted to T3 in tissues.
2. **Triiodothyronine (T3):** The more biologically active thyroid hormone that binds nuclear thyroid receptors to regulate gene transcription.
3. **Hypothalamic–Pituitary–Thyroid (HPT) axis:** Feedback system controlling thyroid hormone

production via TRH (hypothalamus) → TSH (pituitary) → T4/T3 (thyroid).

4. **Deiodinases (D1/D2/D3):** Enzymes that activate or inactivate thyroid hormones; D2 generates T3 from T4 in astrocytes, D3 inactivates hormones in neurons.
5. **MCT8/OATP1C1:** Thyroid hormone transporters at the blood–brain barrier that facilitate entry of T3/T4 into the CNS.
6. **Euthyroidism:** State of normal thyroid hormone function, typically inferred from TSH and free T4 within reference ranges.
7. **Subclinical hypothyroidism:** Elevated TSH with normal free T4; symptoms may be minimal or nonspecific.
8. **Default Mode Network (DMN):** Resting-state brain network involved in self-referential processing and memory; often assessed with resting-state fMRI connectivity.
9. **Diffusion Tensor Imaging (DTI):** MRI method that evaluates white matter microstructure using diffusion metrics (e.g., FA and MD).
10. **Fractional Anisotropy (FA):** DTI measure reflecting directional water diffusion; lower FA often indicates reduced white matter integrity/myelination.
11. **Mean Diffusivity (MD):** DTI measure of overall water diffusion; higher MD may reflect edema or microstructural disruption.
12. **MR Spectroscopy (MRS):** MRI-based technique measuring brain metabolites (e.g., NAA, choline, and myoinositol) as markers of neuronal/glial status.
13. **N-acetylaspartate (NAA):** MRS metabolite associated with neuronal integrity/function; reductions may indicate neuronal dysfunction.
14. **FDG-PET:** Imaging of cerebral glucose metabolism using fluorodeoxyglucose tracer; can show hypo- or hypermetabolism patterns.
15. **P300 (event-related potential):** EEG measure linked to attention and working memory; prolonged latency suggests slower cognitive processing.

Summary points

- Hypothyroidism causes measurable structural and functional brain changes, particularly in hippocampal and cortical regions.
- Levothyroxine restores euthyroidism and improves cognitive and mood symptoms in most patients with overt hypothyroidism.
- Structural MRI demonstrates reduced hippocampal volume in untreated hypothyroidism, often improving after levothyroxine therapy.
- fMRI shows altered functional connectivity and reduced prefrontal activation in hypothyroidism, with partial normalization after treatment.
- DTI reveals white matter microstructural abnormalities in hypothyroidism that correlate with cognitive deficits.
- MR spectroscopy identifies neurochemical alterations during thyroid dysfunction that may normalize with levothyroxine.
- PET and SPECT demonstrate reduced cerebral metabolism and perfusion in hypothyroidism that improve after hormone replacement.
- Some patients experience persistent cognitive or mood symptoms despite biochemical euthyroidism.
- Excess levothyroxine can cause subclinical hyperthyroidism, increasing systemic and cognitive risk in older adults.
- Early diagnosis and treatment of congenital and maternal hypothyroidism are essential to prevent long-term neurodevelopmental impairment.

References

Alaeddin, N., Jongejan, R. M. S., Stingl, J. C., Yolanda, P. R. P., Breteler, M. M. B., et al. (2023). Over- and undertreatment with levothyroxine—findings of the population-based Rhineland study. *Deutsches Ärzteblatt International*. https://doi.org/10.3238/arztebl.m2023.0192

AlAwaji, M. I., & Alhamwy, R. H. (2023). The impact of hypothyroidism on the quality of life of adults in Riyadh, Saudi Arabia. *Cureus, 15*. https://doi.org/10.7759/cureus.37636

Alzoubi, K. H., Gerges, N. Z., Aleisa, A. M., & Alkadhi, K. A. (2009). Levothyroxin restores hypothyroidism-induced impairment of hippocampus-dependent learning and memory: Behavioral, electrophysiological, and molecular studies. *Hippocampus, 19*, 66–78. https://doi.org/10.1002/hipo.20476

Apostu, D., Lucaciu, O., Oltean-Dan, D., Mureşan, A.-D., Moisescu-Pop, C., Maxim, A., et al. (2020). The influence of thyroid pathology on osteoporosis and fracture risk: A review. *Diagnostics, 10*, 149. https://doi.org/10.3390/diagnostics10030149

Bauer, M., Silverman, D. H. S., Schlagenhauf, F., London, E. D., Geist, C. L., van Herle, K., et al. (2009). Brain glucose metabolism in hypothyroidism: A positron emission tomography study before and after thyroid hormone replacement therapy. *Journal of Clinical Endocrinology and Metabolism, 94*, 2922–2929. https://doi.org/10.1210/jc.2008-2235

Bernal, J. (2022). *Thyroid hormones in brain development and function*. Nihgov. https://www.ncbi.nlm.nih.gov/books/NBK285549/.

Bernal, J., Guadaño-Ferraz, A., & Morte, B. (2003). Perspectives in the study of thyroid hormone action on brain development and function. *Thyroid, 13*, 1005–1012. https://doi.org/10.1089/105072503770867174

Bruinstroop, E., van der Spek, A., & Boelen, A. (2023). Role of hepatic deiodinases in thyroid hormone homeostasis and liver metabolism, inflammation and fibrosis. *European Thyroid Journal, 12*. https://doi.org/10.1530/etj-22-0211

Chambers, T., Anney, R., Taylor, P. N., Teumer, A., Peeters, R. P., Medici, M., et al. (2020). Effects of thyroid status on regional brain volumes: A diagnostic and genetic imaging study in UK biobank. *Journal of Clinical Endocrinology and Metabolism, 106*, 688–696. https://doi.org/10.1210/clinem/dgaa903

Cooke, G. E., Mullally, S., Correia, N., O'Mara, S. M., & Gibney, J. (2014). Hippocampal volume is decreased in adults with hypothyroidism. *Thyroid, 24*, 433–440. https://doi.org/10.1089/thy.2013.0058

Cooper, H. E., Kaden, E., Halliday, L. F., Bamiou, D.-E., Mankad, K., Peters, C., et al. (2019). White matter microstructural abnormalities in children with severe congenital hypothyroidism. *NeuroImage: Clinical, 24*, Article 101980. https://doi.org/10.1016/j.nicl.2019.101980

de Escobar, G. M., Obregon, M. J., & del Rey, F. E. (1987). Fetal and maternal thyroid hormones. *Hormone Research, 26*, 12–27. https://doi.org/10.1159/000180681

Drugbank. (2023). *Levothyroxine*. Godrugbankcom. https://go.drugbank.com/drugs/DB00451.

Effraimidis, G., Watt, T., & Feldt-Rasmussen, U. (2021). Levothyroxine therapy in elderly patients with hypothyroidism. *Frontiers in Endocrinology, 12*. https://doi.org/10.3389/fendo.2021.641560

Eghtedari, B., & Correa, R. (2023). *Levothyroxine*. National Library of Medicine. https://www.ncbi.nlm.nih.gov/books/NBK539808/.

El-Akad, R. H., Elosaily, A. H., Gamil, N. M., Merghany, R. M., El-Shiekh, R. A., Almagharbeh, W. T., et al. (2025). Natural product-based interventions for thyroid disorders: Mechanisms and applications. *Naunyn-Schmiedeberg's Archives of Pharmacology*. https://doi.org/10.1007/s00210-025-04604-y

Ettleson, M. D., Karavolos, K., Burnett-Bowie, S.-A. M., Powell, L. H., & Janssen, I. (2024). The association between hypothyroidism and cognitive function change in women across the menopause transition: The study of Women's health across the nation (SWAN). *Thyroid*. https://doi.org/10.1089/thy.2024.0358

Fröhlich, E., & Wahl, R. (2022). Insight into potential interactions of thyroid hormones, sex hormones and their stimulating hormones in the development of non-alcoholic fatty liver disease. *Metabolites, 12*, 718. https://doi.org/10.3390/metabo12080718

Freitas, B. C. G., Gereben, B., Castillo, M., Kalló, I., Zeöld, A., Egri, P., et al. (2010). Paracrine signaling by glial cell–derived triiodothyronine activates neuronal gene expression in the rodent brain and human cells. *Journal of Clinical Investigation, 120*, 2206–2217. https://doi.org/10.1172/jci41977

Gaitonde, D. Y., Rowley, K. D., & Sweeney, L. B. (2012). Hypothyroidism: An update. *American Family Physician, 86*, 244–251.

Garber, J., Cobin, R., Gharib, H., Hennessey, J., Klein, I., Mechanick, J., et al. (2012). Clinical practice guidelines for hypothyroidism in adults: Cosponsored by the American Association of Clinical Endocrinologists and the American Thyroid Association. *Endocrine Practice, 18*, 988–1028. https://doi.org/10.4158/ep12280.gl

Gavigan, C., Abbey, E. J., McGready, J., Simonsick, E. M., & Mammen, J. S. (2023). Levothyroxine dosing in older adults: Recommendations derived from the Baltimore longitudinal study of aging. *Endocrine Practice, 29*, 612–617. https://doi.org/10.1016/j.eprac.2023.05.002

Guadaño-Ferraz, A., María, J. E., Rausell, E., & Bernal, J. (1999). Expression of type 2 iodothyronine deiodinase in hypothyroid rat brain indicates an important role of thyroid hormone in the development of specific primary sensory systems. *Journal of Neuroscience, 19*, 3430–3439. https://doi.org/10.1523/jneurosci.19-09-03430.1999

Hernandez, A. (2018). Cognitive function in hypothyroidism: What is that deiodinase again? *Journal of Clinical Investigation, 129*, 55–57. https://doi.org/10.1172/jci125203

Ishii, S., Amano, I., & Koibuchi, N. (2021). The role of thyroid hormone in the regulation of cerebellar development. *Endocrinology and Metabolism, 36*, 703–716. https://doi.org/10.3803/enm.2021.1150

Jonklaas, J., Bianco, A. C., Bauer, A. J., Burman, K. D., Cappola, A. R., Celi, F. S., et al. (2019). Guidelines for the treatment of hypothyroidism: Prepared by the American Thyroid Association task force on thyroid hormone replacement. *Thyroid, 24*, 1670–1751. https://doi.org/10.1089/thy.2014.0028

Kaur Jhandi, S., Shenoy, S., Randhawa, A. S., & Singh, A. P. (2025). Impaired sustained attention in patients with subclinical hypothyroidism. *Clinical Medicine and Research, 23*, 45–52. https://doi.org/10.3121/cmr.2025.1925

Krausz, Y., Freedman, N., Lester, H., Newman, J. P., Barkai, G., Bocher, M., et al. (2004). Regional cerebral blood flow in patients with mild hypothyroidism. *Journal of Nuclear Medicine: Official Publication, Society of Nuclear Medicine, 45*, 1712–1715.

Kumar, M., Singh, S., Modi, S., Rana, P., D'souza, M., Sekhri, T., et al. (2024). Cortical gray matter thickness and volume changes and their association with memory functions in hyperthyroid patients. *Neuroendocrinology, 114*, 348–355. https://doi.org/10.1159/000536027

Kumar, M., Singh, S., Rana, P., D'souza, M., Kumaran, S. S., Sekhri, T., et al. (2025). Neurometabolite and cognitive changes in hypothyroid patients in response to treatment: In-vivo ^{1}H MRS study. *Neurochemistry International, 182*, Article 105915. https://doi.org/10.1016/j.neuint.2024.105915

Lekurwale, V., Acharya, S., Shukla, S., & Kumar, S. (2023). Neuropsychiatric manifestations of thyroid diseases. *Cureus, 15*. https://doi.org/10.7759/cureus.33987

Li, Y., Du, X., & Lang, X. (2022). Exploration of RCBF and metabolic changes in the brain functional areas of patients with hypothyroidism by ASL and MRS techniques. *Disease Markers, 2022*, 1–6. https://doi.org/10.1155/2022/9204119

Lischinsky, J. E., Skocic, J., Clairman, H., & Rovet, J. (2016). Preliminary findings show maternal hypothyroidism may contribute to abnormal cortical morphology in offspring. *Frontiers in Endocrinology, 7*. https://doi.org/10.3389/fendo.2016.00016

Liu, H., Li, W., Zhang, W., Sun, S., & Chen, C. (2022). Levothyroxine: Conventional and novel drug delivery formulations. *Endocrine Reviews, 44*, Article bnac030. https://doi.org/10.1210/endrev/bnac030

Martin, J. V., & Sarkar, P. K. (2023). Nongenomic roles of thyroid hormones and their derivatives in adult brain: Are these compounds putative neurotransmitters? *Frontiers in Endocrinology, 14*, Article 1210540. https://doi.org/10.3389/fendo.2023.1210540

Nillni, E. A. (2010). Regulation of the hypothalamic thyrotropin releasing hormone (TRH) neuron by neuronal and peripheral inputs. *Frontiers in Neuroendocrinology, 31*, 134–156. https://doi.org/10.1016/j.yfrne.2010.01.001

Nuguru, S., Rachakonda, S., Sripathi, S., Khan, M., Patel, N., & Meda, R. (2022). Hypothyroidism and depression: A narrative review. *Cureus, 14*, Article e28201. https://doi.org/10.7759/cureus.28201

Oppenheimer, J. H., & Schwartz, H. L. (1997). Molecular basis of thyroid hormone-dependent brain development. *Endocrine Reviews, 18*, 462–475. https://doi.org/10.1210/edrv.18.4.0309

Pak, K., Kim, M., Kim, K., Kim, B. H., Kim, S.-J., & Kim, I. J. (2020). Cerebral glucose metabolism and cerebral blood flow in thyroid dysfunction: An activation likelihood estimation meta-analysis. *Scientific Reports, 10*. https://doi.org/10.1038/s41598-020-58255-5

Pearce, E. N. (2019). Levothyroxine dosing after thyroidectomy should be determined by both BMI and body weight. *Clinical Thyroidology, 31*, 470–471. https://doi.org/10.1089/ct.2019;31.470-471

Rastogi, M. V., & LaFranchi, S. H. (2010). Congenital hypothyroidism. *Orphanet Journal of Rare Diseases, 5*. https://doi.org/10.1186/1750-1172-5-17

Remaud, S., Gothié, J.-D., Morvan-Dubois, G., & Demeneix, B. A. (2014). Thyroid hormone signaling and adult neurogenesis in mammals. *Frontiers in Endocrinology, 5*. https://doi.org/10.3389/fendo.2014.00062

Ritchie, M., & Yeap, B. B. (2015). Thyroid hormone: Influences on mood and cognition in adults. *Maturitas, 81*, 266–275. https://doi.org/10.1016/j.maturitas.2015.03.016

Rousset, B., Dupuy, C., Miot, F., & Dumont, J. (2000). *Chapter 2 thyroid hormone synthesis and secretion*. PubMed. https://pubmed.ncbi.nlm.nih.gov/25905405/.

Şorodoc, V., Constantin, M., Asaftei, A., Lionte, C., Ceasovschih, A., Sîrbu, O., et al. (2023). The use of intravenous immunoglobulin in the treatment of Hashimoto's encephalopathy: Case based review. *Frontiers in Neurology, 14*. https://doi.org/10.3389/fneur.2023.1243787

Sabatino, L., Vassalle, C., Del Seppia, C., & Iervasi, G. (2021). Deiodinases and the three types of thyroid hormone deiodination reactions. *Endocrinology and Metabolism, 36*, 952–964. https://doi.org/10.3803/enm.2021.1198

Salas-Lucia, F. (2024). Mapping thyroid hormone action in the human brain. *Thyroid, 34*, 815–826. https://doi.org/10.1089/thy.2024.0120

Samuels, M. H. (2014). Psychiatric and cognitive manifestations of hypothyroidism. *Current Opinion in Endocrinology Diabetes and Obesity, 21*, 377–383. https://doi.org/10.1097/med.0000000000000089

Sharan, P., & Vellapandian, C. (2024). Hypothalamic-pituitary-adrenal (HPA) axis: Unveiling the potential mechanisms involved in stress-induced Alzheimer's disease and depression. *Cureus, 16*. https://doi.org/10.7759/cureus.67595

Singh, S., Trivedi, R., Singh, K., Kumar, P., Shankar, L. R., & Khushu, S. (2014). Diffusion tensor tractography in hypothyroidism and its correlation with memory function. *Journal of Neuroendocrinology, 26*, 825–833. https://doi.org/10.1111/jne.12193

Stott, D. J., Rodondi, N., Kearney, P. M., Ford, I., Westendorp, R. G. J., Mooijaart, S. P., et al. (2017). Thyroid hormone therapy for older adults with subclinical hypothyroidism. *New England Journal of Medicine, 376*, 2534–2544. https://doi.org/10.1056/nejmoa1603825

Taylor, P. N., Medici, M. M., Hubalewska-Dydejczyk, A., & Boelaert, K. (2024). Hypothyroidism. *The Lancet, 404*, 1347–1364. https://doi.org/10.1016/s0140-6736(24)01614-3

Toll, A., Blanco-Hinojo, L., Berge, D., Manzano, A., Abidi, K. E., Perez-Solà, V., et al. (2023). Relationship between thyroid-stimulating hormone, BDNF levels, and hippocampal volume in antipsychotic-naïve first-episode psychosis patients. *Frontiers in Psychiatry, 14*. https://doi.org/10.3389/fpsyt.2023.1301714

Uchida, K., & Suzuki, M. (2021). Congenital hypothyroidism and brain development: Association with other psychiatric disorders. *Frontiers in Neuroscience, 15*. https://doi.org/10.3389/fnins.2021.772382

Uchida, K., Yonezawa, M., Nakamura, S., Kobayashi, T., & Machida, T. (2005). Impaired neurogenesis in the growth-retarded mouse is reversed by T3 treatment. *NeuroReport, 16*, 103–106. https://doi.org/10.1097/00001756-200502080-00005

van Geest, F. S., Gunhanlar, N., Groeneweg, S., & Visser, W. E. (2021). Monocarboxylate transporter 8 deficiency: From pathophysiological understanding to therapy development. *Frontiers in Endocrinology, 12*. https://doi.org/10.3389/fendo.2021.723750

Waliszewska-Prosół, M., Bladowska, J., Budrewicz, S., Sąsiadek, M., Dziadkowiak, E., & Ejma, M. (2021). The evaluation of Hashimoto's thyroiditis with event-related potentials and magnetic resonance spectroscopy and its relation to cognitive function. *Scientific Reports, 11*. https://doi.org/10.1038/s41598-021-82281-6

Wang, Q., Jiang, Y., Lv, H., Lu, Q., Tao, S., Qin, R., et al. (2022). Association of maternal mild hypothyroidism with offspring neurodevelopment in TPOAb-negative women: A prospective cohort study. *Frontiers in Endocrinology, 13*. https://doi.org/10.3389/fendo.2022.884851

Wikipedia Contributors. (2025). *Levothyroxine*. Wikipedia.

Yuen, K. C. J., Chua, E. W. R., Yuen, A. S. M., Martinez-Gill, M., & Verbalis, J. G. (2025). *Physiology and diseases of the hypothalamic-pituitary axis in the elderly*. Nihgov. https://www.ncbi.nlm.nih.gov/books/NBK620038/. (Accessed 17 January 2026) accessed.

Zhang, H., Tong, W., Zeng, W., Luo, H., Zhang, L., Feng, J., et al. (2025). Persistent symptoms in euthyroid Hashimoto's thyroiditis: Current hypotheses and emerging management strategies. *Frontiers in Endocrinology, 16*. https://doi.org/10.3389/fendo.2025.1627787

Zhang, Y., Yang, Y., Tao, B., Lv, Q., Lui, S., & He, L. (2021). Gray matter and regional brain activity abnormalities in subclinical hypothyroidism. *Frontiers in Endocrinology, 12*. https://doi.org/10.3389/fendo.2021.582519

Zhao, C., Wang, Y., Xiao, L., & Li, L. (2022). Effect of levothyroxine on older patients with subclinical hypothyroidism: A systematic review and meta-analysis. *Frontiers in Endocrinology, 13*. https://doi.org/10.3389/fendo.2022.913749

Zinck, J. C., Isaksen Jonas, L., Ahlberg, G., Salling, O. M., Nygaard, B., Ellervik, C., et al. (2023). Interaction between DIO2 polymorphisms and LT4 treatment on psychological well-being outcomes and cardiovascular risk factors: A population-based cross-sectional study in UK biobank. *Endocrine Abstracts*. https://doi.org/10.1530/endoabs.92.op-04-01

Further reading

Aulinger, B. A., & D'Alessio, D. A. (2023). Assessment of the incretin effect in healthy subjects: Concordance between clamp and OGTT methods. *AJP Endocrinology and Metabolism, 325*, E412–E420. https://doi.org/10.1152/ajpendo.00104.2022

Bang, J., Lee, S. A., Koh, G., & Yoo, S. (2023). Association of glucagon to insulin ratio and metabolic syndrome in patients with type 2 diabetes. *Journal of Clinical Medicine, 12*, 5806. https://doi.org/10.3390/jcm12185806

Batterham, R. L., Le Roux, C. W., Cohen, M. A., Park, A. J., Ellis, S. M., Patterson, M., et al. (2003). Pancreatic polypeptide reduces appetite and food intake in humans. *Journal of Clinical Endocrinology and Metabolism, 88*, 3989–3992. https://doi.org/10.1210/jc.2003-030630

Boyle, C. N., Zheng, Y., & Lutz, T. A. (2022). Mediators of amylin action in metabolic control. *Journal of Clinical Medicine, 11*, 2207. https://doi.org/10.3390/jcm11082207

Chan, K., Wong, F. S., & Pearson, J. A. (2022). Circadian rhythms and pancreas physiology: A review. *Frontiers in Endocrinology, 13*. https://doi.org/10.3389/fendo.2022.920261

Che, T., Yan, C., Tian, D., Zhang, X., Liu, X., & Wu, Z. (2021). Time-restricted feeding improves blood glucose and insulin sensitivity in overweight patients with type 2 diabetes: A randomised controlled trial. *Nutrition and Metabolism, 18*. https://doi.org/10.1186/s12986-021-00613-9

Colling, C., Bredella, M. A., Fazeli, P. K., Pachón-Peña, G., Singh, R. J., Rosen, C. J., et al. (2023). Changes in serum cortisol levels after 10 days of overfeeding and fasting. *AJP Endocrinology and Metabolism, 324*, E506–E513. https://doi.org/10.1152/ajpendo.00181.2022

Ellenbroek, J. H., van Dijck, L., Töns, H. A., Rabelink, T. J., Carlotti, F., Ballieux, B. E. P. B., et al. (2014). Long-term ketogenic diet causes glucose intolerance and reduced β- and α-cell mass but no weight loss in mice. *American Journal of Physiology - Endocrinology And Metabolism, 306*, E552–E558. https://doi.org/10.1152/ajpendo.00453.2013

Falkenhain, K. (2025). Ketones and insulin: A paradoxical interplay with implications for glucose metabolism. *Journal of the Endocrine Society, 9*. https://doi.org/10.1210/jendso/bvaf101

Głombik, K., Detka, J., Bobula, B., Bąk, J., Kusek, M., Tokarski, K., et al. (2021). Contribution of hypothyroidism to cognitive impairment and hippocampal synaptic plasticity regulation in an animal model of depression. *International Journal of Molecular Sciences, 22*, 1599. https://doi.org/10.3390/ijms22041599

Geary, N. (2017). Postprandial suppression of glucagon secretion: A puzzlement. *Diabetes, 66*, 1123–1125. https://doi.org/10.2337/dbi16-0075

Guo, L., Xi, Y., Jin, W., Yuan, H., Qin, G., Chen, S., et al. (2024). A 5:2 intermittent fasting meal replacement diet and glycemic control for adults with diabetes. *JAMA Network Open, 7*, e2416786. https://doi.org/10.1001/jamanetworkopen.2024.16786

Horne, B. D., Anderson, J. L., May, H. T., Bair, T. L., Le, V. T., Iverson, L., et al. (2024). Insulin resistance reduction, intermittent fasting, and human growth hormone: Secondary analysis of a randomized trial. *Npj Metabolic Health and Disease, 2*. https://doi.org/10.1038/s44324-024-00025-2

Jamshed, H., Beyl, R. A., Della Manna, D. L., Yang, E. S., Ravussin, E., & Peterson, C. M. (2019). Early time-restricted feeding improves 24-hour glucose levels and affects markers of the circadian clock, aging, and autophagy in humans. *Nutrients, 11*, 1234. https://doi.org/10.3390/nu11061234

Kellard, J. A., Nils, J., Rorsman, G., Hill, T. G., Armour, S. L., van, P. R., et al. (2020). Reduced somatostatin signalling leads to hypersecretion of glucagon in mice fed a high-fat diet. *Molecular Metabolism, 40*, 101021. https://doi.org/10.1016/j.molmet.2020.101021

Kinzig, K. P., Honors, M. A., & Hargrave, S. L. (2010). Insulin sensitivity and glucose tolerance are altered by maintenance on a ketogenic diet. *Endocrinology, 151*, 3105–3114. https://doi.org/10.1210/en.2010-0175

Merovci, A., Finley, B., Hansis-Diarte, A., Neppala, S., Abdul-Ghani, M. A., Cersosimo, E., et al. (2024). Effect of weight-maintaining ketogenic diet on glycemic control and insulin sensitivity in obese T2D subjects. *BMJ Open Diabetes Research and Care, 12*, Article e004199–9. https://doi.org/10.1136/bmjdrc-2024-004199

Paoli, A., Bianco, A., Moro, T., João, F. M., & de, C. (2023). The effects of ketogenic diet on insulin sensitivity and weight loss, which came first: The chicken or the egg? *Nutrients, 15*, 3120. https://doi.org/10.3390/nu15143120

Rix, I., Nexøe-Larsen, C., Bergmann, N. C., Lund, A., & Knop, F. K. (2019). *Glucagon physiology*. National Library of Medicine. https://www.ncbi.nlm.nih.gov/books/NBK279127/.

Stern, J. H., Smith, G. I., Chen, S., Unger, R. H., Klein, S., & Scherer, P. E. (2019). Obesity dysregulates fasting-induced changes in glucagon secretion. *Journal of Endocrinology, 243*, 149–160. https://doi.org/10.1530/JOE-19-0201

Sutton, E. F., Beyl, R., Early, K. S., Cefalu, W. T., Ravussin, E., & Peterson, C. M. (2018). Early time-restricted feeding improves insulin sensitivity, blood pressure, and oxidative stress Even without weight loss in men with prediabetes. *Cell Metabolism, 27*, 1212–1221.e3. https://doi.org/10.1016/j.cmet.2018.04.010

Wiley, J. H., & Leveille, G. A. (1970). Significance of insulin in the metabolic adaptation of rats to meal ingestion. *The Journal of Nutrition, 100*, 1073–1080. https://doi.org/10.1093/jn/100.9.1073

Yang, X., Zhou, J., Shao, H., Huang, B., Kang, X., Wu, R., et al. (2022). Effect of an intermittent calorie-restricted diet on type 2 diabetes remission: A randomized controlled trial. *Journal of Clinical Endocrinology and Metabolism, 108*. https://doi.org/10.1210/clinem/dgac661

Section 5

Nonthyroid treatment regimens

Chapter 21

Impact of epilepsy therapies on thyroid regulation: A narrative review

Anne Rochtus[1] and Brigitte Decallonne[2]

[1]*Department of Pediatrics, Section of Pediatric Endocrinology, University Hospital of Leuven, Leuven, Belgium;* [2]*Department of Endocrinology, University Hospital of Leuven, Leuven, Belgium*

Abbreviations

ACTH adrenocorticotropic hormone
ASMs antiseizure medications
CBZ carbamazepine
CNS central nervous system
CYP450 cytochrome P450
DIO1 type I deiodinase
DIO2 type II deiodinase
DIO3 type III deiodinase
FT3 free triiodothyronine
FT4 free thyroxine
GABA γ-aminobutyric acid
HPT hypothalamic–pituitary–thyroid
KD ketogenic diet
L-T4 levothyroxine
LEV levetiracetam
LTG lamotrigine
NTIS nonthyroidal illness syndrome
OXC oxcarbazepine
PB phenobarbital
PHT phenytoin
rT3 reverse T3
T3 triiodothyronine
T4 thyroxine
TBG thyroxine-binding globulin
TPM topiramate
TPO thyroid peroxidase
TRH thyrotropin-releasing hormone
TSH thyroid-stimulating hormone
UGT uridine 5′-diphospho-glucuronosyltransferase
VPA valproate

Introduction

Epilepsy is one of the most common chronic neurological disorders, affecting approximately 65 million people globally (Moshe et al., 2015). Antiseizure medications (ASMs) are the cornerstone of epilepsy management, often requiring long-term or lifelong use (Perucca et al., 2018). While effective in controlling seizures, ASMs are associated with a range of systemic side effects, including endocrine disturbances (Perucca & Gilliam, 2012). Among these, alterations in thyroid hormone homeostasis are increasingly recognized but remain underappreciated in clinical practice.

The interaction between ASMs and thyroid function was first described in 1961, when phenytoin was shown to influence thyroid hormone levels (Oppenheimer et al., 1961). Since then, numerous studies have explored how ASMs affect the synthesis, release, transport, metabolism, and excretion of thyroid hormones (Paragliola et al., 2015; Verrotti et al., 2008). ASMs are grouped into distinct pharmacological classes according to their specific mechanism of action within the nervous system (Fig. 21.1). The evidence is strongest for older ASMs such as phenytoin, valproate, phenobarbital, carbamazepine, and oxcarbazepine, which belong to pharmacological groups including sodium channel blockers (phenytoin, carbamazepine, oxcarbazepine), GABAergic agents (valproate, phenobarbital), and mixed-mechanism drugs that modulate both excitatory and inhibitory pathways. Data on newer ASMs—including levetiracetam, topiramate, lamotrigine, and gabapentin—are limited and often inconclusive.

Thyroid hormones are essential for neurodevelopment, metabolic regulation, and endocrine balance (Chaker et al., 2017). Their synthesis and release are tightly regulated by the hypothalamic–pituitary–thyroid (HPT) axis. Disruptions in this axis—whether due to illness, medication, or nutritional status—can lead to hypothyroidism, a condition with wide-ranging clinical implications. In epilepsy patients, thyroid dysfunction may not only affect general health but also influence seizure threshold and treatment response.

Hormones in Neuroscience. https://doi.org/10.1016/B978-0-443-30261-9.00021-8

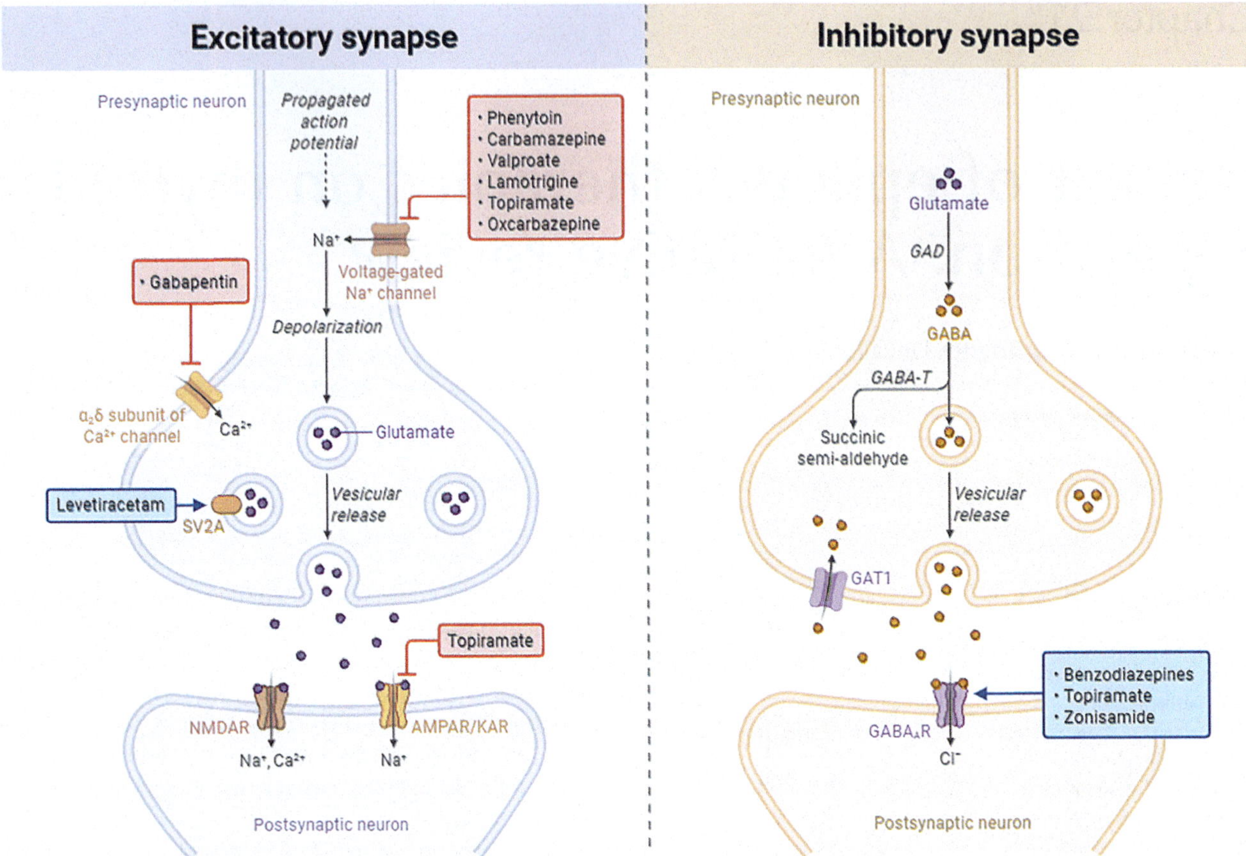

FIGURE 21.1 Mechanisms of action of ASMs. ASMs are classified into several pharmacological groups based on their mechanisms of action. *Figure created by Biorender.*

This chapter provides a comprehensive overview of the relationship between ASMs and thyroid hormone homeostasis. It highlights the mechanisms by which ASMs and ketogenic diet may alter thyroid function, reviews the clinical evidence, and offers practical recommendations for monitoring and managing thyroid health in patients with epilepsy.

Regulation and metabolism of thyroid hormones

Thyroid hormone production and regulation are orchestrated by the HPT axis, a finely tuned negative feedback system essential for maintaining metabolic and neurological homeostasis. The thyroid gland synthesizes two primary hormones: thyroxine (T4) and triiodothyronine (T3). While T4 is the predominant circulating form, T3 is the biologically active hormone with a higher affinity for thyroid hormone receptors.

These hormones are lipophilic and circulate in the bloodstream predominantly bound to transport proteins such as thyroxine-binding globulin (TBG), albumin, and transthyretin. Only a small fraction—approximately 0.2%—exists in the free, unbound form (free T4 (FT4) and free T3 (FT3)), which is available for cellular uptake and biological activity. Upon reaching target tissues, T4 and T3 can dissociate from their carrier proteins and enter cells via passive diffusion or carrier-mediated transport mechanisms.

Thyroid receptors have a higher affinity for T3. As a result, T4 is relatively inactive. Within cells, T4 is converted to either active T3 or inactive reverse T3 (rT3) by a family of enzymes known as deiodinases. Type I (DIO1) and Type II (DIO2) deiodinases, found in the liver, kidneys, muscles, and thyroid gland, catalyze the conversion of T4 to active form T3. Type III (DIO3), primarily located in the central nervous system and placenta, inactivates T4 by converting it to rT3. Thyroid hormones are ultimately metabolized in the liver through sulfation and glucuronidation and excreted in bile.

Thyroid hormone alterations in illness and hypothyroidism

Thyroid hormone levels are sensitive to physiological stress and illness. During periods of acute or chronic illness, particularly in the context of malnutrition or systemic inflammation, serum thyroid hormone concentrations may decline without intrinsic thyroid pathology. In mild illness, this typically manifests as a reduction in T3 levels. However, in more severe or prolonged illness—such as critical care scenarios or prolonged starvation—patients often exhibit a constellation of hormonal changes: reduced T3, elevated rT3, and low-to-normal T4 levels, all occurring without a compensatory rise in thyroid-stimulating hormone (TSH). This adaptive response is known as nonthyroidal illness syndrome (NTIS), or euthyroid sick syndrome, and reflects a shift in peripheral thyroid hormone metabolism rather than primary thyroid dysfunction (Langouche et al., 2019).

In contrast, hypothyroidism refers to a pathological deficiency of thyroid hormones and can range from subclinical (or mild, with increased TSH but normal FT4) to clinical (or overt, with both increased TSH and decreased FT4) forms. Because symptoms are often nonspecific—such as fatigue, weight gain, or cold intolerance—diagnosis typically relies on biochemical markers. The most common cause is chronic autoimmune thyroiditis (Hashimoto's thyroiditis), characterized by the presence of thyroid peroxidase (TPO) antibodies. Hypothyroidism is more prevalent in women, individuals over 65 years of age, White populations, and those with autoimmune comorbidities such as type 1 diabetes or celiac disease.

Clinically, hypothyroidism is categorized as primary (thyroid gland dysfunction), secondary (pituitary TSH deficiency), tertiary (hypothalamic thyrotropin-releasing hormone (TRH) deficiency), or due to peripheral resistance or non-HPT axis causes (Table 21.1). Understanding these distinctions is essential when evaluating thyroid function in patients with epilepsy, particularly those on long-term antiseizure medications.

The possible role of thyroid hormones in epilepsy

Thyroid hormones are fundamental to the development and maintenance of central nervous system (CNS) function. They play a critical role not only in neurodevelopment but also in sustaining normal physiological brain activity and facilitating neurorepair mechanisms. Their influence extends to both inhibitory and excitatory neurotransmission, particularly through modulation of GABAergic and glutamatergic pathways (Tamijani et al., 2015). Although the precise mechanisms remain to be fully elucidated, this dual modulation suggests that thyroid hormones may significantly impact seizure susceptibility and neuronal excitability.

Experimental studies using transcranial magnetic stimulation have provided insights into the effects of thyroid dysfunction on cortical excitability (Terranova et al., 2016). Patients with overt hypothyroidism, as well as those with resistance to thyroid hormone, exhibit altered cortical excitability and impaired inhibitory circuits. These findings underscore the bidirectional influence of both hypothyroid and hyperthyroid states on neuronal activity and seizure thresholds (Fig. 21.2).

Clinically, seizures have been reported in a minority of patients with thyrotoxicosis, further supporting the notion that thyroid hormone excess can disrupt neural stability (Jabbari & Huott, 1980; Song et al., 2010). Conversely, in the context of infantile spasms, adrenocorticotropic hormone (ACTH) therapy—known to suppress thyroxine

TABLE 21.1 Causes of hypothyroidism.

Primary hypothyroidism Chronic auto-immune thyroiditis (Hashimoto's thyroiditis) Iodine deficiency or excess Drugs (e.g., lithium, amiodarone, methimazole) Iatrogenic (e.g., after treatment with radioactive iodine) Subacute thyroiditis (e.g., viral infection) Thyroid gland infiltration (e.g., sarcoidosis) Genetic (e.g., inactivating TPO gene mutation)
Central hypothyroidism Pituitary tumors or dysfunction Hypothalamic tumors or dysfunction Resistance to thyroid-stimulating hormone Drugs (e.g., bexarotene)
Hypothyroidism due to causes outside the hypothalamic–pituitary–thyroid axis Consumptive hypothyroidism (e.g., nephrotic syndrome) Tissue-specific hypothyroidism due to decreased sensitivity to thyroid hormone (RTH)

Adapted with permission from Chaker, L., Bianco, A. C., Jonklaas, J., & Peeters, R. P. (2017). Hypothyroidism. Lancet, 390(10101), 1550-1562. https://doi.org/10.1016/S0140-6736(17)30703-1.

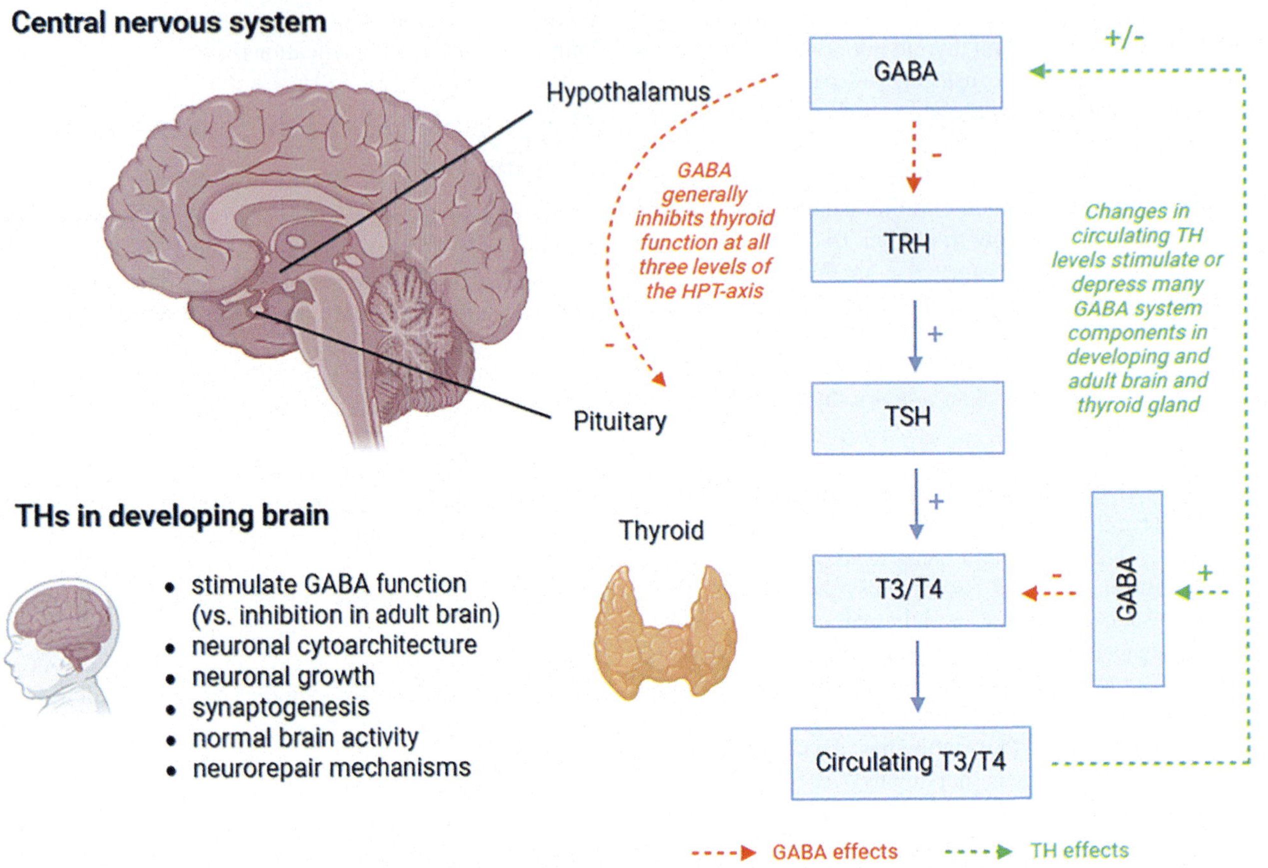

FIGURE 21.2 Altered inhibitory circuits in the developing and adult brain. Schematic representation of the altered inhibitory circuits between thyroid hormones and GABA in the central nervous system. The + and – symbols indicate stimulation and inhibition, respectively. GABA, γ-aminoburtyric acid; *TRH*, thyrotropin releasing hormone; *TSH*, thyroid stimulating hormone; T3, triiodothyronine; T4, thyroxine. *Adapted from Wiens, S. C., & Trudeau, V. L. (2006). Thyroid hormone and gamma-aminobutyric acid (GABA) interactions in neuroendocrine systems. Comparative Biochemistry and Physiology Part A: Molecular & Integrative Physiology, 144(3), 332-344. https://doi.org/10.1016/j.cbpa.2006.01.033. Figure created by Biorender.*

levels—has been associated with favorable outcomes. Notably, patients who responded well to ACTH treatment often exhibited significantly reduced thyroxine concentrations, suggesting a potential therapeutic link between thyroid hormone suppression and seizure control (Izumi & Fukuyama, 1984).

These observations point to a broader neuroendocrine interplay, in which thyroid hormone alterations may reflect or contribute to adaptive responses within the corticotropic axis (hypothalamic-pituitary-adrenal axis). This crosstalk may be particularly relevant in epilepsy, where hormonal fluctuations could influence both seizure activity and treatment outcomes.

In summary, while the exact role of thyroid hormones in epilepsy remains an area of active investigation, current evidence highlights their potential to modulate neuronal excitability and seizure dynamics. Understanding this relationship may open new avenues for therapeutic strategies that integrate endocrine and neurological perspectives.

The impact of antiseizure medications and ketogenic diet on thyroid function

Antiseizure medications (ASMs) are classified into several pharmacological groups based on their mechanisms of action (Fig. 21.1). **Sodium channel blockers** (e.g., carbamazepine, phenytoin, lamotrigine) reduce neuronal excitability by stabilizing inactive sodium channels. **GABAergic agents** (e.g., valproate, benzodiazepines, phenobarbital) enhance inhibitory neurotransmission by increasing GABA levels or receptor activity. **Calcium channel blockers**, such as ethosuximide, target T-type calcium channels, particularly effective in absence seizures. **SV2A modulators** like levetiracetam bind to synaptic vesicle protein 2A, modulating neurotransmitter release. **Glutamate antagonists** (e.g., perampanel) inhibit excitatory transmission. Some newer ASMs, such as topiramate and zonisamide, have **multiple mechanisms**, including sodium channel blockade, GABA enhancement, and carbonic anhydrase inhibition. The choice of ASM depends on seizure type, patient age, comorbidities, and side effect profile.

The possible interactions between ASMs and thyroid homeostasis are shown in Table 21.2, Figs. 21.2 and 21.3, and will be further detailed for each ASM.

Valproate

Valproate (VPA) is one of the most extensively studied antiseizure medications in relation to thyroid hormone alterations. Across 25 studies—primarily retrospective or cross-sectional in design—elevations in TSH were reported in 12%–52% of patients receiving VPA monotherapy, with most studies citing a prevalence around 25% (Rochtus et al., 2022). TSH levels rarely exceeded 10 mIU/L (Eiris-Punal et al., 1999; Hirfanoglu et al., 2007; Kim et al., 2012), and one study reported no significant TSH elevation (Verrotti et al., 2001). Five prospective cohort studies further explored thyroid hormone dynamics during VPA monotherapy, reinforcing the association between treatment duration and TSH elevation (Cansu et al., 2006; Castro-Gago et al., 2007; Kafadar et al., 2015;

TABLE 21.2 Effect of ASMs on thyroid hormone homeostasis.

	T3	FT3	T4	FT4	TSH	TBG Binding	CYP3A4 Induction	UGT induction
PHT	=/↓	↓	↓	↓	=	↓	Yes	Yes
PB	=	=	=/↓	=/↓	=	↑	Yes	Yes
CBZ	=/↓	=/↓	↓	↓	=	↑↑	Yes	Yes
OXC	=	=	=/↓	=/↓	=	↑↑	Yes	Yes
VPA	=/↓	=/↓	=/↓	=/↓	=/↑	↑	No	No
KD	NA	↓	NA	=/↓	=/↑	NA	Yes	NA
LTG	NA	NA	NA	=/↓	=/↑	NA	No	Yes
LEV	=	=	=	=/↓	=	NA	NS	No
TPM	=	=	=	=/↓	=	NA	No	No

T3 = tri-iodothyronine; FT3 = free tri-iodothyronine; T4 = thyroxine; FT4 = free thyroxine; NA = not available; NS = non-significant; TSH = thyroid-stimulating hormone; TBG = thyroxine-binding globulin; UGT = UDP-glucuronyltransferases; CYP3A4 = cytochrome P450 3A4. ASM = antiseizure medication; CBZ = carbamazepine; KD = ketogenic diet; LEV = levetiracetam; LTG = lamotrigine; OXC = oxcarbazepine; PB = phenobarbital; PHT = phenytoin; TPM = topiramate; VPA = valproate.

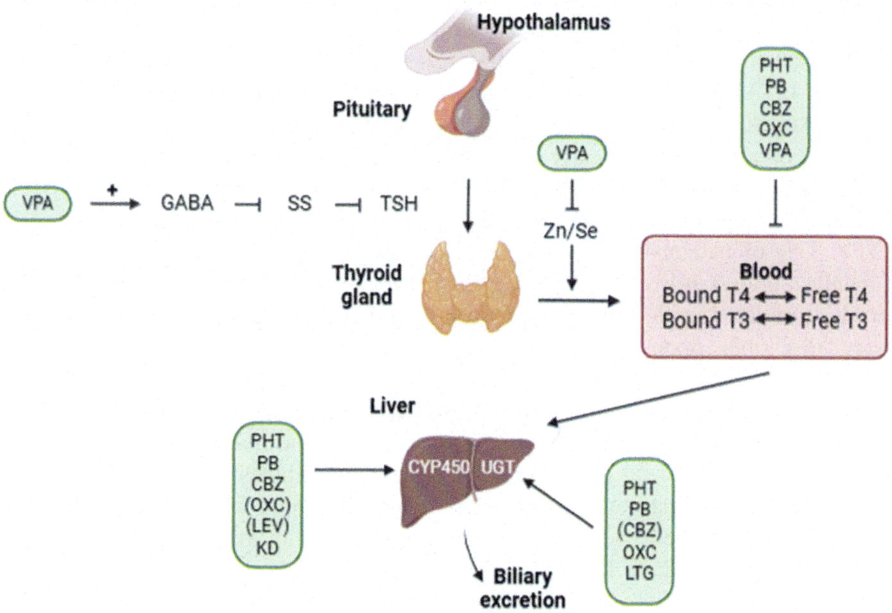

FIGURE 21.3 Overview of the interaction between ASMs and thyroid hormone homeostasis. ASMs = antiseizure medications, CBZ = carbamazepine, CYP3A4 = cytochrome P450 3A4, D1 = iodothyronine deiodinase type 1, D2 = iodothyronine deiodinase type 2, D3 = iodothyronine deiodinase type 3, D4 = iodothyronine deiodinase type 4, GABA = gamma-aminobutyric acid, PB = phenobarbital, rT3 = reverse triiodothyronine, Se = selenium, SS = somatostatin, T2 = diiodothyronine, T3 = triiodothyronine, T4 = thyroxine, TRH = thyrotropin releasing hormone, TSH = thyroid stimulating hormone, UGT = UDP-glucuronosyltransferase, VPA = valproate, Zn = zinc. *Figure created by Biorender.*

Lossius et al., 2009; Verrotti et al., 2009). Notably, therapy exceeding 6 months (OR = 2.98) or 24 months (OR = 4.8) was linked to higher TSH levels (Mikati et al., 2007; Sahu et al., 2012).

Polytherapy, particularly when combining VPA with enzyme-inducing or non-enzyme-inducing ASMs, further increased the risk of TSH elevation. However, findings on FT4, total T4, FT3, and total T3 levels were inconsistent, with studies reporting both reductions and no significant changes. Importantly, some studies demonstrated that thyroid hormone alterations were reversible upon discontinuation of VPA (Lossius et al., 2009).

Mechanistically, VPA appears to influence thyroid function through both central and intrathyroidal pathways. Centrally, VPA enhances γ-aminobutyric acid (GABA) activity, which suppresses somatostatin release—a known inhibitor of TSH secretion—thereby potentially increasing TSH levels. Intrathyroidally, VPA may impair thyroid hormone synthesis by inducing deficiencies in zinc and selenium, essential cofactors for deiodinase enzymes and thyroid peroxidase. Additionally, VPA may reduce TBG levels, alter protein-binding dynamics, and inhibit deiodinase activity (DIO1, DIO2, DIO3), leading to reduced peripheral conversion of T4 to T3 (Kim et al., 2012). Unlike enzyme-inducing ASMs, VPA has minimal hepatic enzyme-inducing properties, and autoimmune mechanisms do not appear to contribute to its thyroid effects (Ilic et al., 2016).

Carbamazepine

Carbamazepine (CBZ) is one of the most frequently studied ASMs in relation to thyroid dysfunction (Rochtus et al., 2022). Across 26 studies—19 of which were retrospective or cross-sectional—CBZ was consistently associated with reductions in circulating thyroid hormones. Seven prospective cohort studies confirmed that monotherapy with CBZ leads to significant decreases in T4, FT4, T3 and FT3 often within the first month of treatment (Attilakos et al., 2007; Castro-Gago et al., 2007; Kafadar et al., 2015; Lossius et al., 2009; Nishiyama et al., 2019; Simko & Horacek, 2007; Verrotti et al., 2009). These reductions persisted with long-term use, with 28%–33% of patients exhibiting decreased FT4 after 24 months. Importantly, thyroid hormone levels were negatively correlated with serum CBZ concentrations and were reversible upon discontinuation of therapy.

TSH elevation was observed in 10%–25% of patients after 12 months of CBZ therapy, although some studies reported no significant or only transient increases. Structural thyroid changes, such as gland enlargement and cyst formation, were inconsistently reported and not confirmed in pediatric cohorts (Hamed et al., 2005; Hegedus et al., 1985; Hirfanoglu et al., 2007).

The mechanisms underlying CBZ-induced thyroid hormone alterations are multifactorial. CBZ is a potent inducer of hepatic cytochrome P450 enzymes, which accelerates the metabolism of thyroid hormones (Connell et al., 1984). It also enhances glucuronidation via upregulation of uridine 5′-diphospho-glucuronosyltransferases (UGTs), particularly UGT1A6, UGT1A9, and UGT1A1 (Benedetti et al., 2005; Surks & DeFesi, 1996). These enzymes not only increase the clearance of T4 but also affect the metabolism of rT3, potentially disrupting the balance of active and inactive thyroid hormone metabolites.

Additionally, CBZ interferes with the binding of TBG, reducing the pool of bioavailable hormone. Interestingly, in some patients with low peripheral thyroid hormone levels, TSH remains within the normal range, suggesting a possible centrally mediated effect. This may involve increased expression of DIO3 in the hippocampus (Baumgartner et al., 1994), which inactivates T4 and T3, further contributing to hypothyroxinaemia without triggering a compensatory TSH response.

Oxcarbazepine

Oxcarbazepine (OXC), a keto-analogue of CBZ, has been studied in eight clinical studies for its effects on thyroid hormone homeostasis. Six of these were retrospective or cross-sectional studies, while two were prospective cohort studies evaluating thyroid hormone changes during OXC monotherapy (Cansu et al., 2006; Garoufi et al., 2014). Although OXC is associated with less hepatic enzyme induction than CBZ (Schmidt & Elger, 2004), its impact on thyroid function is nonetheless clinically relevant.

Most studies reported a significant reduction in T4 and FT4 levels, with prevalence rates ranging from 10% to 32% (Rochtus et al., 2022). These reductions occurred independently of treatment duration or dosage. Interestingly, TSH levels remained within the normal range in nearly all studies, with only one reporting a mild TSH elevation in 9% of patients. The volume of the thyroid gland remained unaffected (Hirfanoglu et al., 2007).

Mechanistically is the enzyme-inducing potential of OXC lower than that of CBZ. Since the effects of these drugs on thyroid function are similar, there must be other mechanisms by which OXC influences thyroid hormone metabolism. One of them is an increased glucuronidation compared to CBZ (Benedetti et al., 2005), which facilitates the hepatic clearance of thyroid hormones. This increased glucuronidation likely contributes to the observed reductions in circulating T4 and FT4. Unlike CBZ, OXC does not appear to interfere with thyroid hormone binding to TBG, as both total and free hormone levels are reduced in OXC treated patients.

Phenobarbital

Phenobarbital (PB), one of the oldest ASMs still in use, has been evaluated in seven studies for its impact on thyroid hormone metabolism. Most of these were retrospective or cross-sectional in design, with two prospective cohort studies specifically investigating thyroid hormone changes during PB monotherapy (Castro-Gago et al., 2007; Lossius et al., 2009). The predominant finding across studies was a significant reduction in FT4 levels, while TSH levels remained largely unchanged (Castro-Gago et al., 2007; Eiris-Punal et al., 1999). A negative correlation between PB serum levels and FT4 was consistently observed, and thyroid hormone levels normalized following discontinuation of PB therapy.

The mechanisms underlying PB-induced thyroid hormone alterations are thought to involve hepatic enzyme induction, particularly of the cytochrome P450 (CYP450) system. However, compared to CBZ, PB exerts a milder effect on thyroid hormone metabolism, suggesting differences in the extent and nature of enzyme activation. PB may preferentially enhance peripheral conversion of T4 to T3, thereby maintaining TSH levels despite reduced FT4 concentrations (Paragliola et al., 2015).

Additionally, PB has only weak protein-binding properties and appears to cause minimal induction of UGTs, which are responsible for the glucuronidation and clearance of thyroid hormones.

Phenytoin

Phenytoin (PHT), a widely used antiseizure medication, has been evaluated in five retrospective studies for its effects on thyroid function. The majority of these studies reported a significant reduction in FT4 levels, with one study noting this in up to 24% of patients (Isojarvi et al., 1990, 1992; Shih et al., 2017; Tiihonen et al., 1995). Despite these reductions, TSH levels typically remained within the normal range, suggesting a dissociation between peripheral thyroid hormone levels and central regulatory feedback.

The mechanisms by which PHT alters thyroid hormone homeostasis are multifaceted. First, PHT is a potent inducer of hepatic cytochrome P450 enzymes, which accelerates the metabolic clearance of thyroid hormones. This increased enzymatic activity enhances the conversion of T4 into inactive metabolites, thereby reducing circulating hormone levels. Second, PHT disrupts the binding of thyroid hormones to their primary transport protein, TBG. By displacing T4 from TBG, PHT increases the proportion of unbound hormone available for metabolism and excretion, further contributing to reduced serum FT4 levels. This displacement may also lead to transient increases in free hormone fractions, which are rapidly cleared due to enhanced hepatic metabolism. Additionally, PHT promotes glucuronidation, a key pathway for thyroid hormone inactivation and biliary excretion (Benedetti et al., 2005). Through upregulation of UGTs, PHT facilitates the conjugation and elimination of T4 and T3, compounding the reduction in hormone availability.

Lamotrigine

Lamotrigine (LTG), a newer-generation antiseizure medication, has been evaluated in only a few studies for its effects on thyroid hormone homeostasis. To date, three studies have explored this relationship, including one prospective cohort study and two retrospective analyses (El-Farahaty et al., 2015; Lossius et al., 2009; Shih et al., 2017). The prospective study, which investigated thyroid hormone changes during LTG monotherapy, was limited by a small sample size and did not yield statistically meaningful conclusions. Among the retrospective studies, one found no significant alterations in thyroid hormone levels, while another reported elevated TSH levels in 9% of patients receiving LTG.

Mechanistically, LTG is considered a weak inducer of the UGT system. It is known to increase the glucuronidation of co-administered VPA, thereby lowering VPA plasma concentrations. However, it remains unclear whether LTG induces the specific UGT isoforms involved in the glucuronidation and clearance of thyroid hormones such as T4 and T3.

Unlike older ASMs such as CBZ or PHT, LTG does not significantly induce hepatic cytochrome P450 enzymes, nor does it appear to interfere with thyroid hormone binding to transport proteins like TBG. The observed TSH elevations in some patients may reflect subtle central effects or individual susceptibility, but the clinical significance remains uncertain.

Given the limited and somewhat conflicting data, further prospective studies are needed to clarify the impact of LTG on thyroid function.

Gabapentin

Gabapentin, a newer-generation antiseizure medication primarily used for focal seizures and neuropathic pain, has not been extensively studied in relation to thyroid hormone homeostasis. To date, no robust clinical studies have systematically evaluated the effects of gabapentin on thyroid function in either pediatric or adult populations. As a result, its endocrine profile remains largely undefined.

Mechanistically, gabapentin differs significantly from older antiseizure medications. It does not induce hepatic cytochrome P450 enzymes and is not known to affect UGT activity. Gabapentin is eliminated unchanged via renal excretion and undergoes minimal hepatic metabolism.

These pharmacokinetic properties suggest a low likelihood of direct interference with thyroid hormone synthesis, metabolism, or clearance. Furthermore, gabapentin does not appear to displace thyroid hormones from their binding proteins, such as TBG, nor does it alter the HPT axis in any known way. However, given the subtle and sometimes delayed endocrine effects observed with other ASMs, the absence of evidence should not be interpreted as evidence of absence.

Levetiracetam

Levetiracetam (LEV), a widely used newer-generation ASM, has been evaluated in six studies for its potential effects on thyroid hormone homeostasis (Attilakos et al., 2019; Aygun et al., 2012; El-Farahaty et al., 2015; Nishiyama et al., 2019; Shih et al., 2017; Yilmaz et al., 2014). Two of these were prospective cohort studies examining thyroid hormone changes during LEV monotherapy, both of which reported no significant alterations in thyroid function. Similarly, most retrospective studies found no meaningful changes in FT4, T4, or TSH levels in patients treated with LEV.

However, some isolated findings suggest that LEV may not be entirely without endocrine effects. A study by El-Farahaty et al. reported significantly lower FT4 levels in pediatric patients treated with LEV compared to both healthy controls and patients receiving valproate or carbamazepine. Notably, TSH levels remained within the normal range in these patients, suggesting a possible central adaptation or altered feedback sensitivity. In a larger adult cohort, Shih et al. found that LEV use was associated with an increased likelihood of low FT4 levels (odds ratio 2.432), although the clinical significance of this finding remains uncertain.

From a pharmacokinetic perspective, LEV is minimally metabolized and is primarily excreted unchanged in the urine. It does not induce or inhibit hepatic CYP450 enzymes or UGTs, which are commonly implicated in thyroid hormone metabolism. This metabolic neutrality likely explains the overall absence of consistent thyroid hormone disturbances in most patients.

Topiramate

Topiramate (TPM), a newer-generation antiseizure medication with broad-spectrum efficacy, has been evaluated in only a few studies for its potential effects on thyroid hormone regulation (Aygun et al., 2012; El-Farahaty et al., 2015; Shih et al., 2017). To date, three studies have addressed this question. A small pediatric study ($n = 5$) by Aygün et al. found no significant changes in thyroid hormone levels during TPM therapy. In contrast, El-Farahaty et al. reported that 33% of patients on TPM monotherapy ($n = 12$) exhibited significantly reduced FT4 levels, although TSH levels remained within the normal range. Similarly, Shih et al. identified a statistically significant association between TPM use and low FT4 levels in a large adult cohort (odds ratio 2.528), suggesting a potential subclinical effect.

Pharmacokinetically, TPM is only minimally metabolized and is primarily excreted unchanged in the urine. Its metabolism involves hydroxylation and hydrolysis, with a portion of its metabolites undergoing glucuronidation. Although TPM is not a strong inducer of hepatic enzymes, its metabolism is inducible, and it remains unclear whether this includes CYP450 isoenzymes or UGTs involved in thyroid hormone clearance.

Ketogenic diet

The ketogenic diet (KD), a high-fat, low-carbohydrate dietary therapy used primarily for drug-resistant epilepsy, has gained renewed interest in recent years. While its anticonvulsant efficacy is well established, emerging evidence suggests that KD may also influence thyroid hormone homeostasis. Three prospective studies have investigated this relationship, focusing on both pediatric and adult populations (Kose et al., 2017; Lee et al., 2017; Molteberg et al., 2020).

In a large pediatric cohort ($n = 120$), Kose et al. reported that 16.7% of children developed subclinical hypothyroidism within the first 6 months of KD initiation. These children exhibited significantly lower FT3 levels compared to baseline, although FT4 and TSH levels remained within or near the normal range. Another pediatric study by Lee et al. found no significant changes in thyroid function overall, but noted that younger age at seizure onset, earlier KD initiation, and higher serum lipid levels were associated with lower FT4 and higher TSH levels. In adults, Molteberg et al. observed a reduction in FT3 and T3 levels and a modest increase in FT4 and TSH after 12 weeks on a modified Atkins diet, a variant of KD.

Mechanistically, KD mimics the metabolic state of fasting, which is known to downregulate the HPT axis. This may lead to reduced peripheral conversion of T4 to T3, mediated by decreased activity of deiodinase enzymes (particularly DIO1 and DIO2). The resulting hormonal profile—low T3, elevated rT3, and relatively stable TSH—is reminiscent of the nonthyroidal illness syndrome (NTIS), a physiological adaptation to conserve energy during metabolic stress (Langouche et al., 2019).

Interestingly, none of the studies found significant differences in thyroid hormone changes between patients on enzyme-inducing ASMs and those on non-inducing ASMs, suggesting that the observed effects are primarily diet-related rather than pharmacologically mediated.

Polytherapy and intractable epilepsy

The use of multiple ASMs, often necessary in patients with refractory or intractable epilepsy, has been associated with a higher risk of thyroid hormone alterations. Five studies have specifically examined the effects of ASM polytherapy on thyroid function (Isojarvi et al., 1990, 1992; Shih et al., 2017; Verrotti et al., 2001; Yildiz et al., 2010). Four smaller studies focused on combined therapy with CBZ and VPA, involving a total of 44 patients. All reported a significant reduction in FT4 levels without a corresponding rise in TSH, suggesting a pattern of subclinical or central hypothyroxinemia.

A larger cross-sectional study by Shih et al. involving 298 adult epilepsy patients provided further insight (Shih et al., 2017). In this cohort, 17.4% of patients had low FT4 levels. Multivariate analysis identified several independent risk factors for FT4 reduction, including older age ($P = .004$), female sex ($P = .014$), longer epilepsy duration ($P = .001$), and intractable epilepsy ($P = .009$). Notably, the use of three or more ASMs was associated with a markedly increased risk (odds ratio 6.346), as was CBZ use specifically (odds ratio 3.152).

The mechanisms underlying these findings are likely multifactorial. Polytherapy may amplify hepatic enzyme induction, particularly of CYP450 and UGTs, leading to accelerated thyroid hormone metabolism and clearance. Additionally, overlapping pharmacodynamic effects may disrupt HPT axis regulation, resulting in a blunted TSH response despite declining peripheral hormone levels.

Practical guideline

Hypothyroxinemia in patients on ASMs

The use of ASMs is frequently associated with mild thyroid hormone abnormalities, most notably hypothyroxinemia—defined as a reduction in FT4 levels with normal or slightly elevated TSH. This biochemical pattern, often subclinical, has been observed in nearly one-third of patients receiving ASMs. Among the various agents studied, VPA shows the highest prevalence of FT4 reduction (up to 25%), followed by CBZ (10%–25%), OXC (9%), LTG (9%), and the ketogenic diet (17%).

The risk of hypothyroxinemia appears to be influenced by several clinical factors. Older age, female sex, longer duration of epilepsy, intractable seizure patterns, and polytherapy with multiple ASMs are all associated with a higher likelihood of thyroid hormone alterations. While most patients in clinical practice achieve seizure control with monotherapy, those requiring polytherapy or with refractory epilepsy represent a subgroup in whom regular thyroid function monitoring is particularly warranted.

Importantly, no studies have demonstrated an association between ASMs and hyperthyroxinemia. Furthermore, thyroid hormone alterations observed during ASM therapy are often reversible upon drug withdrawal, suggesting a functional rather than structural disruption of the HPT axis.

Two main hypotheses have been proposed to explain the paradoxical finding of reduced peripheral thyroid hormone levels without a compensatory rise in TSH. The first posits that ASMs may induce a new setpoint within the HPT axis, allowing lower circulating hormone levels to adequately support cellular function—implying increased tissue sensitivity to thyroid hormones. The second hypothesis suggests that ASMs may blunt the central TSH response, resulting in a form of mild central hypothyroidism. While this may have a stabilizing effect on neuronal excitability and seizure control, it raises concerns about potential long-term consequences for metabolic, cardiovascular, and cognitive health.

Understanding these mechanisms is essential for clinicians managing epilepsy patients, particularly those on long-term or combination ASM therapy. Early recognition and appropriate endocrine follow-up can help mitigate the broader systemic effects of ASM-induced thyroid dysfunction.

The clinical importance of mild hypothyroidism

Mild or subclinical hypothyroidism represents a nuanced yet clinically significant alteration in thyroid function, particularly relevant in both general and specialized populations. Defined by elevated serum TSH levels in the presence of normal FT4 concentrations, this condition is often considered an early or mild form of primary hypothyroidism. In the general adult population, subclinical hypothyroidism is the most prevalent functional thyroid disorder, with an estimated incidence ranging from 3% to 15%. The risk of progression to overt hypothyroidism is approximately 2%–6% annually, especially in individuals with autoimmune thyroiditis, such as Hashimoto's disease (Peeters, 2017).

In pediatric populations, the prevalence of subclinical hypothyroidism is notably lower, estimated at around 1.7% (Gallizzi et al., 2018). Despite often being asymptomatic, affected individuals may report nonspecific symptoms including fatigue, muscle weakness, weight gain, cold intolerance, and constipation. These symptoms, while variable, underscore the potential for subclinical hypothyroidism to impact quality of life and physiological function.

Beyond symptomatic concerns, there is growing apprehension regarding the long-term cardiovascular and metabolic consequences of untreated subclinical hypothyroidism. Associations have been observed with coronary heart disease, congestive heart failure, dyslipidemia, and insulin resistance (Ferrannini et al., 2017; Roef et al., 2012; Roos et al., 2007). The pathophysiological mechanisms underlying these associations remain incompletely understood, though dyslipidemia is a suspected mediator.

Therapeutic intervention with levothyroxine (L-T4) has demonstrated benefit in selected patient groups, particularly those with TSH levels exceeding 10–12 mIU/L. In these individuals, treatment has been associated with improvements in surrogate markers of cardiovascular risk, including reductions in serum cholesterol, decreased carotid intima-media thickness, and enhanced cardiac function.

In contrast, patients receiving ASMs often exhibit a distinct thyroid profile. These individuals more frequently present with reduced FT4 levels, while TSH concentrations remain unchanged or only minimally elevated. This biochemical pattern diverges from classical subclinical hypothyroidism and raises questions about its clinical implications in epilepsy populations. Although the full impact remains to be elucidated, existing evidence suggests that even in euthyroid individuals, thyroid hormone levels correlate with components of the metabolic syndrome, including lipid profiles, insulin sensitivity, and bone mineral density.

Notably, low-normal FT4 levels have been linked to increased insulin resistance and heightened cardiovascular risk. Furthermore, thyroid hormone concentrations appear to be inversely related to bone mineral density, suggesting a broader systemic influence of even subtle thyroid dysfunction.

Guidelines for monitoring and treatment

Despite growing evidence of mild thyroid hormone abnormalities in patients treated with ASMs, there remains a lack of clear, universally accepted guidelines for thyroid function monitoring in this population. The decision to initiate treatment is similarly contentious, with no consensus on thresholds or indications, particularly in asymptomatic individuals.

Nonetheless, clinical judgment suggests that certain patient groups may benefit from closer monitoring and potential intervention. These include individuals on polytherapy regimens and those exhibiting an unfavorable metabolic profile. Additionally, isolated hypothyroxinemia—characterized by low FT4 with normal TSH—has been associated with insulin resistance, supporting the rationale for treatment in selected cases.

An unresolved question is whether mild central hypothyroidism might confer a protective effect in epilepsy patients. This hypothesis draws parallels with the NTIS observed in critically ill patients. In the acute phase of critical illness, NTIS-related hormonal changes are thought to be adaptive, conserving energy to enhance survival. However, in chronic critical illness, these changes may become maladaptive, contributing to muscle wasting, organ dysfunction, and cognitive impairment. Whether a similar dynamic exists in epilepsy patients on long-term ASM therapy remains to be determined.

This uncertainty underscores the ongoing debate about the appropriateness of thyroid hormone replacement in ASM-treated patients. Current evidence supports a more proactive approach in adults, who are at greater risk of progressing to overt hypothyroidism and for whom the metabolic benefits of L-T4 therapy are better established. In contrast, the evidence base for pediatric patients remains limited.

Ultimately, the neurologist must balance the imperative of seizure control with the potential long-term endocrine and metabolic consequences of ASM therapy. This requires a nuanced, individualized approach to monitoring and treatment.

Monitoring recommendations

Our monitoring and treatment recommendations are informed by the guidelines of the European Thyroid Association (ETA), the American Thyroid Association (ATA), and the American Association of Clinical Endocrinologists (AACE) (Rochtus et al., 2022). We advise the following:

- **Baseline Testing:** Measure TSH and FT4 levels prior to initiating any ASM, regardless of the specific drug.
- **Follow-Up Testing:** For patients on older ASMs—such as CBZ, PHT, PB, and VPA—and for those with known risk factors for thyroid dysfunction, repeat thyroid function tests every 6 months.
- **Referral Criteria:** Refer patients to a pediatric or adult endocrinologist if TSH exceeds 10 mIU/L and/or FT4 is decreased.
- **Therapy Adjustments:** When switching ASMs or adding a new drug, assess for clinical signs of hypothyroidism and recheck TSH and FT4 levels after 3 months.

These recommendations are summarized in Tables 21.3 and 21.4, which provide practical guidance for clinicians managing thyroid function in patients receiving ASM therapy.

Conclusions

Epilepsy is a chronic neurological condition that often necessitates long-term treatment with ASMs. While these therapies are essential for seizure control, they are not without systemic effects. Notably, older ASMs such as PB, PHT, CBZ, and VPA are associated with alterations in thyroid hormone metabolism in nearly one-third of

TABLE 21.3 Recommendations for monitoring of thyroid hormones.

Monitor TSH and FT4	
Baseline & every 12 months	Start of ASM therapy (all ASMs) Newer ASMs (LTG, LEV, TPM)
After 3 months	Switch or add-on of ASM Clinical manifestations suggestive of hypothyroidism
Every 6 months	Older ASMs (PB, PHT, CBZ, OXC, VPA) High risk for hypothyroidism - Combined therapy (especially VPA & CBZ) - High serum levels of ASMs - Coexisting hepatic or thyroid abnormality - Pregnant women - Adverse metabolic profile (metabolic syndrome) - Depression - Cognitive impairment

ASM = antiseizure medication; CBZ = carbamazepine; LEV = levetiracetam; LTG = lamotrigine; OXC = oxcarbazepine; PB = phenobarbital; PHT = phenytoin; TPM = topiramate; VPA = valproate; TSH = thyroid-stimulating hormone; FT4 = free thyroxine.

TABLE 21.4 Recommendations for treatment of thyroid hormone abnormalities.

	Children and adolescents	Patients <70 years	Patients >70–75 years
Low FT4/FT3 Normal TSH ≤ 5 mIU/L	Check thyroid function after 1 month. Consult endocrinologist in case of: - Symptoms suggestive for hypothyroidism or - Biochemical progression		
5 < TSH ≤ 10 mIU/L	Consider trial with levothyroxine in case of symptoms suggestive for hypothyroidism or low FT4. Consult with endocrinologist.		Consider observation unless symptoms suggestive for hypothyroidism or low FT4.
TSH > 10 mIU/L	Treatment with levothyroxine is recommended. Follow-up together with endocrinologist.		Consider treatment with levothyroxine in case of symptoms suggestive for hypothyroidism, low FT4 or high CV risk.

Guidelines are based on the American guidelines (American Thyroid Association, ATA, and American Association of Clinical Endocrinologists, AACE) as well as on the European guideline (European Thyroid Association, ETA). CV = cardiovascular; FT4 = free thyroxine; TSH = thyroid-stimulating hormone.
Note: patients already treated with levothyroxine prior to ASM therapy often need to increase their replacement dosage.

patients. The most frequently observed abnormalities include mild hypothyroidism—characterized by elevated TSH with normal FT4—and isolated hypothyroxinemia, where FT4 is reduced despite normal TSH levels.

Given the potential metabolic and endocrine implications of these changes, it is crucial for clinicians to adopt a proactive approach to thyroid function monitoring in patients receiving ASMs. This chapter provides a practical framework for such monitoring and outlines treatment considerations based on current endocrine guidelines.

However, significant knowledge gaps remain, particularly regarding the long-term effects of newer ASMs and the impact of polytherapy on thyroid function. Prospective studies are needed to better understand these relationships and to develop evidence-based strategies for prevention and intervention.

Mini-dictionary of terms

- **Antiseizure Medications.** A class of drugs used to prevent or reduce the frequency of epileptic seizures by modulating neuronal excitability.
- **Autoimmune Thyroiditis.** An immune-mediated condition that can be triggered or exacerbated by certain medications, affecting hormone balance.
- **Clinical Monitoring.** The ongoing assessment of patients for side effects and therapeutic efficacy, essential in managing ASM-related endocrine changes.

- **Cytochrome P450 Enzymes.** A family of liver enzymes involved in the metabolism of drugs and hormones, often affected by ASMs.
- **Drug-Hormone Interaction.** The influence of pharmacological agents on endocrine function, particularly relevant in chronic medication use.
- **Enzyme Induction.** A process by which certain drugs increase the activity of metabolic enzymes, potentially altering hormone levels.
- **Hormone Homeostasis.** The maintenance of stable hormone levels in the body, which can be disrupted by drug interactions.
- **Iodine Uptake.** The process by which iodine is absorbed and utilized for hormone synthesis, potentially disrupted by ASMs.
- **Lamotrigine.** An ASM with a relatively favorable endocrine profile, often used in long-term epilepsy management.
- **Neurodevelopmental Delay.** A condition where the development of the nervous system is slower than expected, possibly linked to hormonal imbalances.
- **Neuroendocrine Axis.** The complex interaction between the nervous system and endocrine glands that regulates physiological processes.
- **Pharmacokinetics.** The study of how drugs are absorbed, distributed, metabolized, and excreted in the body.
- **Sodium Channel Blockers.** A subclass of ASMs that inhibit sodium channels to reduce neuronal firing.
- **Valproate.** A widely used ASM known for its broad-spectrum efficacy and potential endocrine side effects.

Summary points

- Thyroid dysfunction is common in ASM-treated patients, particularly with older agents such as phenobarbital, phenytoin, carbamazepine, and valproate, affecting nearly one-third of individuals.
- The most frequent abnormalities include mild hypothyroidism (elevated TSH with normal FT4) and isolated hypothyroxinemia (low FT4 with normal TSH).
- Routine thyroid function monitoring is not yet standardized, but baseline and periodic testing is advisable, especially in high-risk groups (e.g., polytherapy, older ASMs, metabolic risk).
- Treatment decisions remain controversial, but levothyroxine (L-T4) may be considered in patients with TSH > 10 mIU/L, low FT4, or unfavorable metabolic profiles.
- Thyroid hormones influence neuronal excitability, and their role in epilepsy pathophysiology and treatment response warrants further investigation.
- The balance between seizure control and endocrine health is critical, requiring individualized care and interdisciplinary collaboration.

Artificial intelligence

During the preparation of this work the authors used Copilot from Microsoft 365 strictly to improve readability and language usage. Under no circumstances has it been used to draw insights from data, conclude or produce recommendations. After using this tool/service, the authors reviewed and edited the content as needed and take full responsibility for the content of the publication.

References

Attilakos, A., Dinopoulos, A., Tsirouda, M., Paschalidou, M., Prasouli, A., Stamati, A., Parasxou, N., & Garoufi, A. (2019). Effect of levetiracetam monotherapy on lipid profiles and thyroid hormones in children with epilepsy: A prospective study. *Epilepsy Research, 155*, Article 106162. https://doi.org/10.1016/j.eplepsyres.2019.106162

Attilakos, A., Garoufi, A., Voudris, K., Mastroyianni, S., Fotinou, A., Papadimitriou, D. T., Gavalakis, N., Prassouli, A., & Katsarou, E. (2007). Thyroid dysfunction associated with increased low-density lipoprotein cholesterol in epileptic children treated with carbamazepine monotherapy: A causal relationship? *European Journal of Paediatric Neurology, 11*(6), 358–361. https://doi.org/10.1016/j.ejpn.2007.03.002

Aygun, F., Ekici, B., Aydinli, N., Aydin, B. K., Bas, F., & Tatli, B. (2012). Thyroid hormones in children on antiepileptic therapy. *International Journal of Neuroscience, 122*(2), 69–73. https://doi.org/10.3109/00207454.2011.627486

Baumgartner, A., Campos-Barros, A., Gaio, U., Hessenius, C., Flechner, A., & Meinhold, H. (1994). Carbamazepine affects triiodothyronine production and metabolization in rat hippocampus. *Life Sciences, 54*(23), PL401–PL407. https://doi.org/10.1016/0024-3205(94)90123-6

Benedetti, M. S., Whomsley, R., Baltes, E., & Tonner, F. (2005). Alteration of thyroid hormone homeostasis by antiepileptic drugs in humans: Involvement of glucuronosyltransferase induction. *European Journal of Clinical Pharmacology, 61*(12), 863–872. https://doi.org/10.1007/s00228-005-0056-0

Cansu, A., Serdaroglu, A., Camurdan, O., Hirfanoglu, T., Bideci, A., & Gucuyener, K. (2006). The evaluation of thyroid functions, thyroid antibodies, and thyroid volumes in children with epilepsy during short-term administration of oxcarbazepine and valproate. *Epilepsia, 47*(11), 1855–1859. https://doi.org/10.1111/j.1528-1167.2006.00821.x

Castro-Gago, M., Novo-Rodriguez, M. I., Gomez-Lado, C., Rodriguez-Garcia, J., Rodriguez-Segade, S., & Eiris-Punal, J. (2007). Evolution of subclinical hypothyroidism in children treated with antiepileptic drugs. *Pediatric Neurology, 37*(6), 426–430. https://doi.org/10.1016/j.pediatrneurol.2007.07.003

Chaker, L., Bianco, A. C., Jonklaas, J., & Peeters, R. P. (2017). Hypothyroidism. *Lancet, 390*(10101), 1550–1562. https://doi.org/10.1016/S0140-6736(17)30703-1

Connell, J. M., Rapeport, W. G., Gordon, S., & Brodie, M. J. (1984). Changes in circulating thyroid hormones during short-term hepatic enzyme induction with carbamazepine. *European Journal of Clinical Pharmacology, 26*(4), 453–456. https://doi.org/10.1007/BF00542140

Eiris-Punal, J., Del Rio-Garma, M., Del Rio-Garma, M. C., Lojo-Rocamonde, S., Novo-Rodriguez, I., & Castro-Gago, M. (1999).

Long-term treatment of children with epilepsy with valproate or carbamazepine may cause subclinical hypothyroidism. *Epilepsia, 40*(12), 1761–1766. https://doi.org/10.1111/j.1528-1157.1999.tb01595.x

El-Farahaty, R. M., El-Mitwalli, A., Azzam, H., Wasel, Y., Elrakhawy, M. M., & Hasaneen, B. M. (2015). Atherosclerotic effects of long-term old and new antiepileptic drugs monotherapy: A cross-sectional comparative study. *Journal of Child Neurology, 30*(4), 451–457. https://doi.org/10.1177/0883073814551388

Ferrannini, E., Iervasi, G., Cobb, J., Ndreu, R., & Nannipieri, M. (2017). Insulin resistance and normal thyroid hormone levels: Prospective study and metabolomic analysis. *American Journal of Physiology. Endocrinology and Metabolism, 312*(5), E429–E436. https://doi.org/10.1152/ajpendo.00464.2016

Gallizzi, R., Crisafulli, C., Aversa, T., Salzano, G., De Luca, F., Valenzise, M., & Zirilli, G. (2018). Subclinical hypothyroidism in children: Is it always subclinical? *Italian Journal of Pediatrics, 44*(1), 25. https://doi.org/10.1186/s13052-018-0462-4

Garoufi, A., Koemtzidou, E., Katsarou, E., Dinopoulos, A., Kalimeraki, I., Fotinou, A., Drakatos, A., & Attilakos, A. (2014). Lipid profile and thyroid hormone concentrations in children with epilepsy treated with oxcarbazepine monotherapy: A prospective long-term study. *European Journal of Neurology, 21*(1), 118–123. https://doi.org/10.1111/ene.12262

Hamed, S. A., Hamed, E. A., Kandil, M. R., El-Shereef, H. K., Abdellah, M. M., & Omar, H. (2005). Serum thyroid hormone balance and lipid profile in patients with epilepsy. *Epilepsy Research, 66*(1–3), 173–183. https://doi.org/10.1016/j.eplepsyres.2005.08.004

Hegedus, L., Hansen, J. M., Luhdorf, K., Perrild, H., Feldt-Rasmussen, U., & Kampmann, J. P. (1985). Increased frequency of goitre in epileptic patients on long-term phenytoin or carbamazepine treatment. *Clinical Endocrinology (Oxford), 23*(4), 423–429. https://doi.org/10.1111/j.1365-2265.1985.tb01100.x

Hirfanoglu, T., Serdaroglu, A., Camurdan, O., Cansu, A., Bideci, A., Cinaz, P., & Gucuyener, K. (2007). Thyroid function and volume in epileptic children using carbamazepine, oxcarbazepine and valproate. *Pediatrics International, 49*(6), 822–826. https://doi.org/10.1111/j.1442-200X.2007.02456.x

Ilic, V., Bogicevic, D., Miljkovic, B., Jesic, M., Kovacevic, M., Prostran, M., & Kovacevic, S. V. (2016). Duration of valproic acid monotherapy correlates with subclinical thyroid dysfunction in children with epilepsy. *Epileptic Disorders, 18*(2), 181–186. https://doi.org/10.1684/epd.2016.0821

Isojarvi, J. I., Pakarinen, A. J., & Myllyla, V. V. (1992). Thyroid function with antiepileptic drugs. *Epilepsia, 33*(1), 142–148. https://doi.org/10.1111/j.1528-1157.1992.tb02297.x

Isojarvi, J. I., Pakarinen, A. J., Ylipalosaari, P. J., & Myllyla, V. V. (1990). Serum hormones in male epileptic patients receiving anticonvulsant medication. *Archives of Neurology, 47*(6), 670–676. https://doi.org/10.1001/archneur.1990.00530060082023

Izumi, T., & Fukuyama, Y. (1984). Influence of ACTH on serum hormone content and its anticonvulsant action towards infantile spasms. *Life Sciences, 34*(11), 1023–1028. https://doi.org/10.1016/0024-3205(84)90015-8

Jabbari, B., & Huott, A. D. (1980). Seizures in thyrotoxicosis. *Epilepsia, 21*(1), 91–96. https://doi.org/10.1111/j.1528-1157.1980.tb04048.x

Kafadar, I., Kilic, B. A., Arapoglu, M., Yalcin, K., & Dalgic, N. (2015). Evaluation of thyroid hormones in children receiving carbamazepine or valproate: A prospective study. *Journal of Child Neurology, 30*(1), 63–68. https://doi.org/10.1177/0883073814529822

Kim, S. H., Chung, H. R., Kim, S. H., Kim, H., Lim, B. C., Chae, J. H., Kim, K. J., Hwang, Y. S., & Hwang, H. (2012). Subclinical hypothyroidism during valproic acid therapy in children and adolescents with epilepsy. *Neuropediatrics, 43*(3), 135–139. https://doi.org/10.1055/s-0032-1313913

Kose, E., Guzel, O., Demir, K., & Arslan, N. (2017). Changes of thyroid hormonal status in patients receiving ketogenic diet due to intractable epilepsy. *Journal of Pediatric Endocrinology & Metabolism, 30*(4), 411–416. https://doi.org/10.1515/jpem-2016-0281

Langouche, L., Jacobs, A., & Van den Berghe, G. (2019). Nonthyroidal illness syndrome across the ages. *Journal of the Endocrine Society, 3*(12), 2313–2325. https://doi.org/10.1210/js.2019-00325

Lee, Y. J., Nam, S. O., Kim, K. M., Kim, Y. M., & Yeon, G. M. (2017). Longitudinal change in thyroid hormone levels in children with epilepsy on a ketogenic diet: Prevalence and risk factors. *Journal of Epilepsy Research, 7*(2), 99–105. https://doi.org/10.14581/jer.17015

Lossius, M. I., Tauboll, E., Mowinckel, P., & Gjerstad, L. (2009). Reversible effects of antiepileptic drugs on thyroid hormones in men and women with epilepsy: A prospective randomized double-blind withdrawal study. *Epilepsy and Behavior, 16*(1), 64–68. https://doi.org/10.1016/j.yebeh.2009.07.014

Mikati, M. A., Tarabay, H., Khalil, A., Rahi, A. C., El Banna, D., & Najjar, S. (2007). Risk factors for development of subclinical hypothyroidism during valproic acid therapy. *The Journal of Pediatrics, 151*(2), 178–181. https://doi.org/10.1016/j.jpeds.2007.02.046

Molteberg, E., Thorsby, P. M., Kverneland, M., Iversen, P. O., Selmer, K. K., Nakken, K. O., & Tauboll, E. (2020). Effects of modified atkins diet on thyroid function in adult patients with pharmacoresistant epilepsy. *Epilepsy and Behavior, 111*, Article 107285. https://doi.org/10.1016/j.yebeh.2020.107285

Moshe, S. L., Perucca, E., Ryvlin, P., & Tomson, T. (2015). Epilepsy: New advances. *Lancet, 385*(9971), 884–898. https://doi.org/10.1016/S0140-6736(14)60456-6

Nishiyama, M., Takami, Y., Ishida, Y., Tomioka, K., Tanaka, T., Nagase, H., Nakagawa, T., Tokumoto, S., Yamaguchi, H., Toyoshima, D., Maruyama, A., Nozu, K., Nishimura, N., & Iijima, K. (2019). Lipid and thyroid hormone levels in children with epilepsy treated with levetiracetam or carbamazepine: A prospective observational study. *Epilepsy and Behavior, 90*, 15–19. https://doi.org/10.1016/j.yebeh.2018.11.003

Oppenheimer, J. H., Fisher, L. V., Nelson, K. M., & Jailer, J. W. (1961). Depression of the serum protein-bound iodine level by diphenylhydantion. *The Journal of Clinical Endocrinology & Metabolism, 21*, 252–262. https://doi.org/10.1210/jcem-21-3-252

Paragliola, R. M., Prete, A., Kaplan, P. W., Corsello, S. M., & Salvatori, R. (2015). Treatment of hypopituitarism in patients receiving antiepileptic drugs. *The Lancet Diabetes & Endocrinology, 3*(2), 132–140. https://doi.org/10.1016/S2213-8587(14)70081-6

Peeters, R. P. (2017). Subclinical hypothyroidism. *New England Journal of Medicine, 376*(26), 2556–2565. https://doi.org/10.1056/NEJMcp1611144

Perucca, P., & Gilliam, F. G. (2012). Adverse effects of antiepileptic drugs. *The Lancet Neurology, 11*(9), 792–802. https://doi.org/10.1016/S1474-4422(12)70153-9

Perucca, P., Scheffer, I. E., & Kiley, M. (2018). The management of epilepsy in children and adults. *Medical Journal of Australia, 208*(5), 226–233. https://doi.org/10.5694/mja17.00951

Rochtus, A. M., Herijgers, D., Jansen, K., & Decallonne, B. (2022). Antiseizure medications and thyroid hormone homeostasis: Literature review and practical recommendations. *Epilepsia, 63*(2), 259–270. https://doi.org/10.1111/epi.17117

Roef, G., Lapauw, B., Goemaere, S., Zmierczak, H. G., Toye, K., Kaufman, J. M., & Taes, Y. (2012). Body composition and metabolic parameters are associated with variation in thyroid hormone levels among euthyroid young men. *European Journal of Endocrinology, 167*(5), 719–726. https://doi.org/10.1530/EJE-12-0447

Roos, A., Bakker, S. J., Links, T. P., Gans, R. O., & Wolffenbuttel, B. H. (2007). Thyroid function is associated with components of the metabolic syndrome in euthyroid subjects. *The Journal of Clinical Endocrinology & Metabolism, 92*(2), 491–496. https://doi.org/10.1210/jc.2006-1718

Sahu, J. K., Gulati, S., Kabra, M., Arya, R., Sharma, R., Gupta, N., Kaleekal, T., Reeta, K., & Gupta, Y. K. (2012). Evaluation of subclinical hypothyroidism in ambulatory children with controlled epilepsy on valproate monotherapy. *Journal of Child Neurology, 27*(5), 594–597. https://doi.org/10.1177/0883073811421985

Schmidt, D., & Elger, C. E. (2004). What is the evidence that oxcarbazepine and carbamazepine are distinctly different antiepileptic drugs? *Epilepsy and Behavior, 5*(5), 627–635. https://doi.org/10.1016/j.yebeh.2004.07.004

Shih, F. Y., Chuang, Y. C., Chuang, M. J., Lu, Y. T., Tsai, W. C., Fu, T. Y., & Tsai, M. H. (2017). Effects of antiepileptic drugs on thyroid hormone function in epilepsy patients. *Seizure, 48*, 7–10. https://doi.org/10.1016/j.seizure.2017.03.011

Simko, J., & Horacek, J. (2007). Carbamazepine and risk of hypothyroidism: A prospective study. *Acta Neurologica Scandinavica, 116*(5), 317–321. https://doi.org/10.1111/j.1600-0404.2007.00883.x

Song, T. J., Kim, S. J., Kim, G. S., Choi, Y. C., & Kim, W. J. (2010). The prevalence of thyrotoxicosis-related seizures. *Thyroid, 20*(9), 955–958. https://doi.org/10.1089/thy.2009.0276

Surks, M. I., & DeFesi, C. R. (1996). Normal serum free thyroid hormone concentrations in patients treated with phenytoin or carbamazepine. A paradox resolved. *Journal of the American Medical Association., 275*(19), 1495–1498.

Tamijani, S. M., Karimi, B., Amini, E., Golpich, M., Dargahi, L., Ali, R. A., Ibrahim, N. M., Mohamed, Z., Ghasemi, R., & Ahmadiani, A. (2015). Thyroid hormones: Possible roles in epilepsy pathology. *Seizure, 31*, 155–164. https://doi.org/10.1016/j.seizure.2015.07.021

Terranova, C., Rizzo, V., Calamuneri, A., Bartolone, L., Morgante, F., Bruschetta, D., Roberto, V., Girlanda, P., Benvenga, S., & Quartarone, A. (2016). Cortical excitability in patients with resistance to thyroid hormone compared to patients with hypothyroidism and euthyroid controls: A transcranial magnetic stimulation study. *Archives Italiennes de Biologie, 154*(2–3), 68–77. https://doi.org/10.12871/00039829201624

Tiihonen, M., Liewendahl, K., Waltimo, O., Ojala, M., & Valimaki, M. (1995). Thyroid status of patients receiving long-term anticonvulsant therapy assessed by peripheral parameters: A placebo-controlled thyroxine therapy trial. *Epilepsia, 36*(11), 1118–1125. https://doi.org/10.1111/j.1528-1157.1995.tb00470.x

Verrotti, A., Basciani, F., Morresi, S., Morgese, G., & Chiarelli, F. (2001). Thyroid hormones in epileptic children receiving carbamazepine and valproic acid. *Pediatric Neurology, 25*(1), 43–46. https://doi.org/10.1016/s0887-8994(01)00279-x

Verrotti, A., Laus, M., Scardapane, A., Franzoni, E., & Chiarelli, F. (2009). Thyroid hormones in children with epilepsy during long-term administration of carbamazepine and valproate. *European Journal of Endocrinology, 160*(1), 81–86. https://doi.org/10.1530/EJE-08-0325

Verrotti, A., Scardapane, A., Manco, R., & Chiarelli, F. (2008). Antiepileptic drugs and thyroid function. *Journal of Pediatric Endocrinology & Metabolism, 21*(5), 401–408. https://doi.org/10.1515/jpem.2008.21.5.401

Wiens, S. C., & Trudeau, V. L. (2006). Thyroid hormone and gamma-aminobutyric acid (GABA) interactions in neuroendocrine systems. *Comparative Biochemistry and Physiology Part A: Molecular & Integrative Physiology, 144*(3), 332–344. https://doi.org/10.1016/j.cbpa.2006.01.033

Yildiz, M., Simsek, G., Uzun, H., Uysal, S., Sahin, S., & Balci, H. (2010). Assessment of low-density lipoprotein oxidation, paraoxonase activity, and arterial distensibility in epileptic children who were treated with anti-epileptic drugs. *Cardiology in the Young, 20*(5), 547–554. https://doi.org/10.1017/S1047951110000697

Yilmaz, U., Yilmaz, T. S., Akinci, G., Korkmaz, H. A., & Tekgul, H. (2014). The effect of antiepileptic drugs on thyroid function in children. *Seizure, 23*(1), 29–35. https://doi.org/10.1016/j.seizure.2013.09.006

Chapter 22

Hypothyroidism in patients treated with immune checkpoint inhibitors

N. Laghmouch[1], I. Meeusen[2] and Th.B. Twickler[1,2]
[1]*Antwerp University, Edegem, Belgium;* [2]*AZ Monica, Deurne, Belgium*

List of abbreviations

BMI Body mass index
CTLA-4 Cytotoxic T-Lymphocyte Antigen 4
CXCL10 C-X-C motif chemokine ligand 10
CXCL9 C-X-C motif chemokine ligand 9
FDG Fluorodeoxyglucose
ICIs Immune checkpoint inhibitors
irAEs Immune-related adverse events
PD-1 Programmed Cell Death Protein 1
PDL-1 Programmed Cell Death Protein Ligand 1

Introduction

Immune checkpoint inhibitors (ICIs) have marked a significant milestone in cancer treatment, leading to a substantial increase in their use in everyday clinical practice over the past few years. As they are being used more widely in various tumor treatments, more clinicians are confronted with their notable side effects, with hypothyroidism being one of the most frequently reported (Fig. 22.1).

In this chapter, we will provide clinical information on the use of ICIs in actual practice, with particular focus on the prevalence and the mechanisms behind hypothyroidism. Finally, we offer options for the treatment of ICI-induced hypothyroidism.

ICI therapy, at a glance

Immune checkpoints are surface receptors on immune cells. Checkpoint proteins include Programmed Cell Death Protein 1 (PD-1) and Cytotoxic T-Lymphocyte Antigen 4 (CTLA-4) on T cells, as well as Programmed Cell Death Protein Ligand 1 (PDL-1) on tumor cells. ICI therapy refers to treatment with CTLA-4 inhibitors, PD-1 inhibitors, PDL-1 inhibitors, or a combination of those.

CTLA-4 inhibitors include Ipilimumab and Tremelimumab. In brief, blocking CTLA-4 enhances T cell responses against immunogenic tumor cells. However, in contrast, inhibition of CTLA-4 reduces immune self-tolerance, leading to autoimmunity (Rowshanravan et al., 2018). PD-1 inhibitors, such as Nivolumab, Pembrolizumab, and Cemiplimab, work through different mechanisms by decreasing the interaction between T-cells and tumor cells. Their modes of action are indicative of the side effect profiles they may produce. Many other checkpoint proteins have been discovered as potential targets for future oncology treatments. An updated overview is challenging due to the rapid pace of developments.

In recent years, ICIs have been acknowledged as very effective in the treatment of melanoma and non-small cell lung carcinoma. However, not all patients benefit equally, and organ-specific immune-related adverse events (irAEs), such as rashes or itching, colitis, diarrhea, hyperthyroidism, and hypothyroidism, may occur. Among endocrine irAEs, hypothyroidism is the most common (Hossain, 2024).

Epidemiology of ICI-induced hypothyroidism

Since the introduction of ICIs, the incidence and prevalence of hypothyroidism have risen. A comprehensive retrospective cohort study (TriNetX research network) spanning 2013 to 2024, including 39,749 patients, found that 14.2% of patients over 18 years old developed hypothyroidism within 12 months (Syed et al., 2025). Barroso-Souso et al. (2018) conducted a meta-analysis that reported an overall incidence of hypothyroidism of 6.6% linked to ICI therapy, with rates reaching about 13% when ICIs were combined (e.g., ipilimumab plus nivolumab). Patients treated with PD-1 inhibitors are at a significantly higher

Hormones in Neuroscience. https://doi.org/10.1016/B978-0-443-30261-9.00022-X

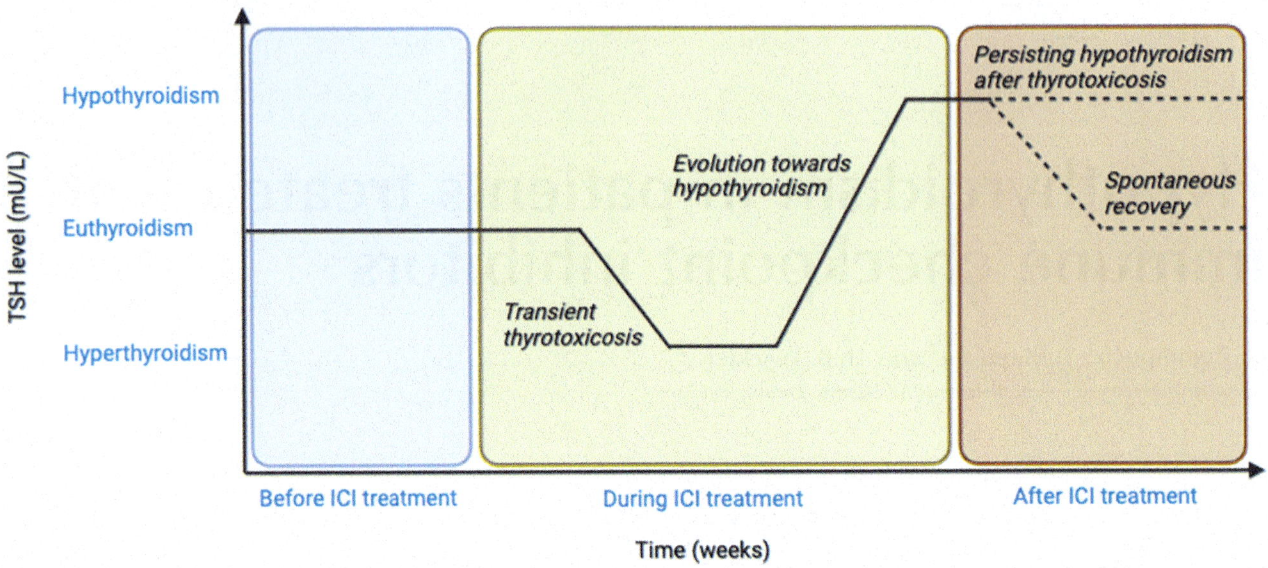

FIGURE 22.1 Time course and evolution of hypothyroidism following thyrotoxicosis. X-axis displays time, Y-axis depicts the TSH level, representing the thyroid function. Usually, the thyrotoxicosis starts 4–6 weeks after the first administration of ICI and lasts about 6 weeks (Iyer et al., 2018). Most often the following hypothyroidism persists. *Figure created using BioRender.com.*

risk of developing hypothyroidism compared to those receiving CTLA-4 inhibitors (odds ratio 1.89; 95% CI 1.17–3.05), as shown in Table 22.1.

Prediction of hypothyroidism along with ICI therapy

Multiple studies have aimed to identify underlying risk factors that could enhance the development of hypothyroidism. There is conflicting evidence regarding whether gender and age influence the risk of developing immune-related hypothyroidism. A retrospective study found that a higher body mass index (BMI) may increase the likelihood of developing endocrine-related immune-related adverse events (irAEs) (Elshafie et al., 2024). On the other hand, such a positive association between BMI and irAEs was not observed in other studies (Zhang et al., 2021). Gender and the prevalence of irAEs is another challenging topic. Female patients have a higher risk of developing irAEs during ICIs (Hu et al., 2022).

Biomarkers have been examined to predict irAEs. In a prospective study by Barlas et al. (2024), baseline serum CXCL10 levels, TSH levels, and thyroid antibody measurements (such as anti-TPO and thyroglobulin antibodies) could predict the likelihood of developing thyroid dysfunction when treated with ICIs. A recent meta-analysis by Quan et al. (2024) indicates that patients treated with ICIs who exhibit higher TSH levels (alongside positive anti-TPO and thyroglobulin antibodies) face a greater risk of adverse events. For example, each 1 mU/L increase in TSH was associated with a 1.3-fold higher risk of irAEs. No significant differences were observed regarding age and

TABLE 22.1 Predicted incidence of hypothyroidism in different types of ICI.

Type of ICI	Predicted incidence of hypothyroidism in % (95% prediction interval)
CTLA-4 inhibitor	3.8 (1.9–7.8)
PD-L1 inhibitor	3.9 (1.7–8.4)
PD-1 inhibitor	7.0 (3.9–12.3)
Combination of ipilimumab (CTLA-4 inhibitor) plus nivolumab (PD-1 inhibitor)	13.2 (6.9–23.8)

Data used from meta-analysis by Barroso-Souso et al. (2018). Created using BioRender.com.

TABLE 22.2 Overview of possible forms of hypothyroidism induced by ICI therapy.

ICI induced hypothyroidism		
Primary hypothyroidism		Secondary hypothyroidism
Direct immune mediated hypothyroidism	Hypothyroidism following thyrotoxicosis	ICI-related hypophysitis

Created using BioRender.com.

gender. Interestingly, PDL-1 did not increase the risk of hypothyroidism compared to PD-1 inhibitors in this meta-analysis. Overall, a patient's profile could be assessed before initiating ICIs to identify those at the highest risk of developing immune therapy-associated hypothyroidism.

Patients exhibiting intense fluorodeoxyglucose (FDG) uptake in the thyroid gland during tumor staging have a higher risk of developing hypothyroidism later (Galligan et al., 2023). This imaging marker can therefore contribute to the patient's risk profile, alongside BMI and thyroid antibodies.

Primary versus secondary hypothyroidism along with ICI therapy

First, it is essential to emphasize that there are two significant forms of hypothyroidism to consider during ICI therapy, as shown in Table 22.2.

The most common form of ICI-induced hypothyroidism is primary hypothyroidism, which originates in the thyroid gland, and the less common form is secondary hypothyroidism, which originates in the pituitary (as part of an ICI-related hypophysitis). The distinction between these two is essential in daily clinical practice because, in the case of secondary hypothyroidism, one should always consider a deficiency in other pituitary-related hormones. A cortisol deficiency may lead to potentially lethal conditions if not treated appropriately, especially when stress from, for instance, infections is present. Clearly, serum TSH and fT4 levels will help differentiate between primary and secondary hypothyroidism.

In the category of primary hypothyroidism, we can distinguish primary hypothyroidism through direct autoimmunity (from now on called "direct immune-mediated hypothyroidism") from hypothyroidism after initial thyrotoxicosis. Primary hypothyroidism during ICI therapy often follows destructive thyroiditis. In this case, it is hyperthyroidism that mostly marks the first step toward hypothyroidism due to local destruction of thyroid tissue and excessive release of thyroid hormone. Ongoing destruction of thyroid tissue later evolves to hypothyroidism. However, it is important to mention that the preceding thyrotoxicosis can remain undetected.

When an ICI-treated patient presents with mild thyrotoxicosis, TSH receptor antibodies may help differentiate direct immune-mediated hyperthyroidism from destructive thyroiditis induced by ICI (Li et al., 2024).

Mechanisms of destructive thyroiditis related to ICI therapy

As mentioned before, primary immune-mediated hypothyroidism is the most common form (Iyer et al., 2018). A recent comprehensive review by Mao et al. (2025) outlined three key features of thyroid tissue that make it susceptible to developing immune-driven hypothyroidism. The first feature is that thyroid-specific proteins, such as TPO (e.g., through anti-TPO antibodies), can act as autoantigens. Therefore, an abnormal release of these proteins following thyroid cell destruction may initiate an autoimmune response. Second, all structures within the thyroid (such as thyrocytes) and their extracellular matrix are highly immunogenic and thus highly vulnerable to destructive immune responses. ICI-enhanced tertiary lymphoid structures could further increase this sensitivity. Finally, thyroid follicular cells contribute to innate immunity by responding to pathogen- and damage-associated molecular patterns and by triggering the production of inflammatory cytokines. This localized pro-inflammatory response results in destruction and impairment of thyroid function.

ICI-induced thyroiditis mainly results from dysregulated T-cell activity because ICIs interfere with peripheral immune tolerance by blocking inhibitory pathways that normally suppress T-cell activation. This disruption leads to autoreactive $CD4^+$ and $CD8^+$ T-cell responses targeting thyroid antigens, causing inflammatory destruction of thyrocytes. Studies show increased infiltration of active T-cell subsets, including cytotoxic $CD8^+$ T cells and various helper T cells, along with upregulation of IFNγ-inducible chemokines like CXCL9 and CXCL10, which draw more T cells into the thyroid (Fig. 22.2) (Iyer et al., 2018). Unlike typical autoimmune thyroid disease, ICI-induced

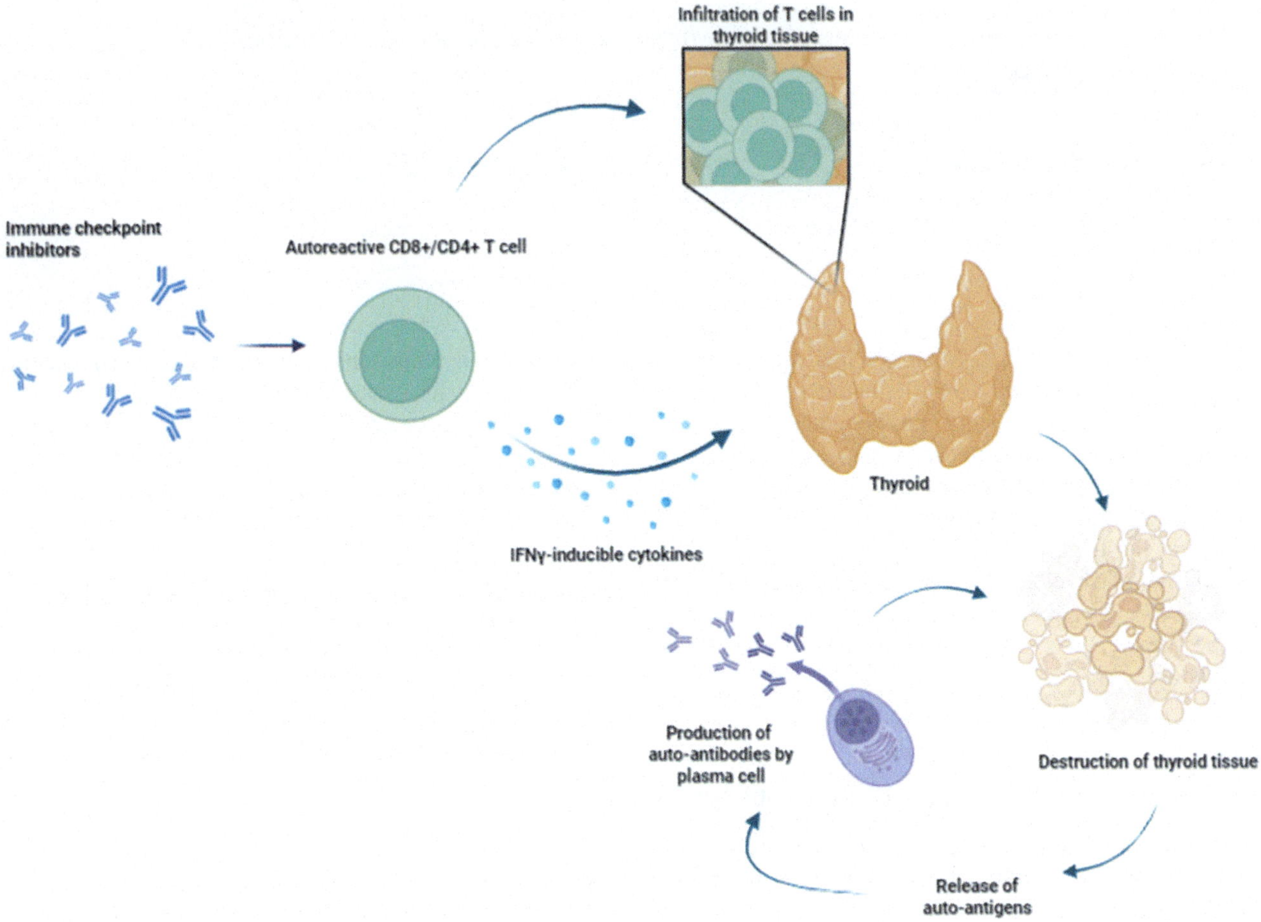

FIGURE 22.2 Mechanisms of destructive thyroiditis related to ICI therapy. There are three: (1) infiltration of autoreactive $CD8^+$ and $CD4^+$ T cells in thyroid tissue, (2) interferon gamma (IFNγ) induced cytokines, and (3) release of auto-antigens, which lead to the production of auto-antibodies. *Created using BioRender.com.*

thyroiditis often occurs without thyroid autoantibodies such as TPO-Abs, indicating a mainly T-cell–mediated process. Additional evidence involves natural killer (NK) cells and changes in PD-L1 and PD-L2 expression on thyroid cells, which worsen tissue damage. Genetic factors such as specific HLA-DR alleles and CTLA-4 polymorphisms may also increase risk (Yamauchi et al., 2017).

Screening for ICI-induced hypothyroidism

Due to the high rate of hypothyroidism in patients treated with ICIs as mentioned before, it is important to screen for hypothyroidism and start treatment early if appropriate. Symptoms of hypothyroidism are usually mild and nonspecific, and their clinical signs may go unnoticed within the overall clinical picture because of the resemblance to oncological symptoms (Duan et al., 2020).

Therefore, screening often includes biochemical analyses. Screening for baseline disturbed thyroid function should occur before ICI therapy is initiated. The European Society for Medical Oncology (ESMO) advises obtaining a baseline thyroid panel including TSH, fT4, and, if indicated, T3. Monitoring during treatment is recommended, too (Haanen et al., 2022). Various monitoring frameworks for hypothyroidism in patients after ICI treatment have been proposed by different endocrinology and oncology societies, depending on the type of ICI.

A TSH level exceeding 10 mU/L is classified as a clinically overt hypothyroidism. Subclinical primary hypothyroidism is defined as a TSH level between 4.2 and 10 mU/L and is usually not an indication to start thyroxine substitution. A low TSH and T4 (and T3) is defined as a secondary hypothyroidism (with its origin in the "inflamed" pituitary).

However, it remains uncertain whether the overall prognosis for non-thyroxine-treated patients with asymptomatic hypothyroidism under ICI therapy is poorer compared to those who receive prompt treatment. Accordingly, some studies have suggested that withholding thyroxine replacement in asymptomatic patients (especially within the range of subclinical hypothyroidism) will not harm the patient. Despite this observation, most health organizations recommend regular screening—at least every 4 weeks after beginning ICI therapy—centered on TSH levels (and T4) during the first 6 months. Because hypothyroidism may still develop after 6 months, regular and long-term screening afterward is highly recommended.

Treatment of ICI-induced hypothyroidism

Starting thyroxine replacement in primary ICI-induced hypothyroidism follows the protocols used for non-AE hypothyroidism. The European Endocrine Society recommends initiating thyroxine at 1 μg/kg of body weight daily in adults and 25 μg/day in the elderly (Husebye et al., 2022). Although no specific target TSH has been established for the ICI-treated population, the most common TSH range is 1–3 mU/L. Ideally, thyroxin should be taken 30 min before breakfast because it has a high affinity for dietary fiber (Wiesner et al., 2021). In oncology settings, malnutrition may be present, and iodide and selenium deficiency, for instance, could impair T4 synthesis. Iodide availability may be limited in regions with few iodide sources, such as seafood, and during seasons with low iodide intake, like winter. Therefore, as the World Health Organisation (n.d.) recommends, a daily intake of 100–150 μg of iodide should be considered in patients at risk for such deficiencies.

In cases of secondary hypothyroidism, cortisol deficiency should be ruled out first before starting thyroxine. Therefore, thyroxine treatment can begin once the adrenal condition is confirmed and hydrocortisone replacement has been given in cases of adrenal deficiency to prevent an Addison's crisis (Husebye et al., 2022).

Despite an optimal thyroxine dose (based on their weight), TSH levels may not reach target levels in a minority of ICI-treated patients. When comorbidities, such as heart disease, are present, maintaining an optimal TSH target is beneficial for reducing long-term morbidity and mortality. Alternatives to thyroxine therapy, such as T3 therapy, are rare but may be considered. In cases of active tumor disease, often associated with a high pro-inflammatory profile, there may be a significant impairment in the peripheral conversion of T4 to T3. T3 is the active thyroid hormone that binds to T3 receptors in the cells of functional tissues. The availability of T4 (and T3) might also be decreased due to low albumin levels, which can occur in catabolic states or malnutrition, thereby affecting the binding of T4 and T3 to circulating proteins such as albumin and thyroxin binding globulin. Indeed, a lower albumin level in ICI-treated patients is linked to a poorer prognosis. To overcome this potential therapeutic obstacle, which may be temporary, T3 monotherapy might be an option for highly vulnerable patients, although still considered experimental (Verelst et al., 2022). In most cases of ICI-induced hypothyroidism, ICI therapy can be continued.

Prognosis of ICI-induced hypothyroidism

Hypothyroidism after ICI therapy is usually not reversible (Iyer et al., 2018), although some cases may show spontaneous recovery of thyroid function (Kotwal et al., 2020).

Development of new-onset hypothyroidism in patients receiving ICIs should not be seen as entirely negative in the course of treatment for the patient and clinician. The onset of more autoimmunity could be considered a necessary side effect when ICIs are highly effective. Recent evidence suggests that the occurrence of thyroid irAEs is associated with improved overall survival and progression-free survival across various cancer types (Albertelli et al., 2024; Cheung et al., 2022). This does not hold for other irAEs, such as pneumonitis and hepatitis (Rong et al., 2024).

Clinical approach (Fig. 22.3)

The clinical relevance of the diagnosis and treatment of ICI-induced hypothyroidism

There are multiple reasons why it is crucial to diagnose and treat (if necessary) ICI-induced hypothyroidism, and those are consistent with expectations.

The patient's overall health and quality of life will be determined by all kinds of comorbidities, including hypothyroidism. One of the treatment goals for cancer patients treated with ICIs - many of whom are relatively young and active in society - is to restore their independence from healthcare and enable them to regain their normal lives. Hospitalizations must subsequently be avoided.

Symptoms of ICI-induced hypothyroidism are relatively non-specific and may resemble oncological or treatment-related symptoms (Hartmann, 2015). For instance, muscle loss (as part of catabolism) is recognized as a less favorable prognostic factor in patients receiving ICI treatment. Muscle quality loss could also be a consequence of ICI-related hypothyroidism, which could be reversible after thyroxine replacement therapy. Physical fitness and nutritional status are important factors associated with survival. An optimal thyroxin substitution empowers the patient in his efforts to achieve optimal physical

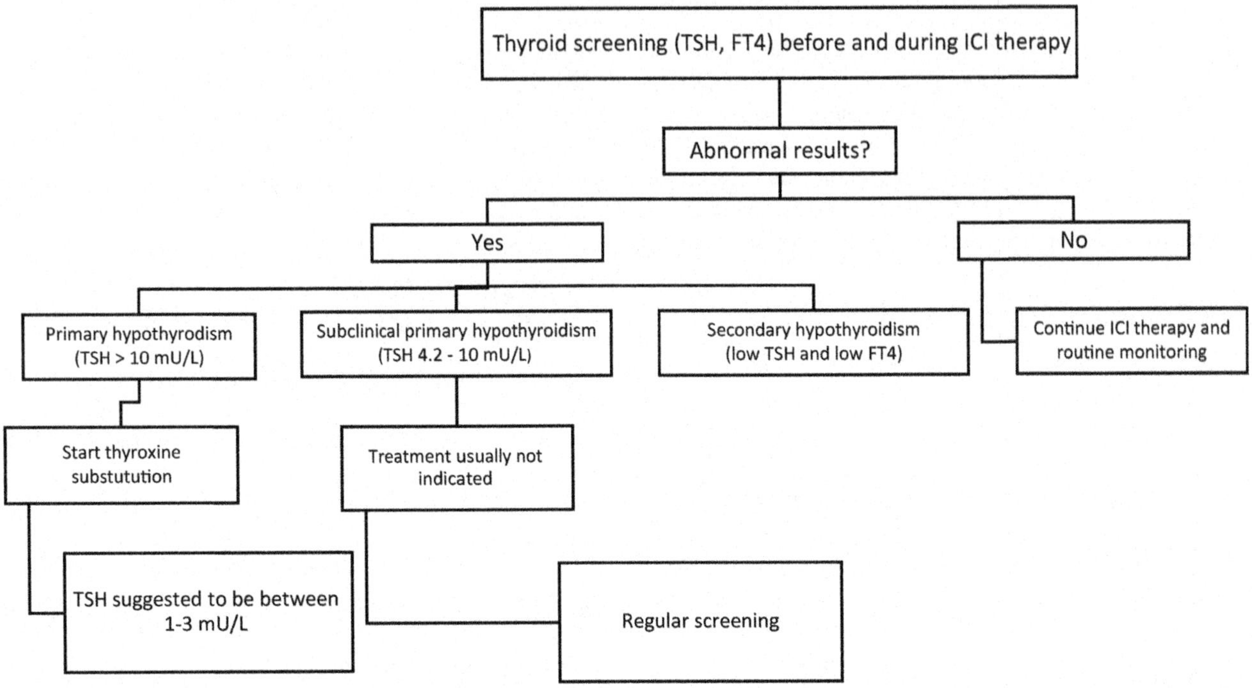

FIGURE 22.3 Suggestion for clinical approach.

fitness. If hypothyroidism is not recognized adequately during ICI treatment, symptoms may not be well delineated, with a possible doctor's delay.

Quality of life is another factor highlighting its relevance. Studies that investigate the impact of ICI-related hypothyroidism on quality of life in this population are lacking at the moment of writing this chapter. However, symptomatic hypothyroidism will certainly impact quality of life, irrespective of its cause. When comparing quality of life in patients with autoimmune thyroiditis (not ICI-induced) before and after 6 months of thyroxine substitution, there is a significant improvement across multiple quality-of-life domains (Winther et al., 2016).

Moreover, it is assumed that patients with ICI-induced hypothyroidism will tolerate the overall oncological treatment better when the co-existing hypothyroidism is treated. However, no blinded randomized trials actually exist proving this assumption.

Summary points

- The use of immune checkpoint inhibitors (ICIs) is rising, increasing the incidence of immune-related adverse events (irAEs).
- Thyroid dysfunction—especially hypothyroidism—is among the most common endocrine irAEs.
- ICI-induced hypothyroidism occurs as either primary hypothyroidism or the less frequent secondary hypothyroidism.
- Primary hypothyroidism results from direct thyroid gland impairment, whereas secondary hypothyroidism arises from pituitary dysfunction due to ICI-related hypophysitis.
- Primary hypothyroidism may present as direct immune-mediated disease or follow a transient thyrotoxic phase caused by destructive thyroiditis.
- Recognizing the distinct clinical patterns of ICI-induced hypothyroidism is essential for timely and accurate diagnosis.
- Management generally mirrors treatment of non-irAE hypothyroidism, relying on levothyroxine replacement therapy.

Conclusion

Hypothyroidism is a common adverse event in treatment with ICIs. As the use of ICIs is expected to increase, the occurrence of ICI-induced hypothyroidism will continue to rise. The risk of developing ICI-induced hypothyroidism is highest when patients are treated with the combination of CTLA-4 inhibitors and PD-1 inhibitors, compared to PD-1 inhibitors alone. The risk is the lowest when CTLA-4 inhibitors are used as monotherapy. Patient characteristics (e.g., BMI, gender, age) are associated with an increased risk of developing ICI-induced hypothyroidism; however, more detailed research on this topic is still needed. Biomarkers such as TSH, thyroid antibodies, and FGD uptake in the thyroid gland (beyond tumor tissue) predict the risk of developing ICI-related hypothyroidism. The distinction between primary and secondary hypothyroidism is essential in daily clinical practice, as the treatment strategy may differ. Clinically overt hypothyroidism in patients under and after ICI treatment demands for thyroxin substitution to improve their physical condition and quality of life. Whereas in the presence of subclinical hypothyroidism, this discussion still exists. The presence of hypothyroidism in patients after ICI therapy predicts the efficiency of ICI therapy and thereby the patient's tumor regression and survival.

Mini dictionary

Immune checkpoint inhibitor	A type of medication (also called immunotherapy) used in the field of oncology which can enhance the immune response against cancer cells.
Immune checkpoint inhibitor induced primary hypothyroidism	Hypothyroidism because of auto-immune processes in the thyroid gland itself.
Immune checkpoint inhibitor induced secondary hypothyroidism	Hypothyroidism that does not originate in the thyroid gland, but in the pituitary (in this context, caused by immune-related hypophysitis).
Immune-related adverse events	Adverse events caused by immunotherapy which stimulates the immune system and therefore can cause inflammation in different organ systems.

References

Albertelli, M., Rossi, G., Nazzari, E., Genova, C., Biello, F., Rijavec, E., Dal Bello, M. G., Patti, L., Tagliamento, M., Barletta, G., Morabito, P., Boschetti, M., Dotto, A., Campana, D., Ferone, D., & Grossi, F. (2024). Improved overall survival in patients developing endocrine toxicity during treatment with nivolumab for advanced non-small cell lung cancer in a prospective study. *Journal of Endocrinological Investigation, 47*(7), 1805–1814. https://doi.org/10.1007/s40618-023-02268-0

Barlas, T., Sutcuoglu, O., Akdogan, O., Balos Toruner, F., Akturk, M., Ozdemir, N., Yazici, O., & Eroglu Altinova, A. (2024). Endocrine immune-related adverse events by immune checkpoint inhibitors and potential predictive markers: A prospective study from a single tertiary center. *Endocrine-Related Cancer, 31*(11), Article e240101. https://doi.org/10.1530/ERC-24-0101

Barroso-Sousa, R., Barry, W. T., Garrido-Castro, A. C., Hodi, F. S., Min, L., Krop, I. E., & Tolaney, S. M. (2018). Incidence of endocrine dysfunction following the use of different immune checkpoint inhibitor regimens: A systematic review and meta-analysis. *JAMA Oncology, 4*(2), 173–182. https://doi.org/10.1001/jamaoncol.2017.3064

Cheung, Y. M., Wang, W., McGregor, B., & Hamnvik, O. R. (2022). Associations between immune-related thyroid dysfunction and efficacy of immune checkpoint inhibitors: A systematic review and meta-analysis. *Cancer Immunology, Immunotherapy, 71*(8), 1795–1812. https://doi.org/10.1007/s00262-021-03128-7

Duan, L., Wang, L., Wang, H., Si, X., Zhang, L., Liu, X., Li, Y., Guo, X., Zhou, J., Zhu, H., & Zhang, L. (2020). Clinical diagnosis and treatment of immune checkpoint inhibitors-related endocrine dysfunction. *Thoracic Cancer, 11*(4), 1099–1104. https://doi.org/10.1111/1759-7714.13347

Elshafie, O., Khalil, A. B., Salman, B., Atabani, A., & Al-Sayegh, H. (2024). Immune checkpoint inhibitors-induced endocrinopathies: Assessment, management and monitoring in a comprehensive cancer centre. *Endocrinology, Diabetes & Metabolism, 7*(4), Article e00505. https://doi.org/10.1002/edm2.505

Galligan, A., Wallace, R., Krishnamurthy, B., Kay, T. W. H., Sachithanandan, N., Chiang, C., Sandhu, S., Hicks, R. J., & Iravani, A. (2023). Increased thyroidal activity on routine FDG-PET/CT after combination immune checkpoint inhibition: Temporal associations with clinical and biochemical thyroiditis. *Cancers, 15*(24), 5803. https://doi.org/10.3390/cancers15245803

Haanen, J., Obeid, M., Spain, L., Carbonnel, F., Wang, Y., Robert, C., Lyon, A. R., Wick, W., Kostine, M., Peters, S., Jordan, K., Larkin, J., & ESMO Guidelines Committee. (2022). Management of toxicities from immunotherapy: ESMO clinical practice guideline for diagnosis, treatment and follow-up. *Annals of Oncology, 33*(12), 1217–1238. https://doi.org/10.1016/j.annonc.2022.10.001

Hartmann, K. (2015). Thyroid disorders in the oncology patient. *Journal of the Advanced Practitioner in Oncology, 6*(2), 99–106. https://doi.org/10.6004/jadpro.2015.6.2.2

Hossain, A. (2024). A comprehensive review of immune checkpoint inhibitors for cancer treatment. *International Immunopharmacology, 143*(Pt 2), Article 113365. https://doi.org/10.1016/j.intimp.2024.113365

Hu, W., Qian, X., Wang, S., Gao, L., Xu, J., & Yan, J. (2022). Sex: A potential factor affecting immune checkpoint inhibitor therapy for cancers. *Frontiers in Immunology, 13*, Article 1024112. https://doi.org/10.3389/fimmu.2022.1024112

Husebye, E. S., Castinetti, F., Criseno, S., Curigliano, G., Decallonne, B., Fleseriu, M., Higham, C. E., Lupi, I., Paschou, S. A., Toth, M., van der Kooij, M., & Dekkers, O. M. (2022). Endocrine-related adverse conditions in patients receiving immune checkpoint inhibition: An ESE clinical practice guideline. *European Journal of Endocrinology, 187*(6), G1–G21. https://doi.org/10.1530/EJE-22-0689

Iyer, P. C., Cabanillas, M. E., Waguespack, S. G., Hu, M. I., Thosani, S., Lavis, V. R., Busaidy, N. L., Subudhi, S. K., Diab, A., & Dadu, R. (2018). Immune-related thyroiditis with immune checkpoint inhibitors. *Thyroid, 28*(10), 1243–1251. https://doi.org/10.1089/thy.2018.0116

Kotwal, A., Kottschade, L., & Ryder, M. (2020). PD-L1 inhibitor-induced thyroiditis is associated with better overall survival in cancer patients. *Thyroid, 30*(2), 177–184. https://doi.org/10.1089/thy.2019.0250

Li, L., Ding, X., Zhang, X., Kong, S., & Chen, M. (2024). Clinical significance of thyroid autoantibodies in differential diagnosis and predicting the course of programmed cell death protein-1 inhibitor-induced thyroid dysfunction. *Endocrine Practice, 30*(12), 1166–1170. https://doi.org/10.1016/j.eprac.2024.08.012

Mao, X., Mao, C., Liu, J., Wang, X., & Mao, Y. (2025). Immune checkpoint inhibitor-induced thyroiditis and its potential mechanisms. *Frontiers in Endocrinology, 16*, Article 1584675. https://doi.org/10.3389/fendo.2025.1584675

Quan, L., Liu, J., Wang, Y., Yang, F., Yang, Z., Ju, J., Shuai, Y., Wei, T., Yue, J., Wang, X., Meng, J., & Yuan, P. (2024). Exploring risk factors for endocrine-related immune-related adverse events: Insights from meta-analysis and Mendelian randomization. *Human Vaccines & Immunotherapeutics, 20*(1), Article 2410557. https://doi.org/10.1080/21645515.2024.2410557

Rong, Y., Ramachandran, S., Bhattacharya, K., Yang, Y., Earl, S., Chang, Y., & Bentley, J. P. (2024). The association between toxicity and efficacy of immune checkpoint inhibitors in older adults with NSCLC. *Immunotherapy, 16*(16–17), 1057–1068. https://doi.org/10.1080/1750743X.2024.2394382

Rowshanravan, B., Halliday, N., & Sansom, D. M. (2018). CTLA-4: A moving target in immunotherapy. *Blood, 131*(1), 58–67. https://doi.org/10.1182/blood-2017-06-741033

Syed, F. F., Jafry, B., Kamran, A., & Collins, J. (2025). Clinical determinants and outcomes of hypothyroidism in immune checkpoint inhibitor (ICI) therapy. *Journal of Clinical Oncology, 43*(16 Suppl. l), Article e14623. https://doi.org/10.1200/JCO.2025.43.16_suppl.e14623

Verelst, F. R., Beyens, M. M. J., Vandenbroucke, E., Forceville, K., & Twickler, M. T. B. (2022). A decrease in peripheral thyroid hormone conversion efficiency in patients treated with immune checkpoint inhibitors and L-T3 as a possible alternative therapeutic escape option. *European Journal of Clinical Investigation, 52*(7), Article e13790. https://doi.org/10.1111/eci.13790

Wiesner, A., Gajewska, D., & Paśko, P. (2021). Levothyroxine interactions with food and dietary supplements: A systematic review. *Pharmaceuticals, 14*(3), 206. https://doi.org/10.3390/ph14030206

Winther, K. H., Cramon, P., Watt, T., Bjorner, J. B., Ekholm, O., Feldt-Rasmussen, U., Groenvold, M., Rasmussen, Å. K., Hegedüs, L., & Bonnema, S. J. (2016). Disease-specific as well as generic quality of life is widely impacted in autoimmune hypothyroidism and improves during the first six months of levothyroxine therapy. *PLoS One, 11*(6), Article e0156925. https://doi.org/10.1371/journal.pone.0156925

World Health Organization. (n.d.). Iodine deficiency. https://www.who.int/data/nutrition/nlis/info/iodine-deficiency.

Yamauchi, I., Sakane, Y., Fukuda, Y., Fujii, T., Taura, D., Hirata, M., Hirota, K., Ueda, Y., Kanai, Y., Yamashita, Y., Kondo, E., Sone, M., Yasoda, A., & Inagaki, N. (2017). Clinical features of nivolumab-induced thyroiditis: A case series study. *Thyroid, 27*(7), 894–901. https://doi.org/10.1089/thy.2016.0562

Zhang, D., Shah, N. J., Cook, M., Blackburn, M., Serzan, M. T., Advani, S., Potosky, A. L., Atkins, M. B., & Braithwaite, D. (2021). Association between body mass index and immune-related adverse events (irAEs) among advanced-stage cancer patients receiving immune checkpoint inhibitors: A pan-cancer analysis. *Cancers, 13*(23), 6109. https://doi.org/10.3390/cancers13236109. Erratum published in Cancers, 14(18), 4525.

Section 6

Diseases of the thyroid or thyroid status and their impact on the brain and behavior

Chapter 23

Focused diseases of the thyroid gland: Graves' hyperthyroidism

Michele Marinò and Elena Sabini
Department of Clinical and Experimental Medicine, Endocrinology Unit I, University of Pisa, Pisa, Italy

List of abbreviations

ACTH Adrenocorticotropic Hormone
ATD Antithyroid Drugs
BAFF B-cell Activating Factor
BMD Bone Mineral Density
cAMP Cyclic AMP
CD25 Cluster of Differentiation 25
CD40 Cluster of Differentiation 40
CD40L Cluster of Differentiation 40 Ligand
CNS Central Nervous System
CRH Corticotropin-Releasing Hormone
CTLA4 Cytotoxic T Lymphocyte–Associated Antigen 4
D2 Type 2 Deiodinase
D3 Type 3 Deiodinase
FcRn Neonatal Fc Receptor
FSH Follicle-Stimulating Hormone
GAG Glycosaminoglycans
GD Graves' Disease
GH Graves' Hyperthyroidism
HLA-DR3 Human Leukocyte Antigen-DR3
HPA Hypothalamic–Pituitary–Adrenal
HPG Hypothalamic–Pituitary–Gonadal
IFN-γ Interferon-γ
IGF-1 Insulin-Like Growth Factor-1
IGF-1R Insulin-Like Growth Factor-1 Receptor
IgG Immunoglobulin G
IL-12 Interleukin-12
IL-1β Interleukin-1β
IL-6 Interleukin-6
IR Insulin Resistance
LH Luteinizing Hormone
MCT8 Monocarboxylate Transporter 8
OATP1C1 Organic Anion Transporting Polypeptide 1C1
OF Orbital Fibroblasts
RAI Radioactive Iodine
SHBG Sex Hormone-Binding Globulin
T3 Triiodothyronine
T4 Thyroxine
TED Thyroid Eye Disease
Tg Thyroglobulin
TH Thyroid Hormones
TNF-α Tumor Necrosis Factor-α
TRAb TSH Receptor Antibody
TRs Thyroid Hormone Receptors
TSH-R Thyrotropin Receptor

Mini-dictionary of terms

Apathetic hyperthyroidism: A clinical presentation of hyperthyroidism, typically observed in older individuals, characterized by the absence of classic hyperadrenergic symptoms. Instead of anxiety, tremor, and weight loss, patients may present with fatigue, apathy, depression, or cognitive impairment, often leading to delayed diagnosis despite significant TH excess.

FcRn: A receptor that regulates IgG homeostasis by recycling immunoglobulins and extending their circulating half-life, contributing to the persistence of pathogenic autoantibodies in autoimmune diseases.

Immune tolerance: Physiological mechanisms that prevent autoreactive immune cells from attacking self-antigens. Loss of immune tolerance is a key mechanism underlying autoimmune diseases.

Iodothyronine deiodinases: A family of intracellular enzymes that regulate local hormone availability by activating or inactivating circulating precursors. Their tissue-specific expression allows local modulation of hormonal signaling independently of systemic levels.

TRAb: Autoantibodies against the thyrotropin receptor that may exert stimulating, blocking, or neutral effects. In GD, stimulating TRAb predominate, drive TH excess, and provide diagnostic and prognostic information.

Hormones in Neuroscience. https://doi.org/10.1016/B978-0-443-30261-9.00023-1

Summary points

- GH is a systemic autoimmune endocrine disorder that affects multiple target tissues beyond the thyroid gland.
- The CNS is a sensitive target of endocrine and immune dysregulation.
- Stimulatory autoantibodies drive sustained receptor activation, leading to widespread hormonal and tissue-specific effects.
- TH excess alters neuronal metabolism, neurotransmitter signaling, synaptic plasticity, and network excitability.
- Immune mechanisms contribute to CNS alterations independent of circulating TH levels.
- Neuropsychiatric symptoms may persist despite restoration of biochemical euthyroidism.

Introduction

GH as a systemic endocrine disorder

GH is the most common cause of autoimmune hyperthyroidism with a global prevalence of 2% in women and 0.5% in men (Taylor et al., 2018). Other causes of hyperthyroidism and thyrotoxicosis include toxic nodules and the thyrotoxic phase of thyroiditis. The disease is caused by the production of stimulatory autoantibodies against the TSH-R, resulting in sustained and inappropriate TH synthesis and secretion (Lanzolla et al., 2024).

While the thyroid gland is the primary target of the autoimmune process, the clinical and biological consequences of GH extend well beyond the thyroid itself.

TH exert pleiotropic effects on virtually all tissues, reflecting their fundamental role in regulating cellular metabolism, energy expenditure, and physiological homeostasis. The CNS represents a major target of TH action, and alterations in TH availability are associated with significant changes in neural function and behavior.

Systemic consequences of hyperthyroidism

TH actions are mediated by nuclear receptors widely expressed in peripheral tissues and in the CNS. Sustained TH excess disrupts metabolic, cardiovascular, and neuroregulatory homeostasis, leading to a hypermetabolic state and multisystem dysfunction.

Clinically, GH manifests through both peripheral and central effects. While weight loss, heat intolerance, and palpitations reflect enhanced peripheral metabolism, neuropsychiatric symptoms such as anxiety, cognitive impairment, and sleep disturbance indicate altered central regulation and may persist despite biochemical euthyroidism.

Neuroendocrine relevance of GH

The CNS is particularly sensitive to changes in TH availability. TH influence brain glucose utilization, neurotransmitter turnover, synaptic plasticity, and network excitability, thus modulating mood, cognition, and behavior. In GH, sustained increases in circulating TH levels are associated with functional alterations in neural circuits. This neuroendocrine interaction provides a conceptual framework for understanding why neurological and neuropsychiatric features are common and clinically relevant components of the disease.

Rationale and scope of the chapter

The aim of this chapter is to approach GH as a systemic autoimmune endocrine disorder, with particular emphasis on its neuroendocrine implications. The following sections examine the immunopathogenesis of GH, the systemic consequences of TH excess, and the impact of the disease on CNS structure and function. Additional contributions from other hormonal systems and signaling pathways are also discussed. Finally, the chapter highlights the translational relevance of GH as a model of endocrine-immune-neural interaction.

Epidemiology and clinical features of GH

Epidemiology

GH is the most frequent cause of hyperthyroidism and represents the leading autoimmune thyroid disorder associated with TH excess. In iodine-sufficient regions, the prevalence of GH in the general population is commonly estimated to range between 0.5% and 1%, with a pronounced female predominance and a female-to-male ratio of approximately 5–10:1. From an epidemiological perspective, GH accounts for roughly 60%–80% of all cases of hyperthyroidism, making it the principal cause of thyrotoxicosis in adult populations. The annual incidence of GH in the general population has been estimated at approximately 20–50 new cases per 100,000 individuals. Although the disease can develop at any age, it most frequently presents during early and middle adulthood, with a peak incidence between the third and fifth decades of life (Lee & Pearce, 2023; Smith & Hegedüs, 2016; Taylor et al., 2018).

The development of GH reflects the combined contribution of genetic susceptibility and environmental influences. Genetic loci associated with disease risk include genes encoding Tg, TSH-R, HLA-DRβ-Arg74, PTPN22, CTLA4, CD25, and CD40, supporting a polygenic architecture underlying disease susceptibility (Brix et al., 2001; Tomer, 2014).

Beyond genetic susceptibility, environmental factors such as iodine intake, tobacco exposure, infections, and psychosocial stress have been implicated in disease onset and progression (Gómez-Roig et al., 2021). In addition, epigenetic alterations, particularly changes in DNA methylation affecting genes involved in autoimmunity and T-cell signaling, including the thyrotropin receptor, have been increasingly recognized (Limbach et al., 2016; Tomer, 2014).

Clinical presentation

The clinical presentation of GH reflects the sustained excess of circulating TH and their effects on multiple organ systems. Common symptoms include weight loss, heat intolerance, excessive sweating, palpitations, and fatigue, although symptom severity and combination vary widely and may change over time.

Cardiovascular involvement represents a major source of morbidity, particularly in older individuals or in those with pre-existing heart disease. Tachycardia and increased cardiac output are frequently observed, and atrial fibrillation occurs in more than 10% of patients aged 60 years or older, while remaining uncommon in younger individuals (Nordyke et al., 1988). Neuromuscular manifestations, including tremor, proximal muscle weakness, hyperreflexia, restlessness, and reduced exercise tolerance, are also common and reflect combined peripheral and central effects of TH excess. Gastrointestinal symptoms, such as increased bowel frequency and dyspepsia, may further impair quality of life and typically improve after restoration of euthyroidism (Gunsar et al., 2003).

In older individuals, GH may present with fewer overt hyperadrenergic features, a pattern known as "apathetic" hyperthyroidism, characterized by nonspecific symptoms such as weight loss or fatigue and associated with delayed diagnosis (Boelaert et al., 2010; Dumitriu & Ursu, 1985).

Thyroid enlargement is a typical feature, particularly in younger patients, whereas less than half of older individuals exhibit a palpable goiter (Nordyke et al., 1988). The gland is usually diffusely enlarged, reflecting hyperplasia and increased vascularity, although nodular features may coexist in iodine-deficient areas. In some cases, thyroid enlargement may cause compressive symptoms, including dysphagia, orthopnea, or voice change.

Extra-thyroidal manifestations

The most common and clinically relevant extra-thyroidal manifestation of GH is TED. Less frequent manifestations include thyroid dermopathy (pretibial myxedema), and thyroid acropachy.

TED affects approximately 25% of patients and may occur before, concurrently with, or independently of thyroid dysfunction. Thyroid dermopathy is observed in 1%–2% of cases, usually in association with TED, whereas acropachy is rare (<1%) and typically coexists with both TED and dermopathy (Bartalena & Fatourechi, 2014).

Extra-thyroidal manifestations are believed to result from autoimmune responses directed against antigens shared by the thyroid and extrathyroidal tissues (Smith, 2025).

In TED, clinical features arise from increased orbital tissue volume due to OF activation, adipose tissue expansion, inflammatory enlargement of extraocular muscles, and GAG accumulation, leading to proptosis, ocular discomfort, and diplopia (Wiersinga et al., 2025).

Thyroid dermopathy is characterized by localized skin thickening, most commonly in the pretibial region. Thyroid acropachy represents the rarest manifestation and is characterized by digital clubbing, soft tissue swelling, and periosteal bone changes. Although typically painless, acropachy may lead to functional impairment in advanced cases and almost always occurs in conjunction with both TED and thyroid dermopathy (Fatourechi et al., 2002).

Disease management

Current therapeutic strategies for GH, including ATD, RAI, and thyroidectomy, are effective in controlling thyrotoxicosis and improving clinical symptoms, but differ substantially in long-term disease control, relapse risk, and overall clinical outcomes.

ATD are commonly used as first-line therapy; a standard 12–18-month course achieves sustained remission in ~50% of patients, while prolonged low-dose regimens may improve long-term disease control in selected cases (Azizi, 2020; Kahaly, 2020; Mohlin et al., 2014).

Definitive therapies are indicated in relapsing disease, intolerance to ATD, or when rapid control is required. RAI markedly reduces relapse risk but frequently results in permanent hypothyroidism. Thyroidectomy provides immediate and definitive disease control and is preferred in patients with large goiters, suspected malignancy, or planned pregnancy, although surgical risks must be considered (Bartalena, 2013).

Beyond conventional therapies, emerging strategies targeting immune pathways, such as CD20, CD40¯CD40L signaling, BAFF, FcRn, HLA-DR3-related mechanisms, and direct modulation of TSH-R signaling, are under investigation and may provide future disease-modifying strategies (Lanzolla et al., 2024; Viola et al., 2025).

Disease course and prognosis

The clinical course of GH is highly heterogeneous and influenced by several factors, including age at onset, sex, smoking status, and the severity and duration of TH excess. Disease expression ranges from rapidly progressive forms to insidious courses with mild or fluctuating symptoms, and relapsing–remitting patterns are common. Overt hyperthyroidism is associated with increased mortality compared with euthyroidism, with population-based studies report a 35% to over 300% increase in all-cause mortality depending on disease severity and duration (Dekkers et al., 2017; Lee & Pearce, 2023). Cardiovascular mortality is increased by approximately 20%, and is primarily driven by endothelial dysfunction, hypercoagulability, and arrhythmogenic effects associated with sustained TH excess (Sohn et al., 2020). Mortality risk correlates strongly with cumulative time spent in the hyperthyroid state, largely independent of treatment modality (Lillevang-Johansen et al., 2017).

Long-term studies highlight the unpredictable trajectory of GH. After ATD, durable remission is achieved in only a subset of patients, while others experience relapse or require definitive treatment. Younger age at diagnosis, large goiter size, persistently elevated TRAb levels, smoking, and thyroid eye disease are consistently associated with increased relapse risk (Shi et al., 2020; Vos et al., 2016). Nearly half of patients developed one or more additional autoimmune conditions over time, supporting the concept of GH as part of a broader autoimmune diathesis (Meling Stokland et al., 2024).

Pathogenesis of GH

Autoantibody-mediated thyroid stimulation in GH

GH is a prototypical autoimmune endocrine disorder resulting from a breakdown of immune tolerance to thyroid antigens in genetically susceptible individuals. Disease onset reflects the interaction between genetic predisposition and environmental factors, leading to a sustained autoimmune response directed primarily against the TSH-R. The hallmark of GH is the production of stimulating IgG autoantibodies directed against the TSH-R. These antibodies activate the receptor, uncoupling TH synthesis and secretion from physiological pituitary regulation. Persistent activation of cAMP–dependent signaling pathways promotes increased iodine uptake, Tg synthesis, TH release, and thyroid growth, thereby driving hyperthyroidism and goiter development. The pathogenic relevance of these autoantibodies is supported by the induction of hyperthyroidism following passive immunoglobulin transfer and by transplacental passage of TSH-R antibodies causing transient neonatal disease (Radetti et al., 1999).

Adaptive immune dysregulation in GH

Adaptive immune dysfunction is central to the pathogenesis of GH and involves coordinated activation of both T and B lymphocytes. Autoreactive $CD4^+$ T helper cells recognizing TSH-R-derived peptides escape central and peripheral tolerance and provide cognate help to B lymphocytes within the thyroid gland. Infiltrating immune cells generate a pro-inflammatory microenvironment through the release of cytokines and costimulatory signals, sustaining local immune activation and amplifying autoantibody production. Immune polarization appears dynamic, with Th1-associated responses predominating during active disease and a relative shift toward humoral immunity in later or inactive stages (Gallo et al., 2020).

The TSH receptor–IGF-1 receptor signaling axis in GH

Beyond TSH-R, increasing evidence supports a pathogenic role for IGF-1R in GH, particularly in mediating extrathyroidal manifestations. Functional crosstalk between TSH-R and IGF-1R results in a signaling complex that amplifies downstream inflammatory and metabolic pathways.

IGF-1R-expressing fibroblasts are overrepresented within connective tissues in GH and, upon interaction with disease-specific Ig, produce T-cell chemoattractants such as IL-16 and RANTES. This response is dependent on intact IGF-1R signaling and supports lymphocyte recruitment to target tissues (Pritchard et al., 2003). In parallel, increased and sustained IGF-1R expression is observed in both T and B lymphocytes, particularly within memory cell compartments, where IGF-1 signaling enhances cell survival, expansion, and immunoglobulin production (Douglas et al., 2008).

Fibroblast-driven tissue remodeling in extrathyroidal manifestations of GH

Extrathyroidal manifestations arise from immune-mediated activation of tissue-resident fibroblasts. In the orbit and skin, fibroblast subsets respond to immune stimulation by producing GAG and differentiating into adipocytes or myofibroblasts, leading to tissue expansion, fibrosis, and organ-specific clinical features. Site-specific differences in fibroblast phenotype and immune activation account for the variable expression of extrathyroidal disease within a shared autoimmune framework.

In the orbit, during active TED, lymphocytes and bone marrow-derived $CD34^+$ fibrocytes that differentiate into $CD34^+$ fibroblasts and coexist with resident $CD34^-$ fibroblasts, generating a heterogeneous stromal cell population with distinct inflammatory and differentiative properties (Douglas et al., 2010). Upon activation, OF undergo lineage-specific differentiation toward adipocytes, contributing to orbital fat expansion, or toward myofibroblasts, promoting fibrosis of extraocular muscles and the development of restrictive myopathy and ocular dysfunction. OF in TED display unique immunological features, including ectopic expression of thyroid-related antigens such as TSH-R, enabling antigen presentation and amplification of local immune responses (Fernando et al., 2012). In addition, overexpression of IGF-1R on OF and infiltrating lymphocytes, together with its functional interaction with TSH-R, potentiates downstream signaling pathways that sustain orbital inflammation and tissue remodeling (Krieger et al., 2015; Tsui et al., 2008).

Pretibial myxedema represents a related fibroblast-mediated manifestation characterized by excessive dermal accumulation of GAG and localized skin thickening. In contrast to TED, inflammatory cell infiltration is typically limited, indicating site-specific differences in immune activation and fibroblast regulation within a shared autoimmune context.

GH and target tissues

Target tissues in GH

GH is a systemic disorder in which sustained TH excess, together with immune-mediated mechanisms, affects multiple target tissues beyond the thyroid gland. While extrathyroidal manifestations such as TED, dermopathy, and acropachy are primarily autoimmune in origin, most systemic features reflect the severity and duration of thyrotoxicosis.

Skeletal and cardiac muscle

Patients with GH may develop diffuse myopathic changes, including muscle fiber atrophy, fatty infiltration, loss of normal striation, vacuolization, and nuclear abnormalities. In some cases, interstitial inflammatory infiltrates composed of plasma cells and macrophages are observed, consistent with a mild myositis. These structural alterations contribute to proximal muscle weakness and reduced exercise tolerance and may explain the persistence of fatigue and weakness even after restoration of biochemical euthyroidism (Malmstroem et al., 2019).

Myocardial involvement has been described in severe thyrotoxicosis, with focal degenerative changes, mononuclear infiltrates, and mucopolysaccharide deposition, resembling alterations observed in skeletal and extraocular muscles (Biondi & Kahaly, 2010).

Pituitary gland

The anterior pituitary represents a hormonally responsive target tissue in GH, reflecting chronic suppression of hypothalamic-pituitary feedback by excess circulating TH. In severe, untreated thyrotoxicosis, TSH-producing cells are markedly reduced, while other pituitary cell populations are preserved. These changes are fully reversible after restoration of euthyroidism (Ezrin et al., 1959; Scheithauer et al., 1992).

Liver

The liver is a major metabolic target of TH excess, and early diagnosis with timely treatment is critical to limit thyrotoxic hepatic injury. While early autopsy studies in severe or untreated GH described hepatocellular necrosis and advanced structural damage, contemporary biopsy-based series in patients diagnosed and treated earlier predominantly report mild, largely reversible alterations (Okosieme et al., 2019). The central role of TH signaling in hepatic metabolism is further supported by the therapeutic efficacy of TH receptor β–selective agonists, such as resmetirom, which have demonstrated improvements in steatohepatitis and liver fibrosis in metabolic dysfunction-associated steatotic liver disease (Harrison et al., 2024).

Bone

Chronic thyrotoxicosis is associated with a marked increase in bone turnover, characterized by enhanced osteoclastic resorption that exceeds compensatory bone formation, leading to loss of bone mass and microarchitectural deterioration, predominantly affecting cortical bone and resulting in osteoporosis and, in some cases, osteitis fibrosa (Follis, 1953).

In untreated hyperthyroidism, BMD reductions of approximately 10%–15% at the spine and hip have been reported, translating into a clinically significant increase in fracture risk, particularly in postmenopausal women and older individuals (Karga et al., 2004).

Patients with euthyroid GD can exhibit persistent reductions in BMD up to 8%–12%, suggesting that ongoing autoimmune activity, particularly persistent TRAb positivity, can impair skeletal recovery despite restoration of euthyroidism (Al Shoufy et al., 2025).

Thyroid gland

The thyroid gland itself is the primary target of autoimmune stimulation in GH and typically manifests as a diffuse toxic goiter. The thyroid may range from minimally enlarged to several-fold its normal volume and may be smooth, lobulated, or, less commonly, nodular. Increased vascularity is a hallmark feature and is often detectable clinically as a thrill or bruit.

Skin, hair, and nails

Cutaneous manifestations in GH reflect increased sympathetic activity and autoimmune involvement, presenting with heat intolerance, warm moist skin, sweating, erythema, pruritus, and occasional urticaria. Autoimmune involvement can extend to the skin, manifesting as vitiligo, while nails and hair may be affected by onycholysis and diffuse thinning.

Central nervous system

The CNS system represents a critical and highly sensitive target of TH excess. Neuropsychiatric manifestations are common and heterogeneous, reflecting both direct hormonal effects on neuronal function and modulation of neurotransmitter systems. Cognitive impairment may also occur, particularly in untreated or longstanding disease. In addition, neuromuscular manifestations are frequently observed and reflect increased neuromuscular excitability. Overall, CNS involvement is largely functional and reversible with restoration of euthyroidism.

The detailed clinical spectrum and the neurobiological mechanisms underlying these manifestations are discussed in the following sections.

Effects of thyroid hormone excess on CNS

TH play a fundamental role in brain development and function. In GH, elevated circulating TH alter neuronal activity, synaptic plasticity, and neuroendocrine regulation, thereby providing the biological substrate for the wide spectrum of neuropsychiatric manifestations observed in affected patients.

CNS involvement in GH likely reflects the combined contribution of intracerebral TH excess, autoimmune mechanisms, and disease-related immune activation (Bunevicius & Prange, 2010). Supporting a role beyond hormone excess alone, anxiety and depressive symptoms have been reported to be more prevalent and more severe in GH than in toxic nodular goiter, and to correlate more closely with disease duration and comorbidity than with TH or TRAb levels (Bové et al., 2014).

Thyroid hormone transport and local availability in the brain

Tightly regulated mechanisms control TH transport and local availability within the brain. T4 and T3 enter the brain through specific transporters, primarily MCT8 and OATP1C1 (Ge et al., 2025).

Once within the CNS, local TH signaling is further modulated by iodothyronine deiodinases. D2, expressed mainly in astrocytes, converts T4 into the biologically active T3, whereas D3 inactivates TH, providing a local protective mechanism against hormone excess. D2 and D3 display region-specific expression with D3 particularly prominent in hippocampal pyramidal neurons, granule cells of the dentate gyrus, and layers II-VI of the cerebral cortex, indicating spatially regulated control of TH signaling. In hyperthyroid states, D2 activity is reduced, while D3 expression is positively regulated by TH levels, leading to complex, region-specific alterations in local T3 availability (Peeters & Visser, 2000).

Thyroid hormone receptors and regional brain vulnerability

TH actions in the CNS are primarily mediated by nuclear TRs, which are widely distributed throughout the brain. Two genes, *TRα* and *TRβ*, encode multiple receptor isoforms with distinct expression patterns and functional roles. TRα is the predominant receptor isoform in the adult brain and is abundantly expressed in neurons, particularly in regions involved in cognition, emotion, and motor control, while TRβ exhibits a more restricted distribution. Regional differences in receptor expression contribute to selective vulnerability of specific brain areas to TH excess, notably the hippocampus, limbic system, prefrontal cortex, and cerebellum. These regions are critically involved in mood regulation, executive function, memory processing, and motor coordination, providing a mechanistic link between altered TH signaling and neuropsychiatric dysfunction (Cook & Koenig, 1990).

TSH receptor signaling and autoimmune interactions in the brain

In addition to classical TRs-mediated pathways, evidences suggest that TSH-R signaling may contribute to CNS effects in GH. Immunohistochemical studies have demonstrated localization of anti-TSH-R immunoglobulins within neuronal cell bodies and axons in multiple cortical regions of the adult human brain (Moodley et al., 2011). Experimental models further suggest a functional role for TSH-R in behavioral regulation: TSH-R-deficient mice exhibit hyperactivity, impulsivity, impaired social behavior, and deficits in short-term memory and object recognition, phenotypes that persist even when TH levels are experimentally normalized (Mouri et al., 2014).

Structural, functional, and neurotransmitter alterations in the hyperthyroid brain

Neuroimaging and functional studies in adults with hyperthyroidism have provided evidence of structural and network-level brain alterations associated with TH excess. Reduced hippocampal volume and decreased gray matter volume have been reported in regions involved in memory, attention, emotion, vision, and motor planning, and these changes correlate negatively with both circulating TH levels and disease duration. In older individuals, higher circulating T4 levels, even within the reference range, have been associated with hippocampal and amygdalar atrophy (Holmberg et al., 2019; Zhang et al., 2014).

At the neurochemical level, TH excess modulates multiple neurotransmitter systems implicated in mood, cognition, and arousal. TH influence central noradrenergic signaling, in part through modulation of β-adrenergic receptor responsiveness to catecholamines, thereby amplifying noradrenergic tone. Serotonergic pathways are also highly sensitive to TH signaling and are closely linked to anxiety and depressive symptoms. In addition, growing evidence implicates disruption of dopaminergic neurotransmission, supported by neuroimaging findings of altered dopamine transporter availability, impaired striatal connectivity, and associations between dopaminergic network dysfunction and cognitive deficits. Oxidative stress and low-grade neuroinflammation may further contribute to dopaminergic neuronal vulnerability (Abdullah et al., 2025).

Neuropsychatric manifestation of GH

Neuropsychiatric manifestations are common in GH and involve affective, cognitive, behavioral, and neuromuscular domains. These features reflect the effects of sustained TH excess on neuronal excitability, neurotransmitter signaling, and central network regulation. Although most symptoms improve after restoration of euthyroidism, they may significantly impair quality of life and can persist in a subset of patients, particularly after prolonged or severe disease.

The clinical spectrum ranges from mild affective and cognitive disturbances to marked behavioral changes and neuromuscular hyperexcitability. Severe or atypical presentations, including acute psychosis or encephalopathic states, are rare and usually reversible with appropriate treatment. A concise overview of the main neuropsychiatric manifestations associated with GH is provided in Table 23.1.

TABLE 23.1 Spectrum of CNS manifestations associated with GH.

Domain	Features
Mood	Anxiety Irritability Emotional lability Restlessness
Behavioral	Hyperreactivity Agitation Hypomanic or manic features Sleep disturbance
Cognitive	Impaired attention Reduced concentration Working memory deficits Executive dysfunction
Neuromuscular	Fine postural tremor Brisk tendon reflexes Proximal weakness Fatigability
Severe/atypical	Acute psychosis Visual or auditory hallucinations Disorientation Aphasia Choreoathetoid movements Hemiparesis or bulbar paralysis Seizures Encephalopathy

Other hormones and endocrine pathways

Although TH excess represents the primary driver of CNS dysfunction in GH, additional endocrine pathways interact with thyroid signaling and may modulate neuropsychiatric manifestations even after restoration of euthyroidism.

HPA axis and stress

GH profoundly affects the HPA axis dynamics through combined alterations in cortisol metabolism and stress responsiveness. TH excess accelerates cortisol clearance via enhanced hepatic metabolism, while simultaneously stimulating adrenal cortisol production and increasing secretory pulse frequency. As a result, circulating total and free cortisol concentrations are typically maintained within the reference range in stable thyrotoxicosis, reflecting a compensated state of HPA axis activation. However, during the early phases of hyperthyroidism or in cases of abrupt disease onset, the increase in cortisol production may temporarily fail to match the accelerated metabolic clearance. This mismatch can result in a transient state of functional adrenal insufficiency, potentially contributing to fatigue, reduced stress tolerance, and neuropsychiatric symptoms observed in untreated patients (Hoshiro et al., 2006).

Beyond endocrine adaptations, psychosocial stress has been extensively investigated as a potential trigger and disease modifier in GH. Retrospective and case–control studies have frequently reported an increased burden of negative life events and chronic stressors preceding disease onset, suggesting a possible role for stress in precipitating autoimmunity in susceptible individuals. In contrast, prospective cohort studies and analyses in high-risk populations have yielded inconsistent results, highlighting substantial methodological limitations in stress research, including recall bias, difficulties in defining stress exposure, and uncertainty in dating disease onset (Fukao et al., 2020).

Despite these controversies, clinical and longitudinal observations consistently indicate that stress may influence disease course rather than initiation. Emotional stress, maladaptive coping styles, and persistent depressive traits have been associated with exacerbations of hyperthyroidism, reduced remission rates, and higher likelihood of relapse during antithyroid drug therapy. Importantly, psychological distress and depressive symptoms may persist even after restoration of euthyroidism, suggesting partial dissociation between biochemical control and central stress regulation.

HPG axis

Alterations of the HPG axis have been reported in GH and may contribute to central and systemic manifestations of the disease. TH excess influences gonadotropin secretion, with potential effects on mood regulation, fatigue, reproductive function, and stress resilience, particularly in women.

Physiological studies in patients with severe, untreated GH indicate that overall HPG axis is largely preserved, as reflected by intact pulsatile secretion of LH and FSH. However, circulating gonadotropin levels, especially LH, are increased, suggesting enhanced pituitary output rather than central axis failure. In men, GH is associated with marked elevations in SHBG, leading to reduced bioavailable androgens despite increased total testosterone levels, consistent with a state of relative functional hypogonadism (Zähringer et al., 2000).

In women, GH is frequently associated with menstrual irregularities, including oligomenorrhea and amenorrhea, and with impaired fertility, effects that are largely reversible following restoration of euthyroidism (Koutras, 1997). These reproductive disturbances likely reflect altered sex steroid bioavailability and disrupted neuroendocrine integration rather than intrinsic gonadal disease.

Metabolic pathway

TH exert potent effects on systemic energy homeostasis, influencing glucose metabolism, lipid turnover, and insulin sensitivity across multiple tissues. In GH, sustained TH excess is associated with loss of both lean and fat mass, increased resting energy expenditure, and alterations in intermediary metabolism.

Despite the catabolic phenotype and reduced body weight typically observed in hyperthyroid patients, accumulating evidence indicates that thyrotoxicosis may paradoxically be associated with impaired glucose tolerance and features of IR. Population-based and longitudinal studies have reported elevated indices of IR in a substantial proportion of patients with overt hyperthyroidism, with abnormalities in fasting glucose, glycated hemoglobin, and insulin levels observed even in the absence of overt diabetes. At the mechanistic level, TH promote hepatic glucose production through upregulation of key gluconeogenic and glycogenolytic enzymes, including phosphoenolpyruvate carboxykinase and glucose-6-phosphatase, as well as increased expression of hepatic glucose transporters. In adipose tissue, TH enhance lipid oxidation and sensitize adipocyte lipolysis to catecholamines, leading to increased circulating free fatty acids, which may further impair insulin action. In skeletal muscle, reduced glycogen synthesis and preferential shunting of glucose toward lactate production contribute to increased hepatic glucose flux via the Cori cycle (Weetman, 2000).

In addition to peripheral IR, altered pancreatic β-cell function has been implicated in the dysregulation of glucose homeostasis in GH. Clinical studies suggest that reduced insulin secretory capacity may contribute to abnormal glucose tolerance in a subset of patients, with partial recovery observed following restoration of euthyroidism. However, in a proportion of individuals, metabolic abnormalities persist despite biochemical control of thyroid function, underscoring the need for ongoing metabolic surveillance (Lakshmana Perumal et al., 2019).

Somatotropin-IGF-1 axis

Bidirectional interactions between TH and the somatotropin-IGF-1 axis have been described and may contribute to systemic and central effects of endocrine dysregulation. Thyroid status influences somatotropin secretion and peripheral IGF-1 signaling, while, conversely, sustained activation of the somatotropin-IGF-1 pathway can modulate thyroid growth and function.

Experimental and clinical studies indicate that hypothyroidism is associated with reduced somatotropin secretion, whereas in hyperthyroid states somatotropin responsiveness appears more variable, reflecting heterogeneity in disease severity, duration, and regulatory feedback mechanisms (Shimatsu et al., 1990). These observations suggest that thyroid hormone excess does not uniformly enhance or suppress somatotropin secretion, but rather alters the dynamic regulation of the axis.

Conversely, chronic elevation of somatotropin and IGF-1 levels is known to promote thyroid cell proliferation and differentiation through activation of IGF-1 receptors expressed on thyroid follicular cells. In this context, thyroid enlargement and nodular disease have been frequently observed in conditions characterized by somatotropin excess (Curtò et al., 2015). Importantly, hyperthyroidism occurring in association with somatotropin excess is most often non-autoimmune, underscoring distinct pathogenetic mechanisms compared with GD.

Although the direct contribution of somatotropin-IGF-1 axis alterations to CNS manifestations in GH remains incompletely defined, IGF-1 signaling plays a recognized role in neuronal metabolism, synaptic plasticity, and neuroprotection. Dysregulation of this axis may therefore act as a modulatory factor, interacting with TH excess and immune-mediated mechanisms to influence CNS function.

Applications of the material

The concepts discussed in this chapter have relevant implications for both clinical practice and research at the interface of endocrinology and neuroscience.

Clinical implications

Recognition of the CNS as a highly sensitive target of TH excess highlights the need for early diagnosis and timely treatment of GH. Clinical evaluation should not be limited to biochemical thyroid function but should systematically include assessment of neuropsychiatric symptoms, cognitive performance, and stress-related factors. The observation that central manifestations may persist despite restoration of euthyroidism supports the need for long-term follow-up and a multidisciplinary management approach, integrating endocrinological, neurological, and psychiatric expertise.

Research and translational perspectives

From a research perspective, the neurobiological mechanisms linking TH excess to neuronal dysfunction, neurotransmitter imbalance, and neuroimmune interactions remain incompletely understood and represent a key area for further investigation. Improved characterization of these pathways may (i) enable the identification of biomarkers of CNS involvement, (ii) support the development of targeted therapeutic strategies addressing neurological and neuropsychiatric manifestations rather than TH levels alone, and (iii) contribute to a broader understanding of endocrine-immune-neural interactions, positioning GH as a valuable human model for translational research in neuroendocrinology.

References

Abdullah, M., Lin, S. H., Huang, L. C., Chiu, N. T., Huang, S. M., & Yang, Y. K. (2025). Graves' disease-associated dopaminergic system dysfunction and its effects on striatocortical functional connectivity and cognitive function. *Journal of the Formosan Medical Association*. https://doi.org/10.1016/j.jfma.2025.10.016

Al Shoufy, M., Nasrallah, J., Al Jebawi, K., El Etry, H., Shafeea, M. S., Hassan, I., & Ali, H. T. (2025). Graves' disease and bone mineral density: A scoping review of the impact of graves' disease on spinal health. *Journal of Clinical Densitometry, 29*(1), Article 101621. https://doi.org/10.1016/j.jocd.2025.101621

Azizi, F. (2020). Long-Term treatment of hyperthyroidism with antithyroid drugs: 35 years of personal clinical experience. *Thyroid, 30*(10), 1451–1457. https://doi.org/10.1089/thy.2019.0814

Bartalena, L. (2013). Diagnosis and management of graves disease: A global overview. *Nature Reviews Endocrinology, 9*(12), 724–734. https://doi.org/10.1038/nrendo.2013.193

Bartalena, L., & Fatourechi, V. (2014). Extrathyroidal manifestations of graves' disease: A 2014 update. *Journal of Endocrinological Investigation, 37*(8), 691–700. https://doi.org/10.1007/s40618-014-0097-2

Biondi, B., & Kahaly, G. J. (2010). Cardiovascular involvement in patients with different causes of hyperthyroidism. *Nature Reviews Endocrinology, 6*(8), 431–443. https://doi.org/10.1038/nrendo.2010.105

Boelaert, K., Torlinska, B., Holder, R. L., & Franklyn, J. A. (2010). Older subjects with hyperthyroidism present with a paucity of symptoms and signs: A large cross-sectional study. *Journal of Clinical Endocrinology & Metabolism, 95*(6), 2715–2726. https://doi.org/10.1210/jc.2009-2495

Bové, K. B., Watt, T., Vogel, A., Hegedüs, L., Bjoerner, J. B., Groenvold, M., … Feldt-Rasmussen, U. (2014). Anxiety and depression are more prevalent in patients with graves' disease than in patients with nodular goitre. *European Thyroid Journal, 3*(3), 173–178. https://doi.org/10.1159/000365211

Brix, T. H., Kyvik, K. O., Christensen, K., & Hegedüs, L. (2001). Evidence for a major role of heredity in graves' disease: A population-based study of two danish twin cohorts. *Journal of Clinical Endocrinology & Metabolism, 86*(2), 930–934. https://doi.org/10.1210/jcem.86.2.7242

Bunevicius, R., & Prange, A. J., Jr. (2010). Thyroid disease and mental disorders: Cause and effect or only comorbidity? *Current Opinion in Psychiatry, 23*(4), 363–368. https://doi.org/10.1097/YCO.0b013e3283387b50

Cook, C. B., & Koenig, R. J. (1990). Expression of erbA alpha and beta mRNAs in regions of adult rat brain. *Molecular and Cellular Endocrinology, 70*(1), 13–20. https://doi.org/10.1016/0303-7207(90)90054-c

Curtò, L., Giovinazzo, S., Alibrandi, A., Campennì, A., Trimarchi, F., Cannavò, S., & Ruggeri, R. M. (2015). Effects of GH replacement therapy on thyroid volume and nodule development in GH deficient adults: A retrospective cohort study. *European Journal of Endocrinology, 172*(5), 543–552. https://doi.org/10.1530/eje-14-0966

Dekkers, O. M., Horváth-Puhó, E., Cannegieter, S. C., Vandenbroucke, J. P., Sørensen, H. T., & Jørgensen, J. O. (2017). Acute cardiovascular events and all-cause mortality in patients with hyperthyroidism: A population-based cohort study. *European Journal of Endocrinology, 176*(1), 1–9. https://doi.org/10.1530/eje-16-0576

Douglas, R. S., Afifiyan, N. F., Hwang, C. J., Chong, K., Haider, U., Richards, P., … Smith, T. J. (2010). Increased generation of fibrocytes in thyroid-associated ophthalmopathy. *Journal of Clinical Endocrinology & Metabolism, 95*(1), 430–438. https://doi.org/10.1210/jc.2009-1614

Douglas, R. S., Naik, V., Hwang, C. J., Afifiyan, N. F., Gianoukakis, A. G., Sand, D., & Smith, T. J. (2008). B cells from patients with graves' disease aberrantly express the IGF-1 receptor: Implications for disease pathogenesis. *Journal of Immunology, 181*(8), 5768–5774. https://doi.org/10.4049/jimmunol.181.8.5768

Dumitriu, L., & Ursu, H. (1985). Hyperthyroidism in the elderly. I. Clinical manifestations. *Endocrinologie, 23*(2), 83–90.

Ezrin, C., Swanson, H. E., Humphrey, J. G., & Dawson Jw Hill, F. M. (1959). The cells of the human adenohypophysis in thyroid disorders.

Journal of Clinical Endocrinology & Metabolism, 19, 958–966. https://doi.org/10.1210/jcem-19-8-958

Fatourechi, V., Ahmed, D. D., & Schwartz, K. M. (2002). Thyroid acropachy: Report of 40 patients treated at a single institution in a 26-year period. *Journal of Clinical Endocrinology & Metabolism, 87*(12), 5435–5441. https://doi.org/10.1210/jc.2002-020746

Fernando, R., Atkins, S., Raychaudhuri, N., Lu, Y., Li, B., Douglas, R. S., & Smith, T. J. (2012). Human fibrocytes coexpress thyroglobulin and thyrotropin receptor. *Proceedings of the National Academy of Sciences of the USA, 109*(19), 7427–7432. https://doi.org/10.1073/pnas.1202064109

Follis, R. H., Jr. (1953). Skeletal changes associated with hyperthyroidism. *Bulletin of the Johns Hopkins Hospital, 92*(6), 405–421.

Fukao, A., Takamatsu, J., Arishima, T., Tanaka, M., Kawai, T., Okamoto, Y., & Imagawa, A. (2020). Graves' disease and mental disorders. *Journal of Clinical & Translational Endocrinology, 19*, Article 100207. https://doi.org/10.1016/j.jcte.2019.100207

Gallo, D., Piantanida, E., Gallazzi, M., Bartalena, L., Tanda, M. L., Bruno, A., & Mortara, L. (2020). Immunological drivers in graves' disease: NK cells as a master switcher. *Frontiers in Endocrinology, 11*, 406. https://doi.org/10.3389/fendo.2020.00406

Ge, Y., Dou, T., Nguyen, T. U., Yadav, G. P., Wensel, T. G., Jiang, J., & Huang, P. (2025). Structural insights into brain thyroid hormone transport via MCT8 and OATP1C1. *Cell, 188*(20), 5576–5588. https://doi.org/10.1016/j.cell.2025.06.032

Gómez-Roig, M. D., Pascal, R., Cahuana, M. J., García-Algar, O., Sebastiani, G., Andreu-Fernández, V., … Vento, M. (2021). Environmental exposure during pregnancy: Influence on prenatal development and early life: A comprehensive review. *Fetal Diagnosis and Therapy, 48*(4), 245–257. https://doi.org/10.1159/000514884

Gunsar, F., Yilmaz, S., Bor, S., Kumanlioğlu, K., Cetinkalp, S., Kabalak, T., & Ozutemiz, O. A. (2003). Effect of hypo- and hyperthyroidism on gastric myoelectrical activity. *Digestive Diseases and Sciences, 48*(4), 706–712. https://doi.org/10.1023/a:1022876423487

Harrison, S. A., Bedossa, P., Guy, C. D., Schattenberg, J. M., Loomba, R., Taub, R., … Ratziu, V. (2024). A phase 3, randomized, controlled trial of resmetirom in NASH with liver fibrosis. *New England Journal of Medicine, 390*(6), 497–509. https://doi.org/10.1056/NEJMoa2309000

Holmberg, M. O., Malmgren, H., Berglund, P., Bunketorp-Käll, L., Heckemann, R. A., Johansson, B., … Nystrom Filipsson, H. (2019). Structural brain changes in hyperthyroid graves' disease: Protocol for an ongoing longitudinal, case-controlled study in Göteborg, Sweden-the CogThy project. *BMJ Open, 9*(11), Article e031168. https://doi.org/10.1136/bmjopen-2019-031168

Hoshiro, M., Ohno, Y., Masaki, H., Iwase, H., & Aoki, N. (2006). Comprehensive study of urinary cortisol metabolites in hyperthyroid and hypothyroid patients. *Clinical Endocrinology, 64*(1), 37–45. https://doi.org/10.1111/j.1365-2265.2005.02412.x

Kahaly, G. J. (2020). Management of graves thyroidal and extrathyroidal disease: An update. *Journal of Clinical Endocrinology & Metabolism, 105*(12), 3704–3720. https://doi.org/10.1210/clinem/dgaa646

Karga, H., Papapetrou, P. D., Korakovouni, A., Papandroulaki, F., Polymeris, A., & Pampouras, G. (2004). Bone mineral density in hyperthyroidism. *Clinical Endocrinology, 61*(4), 466–472. https://doi.org/10.1111/j.1365-2265.2004.02110.x

Koutras, D. A. (1997). Disturbances of menstruation in thyroid disease. *Annals of the New York Academy of Sciences, 816*, 280–284. https://doi.org/10.1111/j.1749-6632.1997.tb52152.x

Krieger, C. C., Neumann, S., Place, R. F., Marcus-Samuels, B., & Gershengorn, M. C. (2015). Bidirectional TSH and IGF-1 receptor cross talk mediates stimulation of hyaluronan secretion by graves' disease immunoglobins. *Journal of Clinical Endocrinology & Metabolism, 100*(3), 1071–1077. https://doi.org/10.1210/jc.2014-3566

Lakshmana Perumal, N., Selvi, J., Sridharan, K., Sahoo, J., & Kamalanathan, S. (2019). Insulin sensitivity and beta-cell function in graves' disease and their changes with the carbimazole-induced euthyroid state. *European Thyroid Journal, 8*(2), 59–63. https://doi.org/10.1159/000496924

Lanzolla, G., Marinò, M., & Menconi, F. (2024). Graves disease: Latest understanding of pathogenesis and treatment options. *Nature Reviews Endocrinology, 20*(11), 647–660. https://doi.org/10.1038/s41574-024-01016-5

Lee, S. Y., & Pearce, E. N. (2023). Hyperthyroidism: A review. *JAMA, 330*(15), 1472–1483. https://doi.org/10.1001/jama.2023.19052

Lillevang-Johansen, M., Abrahamsen, B., Jørgensen, H. L., Brix, T. H., & Hegedüs, L. (2017). Excess mortality in treated and untreated hyperthyroidism is related to cumulative periods of low serum TSH. *Journal of Clinical Endocrinology & Metabolism, 102*(7), 2301–2309. https://doi.org/10.1210/jc.2017-00166

Limbach, M., Saare, M., Tserel, L., Kisand, K., Eglit, T., Sauer, S., … Peterson, P. (2016). Epigenetic profiling in CD4+ and CD8+ T cells from graves' disease patients reveals changes in genes associated with T cell receptor signaling. *Journal of Autoimmunity, 67*, 46–56. https://doi.org/10.1016/j.jaut.2015.09.006

Malmstroem, S., Grove-Laugesen, D., Riis, A. L., Bruun, B. J., Ebbehoj, E., Hansen, K. W., & Rejnmark, L. (2019). Muscle performance and postural stability are reduced in patients with newly diagnosed graves' disease. *Thyroid, 29*(6), 783–789. https://doi.org/10.1089/thy.2018.0318

Meling Stokland, A. E., Austdal, M., Nedrebø, B. G., Carlsen, S., Hetland, H. B., Breivik, L., & Ueland, G. (2024). Outcomes of patients with graves disease 25 years after initiating antithyroid drug therapy. *Journal of Clinical Endocrinology & Metabolism, 109*(3), 827–836. https://doi.org/10.1210/clinem/dgad538

Mohlin, E., Filipsson Nyström, H., & Eliasson, M. (2014). Long-term prognosis after medical treatment of graves' disease in a northern Swedish population 2000-2010. *European Journal of Endocrinology, 170*(3), 419–427. https://doi.org/10.1530/eje-13-0811

Moodley, K., Botha, J., Raidoo, D. M., & Naidoo, S. (2011). Immunolocalisation of anti-thyroid antibodies in adult human cerebral cortex. *Journal of the Neurological Sciences, 302*(1–2), 114–117. https://doi.org/10.1016/j.jns.2010.11.027

Mouri, A., Hoshino, Y., Narusawa, S., Ikegami, K., Mizoguchi, H., Murata, Y., & Nabeshima, T. (2014). Thyrotoropin receptor knockout changes monoaminergic neuronal system and produces methylphenidate-sensitive emotional and cognitive dysfunction. *Psychoneuroendocrinology, 48*, 147–161. https://doi.org/10.1016/j.psyneuen.2014.05.021

Nordyke, R. A., Gilbert, F. I., Jr., & Harada, A. S. (1988). Graves' disease. Influence of age on clinical findings. *Archives of Internal Medicine, 148*(3), 626–631. https://doi.org/10.1001/archinte.148.3.626

Okosieme, O. E., Taylor, P. N., Evans, C., Thayer, D., Chai, A., Khan, I., & Dayan, C. M. (2019). Primary therapy of graves' disease and cardiovascular morbidity and mortality: A linked-record cohort study. *The Lancet Diabetes & Endocrinology, 7*(4), 278–287. https://doi.org/10.1016/s2213-8587(19)30059-2

Peeters, R. P., & Visser, T. J. (2000). Metabolism of thyroid hormone. In K. R. Feingold, R. A. Adler, S. F. Ahmed, B. Anawalt, M. R. Blackman, G. Chrousos, E. Corpas, W. W. de Herder, K. Dhatariya, K. Dungan, E. Hamilton, J. Hofland, S. Jan de Beur, S. Kalra, G. Kaltsas, N. Kapoor, M. Kim, C. Koch, P. Kopp, et al.D. P. Wilson (Eds.), *Endotext.* South Dartmouth (MA): MDText.com, Inc.

Pritchard, J., Han, R., Horst, N., Cruikshank, W. W., & Smith, T. J. (2003). Immunoglobulin activation of T cell chemoattractant expression in fibroblasts from patients with graves' disease is mediated through the insulin-like growth factor I receptor pathway. *Journal of Immunology, 170*(12), 6348–6354. https://doi.org/10.4049/jimmunol.170.12.6348

Radetti, G., Persani, L., Moroder, W., Cortelazzi, D., Gentili, L., & Beck-Peccoz, P. (1999). Transplacental passage of anti-thyroid autoantibodies in a pregnant woman with auto-immune thyroid disease. *Prenatal Diagnosis, 19*(5), 468–471.

Scheithauer, B. W., Kovacs, K. T., Young, W. F., Jr., & Randall, R. V. (1992). The pituitary gland in hyperthyroidism. *Mayo Clinic Proceedings, 67*(1), 22–26. https://doi.org/10.1016/s0025-6196(12)60272-9

Shi, H., Sheng, R., Hu, Y., Liu, X., Jiang, L., Wang, Z., & Cui, D. (2020). Risk factors for the relapse of graves' disease treated with antithyroid drugs: A systematic review and meta-analysis. *Clinical Therapeutics, 42*(4), 662–675.e664. https://doi.org/10.1016/j.clinthera.2020.01.022

Shimatsu, A., Murabe, H., Sasaki, S., Hattori, N., Tanoh, T., & Imura, H. (1990). Secretory dynamics of growth hormone in an acromegalic patient associated with graves' disease. *Journal of Clinical Endocrinology & Metabolism, 71*(4), 1059–1063. https://doi.org/10.1210/jcem-71-4-1059

Smith, T. J. (2025). Thyroid eye disease and the IGF-1 receptor. *The Lancet Diabetes & Endocrinology, 13*(9), 741–742. https://doi.org/10.1016/s2213-8587(25)00218-9

Smith, T. J., & Hegedüs, L. (2016). Graves' disease. *New England Journal of Medicine, 375*(16), 1552–1565. https://doi.org/10.1056/NEJMra1510030

Sohn, S. Y., Lee, E., Lee, M. K., & Lee, J. H. (2020). The association of overt and subclinical hyperthyroidism with the risk of cardiovascular events and cardiovascular mortality: Meta-analysis and systematic review of cohort studies. *Endocrinology and Metabolism, 35*(4), 786–800. https://doi.org/10.3803/EnM.2020.728

Taylor, P. N., Albrecht, D., Scholz, A., Gutierrez-Buey, G., Lazarus, J. H., Dayan, C. M., & Okosieme, O. E. (2018). Global epidemiology of hyperthyroidism and hypothyroidism. *Nature Reviews Endocrinology, 14*(5), 301–316. https://doi.org/10.1038/nrendo.2018.18

Tomer, Y. (2014). Mechanisms of autoimmune thyroid diseases: From genetics to epigenetics. *Annual Review of Pathology: Mechanisms of Disease, 9*, 147–156. https://doi.org/10.1146/annurev-pathol-012513-104713

Tsui, S., Naik, V., Hoa, N., Hwang, C. J., Afifiyan, N. F., Sinha Hikim, A., … Smith, T. J. (2008). Evidence for an association between thyroid-stimulating hormone and insulin-like growth factor 1 receptors: A tale of two antigens implicated in graves' disease. *Journal of Immunology,, 181*(6), 4397–4405. https://doi.org/10.4049/jimmunol.181.6.4397

Viola, N., Colleo, A., Casula, M., Mura, C., Boi, F., & Lanzolla, G. (2025). Graves' disease: Is it time for targeted therapy? A narrative review. *Medicina (Kaunas), 61*(3). https://doi.org/10.3390/medicina61030500

Vos, X. G., Endert, E., Zwinderman, A. H., Tijssen, J. G., & Wiersinga, W. M. (2016). Predicting the risk of recurrence before the start of antithyroid drug therapy in patients with graves' hyperthyroidism. *Journal of Clinical Endocrinology & Metabolism, 101*(4), 1381–1389. https://doi.org/10.1210/jc.2015-3644

Weetman, A. P. (2000). Graves' disease. *New England Journal of Medicine, 343*(17), 1236–1248. https://doi.org/10.1056/nejm200010263431707

Wiersinga, W. M., Eckstein, A. K., & Žarković, M. (2025). Thyroid eye disease (graves' orbitopathy): Clinical presentation, epidemiology, pathogenesis, and management. *Lancet Diabetes Endocrinol, 13*(7), 600–614. https://doi.org/10.1016/s2213-8587(25)00066-x

Zähringer, S., Tomova, A., von Werder, K., Brabant, G., Kumanov, P., & Schopohl, J. (2000). The influence of hyperthyroidism on the hypothalamic-pituitary-gonadal axis. *Experimental and Clinical Endocrinology & Diabetes, 108*(4), 282–289. https://doi.org/10.1055/s-2000-7756

Zhang, W., Song, L., Yin, X., Zhang, J., Liu, C., Wang, J., & Lii, H. (2014). Grey matter abnormalities in untreated hyperthyroidism: A voxel-based morphometry study using the DARTEL approach. *European Journal of Radiology, 83*(1), e43–e48. https://doi.org/10.1016/j.ejrad.2013.09.019

Chapter 24

Graves' disease (autoimmune hyperthyroidism) and fatigue – understandable, identifiable, and manageable

Birgitta Johansson[1], Lars Rönnbäck[1] and Helena Filipsson Nyström[2, 3, 4, 5]

[1]*Institute of Neuroscience and Physiology, Department of Clinical Neuroscience, Sahlgrenska Academy, University of Gothenburg, Gothenburg, Sweden;* [2]*Institute of Medicine, Sahlgrenska Academy, University of Gothenburg, Gothenburg, Sweden;* [3]*Department of Endocrinology, Sahlgrenska University Hospital, Gothenburg, Sweden;* [4]*Wallenberg Center for Molecular and Translational Medicine, Gothenburg, Sweden;* [5]*Gothenburg Centre for Person Centred Care, Gothenburg, Sweden*

Summary points

- Fatigue associated with Graves' disease reduces work capacity and quality of life.
- Fatigue is primarily characterized by an inability to sustain mental activity over extended periods.
- The fatigue is not resolved with conventional recovery strategies.
- Fatigue in Graves' disease presents as a constellation of cognitive, sensory, and emotional symptoms.
- Fatigue can be measured with assessment scales.
- Fatigue can be alleviated, but not cured effectively today.

Understanding fatigue in Graves' disease

Graves' disease (GD) is a well-established autoimmune disorder characterized by elevated thyroid hormone levels and a range of physical and psychological symptoms. While these manifestations are widely recognized, a less acknowledged but equally impactful consequence is the onset of fatigue, particularly in the early stages of the disease. This fatigue is not the ordinary tiredness experienced by healthy individuals, but a profound lack of energy and stamina that impairs one's ability to engage in everyday activities such as reading, socializing, or participating in society as before the illness.

This chapter explores the fatigue associated with GD, emphasizing its potential to persist even after thyroid hormone levels have normalized following treatment. Drawing on clinical experience and research, studies have reported reduced quality of life (Abraham-Nordling et al., 2005; Cramon et al., 2016), emotional distress (Bunevicius & Prange, 2006), and cognitive complaints (Johansson et al., 2023; Lillevang-Johansen et al., 2014; Vogel et al., 2005). However, few studies have specifically aimed to explore fatigue in GD (Fahrenfort et al., 2000; Johansson et al., 2023; Möller et al., 2014). Fatigue is also reported in other autoimmune disorders (Zielinski et al., 2019), and must be treated separately when euthyroidism does not alleviate all symptoms.

There is currently no consensus on the definition or assessment of fatigue. This lack of agreement may contribute to its marginalization in clinical settings, where measurable biomarkers such as hormone levels often take precedence. Adding to the confusion, terms such as fatigue, tiredness, mental fatigue, central fatigue, and cognitive fatigue are used interchangeably. In addition the colloquial term brain fog is frequently invoked to describe the cognitive complaints associated with this condition (Ettleson et al., 2022).

Importantly, the pathological fatigue in this chapter is not alleviated by rest or sleep. It manifests as a persistent and debilitating energy deficit that affects both professional and social functioning. Unlike normal fatigue, pathological

Hormones in Neuroscience. https://doi.org/10.1016/B978-0-443-30261-9.00024-3

fatigue does not resolve with conventional recovery strategies and significantly impairs well-being (Johansson & Rönnbäck, 2014b).

Healthcare professionals may overlook psychosocial consequences, failing to inquire about daily functioning, family life, or work capacity. Instead, assessments are most often narrowly focused on biochemical markers, as less is known about mental consequences. This disconnection between clinical evaluation and patient experience underscores the need for a more holistic approach (Askari et al., 2024).

Fatigue in GD must be understood as a dynamic condition that unfolds over time. It is not simply a momentary lapse in energy, but a persistent inability to sustain mental and physical activity across hours, days, and weeks. Individuals affected by fatigue experience a depletion of energy that is disproportionate to their exertion, and recovery from even minor activities can take an unusually long time. This pattern is often not apparent during routine clinical encounters, and patients themselves may struggle to articulate the nature and extent of their symptoms. As it is also not well-known, it can go undetected. While asking the patient about everyday activities such as reading, conversing, shopping, working, or studying this will be reported as problematic consuming significant energy.

When experiencing fatigue, individuals frequently describe being overwhelmed and distracted by sensory input, unable to filter out background noise or irrelevant stimuli. For example, the sound of a fan may compete with the content of a conversation, indicating a reduced ability to suppress irrelevant information (Hammett et al., 2020). Addition, physical exertion beyond an individual's tolerance can exacerbate fatigue, although cognitive demands tend to be more disruptive in daily life.

The origin of the pathological fatigue, in general is not well understood. However, theories suggest a relation to the brain function and autoimmune diseases (Chaudhuri & Behan, 2004; Ronnback & Hansson, 2004; Zielinski et al., 2019) which are briefly described here. In autoimmune diseases, peripheral inflammation may, due to a disrupted blood-brain barrier, allow foreign inflammatory-inducing particles to reach the brain and disturb its function (Zielinski et al., 2019). In neurological diseases, fatigue is suggested to result from disrupted pathways interconnecting the basal ganglia, thalamus, limbic system, and higher cortical centers (Chaudhuri & Behan, 2004). A dopamine imbalance hypothesis in multiple sclerosis and other neurological disorders suggests altered communication and structural and functional abnormalities in the frontal and striatal regions innervated by dopamine neurons (Dobryakova et al., 2015). The Brain Fatigue Syndrome (BFS) relates fatigue to disease or brain injury, directly or indirectly affecting the brain through inflammatory processes (Johansson & Rönnbäck, 2025; Ronnback & Hansson, 2004; Rönnbäck & Johansson, 2022). This theory proposes astrocyte dysfunction due to initial inflammation, leading to impaired neuronal glutamate signaling, both specific and nonspecific, activating larger networks and reducing glucose uptake from the blood. Ultimately, this results in an energy crisis when the brain is activated and the affected individual experiences profound fatigue with impairments in cognitive, sensory, and emotional systems.

Indirectly, functional neuroimaging studies have revealed altered brain activation patterns in individuals suffering from fatigue after acquired brain injury, mainly mild injuries and in burnout syndrome. These studies showed reduced peak oxygenated hemoglobin levels, indicative of diminished neuronal activity in the frontal cortex and basal ganglia (Berginström et al., 2018; Skau et al., 2019; Skau et al., 2021). However, such findings were not observed in the acute phase of GD (Skau et al., 2025). This raises several questions: Does the severity of fatigue influence these outcomes? Does the origin of fatigue, whether from brain trauma or systemic inflammation play a role? Or are the differences due to study design? Furthermore, using alternative methods to record brain function in relation to fatigue, altered brain activation (Kohl et al., 2009; Wylie et al., 2017), and aberrant connectivity patterns between brain regions have been observed in groups suffering from an acquired brain injury (Engström Nordin et al., 2016; Glavå, Skau, Lövdén, & Johansson, 2025; Ramage et al., 2019), suggesting impaired coordination of neural networks related to fatigue. These findings support the hypothesis that fatigue in brain-related disorders is underpinned by dysfunctional activity, which may explain the persistent energy deficit and resistance to recovery through rest. No such studies have been conducted on patients with GD in relation to fatigue. However, a review suggests that altered brain networks in hyperthyroid individuals may affect cognitive and emotional processing (Tesfaye et al., 2024).

The relationship between prior depression and anxiety may also be significant in the development of fatigue in GD. However, this relationship is complex and requires further investigation (Holmberg et al., 2024).

It is also evident that other factors may influence the severity of fatigue, including poor sleep, emotional disorders, medication, cognitive impairment, and other preexisting conditions. All of these must be evaluated as part of a thorough patient examination, as they are relevant to treatment options.

Identifying fatigue in Graves' disease

Fatigue can be measured using rating scales (Hjollund et al., 2007), although no consensus exists regarding the optimal tool, and no reliable biomarkers have been identified. Some scales are general, while others are disease-specific. Certain instruments assess fatigue as a unidimensional construct, whereas others adopt a multidimensional approach. The design of these scales varies, from those with predefined answer alternatives to those using less specific descriptors such as no problem, mild, slight, or severe problem, or numeric ratings indicating symptom severity. Fatigue is sometimes embedded within broader quality of life questionnaires, such as the ThyPRO, a thyroid-specific quality of life assessment tool comprising 85 items across 13 subscales, including tiredness and energy in daily life (Watt et al., 2010). In both clinical and research settings, the choice of scale often depends on familiarity or perceived appropriateness for the specific context.

In addition to fatigue, patients frequently report cognitive difficulties, particularly with memory and concentration. However, standard neuropsychological tests typically yield results within normal ranges (Johansson et al., 2023; Vogel et al., 2005). This discrepancy suggests that fatigue-related cognitive impairments are not easily captured by conventional cognitive test, even when patients' subjective complaints are substantial.

The few studies, specifically investigating the prevalence of fatigue in GD have employed various tools, including the Fatigue Severity Scale (FSS), the Mental Fatigue Scale (MFS), and fatigue-related items embedded in the Nottingham Health Profile (NHP).

Using the FSS, 56% of participants with untreated GD with hyperthyroidism reported fatigue (Möller et al., 2014). In contrast, the MFS revealed that 89% of participants in the initial phase of GD experienced fatigue, compared to 14% in a control population. After 15 months in long-term euthyroid state, symptoms had improved, but 38% of patients continued to report fatigue (Johansson et al., 2023). Follow-up data showed that 53% of patients still experienced energy deficits according to the NHP, even after successful treatment and normalization of thyroid hormone levels. Notably, one-third of these patients struggled to return to full-time employment, indication that fatigue affect ordinary living (Fahrenfort et al., 2000).

Mood disorders are also commonly reported in GD and often correlate with fatigue (Bunevicius & Prange, 2006; Johansson et al., 2023). However, research has demonstrated that depression and fatigue are distinct separate conditions, although they may co-occur (Cantor et al., 2008; Johansson et al., 2023).

Fatigue in GD presents as a constellation of cognitive, sensory, and emotional symptoms. The MFS captures this symptom complex, including fatigue, cognitive complaints, emotional involvement, and sensory sensitivity. These symptoms tend to vary in synchrony, meaning that an increase in overall fatigue, as reflected in the MFS total score, is often accompanied by increased mental fatigue, concentration difficulties, irritability, and sensitivity to noise (Johansson & Rönnbäck, 2014a). Although the MFS is not disease-specific, it is grounded in extensive clinical experience and research on symptom clusters that co-occur and contribute to the energy deficits reported by patients (Lindqvist & Malmgren, 1993; Rödholm et al., 2001). Originally validated in populations with acquired brain injuries, the MFS has since been applied to patients with GD, Cushing's syndrome (Papakokkinou et al., 2015), hypothyroidism (Johansson, 2022) and burnout syndrome (Skau et al., 2021), all of whom report symptoms strikingly similar to those seen in stroke or traumatic brain injury.

The Mental Fatigue Scale (MFS) has been in use for many years. The term Brain Fatigue Syndrome (BFS) has been proposed to clarify the origin of this symptom complex as brain-based (Johansson & Rönnbäck, 2025). While BFS is commonly associated with neurological disorders, it is increasingly recognized in endocrine conditions such as GD. Despite terminological differences, the MFS detects the BFS (see Table 24.1). A score above 10 on the MFS is indicative of BFS. It is up to clinicians and researchers to further explore the use of the MFS and the BFS concept.

Emotional symptoms such as depression and anxiety may also be present alongside fatigue, adding to the burden of BFS. It is therefore essential to screen for mood disorders to tailor treatment appropriately. In addition, neuropsychological assessments can provide valuable insights, as impaired cognitive function may further complicate daily living. An assessment by an occupational therapist can reveal functional limitations in everyday activities. However, in most cases, cognitive complaints in GD are not associated with measurable cognitive deficits. Instead, there is a distinction between cognitive function and the ability to sustain activity throughout a normal day. Understanding this distinction is crucial for patients, as it can alleviate concerns about cognitive decline and foster insight into the nature of mental energy loss and the need for adaptation.

TABLE 24.1 Brain fatigue syndrome (BFS): Symptom profile.

• Profound energy depletion, especially during everyday cognitive tasks
• Prolonged recovery time following mainly mental exertion
• Cognitive difficulties with concentration, memory, and mental slowness
• Sensory sensitivities, such as heightened sensitivity to light and sound
• Emotional symptoms, including irritability and tearfulness
• Sleep disturbances
• Headaches, often triggered by sustained or intense mental activity

Managing fatigue in Graves' disease

While hyperthyroidism can be effectively treated, residual fatigue may persist and currently lacks a rapid or definitive cure. However, symptoms can be alleviated. Although no pharmacological treatment exists specifically for fatigue, healthcare professionals, including psychologists, occupational therapists, physical therapists, nurses and others, play a vital role in supporting patients who struggle to manage daily life due to fatigue. Their involvement can help prevent deterioration and promote recovery.

Temporary or partial disability leave is often necessary, particularly in the early stages of the disease. It is important to recognize that recovery may take time. Clinical experience shows that exceeding one's energy limits can worsen fatigue and increase the risk of developing depression and anxiety.

Following assessment and planning for treatment options, patients need to be well informed about fatigue and how it can affect everyday life before support and treatment begin. Adjustments in both the home and work environments may be necessary, along with strategies to maintain social engagement and participation in daily activities. These adaptations require planning, self-awareness, and realistic expectations.

Energy-conserving strategies are always important (Vatwani & Margonis, 2019). These include taking regular breaks, both short and long and shortening the duration of activities, and planning tasks in advance. An occupational therapist can provide valuable support in implementing these strategies.

Fatigue also imposes significant psychological stress, in addition to the burden of the disease itself. Emotional responses such as frustration, anxiety, and sadness are common when everyday life and previously effortless activities become disrupted, even after the acute phase has resolved. A psychologist can offer treatment and support (Ponsford et al., 2020; Rezaei et al., 2020). Depression and anxiety should be treated when present, using a combination of medical and psychological interventions tailored to individual needs.

Pharmacological and non-pharmacological interventions, such as methylphenidate (Johansson et al., 2015) and mindfulness-based therapies (Glavå, Rönnbäck, & Johansson, 2025; Grossman et al., 2010; Johansson et al., 2012), have shown promise in alleviating fatigue in patients with acquired brain injuries. Treatment of fatigue in GD needs further exploration.

Patient organizations and peer support groups can also be invaluable. They offer recognition, understanding, and shared experiences that help individuals feel less isolated. Through these communities, patients can gain knowledge, inspiration, and discover new ways to navigate life when old strategies no longer work. This is, of course, a long and gradual process (Table 24.2).

TABLE 24.2 Treatment options for alleviating fatigue. Remember: Recovery takes time.

- Temporary or partial disability leave may be necessary for a period
- Use energy-saving strategies: Take regular breaks (short and long), shorten activity duration, plan activities in advance
- Calm activations as meditation, mindfulness courses, gentle yoga, spending time in nature
- Reduce stress as much as possible
- Engage in physical activity adapted to personal capacity
- Practice good sleep hygiene
- Treat emotional distress such as depression and anxiety
- Reduce light and noise exposure when necessary
- Join patient organizations and peer support groups

When fatigue is present, treatment must be adapted to the patient's available energy. This does not imply that patients are incapable of engaging in therapy or activities, but rather that longer treatment periods at a slower pace may be necessary. All therapeutic approaches must be tailored to the patient's energy limitations. Regardless of the treatment modality, regular follow-up is essential to monitor progress.

Recovery is a biological and emotional process that takes time, and both the body and brain require sustained support. Establishing a daily routine that reduces stress and cognitive load, while still offering meaningful stimulation, is key. Finding joy in everyday life, even in small ways is an important part of the healing process.

Although the underlying mechanisms of fatigue remain poorly understood and effective treatments are lacking, healthcare providers must not remain passive. The responsibility of care includes alleviating suffering wherever possible, offering support, and providing individualized treatment to improve quality of life.

References

Abraham-Nordling, M., Törring, O., Hamberger, B., Lundell, G., Tallstedt, L., Calissendorff, J., & Wallin, G. (2005). Graves'disease: A long-term quality-of-life follow up of patients randomized to treatment with antithyroid drugs, radioiodine, or sugery. *Thyroid, 15* (11), 1279–1286.

Askari, S., Harvey, K., & Sam-Odutola, M. (2024). Patients experience of cognitive fatigue post-stroke: An exploratory study. *Disability & Rehabilitation, 46*(19), 4514–4521. https://doi.org/10.1080/09638288.2023.2280065

Berginström, N., Nordström, P., Ekman, U., Eriksson, J., Andersson, M., Nyberg, L., & Nordström, A. (2018). Using functional magnetic resonance imaging to detect chronic fatigue in patients with previous traumatic brain injury: Changes linked to altered striato-thalamic-cortical functioning. *The Journal of Head Trauma Rehabilitation, 33* (4), 266–274.

Bunevicius, R., & Prange, A. J. (2006). Psychiatric manifestations of graves' hyperthyriodism. *Theraphy In Practice, 20*(11), 897–909.

Cantor, J. B., Ashman, T., Gordon, W., Ginsberg, A., Engmann, C., Egan, M., … Flanagan, S. (2008). Fatigue after traumatic brain injury and its impact on participation and quality of life. *The Journal of Head Trauma Rehabilitation, 23*(1), 41–51.

Chaudhuri, A., & Behan, P. O. (2004). Fatigue in neurological disorders. *Lancet, 363*, 978–988.

Cramon, P., Winther, K. H., Watt, T., Bonnema, S. J., Bue Bjorner, J., Ekholm, O., … Krogh Rasmussen, Å. (2016). Quality-of-Life impairments persist six months after treatment of graves' hyperthyroidism and toxic nodular goiter: A prospective cohort study. *Thyroid, Thyroid(Aug)*, 1010–1018. https://doi.org/10.1089/thy.2016.0044

Dobryakova, E., Genova, H. M., DeLuca, J., & Wylie, G. R. (2015). The dopamine imbalance hypothesis of fatigue in multiple sclerosis and other neurological disorders. *Frontiers in Neurology, 12*(6), 52. https://doi.org/10.3389/fneur.2015.0005

Engström Nordin, L., Möller, M. C., Julin, P., Bartfai, A., Hashim, F., & Li, T.-Q. (2016). Post mTBI fatigue is associated with abnormal brain functional connectivity. *Scientific Reports, 6*(21183). https://doi.org/10.1038/srep21183

Ettleson, M. D., Raine, A., Batistuzzo, A., Batista, S. P., McAninch, E., Teixeira, M. C. T. V., … Bianco, A. C. (2022). Brain fog in hypothyroidism: Understanding the patient's perspective. *Endocrine Practice, 28*(3), 257–264.

Fahrenfort, J. J., Wilterdink, A. M. L., & van der Veen, E. A. (2000). Long-term residual complaints and psychosocial sequelae after remission of hyperthyroidism. *Psychoneuroendocrinology, 25*, 201–211.

Glavå, G., Rönnbäck, L., & Johansson, B. (2025). A new mindfulness and psycho-educative program for treatment of brain fatigue, evaluated after an acquired brain injury and multiple sclerosis. *Health Psychology and Behavioral Medicine, 13*(1), Article 2502039. https://doi.org/10.1080/21642850.2025.2502039

Glavå, G., Skau, S., Lövdén, M., & Johansson, B. (2025). Differences in frontal cortical brain function between individuals suffering from pathological mental fatigue following acquired brain injury and healthy individuals. *Behavioural Brain Research, 490*(115631). https://doi.org/10.1016/j.bbr.2025.115631

Grossman, P., Kappos, L., Gensicke, H., D'Souza, M., Mohr, D. C., Penner, I. K., & Steiner, C. (2010). MS quality of life, depression, and fatigue improve after mindfulness training: A randomized trial. *Neurology, 75*(13), 1141–1149.

Hammett, S. T., Cook, E., Hassan, O., Hughes, C.-A., Rooslien, H., Tizkar, R., & Larsson, J. (2020). GABA, noise and gain in human visual cortex. *Neuroscience Letters, 736*(135294). https://doi.org/10.1016/j.neulet.2020.135294. PMCID: PMC7511597 PMID: 32777347.

Hjollund, N. H., Andersen, J. H., & Bech, P. (2007). Assessment of fatigue in chronic disease: A bibliographic study of fatigue measurement scales. *Health Qual Life Outcomes, 27*(5), 12. https://doi.org/10.1186/1477-7525-5-12

Holmberg, M., Malmgren, H., Berglund, P. F., Birgitta Johansson, B., & Filipsson Nyström, H. (2024). Psychiatric complications in Graves' disease. *European Thyroid Journal, 13*, Article e230247. https://doi.org/10.1530/ETJ-23-0247

Johansson, B. (2022). Screening method for assessment of work ability for patients suffering from mental fatigue. *Frontiers in Behavioral Neuroscience, 16*, Article 869377. https://doi.org/10.3389/fnbeh.2022.869377

Johansson, B., Bjuhr, H., & Rönnbäck, L. (2012). Mindfulness based stress reduction improves long-term mental fatigue after stroke or traumatic brain injury. *Brain Injury, 26*(13–14), 1621–1628.

Johansson, B., Holmberg, M., Skau, S., Malmgren, H., & Filipsson Nyström, H. (2023). The relationship between mental fatigue, depression, and cognition in Graves' disease. *European Thyroid Journal, 12*(4), Article e230040. https://doi.org/10.1530/ETJ-23-0040

Johansson, B., & Rönnbäck, L. (2014a). Evaluation of the mental fatigue scale and its relation to cognitive and emotional functioning after traumatic brain injury or stroke. *International Journal of Physical Medicine and Rehabilitation, 2*, 182. https://doi.org/10.4172/2329-9096.1000182

Johansson, B., & Rönnbäck, L. (2014b). Long-lasting mental fatigue after traumatic brain injury –a major problem most often neglected diagnostic criteria, assessment, relation to emotional and cognitive problems, cellular background, and aspects on treatment. In F. Sadaka (Ed.), *Traumatic brain injury* (pp. 1–20). Rijeka, Croatia: INTECH.

Johansson, B., & Rönnbäck, L. (2025). The brain fatigue syndrome—symptoms, probable definition, and pathophysiological mechanisms. *Journal of Clinical Medicine, 14*(3271). https://doi.org/10.3390/jcm14103271

Johansson, B., Wentzel, A.-P., Andréll, P., Mannheimer, C., & Rönnbäck, L. (2015). Methylphenidate reduces mental fatigue and improves processing speed in persons suffered a traumatic brain injury. *Brain Injury, 29*(6), 758–765.

Kohl, A. D., Wylie, G. R., Genova, H. M., Hillary, F., & DeLuca, J. (2009). The neural correlates of cognitive fatigue in traumatic brain injury using functional MRI. *Brain Injury, 23*(5), 420–432.

Lillevang-Johansen, M., Petersen, I., Christensen, K., Hegedüs, L., & Heiberg Brix, T. (2014). Is previous hyperthyroidism associated with long-term cognitive dysfunction? A twin study. *Clinical Endocrinology, 80*(2), 290–295. https://doi.org/10.1111/cen.12255

Lindqvist, G., & Malmgren, H. (1993). Organic mental disorders as hypothetical pathogenetic processes. *Acta Psychiatrica Scandinavica, 88*(Suppl. 373), 5–17.

Möller, M. C., Bartfai, A., Nygren de Boussard, C., Flöter Rådestad, A., & Calissendorff, J. (2014). High rates of fatigue in newly diagnosed Graves' disease. *Fatigue: Biomedicine, Health & Behavior, 2*(3), 153–162.

Papakokkinou, E., Johansson, B., Berglund, P., & Ragnarsson, O. (2015). Mental fatigue and executive dysfunction in patients with Cushing's syndrome in remission. *Behavioural Neurology*, 6. https://doi.org/10.1155/2015/173653

Ponsford, J., Lee, N. K., Wong, D., McKay, A., Haines, K., Downing, M., … O'Donnell, M. L. (2020). Factors associated with response to adapted cognitive behavioral therapy for anxiety and depression following traumatic brain injury. *The Journal of Head Trauma Rehabilitation, 35*(2), 117–126. https://doi.org/10.1097/HTR.0000000000000510

Ramage, A. E., Tate, D. F., New, A. B., Lewis, J. D., & Robin, D. A. (2019). Effort and fatigue-related functional connectivity in mild traumatic brain injury. *Frontiers in Neurology, 9*, 1165. https://doi.org/10.3389/fneur.2018.01165

Rezaei, S., Abedi, P., Maraghi, E., Hamid, N., & Rashidi, H. (2020). The effectiveness of cognitive- behavioral therapy on quality of life in women with hypothyroidism in the reproductive age: A randomized controlled trial. *Thyroid Research, 13*(6).

Rödholm, M., Starmark, J.-E., Svensson, E., & von Essen, C. (2001). Asteno-emotional disorder after aneurysmal SAH: Reliability, symptomatology and relation to outcome. *Acta Neurologica Scandinavica, 103*, 379–385.

Ronnback, L., & Hansson, E. (2004). On the potential role of glutamate transport in mental fatigue. *Journal of Neuroinflammation, 1*, 22.

Rönnbäck, L., & Johansson, B. (2022). Long-lasting pathological mental fatigue after Brain Injury–A dysfunction in glutamate neurotransmission? *Frontiers in Behavioral Neuroscience, 15*, Article 791984. https://doi.org/10.3389/fnbeh.2021.791984

Skau, S., Bunketorp-Käll, L., Kuhn, H. G., & Johansson, B. (2019). Mental fatigue and functional near-infrared spectroscopy (fNIRS) –based assessment of cognitive performance after mild traumatic brain injury. *Frontiers in Human Neuroscience, 13*(145). https://doi.org/10.3389/fnhum.2019.00145

Skau, S., Holmberg, M., Johansson, B., Bunketorp Käl, I. L., Malmgren, H., Kuhn, H.-G., & Filipsson Nyström, H. (2025). Assessing prefrontal cortex activity in Graves' disease: A functional near-infrared spectroscopy study. *Frontiers in Human Neuroscience, 19*, Article 1559914. https://doi.org/10.3389/fnhum.2025.1559914

Skau, S., Jonsdottir, I. H., Sjörs Dahlman, A., Johansson, B., & Kuhn, H. G. (2021). Exhaustion disorder and altered brain activity in frontal cortex detected with fNIRS. *Stress and Health, 24*(1), 64–75.

Tesfaye, E., Getnet, M., Anmut Bitew, D., Adugna, D. G., & Maru, L. (2024). Brain functional connectivity in hyperthyroid patients: Systematic review. *Frontiers in Neuroscience, 18*, Article 1383355. https://doi.org/10.3389/fnins.2024.1383355

Vatwani, A., & Margonis, R. (2019). *Energy conservation techniques to decrease fatigue*. Retrieved from.

Vogel, E. K., McCollough, A. W., & Machizawa, M. G. (2005). Neural measures reveal individual differences in controlling access to working memory. *Nature, 438*, 500–503.

Watt, T., Hegedus, L., Groenvold, M., Bjorner, J. B., Rasmussen, A. K., Bonnema, S. J., & Feldt-Rasmussen, U. (2010). Validity and reliability of the novel thyroid-specific quality of life questionnaire, ThyPRO. *European Journal of Endocrinology, 162*(1), 161–167.

Wylie, G. R., Dobryakova, E., DeLuca, J., Chiaravalloti, N., Essad, K., & Genova, H. (2017). Cognitive fatigue in individuals with traumatic brain injury is associated with caudate activation. *Scientific Reports, 7* (8973). https://doi.org/10.1038/s41598-41017-08846-41596

Zielinski, M. R., Systrom, D. M., & Rose, N. R. (2019). Fatigue, sleep, and autoimmune and related disorders. *Frontiers in Immunology, 10*, Article 1827. https://doi.org/10.3389/fimmu.2019.01827

Section 7

Diet and nutrition and their impact on the thyroid status and neuroscientific domains or behavior

Chapter 25

The thyroid–brain axis in hyperthyroidism: Neurotoxic mechanisms and nutraceutical interventions

Mohammed A. Abdel-Rasol[1], Mariam A. Tammam[2], Yaser M. Hassan[3] and Wael M. El-Sayed[1]

[1]*Department of Zoology, Faculty of Science, Ain Shams University, Cairo, Egypt;* [2]*Biotechnology Program, Faculty of Agriculture, Beni-Suef University, Beni-Suef, Egypt;* [3]*Biotechnology Program, Faculty of Science, Ain Shams University, Cairo, Egypt*

List of abbreviations

BDNF Brain-Derived Neurotrophic Factor
CoQ10 Coenzyme Q10
DHA Docosahexaenoic Acid
EPA Eicosapentaenoic Acid
FADS2 Fatty Acid Desaturase 2
GABA Gamma-Aminobutyric Acid
GPx Glutathione Peroxidase
HDAC Histone Deacetylase
HPA Hypothalamic-Pituitary-Adrenal
IL-1β Interleukin-1 Beta
IL-6 Interleukin-6
iNOS Inducible Nitric Oxide Synthase
LTP Long-Term Potentiation
MMP Matrix Metalloproteinase
NPD1 Neuroprotectin D1
ROS Reactive Oxygen Species
SCFA Short Chain Fatty Acids
SOD Superoxide Dismutase
T3 Triiodothyronine
T4 Thyroxine
THRs Thyroid Hormone Receptors
TNF-α Tumor Necrosis Factor Alpha

Introduction

Hyperthyroidism is characterized by overproduction of thyroxine (T4) and triiodothyronine (T3), which regulate basal metabolism in nearly all tissues. Common causes include Graves' disease, toxic nodular goiter, and thyroid adenomas. While its effects on cardiovascular, skeletal, and gastrointestinal systems are well documented, its neurological consequences receive less clinical attention (Sawicka-Gutaj et al., 2022; Yorke, 2022).

Evidence shows that hyperthyroidism can impair brain function early in its course. Cognitive deficits, mood disturbances, and structural brain changes may appear before overt endocrine symptoms, underscoring the brain's vulnerability to thyroid hormone excess (Bernardes et al., 2024; Lou et al., 2023; Tesfaye et al., 2024).

The brain consumes ~20% of the body's oxygen and glucose and relies on tight metabolic control. Excess thyroid hormones accelerate neuronal firing and energy demand, leading to excitotoxicity, mitochondrial overload, and elevated ROS (Somayaji et al., 2024). These stressors disrupt neural homeostasis and activate inflammatory pathways, driving memory loss, mood instability, and possibly long-term neurodegeneration (Shahat et al., 2022; Altunkaya et al., 2024; Z. Li & Liu, 2024).

Nutraceuticals—bioactive dietary compounds—are gaining interest as adjunctive neuroprotective therapies. Nutrients such as omega-3 fatty acids, curcumin, resveratrol, and vitamins C and E exhibit antioxidant and anti-inflammatory properties, buffering the CNS against oxidative and metabolic stress by neutralizing ROS, modulating immune signaling, and preserving mitochondrial function.

This chapter examines how hyperthyroidism disrupts brain function through excitotoxicity, oxidative stress, and neuroinflammation, and explores nutraceuticals as complementary interventions to preserve cognitive and emotional health in hyperthyroid states.

Neurological and cognitive effects of hyperthyroidism

Excess thyroid hormones disrupt brain function by overstimulating pathways that regulate neuronal development,

Hormones in Neuroscience. https://doi.org/10.1016/B978-0-443-30261-9.00025-5

synaptic activity, metabolism, and neurotransmitter systems, leading to diverse cognitive, psychiatric, and neuroanatomical changes.

Cognitive dysfunctions in hyperthyroidism

Cognitive deficits are among the most common neurological symptoms in hyperthyroidism, including memory loss, reduced attention, and impaired learning. The hippocampus, which governs memory and spatial processing, is highly vulnerable due to its metabolic demand and sensitivity to oxidative stress.

Excess thyroid hormones overstimulate THRs in hippocampal neurons, altering gene expression for synaptic organization, ion channels, and mitochondrial activity. This initially boosts energy output but generates excessive ROS, causing oxidative stress and synaptic dysfunction (Alšauskė et al., 2024).

Neurotransmitter imbalances further drive decline. Elevated glutamate and reduced GABAergic inhibition disrupt excitatory–inhibitory balance and impair LTP, essential for memory formation. Serotonin dysregulation weakens hippocampal–prefrontal connectivity, reducing executive function and attention (Sepúlveda et al., 2024). Neuroinflammation compounds these effects: activated microglia and elevated cytokines (IL-1β, TNF-α) destabilize synapses and promote neuronal injury (Lou et al., 2023; Sepúlveda et al., 2024).

In summary, cognitive impairment arises from ROS-driven synaptic damage, neurotransmitter imbalance, inflammation, and disrupted plasticity—centered on hippocampal vulnerability (Table 25.1).

Mood and psychiatric disturbances

Hyperthyroidism is often accompanied by mood symptoms such as anxiety, irritability, emotional lability, and, in severe cases, psychosis. These features arise from combined neurochemical and hormonal disruptions.

Monoaminergic dysregulation is central: excess thyroid hormones alter synthesis, release, and receptor responsiveness of serotonin, dopamine, and norepinephrine. Reduced serotonergic tone contributes to anxiety and depression, while dopaminergic changes may drive agitation or psychotic symptoms (Bakalov et al., 2023; Bernardes et al., 2024).

Thyroid hormone excess also sensitizes the CNS to catecholamines, increasing sympathetic activity and causing restlessness, palpitations, and hyperarousal.

HPA axis overactivation raises cortisol, exerting neurotoxic effects on the hippocampus and amygdala and weakening stress resilience. Chronic THR overstimulation in the limbic system and prefrontal cortex disrupts synaptic plasticity and emotional processing, destabilizing mood regulation and sometimes mimicking primary psychiatric disorders (Sawicka-Gutaj et al., 2022; Spaggiari et al., 2021) (Table 25.1).

Case Study: A 33-year-old woman presented with restlessness, irritability, and poor concentration. Initially diagnosed with GAD, she failed to respond to anxiolytics. Thyroid testing revealed hyperthyroidism, clarifying the cause of her symptoms. This case highlights how hyperthyroidism can mimic psychiatric disorders and underscores the need for thyroid screening in patients with unexplained mood disturbances (Moideen Sheriff et al., 2023).

Congenital hyperthyroidism and neurodevelopmental consequences

Congenital hyperthyroidism, though rare, can significantly impair neurodevelopment. Thyroid hormones are crucial for fetal and neonatal brain maturation, regulating neurogenesis, synaptogenesis, axonal guidance, and myelination. Excess hormone during these critical windows disrupts normal trajectories, often causing lasting effects (Grob et al., 2024).

Overactivation of THRs in neural progenitor cells alters gene expression controlling proliferation and differentiation, leading to premature cell cycle exit and imbalance between excitatory and inhibitory neurons (Salloum-Asfar et al., 2024). Elevated metabolic activity increases ROS, damaging developing neurons and glia, destabilizing cytoskeleton, and impairing synapse formation.

Excess thyroid hormone may accelerate synaptic maturation, mistime critical periods, and impair synaptic pruning, reducing long-term plasticity. Disrupted neurotrophic signaling, including reduced BDNF, limits dendritic arborization and synaptic connectivity. Myelination may also be abnormal due to altered oligodendrocyte maturation and myelin protein expression (Chincarini et al., 2024; Klosinska et al., 2022).

These changes can result in delayed milestones, attentional deficits, and learning disabilities, with severity determined by timing and duration of hormonal imbalance (Table 25.1).

Case Study: A 36-year-old woman with Graves' disease developed thyrotoxicosis during pregnancy. At 26 weeks, she reported decreased fetal movements and tachycardia. Labs confirmed elevated thyroid hormones and TRAB levels. Ultrasound revealed fetal hydrops and goiter; MRI showed mild ventriculomegaly. Amniocentesis ruled out chromosomal abnormalities and infections. This case highlights the neurodevelopmental risk of maternal Graves' disease and the need for early thyroid screening and treatment in pregnancy to prevent adverse maternal and fetal outcomes (Doucette et al., 2018).

TABLE 25.1 Cognitive, mood, neurodevelopmental, brain changes, and molecular mechanisms in hyperthyroidism.

Category	Key mechanisms	Main effects	Regions affected	Consequences	References
Cognitive dysfunctions	THR overstimulation, oxidative stress, neurotransmitter imbalance	Memory loss, poor attention, learning deficits; ROS-induced synaptic dysfunction	Hippocampus, prefrontal cortex	Cognitive decline, hippocampal vulnerability	(Alšauskė et al., 2024; Lou et al., 2023; Sepúlveda et al., 2024)
Mood & psychiatric disturbances	Monoamine dysregulation (serotonin, dopamine, norepinephrine), SNS hyperactivity, HPA overactivation	Anxiety, irritability, emotional lability, psychosis, palpitations	Prefrontal cortex, amygdala, hippocampus	Psychiatric-like symptoms; frequent misdiagnosis as GAD or bipolar disorder	(Bakalov et al., 2023; Bernardes et al., 2024; Sawicka-Gutaj et al., 2022; Spaggiari et al., 2021)
Congenital hyperthyroidism	Disrupted neurogenesis, synaptogenesis, axonal guidance, myelination	Premature cell cycle exit, imbalanced neuronal populations, altered synapse maturation	Fetal brain, neural progenitors	Developmental delay, learning disability, attention deficits (severity time-dependent)	(Chincarini et al., 2024; Grob et al., 2024; Klosinska et al., 2022; Salloum-Asfar et al., 2024)
Structural/Functional brain changes	Oxidative stress, excitotoxicity, mitochondrial dysfunction, BBB disruption	Hippocampal atrophy, cortical thinning, network disconnection	Hippocampus, prefrontal cortex, brainstem	Impaired memory, emotional regulation, chronic neurocognitive symptoms	(Lou et al., 2023; Tan et al., 2021; Tesfaye et al., 2024; Ullrich et al., 2025; Zhu et al., 2022)
Molecular mechanisms	NMDA overactivation, ROS buildup, mitochondrial damage, neuroinflammation	Synaptic dysfunction, impaired LTP, neuronal injury	Hippocampus, cortex	Reduced plasticity, worsened learning/memory	(Lou et al., 2023; Zhu et al., 2022)

BBB, Blood-Brain Barrier; *fMRI*, Functional Magnetic Resonance Imaging; *HPA*, Hypothalamic-Pituitary-Adrenal; *IL-1β*, Interleukin-1 Beta; *LTP*, Long-Term Potentiation; *ROS*, Reactive Oxygen Species; *THR*, Thyroid Hormone Receptor; *TNF-α*, Tumor Necrosis Factor Alpha.

Structural and functional brain changes

Neuroimaging studies reveal hippocampal atrophy, cortical thinning, and disrupted connectivity between cognitive and emotional networks in hyperthyroidism.

Hippocampal shrinkage correlates with memory loss and likely reflects chronic excitotoxic damage from glutamate overstimulation and metabolic overload. Prefrontal cortex changes impair executive function and emotional regulation.

fMRI studies show altered connectivity across memory, attention, and affective circuits, likely due to hormone-induced mitochondrial stress, neurotransmitter imbalance, and synaptic dysfunction (Tesfaye et al., 2024).

At the molecular level, hyperthyroidism upregulates genes linked to synaptic signaling and neurotransmitter homeostasis. Excessive NMDA receptor activation causes calcium influx, mitochondrial impairment, and neuronal injury, especially in hippocampal and cortical areas (Zhu et al., 2022).

ROS accumulation further damages neurons, activating NF-κB–mediated inflammatory pathways that suppress synaptic plasticity and promote cell death. These processes impair LTP and undermine memory formation (Lou et al., 2023; Tan et al., 2021).

BBB integrity is also compromised, increasing permeability to circulating hormones and ROS, which amplify neuroinflammation. HPA axis dysregulation and elevated cortisol worsen hippocampal and prefrontal dysfunction, contributing to mood instability (Ullrich et al., 2025).

Together, excitotoxicity, oxidative stress, inflammation, and metabolic overload drive structural degeneration and functional disruption, producing the cognitive and emotional symptoms observed in hyperthyroidism (Table 25.1).

Key Takeaways:

1. Cognitive and executive dysfunctions in hyperthyroidism arise from hippocampal vulnerability, oxidative stress, neurotransmitter imbalance, and synaptic disruption.
2. Mood disturbances such as anxiety, irritability, and psychosis are rooted in monoaminergic dysregulation, HPA axis overactivation, and increased sympathetic drive.
3. Congenital hyperthyroidism, though rare, can have profound and lasting effects on brain development—impacting neurogenesis, synaptogenesis, myelination, and plasticity.
4. Structural and functional brain changes, including hippocampal atrophy and altered cortical connectivity, are visible through neuroimaging and stem from excitotoxicity, mitochondrial dysfunction, and inflammatory responses.
5. These neurological manifestations may persist even after achieving hormonal normalization, underscoring the need for early intervention and adjunctive neuroprotective strategies.

Pathophysiological mechanisms of brain damage

Hyperthyroidism-induced neurotoxicity arises from a convergence of metabolic and cellular disruptions driven by excessive thyroid hormone activity. The brain, given its high metabolic demand and limited antioxidant capacity, becomes especially vulnerable. Key mechanisms include oxidative stress, neuroinflammation, excitotoxicity, BBB dysfunction, and transcriptional dysregulation via thyroid hormone receptor (THR) overactivation.

Oxidative stress

Excess thyroid hormones (T3, T4) markedly elevate brain metabolic rate and oxygen consumption, especially in mitochondria where they accelerate oxidative phosphorylation. Increased activity at complexes I and III leads to electron leakage and superoxide formation, which is converted to hydrogen peroxide and hydroxyl radicals (Kochman et al., 2021; Vujovic & Farahani, 2025).

This ROS surge damages mitochondria early: membrane integrity is lost, permeability transition pores open, ATP synthesis falls, and cytochrome *c* release activates caspase-dependent apoptosis, driving neuronal loss (Wei et al., 2022).

ROS also trigger lipid peroxidation, generating malondialdehyde and 4-hydroxynonenal, which perpetuate membrane injury. DNA and protein oxidation disrupt gene transcription, enzymatic function, and synaptic signaling (Kochman et al., 2021; Macvanin et al., 2023).

Endogenous antioxidants (glutathione, SOD, catalase) become depleted, shifting the redox state toward chronic oxidative stress. NF-κB activation amplifies inflammation via cytokine induction and microglial activation, compounding mitochondrial injury (Shahat et al., 2022).

ROS also enhance NMDA receptor activity, increasing calcium influx and activating degradative enzymes such as calpains and phospholipases (Zhu et al., 2022).

Together, mitochondrial dysfunction, antioxidant depletion, lipid and protein oxidation, and calcium dysregulation create a self-perpetuating cycle of oxidative injury central to hyperthyroidism-induced neurodegeneration (Fig. 25.1).

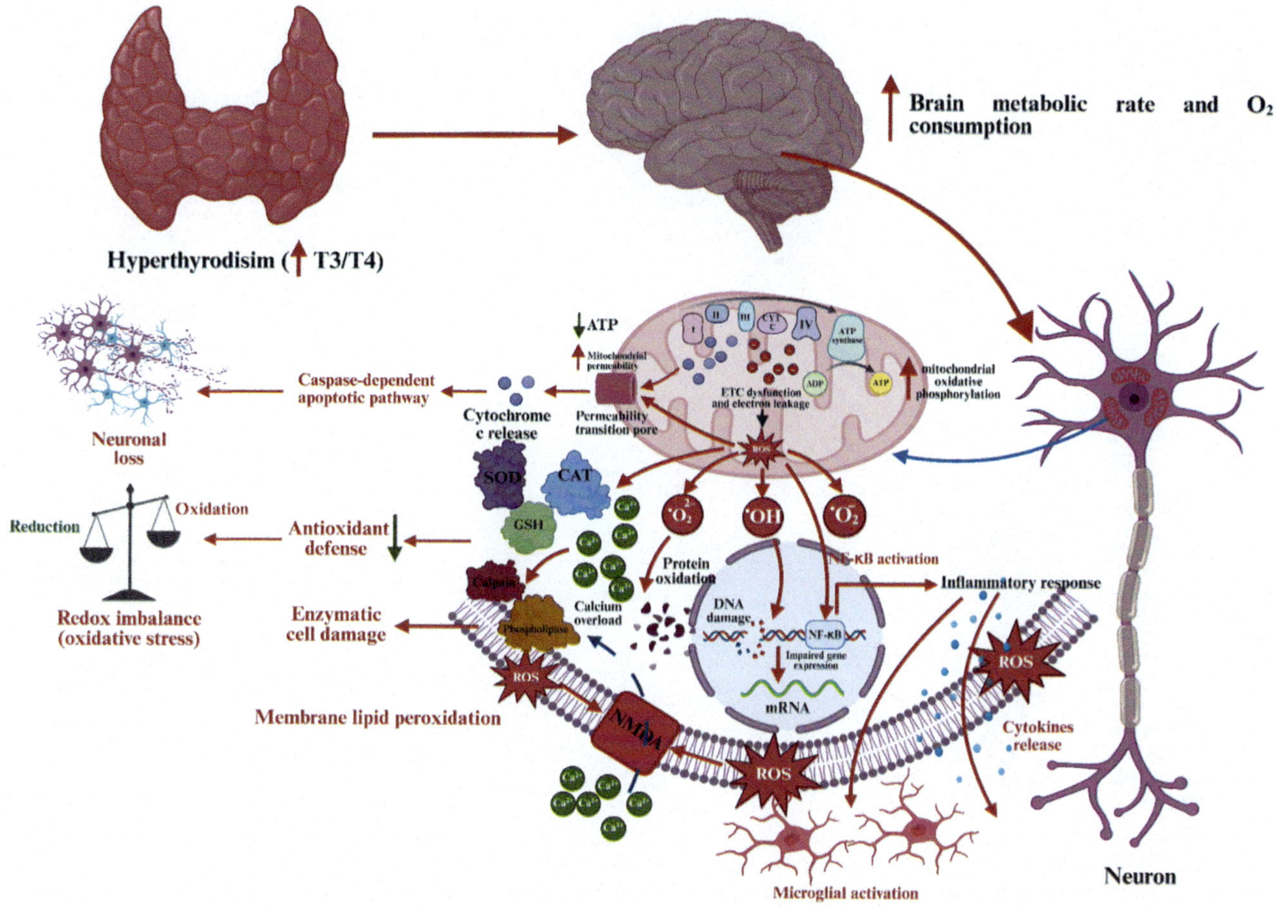

FIGURE 25.1 Hyperthyroidism-induced oxidative stress pathway in the brain. **Abbreviations**: *ATP*: Adenosine triphosphate, *ADP*: Adenosine triphosphate, *Ca2+*: Calcium ion, *CAT*: Catalase, *cyt c*: Cytochrome c, *DNA*: Deoxyribonucleic acid, *ETC*: Electron transport chain, *GSH*: Glutathione, *iNOS*: Inducible nitric oxide synthase, *NF-κB*: Nuclear factor kappa-light-chain-enhancer of activated B cells, *NMDA*: N-methyl-d-aspartate, *ROS*: Reactive oxygen species, *SOD*: Superoxide dismutase, *T3*: Triiodothyronine, *T4*: Thyroxine.x2

Neuroinflammation

Oxidative stress and metabolic overload in hyperthyroidism activate microglia, the CNS's immune cells, which release IL-1β, TNF-α, and IL-6 (Sepúlveda et al., 2024).

These cytokines disrupt synaptic signaling, impair plasticity, and promote glutamate buildup, raising excitotoxic risk. They also induce iNOS, elevating NO, which reacts with superoxide to form peroxynitrite—damaging proteins, lipids, and mitochondria (Baev et al., 2022; Davidson & Stevenson, 2024).

NF-κB activation perpetuates cytokine release and microglial reactivity. BBB disruption follows, permitting peripheral immune cell entry and amplifying neuroinflammation (Davidson & Stevenson, 2024; Sawicka-Gutaj et al., 2022).

Cytokines further impair glutamate transporters, reduce astrocytic clearance, and sensitize NMDA receptors, leading to calcium overload, calpain activation, cytoskeletal degradation, and neuronal injury. Mitochondrial dysfunction worsens energy failure and triggers apoptosis (Fernández-Serra et al., 2022; Sawicka-Gutaj et al., 2022).

Chronic microglial activation thus creates a toxic neural milieu that disrupts synapses, accelerates neurodegeneration, and drives the cognitive and psychiatric symptoms of hyperthyroidism (Fig. 25.2).

Excitotoxicity and neurotransmitter imbalance

Hyperthyroidism disrupts excitatory–inhibitory balance, driving excitotoxic stress. Glutamate becomes pathologically elevated through increased synthesis, excessive release, and impaired astrocytic uptake, causing persistent extracellular accumulation in synaptically dense regions (Zhu et al., 2022).

Chronic glutamate excess overstimulates NMDA and AMPA receptors, leading to calcium influx. Elevated calcium activates calpains, endonucleases, and phospholipases, damaging membranes, fragmenting DNA, and

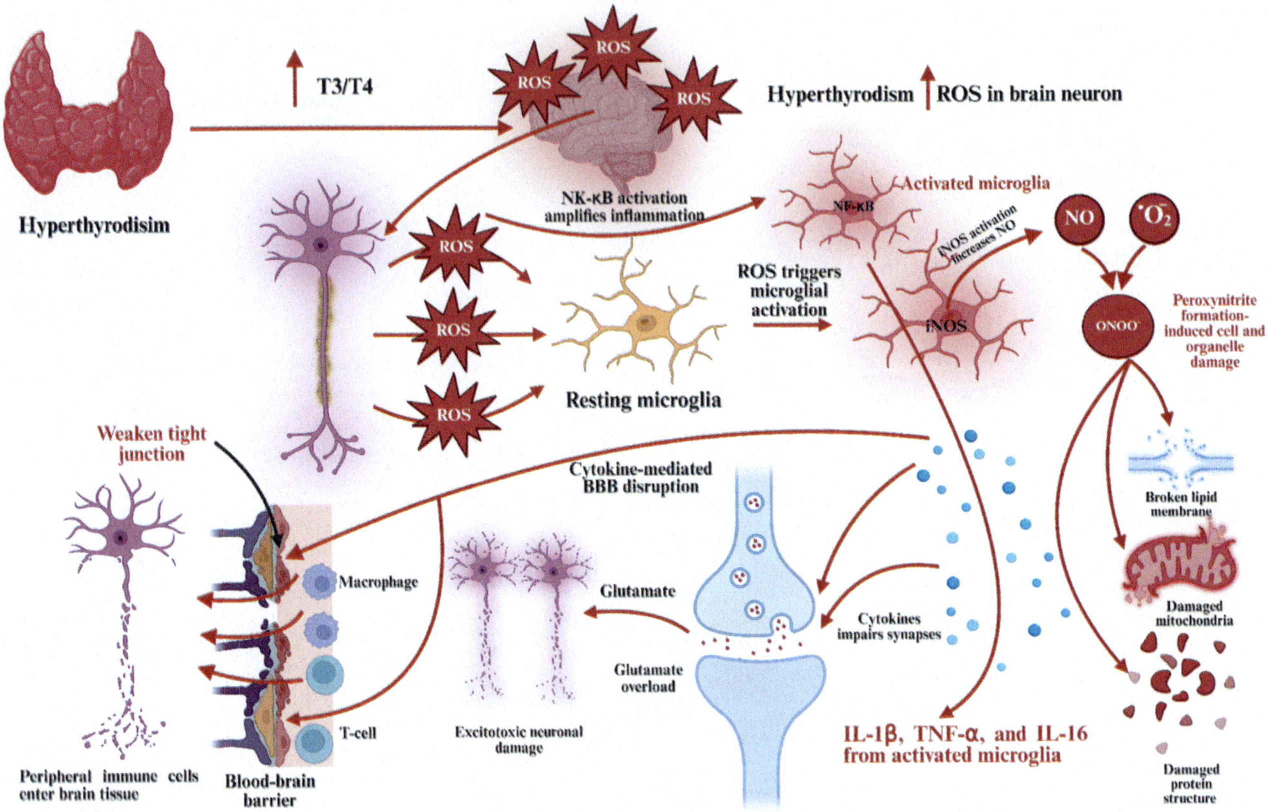

FIGURE 25.2 Neuroinflammation cascade in hyperthyroidism-induced brain dysfunction. **Abbreviations**: *ATP*: Adenosine triphosphate, *BBB*: Blood-brain barrier, *Ca2+*: Calcium ion, *IL-1β*: Interleukin-1 beta, *IL-6*: Interleukin-6, *iNOS*: Inducible nitric oxide synthase, *NF-κB*: Nuclear factor kappa-light-chain-enhancer of activated B cells, *NO*: Nitric oxide, *ONOO−*: Peroxynitrite, *ROS*: Reactive oxygen Species, *TNF-α*: Tumor necrosis factor-alpha.fx3

disrupting the cytoskeleton. Mitochondrial calcium overload further impairs ATP production, boosts ROS generation, and triggers apoptosis (Altunkaya et al., 2024; Li & Liu, 2024).

Inhibitory signaling is also reduced: GABA levels and receptor expression decline, removing a key brake on excitatory activity. Serotonin synthesis and receptor responsiveness are impaired, weakening modulation of excitatory transmission and neuronal resilience (Bogielski et al., 2024; Flamant & Richard, 2024).

These imbalances—elevated glutamate, reduced GABA, and impaired serotonin—create a self-reinforcing cycle of calcium dysregulation, oxidative injury, and neuronal loss, contributing to both acute symptoms and long-term neurodegeneration in hyperthyroidism.

Blood brain barrier disruption

In hyperthyroidism, BBB integrity is compromised, increasing the brain's vulnerability (Sawicka-Gutaj et al., 2022). Normally, the BBB—composed of endothelial cells, astrocytic end-feet, and pericytes—restricts harmful substances and maintains CNS homeostasis. Excess thyroid hormones destabilize this barrier by downregulating tight junction proteins such as claudin-5, occludin, and ZO-1, weakening endothelial cohesion and increasing paracellular permeability (Davidson & Stevenson, 2024). Matrix metalloproteinases, notably MMP-9, degrade the extracellular matrix and disrupt tight junctions (Sunny et al., 2024). The compromised BBB allows infiltration of peripheral immune cells, cytokines, and plasma toxins, triggering microglial activation and sustained inflammation, which accelerates oxidative stress, impairs synaptic signaling, and exacerbates neuronal injury (Gryka-Marton et al., 2025). BBB endothelial cells are also damaged by ROS from mitochondrial dysfunction and NADPH oxidase activity, impairing nitric oxide signaling and reducing cerebral perfusion (Shahat et al., 2022). Resulting hypoxia and vascular instability increase neuronal sensitivity to injury (Dunn & Isaacs, 2021). Altered BBB permeability also disrupts ion and neurotransmitter gradients; elevated extracellular glutamate and impaired astrocytic clearance promote excitotoxicity, while calcium and potassium dysregulation destabilizes neuronal firing (Gruenbaum et al., 2024). Thus, BBB disruption amplifies neurotoxicity in hyperthyroidism, facilitating the interplay of oxidative stress, inflammation, and excitotoxicity that drives neurodegeneration (Fig. 25.3).

Thyroid hormone receptor overactivation

THRs—primarily THRα and THRβ—are expressed in neurons, astrocytes, and oligodendrocytes, regulating genes related to metabolism, plasticity, and cell survival. In hyperthyroidism, overactivation of these receptors causes transcriptional dysregulation, transforming normal thyroid signaling into a driver of cellular dysfunction (Ariyani et al., 2022; Sawicka-Gutaj et al., 2022).

In neurons, excessive THR activity upregulates genes for mitochondrial biogenesis and oxidative phosphorylation, increasing ROS production, damaging mitochondria, and impairing ATP synthesis. Concurrently, elevated expression of voltage-gated calcium channels and excitatory receptor subunits enhances calcium influx, raising susceptibility to excitotoxic injury (Ollivry & Joubert, 2023).

Glial cells are also affected. In astrocytes, THR overactivation downregulates glutamate transporter genes (EAAT1, EAAT2), reducing glutamate clearance and promoting extracellular accumulation, while suppressing antioxidant enzymes and weakening ROS defense. In oligodendrocytes, disrupted THR signaling impairs myelin-related gene expression, compromising axonal conduction and structural integrity (Jurga et al., 2021; Li & Liu, 2024).

THRs further modulate inflammation. Overactivation in glial cells increases pro-inflammatory cytokines and chemokines, fueling microglial activation and chronic neuroinflammation. At the systems level, dysregulated THR signaling impairs synaptic protein expression, neurotrophic support, and cytoskeletal dynamics, leading to reduced plasticity, dendritic retraction, and impaired network connectivity (Lou et al., 2023; O'Shaughnessy et al., 2023).

Overall, THR overactivation provides a genomic mechanism by which thyroid hormone excess drives neurotoxicity, affecting neuronal excitability, glial homeostasis, and synaptic architecture (Fig. 25.4).

Key Takeaways:

1. Hyperthyroidism disrupts brain homeostasis via multiple overlapping mechanisms—oxidative stress, inflammation, excitotoxicity, BBB dysfunction, and transcriptional changes.

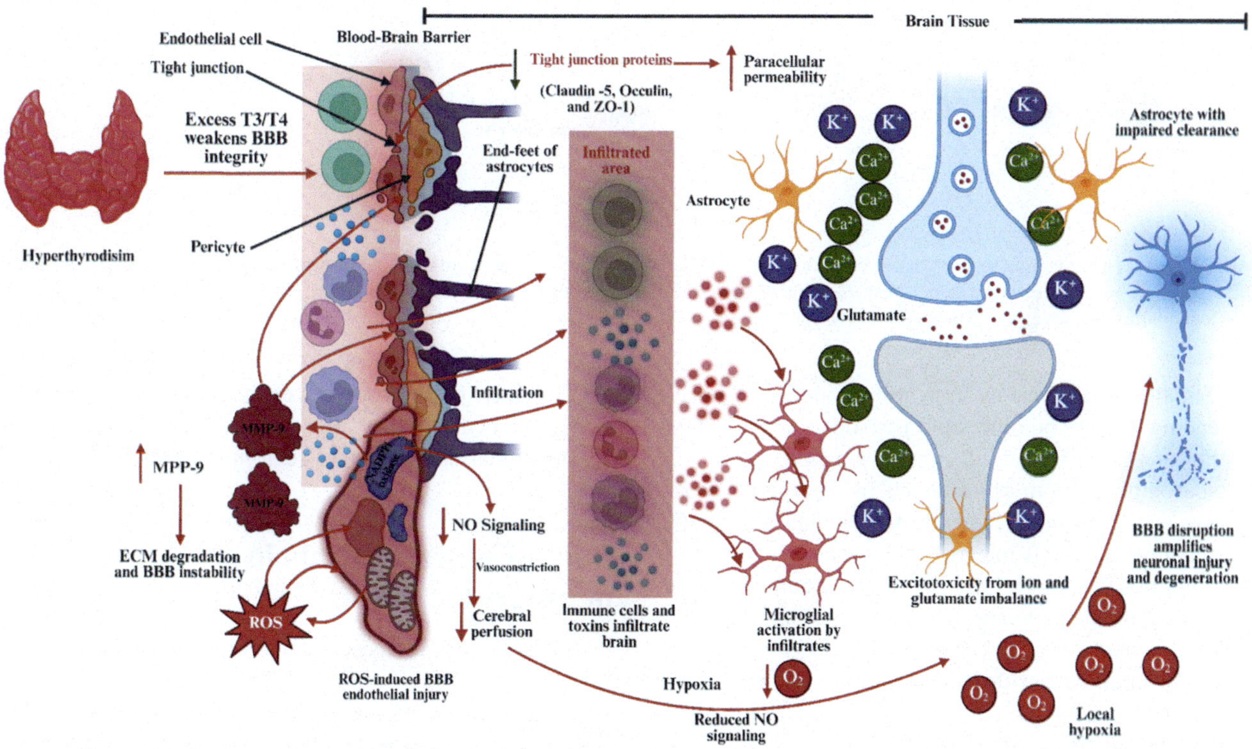

FIGURE 25.3 Blood-brain barrier (BBB) disruption and its role in hyperthyroidism-induced neurotoxicity. **Abbreviations**: *BBB*: Blood-brain barrier, Ca^{2+}: Calcium ion, *ECM*: Extracellular matrix, K^{+}: Potassium ion, *MMP-9*: Matrix Metalloproteinase-9, *NADPH*: Nicotinamide adenine dinucleotide phosphate, *NO*: Nitric oxide, *ROS*: Reactive oxygen species, *T3*: Triiodothyronine, *T4*: Thyroxine, *TNF-α*: Tumor necrosis factor-alpha, *ZO-1*: Zonula Occludens-1.fx4

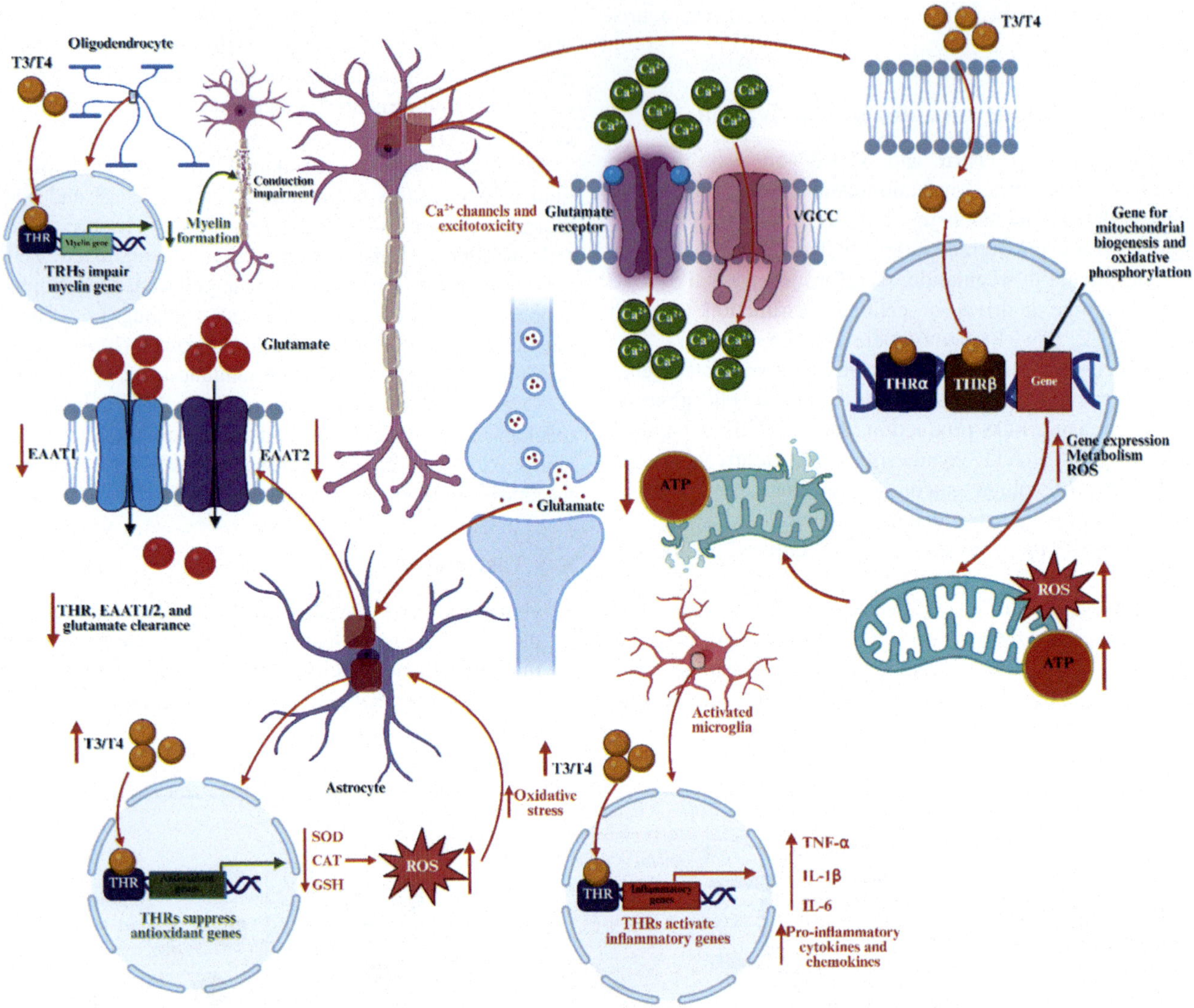

FIGURE 25.4 Thyroid hormone receptor overactivation and genomic neurotoxicity in hyperthyroidism. **Abbreviations**: *ATP*: Adenosine triphosphate, *Ca2+*: Calcium ion, *IL-1β*: Interleukin-1 beta, *IL-6*: Interleukin-6, *EAAT1*: Excitatory amino acid transporter, *ROS*: Reactive oxygen species, *T3*: Triiodothyronine, *T4*: Thyroxine, *THRs*: Thyroid hormone receptors, *TNF-α*: Tumor necrosis factor-alpha, *VGCCs*: Voltage-gated calcium channel.

2. Oxidative stress results from mitochondrial overdrive, leading to ROS buildup, lipid peroxidation, and activation of apoptosis and inflammatory pathways.
3. Neuroinflammation is sustained by microglial activation and cytokine release, further impairing synaptic signaling and BBB integrity.
4. Excitotoxicity arises from glutamate accumulation and calcium overload, worsened by reduced GABAergic inhibition and serotonergic imbalance.
5. Blood–brain barrier disruption allows peripheral toxins and immune cells to enter the brain, amplifying oxidative and inflammatory injury.
6. Thyroid hormone receptor overactivation dysregulates gene expression in neurons and glia, altering metabolism, synaptic architecture, and neuroimmune signaling.
7. Together, these processes create a self-reinforcing cycle of neurotoxicity that underlies the structural and functional brain changes seen in hyperthyroidism.

Role of nutraceuticals in neuroprotection

Hyperthyroidism increases brain vulnerability to oxidative stress, inflammation, and excitotoxicity, prompting interest in nutraceuticals as adjunctive neuroprotective agents. Antioxidants, anti-inflammatory compounds, adaptogens, micronutrients, and gut–brain modulators target multiple pathways to limit oxidative injury, regulate inflammation, support mitochondrial function, and preserve cognition (Shahat et al., 2022).

Antioxidants

Excess thyroid hormones accelerate metabolism and ROS production, overwhelming neural defenses and causing mitochondrial dysfunction. Nutraceutical antioxidants—including vitamin E, vitamin C, CoQ10, resveratrol, and curcumin—neutralize ROS, stabilize mitochondria, and protect against hormone-induced neurotoxicity (Sultana et al., 2022).

Vitamin E, vitamin C, and coenzyme Q10

Vitamin E integrates into neuronal membranes, preventing lipid peroxidation and preserving synaptic function. Through Nrf2/ARE activation, it upregulates HO-1, SOD, and catalase, reducing ROS generation. γ-Tocopherol scavenges peroxynitrite, elevated in hyperthyroidism (Andrés et al., 2024).

Vitamin C regenerates oxidized vitamin E, detoxifies ROS/RNS, and supports mitochondrial function and neurotransmitter regulation, improving mood and cognition (Maliki et al., 2021).

CoQ10 supports oxidative phosphorylation, reduces ROS, stabilizes membranes, and promotes mitochondrial biogenesis via PGC-1α (Bagheri et al., 2023; Pradhan et al., 2021). Combined, these antioxidants synergistically maintain mitochondrial integrity and neuronal energy balance (Table 25.2).

Polyphenols

Resveratrol and curcumin provide complementary neuroprotection by reducing oxidative stress, excitotoxicity, and mitochondrial dysfunction. Resveratrol activates the SIRT1–PGC-1α axis, enhances mitochondrial biogenesis, modulates NMDA receptors, and promotes autophagy (Bhandari et al., 2025).

Curcumin activates Nrf2/ARE signaling, increases antioxidant enzymes, prevents lipid peroxidation, and regulates ferroptosis (Foroutan et al., 2024). Together, they represent a promising polyphenolic approach for neuroprotection in hyperthyroidism (Table 25.2).

Anti-inflammatory agents

Omega-3 fatty acids and quercetin counter neuroinflammation. DHA and EPA inhibit NF-κB, downregulate TNF-α and IL-6, and promote NPD1 production, restoring anti-inflammatory signaling suppressed by FADS2 downregulation (Korbecki et al., 2023).

Quercetin inhibits COX-2, iNOS, and NF-κB while modulating TLR4/MyD88 signaling, reducing glial activation and supporting a balanced neuroimmune state (Yu et al., 2022).

Adaptogens and neuroprotective herbs

Bacopa monnieri, Ginkgo biloba, and *Withania somnifera* provide multidimensional protection. *Bacopa* enhances neuroplasticity, boosts BDNF, supports cholinergic signaling, and regulates HPA activity (Sekhar et al., 2019).

Ginkgo improves cerebral blood flow, supports mitochondrial integrity, and maintains synaptic function under metabolic strain (Bagheri et al., 2023).

Withania modulates oxidative stress, stabilizes mitochondria, calms the HPA axis, and interacts with GABA-A receptors, potentially reducing anxiety (Anand et al., 2025).

Micronutrients

Selenium, zinc, magnesium, and iodine support thyroid–brain homeostasis.

Selenium serves as a cofactor for GPx enzymes, detoxifying peroxides and supporting Dio2 activity to stabilize intracellular thyroid signaling (Bryliński et al., 2025).

Zinc modulates NMDA receptors, limits excitotoxicity, and supports SOD activity (Mitchell et al., 2023).

Magnesium facilitates T4-to-T3 conversion, prevents calcium overload, and dampens microglial activation (Ollivry & Joubert, 2023; Patel et al., 2024).

Iodine is essential for hormone synthesis but excess may increase ROS and thyroid antigenicity, exacerbating autoimmunity (Khudair et al., 2025).

Gut–brain axis modulators

Dysbiosis in hyperthyroidism worsens systemic inflammation and neurotransmitter imbalance. Probiotics (*Lactobacillus, Bifidobacterium*) enhance gut barrier integrity, regulate immune signaling, and modulate the kynurenine–tryptophan pathway to favor neuroprotection. Their SCFAs, particularly butyrate, cross the BBB, upregulate BDNF, and support synaptic plasticity, while also enhancing GABAergic tone (Gou et al., 2022; Shu et al., 2024).

Prebiotics (FOS, GOS) restore microbial diversity, promote SCFA production, fortify gut barriers, and reduce systemic inflammation. Butyrate enhances mitochondrial function and suppresses pro-inflammatory gene expression. Some prebiotics may also reduce thyroid autoantibody levels in Graves' disease (Shu et al., 2024; Zawadzka et al., 2023).

Limitations of nutraceuticals

Clinical translation is limited by dosing discrepancies, poor bioavailability, variable host response, and drug

TABLE 25.2 Role of nutraceuticals in neuroprotection in hyperthyroidism-induced neurotoxicity.

Nutraceutical	Key mechanisms	Main effects & outcomes	References
Vitamin E	Integrates into membranes, scavenges lipid radicals, activates Nrf2/ARE (↑HO-1, SOD, catalase)	Prevents peroxidation, preserves synaptic function, stabilizes mitochondria & energy production	(Andrés et al., 2024; Macvanin et al., 2023; Sultana et al., 2022)
Vitamin C	Regenerates vitamin E, detoxifies ROS/RNS, supports mitochondrial & neurotransmitter function	Reduces oxidative stress, maintains synaptic plasticity, supports mood & cognition	Maliki et al. (2021)
CoQ10	Supports oxidative phosphorylation, reduces ROS, activates PGC-1α	Enhances ATP production, improves mitochondrial resilience, preserves neuronal energy balance	(Bagheri et al., 2023; Macvanin et al., 2023; Pradhan et al., 2021)
Resveratrol	Activates SIRT1–PGC-1α, promotes autophagy, modulates NMDA receptors	Improves mitochondrial biogenesis, reduces oxidative injury, preserves synaptic integrity	(Bhandari et al., 2025; Kung et al., 2021)
Curcumin	Activates Nrf2/ARE, prevents lipid peroxidation, protects mitochondria	Boosts antioxidant enzymes, stabilizes mitochondria, reduces oxidative stress	(Foroutan et al., 2024; Shakeri et al., 2022)
Omega-3 (DHA, EPA)	Inhibit NF-κB, ↓TNF-α/IL-6, restore FADS2 activity	Reduce neuroinflammation, protect neurons, improve cognition & mood	(Korbecki et al., 2023; Ruggeri et al., 2023)
Quercetin	Inhibits COX-2, iNOS, NF-κB; modulates TLR4/MyD88	Limits glial activation, supports balanced neuroimmune signaling	Yu et al. (2022)
Bacopa monnieri	Enhances BDNF, cholinergic signaling, regulates HPA axis	Improves memory, reduces stress, stabilizes neuroendocrine function	(Sekhar et al., 2019; Tesse et al., 2018)
Ginkgo biloba	Improves cerebral blood flow, preserves mitochondria, reduces oxidative/inflammatory stress	Maintains synaptic and cognitive function under metabolic strain	(Aritonang, 2021; Bagheri et al., 2023)
Withania somnifera	Modulates oxidative stress, stabilizes mitochondria, interacts with GABA-A receptors	Reduces anxiety, calms HPA axis, supports cognitive stability	Anand et al. (2025)
Selenium	Cofactor for GPx, detoxifies peroxides, supports Dio2 activity	Prevents ferroptosis, stabilizes thyroid hormone levels, protects neurons	Bryliński et al. (2025)
Zinc	Modulates NMDA receptors, supports SOD activity	Reduces excitotoxicity, preserves neuronal integrity & emotional stability	Mitchell et al. (2023)
Magnesium	Stabilizes mitochondria, converts T4→T3, NMDA antagonism	Prevents calcium overload, reduces excitotoxicity, protects neurons	(Ollivry & Joubert, 2023; Patel et al., 2024)
Iodine	Excess increases ROS, immune reactivity, impairs thyroid peroxidase	May worsen neuroinflammation and cognitive dysfunction in autoimmune thyroid disease	Khudair et al. (2025)
Probiotics (*Lactobacillus*, *Bifidobacterium*)	Enhance gut barrier, modulate neurotransmitters & GABA signaling	Reduce neuroinflammation, improve cognition & emotional stability via gut–brain axis	(Gou et al., 2022; Shu et al., 2024)
Prebiotics (FOS, GOS)	Promote SCFA-producing bacteria, restore microbiota balance	Enhance gut integrity, modulate immune responses, support brain function	Zawadzka et al. (2023)

BDNF, Brain-Derived Neurotrophic Factor; CoQ10, Coenzyme Q10; COX-2, Cyclooxygenase-2; FOS, Fructooligosaccharides; GOS, Galactooligosaccharides; GPx, Glutathione Peroxidase; HPA, Hypothalamic-Pituitary-Adrenal; iNOS, Inducible Nitric Oxide Synthase; NF-κB, Nuclear Factor Kappa-Light-Chain-Enhancer of Activated B Cells; NMDA, N-Methyl-D-Aspartate; NPD1, Neuroprotectin D1; PGC-1α, Peroxisome Proliferator-Activated Receptor Gamma Coactivator 1 Alpha; RNS, Reactive Nitrogen Species; ROS, Reactive Oxygen Species; SCFAs, Short-Chain Fatty Acids; SIRT1, Sirtuin 1; TLR4/MyD88, Toll-Like Receptor 4/Myeloid Differentiation Factor 88.fx1 interactions. Rodent studies often use doses impractical for humans, and many compounds (e.g., resveratrol, curcumin) undergo rapid metabolism, requiring delivery systems like nanoparticles or piperine. Fat-soluble antioxidants depend on dietary fat absorption, which may be impaired in hyperthyroidism.

Micronutrients must be carefully dosed—excess selenium may disrupt thyroid metabolism, and zinc can upset copper balance. Probiotic efficacy varies with strain survival and colonization.

Potential drug interactions include vitamin E and CoQ10 interfering with oxidative pathways targeted by methimazole, resveratrol affecting CYP3A4 metabolism, and adaptogens altering thyroid–adrenal signaling (Macvanin et al., 2023). Thus, nutraceuticals should be considered complementary and used under medical supervision.

Key Takeaways:

1. Hyperthyroidism impairs brain function through oxidative stress, inflammation, mitochondrial dysfunction, and excitotoxicity—key drivers of cognitive and emotional symptoms.
2. Antioxidants and polyphenols (e.g., vitamin E, CoQ10, resveratrol, curcumin) offer neuroprotection by reducing ROS, preserving mitochondria, and modulating inflammatory pathways.
3. Anti-inflammatory and adaptogenic agents such as omega-3s, quercetin, *Bacopa monnieri,* and *Withania somnifera* improve neuroimmune balance, stress resilience, and synaptic health.
4. Micronutrients like selenium, zinc, and magnesium support thyroid hormone regulation and neurotransmitter stability but require careful dosing to avoid adverse effects.
5. Probiotics, prebiotics, and butyrate enhance gut–brain signaling, reduce systemic inflammation, and promote neurotrophic factors—indirectly benefiting cognition and mood.
6. Clinical use of nutraceuticals must account for bioavailability, individual variability, and potential drug interactions; they should complement—not replace—standard therapy.

Other hormones, endocrine tissues, and pathways

While thyroid hormones play the central role in hyperthyroidism, their effects reverberate throughout the endocrine network. Understanding these interactions is critical for translating mechanistic insights into clinical strategies that address the full spectrum of systemic and neurological consequences.

Crosstalk with the hypothalamic–pituitary–adrenal (HPA) axis

Excess thyroid hormones activate the HPA axis, leading to elevated cortisol levels and heightened stress responses. Chronically high cortisol can damage hippocampal neurons, impair synaptic plasticity, and exacerbate mood disturbances commonly reported in hyperthyroid patients. Therapeutically, interventions that stabilize cortisol output—such as stress-reduction programs or adaptogenic botanicals—may help restore emotional balance and protect vulnerable brain regions (Ring, 2025).

Interactions with sex hormones

Hyperthyroidism often alters gonadal steroid levels, producing menstrual irregularities, decreased fertility, and changes in libido. In women, elevated estrogen metabolism can worsen mood instability, while in men, suppressed testosterone may contribute to fatigue and muscle wasting. These hormonal shifts may indirectly influence brain function by altering serotonergic and dopaminergic signaling, providing a potential rationale for combined endocrine and psychiatric management strategies (Liu et al., 2024).

Metabolic and pancreatic effects

Thyroid hormone excess accelerates glucose turnover, increases insulin clearance, and may precipitate impaired glucose tolerance or overt diabetes in susceptible individuals. Dysregulated insulin signaling has been linked to cognitive decline and neuroinflammation, suggesting that monitoring and optimizing glucose control could have neuroprotective benefits (Choi et al., 2021).

Bone–endocrine interactions

The hypermetabolic state also affects the parathyroid–vitamin D axis. Hyperthyroidism accelerates bone resorption, increasing fracture risk and releasing calcium into circulation. Elevated calcium levels can affect neuronal excitability and synaptic signaling, potentially compounding excitotoxicity in the brain. Early detection and management of bone loss—through antiresorptive agents or vitamin D repletion—represent important translational steps to protect both skeletal and neural health (Daya et al., 2022).

Catecholaminergic and sympathetic pathways

Excess thyroid hormone sensitizes tissues to catecholamines by upregulating β-adrenergic receptors. Clinically, this manifests as tachycardia, tremor, and heightened anxiety. β-Blockers are already a mainstay of symptomatic treatment, but emerging evidence suggests that modulating adrenergic signaling may also reduce excitotoxicity and improve cognitive outcomes (Szybiak-Skora et al., 2025).

Translational opportunities

These interlinked endocrine pathways highlight opportunities for integrative care. For example, combining antithyroid therapy with interventions that normalize cortisol, support sex hormone balance, stabilize glucose metabolism, and protect bone health may produce better neurological and systemic outcomes than focusing solely on thyroid hormone levels. This systems-based approach aligns with precision medicine principles, encouraging clinicians to consider each patient's broader endocrine and metabolic profile when designing treatment strategies (Das et al., 2022).

Applications of the material

The mechanistic and therapeutic insights described in this chapter have several practical applications across clinical care, research, and public health. Understanding the thyroid–brain axis and its interaction with other endocrine systems offers a framework for more comprehensive patient management and for developing novel strategies to mitigate hyperthyroidism-related neurotoxicity.

Clinical practice

One key application is the early recognition of neuropsychiatric symptoms in patients with hyperthyroidism. Many individuals present first with anxiety, irritability, or cognitive complaints before overt endocrine signs are apparent. Incorporating cognitive screening tools and mood assessments into routine evaluation of hyperthyroid patients may facilitate timely intervention and prevent prolonged neurotoxicity.

Another application is the use of adjunctive nutraceutical strategies alongside antithyroid medications. For example, antioxidants such as vitamin E and CoQ10 may help preserve mitochondrial function, while omega-3 fatty acids and adaptogens can modulate neuroinflammation and HPA axis overactivity. Although more clinical trials are needed, these approaches have the potential to improve quality of life by addressing residual neurological symptoms that persist even after euthyroidism is achieved (Li et al., 2017).

Attention to broader endocrine interactions can also inform treatment. For instance, monitoring cortisol, glucose, and bone turnover markers can guide interventions that protect brain health, such as stress management programs, insulin-sensitizing therapies, and vitamin D repletion (Mantle & Hargreaves, 2023).

Research and innovation

The mechanistic framework presented here can guide biomarker discovery and precision medicine approaches. Oxidative stress markers, cytokine profiles, and neuroimaging findings could be incorporated into future studies to stratify patients by risk of cognitive decline and track treatment response.

Similarly, preclinical work on nutraceutical combinations—such as CoQ10 with lipoic acid or psychobiotic probiotics—can be translated into pilot clinical trials. These trials should integrate cognitive endpoints and neuroinflammatory biomarkers to evaluate whether adjunctive therapy reduces long-term neurological burden.

Emerging computational tools, including AI-based predictive modeling, can leverage genetic, metabolic, and microbiome data to personalize therapy. Such approaches may ultimately allow clinicians to identify which patients will benefit most from specific nutraceutical regimens or lifestyle interventions (Lu et al., 2025).

Public health and education

Finally, this material has implications for medical education and awareness campaigns. Teaching clinicians to consider thyroid function testing in patients with unexplained mood disorders, psychosis, or cognitive complaints can reduce misdiagnosis and expedite appropriate treatment. Public health initiatives that promote diets rich in neuroprotective nutrients—such as omega-3 fatty acids and polyphenols—may support brain health in populations at risk for thyroid dysfunction (Tur et al., 2012).

Therapeutic perspectives and future directions

Several strategies hold promise for advancing the management of hyperthyroidism-related neurotoxicity.

Personalized nutraceutical strategies

Redox phenotyping and genomic screening may allow tailoring of antioxidant or micronutrient regimens to individual profiles. Patients with low glutathione or SOD2 polymorphisms may benefit most from CoQ10 or vitamin E supplementation.

Synergistic combinations

Combining CoQ10, lipoic acid, and creatine improved oxidative stress markers and muscle function in mitochondrial disorders, suggesting potential for hyperthyroid-related mitochondrial deficits. Future studies should test this combination to preserve neural energy metabolism and reduce cognitive symptoms.

Vitamin E for mitochondrial protection

Animal studies show vitamin E reduces mitochondrial ROS and preserves integrity in hyperthyroidism. Early-phase trials could assess its impact on mitochondrial biomarkers and cognition in newly diagnosed patients.

Gut–brain axis interventions

Psychobiotic probiotics such as *Lactobacillus rhamnosus* GG have improved cognition in clinical trials outside hyperthyroidism. Translational studies could test targeted probiotics in Graves' disease or subclinical hyperthyroidism to improve emotional and cognitive outcomes.

Microbiome modulation

Some probiotic formulations may worsen Graves' orbitopathy, highlighting the need for carefully selected synbiotics (probiotics + prebiotics) and immune-profile–based stratification to maximize safety and efficacy.

Systems biology and AI-guided design

Metabolomics and microbiome data can feed AI models to generate personalized nutraceutical blends—e.g., curcumin + omega-3 or CoQ10 + selenium—optimized for molecular signatures and symptom profiles.

Biomarker-led trials

Future trials should include objective biomarkers of neurotoxicity—oxidative damage markers, imaging of hippocampal integrity, and cognitive measures—to improve mechanistic insight and guide personalized interventions.

Key Takeaways:

1. Personalized nutraceutical therapy using redox biomarkers and genetic profiling may optimize neuroprotective outcomes in hyperthyroidism.
2. Synergistic combinations (e.g., CoQ10 + lipoic acid + creatine) and targeted agents like vitamin E show translational promise for protecting mitochondrial and cognitive health.
3. Gut–brain interventions, including psychobiotics and synbiotics, may support emotional and cognitive stability but must be tailored to avoid autoimmune complications.
4. AI-guided and systems biology approaches offer future potential for individualized nutraceutical design based on molecular, metabolic, and microbiome data.
5. Biomarker-based clinical trials are needed to validate nutraceutical efficacy in hyperthyroid neurotoxicity, using objective measures of neuronal damage and cognitive function.

Summary points

- Hyperthyroidism disrupts brain function via oxidative stress, neuroinflammation, excitotoxicity, blood–brain barrier breakdown, and thyroid hormone receptor overactivation.
- Cognitive and psychiatric symptoms (memory loss, anxiety, irritability, psychosis) often appear early and may persist after hormone normalization.
- Neuroimaging shows hippocampal atrophy and altered cortical connectivity, correlating with cognitive impairment.
- Hyperthyroidism also impacts adrenal, gonadal, pancreatic, and bone pathways, affecting cortisol, sex hormones, glucose, and calcium balance.
- Nutraceuticals (antioxidants, omega-3s, adaptogens, probiotics) target key neurotoxic mechanisms and may improve neurological outcomes.
- Bioavailability issues, dosing variability, and drug–nutrient interactions limit clinical application and require individualized approaches.
- Future work should focus on biomarker-guided interventions and combined endocrine–nutraceutical strategies for precision care.

Conclusion

Hyperthyroidism causes neurotoxicity through oxidative stress, mitochondrial dysfunction, neuroinflammation, and excitotoxicity, leading to cognitive and emotional impairments. Conventional therapies restore thyroid hormone levels but often do not fully address neurological symptoms. Nutraceuticals—including antioxidants, anti-inflammatory agents, adaptogens, micronutrients, and microbiota-targeted compounds—offer a multifaceted adjunctive approach. They support mitochondrial resilience, modulate immune signaling, and stabilize neurotransmission. Clinical application is limited by formulation challenges, absorption issues, potential drug-nutrient interactions, and interindividual variability related to genetics, metabolism, and microbiota. Future research should focus on rigorously designed trials, optimized delivery systems, and integrative strategies combining dietary, pharmacologic, and lifestyle approaches. Within precision medicine, nutraceuticals may expand therapeutic options for managing neuropsychiatric consequences of thyroid dysfunction.

Mini-dictionary of terms

Thyrotoxicosis: A clinical state caused by excessive thyroid hormone activity, leading to symptoms such as weight loss, heat intolerance, and palpitations.

Hyperthyroidism: Overproduction of thyroid hormones (T3 and T4) by the thyroid gland, most often due to Graves' disease, toxic multinodular goiter, or thyroid adenomas.

Thyroid Hormone Receptors (THRs): Nuclear receptors that bind thyroid hormones and regulate the expression of genes controlling metabolism, brain development, and cellular energy use.

Excitotoxicity: Damage to nerve cells caused by excessive stimulation by excitatory neurotransmitters, especially glutamate, resulting in calcium overload and cell death.

Oxidative Stress: An imbalance between free radical production and antioxidant defenses, causing damage to proteins, lipids, and DNA.

Neuroinflammation: Activation of immune responses in the brain, primarily involving microglia, leading to cytokine release and potential damage to neural circuits.

HPA Axis (Hypothalamic–Pituitary–Adrenal Axis): The hormonal system controlling the stress response; chronic activation can raise cortisol levels and affect brain health.

Blood--Brain Barrier (BBB): A selective barrier that separates the brain from circulating blood, protecting neural tissue from toxins and pathogens.

Nutraceuticals: Natural compounds from foods or plants that provide health benefits beyond basic nutrition and may help counteract disease processes.

Adaptogens: Substances that help the body adapt to stress and restore balance to hormonal and immune systems.

Gut–Brain Axis: The communication network linking the gut microbiome, immune system, and nervous system, influencing mood, cognition, and stress responses.

Ethics approval

Not applicable.

Availability of data and materials

Manuscript has no associated data.

Acknowledgments

Not applicable.

Competing interests

The authors have no conflict of interest to disclose.

Funding

The authors declare that no funds were received during the preparation of this manuscript.

Authors' contributions

Conceptualization and Design of the Study: WME. Supervision: WME. Writing original draft: MAA, MAT, YMH. Design of figures: MAA. Review and editing: WME. All authors read and approved the final manuscript.

References

Alšauskė, S. V., Liseckienė, I., & Verkauskienė, R. (2024). The association of thyroid disease with risk of dementia and cognitive impairment: A systematic review. *Medicina, 60*(12), 1917. https://doi.org/10.3390/medicina60121917

Altunkaya, M., Babur, E., Tufan, E., Süer, C., & Dursun, N. (2024). How thyroid hormone disorder affect the expression of genes related to neurodegeneration in the hippocampus tissue of rats: An experimental study. *Journal of Clinical Sciences, 21*(4), 175–179. https://doi.org/10.4103/jcls.jcls_34_24

Anand, A., Shrivastava, A., Singh, K., Barik, R., Gayakwad, D.,S.,J., Shamim, & Dwivedi, S. (2025). Neuroprotective efficacy and complementary treatment with medicinal herbs: A comprehensive review of recent therapeutic approaches in epilepsy management. *CNS & Neurological Disorders - Drug Targets, 24*(1), 60–73. https://doi.org/10.2174/0118715273332140240724093837

Andrés, C. M. C., Pérez De La Lastra, J. M., Juan, C. A., Plou, F. J., & Pérez-Lebeña, E. (2024). Antioxidant metabolism pathways in vitamins, polyphenols, and selenium: Parallels and divergences. *International Journal of Molecular Sciences, 25*(5), 2600. https://doi.org/10.3390/ijms25052600

Aritonang, D. Y. E. (2021). The potency of ginkgo biloba in treating tinnitus: A review. *Eureka Herba Indonesia, 3*(2), 145–150. https://doi.org/10.37275/ehi.v3i2.34

Ariyani, W., Miyazaki, W., Amano, I., & Koibuchi, N. (2022). Involvement of integrin αvβ3 in thyroid hormone-induced dendritogenesis. *Frontiers in Endocrinology, 13*, Article 938596. https://doi.org/10.3389/fendo.2022.938596

Baev, A. Y., Vinokurov, A. Y., Novikova, I. N., Dremin, V. V., Potapova, E. V., & Abramov, A. Y. (2022). Interaction of mitochondrial calcium and ROS in neurodegeneration. *Cells, 11*(4), 706. https://doi.org/10.3390/cells11040706

Bagheri, S., Haddadi, R., Saki, S., Kourosh-Arami, M., Rashno, M., Mojaver, A., & Komaki, A. (2023). Neuroprotective effects of coenzyme Q10 on neurological diseases: A review article. *Frontiers in Neuroscience, 17*, Article 1188839. https://doi.org/10.3389/fnins.2023.1188839

Bakalov, D., Iliev, P., Sabit, Z., Tafradjiiska-Hadjiolova, R., & Bocheva, G. (2023). Attenuation of hypothyroidism-induced cognitive impairment by modulating serotonin mediation. *Veterinary Sciences, 10*(2), 122. https://doi.org/10.3390/vetsci10020122

Bernardes, B. M., Borba, M. C. D. O., Moreira, S., Sartori, C. S., Júnior, I. F. D. S., Souza, M. T. L. D., Câmara, A. C. B., Bastos, W. L., Silveira, L. G., Rezende, J. M. E., Folgosi, F., Franzoni, I., Berquo, J. M., & Moreira, A. K. L. G. (2024). Relationship between mood disorders and thyroid changes. *Brazilian Journal of Implantology and Health Sciences, 6*(2), 2241–2256. https://doi.org/10.36557/2674-8169.2024v6n2p 2241-2256

Bhandari, S., Trivedi, R., Garg, V., Saini, A., Gupta, M., & Singh, R. (2025). Exploring the role of phytoconstituents for endocrine disorder: A review. *Pharmacognosy Research, 17*(2), 402–410. https://doi.org/10.5530/pres.20252070

Bogielski, B., Michalczyk, K., Głodek, P., Tempka, B., Gębski, W., & Stygar, D. (2024). Association between small intestine bacterial overgrowth and psychiatric disorders. *Frontiers in Endocrinology, 15*, Article 1438066. https://doi.org/10.3389/fendo.2024.1438066

Byliński, Ł., Kostelecka, K., Woliński, F., Komar, O., Miłosz, A., Michalczyk, J., Biłogras, J., Machrowska, A., Karpiński, R., Maciejewski, M., Maciejewski, R., Garruti, G., Flieger, J., & Baj, J. (2025). Effects of trace elements on endocrine function and pathogenesis of thyroid diseases—A literature review. *Nutrients, 17*(3), 398. https://doi.org/10.3390/nu17030398

Chincarini, G., Walker, D. W., Wong, F., Richardson, S. J., Cumberland, A., & Tolcos, M. (2024). Thyroid hormone analogues: Promising therapeutic avenues to improve the neurodevelopmental outcomes of intrauterine growth restriction. *Journal of Neurochemistry, 168*(9), 2335–2350. https://doi.org/10.1111/jnc.16124

Choi, Y. M., Kim, M. K., Kwak, M. K., Kim, D., & Hong, E.-G. (2021). Association between thyroid hormones and insulin resistance indices based on the Korean national health and nutrition examination survey. *Scientific Reports, 11*(1), Article 21738. https://doi.org/10.1038/s41598-021-01101-z

Das, D., Banerjee, A., Jena, A. B., Duttaroy, A. K., & Pathak, S. (2022). Essentiality, relevance, and efficacy of adjuvant/combinational therapy in the management of thyroid dysfunctions. *Biomedicine & Pharmacotherapy, 146*, Article 112613. https://doi.org/10.1016/j.biopha.2022.112613

Davidson, T. L., & Stevenson, R. J. (2024). Vulnerability of the hippocampus to insults: Links to blood–brain barrier dysfunction. *International Journal of Molecular Sciences, 25*(4), 1991. https://doi.org/10.3390/ijms25041991

Daya, N. R., Fretz, A., Martin, S. S., Lutsey, P. L., Echouffo-Tcheugui, J. B., Selvin, E., & Juraschek, S. P. (2022). Association between subclinical thyroid dysfunction and fracture risk. *JAMA Network Open, 5*(11), Article e2240823. https://doi.org/10.1001/jamanetworkopen.2022.40823

Doucette, S., Tierney, A., Roggensack, A., & Yusuf, K. (2018). Neonatal thyrotoxicosis with tricuspid valve regurgitation and hydrops in a preterm infant born to a mother with graves' disease. *American Journal of Perinatology Reports, 08*(02), e85–e88. https://doi.org/10.1055/s-0038-1645879

Dunn, J. F., & Isaacs, A. M. (2021). The impact of hypoxia on blood-brain, blood-CSF, and CSF-brain barriers. *Journal of Applied Physiology, 131*(3), 977–985. https://doi.org/10.1152/japplphysiol.00108.2020

Fernández-Serra, R., Martínez-Alonso, E., Alcázar, A., Chioua, M., Marco-Contelles, J., Martínez-Murillo, R., Ramos, M., Guinea, G. V., & González-Nieto, D. (2022). Postischemic neuroprotection of aminoethoxydiphenyl borate associates shortening of peri-infarct depolarizations. *International Journal of Molecular Sciences, 23*(13), 7449. https://doi.org/10.3390/ijms23137449

Flamant, F., & Richard, S. (2024). Thyroid hormone receptors function in GABAergic neurons during development and in adults. *Endocrinology, 165*(9), Article bqae101. https://doi.org/10.1210/endocr/bqae101

Foroutan, Z., Butler, A. E., Zengin, G., & Sahebkar, A. (2024). Curcumin and ferroptosis: A promising target for disease prevention and treatment. *Cell Biochemistry and Biophysics, 82*(2), 343–349. https://doi.org/10.1007/s12013-023-01212-6

Gou, H.-Z., Zhang, Y.-L., Ren, L.-F., Li, Z.-J., & Zhang, L. (2022). How do intestinal probiotics restore the intestinal barrier? *Frontiers in Microbiology, 13*, Article 929346. https://doi.org/10.3389/fmicb.2022.929346

Grob, F., Brown, A., & Zacharin, M. (2024). Neurodevelopmental Follow-Up of children born to mothers with graves' disease and neonatal hyperthyroidism. *Hormone Research in Paediatrícs*, 1–8. https://doi.org/10.1159/000539268

Gruenbaum, B. F., Schonwald, A., Boyko, M., & Zlotnik, A. (2024). The role of glutamate and blood–brain barrier disruption as a mechanistic link between epilepsy and depression. *Cells, 13*(14), 1228. https://doi.org/10.3390/cells13141228

Gryka-Marton, M., Grabowska, A. D., & Szukiewicz, D. (2025). Breaking the barrier: The role of proinflammatory cytokines in BBB dysfunction. *International Journal of Molecular Sciences, 26*(8), 3532. https://doi.org/10.3390/ijms26083532

Jurga, A. M., Paleczna, M., Kadluczka, J., & Kuter, K. Z. (2021). Beyond the GFAP-astrocyte protein markers in the brain. *Biomolecules, 11*(9), 1361. https://doi.org/10.3390/biom11091361

Khudair, A., Khudair, A., Niinuma, S. A., Habib, H., & Butler, A. E. (2025). Beyond thyroid dysfunction: The systemic impact of iodine excess. *Frontiers in Endocrinology, 16*, Article 1568807. https://doi.org/10.3389/fendo.2025.1568807

Klosinska, M., Kaczynska, A., & Ben-Skowronek, I. (2022). Congenital hypothyroidism in preterm newborns – The challenges of diagnostics and treatment: A review. *Frontiers in Endocrinology, 13*, Article 860862. https://doi.org/10.3389/fendo.2022.860862

Kochman, J., Jakubczyk, K., Bargiel, P., & Janda-Milczarek, K. (2021). The influence of oxidative stress on thyroid diseases. *Antioxidants, 10*(9), 1442. https://doi.org/10.3390/antiox10091442

Korbecki, J., Rębacz-Maron, E., Kupnicka, P., Chlubek, D., & Baranowska-Bosiacka, I. (2023). Synthesis and significance of arachidonic acid, a substrate for cyclooxygenases, lipoxygenases, and cytochrome P450 pathways in the tumorigenesis of glioblastoma multiforme, including a pan-cancer comparative analysis. *Cancers, 15*(3), 946. https://doi.org/10.3390/cancers15030946

Kung, H.-C., Lin, K.-J., Kung, C.-T., & Lin, T.-K. (2021). Oxidative stress, mitochondrial dysfunction, and neuroprotection of polyphenols with respect to resveratrol in parkinson's disease. *Biomedicines, 9*(8), 918. https://doi.org/10.3390/biomedicines9080918

Li, Z., & Liu, J. (2024). Thyroid dysfunction and alzheimer's disease, a vicious circle. *Frontiers in Endocrinology, 15*, Article 1354372. https://doi.org/10.3389/fendo.2024.1354372

Li, L., Zhi, M., Hou, Z., Zhang, Y., Yue, Y., & Yuan, Y. (2017). Abnormal brain functional connectivity leads to impaired mood and cognition in hyperthyroidism: A resting-state functional MRI study.

Oncotarget, 8(4), 6283–6294. https://doi.org/10.18632/oncotarget.14060

Liu, X., Wang, Y., Ma, L., Wang, D., Peng, Z., & Mao, Z. (2024). High prevalence of erectile dysfunction in men with hyperthyroidism: A meta-analysis. *BMC Endocrine Disorders, 24*(1), 58. https://doi.org/10.1186/s12902-024-01585-6

Lou, K., Liu, S., Zhang, F., Sun, W., Su, X., Bi, W., Yin, Q., Qiu, Y., Zhang, Z., Jing, M., & Ma, S. (2023). The effect of hyperthyroidism on cognitive function, neuroinflammation, and necroptosis in APP/PS1 mice. *Journal of Translational Medicine, 21*(1), 657. https://doi.org/10.1186/s12967-023-04511-x

Lu, Q., Wu, Y., Chang, J., Zhang, L., Lv, Q., & Sun, H. (2025). Application progress of artificial intelligence in managing thyroid disease. *Frontiers in Endocrinology, 16*, Article 1578455. https://doi.org/10.3389/fendo.2025.1578455

Macvanin, M. T., Gluvic, Z., Zafirovic, S., Gao, X., Essack, M., & Isenovic, E. R. (2023). The protective role of nutritional antioxidants against oxidative stress in thyroid disorders. *Frontiers in Endocrinology, 13*, Article 1092837. https://doi.org/10.3389/fendo.2022.1092837

Maliki, O. O., Alagbonsi, A. I., Ibitoye, C. M., & Olayaki, L. A. (2021). Melatonin and vitamin C modulate cassava diet-induced alteration in reproductive and thyroid functions. *Nigerian Journal of Experimental and Clinical Biosciences, 9*(3), 133–143. https://doi.org/10.4103/njecp.njecp_9_21

Mantle, D., & Hargreaves, I. P. (2023). Coenzyme Q10 and endocrine disorders: An overview. *Antioxidants, 12*(2), 514. https://doi.org/10.3390/antiox12020514

Mitchell, J. S., Hermens, D. F., Bennett, M. R., Can, A. T., & Lagopoulos, J. (2023). Ketamine and zinc: Treatment of anorexia nervosa via dual NMDA receptor modulation. *CNS Drugs, 37*(2), 159–180. https://doi.org/10.1007/s40263-022-00984-4

Moideen Sheriff, S., Singh, G., Chukwunyelu, N. K., Ezeafulukwe, C. J., & Hassan, O. A. (2023). Hyperthyroidism masquerading as an anxiety disorder: A report on a misdiagnosed case. *Cureus, 15*(8), e44071. https://doi.org/10.7759/cureus.44071. PMID: 37750133; PMCID: PMC10517880.

Ollivry, A., & Joubert, Z. (2023). Thyrotoxic periodic paralysis: A presentation of hyperthyroidism increasing in frequency around the world. *Journal of Endocrinology, Metabolism and Diabetes of South Africa, 28*(2), 72–75. https://doi.org/10.1080/16089677.2023.2170955

O'Shaughnessy, K. L., McMichael, B. D., Sasser, A. L., Bell, K. S., Riutta, C., Ford, J. L., Stoker, T. E., Grindstaff, R. D., Pandiri, A. R., & Gilbert, M. E. (2023). Thyroid hormone action controls multiple components of cell junctions at the ventricular zone in the newborn rat brain. *Frontiers in Endocrinology, 14*, Article 1090081. https://doi.org/10.3389/fendo.2023.1090081

Patel, V., Akimbekov, N. S., Grant, W. B., Dean, C., Fang, X., & Razzaque, M. S. (2024). Neuroprotective effects of magnesium: Implications for neuroinflammation and cognitive decline. *Frontiers in Endocrinology, 15*, Article 1406455. https://doi.org/10.3389/fendo.2024.1406455

Pradhan, N., Singh, C., & Singh, A. (2021). Coenzyme Q10 a mitochondrial restorer for various brain disorders. *Naunyn-Schmiedeberg's Archives of Pharmacology, 394*(11), 2197–2222. https://doi.org/10.1007/s00210-021-02161-8

Ring, M. (2025). An integrative approach to HPA axis dysfunction: From recognition to recovery. *The American Journal of Medicine, 138*(10), 1451–1463. https://doi.org/10.1016/j.amjmed.2025.05.044. S0002934325003535.

Ruggeri, R. M., Barbalace, M. C., Croce, L., Malaguti, M., Campennì, A., Rotondi, M., Cannavò, S., & Hrelia, S. (2023). Autoimmune thyroid disorders: The mediterranean diet as a protective choice. *Nutrients, 15*(18), 3953. https://doi.org/10.3390/nu15183953

Salloum-Asfar, S., Shin, K. C., Taha, R. Z., Khattak, S., Park, Y., & Abdulla, S. A. (2024). The potential role of thyroid hormone therapy in neural progenitor cell differentiation and its impact on neurodevelopmental disorders. *Molecular Neurobiology, 61*(6), 3330–3342. https://doi.org/10.1007/s12035-023-03751-8

Sawicka-Gutaj, N., Zawalna, N., Gut, P., & Ruchała, M. (2022). Relationship between thyroid hormones and central nervous system metabolism in physiological and pathological conditions. *Pharmacological Reports, 74*(5), 847–858. https://doi.org/10.1007/s43440-022-00377-w

Sekhar, V. C., Viswanathan, G., & Baby, S. (2019). Insights into the molecular aspects of neuroprotective bacoside A and bacopaside I. *Current Neuropharmacology, 17*(5), 438–446. https://doi.org/10.2174/1570159X16666180419123022

Sepúlveda, P., Ferreira, A. F. F., Sandoval, C., Bergoc, G., Moreno, A. C. R., Nunes, M. T., & Torrão, A. D. S. (2024). Thyroid hormone supplementation restores cognitive deficit, insulin signaling, and neuroinflammation in the hippocampus of a sporadic alzheimer's-like disease rat model. *Cells, 13*(21), 1793. https://doi.org/10.3390/cells13211793

Shahat, A. S., Hassan, W. A., & El-Sayed, W. M. (2022). N-Acetylcysteine and safranal prevented the brain damage induced by hyperthyroidism in adult male rats. *Nutritional Neuroscience, 25*(2), 231–245. https://doi.org/10.1080/1028415X.2020.1743917

Shakeri, F., Bibak, B., Safdari, M. R., Keshavarzi, Z., Jamialahmadi, T., Sathyapalan, T., & Sahebkar, A. (2022). Cellular and molecular mechanisms of curcumin in thyroid gland disorders. *Current Medicinal Chemistry, 29*(16), 2878–2890. https://doi.org/10.2174/0929867329666220210145033

Shu, Q., Kang, C., Li, J., Hou, Z., Xiong, M., Wang, X., & Peng, H. (2024). Effect of probiotics or prebiotics on thyroid function: A meta-analysis of eight randomized controlled trials. *PLoS One, 19*(1), Article e0296733. https://doi.org/10.1371/journal.pone.0296733

Somayaji, H., Shetty, S., Hegde, P., & Shetty, V. (2024). Association of antioxidants in thyroid dysfunction. *Chinese Journal of Laboratory Medicine, 6*, 203–211. https://doi.org/10.0711/Cjlm.2024658474

Spaggiari, G., Setti, M., Tagliavini, S., Roli, L., De Santis, M. C., Trenti, T., Rochira, V., & Santi, D. (2021). The hypothalamic–pituitary–adrenal and -thyroid axes activation lasting one year after an earthquake swarm: Results from a big data analysis. *Journal of Endocrinological Investigation, 44*(7), 1501–1513. https://doi.org/10.1007/s40618-020-01457-5

Sultana, D. R., Shahin, A. D., Md, & Jawadul, H. (2022). Measurement of oxidative stress and total antioxidant capacity in hyperthyroid patients following treatment with carbimazole and antioxidant. *Heliyon, 8*(1), Article e08651. https://doi.org/10.1016/j.heliyon.2021.e08651

Sunny, A., James, R. R., Menon, S. R., Rayaroth, S., Daniel, A., Thompson, N. A., & Tharakan, B. (2024). Matrix Metalloproteinase-9 inhibitors as therapeutic drugs for traumatic brain injury.

Neurochemistry International, 172, Article 105642. https://doi.org/10.1016/j.neuint.2023.105642

Szybiak-Skora, W., Miedziaszczyk, M., Szałek, E., & Lacka, K. (2025). The therapeutic potential of propranolol and other beta-blockers in hyperthyroidism. *International Journal of Molecular Sciences, 26*(17), 8322. https://doi.org/10.3390/ijms26178322

Tan, B., Babur, E., Aşçıoğlu, M., & Süer, C. (2021). Effect of L-thyroxine administration on long-term potentiation and accompanying mitogen-activated protein kinases in rats. *International Journal of Developmental Neuroscience, 81*(3), 259–269. https://doi.org/10.1002/jdn.10097

Tesfaye, E., Getnet, M., Anmut Bitew, D., Adugna, D. G., & Maru, L. (2024). Brain functional connectivity in hyperthyroid patients: Systematic review. *Frontiers in Neuroscience, 18*, Article 1383355. https://doi.org/10.3389/fnins.2024.1383355

Tesse, A., Grossini, E., Tamma, G., Brenner, C., Portincasa, P., Marinelli, R. A., & Calamita, G. (2018). Aquaporins as targets of dietary bioactive phytocompounds. *Frontiers in Molecular Biosciences, 5*, 30. https://doi.org/10.3389/fmolb.2018.00030

Tur, J. A., Bibiloni, M. M., Sureda, A., & Pons, A. (2012). Dietary sources of omega 3 fatty acids: Public health risks and benefits. *British Journal of Nutrition, 107*(S2), S23–S52. https://doi.org/10.1017/S0007114512001456

Ullrich, D., Führer, D., Heuer, H., Mayerl, S., Haupeltshofer, S., Schmitt, L.-I., Leo, M., Szepanowski, R. D., Hagenacker, T., Schwaninger, M., Kleinschnitz, C., & Langhauser, F. (2025). Triiodothyronine treatment in mice improves stroke outcome and reduces blood–brain barrier damage. *European Thyroid Journal, 14*(1), Article e240143. https://doi.org/10.1530/ETJ-24-0143

Vujovic, F., & Farahani, R. M. (2025). Thyroid hormones and brain development: A focus on the role of mitochondria as regulators of developmental time. *Cells, 14*(3), 150. https://doi.org/10.3390/cells14030150

Wei, X., Tan, J., & Gao, H. (2022). Role of sirtuin 1 in the brain development in congenital hypothyroidism rats via the regulation of p53 signaling pathway. *Bioengineered, 13*(4), 9455–9466. https://doi.org/10.1080/21655979.2022.2060626

Yorke, E. (2022). Hyperthyroidism and liver dysfunction: A review of a common comorbidity. *Clinical Medicine Insights: Endocrinology and Diabetes, 15*, Article 11795514221074672. https://doi.org/10.1177/11795514221074672

Yu, C., Wang, D., Yang, Z., & Wang, T. (2022). Pharmacological effects of polyphenol phytochemicals on the intestinal inflammation via targeting TLR4/NF-κB signaling pathway. *International Journal of Molecular Sciences, 23*(13), 6939. https://doi.org/10.3390/ijms23136939

Zawadzka, K., Kałuzińska, K., Świerz, M. J., Sawiec, Z., Antonowicz, E., Leończyk-Spórna, M., Abadi, A. K., Trofimiuk-Müldner, M., & Bała, M. M. (2023). Are probiotics, prebiotics, and synbiotics beneficial in primary thyroid diseases? A systematic review with meta-analysis. *Annals of Agricultural and Environmental Medicine, 30*(2), 217–223. https://doi.org/10.26444/aaem/162732

Zhu, W., Wu, F., Li, J., Meng, L., Zhang, W., Zhang, H., Cha, S., Zhang, J., & Guo, G. (2022). Impaired learning and memory generated by hyperthyroidism is rescued by restoration of AMPA and NMDA receptors function. *Neurobiology of Disease, 171*, Article 105807. https://doi.org/10.1016/j.nbd.2022.105807

Chapter 26

Excess iodine intake and thyroid dysfunction

Hernando Vargas-Uricoechea
Metabolic Diseases Study Group, Department of Internal Medicine, Universidad del Cauca, Popayán, Colombia

List of abbreviations

AIT Amiodarone-induced thyrotoxicosis
CT Computed tomography
D1 Type 1 deiodinase
D2 Type 2 deiodinase
D3 Type 3 deiodinase
DEHAL1 Iodotyrosine Dehalogenase 1
DIT Diiodothyronine
DUOX Dual oxidases
DUOXA1 Dual Oxidase Maturation Factor 1
DUOXA2 Dual Oxidase Maturation Factor 2
EAR Estimated Average Requirements
FT3 Free T3
FT4 Free T4
H_2O_2 Hydrogen peroxide
ICAM-1 Intercellular adhesion molecule 1
IL-6 Interleukin-6
KCLO4 Potassium perchlorate
KCNE2 Potassium voltage-gated channel subfamily E regulatory subunit 2
KCNQ1 Potassium voltage-gated channel subfamily Q member 1
KI Potassium iodide
MCT8, SLC16A2 Monocarboxylate transporter type 8
MHC Major histocompatibility complex
MIT Monoiodothyronine
mRNA Messenger RNA
Na^+/K^+/ATPase Sodium–potassium adenosine triphosphatase
NADPH Nicotinamide adenine dinucleotide phosphate
NIS Sodium/iodine symporter
PTC Papillary thyroid cancer
RDA Recommended Daily Allowances
ROS Reactive oxygen species
rT3 Reverse T3
SLC26A Solute carrier 26, member A
T3 Triiodothyronine
T4 Thyroxine
Tg Thyroglobulin
TPO Thyroid peroxidase
TRAb Antibodies against the TSH receptor
TRH Thyroid-releasing hormone
TSH Thyrotropin
UL Tolerable Upper Intake Level
USI Universal salt iodization
WDTC Well-differentiated thyroid cancer

Introduction

Iodine intake is necessary for the synthesis of thyroid hormones, and the effects of iodine on the thyroid are complex, maintaining a "U"-shaped relationship between iodine intake and the risk of thyroid diseases; therefore, low or high intake is associated with a significant increase in the risk of functional and/or anatomical alterations in the gland (Khudair et al., 2025; Laurberg et al., 2010; Armstrong et al., 2025; Shahid et al., 2025).

Under normal conditions, an adult can tolerate a high iodine intake (up to 600–1100 μg/day) without adverse effects; however, the optimal intake level for preventing thyroid disease is approximately 150 μg/day. Similarly, in areas where most of the population has sufficient iodine, they may be tolerant of high iodine intakes. Globally, disorders associated with iodine intake remain highly prevalent, and the most effective way to correct population iodine deficiency is through USI programs (WHO/UNICEF/ICCIDD, 2007; Hatch-McChesney & Lieberman, 2022).

Therefore, if USI programs (or those focused on evaluating and regulating micronutrient intake) are inadequate, or if dietary salt intake is excessive, the population is at risk of developing disorders due to excessive iodine intake (e.g., hyperthyroidism, hypothyroidism, thyroid cancer, and thyroid autoimmunity, among other conditions) (WHO/UNICEF/ICCIDD, 2007; Hatch-McChesney & Lieberman, 2022; Sohn et al., 2024).

Hormones in Neuroscience. https://doi.org/10.1016/B978-0-443-30261-9.00026-7

This chapter describes the main effects of excess iodine on the thyroid gland.

Iodine metabolism

The primary function of the thyroid gland is to synthesize and secrete thyroid hormones (T4 and T3) (Armstrong et al., 2025; Shahid et al., 2025). These hormones are the only hormones that require iodine for their synthesis, and their secretion is determined by TSH, which is secreted by the anterior pituitary gland. TSH secretion is regulated by TRH (Armstrong et al., 2025; Shahid et al., 2025; Contreras-Jurado, 2025).

Both, T4 and T3, act on multiple tissues and are essential for normal development, growth, and metabolism. Consequently, in the presence of conditions that stimulate excess or deficiency of T4 and T3, various metabolic alterations occur (Aranda, 2025; Contreras-Jurado, 2025) (Fig. 26.1).

Iodine metabolism and thyroid hormone synthesis

Iodine is the main component of thyroid hormones; more than 95% of the body's iodine is metabolized in the thyroid gland, with the remainder being used by other organs and systems. The biosynthesis and secretion of T4 and T3 are complex processes that require several sequential steps, as described in Table 26.1 (Contreras-Jurado, 2025; Aranda, 2025; Leung et al., 2010).

Iodine represents 65% and 58% of the net weights of T4 and T3, respectively. When ingested, iodine is absorbed through the small intestine and transported in the plasma to the thyroid gland, where it is concentrated, oxidized, and incorporated into Tg to form MIT and DIT, which subsequently leads to the formation of T3 and T4 (Cavalieri, 1997; Leung et al., 2010).

Of the total iodine used by the body, 96%–99% is metabolized in the thyroid, while the remainder is used by organs such as the retina and choroid plexus. The thyroid

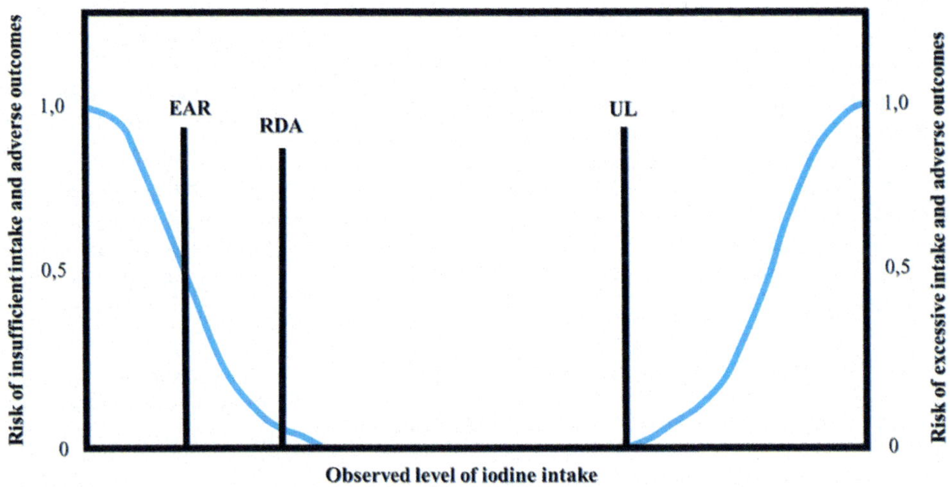

FIGURE 26.1 U-shaped relationship between iodine intake and adverse outcomes; both "low" and "excessive" intake lead to functional and anatomical alterations of the thyroid.

TABLE 26.1 Sequential steps in the biosynthesis and secretion of thyroid hormones.

Absorption of iodide from the gastrointestinal tract into the bloodstream
Iodide uptake in the thyroid by the NIS
Iodide uptake in the thyroid by the NIS. This transporter is a membrane glycoprotein that modulates iodine transport in the thyroid and other tissues
Iodide oxidation, this process involves TPO and H_2O_2, mediated by DUOX
Iodization of Tg
Storage of thyroid hormones (in their Tg-bound form)
Reabsorption and hydrolysis of Tg (in follicular cells) and secretion of T4 and T3

and kidneys eliminate most iodine from the blood, regardless of the amount and presence of other anions (iodine clearance depends mainly on the glomerular filtration rate) (Cavalieri, 1997; Contreras-Jurado, 2025) (Fig. 26.2).

The transport of iodide into thyroid follicular cells is the first step in T4 and T3 biosynthesis. This transport from the extracellular space to the follicular lumen requires two steps: transport into the cell (on the basal side), and into the lumen (on the apical side). The first step is mediated by NIS, and the second step is mediated by pendrin (Leung et al., 2010; Cavalieri, 1997; Dunn & Dunn, 2001; Bizhanova & Kopp, 2009).

NIS is located in the basolateral plasma membrane of thyroid follicular cells and actively transports iodide into the gland using an electrochemical gradient generated by the Na^+/K^+/ATPase pump. This pump regulates the active transport of sodium and potassium across the cell membrane against their respective concentration gradients. Thus, for every ATP molecule broken down, the pump transports two potassium ions into the cell and three sodium ions out of the cell (Bizhanova & Kopp, 2009; Riesco-Eizaguirre et al., 2021).

Furthermore, iodine efflux is mediated (at least in part) by pendrin, a highly hydrophobic membrane protein located in the apical membrane of thyrocytes, although it is also expressed in the kidneys and inner ear. Pendrin belongs to the SLC26A family and includes several anion transporters and a motor protein called prestin, which is expressed in outer hair cells (Bizhanova & Kopp, 2009; Geertsma & Oliver, 2024).

On the intraluminal side, iodide is oxidized by a reaction requiring H_2O_2 (the final electron acceptor for thyroid hormone biosynthesis), catalyzed by TPO on the apical surface of thyrocytes (Aranda, 2025; Contreras-Jurado, 2025).

The KCNQ1–KCNE2 channels, on the other hand, are essential for the physiology of gastric parietal cells and

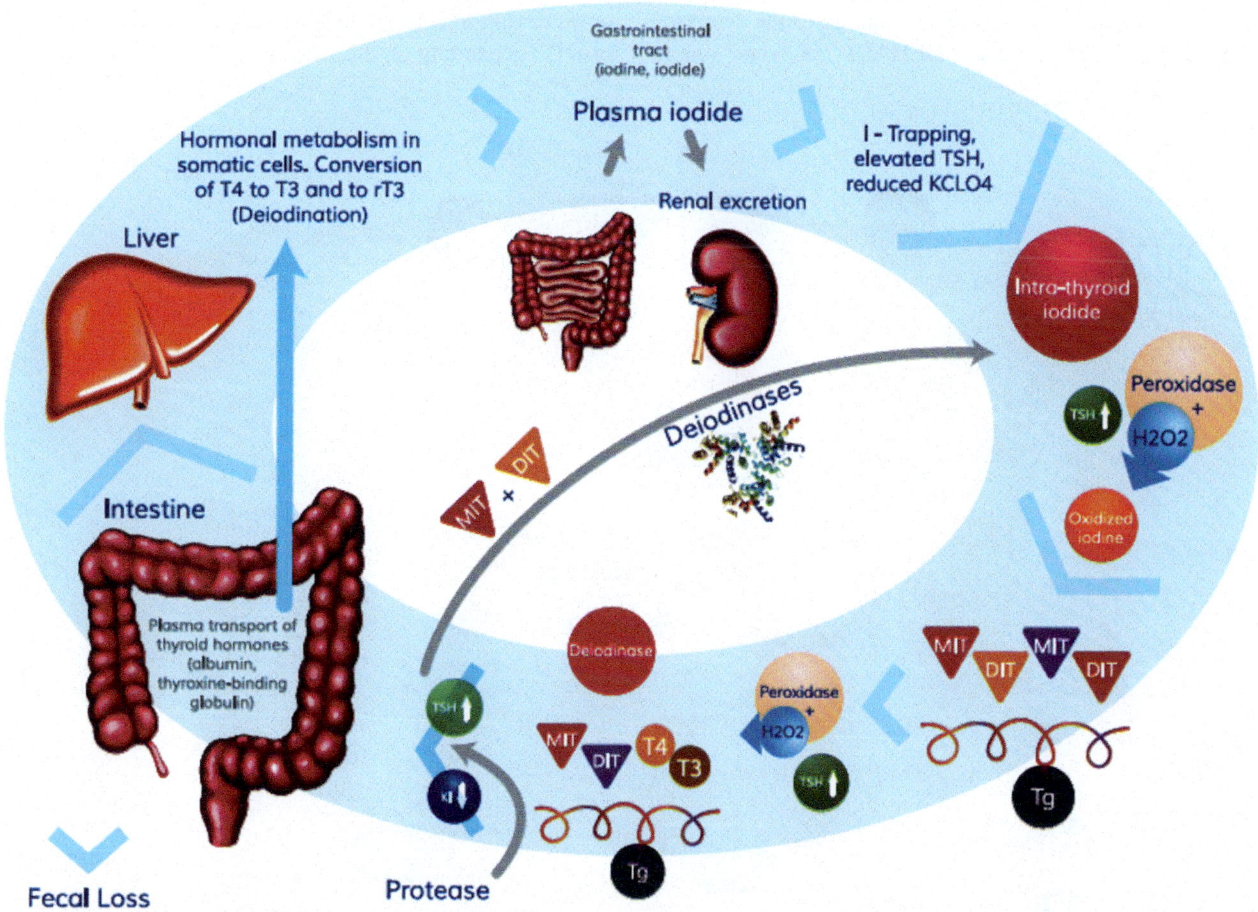

FIGURE 26.2 Iodine metabolism. After ingestion, iodide is captured and oxidized in the thyroid gland, subsequently binding to tyrosine to form iodotyrosines. In Tg, the binding of iodotyrosyl residues forms T4 and T3. Both hormones are secreted and transported in the serum (and some T4 is deiodinated to T3). Thyroid hormones (especially T3) exert their metabolic effects at the cellular level and eventually undergo a deiodination process; the resulting iodide is either reused or excreted. A second cycle occurs within the thyroid gland through the deiodination of iodotyrosines, leading to the generation of iodides, some of which are reused without leaving the gland.

thyroid epithelial cells. Their role in iodine uptake in the thyroid indicates that they are necessary for proper NIS function; therefore, iodine uptake requires a constitutively active potassium channel that promotes potassium efflux (Thompson et al., 2021).

H_2O_2 is produced by two enzyme isoforms (DUOX1 and DUOX2), which are calcium-dependent NADPH-flavoprotein oxidases. These require maturation factors, DUOXA1 and DUOXA2, to position DUOX1 and DUOX2 in the cell membrane, respectively (Contreras-Jurado, 2025; Ohye & Sugawara, 2010).

The H_2O_2 produced by DUOX1 and DUOX2 is used by TPO to oxidize iodide (thus iodinating the tyrosyl residues in Tg to form MIT and DIT); subsequently, it combines with MIT and DIT to form T3 and T4 (Contreras-Jurado, 2025; Mullur et al., 2014; Ohye & Sugawara, 2010).

To ensure the necessary iodine supply for thyroid hormone production, the enzyme DEHAL1 deiodinates MIT and DIT (both of which are formed in excess within the "prehormonal" matrix of Tg and are released again after Tg hydrolysis by cathepsins, leading to the secretion of T3 and T4). Subsequently, DEHAL1 "recovers" one to two iodine molecules contained in MIT and DIT (this function represents an intrathyroidal iodine recycling mechanism and, therefore, efficient hormone production, allowing the reuse of iodine within the thyroid) (Contreras-Jurado, 2025; Krause et al., 2007).

Iodinated Tg enters the cell through micro- and macropinocytosis, where it subsequently undergoes digestion in lysosomes. MCT8 (SLC16A2) also plays a role in this process. MCT belongs to a superfamily that independently facilitates the bidirectional transport of MIT and DIT across the plasma membrane. In this sense, T4 and T3 are excreted through MCT8 and other transporters, respectively. MIT and DIT are dehalogenated by DEHAL1, and the released iodide is recycled (Markova et al., 2024; Tan et al., 2025) (Fig. 26.3).

Deiodinases, for their part, exert important metabolic control over intracellular T4 and T3 concentrations, leading to the bioavailability of thyroid hormones in the tissue. Three deiodinases have been described:

Type 1 deiodinase (D1), located in the plasma membrane, catalyzes the removal of iodine atoms from the inner or outer ring in equimolar proportions to generate T3, rT3, or DIT, depending on the substrate. Most circulating T3 is

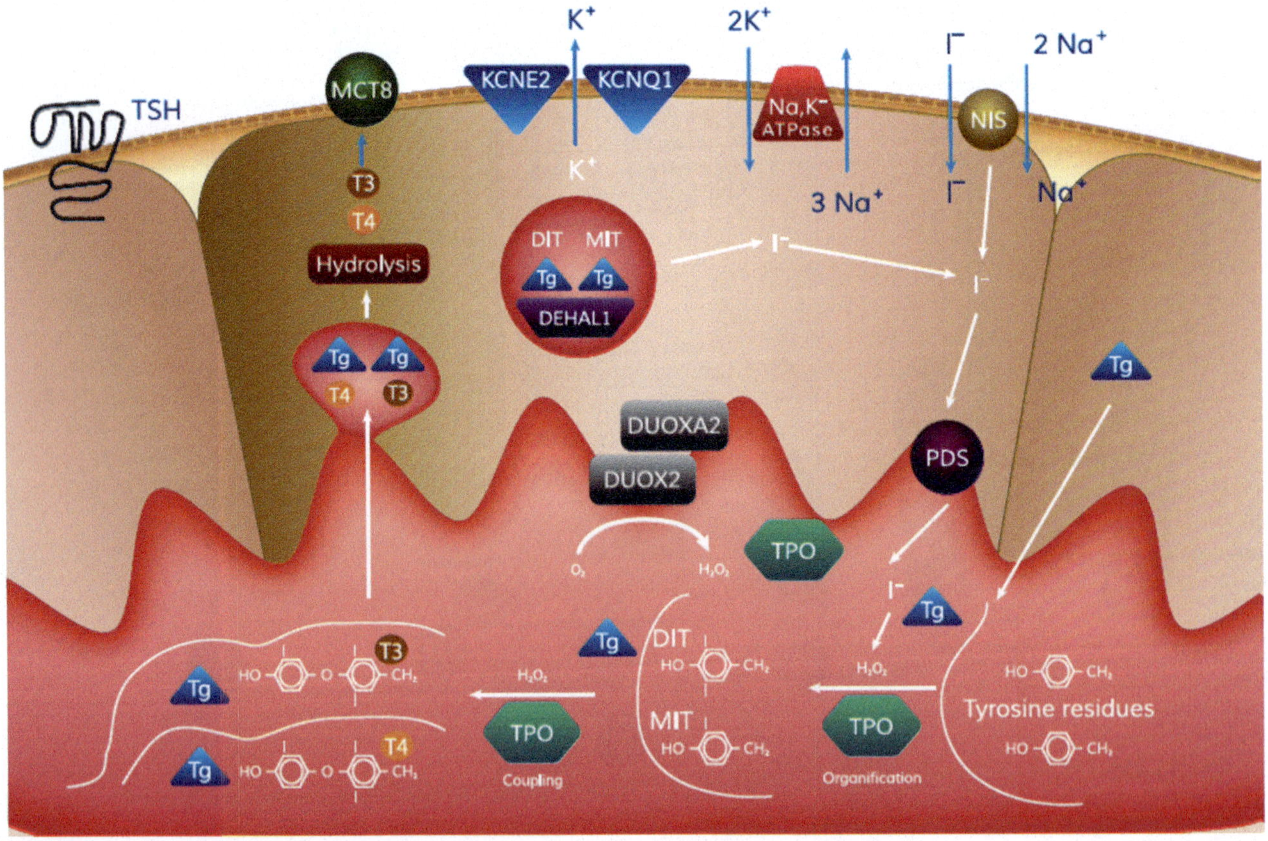

FIGURE 26.3 Iodine absorption and thyroid hormone synthesis. The first step in iodine absorption is mediated by the NIS. In the apical membrane, pendrin mediate iodine efflux. TPO, through H_2O_2, mediates the oxidation, organification, and coupling reaction that leads to the synthesis of thyroid hormones. MIT and DIT are dehalogenated by DEHAL1, and finally, the released iodide is recycled.

derived from the conversion of T4 to T3 by D1 (Giri & Mugesh, 2025; Sabatino et al., 2021).

Type 2 deiodinase (D2), which is considerably more efficient than D1, catalyzes the removal of an iodine atom from the outer ring of T4, generating its active product (T3) (Giri & Mugesh, 2025; Sabatino et al., 2021).

Type 3 deiodinase (D3) is expressed in the brain and other tissues. It irreversibly inactivates T3 or prevents the activation of T4, catalyzing the removal of an iodine atom from the inner ring to generate DIT or rT3, respectively (Giri & Mugesh, 2025; Sabatino et al., 2021).

Integrated physiological model between iodine metabolism and the thyroid hormones

Thyrotroph cells (in the adenohypophysis) produce TSH, which regulates NIS activity, the synthesis, storage and iodination of Tg, TPO activity, the reentry of Tg into follicular cells and the subsequent release and secretion of T4 and T3. TSH is produced under the control of hypothalamic secretion of TRH and somatostatin, which stimulate and inhibit the release of TSH, respectively (Aranda, 2025; Contreras-Jurado, 2025; Jing & Zhang, 2022).

Hypothalamic TRH production is sensitive to feedback from FT3 and FT4; therefore, low levels of both thyroid hormones stimulate the production of TRH, whereas high levels of FT3 and FT4 induce the hypothalamus to secrete somatostatin (which inhibits the release of TSH) (Jing & Zhang, 2022; Mariotti et al., 2000).

However, the components that determine adaptation to inadequate iodine intake are only partially understood, and there is insufficient data to construct a response curve to a precise dose or amount of iodine to explain the primary events that control and affect iodine homeostasis and thyroid function; for example, if the iodine supply is adequate, the thyroid captures sufficient iodine for the synthesis of T4 and T3, whereas excess iodine stimulates its renal excretion (Khudair et al., 2025; Armstrong et al., 2025; Shahid et al., 2025).

The stimulus for iodine absorption and retention depends on the degree of need for T3 or T4; therefore, if the amount of endogenous iodine recycled and that which comes from the diet does not satisfy the need required for the synthesis of thyroid hormones, insufficient feedback (mediated by FT3 and FT4) to the hypothalamus stimulates the production of TRH (increasing the secretion of TSH). Thus, adequate iodine intake allows the adequate production of FT3 and FT4 (regulating TRH and TSH, thereby modulating intestinal uptake and renal retention of iodine) (Armstrong et al., 2025; Shahid et al., 2025; Mariotti et al., 2000; Lechan et al., 2009).

When thyroid iodine content is depleted (as a result of inadequate iodine supply), FT3 and FT4 levels decrease, and TRH and TSH production increases. The above induces an increase in intestinal uptake and iodine transfer through NIS (present in enterocytes); additionally, iodine uptake in the thyroid and renal activity of the NIS also increases (causing greater renal iodine retention). Consequently, the thyroid produces a greater amount of Tg, and therefore, simple diffuse goiter occurs (in an early form) (Armstrong et al., 2025; Shahid et al., 2025; Leung et al., 2010).

In summary, ingested iodine is absorbed and transported in the circulation (as iodide); the thyroid concentrates iodide along the basal-lateral membrane of the thyrocytes through the NIS, and the intracellular iodide is transported to the lumen of the thyroid follicles. Meanwhile, proteins such as Tg and TPO are synthesized; in Tg, tyrosyl residues serve as a substrate for the iodination and formation of T3 and T4.

TPO is located on the apical cell membrane, where it stimulates iodide oxidation and attaches iodine to tyrosyl residues on Tg. Iodination of Tg produces MIT and DIT, and subsequent iodination couples two DIT residues to produce thyroxine (T4). When a certain amount of thyroid hormones is needed, Tg is internalized into the apical pool of thyrocytes and undergoes protease-mediated digestion; after digestion, T4 and T3 are released into the circulation (Armstrong et al., 2025; Shahid et al., 2025; Carvalho & Dupuy, 2017).

Iodine-induced hyperthyroidism

The main consequence of excessive iodine intake is iodine-induced hyperthyroidism, which usually occurs in people with endogenous goiter (induced by population-level iodine deficiency) who have been supplemented with this trace element (e.g., through USI programs) (Sohn et al., 2024).

This iodine-induced hyperthyroidism is known as the Jod-Basedow effect and occurs in a proportion of patients at risk, such as individuals with autoimmune thyroid disease (including Hashimoto's disease and Graves-Basedow disease), those with nontoxic nodular or diffuse goiter, and patients with a history of thyroidectomy. Furthermore, since iodine excretion occurs at the renal level, chronic kidney disease and end-stage renal disease can be considered risk factors for the development of the Jod-Basedow effect (Khudair et al., 2025; Sohn et al., 2024).

Additionally, iodinated antiseptic solutions, oral iodine supplements, and medications such as amiodarone have been associated with an increased risk of developing the Jod-Basedow effect. Also, iodinated contrast media are significant exogenous sources of iodide.

Both ionic and nonionic contrast media contain iodine in amounts ranging from 300 to 370 μg/mL. Typically, the contrast medium doses used in imaging procedures, such as computed tomography (CT) scans, are 50–100 mL, which means that a patient undergoing a contrast-enhanced CT scan receives approximately 15–37 gm of total iodine and 2500–5000 μg of free iodine (Carlqvist et al., 2024; Khudair et al., 2025; Sohn et al., 2024, 2025).

The best-documented experience of the Jod-Basedow effect was described in Tasmania, where a temporary increase in the diagnosis of thyrotoxicosis occurred shortly after the addition of small amounts of iodine to daily bread (as the chosen method to correct iodine deficiency) (Connolly, 1973).

Analysis of this population revealed two patterns of underlying thyroid dysfunction. The first pattern documented the presence of a nodular goiter with areas of autonomous function, especially in elderly individuals, without the presence of TRAb. The second pattern occurs in a younger population with diffuse goiter (and positive TRAb) (Connolly, 1973).

Moreover, it has been suggested that the development of the Jod-Basedow effect (following iodine fortification and/or supplementation) is due to iodine deficiency, which increases the proliferation and mutation rate of thyrocytes, triggering the development of multifocal autonomous growth with scattered cell clones harboring TSH receptor activation mutations (Farebrother et al., 2019; Khudair et al., 2025; Sohn et al., 2024).

One way to confirm the Jod-Basedow effect is by measuring urinary iodine (which increases) and using imaging techniques such as thyroid scintigraphy with I-131 uptake (which shows low or absent uptake of the radiotracer) (Bürgi, 2010; Farebrother et al., 2019; Sohn et al., 2024).

Some commonly used medications and substances with significant iodine content that may be associated with an increased risk of Jod-Basedow syndrome are listed in Table 26.2.

Iodine-induced hypothyroidism

Depending on the excess amount of iodine supplied, organification can be carried out in the same way; this organification phenomenon exhibits a biphasic response to excess iodine: a first phase of increased organification and a second phase of reduction (as a response to a relative blockage of this phenomenon) (Bürgi, 2010; Markou et al., 2001).

The decrease in iodine organification (due to the increase in the amount of iodine supplemented) is called the Wolff–Chaikoff effect; this "self-regulated" block is independent of TSH and is based on an intracellular molar concentration of iodine $\geq 10^{-3}$) (Markou et al., 2001; Vargas-Uricoechea et al., 2016).

Susceptibility to this phenomenon may be due to the stimulation of the mechanisms that "trap" iodine, as occurs in Graves-Basedow disease, or to the inability to form organic iodine, as occurs after administering radioactive iodine therapies or during treatment with thionamides (methimazole, carbimazole, propylthiouracil) or in patients with Hashimoto's disease (Vargas-Uricoechea et al., 2016).

In such situations, goiter or hypothyroidism may develop if iodine administration is continued for prolonged periods. The exact biochemical mechanism explaining the Wolff–Chaikoff effect remains controversial, but it could be explained by the triiodide reaction ($I^- + I_2 \leftrightarrow I^-_3$). The above indicates that at a high concentration of iodide, this reaction shifts to the right due to a mass action phenomenon, thus sequestering I_2, which is supposedly an intermediate in iodide organification (Leung & Braverman, 2012; Vargas-Uricoechea et al., 2016).

When moderate or high doses of iodine are administered repeatedly, the inhibition of organification and thyroid hormone formation is partially relieved; this "escape or adaptation" mechanism occurs because iodine transport decreases and the thyroid iodine concentration is insufficient to maintain a full Wolff–Chaikoff effect (Khudair et al., 2025; Vargas-Uricoechea et al., 2016; Leung & Braverman, 2012; Koukkou et al., 2017).

This response is a manifestation of the thyroid's autoregulatory inhibition of iodine transport (which prevents the development of hypothyroid goiter in the normal population). In some cases, the "escape or adaptation" mechanism does not occur, resulting in persistent inhibition of thyroid hormone formation, leading to hypothyroidism and goiter (iodine-induced myxedema).

The Wolff–Chaikoff effect is short-lived because the biosynthesis of the thyroid gland quickly ceases, and the intracellular iodine concentration falls below 10^{-3} M (and iodine organification resumes). This marked decrease in thyroid gland activity occurs through autoregulation, independent of TSH (Vargas-Uricoechea et al., 2016) (Fig. 26.4).

In general, the mechanism by which excess iodine causes thyroid dysfunction is not well established; although some explanations can be given; for example, excess iodine induces the production of cytokines and chemokines, which can recruit immunocompetent cells to the thyroid, and together with the processing of excess iodine at the intrathyroidal level, causes an increase in oxidative stress; the incorporation of iodine into Tg induces greater antigenicity of the same, and therefore, a greater risk of thyroid autoimmunity (Hashimoto's disease) (Khudair et al., 2025; Zimmermann & Galetti, 2015).

Similarly, the Wolff–Chaikoff effect can be explained by the generation of inhibitory substances (iodolactones, iodoaldehydes, and iodolipids) that act on TPO activity, with a subsequent decrease in the activity of intrathyroidal deiodinases, thus reducing the synthesis of T4 and T3 (Khudair et al., 2025; Sohn et al., 2024; Vargas-Uricoechea et al., 2016).

TABLE 26.2 Commonly used medicines and substances that contain a significant amount of iodine.

Oral use	
Amiodarone	75 mg/tablet
Benziodarone	49 mg/100 mg/tablet
Calcium iodide (syrup)	26 mg/mL
Diiodohydroxyquin	134 mg/tablet
Echothiophate iodide ophthalmic solution	5–41 mg/drop
Hydroiodic acid (syrup)	13–15 mg/mL
Iodochlorhydroxyquin	104 mg/tablet
Vitamins that contain iodine	0.15 mg/tablet
Iodized glycerol	15 mg/tablet or 25 mg/mL
Idoxuridine (ophthalmic solution)	18 μg/drop
Isopropamide iodide	1.8 mg/tablet
Kelp (seaweed)	0.15 mg/tablet
KI	145 mg/tablet or 24 mg/mL
Lugol's solution	6.3 mg/drop
Niacinamide hydroiodide + KI	115 mg/tablet
Iodinated nasal emollient	5 mg/0.8 mL
Saturated KI solution	38 mg/drop
Parenteral preparations	
Sodium iodide, 10% solution	85 mg/mL
Topical antiseptics	
Diiodohydroxyquinoline cream	6 mg/gm
Iodine tincture	40 mg/mL
Iodochlorhydroxyquinoline cream	12 mg/gm
Iodized gauze	4.8 mg/100 mg of gauze
Povidone-iodine	10 mg/mL
Iodinated contrast media	
Sodium meglumine diatrizoate	370 mg/mL
Iodized oil	380 mg/mL
Iopanoic acid	333 mg/tablet
Sodium ipodate	308 mg/capsule
Iotalamato	480 mg/mL
Metrizamide	483 mg/mL (before dilution)

The decrease in the expression of the thyroid gland occurs approximately 24 h after the acute excess of iodine, causing a decrease in the concentration of intrathyroidal iodine and in the levels of iodinated substances, thus inhibiting the synthesis of thyroid hormones and subsequently inducing (under normal conditions) the restart of their production (Khudair et al., 2025; Sohn et al., 2024; Vargas-Uricoechea et al., 2016; Zimmermann & Galetti, 2015).

In vulnerable individuals (such as those with thyroid antibodies, a history of thyroiditis, prior use of radioactive iodine, or those using medications such as amiodarone, interferon-α, and lithium), the "escape" mechanism to the acute Wolff–Chaikoff effect may fail, and they have an

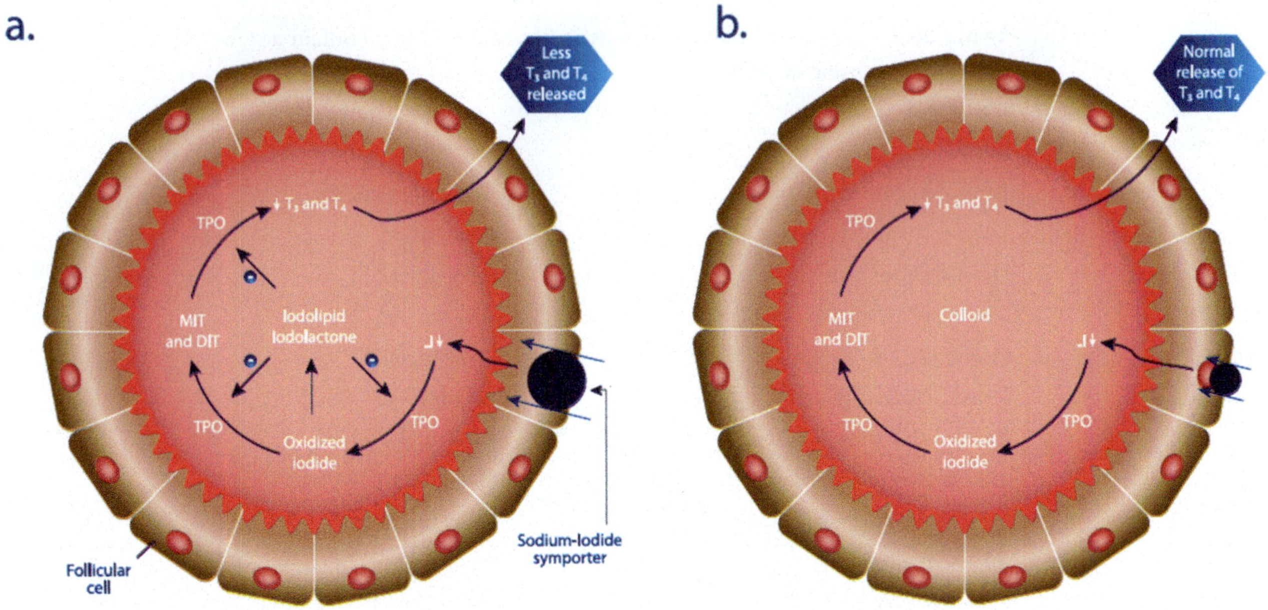

FIGURE 26.4 The Wolff–Chaikoff effect and escape mechanism. The Wolff–Chaikoff effect is a decrease in thyroid hormone synthesis caused by the ingestion of a large amount of iodine. It is an autoregulatory phenomenon that inhibits iodine organification, the formation and the release of thyroid hormones into the bloodstream (a). When moderate or high doses of iodine are administered (repeatedly), the inhibition of organification and thyroid hormone formation is partially corrected. This "escape" or "adaptation" mechanism occurs because iodine transport decreases, and the thyroid iodine concentration is insufficient to maintain the full Wolff–Chaikoff effect (b).

increased risk of developing permanent hypothyroidism (Sohn et al., 2024; Vargas-Uricoechea et al., 2016).

Excessive iodine intake and thyroid cancer

An increase in the frequency of well-differentiated thyroid cancer (WDTC), especially papillary thyroid cancer (PTC), has been described in areas where USI programs have been implemented; however, despite the association found in some countries between WDTC and iodized salt consumption, it is also true that in other countries where the intake of this type of salt is low, an increase in the frequency of this type of cancer has been observed (Zimmermann & Galetti, 2015).

Similarly, other studies have described an increase in the frequency of PTC (BRAF positive) in relation to the presence of microcarcinomas after iodine supplementation. In this regard, a case-control study found that, in the presence of a high concentration of urinary iodine, there was an association with a higher risk of PTC, especially in individuals aged <45 years (Kim et al., 2021).

Similarly, in Korea, where BRAF mutations have a prevalence $>80\%$ (in individuals with PTC), it has been intriguingly documented that both low and excessively high iodine levels are associated with a higher probability of having BRAF mutations, suggesting that both relatively low and excessive iodine intake appear to be risk factors for the occurrence of BRAF mutations in the thyroid and, therefore, may be risk factors for the development of PTC in iodine-rich areas (U-shaped relationship) (Kim et al., 2017).

Additionally, a meta-analysis found that excessive iodine intake increases the risk of PTC, whereas adequate intake has a protective effect, although no robust association was found between iodine levels and BRAF mutations or lymph node metastasis in patients with PTC (Zhang et al., 2022).

Several mechanisms have been proposed to explain the complex relationship between excessive iodine intake and thyroid cancer; for example, some studies have documented that excessive iodine intake can accelerate cell cycle progression in thyroid cancer cells, promoting cell proliferation through pathways mediated by the AKT protein, and that blocking its phosphorylation can reverse these effects (Barrea et al., 2021; Hu et al., 2021).

Also, a probable association has been suggested between the presence of thyroid autoantibodies (especially TgAb), elevated TSH levels, and PTC incidence. In this regard, animal models have documented that both iodine excess and deficiency increase thyroid cell proliferation and the formation of thyroid adenomas but not the risk of carcinogenesis. However, exposure of these animals to "radiation" has shown that both iodine deficiency and excess increase the risk of thyroid cancer, leading to the hypothesis that both excess and deficiency of iodine can be considered "promoters" of thyroid cancer, rather than "inducers" (Barrea et al., 2021; Hu et al., 2021; Vargas-Uricoechea, 2024).

In summary, excessive iodine intake appears to be associated with an increased risk of WDTC development. However, the magnitude and impact on carcinogenesis seem to depend on other factors, such as genetic predisposition and environmental exposure.

Amiodarone-induced thyroiditis (AIT)

Amiodarone is an antiarrhythmic medication (class III) that contains 37.3% iodine (of its molecular weight), has a structural similarity to thyroid hormones, and has a half-life of approximately 100 days. As usual, doses range between 100 and 600 mg/day, and individuals with indications for its use can receive between 3 and 21 mg of iodine/day as a result of treatment (Gelman et al., 2024; Trohman et al., 2019).

Amiodarone-induced thyroid disturbances are of particular concern in individuals with underlying thyroid disease, and it is estimated that it may cause thyroid dysfunction in 15%–20% of individuals receiving it; even low doses of amiodarone (200 mg/day) can increase the load on daily iodine intake by 50–100 times. The frequency of amiodarone-induced hypothyroidism and hyperthyroidism is estimated to occur in 5%–10% and 0.9%–10%, respectively (in patients receiving the medication) (Medić et al., 2022).

The most frequent alterations in thyroid function tests induced by amiodarone include a decrease in the serum concentration of T3, along with an increase in T4 and rT3, with normal TSH values (although it may also be slightly elevated). Early elevations in serum TSH concentration depend on the dose and time of use, which usually normalizes over weeks to months (Medić et al., 2022; Ylli et al., 2021; Elnaggar et al., 2018).

However, the high iodine content of amiodarone is insufficient to explain the full spectrum of thyroid abnormalities associated with its use. The effects of this medication on thyroid function can be divided into those produced by the intrinsic properties of the drug and those due to iodine (Trohman et al., 2019; Medić et al., 2022; Ylli et al., 2021; Elnaggar et al., 2018).

An increase in the positivity of thyroid autoantibodies in individuals exposed to amiodarone has been documented (and although it is rare for such autoantibodies to become positive in individuals with negative autoantibodies before starting the drug), its use has been associated with an increase in some lymphocyte subpopulations, which suggests that, in susceptible individuals, amiodarone may precipitate or exacerbate thyroid autoimmunity (Trohman et al., 2019; Medić et al., 2022; Ylli et al., 2021; Elnaggar et al., 2018).

In this sense, amiodarone has been found to reduce the expression of NIS mRNA (a finding i.e. reversible upon discontinuation of the medication). Amiodarone-induced persistent hypothyroidism is attributed to a subtle defect, resulting in increased susceptibility to the inhibitory effect of iodine on hormone synthesis or the inability to escape the Wolff–Chaikoff effect (or both) (Elnaggar et al., 2018; Vargas-Uricoechea et al., 2016; Ylli et al., 2021).

Hypothyroidism may resolve or persist after discontinuation of amiodarone; resolution may have a typical time frame of 2–4 months (which is consistent with the half-life of amiodarone), persistent hypothyroidism despite medication discontinuation is almost always associated with underlying autoimmune thyroid disease (the most likely pathogenetic mechanism is the preexistence of Hashimoto's disease and the inability to escape the Wolff–Chaikoff effect) (Elnaggar et al., 2018; Vargas-Uricoechea et al., 2016; Ylli et al., 2021).

However, AIT has been classified into two types: AIT type 1 and AIT type 2. AIT type 1 is a form of iodine-induced thyrotoxicosis that is more prevalent among people with preexisting thyroid disease (multinodular toxic goiter and Graves-Basedow disease) who live in regions of low iodine consumption and is secondary to the Jod-Basedow effect. Treatment is based on a combination of thionamides, beta blockers, and perchlorate (to reduce the entry of iodine into the thyroid), if available (Macchia & Feingold, 2000; Oktaviono et al., 2025).

AIT type 2 is a destructive thyroiditis in which thyrotoxicosis is caused by the excessive release of thyroid hormones. It usually appears in patients without a history of thyroid disease; thyroid autoimmunity (humoral) appears to play a role in the development of AIT type 2 in patients without underlying thyroid disorders; this thyroiditis persists for 1–3 weeks (until thyroid hormone stores are depleted) but resolves more quickly after treatment with glucocorticoids. The prevalence of AIT type 2 in iodine-deficient regions is estimated to be 5%–10% and occurs in a male-to-female ratio of 3:1 (Oktaviono et al., 2025; Souza et al., 2021).

Treatment consists mainly of the use of glucocorticoids and relief of adrenergic symptoms. The ability to distinguish between the two types of AIT is a real challenge, since some patients may present findings of both; for example, images with thyroid scintigraphy with I-131 uptake can help in the differential diagnosis, since radiotracer uptake is increased in type 1 AIT, while it is decreased or absent in type 2 (due to the destruction or damage of the thyroid tissue) (Macchia & Feingold, 2000; Oktaviono et al., 2025; Souza et al., 2021).

However, some patients with AIT type 1 may have low radiotracer uptake (less than 3%) because such uptake can be inhibited by high concentrations of intrathyroidal iodine, even in the presence of increased thyroid activity.

Biochemical markers to detect thyroid gland destruction can provide additional information, but do not necessarily have a high predictive value in the differential diagnosis of AIT; for example, the serum Tg concentration may be elevated in both AIT (in type 2, due to the tissue destruction present, and in type 1 due to the potential underlying presence of goiter, regardless of the existence of hyperthyroidism or thyroid autoimmunity) (Ylli et al., 2021; Elnaggar et al., 2018; Macchia & Feingold, 2000; Oktaviono et al., 2025; Souza et al., 2021).

Otherwise, the presence of TRAb suggests Graves-Basedow disease, and the positivity of this autoantibody should always be considered in the differential diagnosis (Oktaviono et al., 2025). Interleukin-6 (IL-6) levels could be considered a better marker of AIT, since a normal or slightly elevated serum concentration has been documented in patients with AIT type 1, while in AIT type 2, it is significantly elevated (Elnaggar et al., 2018; Macchia & Feingold, 2000; Oktaviono et al., 2025; Souza et al., 2021; Cappellani et al., 2024).

Another tool that is considered very useful in the differential and early diagnosis of AIT is ultrasound with flow evaluation using Doppler (color), which determines the amount of blood flow within the thyroid and provides additional information on thyroid morphology. Four main patterns have been described to help differentiate AIT (Oktaviono et al., 2025; Souza et al., 2021; Cappellani et al., 2024; Krausová, 2025). The characteristics are summarized in Table 26.3.

In summary, excess iodine intake (including the use of drugs such as amiodarone) can cause both hyperthyroidism and hypothyroidism, either due to a higher frequency of Jod-Basedow or the Wolff–Chaikoff effect, respectively (Fig. 26.5).

Iodine intake and autoimmunity

Excess iodine is an environmental risk factor for developing thyroid autoimmunity in both genetically predisposed animal models (nonobese diabetic mice) and humans, and iodine appears to stimulate thyroid autoimmunity in a dose-dependent manner (Vargas-Uricoechea et al., 2025).

Among the proposed mechanisms, the fact that excess iodine can increase oxidative stress in thyrocytes through the activation of ROS is noteworthy. This is likely caused

TABLE 26.3 Characteristics of AIT.

Characteristics	AIT type 1	AIT type 2
Underlying thyroid disorders	Yes (multinodular goiter, Graves-Basedow disease)	No
Time of onset from the start of amiodarone	Short (average 3 months)	Length (average 30 months)
Thyroid scintigraphy (I-131 uptake)	Low, normal, or high	Low
IL-6 concentration	Normal or slightly elevated	Markedly elevated
T4/T3 ratio	Usually <4	Usually >4
Thyroid antibodies (TgAb, TPOAb, TRAb)	They may be present	Usually absent
Pathogenic mechanisms	Excessive synthesis of thyroid hormones	Destructive thyroiditis
Response to thionamides	Likely	No
Response to perchlorate	Yes	No
Response to glucocorticoids	Poor	Yes
Subsequent hypothyroidism	Unlikely	Likely
Probable effects following the thyrotoxic phase due to iodine	Iodine-induced hyperthyroidism	Iodine-induced hypothyroidism
Ultrasound patterns (evaluation of thyroid blood flow with color Doppler)		
Pattern 0	Vascularity, glandular destruction	
Pattern 1	Irregular and uneven intraparenchymal blood flow	
Pattern 2	Increased intraparenchymal blood flow, homogeneous and diffuse, similar to that found in Graves-Basedow disease.	
Pattern 3	Homogeneous, diffuse, and noticeably increased distribution	
Pattern 0 is associated with AIT type 2, while patterns 1 through 3 are associated with AIT type 1		

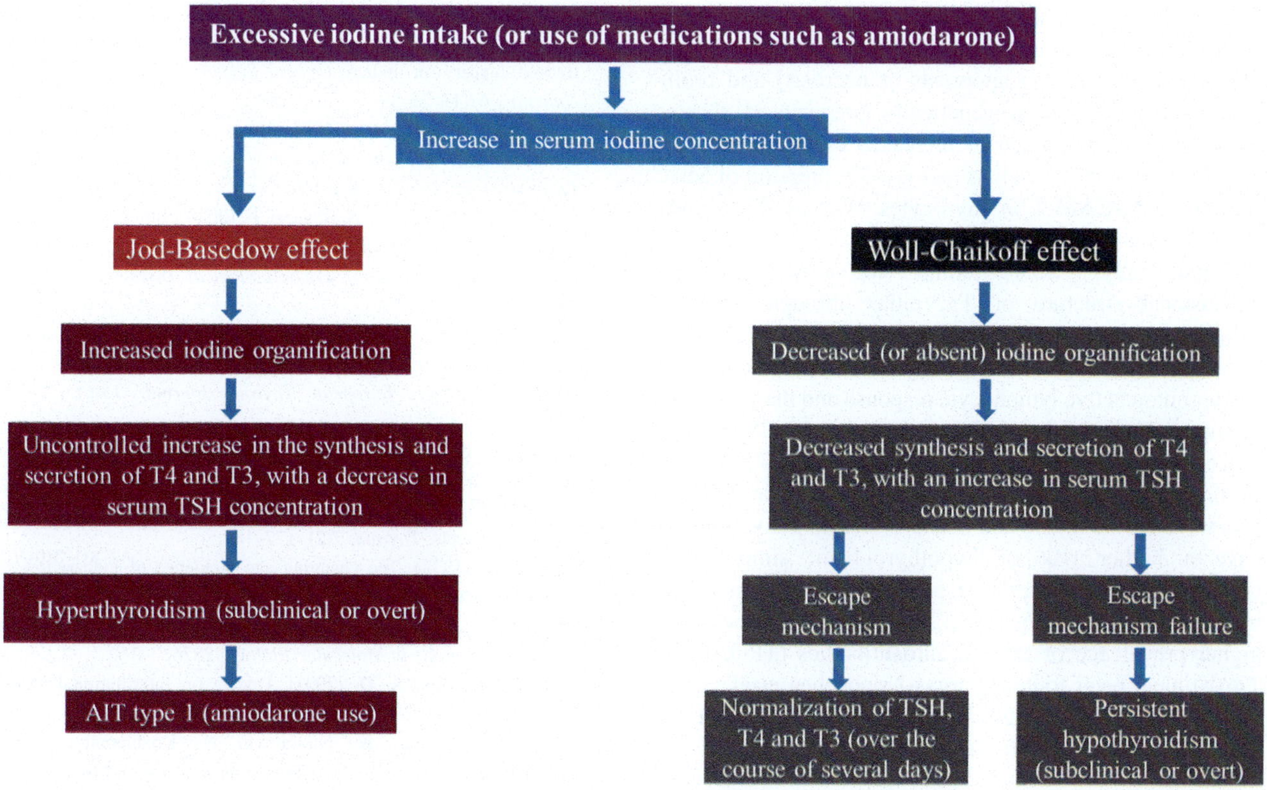

FIGURE 26.5 Functional thyroid disorders induced by excessive iodine intake (including the use of medications such as amiodarone).

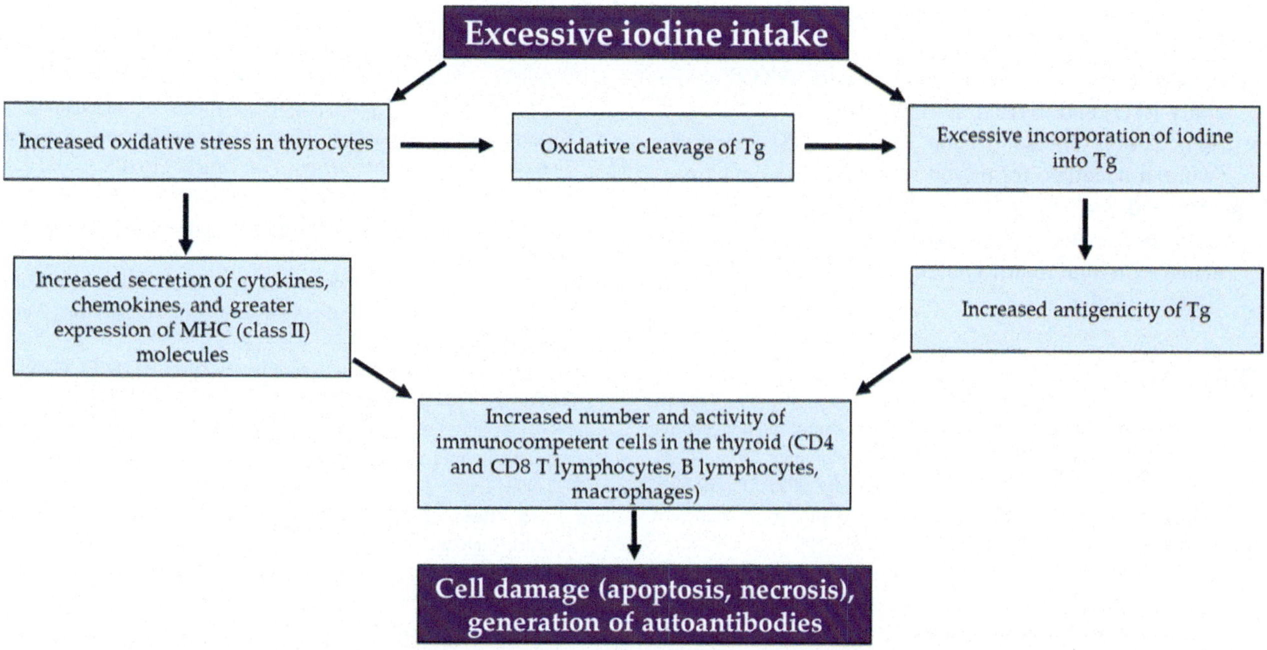

FIGURE 26.6 Iodine excess and thyroid autoimmunity. Iodine excess is a risk factor for thyroid autoimmunity. Among the proposed mechanisms, the increase in oxidative stress of thyrocytes through the activation of ROS stands out, causing cell damage and proinflammatory changes in the gland. Additionally, excess iodine can change the three-dimensional structure of Tg, with a greater response of autoreactive lymphocytes and an increase in the synthesis and secretion of autoantibodies.

by a (latent) alteration in intracellular iodine metabolism, as evidenced by iodine overload. This overload can lead to cell damage (including apoptosis or necrosis) and proinflammatory changes in genetically predisposed thyroid tissue (as a consequence of increased cytokine and chemokine secretion), as well as increased expression of MHC class II and ICAM-1 in thyrocytes (Vargas-Uricoechea, 2023; Vargas-Uricoechea et al., 2025).

Additionally, excess iodine can alter the three-dimensional structure of Tg, either through increased iodine organization or ROS-mediated Tg proteolysis. Consequently, Tg autoantigenicity increases, leading to a greater autoreactive lymphocyte response and the generation of autoantibodies (Vargas-Uricoechea et al., 2025; Vargas-Uricoechea, 2023; Petranović Ovčariček et al., 2024).

In fact, it has been documented that populations with excessive iodine intake and positive thyroid autoantibodies have a higher risk of hypothyroidism; similarly, in geographic areas where USI programs have been implemented, a higher frequency of Hashimoto's disease and a higher prevalence of thyroid autoantibodies (TPOAb and TgAb) have been found (Vargas-Uricoechea et al., 2025; Vargas-Uricoechea, 2023; Petranović Ovčariček et al., 2024; Atapattu et al., 2024) (Fig. 26.6).

Summary points

- Iodine is an essential component of thyroid hormone synthesis.
- Excess iodine is associated with either hyperthyroidism (Jod-Basedow effect) or hypothyroidism (Wolff–Chaikoff effect).
- Excess iodine has been associated with WTDC (especially PTC) and thyroid autoimmunity.
- Sources of excess iodine include USI programs (without adequate follow-up or monitoring over time), diets with excessive intake of iodized salt, the use of medications such as amiodarone, nutritional supplements with high iodine content, and the use of iodinated contrast media.

References

Aranda, A. (2025). Thyroid hormone action by genomic and nongenomic molecular mechanisms. *Methods in Molecular Biology, 2876*, 17–34.

Armstrong, M., Asuka, E., & Fingeret, A. (January 2025). Physiology, thyroid function. In *StatPearls*. Treasure Island (FL): StatPearls Publishing.

Atapattu, N., Jayatissa, R., de Silva, H., Adlan, M. A., Obuobie, E. K., & Premawardhana, L. D. (2024). Thyroid autoimmunity during universal salt iodisation-possible short-term modulation with longer-term stability. *Nutrients, 16*(24), 4299.

Bürgi, H. (2010). Iodine excess. *Best Practice & Research Clinical Endocrinology & Metabolism, 24*(1), 107–115.

Barrea, L., Gallo, M., Ruggeri, R. M., Giacinto, P. D., Sesti, F., Prinzi, N., & Group, E. O. L. O. (2021). Nutritional status and follicular-derived thyroid cancer: An update. *Critical Reviews in Food Science and Nutrition, 61*(1), 25–59.

Bizhanova, A., & Kopp, P. (2009). Minireview: The sodium-iodide symporter NIS and pendrin in iodide homeostasis of the thyroid. *Endocrinology, 150*(3), 1084–1090.

Cappellani, D., Bartalena, L., & Bogazzi, F. (2024). Short review: Novel concepts in the approach to patients with amiodarone-induced thyrotoxicosis. *Journal of Endocrinological Investigation, 47*(2), 275–283.

Carlqvist, J., Nyman, U., Brandberg, J., Nyström, H. F., & Hellström, M. (2024). Effects of iodine contrast media on thyroid function—A prospective study. *European Thyroid Journal, 13*(6), Article e240244.

Carvalho, D. P., & Dupuy, C. (2017). Thyroid hormone biosynthesis and release. *Molecular and Cellular Endocrinology, 458*, 6–15.

Cavalieri, R. R. (1997). Iodine metabolism and thyroid physiology: Current concepts. *Thyroid, 7*(2), 177–181.

Connolly, R. J. (1973). The changing age incidence of jodbasedow in Tasmania. *Medical Journal of Australia, 2*(4), 171–174.

Contreras-Jurado, C. (2025). Thyroid hormones and co-workers: An overview. *Methods in Molecular Biology, 2876*, 3–16.

Dunn, J. T., & Dunn, A. D. (2001). Update on intrathyroidal iodine metabolism. *Thyroid, 11*(5), 407–414.

Elnaggar, M. N., Jbeili, K., Nik-Hussin, N., Kozhippally, M., & Pappachan, J. M. (2018). Amiodarone-induced thyroid dysfunction: A clinical update. *Experimental and Clinical Endocrinology & Diabetes, 126*(6), 333–341.

Farebrother, J., Zimmermann, M. B., & Andersson, M. (2019). Excess iodine intake: Sources, assessment, and effects on thyroid function. *Annals of the New York Academy of Sciences, 1446*(1), 44–65.

Geertsma, E. R., & Oliver, D. (2024). SLC26 anion transporters. *Handbook of Experimental Pharmacology, 283*, 319–360.

Gelman, I., Sharma, N., Mckeeman, O., Lee, P., Campagna, N., Tomei, N., Baranchuk, A., Zhang, S., & El-Diasty, M. (2024). The ion channel basis of pharmacological effects of amiodarone on myocardial electrophysiological properties, a comprehensive review. *Biomedicine & Pharmacotherapy, 174*, Article 116513.

Giri, D., & Mugesh, G. (2025). Thyroid hormone deiodination-mechanisms and small molecule enzyme mimics. *Biomolecules, 15*(4), 529.

Hatch-McChesney, A., & Lieberman, H. R. (2022). Iodine and iodine deficiency: A comprehensive review of a re-emerging issue. *Nutrients, 14*(17), 3474.

Hu, M. J., He, J. L., Tong, X. R., Yang, W. J., Zhao, H. H., Li, G. A., & Huang, F. (2021). Associations between essential microelements exposure and the aggressive clinicopathologic characteristics of papillary thyroid cancer. *Biometals*. https://doi.org/10.1007/s10534-021-00317-w

Jing, L., & Zhang, Q. (2022). Intrathyroidal feedforward and feedback network regulating thyroid hormone synthesis and secretion. *Frontiers in Endocrinology, 13*, Article 992883.

Khudair, A., Khudair, A., Niinuma, S. A., Habib, H., & Butler, A. E. (2025). Beyond thyroid dysfunction: The systemic impact of iodine excess. *Frontiers in Endocrinology, 16*, Article 1568807.

Kim, K., Cho, S. W., Park, Y. J., Lee, K. E., Lee, D. W., & Park, S. K. (2021). Association between iodine intake, thyroid function, and papillary thyroid cancer: A case-control study. *Endocrinology and Metabolism, 36*(4), 790–799.

Kim, S. J., Lee, J., & Soh, E. Y. (2017). The clinical significance of the BRAF mutation in patients with papillary thyroid cancer. *Journal of Endocrine Surgery, 17*(4), 175–183.

Koukkou, E. G., Roupas, N. D., & Markou, K. B. (2017). Effect of excess iodine intake on thyroid on human health. *Minerva Medica, 108*(2), 136–146.

Krausová, A. (2025). Amiodarone-induced thyreopaties. *Casopis Lékaru Ceských, 164*(3), 94–99.

Krause, K., Karger, S., Gimm, O., Sheu, S. Y., Dralle, H., Tannapfel, A., Schmid, K. W., Dupuy, C., & Fuhrer, D. (2007). Characterisation of DEHAL1 expression in thyroid pathologies. *European Journal of Endocrinology, 156*(3), 295–301.

Laurberg, P., Cerqueira, C., Ovesen, L., Rasmussen, L. B., Perrild, H., Andersen, S., Pedersen, I. B., & Carlé, A. (2010). Iodine intake as a determinant of thyroid disorders in populations. *Best Practice & Research Clinical Endocrinology & Metabolism, 24*(1), 13–27.

Lechan, R. M., Hollenberg, A., & Fekete, C. (2009). Hypothalamic-pituitary-thyroid axis: Organization, neural/endocrine control of TRH. In L. R. Squire (Ed.), *Encyclopedia of neuroscience* (pp. 75–87). Oxford, UK: Academic Press.

Leung, A. M., & Braverman, L. E. (2012). Iodine-induced thyroid dysfunction. *Current Opinion in Endocrinology Diabetes and Obesity, 19*(5), 414–419.

Leung, A., Pearce, E. N., & Braverman, L. E. (2010). Role of iodine in thyroid physiology. *Expert Review of Endocrinology and Metabolism, 5*(4), 593–602.

Macchia, P. E., & Feingold, K. R. (2000). Amiodarone induced thyrotoxicosis. In K. R. Feingold, S. F. Ahmed, B. Anawalt, M. R. Blackman, A. Boyce, G. Chrousos, E. Corpas, W. W. de Herder, K. Dhatariya, K. Dungan, J. Hofland, S. Kalra, G. Kaltsas, N. Kapoor, C. Koch, P. Kopp, M. Korbonits, C. S. Kovacs, W. Kuohung, & D. P. Wilson (Eds.), *Endotext*. South Dartmouth (MA): MDText.com, Inc.

Mariotti, S., & Beck-Peccoz, P. (2000). Physiology of the hypothalamic-pituitary-thyroid axis. In K. R. Feingold, S. F. Ahmed, B. Anawalt, M. R. Blackman, A. Boyce, G. Chrousos, E. Corpas, W. W. de Herder, K. Dhatariya, K. Dungan, J. Hofland, S. Kalra, G. Kaltsas, N. Kapoor, C. Koch, P. Kopp, M. Korbonits, C. S. Kovacs, W. Kuohung, & D. P. Wilson (Eds.), *Endotext*. South Dartmouth (MA): MDText.com, Inc.

Markou, K., Georgopoulos, N., Kyriazopoulou, V., & Vagenakis, A. G. (2001). Iodine-induced hypothyroidism. *Thyroid, 11*(5), 501–510.

Markova, B., Mayerl, S., & Heuer, H. (2024). Toward a treatment for thyroid hormone transporter MCT8 deficiency - Achievements and challenges. *European Thyroid Journal, 13*(6), Article e240286.

Medić, F., Bakula, M., Alfirević, M., Bakula, M., Mucić, K., & Marić, N. (2022). Amiodarone and thyroid dysfunction. *Acta Clinica Croatica, 61*(2), 327–341.

Mullur, R., Liu, Y. Y., & Brent, G. A. (2014). Thyroid hormone regulation of metabolism. *Physiological Reviews, 94*(2), 355–382.

Ohye, H., & Sugawara, M. (2010). Dual oxidase, hydrogen peroxide and thyroid diseases. *Experimental Biology and Medicine, 235*(4), 424–433.

Oktaviono, Y. H., Mustofa, A., Saputra, P. B. T., Widiarti, W., Faizah, N. N., & Alkaff, F. F. (2025). Exploring current diagnosis and management of amiodarone-induced thyrotoxicosis. *The American Journal of Cardiology, 242*, 75–81.

Petranović Ovčariček, P., Görges, R., & Giovanella, L. (2024). Autoimmune thyroid diseases. *Seminars in Nuclear Medicine, 54*(2), 219–236.

Riesco-Eizaguirre, G., Santisteban, P., & De la Vieja, A. (2021). The complex regulation of NIS expression and activity in thyroid and extrathyroidal tissues. *Endocrine-Related Cancer, 28*(10), T141–T165.

Sabatino, L., Vassalle, C., Del Seppia, C., & Iervasi, G. (2021). Deiodinases and the three types of thyroid hormone deiodination reactions. *Endocrinology and Metabolism, 36*(5), 952–964.

Shahid, M. A., Ashraf, M. A., & Sharma, S. (January 2025). Physiology, thyroid hormone. In *StatPearls*. Treasure Island (FL): StatPearls Publishing.

Sohn, S. Y., Inoue, K., Bashir, M. T., Currier, J. W., Neverova, N. V., Ebrahimi, R., Rhee, C. M., Lee, M. L., & Leung, A. M. (2025). Thyroid dysfunction risk after iodinated contrast media administration: A prospective longitudinal cohort analysis. *The Journal of Cinical Endocrinology and Metabolism, 110*(4), e1204–e1210.

Sohn, S. Y., Inoue, K., Rhee, C. M., & Leung, A. M. (2024). Risks of iodine excess. *Endocrine Reviews, 45*(6), 858–879.

Souza, L. V. F., Campagnolo, M. T., Martins, L. C. B., & Scanavacca, M. I. (2021). Amiodarone-induced thyrotoxicosis - Literature review and clinical update. *Arquivos Brasileiros de Cardiologia, 117*(5), 1038–1044.

Tan, J., Xiao, Y., Kong, F., Qian, J., Zhu, A., & Yan, C. (2025). Structural insights into thyroid hormone transporter MCT8. *Nature Communications, 16*(1), 2958.

Thompson, E., Eldstrom, J., & Fedida, D. (2021). Hormonal signaling actions on Kv7.1 (KCNQ1) channels. *Annual Review of Pharmacology and Toxicology, 61*, 381–400.

Trohman, R. G., Sharma, P. S., McAninch, E. A., & Bianco, A. C. (2019). Amiodarone and thyroid physiology, pathophysiology, diagnosis and management. *Trends in Cardiovascular Medicine, 29*(5), 285–295.

Vargas-Uricoechea, H. (2023). Molecular mechanisms in autoimmune thyroid disease. *Cells, 12*(6), 918.

Vargas-Uricoechea, H. (2024). Autoimmune thyroid disease and differentiated thyroid carcinoma: A review of the mechanisms that explain an intriguing and exciting relationship. *World Journal of Oncology, 15*(1), 14–27.

Vargas-Uricoechea, H., Bonelo-Perdomo, A., & Sierra-Torres, C. H. (2016). Iodine and the thyroid. In S. Imam, & S. I. Ahmad (Eds.), *Thyroid disorders: Basic science and clinical practice* (pp. 27–48). https://doi.org/10.1007/978-3-319-25871-3

Vargas-Uricoechea, H., Castellanos-Pinedo, A., Meza-Cabrera, I. A., Pinzón-Fernández, M. V., Urrego-Noguera, K., & Vargas-Sierra, H. (2025). Iodine intake from universal salt iodization programs and hashimoto's thyroiditis: A systematic review. *Diseases, 13*(6), 166.

WHO/UNICEF/ICCIDD. Assessment of iodine deficiency disorders and monitoring their elimination: A guide for programme managers. 3rd ed. Geneva, Switzerland: World Health Organization; 2007.

Ylli, D., Wartofsky, L., & Burman, K. D. (2021). Evaluation and treatment of amiodarone-induced thyroid disorders. *The Journal of Cinical Endocrinology and Metabolism, 106*(1), 226–236.

Zhang, X., Zhang, F., Li, Q., Aihaiti, R., Feng, C., Chen, D., Zhao, X., & Teng, W. (2022). The relationship between urinary iodine concentration and papillary thyroid cancer: A systematic review and meta-analysis. *Frontiers in Endocrinology, 13*, Article 1049423.

Zimmermann, M. B., & Galetti, V. (2015). Iodine intake as a risk factor for thyroid cancer: A comprehensive review of animal and human studies. *Thyroid Research, 8*, 8.

Section 8

Resources

Chapter 27

Recommended resources for the study of hormones in neuroscience: The thyroid gland and thyroid hormones

Rajkumar Rajendram[1, 2, 3], M. Aadam Siraj[4], Vinood B. Patel[5] and Victor R. Preedy[6]

[1]*Research Unit, Department of Medical Education, College of Medicine, King Saud bin Abdulaziz University for Health Sciences, Riyadh, Saudi Arabia;* [2]*Department of Medicine, King Abdulaziz Medical City, Ministry of National Guard Health Affairs, Riyadh, Saudi Arabia;* [3]*Research Offices, King Abdullah International Medical Research Center, Riyadh, Saudi Arabia;* [4]*Department of Clinical and Biomedical Sciences, University of Exeter Medical School, University of Exeter, Exeter, United Kingdom;* [5]*School of Life Sciences, University of Westminster, London, United Kingdom;* [6]*School of Life Course and Population Sciences, Faculty of Life Science and Medicine, King's College London, London, United Kingdom*

Introduction

In the 19th century research defined the role of thyroid hormone deficiency in the developmental disorder formerly known as cretinism. This severe disease characterized by intellectual disability is now called congenital hypothyroidism. Scientists subsequently uncovered some of the intricate mechanisms through which thyroid hormones orchestrate normal brain development and maintain adult neurological function (Preedy et al., 2009; Stepien & Huttner, 2019).

Thyroid hormones regulate metabolism and drive neural progenitor proliferation, neuronal migration, functional maturation, and neuronal survival in the developing brain. Neuroendocrine interactions are very complicated. Thyroid hormones, particularly T3 (triiodothyronine) and T4 (thyroxine), influence several metabolic pathways. These involve specific cell membrane transporters, intracellular deiodinase enzymes, and nuclear hormone receptors present in the brain (Sabatino et al., 2024). Investigating the endocrine aspects of neuroscience requires understanding of both classical genomic mechanisms and emerging non-genomic pathways that operate at the cell membrane level (Schiera et al., 2021).

Historical development of key mechanistic insights

These interlinking fields of thyroid hormone biology have progressively evolved and expanded through distinct conceptual phases. Early studies focused on whole-organism effects of thyroid hormone deprivation on brain structure and function. These foundational investigations established that hypothyroidism results in serious anomalies in nervous system organization and function (Schiera et al., 2021). By the late twentieth century, research advanced to examine specific cellular populations and brain regions, revealing differential sensitivity to thyroid hormones across development and revealing the importance of local thyroid hormone metabolism.

Advancing the scientific fields of endocrinology and neuroscience

Those active in the fields of endocrinology and neuroscience must keep up to date and abreast of current research. To address this, we have compiled a list of resources as recommended by researchers, physicians, practitioners, contributing authors, and others who actively work in these fields. By providing a curated collection of impactful web-based links, we enhance the access to current knowledge in these important areas. Information on over 150 organizations, scientific societies, regulatory bodies and other web-based resources are included.

The lists of material in these tables is to provide general information only. It does not constitute any recommendation or endorsement of the activities of these sites, facilities, or other resources listed in this chapter, by the authors or editors of this book. Readers should visit the listed websites sites at their own discretion and ensure they have appropriate security measures, such as up-to-date antivirus and anti-malware protection, in place.

Tables 27.1–27.4 list the most up-to-date information on the regulatory bodies (Table 27.1), professional societies (Table 27.2), emerging technologies and platforms

Hormones in Neuroscience. https://doi.org/10.1016/B978-0-443-30261-9.00027-9

TABLE 27.1 Websites, regulatory bodies or organizations associated with the study of hormones in neuroscience, the thyroid gland, related hormones and associated clinical conditions.

Regulatory body or organisation	Web address
Alzheimer's Drug Discovery Foundation – Biomarkers Program	https://www.alzdiscovery.org
American Association of Clinical Endocrinologists	https://www.aace.com/
American Association of Clinical Endocrinologists (AACE)	https://www.aace.com/
American Brain Coalition	https://www.americanbraincoalition.org/
American Thyroid Association (ATA)	https://www.thyroid.org
Association for Nutrition	https://www.associationfornutrition.org/
Association of British Neurologists	https://www.theabn.org/
British Association of Endocrine and Thyroid Surgeons	https://baets.org.uk/
British Neuroscience Association	https://www.bna.org.uk/
British Thyroid Association (BTA)	https://british-thyroid-association.org/
British Thyroid Foundation	https://www.btf-thyroid.org/
Centers for Disease Control and Prevention (CDC)	https://www.cdc.gov
ClinicalTrials.gov (National Institute of Health Trials Registry)	https://clinicaltrials.gov/
Endocrine Society	https://www.endocrine.org/
European Medicines Agency (EMA)	https://www.ema.europa.eu/
European Society of Endocrinology	https://www.ese-hormones.org/
European Thyroid Association (ETA)	https://www.eurothyroid.com
Food and Drug Administration (FDA)	https://www.fda.gov/
Graves' Disease and Thyroid Foundation	https://www.gdatf.org/
Health and Care Professions Council (HCPC)	https://www.hcpc-uk.org/
Health Canada (regulatory body)	https://www.canada.ca/en/health-canada.html
Health Research Authority (HRA)	https://www.hra.nhs.uk/
Hormone Foundation	https://www.endocrine.org/hormone-foundation
Human Tissue Authority (HTA)	https://www.hta.gov.uk/
International Neuroendocrine Federation (INF)	https://inf-neuroendocrinology.org/
International Thyroid Oncology Group	https://itog.org/
Iodine Global Network (IGN)	https://ign.org/
Medicines and Healthcare products Regulatory Agency (MHRA, UK)	https://www.gov.uk/government/organisations/medicines-and-healthcare-products-regulatory-agency
Mental Health America	https://mhanational.org/
National Institute for Health and Care Excellence (NICE, United Kingdom)	https://www.nice.org.uk/
National Institute of Diabetes and Digestive and Kidney diseases (NIDDK, NIH)	https://www.niddk.nih.gov/
National Institute of Neurological Disorders and Stroke (NINDS)	https://www.ninds.nih.gov/
National Institutes of Health (NIH)	https://www.nih.gov/
National Library of Medicine (NLM)	https://www.nlm.nih.gov/
Society for Endocrinology (United Kingdom)	https://www.endocrinology.org/
Thyroid Cancer Survivors' Association (ThyCa)	https://www.thyca.org/

Continued

TABLE 27.1 Websites, regulatory bodies or organizations associated with the study of hormones in neuroscience, the thyroid gland, related hormones and associated clinical conditions.—cont'd

Regulatory body or organisation	Web address
Thyroid Federation Australia	https://thyroidfoundation.org.au/
Thyroid Federation International (TFI)	https://www.thyroid-federation.org
Thyroid Foundation of America	http://www.allthyroid.org/
Thyroid Foundation of Canada	https://thyroid.ca
Thyroid Trust	https://www.thyroidtrust.org/
Turkish Endocrinology and Metabolism Association	https://temd.org.tr/
Veterans Affairs (VA)	https://www.va.gov/
World Health Organization (WHO)	https://www.who.int
World Iodine Association	https://www.worldiodineassociation.com/

This table lists the regulatory bodies and organizations relevant to the study of hormones in neuroscience, the thyroid gland, related hormones and associated clinical conditions. The links were accurate at the time of going to press but may move or alter. In these cases, the use of the "Search" tabs should be explored at the parent address or site. In some cases, links direct the reader to relevant pages within parent sites. Some societies and organizations prefer shortened terms, such as acronyms and abbreviations. Some organizations provide a dual role of imparting information on specific conditions and acting as professional or regulatory bodies. Where possible we provided translations of the name of the site if not already provided by the invited authors. See also Table 27.2.

TABLE 27.2 Professional societies that are associated with the study of hormones in neuroscience, the thyroid gland, related hormones and associated clinical conditions.

Society name	Web address
Alzheimer's Association International Society to Advance Alzheimer's Research and Treatment (ISTAART)	https://www.alz.org/istaart
American Association of Clinical Endocrinologists (AACE)	https://www.aace.com/
American Medical Association	https://www.ama-assn.org/
American Society for Nutrition	https://nutrition.org/
American Thyroid Association	https://www.thyroid.org/
Asia & Oceania Thyroid Association (AOTA)	http://aota.kr/
Brazilian Society of Endocrinology and Metabolism	https://endocrino.org.br/en/
British Association of Endocrine and Thyroid Surgeons (BAETS)	https://baets.org.uk/
British Society for Neuroendocrinology (BSN)	https://www.neuroendo.org.uk/
British Thyroid Association (BTA)	https://www.british-thyroid-association.org
Endocrine Society	https://www.endocrine.org/
European Association of Nuclear Medicine (EANM)	https://www.eanm.org/
European Society for Clinical Nutrition and Metabolism (ESPEN)	https://www.espen.org/
European Society for Pediatric Endocrinology (ESPE)	https://www.eurospe.org/
European Society of Endocrinology (ESE)	https://www.ese-hormones.org/
European Thyroid Association (ETA)	https://www.eurothyroid.com/
French Society of Nutrition	https://sf-nutrition.fr/
International Society for Neurochemistry (ISN)	https://www.neurochemistry.org
International Thyroid Oncology Group (ITOG)	https://www.itog.org/

Continued

TABLE 27.2 Professional societies that are associated with the study of hormones in neuroscience, the thyroid gland, related hormones and associated clinical conditions.—cont'd

Society name	Web address
Japanese Thyroid Association	https://www.japanthyroid.jp
Latin America Thyroid Society	http://www.lats.org/
National Institute for Health and Care Excellence	https://www.nice.org.uk/
National Neurological Societies	https://www.ean.org/home/members/national-neurological-societies
Royal Society of Medicine	https://www.rsm.ac.uk/
Society for Endocrinology (UK)	https://www.endocrinology.org/
Society for Neuroscience	https://www.sfn.org/
Society of Nuclear Medicine and Molecular Imaging (SNMMI)	https://www.snmmi.org/
ThyCa (Thyroid Cancer Survivors' Association)	https://thyca.org/
Thyroid & Hormones Endocrinology Center	https://www.thyroidandhormones.com/associations

This table lists the professional societies relevant to the study of hormones in neuroscience, the thyroid gland, related hormones and associated clinical conditions. See also Table 27.1.

TABLE 27.3 Emerging techniques, instruments, tools and platforms associated with hormones in neuroscience, the thyroid gland, related hormones and associated clinical conditions.

Technique/Instrument/Platform	Organisation or company Name	Web address
3D organoid cultures for tissue modeling and drug screening	Stem Cell Technologies	https://www.stemcell.com/
3T magnetic resonance imaging (MRI) systems for clinical/research neuroimaging	GE HealthCare	https://www.gehealthcare.com/
3T/7T MRI systems for structural MRI, fMRI, DTI, MRS	Siemens Healthineers	https://www.siemens-healthineers.com/
AI-powered thyroid nodule classification tools	aidoc	https://www.aidoc.com/
Analytical platforms and various measuring devices	Mettler Toledo	https://www.mt.com/
Antibodies for immunohistochemistry and northern blotting	ThermoFisher	https://www.thermofisher.com/
Blood-based multiplex assays for T_3/T_4 and neurodegeneration markers	Quanterix	https://www.quanterix.com/
Cambridge neuropsychological test automated battery (CANTAB) test battery: Digital cognitive assessment	Cambridge cognition	https://cambridgecognition.com/executive-function/
Cerebrospinal fluid (CSF) TH & amyloid biomarker assays	Fujirebio	https://www.fujirebio.com/
Clustered regularly interspaced short palindromic repeats (CRISPR)-Cas9 gene editing technology	Horizon Discovery, Thermo Fisher	https://horizondiscovery.com/en/applications/gene-editing https://www.thermofisher.com/in/en/home.html
Diffusion tensor imaging (DTI) tractography and diffusion processing toolkit	MRtrix3	https://www.mrtrix.org/

Continued

TABLE 27.3 Emerging techniques, instruments, tools and platforms associated with hormones in neuroscience, the thyroid gland, related hormones and associated clinical conditions.—cont'd

Technique/Instrument/Platform	Organisation or company Name	Web address
Digital and imaging pathology platforms for thyroid histology	Leica Biosystems	https://www.leicabiosystems.com/
Digital enzyme-linked immunosorbent assays (ELISA) single molecule arrays (simoa) for ultra-sensitive hormone detection	Quanterix	https://www.quanterix.com
Elastosonography for preoperative risk assessment of thyroid nodules	Siemens	https://www.siemens-healthineers.com/tr/ultrasoncihazlari/advanced-applications/ultrasound-elastography-systems
Electroencephalogram (EEG) amplifiers and acquisition systems	Brain Products	https://www.brainproducts.com/
Functional MRI (fMRI) for thyroid–Brain axis Studies	Siemens Healthineers	https://www.siemens-healthineers.com
High-resolution Mass spectrometry for thyroid hormone metabolomics	Thermo Fisher Scientific	https://www.thermofisher.com
High-resolution thyroid ultrasonography systems	GE HealthCare	https://www.gehealthcare.com/
Integrated lab information systems for thyroid test data analysis	Oracle Health (formerly Cerner Corporation)	https://www.oracle.com/health/
Liquid biopsy platforms for thyroid cancer monitoring	Guardant Health	https://guardanthealth.com/
Liquid chromatography–mass spectrometry (LC-MS) and ultra high performance liquid chromatography (UHPLC) for hormone analysis	Agilent, Waters	https://www.agilent.com/ https://www.waters.com/
Machine learning diagnostic platforms for thyroid disorders	Google DeepMind Health	https://deepmind.google/
Magnetic resonance imaging (MRI) systems and neuroimaging workflows	Philips	https://www.philips.com/healthcare
Medical E-Governance (MEG) systems and clinical MEG workflow platforms	MEGIN	https://megin.com/
Microfluidic platforms for thyroid hormone detection	Dolomite Microfluidics	https://www.unchainedlabs.com/dolomite-microfluidics-systems/
Nuclear magnetic resonance (NMR) spectroscopy and associated instrumentation. Can be applied for assessing molecular interactions and metabolic changes in thyroid disorders.	Bruker	https://www.bruker.com/en.html
Optical biosensors for thyroid stimulating hormone (TSH) detection	HORIBA Scientific	https://www.horiba.com/int/scientific/
Point-of-care (POC) diagnostic devices for rapid testing	Siemens Healthineers, Abbott	https://www.siemens-healthineers.com/ https://www.abbott.com/
Positron emission tomography/Computed tomography (PET/CT), fluorodeoxyglucose (FDG)-PET imaging and PET/MR hybrid imaging for thyroid cancer metabolism and brain metabolism	Siemens Healthineers	https://www.siemens-healthineers.com/
Proteomics and mass spectrometry platforms for protein profiling	Bruker Thermo Fisher	https://www.bruker.com https://www.thermofisher.com/
Real-time PCR systems for thyroid gene expression profiling	Thermo Fisher	https://www.thermofisher.com/

Continued

TABLE 27.3 Emerging techniques, instruments, tools and platforms associated with hormones in neuroscience, the thyroid gland, related hormones and associated clinical conditions.—cont'd

Technique/Instrument/Platform	Organisation or company Name	Web address
Real-time polymerase chain reaction (PCR) for thyroid gene expression	Bio-Rad Laboratories	https://www.bio-rad.com/
Single photon emission computed tomography (SPECT) systems for cerebral perfusion imaging	GE HealthCare	https://www.gehealthcare.com/
Single-cell ribonucleic acid (RNA) sequencing for gene expression analysis	Illumina, Thermo Fisher 10x Genomics	https://www.illumina.com/ https://www.thermofisher.com/ https://www.10xgenomics.com
Thyroid hormone metabolomics profiling tools	Metabolon	https://www.metabolon.com/
Thyroid scintigraphy imaging systems	Philips Healthcare	https://www.philips.co.uk/healthcare
Thyroid ultrasound elastography for tissue stiffness analysis	GE Healthcare	https://www.gehealthcare.co.uk/

This table lists some emerging technologies and online resources relevant to hormones in neuroscience, the thyroid gland, related hormones and associated clinical conditions. A cautionary note is needed as some platforms and methods are still at the experimental stage and used in research or they have not been used for routine clinical use. Some companies are mentioned more than once and, in some cases, we have merged suggestions into a single entry. The links were accurate at the time of going to press but may move or alter. In these cases, the use of the "Search" tabs should be explored at the parent address or site. In some cases, links direct the reader to relevant pages within parent sites. Some material suggested by authors is no longer available at specific sites, so we have inserted the parent url.

TABLE 27.4 Other resources of interest or relevance for health care professionals or patients associated with the study of hormones in neuroscience, the thyroid gland, related hormones and clinical conditions.

Type of material/Platform/Resource	Organisation or company Name	Web address
Alzheimer's Disease neuroimaging initiative (ADNI) data	University of California, San Diego	https://adni.loni.usc.edu
Clinical practice guidelines on thyroid disorders and disclaimer	American Thyroid Association	https://www.thyroid.org/professionals/ata-professional-guidelines/ https://www.thyroid.org/professionals/ata-guidelines/
Clinical trials on Thyroid Hormone and cognitive Function. Ongoing clinical trials assessing the neuroprotective effects of nutraceuticals and standard therapies in thyroid disorders. Also material on thyroid–neurological studies	ClinicalTrials.gov	https://www.clinicaltrials.gov/
Drug interaction/absorption considerations for levothyroxine	DrugBank	https://go.drugbank.com/drugs/DB00451
Educational e-modules on thyroid function	Endocrine Society Learning Center	https://education.endocrine.org/Public/Catalog/ChefView.aspx
Endocrinology clinical decision support tools	UpToDate	https://www.uptodate.com
European Thyroid Association Clinical Guidelines	European Thyroid Association	https://www.eurothyroid.com/guidelines/eta_guidelines.html
Genetic and molecular thyroid data repository	NCBI Gene/National Institute for Health (NIH)	https://www.ncbi.nlm.nih.gov/gene
Guidelines on health care (various areas and UK based)	National Institute for Health and Clinical Excellence - NICE	https://www.nice.org.uk

Continued

TABLE 27.4 Other resources of interest or relevance for health care professionals or patients associated with the study of hormones in neuroscience, the thyroid gland, related hormones and clinical conditions.—cont'd

Type of material/Platform/Resource	Organisation or company Name	Web address
Guidelines on management of various thyroid disorders (American based)	American Thyroid Association	https://www.thyroid.org/
Guidelines on management of various thyroid disorders (European based)	European Thyroid Association	https://etj.bioscientifica.com/page/993
Guidelines on Thyroid Disease Management (Endocrinology). Updated guidelines and protocols for diagnosing and treating thyroid dysfunction and related neurological impacts.	American Thyroid Association	https://www.thyroid.org/professionals/
Human Protein Atlas Also open-access hormone reference data	Brain and Endocrine Expression Profiles	https://www.proteinatlas.org
Large neuroimaging/behavior dataset + pipelines (connectomics)	Human Connectome Project	https://www.humanconnectome.org/
Magnetic resonance (MR) spectroscopy quantitation (LCModel manual/tools)	S Provencher	https://www.s-provencher.com/pub/LCModel/manual/manual.pdf
Measuring alcohol units (UK relevance from the Department of Health)	The Learning Exchange	https://lx.iriss.org.uk/sites/default/files/resources/Alcohol%20Units%20a%20brief%20guide.pdf
Montreal Cognitive Assessment (MoCA) cognitive screening (official access/licensing information)	Mapi Research Trust (ePROVIDE)	https://eprovide.mapi-trust.org/instruments/montreal-cognitive-assessment
National Center for Biotechnology Information (NCBI) Bookshelf. A resource for biomedical literature	National Center for Biotechnology Information	https://www.ncbi.nlm.nih.gov/books/
Open neuroimaging data repository (Brain Imaging Data Structure (BIDS)-compatible sharing)	OpenNeuro	https://openneuro.org/
Patient and professional thyroid information.	British Thyroid Foundation	https://www.btf-thyroid.org/
Health information—thyroid	MedlinePlus	https://medlineplus.gov/thyroiddiseases.html
Patient Health Questionnaire 9 (PHQ-9) depression questionnaire	American Psychological Association	https://www.apa.org/depression-guideline/patient-health-questionnaire.pdf
Patient information on hypothyroidism and thyroid disease	MedlinePlus	https://medlineplus.gov/thyroiddiseases.html
Patient support + education + advocacy network	Thyroid Federation International	https://www.thyroid-federation.org/
PubChem (Chemical Database for Thyroid Hormone Analogs)	National Institute of Health (NIH)	https://pubchem.ncbi.nlm.nih.gov
Quality of Life Questionnaire	World Health Organization	https://iris.who.int/bitstream/handle/10665/63529/WHOQOL-BREF.pdf
Resources on Endocrine Disorders and Research	The Endocrine Society	https://www.endocrine.org/
ThyPRO (thyroid-specific quality-of-life instrument)	ThyPRO	https://eprovide.mapi-trust.org/instruments/thyroid-specific-patient-reported-outcome
Thyroid cancer staging and treatment guidelines	National Comprehensive Cancer Network (NCCN)	https://www.nccn.org

Continued

TABLE 27.4 Other resources of interest or relevance for health care professionals or patients associated with the study of hormones in neuroscience, the thyroid gland, related hormones and clinical conditions.—cont'd

Type of material/Platform/Resource	Organisation or company Name	Web address
Thyroid disease coding and classification	International Classification of Diseases (ICD)-11/WHO	https://icd.who.int/en
Thyroid disease patient support and education	ThyCa	https://thyca.org/
Thyroid Diseases Diagnosis and Treatment Guide 2025	Turkish Endocrinology and Metabolism Association	https://file.temd.org.tr/Uploads/publications/guides/documents/tiroid2025_11042025.pdf
Thyroid hormone assay protocols	Clinical and Laboratory Standards Institute	https://clsi.org
Thyroid imaging and case library		https://radiopaedia.org/
Thyroid Trust's patient support and research references	The Thyroid Trust	https://www.thyroidtrust.org/
United Kingdom clinical guideline hub (thyroid disease, imaging appropriateness)	NICE	https://www.nice.org.uk/
Working with clinical samples	Guidelines for Working with Human Tissue and Patient Information (University College London; UCL)	https://www.ucl.ac.uk/human-tissue/sites/human_tissue/files/1i_working_with_human_tissue_and_patient_information_-_ucl_guidelines_20231107v.4-a.pdf
World Health Organization (WHO) iodine nutrition and thyroid resources	World Health Organization	https://www.who.int/health-topics/iodine

This table lists some other resources relevant to hormones in neuroscience, the thyroid gland, related hormones and associated clinical conditions. The links were accurate at the time of going to press but may move or alter. In these cases, the use of the "Search" tabs should be explored at the parent address or site. In some cases, links direct the reader to relevant pages within parent sites. As a cautionary note, some resources are still at the experimental stage and used in research or they have not been used for routine clinical use.TUD QC

(Table 27.3), and other resources of interest (Table 27.4) that are relevant to an evidence-based approach to the investigation of the endocrine aspects of neuroscience, the thyroid gland, related hormones and clinical conditions. Some organizations are listed in more than one table as they occasionally fulfill multiple roles.

Other resources

The Welcome Collection (https://wellcomecollection.org/collections) and The British Library (www.bl.uk/) also hold material on topics related to Hormones in Neuroscience: The Thyroid and Thyroid Hormones.

Other chapters on resources (recommended by authors and practitioners) may also be relevant to the study of the thyroid gland, related hormones and associated clinical conditions in neuroscience (Rajendram & Preedy, 2019; Rajendram et al., 2013a; Rajendram et al., 2013b; Rajendram et al., 2017; Rajendram et al., 2019a; Rajendram et al., 2019b; Rajendram et al., 2022a; Rajendram et al., 2022b; Rajendram et al., 2023; Rajendram et al., 2025a; Rajendram et al., 2025b; Rajendram et al., 2025c; Rajendram et al., 2025d; Rajendram et al., 2025e; Rajendram et al., 2025f; Rajendram et al., 2026a; Rajendram et al., 2026b).

Acknowledgments(in alphabetical order)

We thank the following authors for their contributions to the development of this resource. We apologise if some of the suggested material was not included in this chapter (e.g. for being expired sites, incomplete or flagged as potential security threats by university cybersecurity software) or has been moved to, or merged into, different tables. We are also grateful for the anonymous colleagues who provided useful information.

Alizadeh A
Arata N
Deravi N
Dolu N
EL-Sayed W
Ghorbani R
Iranfar A
Jabbarpour A
Jhandi SK.
Johansson B
Minadi M
Rahmanian M
Ressler K
Sahu S
Sankoda A
Singh A

Varjovi MN
Watanabe N
Yoshihara A
Yu OHY.

References

Preedy, V. R., Burrows, G. N., & Watson, R. R. (Eds.). (2009). *Comprehensive handbook of iodine: Nutritional, biochemical, pathological and therapeutic aspects*. Academic Press.

Rajendram, R., Gyamfi, D., Patel, V. B., & Preedy, V. R. (2022a). Recommended resources for biomarkers of nutrition. In V. R. Preedy, & V. B. Patel (Eds.), *Biomarkers of nutrition*. New York, USA: Elsevier.

Rajendram, R., Gyamfi, D., Patel, V. B., & Preedy, V. R. (2022b). Recommended resources for biomarkers in diabetes: Methods, discoveries, and applications. In V. B. Patel, & V. R. Preedy (Eds.), *Biomarkers in diabetes: Biomarkers in disease: Methods, discoveries and applications*. Berlin, Germany: Springer.

Rajendram, R., Gyamfi, D., Patel, V. B., & Preedy, V. R. (2023). Recommended resources on diabetes in relation to foods, plants, herbs and spices in human health. In R. Rajendram, V. B. Patel, & V. R. Preedy (Eds.), *Ancient and traditional foods, plants, herbs and spices used in diabetes*. Boca Raton, USA: CRC Press.

Rajendram, R., Patel, V. B., & Preedy, V. R. (2017). Recommended resources on maternal nutrition. In R. Rajendram, V. B. Patel, & V. R. Preedy (Eds.), *Nutrition and diet in maternal diabetes* (pp. 495–500). New York, USA: Springer.

Rajendram, R., Patel, V. B., & Preedy, V. R. (2019a). Resources in famine, starvation, and nutrient deprivation. In V. B. Patel, & V. R. Preedy (Eds.), *Famine, starvation, and nutrient deprivation* (pp. 2399–2406). New York, USA: Springer.

Rajendram, R., Patel, V. B., & Preedy, V. R. (2019b). Resources in diet, nutrition and epigenetics. In V. B. Patel, & V. R. Preedy (Eds.), *Nutrition and epigenetics* (pp. 2309–2314). New York, USA: Springer.

Rajendram, R., Patel, V. B., & Preedy, V. R. (2025a). Recommended resources on protein intake in health and disease. In V. R. Preedy (Ed.), *Protein intake in health and disease*. Boca Raton: Taylor and Francis.

Rajendram, R., Patel, V. B., & Preedy, V. R. (2025b). Recommended resources on plant and animal proteins in health and disease prevention. In V. R. Preedy (Ed.), *Plant and animal proteins in health and disease prevention*. Boca Raton: Taylor and Francis.

Rajendram, R., Patel, V. B., & Preedy, V. R. (2025c). Recommended resources for the scientific basis of fatigue. In C. R. Martin, V. R. Preedy, V. B. Patel, & R. Rajendram (Eds.), *The scientific basis of fatigue* (pp. 585–594). Academic Press.

Rajendram, R., Patel, V. B., & Preedy, V. R. (2025d). Recommended resources for the behavior and psychology of disease. In V. R. Preedy (Ed.), *Handbook of the behavior and psychology of disease* (pp. 2927–2950). Springer.

Rajendram, R., Patel, V. B., & Preedy, V. R. (2025e). Recommended resources for the neuroscience of disease. In V. R. Preedy (Ed.), *Handbook of the neuroscience of disease* (pp. 1745–1760). Springer.

Rajendram, R., Patel, V. B., & Preedy, V. R. (2025f). Recommended resources for the biology and pathology of mental disorders. In C. R. Martin, V. R. Preedy, V. B. Patel, & R. Rajendram (Eds.), *Handbook of the biology and pathology of mental disorders* (pp. 1–15). Springer Nature.

Rajendram, R., Patel, V. B., & Preedy, V. R. (2026a). Recommended resources for hormones in neuroscience: The pancreas and pancreatic hormones. In V. R. Preedy, V. B. Patel, & R. Rajendram (Eds.), *Hormones in neuroscience: The pancreas and pancreatic hormones*. New York, USA: Elsevier.

Rajendram, R., & Preedy, V. R. (2019). Resources for the neuroscience of alcohol. In V. R. Preedy (Ed.), *Neuroscience of alcohol: Mechanisms and treatment* (pp. 853–862). Elsevier.

Rajendram, R., Rajendram, R., Patel, V. B., & Preedy, V. R. (2013a). Interlinking diet, nutrition, the menopause and recommended resources. In C. J. Hollins-Martin, R. R. Watson, & V. R. Preedy (Eds.), *Nutrition and diet in menopause*. Berlin, Germany: Springer.

Rajendram, R., Rajendram, R., Patel, V. B., & Preedy, V. R. (2013b). Diet quality: What more is there to know? In V. R. Preedy, L.-A. Hunter, & V. B. Patel (Eds.), *Diet quality: An evidence-based approach* (pp. 397–401). Berlin, Germany: Springer.

Rajendram, R., Siraj, M. A., Patel, V. B., & Preedy, V. R. (2026b). Recommended resources for the biological measures of well-being. In C. R. Martin, V. R. Preedy, V. B. Patel, & R. Rajendram (Eds.), *Biological measures of well-being*. New York, USA: Elsevier.

Sabatino, L., Lapi, D., & Del Seppia, C. (2024). Factors and mechanisms of thyroid hormone activity in the brain: Possible role in recovery and protection. *Biomolecules, 14*(2), 198. https://doi.org/10.3390/biom14020198

Schiera, G., Di Liegro, C. M., & Di Liegro, I. (2021). Involvement of thyroid hormones in brain development and cancer. *Cancers, 13*(11), 2693. https://doi.org/10.3390/cancers13112693

Stepien, B. K., & Huttner, W. B. (2019). Transport, metabolism, and function of thyroid hormones in the developing mammalian brain. *Frontiers in Endocrinology, 10*, 209. https://doi.org/10.3389/fendo.2019.00209

Index

'Note: Page numbers followed by "f" indicate figures and "t" indicate tables.'

A

Acetylcholine, 45–46
ACTH. *See* Adrenocorticotropic hormone (ACTH)
Action of hCG on TSH receptors, 225, 226t
Active surveillance, 186
Adaptive immune dysregulation, 296
Adaptogens, 320–323, 325–326
  Bacopa monnieri, 321, 322t, 323
  Ginkgo biloba, 321, 322t
  Withania somnifera, 321, 322t, 323
Adipose tissue, 180
  adipokines, 180–181
  dysfunction, 180
  metabolic signaling, 180
Adjuvant therapy in thyroid cancer, 186
Adrenocorticotropic hormone (ACTH), 59
Advanced cancer treatment, 186
Adverse outcome pathway (AOP), 15f, 16t, 23
  mechanistic mapping, 16t
  neurodevelopmental assessment pipeline, 15f
AHDS. *See* Allan–Herndon–Dudley syndrome (AHDS)
AIT. *See* Amiodarone-induced thyroiditis (AIT)
AJCC. *See* American Joint Committee on Cancer (AJCC)
Allan–Herndon–Dudley syndrome (AHDS), 22, 35, 37, 80–81, 84, 87, 157, 164, 237–238
  clinical manifestation, 81
  MCT8 mutations, 22, 80–81
  neurodevelopmental abnormalities, 22
  neurological phenotype, 80–81, 86
Allopregnanolone, 216–218
  GABA-A modulation, 217–218
  synthesis, 217
5-Alpha reductase, 217
Alzheimer's disease (AD), 4–5, 7–8, 94t, 96–99, 97f–98f, 100t, 101–102, 107–108, 111–119, 196, 202, 207
  amyloid-beta plaque formation, 111
  amyloid-PET, 7–8
  neuroinflammation and oxidative stress, 111
  tau hyperphosphorylation, 111
  tau-PET, 7–8
  tau protein hyperphosphorylation, 96, 98–99, 100t, 101–102
  thyroid signaling in, 96, 101–102
American College of Obstetricians and Gynecologists (ACOG), 43–44
American Joint Committee on Cancer (AJCC), 186
American Thyroid Association (ATA), 43–44, 63–64, 66–67, 70, 186, 348t–350t
Amiodarone, 335, 337–340, 337t, 341f, 342
  effects on thyroid function, 339
  iodine content, 339
  thyroid disturbances, 339
Amiodarone-induced thyroiditis (AIT), 339–340, 340t
  type 1, 339–340, 340t
  type 2, 339–340, 340t
  ultrasound patterns, 340, 340t
Amygdala, 214–216, 214f, 217t, 218
  mRNA after fear learning, 217t
  thyroid signaling in, 216, 218
Amyloid, 188, 191
Amyloid-PET, 7–8
Amyloid precursor protein (APP), 107, 109f, 111–114
Amyloid-β (Aβ), 96, 99, 100t, 101–102, 107–108, 109f, 111–116, 118–119
Amyotrophic lateral sclerosis (ALS), 96–97, 98f, 100t
Anabas, 142, 144f
Antioxidants, 313, 316, 320–325
  vitamin C, 321, 322t, 323
  vitamin E, 313, 321–325, 322t
Antipsychotics, 172–176, 178, 180
  clozapine, 175–176
  quetiapine, 175–176
  treatment-related changes, 174
Antiseizure medications (ASMs), 269–270, 270f, 272–280, 273f, 273t, 279t
  GABAergic agents, 269, 272–274
  glutamate antagonists, 272–273
  sodium channel blockers, 269, 272–273, 280
Antithyroid drugs (ATDs), 66, 68, 125, 128, 131–132, 134, 136–137, 295–296, 300
Apathetic hyperthyroidism, 293, 295
Aphasia, 189–191
Aquaporin 4 (AQP4), 155–156, 160–162
Artificial intelligence (AI), 350t–352t
Assisted reproductive technologies (ART), 68–70
Astrocytes, 77, 79t, 80, 82t, 83–84, 85f, 86, 155–166, 156f–161f, 163f, 236, 240, 250–251, 262
  defects associated with thyroid hormone alterations, 158–160
  DIO2 expression, 84
  functions in health and disease, 156, 157f
  in neuroinflammation, 162–163, 163f
  OATP1C1 immunoreactivity, 83
  thyroid hormone regulation, 236
  T4 to T3 conversion, 84, 85f
Astrogliosis, 156, 161–162, 164–165
  reactive, 161–162, 164–165
ATA. *See* American Thyroid Association (ATA)
Atrial fibrillation, 188–189, 191
Auditory fear conditioning, 213, 214f, 215–217
Autoimmune diseases, 126–127, 135–137
Autoimmune disorders, 305–306
Autoimmune response, 285
Autoimmune thyroid diseases, 30, 37, 40–41, 43–44, 46–47, 63, 66–67, 70, 171, 175
  Graves' disease, 28–30, 37
  Hashimoto's thyroiditis, 30, 37
  immunoendocrine marker, 175
  increased psychosis risk, 175
  suicide mortality, 175
Autoimmune thyroiditis, 271, 271t, 277, 279
Autoimmunity, 294–301
  immune tolerance, 293, 296
  neuroimmune interactions, 301
Autonomic nervous system (ANS), 139–140, 140f, 148–149, 148f

B

Bacopa monnieri. *See* Adaptogens; *Bacopa monnieri*
BBB. *See* Blood–brain barrier (BBB)
Bergmann glia, 155–156, 158–159, 164–165
BFS. *See* Brain Fatigue Syndrome (BFS)
B-G axis. *See* Brain-gut axis (B-G axis)
Bifidobacterium. *See* Probiotics; Bifidobacterium
Bioavailability, 238t
  increased, 238t
  low, 238t
Biomarkers, 172, 176–179, 181
  prognostic, 172, 178

Biomarkers (*Continued*)
translational, 181
treatment response, 177
Blood–Brain Barrier (BBB), 78, 80f, 81–87, 85f, 92–93, 94t, 95–96, 102, 113, 115–116, 115t, 118–119, 156–158, 157f–158f, 160–166, 160f–161f, 235, 237–240, 238t, 239f, 250, 262, 306
dysfunction, 156, 161f, 162, 164, 166
enhanced penetration, 238t
maintenance, 156, 160–162
MCT8 at, 78, 80f, 84–86
OATP1C1, 81–84, 85f
permeability, 235, 237–238
stabilization, 235, 240
thyroid hormone transport across, 78, 84–86, 85f
thyroid hormone transporters, 250, 262
Blood–brain barrier (BBB) disruption, 315t, 316, 318–320, 318f–319f, 325–326
Bone, 297
bone mineral density, 297
osteoporosis, 297
BRAF mutations, 338–339
Brain, 250–259, 251f, 251t, 254f, 261
activation patterns, 306
network analysis, 8
thyroid axis, 249–250, 251t
Brain-derived neurotrophic factor (BDNF), 197f, 199–202, 207
Brain development, 13–14, 15f, 16–23, 16t, 17f, 18t, 19f–20f, 77, 86–87, 195–202, 197f, 200t, 207, 347
critical period, 13
fetal brain development, 199
maternal TH reliance, 13
MCT8 deficiency effects, 86
neurodevelopmental processes, 197–199, 200t
neuronal migration, 86
thyroid hormone role, 77, 86
timing signal, 13
Brain Fatigue Syndrome (BFS), 306–307, 307t
symptom profile, 307t
Brain fog, 202, 205, 208
Brain-gut axis (B-G axis), 139–149
bidirectional communication, 139–140, 140f, 145, 147–149, 148f
in fish, 139, 140f, 141–142, 143f–144f, 145–149, 146f, 148f
in humans/mammals, 139, 140f, 141–142, 145–149, 148f
Brainstem, 214, 214f
Brain structures, 4–8
gray matter, 4–6, 9
hippocampus, 3–5, 7
white matter, 5–7, 9
Brain-sympathetic-chromaffin (BSC) axis, 140f, 145, 148–149, 148f
British Thyroid Association (BTA), 348t–350t
BSC axis. *See* Brain-sympathetic-chromaffin (BSC) axis
Burch and Wartofsky scale, 129, 129t

C

Cajal–retzius cells, 78–80, 79t, 82t, 83
Calcitonin, 123, 134, 136
Cambridge Neuropsychological Test Automated Battery (CANTAB), 245–246
computerized test batteries, 246
executive functions, 246
spatial working memory, 245–246
CANTAB. *See* Cambridge Neuropsychological Test Automated Battery (CANTAB)
Carbamazepine (CBZ), 269, 273f, 273t, 274–275, 277–278, 279t, 280
mechanisms of action, 274
Cardiovascular outcomes, 39–43, 46
coronary heart disease, 41–43
mortality, 41–43, 46
risk factors, 41, 43
Carrier proteins, 27, 33, 37
albumin, 27, 33–35
thyroxine-binding globulin (TBG), 27, 33–35
transthyretin (TTR), 27, 33–35
Central nervous system (CNS), 139–142, 149, 155–157, 160–162, 164, 166, 195–202, 208, 235–236, 238–240, 244, 294, 298–301, 347
development, 244
function and structure, 236
hypothyroidism, 240
neuroendocrine regulation, 294, 298
thyroid hormones effects, 244
thyroid hormone transport, 298
Central nervous system development, 223, 228–230, 229f
neuronal differentiation, 228–230
Cerebral cortex, 197, 202, 205, 207
Cerebral plasticity, 5–6
Cerebral small vessel disease, 5
Cerebrospinal fluid (CSF), 113, 118–119
Chemotherapy, 190–191
Children, 41, 46
autoimmune disease, 41
TSH cutoff, 41
Chitosan nanoparticles, 240
engineered, 240
Cholinergic neurons, 14, 16, 17f, 18t, 23
acetylcholine production, 17f, 18t
basal forebrain, 14, 17f
memory and learning, 14, 17f, 18t
Clinical staging and risk classification, 186
ClinicalTrials.gov, 348t–349t
Clustered regularly interspaced short palindromic repeats (CRISPR)-Cas9 gene editing, 350t–352t
CNS. *See* Central nervous system (CNS)
CNS hypothyroidism, 240
Coenzyme Q10 (CoQ10), 321, 322t, 323–325
Cognition, 171–172, 176–178, 177t, 185–191, 195–205, 197f, 198t, 203t–204t, 206t, 207, 223, 228–232, 243–245, 305–308
cognitive complaints, 305, 307
cognitive decline, 202–205, 243
cognitive deficit, 245
cognitive development, 228–230
cognitive difficulties, 306–307
cognitive impairments, 176–178, 180, 196, 199–202, 205, 244–245, 306–307
cognitive performance, 228–230
cognitive remediation therapy (CRT), 177–178
dysfunction, 185–191
recovery, 190–191
before thyroid cancer treatment, 186–187
thyroid hormones and, 177
Cognitive decline, 107–108, 111–118
Cognitive dysfunctions, 314, 315t, 316
Cognitive impairment, 3–8, 91–92, 96–102, 100t, 293–294, 298–299, 301
dementia, 3, 7–8
in hyperthyroidism, 97–99, 101–102
in hypothyroidism, 97–102
mild cognitive impairment (MCI), 4–5, 99, 102
Cognitive outcomes, 39–41, 45–46, 253, 256t, 257–261
cognitive impairment, 39–40, 45
neurotransmitter activity, 45
Computed tomography (CT), 350t–352t
Computerized test batteries, 246
advantages, 246
Cambridge Neuropsychological Test Automated Battery (CANTAB), 246
Congenital hyperthyroidism, 314, 315t, 316
neurodevelopmental consequences, 314, 316
Congenital hypothyroidism, 30, 37, 347
Corticotropin-releasing hormone (CRH), 59
Cortisol deficiency, 285, 287
hydrocortisone replacement, 287
C oxidase, 21
subunit Va, 21
CRH. *See* Corticotropin-releasing hormone (CRH)
CTLA-4 inhibitors, 283–284, 284t, 288–289
ipilimumab, 283–284, 284t
tremelimumab, 283
Curcumin. *See* Nutraceuticals; curcumin
Cytochrome P450 enzymes, 273f, 273t, 274–277, 280

D

D1. *See* Deiodinases
D2. *See* Deiodinases
D3. *See* Deiodinases
Danio rerio, 141
Default mode network (DMN), 256–257, 261–262
DEHAL1, 334, 334f

Deiodinases (Dio), 32–37, 32t, 77, 87, 92–93, 94t, 110, 115–116, 115t, 118–119, 123–124, 136, 140–145, 148–149, 156–158, 158f–159f, 163–166, 196, 205–207, 214–217, 250–252, 259–262, 270, 274, 276, 334, 336
Dio1, 92–93, 140–142, 143f–144f, 146f, 196
Dio2, 84, 85f, 92–93, 94t, 140–142, 143f–144f, 146f, 156–158, 158f–159f, 163–166, 196, 205, 216, 217t
Dio3, 92–93, 94t, 140–142, 143f–144f, 146f, 157–158, 159f, 163–165, 196, 205, 216–217
selenium, 32–33, 37
type 1 (D1), 32t, 33, 334–335
type 2 (D2), 32t, 33, 335
type 3 (D3), 335
Dementia, 92, 97–102, 100t, 101f, 202–205, 207–208
risk, 108, 111–113, 117–119
risk factors, 92, 98–101, 100t, 101f
Demyelinating diseases, 240
treatment, 240
Dendrimers, 239t, 240
Depression, 245
depressive symptoms, 245–246
Hamilton Depression Rating Scale, 245
Destructive thyroiditis, 285–286, 286f, 288
mechanisms of, 285–286, 286f
Differentiated thyroid carcinoma (DTC), 185–186, 188–190
Diffusion MRI, 4, 8
diffusion tensor imaging, 6
Diffusion tensor imaging (DTI), 257–258, 261–262
fractional anisotropy (FA), 257–258, 262
mean diffusivity (MD), 257–258, 262
white matter integrity, 257–258, 262
Digital enzyme-linked immunosorbent assays (ELISA), 350t–352t
Dio. *See* Deiodinases (Dio)
Disability leave, 308
DITPA, 20f, 22–23
MCT8-independent, 22
oligodendrocyte differentiation, 22
Wnt-Notch inhibition, 20f, 22
Dopamine imbalance hypothesis, 306
Dopaminergic neurons, 16, 17f, 18t, 23
ADHD-like behaviors, 18t
movement disorders, 16, 17f, 18t
substantia nigra pars compacta (SNc), 16, 17f
DTC. *See* Differentiated thyroid carcinoma (DTC)
Dual oxidase 2, 41, 46
DUOX1, 30–31, 37, 334
DUOX2, 30–31, 37, 334
Dysphagia, 56, 59–60
assessment, 56
post-operative, 56, 59–60
Dysphonia, 55–56, 59–60
assessment, 55–56
non-nerve related, 56
post-operative, 55–56, 59–60

E

Ease response, 148–149
EBSLN. *See* External branch of the superior laryngeal nerve (EBSLN)
Efflux pumps, 237–238
breast cancer resistance protein, 237–238
P-glycoprotein, 237–238
Electroencephalography (EEG), 258–259, 261–262
event-related potential (P300), 258, 262
Emotional distress, 305–308
anxiety, 306–308
depression, 306–308
Endocrine-disrupting chemicals (EDC), 164
Endocrine Society (ES), 348t–350t
Endocrinology, 347–354
Endoscopic thyroidectomy, 58
CO_2 insufflation, 58
gasless procedures, 58
Endothelial cells, 78, 79t, 80f, 81–87, 82t
MCT8 expression, 78, 79t, 80f
OATP1C1 expression, 81–83, 82t
Energy-conserving strategies, 308
ENS. *See* Enteric nervous system (ENS)
Enteric nervous system (ENS), 139–140, 140f, 148–149, 148f
Enzyme induction, 274–277, 280
Ependymal cells, 78, 79t, 82t, 83
Epigenetics, 207
Epilepsy, 269–272, 277–280
intractable epilepsy, 277
ESS. *See* Euthyroid sick syndrome (ESS)
Estrogens, 86–87, 190–191
effects on TBG, 225, 226t
estrogen receptors (ER), 86
mutual modulation with THs, 86
ETA. *See* European Thyroid Association (ETA)
European Endocrine Society, 287
European Society for Medical Oncology (ESMO), 286
European Thyroid Association (ETA), 63–64, 66–67, 70, 186, 348t–350t
Euthyroidism, 131–132, 134, 250, 253–255, 257–258, 260–262
Euthyroid sick syndrome (ESS), 188
Excitation/inhibition (E/I) balance, 14, 17f, 18, 18t, 20f, 23
GABAergic pathway, 20f
glutamatergic pathway, 20f
hypothyroidism-induced, 18
Excitotoxicity, 313, 316–321, 323, 325–326
neurotransmitter imbalance, 314, 315t, 316–318, 320–321, 323
N-methyl-d-aspartate (NMDA) receptors, 315t, 316–318, 317f, 320f, 321, 322t
Executive functions, 244–246
frontal, 245
spatial working memory, 244
Exosomes, 240
delivery platforms, 236, 240
External branch of the superior laryngeal nerve (EBSLN), 54–55, 60
injury, 54–55, 60
surgical anatomy, 54–55
Extracellular matrix, 156, 162, 165
Extra-thyroidal manifestations, 295–297
pretibial myxedema, 295, 297
thyroid acropachy, 295, 297
thyroid eye disease (TED), 295–297

F

Fatigue, 305–308
assessment of, 305–307
in Graves' disease, 305–308
pathological, 305–306
Fatigue Severity Scale (FSS), 307
FcRn, 293, 295
FDG uptake. *See* Fluorodeoxyglucose (FDG) uptake
Fear learning, 213–217, 214f–215f, 217t
brain regions involved, 214, 214f
disorders of, 214, 215f
extinction, 213–216, 214f, 218
Fear memory, 213–217
consolidation, 213, 216, 218
extinction. *See* Fear learning; extinction
Fetal brain development, 224–225, 228, 230
Fetal development, 63–65, 67–71
18F-Fluorodeoxyglucose-Positron emission tomography (FDG-PET), 6–8
Fibroblast-driven tissue remodeling, 296–297
Fine-needle aspiration biopsy (FNAB), 126–127, 130–131, 133–137
First-episode psychosis (FEP), 171–172, 174, 176–177, 179–181
HPT axis dysregulation, 171–172, 174
metabolic dysregulation, 180
thyroid markers in, 176
Fluorodeoxyglucose (FDG) uptake, 285, 288–289
FNAB. *See* Fine-needle aspiration biopsy (FNAB)
Free thyroxine (FT4), 3–7, 39–40, 44, 46, 171–178, 180–181
cognitive performance, 177–178
dimensional marker, 176
phenotype associations, 171–172, 174
Free triiodothyronine (fT3), 3–5, 7
Freezing behavior, 213, 215–216, 218
FSS. *See* Fatigue Severity Scale (FSS)
FT4. *See* Free thyroxine (FT4)
Functional connectivity, 6, 8–9
Functional magnetic resonance imaging (fMRI), 256–257, 259–262, 350t–352t
resting-state connectivity, 256–257
task-based fMRI, 257, 257t

G

GABA-A signaling, 217–218
GABAergic interneurons, 14, 16, 17f, 18t, 23, 78–80, 79t, 82t, 83
  differentiation, 14, 17f, 20f
  Nkx2.1-mediated, 17f, 20f
  synaptic connectivity, 20f
Gabapentin, 269, 275–276
Gastrointestinal tract (GIT), 142, 147
GD. *See* Graves' disease (GD)
Gene expression, 216–218
Gene therapy, 87
  AAV vectors, 86–87
  molecular correction, 165
Genomic signaling, 13–14, 16–21, 19f, 23, 196–197, 197f
  gene regulation, 16, 18, 19f, 21–22
  nuclear receptors, 13–14, 18, 19f, 21
  TR-mediated, 18, 21
Gestational hypothyroidism, 225–226, 228–231
  complications in offspring, 228–230
  diagnosis, 225–227
  prevalence, 226–227
Gestational transient thyrotoxicosis (GTT/GTH), 66–68, 70–71
Ginkgo biloba. *See* Adaptogens; *Ginkgo biloba*
GIT. *See* Gastrointestinal tract (GIT)
Glial fibrillary acidic protein (GFAP), 155–156, 158–160, 160f, 162–165
Glucocorticoid receptor (GR), 116–118
Glucose metabolism, 244–245
  limbic system, 244–245
Glutamatergic neurons, 14, 16, 17f, 18t, 23
  cortical thinning, 14, 17f, 18t
  dendritogenesis, 17f, 20f
  pyramidal cells, 14, 17f
Glycogen synthase kinase-3 (GSK-3), 109f, 111–112, 115, 119
Goiter, 126–131, 131t, 133–136
  nodular goiter, 128, 133–134, 136
  toxic nodular goiter, 128, 131t, 134, 136
Gonadal steroids, 179–180
  estrogens, 179–180
  prolactin, 180
  testosterone, 179–180
Graves–Basedow disease, 335–336, 339–340, 340t
  risk factor for Jod–Basedow effect, 335
  susceptibility to Wolff–Chaikoff effect, 336
Graves' disease (GD), 96–97, 99, 100t, 128, 130–132, 131t, 136, 305–308
  autoimmune disorder, 305
  hyperthyroidism, 305–308
  orbitopathy, 130–132
  treatment, 306, 308, 308t
Graves' hyperthyroidism (GH), 294–301
  disease management, 295
  pathogenesis, 296–297
  systemic consequences, 294, 297–298
Gut–brain axis, 320–323, 325–326
Gut microbiota, 139–140, 142–145, 147–148

H

Half-life, 236–238, 238t
  prolonged circulation, 238t
  short, 238t
Hashimoto's disease, 335–336, 339, 342
  prevalence with USI programs, 342
  risk factor for Jod–Basedow effect, 335
  susceptibility to Wolff–Chaikoff effect, 336, 339
Hashimoto's encephalopathy (HE), 171–173, 173t, 178
  autoimmune nature, 173
  corticosteroid therapy, 173, 178
  psychosis in, 172–173
Hashimoto's thyroiditis, 96–97, 100t, 252, 255, 256t, 258, 260–261
HCG. *See* Human chorionic gonadotropin (hCG)
Hematoma, 51–53, 58–60
  airway obstruction, 51, 59–60
  post-operative, 51–53, 58–60
Hemi-thyroidectomy, 53, 57, 59–60
  hypothyroidism, 53, 57, 59–60
Heterochrony, 20–21
  developmental processes, 20
  dual function, 20
Hippocampal and cortical alterations, 253, 256, 262
Hippocampus, 96–99, 100t, 101–102, 195–205, 197f, 207–208, 214, 214f, 216, 218, 228–232, 244–245
  atrophy, 244
  cerebral blood flow, 245
  hippocampal volume, 196, 201
  memory deficits, 244
  thyroid signaling in, 216, 218
Hormone homeostasis, 269–270, 273–280
  drug-hormone interaction, 280
Hormones in neuroscience, 348t–352t
HPA axis. *See* Hypothalamic–pituitary–adrenal (HPA) axis
HPG axis, 300
HPI axis. *See* Hypothalamic–pituitary–interrenal (HPI) axis
HPT axis. *See* Hypothalamic–pituitary–thyroid (HPT) axis
Human chorionic gonadotropin (hCG), 59, 63–64, 66–69, 71, 224–225, 226t
Hydrogen peroxide (H_2O_2), 27, 29–31, 37
Hyperthyroidism, 3–8, 63–64, 66, 68, 93–99, 94t, 100t, 101–102, 101f, 108, 112–113, 115, 118–119, 127–132, 135–137, 171–173, 173t, 176–180, 185–190, 195–196, 197f, 199–205, 203t–204t, 206t, 207–208, 313–325, 331, 335, 339–340, 342
  causes, 127–128
  cognitive impairment, 199–202
  diagnostic approach, 129–130, 130f, 131t
  Graves' disease, 4–5, 172, 173t, 178
  iodine-induced, 335–336, 339–340
  Jod–Basedow effect, 335–336, 339–340, 342
  overt, 66, 68, 97–99
  psychosis in, 172
  subclinical, 3, 99, 100t
  symptoms, 172, 176–177
  treatment, 131–132, 202, 207
Hypocalcemia, 53–54, 57, 59–60
  biochemical predictors, 54
  management, 54, 57
  post-operative, 53–54, 57, 59–60
Hypometabolic coma, 124, 136
Hypophysitis, 285, 285t, 288, 289t
Hypothalamic–pituitary–adrenal (HPA) axis, 59, 116, 118, 140f, 145–149, 148f, 171, 173–175, 179–181, 214–215, 314, 315t, 316, 321, 323, 326
  cortisol awakening response (CAR), 179, 181
  dysregulation, 59, 179
  stress response, 181
Hypothalamic–pituitary–interrenal (HPI) axis, 140f, 145, 148–149, 148f
Hypothalamic–pituitary–thyroid (HPT) axis, 35–37, 36f, 110, 115t, 116–119, 123–124, 124f, 136, 139, 148f, 149, 171–175, 214–215, 250, 260–262, 269–271, 275–277
  dysregulation, 171–172, 174
  modulation potential, 116, 118
  phase-dependent endocrine pattern, 174
  regulation, 110
  thyroid-stimulating hormone (TSH), 110
Hypothalamus, 214, 214f
Hypothalamus–pituitary–thyroid (HPT) axis, 207
Hypothyroidism. *See also* ICI-induced hypothyroidism
  amiodarone-induced, 339–340
  animal models, 243–245
  causes, 124–125, 243
  central hypothyroidism, 271, 271t, 277–278
  clinical, 225–230, 226t
  cognition in, 176–177
  cognitive impairment, 196, 199–202, 205, 244–245
  congenital, 5–6, 195–196, 199, 207, 237, 251–252, 255, 257–258, 261–262
  diagnostic approach, 126–127, 126f
  gestational, 5
  iodine-induced, 336–338, 340
  mild hypothyroidism, 277–279
  myxedema psychosis, 171–173, 173t, 178
  neuroinflammation, 235
  overt, 63, 65, 67, 70–71, 97–99
  in pediatric population, 57
  post-operative, 53, 57, 59–60
  pregnancy-related, 252, 255t
  psychosis in, 172
  subclinical, 3–6, 63, 65, 67, 70–71, 99–101, 100t, 111–112, 117, 119, 225–230, 226t, 253, 255–257, 260–262, 271, 276–278
  treatment, 127, 196, 199, 201–202, 207
  Wolff–Chaikoff effect, 336–340, 342
Hypothyroxinemia, 277–280

I

ICI-induced hypothyroidism, 283–289, 285t, 289t
clinical management, 286–288
epidemiology and prediction, 283–285, 284t, 288–289
prognosis, 287–289
ICIs. *See* Immune checkpoint inhibitors (ICIs)
IFN-γ. *See* Interferon-γ (IFN-γ)
IGF-1 receptor (IGF-1R), 296–297, 300
Immune checkpoint inhibitors (ICIs), 283–289, 289t
combination therapy, 283–284, 284t, 288–289
modes of action, 283
use in cancer treatment, 283
Immune-related adverse events (irAEs), 283–285, 287–288, 289t
endocrine irAEs, 283, 288
thyroid irAEs, 287
Immunotherapy, 40, 46
Induced pluripotent stem cells (iPSCs), 86
cerebral organoids, 86
MCT8 deficiency model, 86
Insulin resistance (IR), 300
Interferon-γ (IFN-γ), 285–286, 286f
Intracellular signaling, 236
mitogen-activated protein kinase, 236
phosphatidylinositol-3-kinase, 236
Intraoperative nerve monitoring (IONM), 54–55, 57–60
nerve preservation, 54–55, 57–60
Intrauterine growth restriction (IUGR), 21–23
cerebral TH signaling, 21–22
neurodevelopmental consequences, 21
white matter tracts, 21
Iodide, 224–225, 231
plasma iodide, 224–225, 226t
reduction in plasma iodide, 224–225, 226t
Iodinated contrast media, 335–336, 337t, 342
iodine content, 335–336, 337t
risk factor for Jod–Basedow effect, 335
Iodine. *See also* Micronutrients; iodine
daily requirement, 28
deficiency, 63–64, 67, 69–70, 195–196, 199, 205, 207, 224–227, 231
iodide ions, 27–29, 37
organification, 29–30
recycling, 110
Iodine intake, 331–332, 332f, 342
excessive, 331, 335, 338–342
optimal level, 331
U-shaped relationship, 331, 332f, 338
Iodine metabolism, 332–335, 333f, 340–342
biosynthesis and secretion of thyroid hormones, 332, 332t
intrathyroidal recycling, 334
transport, 332–335
Iodothyronine deiodinases (D2/D3), 232, 293, 298
type 2 deiodinase (D2), 225, 232
type 3 deiodinase (D3), 224–225, 226t, 232
Iodotyrosine conjugation, 30
IONM. *See* Intraoperative nerve monitoring (IONM)
IPSCs. *See* Induced pluripotent stem cells (iPSCs)
IrAEs. *See* Immune-related adverse events (irAEs)
Ischemic stroke, 235–236, 238–239
Isolated hypothyroxinemia (IH), 65, 67–68, 71

J

Jod–Basedow effect, 335–336, 339–340, 342
diagnosis, 336
risk factors, 335
underlying thyroid dysfunction patterns, 336

K

Ketogenic diet (KD), 270, 273t, 276–277
Kocher incision, 56

L

Lactobacillus. *See* Probiotics; *Lactobacillus*
Lactobacillus rhamnosus. *See* Probiotics; *Lactobacillus rhamnosus*
L-amino acid transporters (LAT1, LAT2), 35
Lamotrigine (LTG), 269, 273t, 275, 277, 279t, 280
LAT1 (SLC7A5), 83–84
expression, 83–84
T3 transport, 83–84
LAT2 (SLC7A8), 83–84
expression, 83–84
T3 transport, 83–84
Learning and memory, 195–205, 207–208
deficits, 195–196, 199–202
Levetiracetam (LEV), 269, 273t, 276, 279t
Levothyroxine (LT-4), 41–47, 127, 132–136, 171–173, 173t, 178, 185, 188–190, 216, 218, 227–232, 244–246, 249–250, 252–262, 254f, 278, 280
absorption, 41
cognitive impairment treatment, 245
dose, 41
dose adjustment, 227–228
dosing considerations, 253, 255t
indications for therapy, 252–253
memory improvement, 245
pharmacokinetics and pharmacodynamics, 252–253
side effects, 41
spatial working memory, 245–246
therapy, 65–70, 196, 201–202, 207
treatment in pregnancy, 227–228, 231
Lipid nanoparticles, 240
Liposomes, 235, 238, 239t
Liquid chromatography–mass spectrometry (LC-MS), 350t–352t
Liver, 32–35, 37, 297
hepatic metabolism, 299–300
LT-4. *See* Levothyroxine (LT-4)

M

Magnesium. *See* Micronutrients; magnesium
Magnetic resonance imaging (MRI), 4–8, 118, 350t–352t
resting-state MRI, 4, 6, 8
structural MRI, 8
task-based functional MRI, 4, 6, 8
Maternal-fetal-placental thyroid axis, 63–65, 71
Maternal thyroid dysfunction, 5
MCT8. *See* Monocarboxylate transporters (MCT8)
MCT10 (SLC16A10), 83–84
Expression, 83
iodothyronine transport, 83
MCT8 deficiency, 157, 162, 164
Mct8/Dio2-KO mice, 160, 162, 164
MCT8/OATP1C1, 250, 262
MCTs. *See* Monocarboxylate transporters (MCTs)
MDADI. *See* MD Anderson Dysphagia Index (MDADI)
MD Anderson Dysphagia Index (MDADI), 56
Medium-sized spiny neurons (MSNs), 78–80, 79t, 82t, 83
Medullary thyroid carcinoma, 185–186
Memory, 243–245
deficits, 244–245
logical, 245
memory quotient, 245
Mental fatigue, 305–307
Mental Fatigue Scale (MFS), 306–307
Metabolic dysfunction-associated steatotic liver disease (MASLD), 39–40, 46–47
liver fibrosis, 46
pathogenesis, 46
risk factors, 46
Metabolic hormones, 180
adiponectin, 180
insulin, 180
leptin, 180
Metabolic pathway, 300
Methylphenidate, 308
MFS. *See* Mental Fatigue Scale (MFS)
Micronutrients, 320–325
iodine, 321, 322t
magnesium, 321, 322t, 323
selenium, 321, 322t, 323, 325
Mild cognitive impairment (MCI). *See* Cognitive impairment; mild cognitive impairment (MCI)
Mindfulness-based therapies, 308
Miscarriage, 65–71
Mitochondrial dysfunction, 94t, 95–96, 100t, 102, 315t, 316–320, 323, 325

Modulating extracellular matrix and/or neurotrophic factors, 165
Monitoring recommendations, 278
baseline testing, 278, 279t
referral criteria, 278
therapy adjustments, 278
Monocarboxylate Transporter 8 (MCT8), 15f, 17f, 18, 20f, 21–23, 35, 37, 77–81, 78f, 79t, 80f, 83–87, 85f, 92–93, 94t, 235, 237–238, 240
Allan–Herndon–Dudley syndrome, 22
functional features, 78
localization in CNS, 78–80
mutations, 80–81
TH transport, 15f, 21–23
transport deficiency, 22–23
Monocarboxylate transporter 10 (MCT10), 35, 37, 92–93, 94t
Monocarboxylate transporters (MCTs), 141–142, 149
MCT8, 141–142, 144f, 146f
MCT10, 141–142, 144f, 146f
Mood, 39, 45–46
and affective disorders, 253–257, 256t, 259–261
depression, 45–46
and psychiatric disturbances, 314, 315t, 316
MR spectroscopy (MRS), 258, 261–262
N-acetylaspartate (NAA), 258, 262
neurochemical alterations, 258, 262
MSNs. *See* Medium-sized spiny neurons (MSNs)
Multiple sclerosis (MS), 240
Multisensory stimulation, 231
Myelination, 15f, 16t, 19f, 21–23, 197–199, 205–208
defects, 15f
delayed, 22
reduced, 21
Myxedema madness. *See* Myxedema psychosis (MP)
Myxedema psychosis (MP), 171–173, 173t, 178
biological mechanisms, 172
levothyroxine treatment, 172, 178
symptoms, 172

N

Nanocarriers, 235–236, 238–241, 238t–239t
nanoparticles, 235, 238–240
systems, 239–240
T3-loaded, 239f
Nanomedicine, 235–236, 238–241
personalized, 240
platforms, 238, 240
National Institutes of Health (NIH), 348t–349t
Nervous system, 243–244
development, 243–244
thyroid hormones, 243
Neural cells, 77, 79t, 82t, 83–86, 85f
immature neurons, 82t, 83
OATP1C1 expression, 83
thyroid hormone availability, 84, 85f
Neural migration, 243–244
Neurocognitive changes, 186–187, 189–191
Neurodegeneration, 91–97, 107–108, 111–113, 115–119, 235–236, 238–240
age-related, 93–96
mechanisms of, 93–97, 94t, 99, 100t, 101–102
Neurodevelopment, 91–93, 94t, 95–96, 157–158, 162, 165–166, 171, 173–174, 223–225, 227–232
fetal, 223–225, 227–228
glia-neuron crosstalk, 174
offspring, 223, 228–232
thyroid hormones and, 174
Neurodevelopmental consequences. *See* Congenital hyperthyroidism; neurodevelopmental consequences
Neurodevelopmental disorders, 13, 18, 21–23
ADHD, 16, 18t, 20f, 21, 23
cretinism, 13
epilepsy, 23
Neurodevelopmental implications, 250–252, 255, 260–262
myelination, 250–252, 256–258, 261
neurogenesis, 250, 252, 256
Neurodevelopmental outcomes, 64–65, 67–71
Neuroendocrine axis, 280
Neuroendocrine interactions, 347
Neuroendocrine system, 207
Neurogenesis, 94t, 96–99, 98f, 100t, 101f, 195–202, 207–208, 236–237, 241
hippocampus, 238
Neuroimaging, 4, 7–8, 250, 253, 256–261
functional neuroimaging, 6
molecular neuroimaging, 6–7
Neuroinflammation, 98–99, 100t, 102, 108, 111–119, 115t, 156, 162–166, 299, 313–314, 316–321, 323, 325–326
astrocyte dysfunction during, 163–164
homeostatic role and reactive responses, 162–163
Neuronal migration, 197–199, 208
Neurons, 77–80, 79t, 82t, 83–87, 250–252, 258, 262
MCT8 expression, 78–80
OATP1C1 expression, 83
thyroid hormone uptake, 77
Neuroplasticity, 195–196, 199–205, 207–208, 231–232
Neuroprotection, 235–236, 238–239
Neuropsychiatric manifestations, 294, 298–301
affective disturbances, 299
behavioral changes, 299
neuromuscular hyperexcitability, 299
Neuropsychological assessments, 307
Neuropsychological symptoms, 245–246
depressive symptoms, 245–246
forgetfulness, 245–246
insomnia, 246
Neuropsychological tests, 307
Neuroscience, 347–354
neurological function, 347
Neurotransmitter imbalance. *See* Excitotoxicity; neurotransmitter imbalance
Neurotransmitters, 142, 145–147, 197–205
Neurovascular unit (NVU), 156, 157f, 160–162, 161f, 164–165
alterations, 161–162, 161f
components, 160, 165
NF-κB pathways, 97–99, 100t
NHP. *See* Nottingham Health Profile (NHP)
NIS. *See* Sodium-iodide symporter (NIS)
Nitric oxide (NO), 142, 143f, 147
N-methyl-d-aspartate (NMDA) receptors. *See* Excitotoxicity; N-methyl-d-aspartate (NMDA) receptors
NO. *See* Nitric oxide (NO)
Non-genomic signaling, 13–14, 16–21, 19f, 23, 196–197, 197f
integrin αvβ3-mediated, 18, 19f, 23
PI3K/Akt/MAPK activation, 13–14, 16–20, 19f
reactive oxygen species (ROS) production, 16–18, 19f, 21, 23
Nonthyroidal illness syndrome (NTIS), 270–271, 276, 278
Nottingham Health Profile (NHP), 307
NPs. *See* Nanocarriers; nanoparticles
Nucleophagy, 23
Nutraceuticals, 313, 320–326, 322t
curcumin, 313, 321–323, 322t, 325
omega-3 fatty acids, 313, 321, 322t, 323–325
resveratrol, 313, 321–323, 322t

O

OATP1C1. *See* Organic anion-transporting polypeptide 1C1 (OATP1C1)
OECD guidelines, 15f, 16t, 23
TG 426, 15f, 16t
TG 443, 15f, 16t
Omega-3 fatty acids. *See* Nutraceuticals; omega-3 fatty acids
Oral use, 337t
amiodarone, 337t
iodine supplements, 335
kelp, 337t
Organic anion transporters (OATPs), 35, 37
Organic anion-transporting polypeptide 1C1 (OATP1C1), 77, 81–84, 81f, 82t, 85f, 87
deficiency, 159–160
functional features, 81
localization in CNS, 81–83
mutations, 83
Slco1c1, 159f, 164–165
Organic anion transporting polypeptides (OATP1C1), 92–93, 94t
Osteoporosis, 57
post-thyroidectomy, 57

Oxcarbazepine (OXC), 269, 273t, 274, 277–278, 279t
Oxidative stress, 93–97, 94t, 101–102, 101f, 108, 109f, 111–112, 114–116, 118–119, 313–314, 316–320, 323–326, 336, 340–342, 341f
  aging and, 93–96, 94t
  increase by excess iodine, 336, 340–342
  reactive oxygen species (ROS), 313–314, 315t, 316, 317f–320f, 318–321, 322t, 323, 325
  thyrocyte damage, 340–342
  thyroid hormones and, 93–96, 94t, 99

P

Papillary thyroid cancer (PTC), 338–339, 342
  BRAF mutations, 338–339
  excess iodine association, 338
  microcarcinomas, 338
Parathyroid glands, 51, 53–54, 58
  injury, 51, 53
  preservation, 53, 58
Parenteral preparations, 337t
  sodium iodide solution, 337t
Parkinson's disease (PD), 94t, 96–97, 97f–98f, 100t
Patient organizations, 308
PD-1 inhibitors, 283–285, 284t, 288–289
  cemiplimab, 283
  nivolumab, 283–284, 284t
  pembrolizumab, 283
PD-L1 inhibitors, 283–285, 284t
Pediatric population, 57–60
  hypothyroidism, 57
  post-operative complications, 57
Peer support groups, 308
Pendred syndrome, 29, 37
Pendrin, 28–29, 36–37, 333, 334f
  anion transporter, 333
  autoantibodies against, 29
  iodine efflux, 333, 334f
  PDS gene, 28–29
Pericytes, 78, 79t, 80, 87
Personalized medicine, 259–260
Pharmacokinetics, 235–238, 240, 280
Phenobarbital (PB), 269, 273f, 273t, 275, 277–278, 279t, 280
Phenytoin (PHT), 269, 273t, 275, 277–278, 279t, 280
Pituitary gland, 297
Placenta, 63–66, 68–71
  development, 63–65, 69–71
Pneumomediastinum, 58, 60
  CO_2 insufflation, 58
Polymeric nanoparticles, 235, 238, 239t, 241
Polyphenols, 321, 323–324
Polytherapy, 274, 277–280
Positron emission tomography (PET), 4, 6–8, 258–262, 350t–352t
  glucose metabolism, 6–7, 9
Posttraumatic stress disorder (PTSD), 213–218, 215f
  biomarker, 215, 217
  clinical associations with thyroid status, 215f
  thyroid dysfunction, 213–215, 215f
Preeclampsia, 65
Prefrontal cortex (PFC), 214, 214f, 217
Pregnancy, 63–71, 223–232
  adverse outcomes, 63–71
  complications, 228, 231–232
  maternal thyroid adaptation, 63–65, 70–71
  thyroid function during, 224–232
Pregnancy outcomes, 39–40, 43–44, 46
  maternal-fetal outcomes, 43–44
  miscarriage, 39–40, 44, 46
  neonatal death, 39–40
Pregnancy-specific reference ranges, 63–66, 68, 70–71
Presence of D3 in the placenta, 225, 226t
Preterm birth, 63–66, 68–69, 71
Primary hypothyroidism, 285–289, 285t, 289t
  direct immune-mediated, 285, 285t, 288, 289t
  following thyrotoxicosis, 284f, 285, 285t, 288
Probiotics, 321, 322t, 323–325
  Bifidobacterium, 321, 322t
  Lactobacillus, 321, 322t
  Lactobacillus rhamnosus, 325
Projection neurons, 78–80, 79t, 82t, 83
Propylthiouracil, 32t, 37
Proteomics, 350t–352t
Proton-magnetic resonance spectroscopy (1H-MRS), 4, 7–8
  choline, 7
  glutamate, 7
  myo-inositol, 7
Psychiatric disorders, 4–5, 7–8
  anxiety, 4, 7–8
  depression, 4–5, 7–8
  psychosis, 3–5
Psychosis, 171–181
  affective, 171–172, 174–175
  immunoendocrine model, 175
  primary psychotic disorders, 171–174, 178–181
PTC. *See* Papillary thyroid cancer (PTC)
Purkinje cells, 78–80, 79t, 86
Pyramidal neurons, 78–80, 79t, 82t, 83, 86

Q

Quality of life, 54–57, 60, 305–308
  psychological well-being, 57, 60
  swallowing symptoms, 56
  voice impairment, 54–56
Quantitative susceptibility mapping (QSM), 7–8
  brain iron levels, 7, 9

R

Radial glia cells, 79t, 80, 82t, 83, 87
Radioactive iodine (RAI) therapy, 186, 188–191
Radioiodine therapy, 124–125, 131–132, 134–137
RAI. *See* Radioactive iodine (RAI) therapy
Reactive oxygen species (ROS). *See also* Oxidative stress; reactive oxygen species (ROS)
  activation by excess iodine, 340–342
  production, 93–96, 94t, 97f, 99
  regulation, 93–95
  Tg proteolysis, 342
Recovery, 305–306, 308
Recurrent laryngeal nerve (RLN), 51, 54–58, 60
  injury, 51, 54–58, 60
  surgical anatomy, 54
Recurrent pregnancy loss (RPL), 67, 69–70
Renal outcomes, 39–40, 45–46
  chronic kidney disease (CKD), 39–40, 45–46
  glomerular filtration rate (eGFR), 45
  proteinuria, 45
Replacement therapy, 165
Resting-state MRI, 4, 6, 8
Resveratrol. *See* Nutraceuticals; resveratrol
RLN. *See* Recurrent laryngeal nerve (RLN)
Robotic thyroidectomy, 56, 58
  complications, 58
  cosmetic outcomes, 58
ROS. *See* Reactive oxygen species (ROS)

S

SAM. *See* Sympatho-adrenomedullary system (SAM)
Schizophrenia, 171–180, 176t–177t
  cognitive impairment, 176
  HPT axis activity, 174
  thyroid abnormalities in, 173–176
Secondary hypothyroidism, 285–289, 285t, 289t
  ICI-related hypophysitis, 285, 285t, 288, 289t
Secondary psychoses, 172–174, 173t, 176, 178, 180–181
  clinical implications, 173, 178
  differential diagnosis, 173
  management, 178
Selenium. *See* Micronutrients; selenium
Sensory sensitivities, 305–307
Seroma, 56, 59–60
  post-operative, 56, 59–60
Serotonin, 96, 98f, 100t
Sex-specific effects, 93, 94t, 95f, 96, 100t
Single-cell ribonucleic acid (RNA) sequencing, 350t–352t
SIS. *See* Swallow Impairment Score (SIS)
Skeletal and cardiac muscle, 297
  myocardial involvement, 297
Skin, hair, and nails, 297–298
Sleep disturbances, 306
Sodium-iodide symporter (NIS), 27–29, 36–37, 332–335, 332t, 334f, 339
  abnormalities, 28
  activity regulation by TSH, 335

Sodium-iodide symporter (NIS) (*Continued*)
amiodarone effect on mRNA expression, 339
expression, 28–29
iodide transport, 332t, 333, 334f, 335
Somatotropin-IGF-1 axis, 300–301
Spatial working memory, 244–246
Cambridge Neuropsychological Test Automated Battery (CANTAB), 245–246
executive function, 244
screening, 246
SPECT, 258–259, 261–262
ST. *See* TSH suppression therapy (ST)
Steroid-responsive encephalopathy associated with autoimmune thyroiditis. *See* Hashimoto's encephalopathy (HE)
Stress, 295, 299–301
acclimation, 142, 147–148
reduction, 308
Stress response, 145–149
acute stress, 147
chronic stress, 145, 147
Structural and functional brain changes, 315t, 316, 320, 325
Structural MRI, 8, 256, 261–262
gray and white matter changes, 256
Subclinical hypothyroidism (SCH), 39–47, 250, 256, 261–262
causes, 40
diagnosis, 39–40, 46
prevalence, 40–41, 43–45
Surgery, 186–188, 190–191
Swallow Impairment Score (SIS), 56
Sympatho-adrenomedullary system (SAM), 140f, 145, 148–149, 148f
Synapse, 199–202, 207
synaptic plasticity, 195–196, 199–202
synaptic transmission, 207–208
Synaptic plasticity, 94t, 98–99, 100t, 102, 243
Synaptogenesis, 13, 20f, 21, 23, 197–202, 205–208
Systemic toxicity, 236–239, 238t
reduced, 238t

T

T3. *See* Thyroid hormones; triiodothyronine (T3)
T4. *See* Thyroid hormones; thyroxine (T4)
Tanycytes, 78, 79t, 80, 82t, 83, 87, 155–157, 164–165
Task-based functional MRI, 4, 6, 8
Tau pathology, 107–108, 111–116, 114t, 118–119
Tau-PET, 7–8
TBG. *See* Thyroxine-binding globulin (TBG)
T-cell activity, 285–286
CD4+ T cells, 285–286, 286f
CD8+ T cells, 285–286, 286f
T-cell responses, 283, 285–286
Testosterone, 190–191
TETRAC, 20f, 22–23
pro-hormone of TRIAC, 22
T4 derivative, 22
Tg. *See* Thyroglobulin (Tg)
THs. *See* Thyroid hormones (THs)
ThyPRO, 306–307
Thyroglobulin (Tg), 27–31, 37, 108–110, 119, 123, 126, 135–136, 223–224, 228–230, 332–336, 332t, 333f, 338–340, 342
autoantigenicity, 342
degradation, 31
gene, 31
internalization and degradation, 110
iodination, 28, 31, 334–335
storage of thyroid hormones, 332t
Thyroid antibodies, 284–286, 288–289
thyroglobulin antibodies, 284–285
thyroid peroxidase (TPO) antibodies, 284–286
TSH receptor antibodies, 285
Thyroid autoantibodies, 171–173, 175–176, 178, 181
immunoendocrine marker, 175
negative symptoms, 171–172, 175, 178
psychosocial functioning, 175, 178
Thyroid autoimmunity, 63, 65–71, 331, 336, 339–342, 341f
environmental risk factor, 340
immune dysregulation, 65–66, 69
mechanisms, 336, 340–342, 341f
prevalence with USI programs, 342
thyroidal reserve, 65–67, 69
TPOAb positivity, 63, 65–71
Thyroid cancers, 133–135, 137, 185–191, 350t–352t
anaplastic carcinoma, 134–135
diagnosis, 185–186
medullary carcinoma, 134–135
papillary carcinoma, 134–135
treatment, 186
types, 185
Thyroid disease, 171–174, 173t, 176–178
autoimmune, 171, 175
cognitive dysfunction in, 176–177
psychotic symptoms in, 173t
Thyroid disorders, 63–70, 350t–352t
Thyroid dysfunction. *See also* ICI-induced hypothyroidism
amiodarone-induced, 339
excess iodine-induced, 331, 340
functional and anatomical alterations, 331, 332f
neurocognitive impairment, 3–8
sex-specific effects, 4
Thyroidectomy, 51–60, 185–191, 295
complications, 51–60
pediatric population, 57–60
surgical techniques, 51, 53, 56, 58–60
Thyroid function, 171–179, 181
abnormalities in schizophrenia, 173–176
assessment, 172–173, 178, 181
psychosis and, 171–172, 179
tests, 66–68, 71
Thyroid gland, 123–137, 223–225, 224f, 294–295, 297, 348t–352t
development, 223–224, 224f
function, 123–124, 136
hormone synthesis, 123–124, 128, 132–133, 136
organogenesis, 223–224
structure, 123–124
Thyroid hormone (TH), 13–14, 16–23, 39, 41, 45–46, 139–149, 156–166, 158f–159f
availability, 156–157, 160, 162–166
deficiency, 156, 158–160, 161f, 162–165, 163f
deiodinase enzymes, 13–14, 16, 21–22
disruption targets, 15f
gene expression, 41, 43f
maternal hypothyroidism, 13–14, 23
physiologic roles, 41
receptors (TR), 41, 42t, 46, 157–158, 158f–159f, 165
T3, 139–145, 144f, 147, 148f
T4, 139–145, 148f
transport, 141–142, 149
Thyroid hormone analogs, 22–23
MCT8 avoidance, 22
therapeutic potential, 22
thyrotoxicosis evasion, 22
Thyroid hormone excess (TH excess), 293–301
neuronal metabolism, 294, 300–301
neurotransmitter signaling, 294, 298–299
synaptic plasticity, 294, 298–301
Thyroid hormone levels, 112–113, 117–119
free thyroxine (FT4), 112–113, 117–118
free triiodothyronine (FT3), 113, 117–118
total thyroxine (TT4), 113
Thyroid hormone receptor (TR), 216, 218, 223
alpha (Thra), 216, 217t
beta (Thrb), 216, 217t
Thyroid Hormone Receptor and Regulator (TR-R) system, 142, 143f–144f, 145, 146f, 148–149, 148f
Thyroid hormone receptors (THRs), 41, 42t, 46, 92–93, 94t, 97, 101, 110–111, 113, 115t, 117–119, 141–142, 146f, 148–149, 196–197, 199, 205, 250, 256, 261, 298, 314, 315t, 319–320, 320f, 325–326
alpha isoform, 42t
beta isoform, 42t, 46
THRα, 92–93, 94t
THRβ, 141–142, 144f, 146f
Thyroid hormone replacement, 57, 59–60
levothyroxine, 57, 59
optimal levels, 57

Thyroid hormones (THs), 3–8, 27–37, 30f, 63–71, 77–81, 84–87, 91–102, 94t, 95f, 97f–98f, 100t, 107–119, 109f, 123–137, 171–181, 185, 187–191, 195–208, 197f, 198t, 200t, 203t–204t, 206t, 223–225, 227–232, 235–239, 241, 243–245, 249–262, 269–280, 332–340, 332t, 333f–334f, 342, 347
biological activity, 27, 31, 33, 36
biosynthesis, 332–335, 332t, 334f
biosynthesis and regulation, 108–111
brain availability, 84–87
brain-derived neurotrophic factor (BDNF) levels, 3
brain development and function, 243
brain development role, 77, 86
CNS development and function, 110–111
cognition in psychosis, 177
dysregulation, 237
free T3 (FT3), 270, 273t, 274, 276, 279t
free T4 (FT4), 63–69, 71, 270–271, 273t, 274–280, 279t
maternal, 223–225, 227–232
mechanisms of action, 92, 94t
metabolic effects, 332, 333f
metabolism, 32–33, 34f, 37
mitochondrial function, 112
neurobiological actions, 171
neurodevelopmental stages, 3
neurogenesis, 3
physiological roles, 250–252, 251t
production and transport, 92–93, 94t
receptors, 223–224, 235–236
secretion, 332, 332t, 333f, 335
synthesis, 27–32, 30f, 36–37
T3, 195–197, 197f, 198t, 199–202, 200t, 203t–204t, 205–208, 206t
T4, 91–93, 94t, 95f, 96–97, 98f, 100t, 101, 195–196, 197f, 198t, 199–202, 200t, 203t–204t, 205–208, 206t
target genes, 243
therapeutic implications, 178–179
thyroid-stimulating hormone (TSH), 63–69, 71
thyroxine (T4), 123–124, 126–130, 132, 135–136, 243, 249–255, 259–262
transporter, 157–160, 158f–159f, 164–165
transport mechanisms, 77–78, 81, 83–87, 85f
triiodothyronine (T3), 123–124, 126–128, 130, 132, 135–136, 249–255, 259–262, 270–271, 272f–273f, 273t, 274–276
Thyroid hormone signaling, 213–217
in amygdala, 216, 218
in hippocampus, 216, 218
in PTSD, 213–215, 215f
Thyroiditis, 124–128, 130, 135–136
chronic autoimmune thyroiditis, 124–127, 133, 136
postpartum thyroiditis, 124–125, 128, 135–136
subacute thyroiditis, 124–125, 128, 130, 131t, 135–136
Thyroid-NDT-TAS, 14, 16t, 23
framework, 16t
neurodevelopmental assessment, 14, 16t
regulatory application, 16t
Thyroid peroxidase (TPO), 27, 29–30, 36–37, 109–110, 123, 126, 135, 223–224, 227–230, 332t, 333–336, 334f
activity inhibition, 336
autoantibodies, 30, 37
DUOX1 and DUOX2, 30–31, 37
gene, 29–30
iodide oxidation, 332t, 333–335, 334f
location, 333, 335
TPO antibodies, 227–230
Thyroid peroxidase antibody (TPOAb), 63, 65–71
Thyroid receptors (TRs), 13–14, 15f, 16–23, 17f, 19f–20f
TR-mediated genomic effects, 18, 21
Thyroid stimulating hormone (TSH), normalization, 3–7, 27–31, 33, 35–36, 39–47, 53, 57, 59, 91–92, 94t, 97–99, 100t, 101–102, 108–110, 109f, 112–113, 115–119, 115t, 124–128, 130, 132–136, 171–176, 173t, 176t, 180–181, 214–215, 224–230, 226t, 243–246, 269–271, 272f–273f, 273–280, 273t, 279t, 284–289, 284f, 350t–352t
elevated levels, 243–246
HPT axis regulation, 181
indicator for thyroid dysfunction, 243
levels in FEP, 171–172, 174, 176
normalization, 40
PANSS severity, 175
pregnancy-specific reference ranges, 43–44
reference values, 224–227, 226t
secretion, 33, 35–36
suppression, 59
target range, 57
TSH receptor (TSH-R), 27
TSH-receptor mutations, 41
Thyroid storm, 53, 59, 127, 129, 132, 136–137
management, 53, 59, 132
thyrotoxic crisis, 53, 59
Thyroid ultrasonography, 350t–352t
Thyrotoxicosis, 53–54, 59, 127–130, 132, 135–137, 314, 326
hungry bone syndrome, 53–54
pre-operative assessment, 53, 59
Thyrotropin (TSH), 332, 335–336, 338–339
feedback regulation, 335
production, 335
regulation of NIS activity, 335
Thyrotropin receptor (TSH-R), 294–298
Thyrotropin receptor antibodies (TRAb), 66–68
Thyrotropin-releasing hormone (Trh), 35–36, 216, 217t
Thyrotropin-releasing hormone receptor (Trhr), 216, 217t
Thyroxine (T4), 77–78, 81, 83–87, 85f, 108, 109f, 110–113, 115–119, 214–218, 215f
conversion to T3, 84, 85f
transport by MCT8, 78, 87
transport by OATP1C1, 81, 87
Thyroxine-binding globulin (TBG), 63–64, 224–226, 226t, 270, 273t, 274–276
Thyroxine replacement therapy, 286–288
iodide intake, 287
T3 therapy, 287
Topical antiseptics, 337t
diiodohydroxyquinoline cream, 337t
iodine tincture, 337t
povidone-iodine, 337t
Topiramate (TPM), 269, 273t, 276–277, 279t
TPO. *See* Thyroid peroxidase (TPO)
TR. *See* Thyroid hormone receptor (TR)
TRAb. *See* TSH receptor antibodies (TRAb)
Transgenic AD models, 114, 114t
Okadaic acid (OA) mouse model, 114, 114t
PDAPP mouse model, 114, 114t
3xTg-AD mouse model, 114–115, 114t
Transthyretin (Ttr), 216, 217t
Trauma-related disorders, 213–217
TRIAC, 20f, 22–23
clinical trials, 22
congenital MCT8 deficiency, 22
T3 derivative, 22
Triiodothyronine (T3), 41, 43f, 46, 77–78, 81, 83–87, 85f, 108, 109f, 110–119, 171–172, 174–175, 177–179, 214–218, 215f, 235–241
augmentation, 172, 179
brain availability, 84–86
delivery, 235–238, 240
dysregulation, 237
nanoformulated, 238, 240
pro-cognitive effects, 178
total T3 levels, 174
transport by MCT8, 78, 87
uptake in cerebral organoids, 86
TR-R system. *See* Thyroid Hormone Receptor and Regulator (TR-R) system
TRs. *See* Thyroid hormone receptors (TRs)
TSH. *See* Thyrotropin (TSH)
TSHR. *See* TSH receptor (TSHR)
TSH receptor (TSHR), 223–225, 226t
TSH receptor antibodies (TRAb), 130–132, 131t, 135–136, 336, 340, 340t
TSH suppression therapy (ST), 185–186, 188–190
beneficial effect, 189
no effect, 189
reducing effect, 189

U

UDP-glucuronosyltransferases (UGTs), 273f, 273t, 274–277
Ultrasonography (US), 130–135, 137
Universal salt iodization (USI) programs, 331, 335, 338, 342
  correction of iodine deficiency, 331
  risk of disorders, 331, 335
  thyroid cancer increase, 338
Universal screening, 67, 70
US. *See* Ultrasonography (US)
Use of TH analogs, 165
USI. *See* Universal salt iodization (USI) programs

V

Vagus nerve (VN), 139–140, 140f, 145, 147–149, 148f
Valproate (VPA), 269, 273–278, 273f, 273t, 279t, 280
VHI. *See* Voice Handicap Index (VHI)
Vitamin C. *See* Antioxidants; vitamin C
Vitamin E. *See* Antioxidants; vitamin E
VN. *See* Vagus nerve (VN)
Voice Handicap Index (VHI), 56
Volumetric analysis, 4–5

W

WDTC. *See* Well-differentiated thyroid cancer (WDTC)
Well-differentiated thyroid cancer (WDTC), 338–339, 342
Withania somnifera. See Adaptogens; *Withania somnifera*
Wolff–Chaikoff effect, 29, 36–37, 45–46, 336–340, 338f, 342
  adaptation mechanism, 336–338, 338f
  biochemical mechanism, 336
  susceptibility, 336–339
Work capacity, 306–307
Working memory, 185–187, 189–191, 244–245
  affected domain in hypothyroidism, 244
  impairment, 244–245
  spatial, 244–246
World Health Organization (WHO), 348t–349t

Z

Zebrafish, 139, 141–142, 143f–144f, 148

Printed and bound by CPI Group (UK) Ltd, Croydon, CR0 4YY
24/09/2026
02222288-0006